BIOLOGICAL SCIENCE

AN INTRODUCTORY STUDY

PRENTICE-HALL INTERMEDIATE SCIENCE SERIES

SERIES EDITOR:

William A. Andrews
Professor of Science Education
Faculty of Education, University of Toronto

Understanding Science 1
Understanding Science 2
Physical Science: An Introductory Study
Biological Science: An Introductory Study

William A. Andrews

Nancy J. Purcell

David A. Balconi

Nancy D. Davies

Donna K. Moore

BIOLOGICAL SCIENCE

AN INTRODUCTORY STUDY

PRENTICE-HALL OF CANADA, LTD.,
Scarborough, Ontario

Canadian Cataloguing in Publication Data

Main entry under title:
Biological science, an introductory study

ISBN 0-13-076562-7

1. Biology   I. Andrews, William A., 1930-

QH45.2.B56   574   C79-094723-4

Prentice-Hall Inc., Englewood Cliffs, New Jersey
Prentice-Hall International, Inc., London
Prentice-Hall of Australia, Pty., Ltd., Sydney
Prentice-Hall of India, Pvt., Ltd., New Delhi
Prentice-Hall of Japan, Inc., Tokyo
Prentice-Hall of Southeast Asia (PTE) Ltd., Singapore

ISBN 0-13-076562-7

 Metric Commission Canada has granted permission for
use of the National Symbol for Metric Conversion.

Production Editors: Janet MacLean/Beth Burgess
Designer: John Zehethofer
Illustrators: James Loates Illustrating/Julian Mulock/
Peter Van Gulik/Jean Miller and Audra Geras of Biomedical
Productions
Compositor: CompuScreen Typesetting Limited

Printed and bound in Canada by Bryant Press

1 2 3 4 5 6 7 8 9   88 87 86 85 84 83 82 81 80

Contents

Acknowledgements

The authors wish to acknowledge the competent professional help received from the staff of Prentice-Hall of Canada, Ltd. in the production of this text. In particular, we extend our thanks to Janet MacLean and Beth Burgess for their skillful editorial work and to Barb Steel for her dedicated coordination of the editorial aspect of the production of this text. We also thank Steve Lane for his untiring and valuable assistance in the planning and development of this text. This book owes its final shape and form largely to the efforts of John Zehethofer and Rand Paterson.

We wish, further, to thank the many teachers who reviewed the manuscript and offered their constructive criticisms. Also, we would be remiss if we did not express our appreciation of the imaginative, attractive, and accurate art work of James Loates, Julian Mulock, Peter Van Gulik, Jean Miller and Audra Geras. Finally, we extend a special word of appreciation to Brenda Andrews for her preparation of an excellent index and to Lois Andrews for her careful and dedicated preparation of the final manuscript.

The writing of a text is a time-consuming undertaking that leaves little time for one's family and friends. Thus we thank our families and friends for their understanding and support during the many months that we were writing this text.

W. A. Andrews,
Editor

Photo Credits

Cover photo: Miller Services. Unit Introductory Photos: Unit One: Helen Sutton. Unit Two: P. Moens, Department of Biology, York University. Units Three and Six: courtesy of Federation of Ontario Naturalists. Unit Five: Eric S. Grace. Unit Seven: Bob Wood. Figs. 1-1; 2-7,A; 2-7,B; 2-8; 11-2,A; 11-2,B; 11-2,C; 11-25; 11-29; 12-1,C; 13-26; 23-4; 23-22; 23-25; 25-22; 25-30; 25-34: courtesy of Ontario Ministry of Natural Resources. Figs. 2-4,A: 2-4,B: courtesy of Manfred C. Schmid. Figs. 5-15; 5-16; 5-18; 5-19; 5-20; 5-22: courtesy of Dr. K. Kovacs and Dr. E. Howarth, Department of Pathology, St. Michael's Hospital, Toronto and Ena Ilse, Sandra Briggs, Donna McComb and Sandra Cohen of the Electron Microscopy Unit of St. Michael's Hospital, Toronto. Figs. 6-1; 9-1,A; 9-1,B; 10-1,A; 10-1,B; 10-1,C; 12-1,A; 12-1,B; 13-2,A: courtesy of Dr. Pamela Stokes, Botany Department, University of Toronto. Figs. 6-10; 10-1,D; 10-1,E; 19-4; 20-14,A: Gilbert L. Twiest, excerpted from *Fresh-water Plants and Animals,* sound filmstrip series, Prentice Hall Media. Figs. 9-1,C; 9-1,D: courtesy of Ontario Ministry of Health, Laboratory Services. Fig. 9-11: courtesy of Dr. P.C. Fitz-James and K. Ebisuzaki. Figs. 9-12; 9-17; 9-19; 9-23: courtesy of Professor Frances Deane, Department of Microbiology and Parasitology, University of Toronto. Figs. 9-15; 11-2,D; 11-6; 11-8; 11-19; 11-21; 26-1: courtesy of Ontario Ministry of Agriculture and Food. Fig. 14-1: Nancy Purcell. Fig. 25-3: courtesy of J.C. Ritchie. Figs. 25-4; 25-40: Duncan M. Cameron, Jr. Figs. 25-5; 25-12: Donna K. Moore. Figs. 25-6,A; 25-6,B: T. C. Dauphiné, Jr. Fig. 25-10: Eric S. Grace. Fig. 25-14: courtesy of Biology Department, York University. Fig. 25-16: courtesy of Ontario Ministry of Natural Resources and Information Canada. Fig. 25-20: courtesy of Ontario Ministry of Natural Resources and Dr. D. Gunn. Fig. 25-29: Norman R. Lightfoot. Fig. 25-44: courtesy of Jon C. Barlow. Figs. 27-2; 27-3; 27-8; 27-10; 27-13; 18-1: courtesy of Cytogenetics Laboratory, Queen's University. All other photos by W.A. Andrews.

Preface

This is an introductory course in biological science for secondary school students. As such, it has two major aims. One aim is to provide some information about the living things that share the earth with you. Toward this end, you will spend much of your time studying the form and function of a wide variety of living things. The book begins with an overview of the various forms of life. It then investigates the most basic unit of life, the cell. After that, it deals with the form and function of simple living things such as bacteria, protozoans, algae, and fungi. From there it proceeds to a study of the form and function of plants and animals. You, the human, are one of the animals included in this study. The final units deal with the important and exciting concepts of ecology, genetics, and evolution.

The second major purpose of this course is to familiarize you with the scientific method of inquiry. Although there are many scientific methods, there is a common underlying pattern of logical thought in every scientific study. After you have done several of the investigations in this book, you should be familiar with this pattern and you should be able to approach problems in a manner similar to that used by scientists. More important, you should understand what science is. You should be able to appreciate its methods and be aware of its limitations. Often in our daily lives we encounter problems that can be solved by using the scientific method. Learn it well!

This book contains much more information than can be studied in one year. Your teacher will select those chapters and sections that will be most useful to you and your classmates. You will study the topics in this book through investigations and through narratives. Investigations are experiments carried out in the laboratory. They are included to give you practical experience in using the scientific method. Scientists, however, do not simply conduct experiments; they also read a great deal. Therefore, this book includes narratives, or reading sections, which contain important information to add to the investigations.

The combination of investigations and narratives should provide you with the knowledge and skills needed to solve many biological problems and to answer many questions on biological topics. More important, it should open your eyes to many new and interesting things. It may even start you asking questions about life around you. If that happens, this course will have been truly successful.

W.A.A.
Editor

UNIT ONE

Life And Its Diversity

All living things have basic similarities that distinguish them from nonliving things. However, each living thing differs from all other living things in some ways. In this unit you will study the characteristics of living things and explore the wide variety of life forms on earth.

1

The Science Of Biology

Science has two aspects. It is both a body of knowledge and a method used for discovering new knowledge. The main purpose of this chapter is to acquaint you with the second aspect, which is commonly called the **scientific method**. Although there is no one scientific method that every scientist follows step by step, there is an underlying pattern of logical thought in every scientific study. It is this pattern that is called the scientific method. You will be using it throughout this course to gain new knowledge about the science of **biology**.

1.1 What is Biology?

The word *biology* comes from the Greek words *bios,* which means "life", and *logos,* which means "thought". Thus, **biology** *is the science that deals with the study of life*.

A Brief History

Biology is a very old science. In fact, it probably began in prehistoric times. Over thirty thousand years ago, cave dwellers in France and Spain decorated the walls and ceilings of their caves with paintings of animals. These paintings show that the cave dwellers had closely observed the structure and behaviour of many animals. When humans were mainly hunters of animals and gatherers of edible plants, they studied biology at a simple level. They had to know the behaviour of their prey and, of course, the behaviour of animals that might prey on them. They also had to observe the growing conditions required by the various plants as they collected them for food.

When humans began an agricultural way of life about ten thousand years ago, they had to know much more about plants and animals. For example, they had to know when to collect seeds for the next year's crop, when to plant the seeds, and how much moisture the various types of plants needed. They also had to know what to feed their animals, how long the animals would live, and what might kill the animals. In a sense then, they were biologists because they studied living things. From such early beginnings came the science of biology (Fig. 1-1).

Prehistoric humans had to rely on their senses alone to make observations. They had no laboratories, microscopes, dissecting kits, or other equipment. As a result, they could only ask simple questions such as: What is it? and, What other living things is it like? While seeking answers to such questions, they probably discovered that the wide variety of living things had much in common, as well as many differences. They probably grouped similar living things and named each group. For example, they might have noted that all flying animals with feathers had much in common. They may have grouped them under one name such as "birds". This process of grouping similar things is called **classification**.

These early "biologists" also would have observed other patterns in nature, such as regular migration of birds and life cycles of plants and animals.

As the centuries passed, humans became more knowledgeable and curious. They began to ask more complex questions such as: What is it made of? and, How is it put together? Modern biology began during the 17th century when humans finally had the knowledge, skill, and equipment to seek answers to such questions. During that century Robert Hooke (1635–1703) and Anton van Leeuwenhoek (1632–1723) introduced a new tool, the microscope, to the scientific world. Another pioneer of modern biology was William Harvey (1578–1657), an English physician. He showed the importance in biological studies of well-designed experiments and careful observations. He traced the pattern of circulation of blood in humans, and showed that it travelled in one direction through the arteries and veins in a circular path.

The search for still more knowledge by curious scientists led them to ask even more complex questions: What do the parts of a living thing do? How do the parts work? With the asking of such questions, biology truly came of age.

Fig. 1-1 This Indian petroglyph shows that the early residents of Canada had an interest in plants and animals.

Review

1. **a)** State the two main aspects of every science.
   b) Explain the origin of the word "biology".
   c) Give a definition of biology.

2. **a)** What is classification?
   b) Describe two examples of classification that you have observed.

1.2 Biology Today

Modern biology is a vast science. Over 1 500 000 different kinds (species) of organisms have been identified and new ones are still being discovered. Biologists think that there may be over 2 000 000 different species on earth. They range in size and complexity from tiny bacteria to trees and humans. Because biology is such a large field, it is broken down into several subdivisions for easier study. These subdivisions are formed according to the group of organisms being studied or the approach taken to the study of the organisms.

Group of Organisms Being Studied

Examples of some of the main fields of biology formed using this method of subdivision are: **botany**, the study of plants; **zoology**, the study of animals; **microbiology**, the study of microscopic organisms; **bacteriology**, the study of bacteria; **mycology**, the study of moulds and mushrooms; **entomology**, the study of insects; and **ornithology**, the study of birds.

Approach Taken to the Study of Organisms

Examples of some of the main fields of biology formed using this second method of subdivision are: **taxonomy**, the classification of organisms; **morphology**, the study of the external form and structure of organisms; **anatomy**, the study of the internal structure of organisms; **physiology**, the study of the function of organisms; **cytology**, the study of cells; **ecology**, the study of the relationship of organisms to their environment; **genetics**, the study of inheritance; and **pathology**, the study of diseases.

A New Definition of Biology

Biology was first defined as the science that deals with the study of life. However, as we learn more, we see that biology involves many other things. It is also a study of all those things that affect life. For example, if botanists wish to study plant growth, they must not only study the morphology and physiology of plants but also energy input from the sun and nutrient input from the soil. If zoologists wish to study the ecology of fish in a river, they must study the physical and chemical properties of the water as well as the morphology and physiology of the fish (Fig. 1-2). Thus the following is a more accurate definition of biology: **Biology** *is the study of living things and things that were once alive, together with the matter and energy that surround them.*

Fig. 1-2   This zoologist is doing chemical tests on the water as part of an ecological study of a river.

Biology used to be a simple science that dealt largely with the identification and description of organisms. These aspects are still important and they will be a major part of this course. However, modern biology moves beyond that stage into exciting new areas that involve many other sciences and disciplines. For example, many biologists find that they have to be skilled in mathematics and computer programming in order to solve their research problems. Biologists who have specialized in biophysics must be very knowledgeable in physics. This knowledge is needed to operate the complex instruments used in such work as cancer research. Biologists who work in the relatively new field of molecular biology (the study of the structure and function of the parts of cells) must know a great deal of chemistry. Biologists are even involved in space science as they seek evidence of life on distant planets.

Review

1. **a)** What field of biology seeks answers to these types of questions: What is it? What other living thing is it like?
   b) What field of biology seeks answers to these types of questions: What is it made of? How is it put together?
   c) What field of biology seeks answers to these types of questions: What do the parts of a living thing do? How do the parts function?
2. **a)** The names of some of the subdivisions of biology may indicate the group of organisms being studied. For example, zoology indicates the study of animals. Name three subdivisions of this type that are not described in this text.
   b) The names of other subdivisions of biology may indicate the approach taken to the study of organisms. For example, pathology indicates the study of diseases. Name three subdivisions of this type that are not described in the text.
3. **a)** State the "new" definition of biology.
   b) Explain why this definition is better than the first one.

1.3 The Scientific Method

When most people hear the term *scientific method* they assume that it involves a formula too complicated for ordinary people to understand. This is not true. In fact, the scientific method is used to some extent by most people every day. A mechanic trying to find out why a car will not start, a television repair-person trying to determine why a television set will not produce a picture, a person shopping for a new car, and a person trying to figure out why a plant will not grow, all approach their problems in a general step-by-step manner that can be called the scientific method (Fig. 1-3). Each of these people uses a method that is unique to his or her particular problem. Yet all their methods show an **underlying pattern of logical thought** that is very similar to that used by scientists:

Fig. 1-3 Each of these persons is approaching a problem by using a form of scientific method.

A

B

C

D

1. Each person recognizes that a problem exists that needs to be solved.
2. Each person knows or collects some background information on the problem.
3. Each person makes up a hypothesis or prediction regarding the problem.
4. Each person experiments or tries various things in some step-by-step way to try to solve the problem.
5. Each person observes the results of the experiments and records them in a useful way.
6. Each person comes to a conclusion regarding the problem and its solution.

While seeking solutions to their problems, these people must always maintain a certain **attitude**. They must keep an open mind, work hard and honestly, question their results, and always remain unbiased. This attitude will save time and money. This same attitude is characteristic of scientists.

Remember then, that there is no one scientific method that can be followed step by step to solve all problems. Instead, it is the attitude and the underlying pattern of logical thought that make up what is called the scientific method.

The scientific method has become a powerful tool in helping scientists solve their problems. They usually follow, consciously or unconsciously, a series of steps that are called **processes of science**. Together these form a pattern of thought and action that is called the **process of scientific inquiry** or the **scientific method**. Let us look more closely at these steps. As we do so, keep in mind that the scientific method is really just common sense and an orderly, logical way to solve problems. Remember, too, that we all use, or should, this method in our daily lives.

Recognition of a Problem: Curiosity

All scientific studies must begin with the **recognition of a problem**. The scientist and you, the student of science, must see something worth studying. Usually **curiosity** helps us recognize a problem. Most of us always wonder about the hows and whys of things that happen around us. Scientists are different from us only because they have made this same curiosity their profession. They are always asking questions about the observations they make. In fact, they usually produce more questions than answers. If this were not so, the wheels of science would grind to a halt because no new frontiers would be opened for study. However, simple curiosity will not put a person on the moon or produce a new medicine. Scientists learn to direct their curiosity. Once they have recognized a problem, they read as much as they can find about it, discuss it with others, and plan a course of action. The following story about the discovery of penicillin illustrates this point.

In 1928 a British bacteriologist, Sir Alexander Fleming, was culturing (growing) bacteria in petri dishes as part of an experiment. One day Fleming's assistant noticed that a blue-green mould was growing in some of the cultures (Fig. 1-4). He also noticed that the bacteria did not grow near the mould. Fleming became curious; he recognized a problem. Why would the bacteria not grow near this mould? Because Fleming asked this question, research was begun which proved that certain blue-green moulds give off a substance that kills bacteria or slows down their growth. That substance is called **penicillin**. It was the first **antibiotic** to be produced and it has proved effective in treating many diseases caused by bacteria. Millions of lives were saved during World War II and the following decades because of this discovery. It all began because Fleming became curious and recognized a problem. Just think for a moment. Another person might have thrown out the cultures because they were mouldy, just as you and I would throw out a mouldy orange!

Fig. 1-4   Blue and green moulds often grow on oranges, cantaloupes, jams, and damp leather. Many of these belong to the genus *Penicillium*, as did the mould found in Fleming's bacteria cultures.

Fig. 1-5 Many people think that scientists spend all their time in the laboratory. Actually, most scientists spend more time reading, thinking, and discussing than they do experimenting.

Fig. 1-6 Do you think all insects have three pairs of legs?

Collection of Information on the Problem

Once scientists have recognized a problem, they collect all the related information they can. This allows them to build on the work of preceding scientists. If they did not do this, science would never make any progress. We would be discovering the same things over and over again. Therefore, before scientists begin experimenting, they visit the library and **collect information** from books, journals, and other sources (Fig. 1-5). You, as a student of science, should also do some reading. Before you do any experiment in this book, read the related material.

Formation of a Hypothesis

Sometimes the information a scientist collects is enough to solve the problem and the investigation ends. However, if the problem is still not solved, the scientist must proceed with the scientific method in order to try to solve the problem.

Before doing an experiment, scientists usually make a **hypothesis**. A hypothesis is a prediction of the results of the experiment. It gives the scientists something to work toward. It directs the course of their experiments. Yet, no matter how reasonable a hypothesis may seem to be, it is not "true" until experiments have shown it to be so. Often scientists have to change or completely discard a hypothesis after some experiments have been done. Consider this example:

Nancy noticed that six kinds of flies, two kinds of bugs, five kinds of beetles, and two kinds of grasshoppers have three pairs of legs. She became curious and recognized a problem to be explored. Why do animals that are so different all have three pairs of legs? She did a little reading and made up this hypothesis: All insects have three pairs of legs (Fig. 1-6). In making this generalization, she reasoned **inductively**. (Inductive reasoning involves making a generalization from specific observations or facts.) Is the hypothesis true? Until Nancy does an experiment, namely the examination of many more types of insects, the hypothesis remains a prediction, or an educated guess.

The Experiment

The real value of a hypothesis is that it aids the scientist in designing the experiment. For example, after Nancy made the prediction, or hypothesis, that all insects have three pairs of legs, she reasoned **deductively** and worked in reverse. (Deductive reasoning involves going from a generalization to specifics.) Nancy reasoned that, "Since all insects have three pairs of legs and since moths, butterflies, and bees are insects, then moths, butterflies, and bees must have three pairs of legs." The first hypothesis led Nancy to make a second hypothesis about things she had not yet seen. Further, this combination of inductive and deductive reasoning suggested some experiments that should be done. Clearly Nancy should design and conduct an experiment that explored whether or not moths, butterflies, and bees have three pairs of legs.

An experiment must always be conducted under **controlled conditions**. Scientists must be certain that they have controlled or kept constant all **variables** (factors that can affect the result) except the one being studied. For example, Nancy should ask herself, "What variables could affect the number of legs on an insect?" Clearly age is one such variable. Immature flies (grubs) and immature butterflies (caterpillars) do not have three pairs of legs. To control this variable, Nancy should examine only mature or adult insects. Her first hypothesis should now be changed to, "All adult insects have three pairs of legs." What other variables should be controlled in her experiment?

Figure 1-7 shows another example of a **controlled experiment**. This experiment was done to discover the effects of lack of light on the growth of a plant. Plant A was the **control**. It was grown in light. Plant B was grown in darkness. The variable being tested was light; therefore, all variables such as soil nutrients, soil type, soil moisture, and temperature, were kept the same in B as in the control A. If this were not done, we would not know for sure what caused the difference in growth of the plants.

Observation and Recording of Results

A scientist learns to record everything that is observed during an experiment. This may be done in a table, graph, diagram, or written paragraph. Such information is called the **data** of the experiment.

Observations may be both qualitative and quantitative. **Qualitative observations** describe such characteristics as colour, odour, shape, taste, and pattern of movement. **Quantitative observations** deal with measurements such as length, volume, mass, density, and temperature. Both types of observations are made by scientists. They usually make qualitative observations first, since they are the easiest to make. However, quantitative observations are often more useful. For example, if you were using the microscope to try to identify an organism, you might notice that it was green, oval in shape, had a whip-like tail and moved in a jerky fashion. These are all qualitative observations. However, you probably could not identify

A   B

Fig. 1-7 In a control experiment, all variables are kept constant except one. In this case, the variable changed was light.

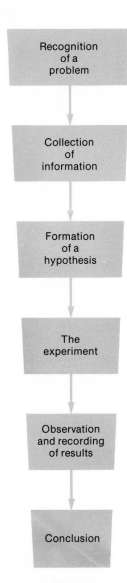

Fig. 1-8 The six main steps that are normally used in the scientific method for solving a problem.

Recognition of a problem

Collection of information

Formation of a hypothesis

The experiment

Observation and recording of results

Conclusion

the organism for sure until you made a quantitative observation such as a measurement of its length. Most identification manuals give the average lengths of organisms.

The Conclusion

Once a scientist has organized the data, she may, through inductive reasoning, make a generalization or **conclusion**. If the original hypothesis was correct, this conclusion should be the same as the hypothesis. If further experiments are done and they support the original conclusion, the hypothesis may be called a **theory**. However, if further experiments do not support the conclusion, the original hypothesis must be changed or discarded and the whole procedure repeated.

Summary of Scientific Method

Figure 1-8 shows the six main processes of science that usually make up the scientific method. Scientific inquiry, or the scientific method, usually begins with the **recognition of a problem**. Of course, this can only happen if a person's curiosity is aroused. In turn, curiosity is only aroused when a person already has some knowledge about the problem. The next step is the **collection of information** about the problem by reading and discussing with others. This is followed by the formation of a **hypothesis**, or prediction, about the solution to the problem. Next comes an **experiment** to test the hypothesis. If the experiment is done with proper controls, **observations** can be made and recorded that may lead to a **conclusion**. If the conclusion is not the same as the hypothesis, the hypothesis must be changed and the entire process repeated.

Review

1. **a)** List the six main steps that are usually part of the scientific method.
   b) Figure 1-3 on page 6 shows four different people who are attempting to solve a problem using the scientific method. Select one person and explain in detail how you think that person might have gone through the six main steps in the scientific method.
2. **a)** What is the difference between a hypothesis (prediction) and a guess?
   b) What values are there in predicting before you do an experiment?
3. **a)** What is a variable in an experiment?
   b) What is a control in an experiment?
   c) Why is it necessary to use controls in an experiment?
4. **a)** What is the difference between a qualitative and a quantitative observation? Illustrate your answer with an example of each.
   b) What is the difference between a hypothesis and a theory?
   c) What is the difference between inductive and deductive reasoning?
5. **a)** Describe how you would conduct a controlled experiment to find out if cats like a new cat food better than other brands.
   b) Describe an experiment to test the hypothesis that moths are attracted only to certain colours of light. Be sure to describe your control.

1.4 Investigation

Using the Scientific Method

In this investigation you use the scientific method to study the growth of bean seedlings. Be sure to read and follow the steps outlined in Section 1.3.

You are to do this experiment at home and submit a report to your teacher about three or four weeks from today. Your report is to be written up under these headings:

1. Problem
2. Background Information
3. Hypothesis
4. Experiment
5. Observations
6. Conclusion
7. Further Studies

Materials

jars (2)
absorbent paper (e.g. blotting paper, paper towels, or newspaper)
absorbent material (e.g. vermiculite or granular styrofoam)
bean seeds (10)

Procedure A Obtaining Bean Seedlings

Before you can begin your experiment, you must germinate some bean seeds to obtain seedlings. Use the set-up of Figure 1-9 and follow these instructions.

a. Soak 10 bean seeds overnight in water.
b. Line a glass jar with absorbent paper such as blotting paper, paper towels, or newspaper.
c. Fill the remainder of the jar with an absorbent material such as vermiculite or granular styrofoam. Torn-up pieces of paper towelling or blotting paper are acceptable if the other materials are not available.
d. Place 5 seeds two thirds of the way up the jar, between the absorbent paper and the wall of the jar.
e. Add water to the jar to moisten the entire contents. However, do not cover the seeds with water.
f. Set up a second jar with identical contents to the first one.
g. Place both jars in a warm location. Add water from time to time, but do not cover the seeds with water. Do not let the contents dry out. Observe them each day until seedlings (small plants) appear. Then move to Procedure B.

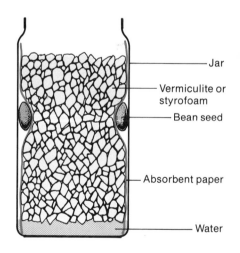

Fig. 1-9 A method of germinating bean seeds.

Procedure B   Using the Scientific Method

Once seedlings have appeared from the seeds in the two jars, use the scientific method to study the effect of one factor, light, on the growth of the seedlings. Since this is the first time you have used the scientific method in this course, we will give you some help.

Problem

You probably know that light is needed for the normal growth of green plants. If you now wonder if light is needed for the normal growth of bean seedlings, you have recognized a problem. The problem may be stated as: *Is light needed for the normal growth of bean seedlings?* Include a statement of this problem in your report.

Background Information

You should now refer to the index at the back of this book to see if you can find any information relating to the problem. Look under "light" and "growth". A visit to the library may also be helpful. Summarize any relevant information under the heading "Background Information" in your report.

Hypothesis

Several hypotheses are possible. However, the most likely one, in view of our past observations on the growth of plants, is: *Light is needed for the normal growth of bean seedlings.* Include this hypothesis in your report.

Experiment

Using the two jars with seedlings, design and carry out an experiment to test the hypothesis. Think about these questions as you do so: What variable is being tested? What other variables could affect the results? How can those variables be kept the same in both jars? How can a proper control be set up? Remember, if a control is not established, you will not know which variable caused any differences that you observe between the two jars.

Write out your procedure and include it in your report under the heading "Experiment".

Observations

Record the data in a table similar to Table 1-1. Continue to note and record data until you have enough information to support or disprove the hypothesis.

	CONTROL JAR					EXPERIMENTAL JAR				
Date	Stem length	Stem diameter	Leaf colour	Number of leaves	Other obser-vations	Stem length	Stem diameter	Leaf colour	Number of leaves	Other obser-vations

TABLE 1-1 **Light and the Growth of Bean Seedlings**

Conclusion

Use your data to draw a conclusion about the effect of light on the normal growth of bean seedlings. Record your conclusion in your report. Compare your conclusions to the original hypothesis.

Further Studies

Most scientific experiments raise more questions than they answer. Now that this experiment is complete, ideas for further problems to study may come to mind. For example, we may wonder what would happen if the positions of the control jar and the experimental jar were interchanged at the end of the experiment and the plants were allowed to continue growing. Make a list of other problems that you have recognized, and for each, state a hypothesis that could be studied. Include the list and the hypothesis in your report under the heading "Further Studies".

Highlights

Science has two aspects. It is both a body of knowledge and a method for discovering new knowledge.

Biology is a vast science that deals with the study of living and once-living things, together with the matter and energy that surround them.

There is no one scientific method that scientists follow step by step. Instead, the scientific method consists of an underlying pattern of logical thought and a certain attitude. However, most scientific studies follow six main steps: recognition of a problem; collection of information; formation of a hypothesis; the experiment; observations; conclusions. You will follow these steps many times in this course. Hopefully, you will also follow them in your daily life.

2

What Is Life?

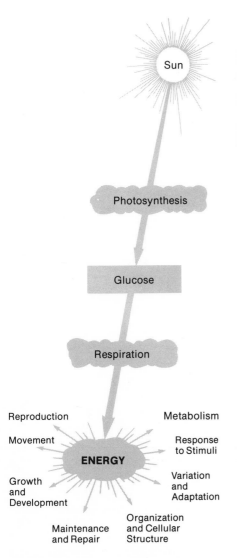

Fig. 2-1 All living things have common characteristics. These characteristics are functions that require energy.

What is life? What is the difference between a *living* and a *nonliving* thing? You would have no trouble deciding that a dog running down the street was alive; nor would you have any trouble deciding that a stone was nonliving. However, if you ask yourself whether a bean seed, an apple, or a potato is living or nonliving, you may have problems deciding on the answer. All these appear just as nonliving as a stone. Yet we know that all three can produce a living plant. Since it seems unlikely that a nonliving thing can produce a living thing, we can assume that the bean seed, apple, and potato are living. What, then, are the characteristics of living things?

2.1 Characteristics of Living Things

This section briefly discusses the main characteristics of living things. The rest of this book explains them more fully and allows you to do some experiments to study them.

1. The Need for Energy

The most important, although not the most obvious, characteristic of all living things is the need for energy. All living things require a continuous supply of energy to support their more obvious characteristics such as movement, growth, and reproduction.

Almost all the energy used by living things comes originally from the sun (Fig. 2-1). Green plants, through photosynthesis, store some of the sun's energy in compounds such as glucose (a sugar). These plants, through respiration, then "burn" or break down the glucose, releasing the energy needed to support their life processes. Animals get their supply of energy by eating the plants or by eating other animals that have eaten plants. By doing this, they obtain glucose and other compounds which they too, break down through respiration to release energy to support their life processes.

2. Movement

One of the most obvious characteristics of living things is movement. If you came across an animal lying on the roadside as a result of being struck by a car, you would probably look for movement of the eyes and chest to see if it was still alive. Most animals show obvious signs of movement when they are alive.

Although movement in plants is not as obvious, it does occur. This movement can be very slow, such as the opening of buds on a tree or the turning of leaves of a plant toward the sun (Fig. 2-2). In contrast, the tiny sundew of northern bogs and the Venus' flytrap of Carolinian bogs show much more rapid motion (Fig. 2-3). One of the most interesting examples of motion in plants is shown by the *Mimosa pudica,* commonly called the Sensitive Plant. If this plant is touched, its leaves quickly fold up (Fig. 2-4).

As you will see later, many animals, plants, and microscopic organisms show few or no outward signs of movement. Yet under the microscope, you can see that the cell contents of these organisms are in continuous motion. This proves that in one way or another, all living things show movement.

Some organisms show a special type of movement called locomotion. **Locomotion** *is the movement of an organism from one place to another.* Most animals can carry out locomotion but very few plants can. Remember that both movement and locomotion, in a biological sense, must be initiated or caused by the organism itself. Locomotion does not occur when the wind blows a plant from one place to another; nor does movement occur when the wind moves the branches of a tree.

3. Cellular Structure and Organization

All living things are made up of cells. Some have only one cell; others have millions of cells. Some cells are very simple and others are very complex. However, from bacteria and amoebas to trees and humans, the cell is the basic unit in which substances are organized to produce a living thing.

Sometimes cells may remain visibly together after an organism dies. Therefore the simple presence of cells is not proof of life. It only shows that the organism was alive at one time. Only living things can organize materials into cells in the first place.

Living cells contain a complex mixture of substances that is called **protoplasm**. This mixture is found only in living cells. The protoplasm itself,

Fig. 2-2   A geranium plant in a window will turn its leaves toward the light.

A   B

Fig. 2-3   The sundew (A) and Venus' fly-trap (B) are insectivorous (insect-eating). The tentacles on the leaves of the sundew quickly bend inward to trap an insect that triggers them. The two lobes of a leaf of the Venus' flytrap move together to trap an insect.

Fig. 2-4 In just seconds a normal *Mimosa* plant (A) will turn limp (B) if it is touched.

however, is not alive. None of the materials of which it is made—carbohydrates, fats, proteins, water, and other compounds—are alive. Yet, living cells have the ability to organize all these materials into what biologists call a **living condition**. Protoplasm differs from one kind of organism to another and even from one individual to another of the same kind. It even differs from one part of an individual to another part of the same individual. In fact, the composition, or makeup, of any particular sample of protoplasm is always changing.

Living things have the ability to organize materials into protoplasm and to organize protoplasm and other substances into cells. Living things are therefore called *organisms* because of this ability to organize substances.

4. Growth and Development

All living things grow at some time during their lives. The total growth may be very small, as in the case of a bacterium or an amoeba. Total growth can also be quite extensive, as in the case of a whale or a large tree. Yet, whether great or small, growth is a characteristic of all living things. However, many nonliving things can also grow. For example, crystals of sugar, salt, and bluestone can be made to grow larger. You probably have seen an icicle grow. How, then, can we say that growth is a characteristic of living things? What kind of growth are we referring to?

The crystals and the icicle grow larger by adding more material of the same kind to their surfaces. The growth of living things is quite different from this. A dog does not grow by the collection of more dog on its surface; nor does a geranium plant grow by the collection of more geranium on its surface. Yet, neither of these organisms grows simply by taking in food. They must organize the food, along with water, minerals, and other chemicals, into the complex materials that make up protoplasm and the other parts of living cells. Living things grow, not by adding more of their own material to their surfaces, but by organizing materials that they take in to form their own special kind of protoplasm.

If you plant a bean seed, it will become a bean plant. It never becomes a potato plant or a tree. It becomes a unique living thing with specialized parts that make it different from other living things. *The series of changes that take place as an organism grows toward its final form is called* **development**. All living things undergo development (Fig. 2-5).

5. Maintenance and Repair

Most living things live long after growth appears to have stopped. Yet, in one sense, they continue to grow as long as they are alive. They may not grow any larger but they must continually maintain and repair the materials of which they are made. For example, skin cells on your body wear away and must be replaced by new ones. A cut on your finger heals because a blood clot covers it with a scab to prevent infection while new tissue is produced to

Fig. 2-5 Stages in the development of a bean plant.

cover the cut. Some organisms, such as the salamander and crayfish, can even grow new limbs to replace lost ones by reorganizing old, and adding new, material.

Living things use a great deal of energy in the maintenance and repair of worn-out and damaged parts. This is a characteristic of all living things.

6. Reproduction

Only living things can produce offspring similar to themselves. Trout lay eggs that hatch and develop into trout; bluebirds lay eggs that eventually produce bluebirds; horses give birth to horses; corn seeds grow into corn plants; and maple seeds become maple trees. *It is a basic law of biology that only life can produce life and like produces like*.

Organisms must be able to reproduce themselves because they have a limited life span. After most organisms are formed, they go through a period of rapid growth. They eventually reach a stage called maturity at which growth in size usually ceases but maintenance and repair may continue. They then enter a period of decline in which maintenance and repair of worn-out and damaged materials is no longer fast enough to keep the organisms in a stable state. Finally, death occurs.

Life spans vary considerably from one type of organism to another. Some insects live only a few weeks. A person in this country can expect to live, on the average, about seventy years. A horse lives about thirty years. Some trees live for a few decades and others for hundreds of years. Some redwood trees in California have lived for several thousand years. Some simple organisms such as bacteria and amoebas appear to have an indefinite life span. In a sense, they live forever, because they reproduce by splitting in two. The offspring repeat this process. Clearly such organisms never die of old age!

Organisms use a great deal of energy in the reproduction of offspring. This also is a characteristic of all living things.

7. Response to Stimuli

All living things are able to respond to certain **stimuli** or changes in their environment (surroundings). A dog comes when you whistle. A fly moves when you try to swat it. A *Mimosa* plant folds its leaves in response to darkness, touch, and heat. A plant in a window turns its leaves toward the light. Earthworms seek out moist soil containing decaying vegetation. In all these examples a **stimulus**—sound, touch, heat, light, moisture—causes a **response** by a living thing. *A living thing's response to a stimulus is called* **irritability**.

Irritability is valuable to animals in many ways. It helps them obtain food and avoid predators. It is most highly developed in those animals that have nervous systems and keen sense organs such as eyes, nose, and ears.

Plants usually respond slowly to stimuli because they lack sense organs, muscles, and other parts needed for a quick response. However, they usually respond to light by turning their leaves toward it. They also respond to gravity by sending roots downward into the soil. Many homeowners have discovered, to their dismay, that poplar and willow trees often respond to the presence of water around a home by clogging the water drains with roots.

Even single-celled organisms such as amoebas show irritability. As you will see later, such organisms respond to touch, light, heat, and other environmental stimuli.

Responses to stimuli must be coordinated if they are to be effective. Even simple organisms have many parts and each part must do the right thing at the right time if the proper response is to be carried out. For example, when you call a dog to supper, stimuli will be received by one or more of the eyes, ears, and nose. The responses to these stimuli must be coordinated within the dog before it can respond properly. Some muscles must contract; others must relax; digestive juices must be secreted (Fig. 2-6). A system of **nerves** and a system of chemical regulators called **hormones** coordinate these responses in a dog and many other animals. In plants, only hormones are involved in the coordination of responses.

Organisms respond to stimuli by changing their relationship to it. For example, a dog usually comes when you whistle. It changes its location in response to the stimulus. *Such responses, which often occur in definite patterns, are called* **behaviour**. Remember that behaviour must begin with the organism. A ball rolling down a hill is not showing behaviour. It is simply being pulled along by the force of gravity. However, a dog that responds to a whistle creates a change in its relationship to its environment. Your whistle does not pull the dog to you.

Organisms use a considerable amount of energy as they respond to stimuli within their environments.

8. Variation and Adaptation

As you have just learned, organisms show irritability; they respond to changes in their environment. These changes include temperature, moisture, light, availability of food and the presence of disease organisms. Often the changes are sudden. For example, a storm might bring unusual amounts of water to an area. In this case an animal might respond by seeking shelter on higher ground. If its food supply in an area is suddenly destroyed by fire, an animal might respond by moving to a new area.

In contrast, other changes occur over a long period of time. The climate of an area may gradually change over thousands of years, or soil may gradually build up on a rocky landscape. Some organisms may no longer be able to survive in that area because they are not suited to the new conditions. They can either leave the area, or they can change so that they are suited to the new conditions. The first option can be done quickly. However, basic change in an organism can occur only over many generations.

Change occurs as a result of a characteristic called **variation**. Offspring always differ in some ways from one another and from their parents. These differences are called variations. Most variations do not affect an organism's chances of survival. For example, the fact that your hair is a different colour from your parents' will not likely affect your chances of survival. However, now and then a variation occurs that does give an organism a better chance of surviving in a changing environment. Suppose that the climate of an area

Fig. 2-7 What adaptations do these organisms show? How would each adaptation aid survival?

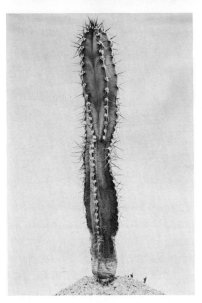

is changing and deeper snow is produced each winter. Clearly a variation that produced longer legs in a deer would increase that deer's chances of surviving in that area. If this variation is passed on to the offspring of that deer, they, also, would have an increased chance of survival. Gradually the only deer of that type to be found in the deep snow area may be the long-legged types. The others would have moved away or died. *The process by which a certain type of organism becomes better suited to survive in its environment is called* **adaptation**.

Keep in mind that organisms do not change in order to survive in a changing environment. The deer in our example did not grow long legs because they needed them to survive in the deep snow. Organisms do not change to survive; they survive because they change.

Variation and adaptation are characteristics of most living things (Fig. 2-7). The exceptions occur for those organisms that reproduce asexually (see Chapter 8).

9. Metabolism

Metabolism *is the exchange of matter and energy between an organism and its environment, and the changes that occur in this matter and energy when they are within the organism.* In effect, metabolism is the sum of all the processes occurring in an organism. It includes taking in food, or **ingestion**, as well as taking in water and air. It also includes all the changes in food materials that occur within organisms during **digestion**. It includes all changes that occur as the products of digestion are **assimilated**, or put together, during growth, maintenance, and repair. Finally, metabolism includes the release of energy through **respiration** and the elimination of by-products through **excretion**.

A horse is always exchanging materials with its surroundings. It consumes vegetation and water. It returns to the environment fecal matter and urine. It breathes in air and breathes out air that contains less oxygen and more carbon dioxide. All living things, like the horse, take in substances from the environment and return substances to the environment. However, the substances returned are not the same as those taken in. The air breathed out is not the same as the air breathed in; the fecal matter and urine contain substances that were not present in the original food. While the materials were inside the horse, they underwent great changes. As this happened, energy was released. This energy is used to drive the other life processes. However, much of it eventually ends up as heat which keeps the horse's body warm.

Metabolism has two distinct phases, anabolism and catabolism. **Anabolism** *is a constructive or building-up phase.* It includes **assimilation**, or the building of protoplasm from simple compounds and elements that were obtained as a result of ingestion and digestion. It also includes the process of photosynthesis. **Catabolism** *is a destructive or breaking-down phase.* It involves the release of energy by the breakdown of food materials through respiration.

Review

1. **a)** Why do all living things require energy?
   b) Energy is needed for you to move your hand. Show that this energy originally came from the sun.
2. **a)** Describe an example of movement in a plant that is not described on page 15 of this text.
   b) What is locomotion? Name an animal that has movement but not locomotion.
3. **a)** What is protoplasm?
   b) Why is a living thing called an organism?
4. **a)** Describe the difference between the growth of a living and a nonliving thing.
   b) What is development?
   c) What is the difference between development and growth?
5. **a)** How do maintenance and repair differ from growth?

Fig. 2-8   What important adaptations does a white-tailed deer show?

b) A salamander and a crayfish can grow new limbs but humans cannot. Why do you suppose this is so?

6. a) What is reproduction?
   b) Why is reproduction necessary?

7. a) What is a stimulus?
   b) What is irritability?
   c) Describe an example of the importance of irritability to an animal.
   d) Describe an example of the importance of irritability to a plant.
   e) What is behaviour? Use an example to illustrate your answer.

8. a) What are variations?
   b) What is adaptation?
   c) Explain the relationship of variations to adaptations.

9. a) White-tailed deer do have long legs (Fig. 2-8). Our discussion on

page 20 stated that this adaptation is related to survival in deep snow. What is a more likely explanation for the fact that all white-tailed deer show this adaptation?

    b) Look back to Figure 2-7 on page 20. Answer the questions in the caption.

10. a) What is metabolism?

    b) Name and define the two phases of metabolism.

    c) Name and briefly describe five processes that can occur during metabolism.

    d) Why is metabolism a characteristic of all living things?

2.2 Biogenesis and Spontaneous Generation

Biogenesis

Only life can produce life. This is a basic principle of biology called the **Theory of Biogenesis**. This theory probably seems obvious to you. All living things seem to come from a living parent or parents. But does life ever come from nonliving matter? Has it ever done so in the past? Might it do so some time in the future when conditions on the earth are different? We do not know the answers to these questions. However, to the best of our knowledge, life comes only from life. The principle of biogenesis is usually called a theory because we are reasonably sure but not absolutely certain that it is true. Because of this uncertainty, the theory of biogenesis would be better stated as: *Under the present conditions on earth and at the present time, only life can produce life.*

Spontaneous Generation

Although the theory of biogenesis may seem obvious to you, it is relatively new. Until about one hundred years ago, most people, including many scientists, believed that some nonliving things could change directly into living things. This concept is called **spontaneous generation**. Some of the beliefs associated with spontaneous generation are quite interesting. The following are a few of them: "Eels are made out of the ooze at the bottom of rivers, lakes, and oceans." "Many kinds of worms are produced from horse hair in water." "Frogs and fish are formed in clouds and drop to the earth with the rain." "Maggots (fly larvae) are produced from decaying meat." Some people even believed that adult flies were made directly from decaying animal bodies. One of the most interesting statements about spontaneous generation came from a Belgian scientist named Johann van Helmont. About three hundred years ago, he stated that if you placed a dirty shirt in a container of wheat, mice would be produced in 21 d. What do you suppose led van Helmont to this strange conclusion?

Since the early scientists knew little about the life cycles of most organisms, their conclusions are not surprising. However, during the 17th,

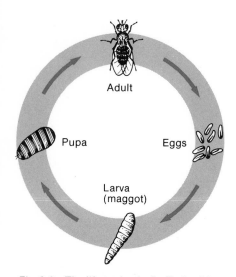

Fig. 2-9 The life cycle of a fly. Today it is well-known that maggots develop from eggs that are laid by an adult fly.

18th, and 19th centuries some scientists began to question the theory of spontaneous generation. Naturally, other scientists defended it. Both groups of scientists made hypotheses related to the problem of spontaneous generation and experimented to test their hypotheses. The rest of this section presents some of this work.

Redi's Investigations

Francesco Redi, a 17th century Italian scientist, was the first person to seriously challenge the theory of spontaneous generation. He felt that spontaneous generation was not possible and decided to do experiments to prove this. He focused his attention on the belief that maggots were formed directly from decaying meat. He had noticed flies hovering over decaying meat. He therefore hypothesized that the maggots came from eggs laid by the flies instead of directly from the decaying meat (Fig. 2-9). The meat, he reasoned, was simply food for the maggots.

To test his hypothesis, Redi had to prove that no maggots would form if adult flies were kept away from decaying meat. In 1668 he tested his hypothesis with the following experiment. He placed pieces of veal, snake, eel, and fish each in separate clean jars and left the jars open (Fig. 2-10,A). He then prepared an identical set of four jars which he sealed tightly (Fig. 2-10,B). After a few days, the open jars were swarming with adult flies and all four types of meat contained maggots. However, the sealed jars contained no maggots, and even after several weeks, they still contained only decaying meat. Therefore, Redi concluded that his hypothesis was correct.

This did not convince Redi's opponents however. They were quick to point out that air was needed to make maggots from decaying meat. They felt that sealed jars had kept the air out and prevented the maggots from forming. (Today we might have accused Redi of having a poor control in his experiment.) Perhaps air was needed to produce maggots from decaying meat. Redi soon added another step to his experiment. He set up four jars similar to the first two sets. This time, however, he covered the jars with a

Fig. 2-10 Redi's experiment to disprove the theory of spontaneous generation.

A. Open jars

Flies present

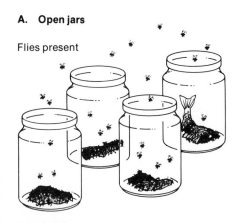

Maggots on meat

B. Sealed jars

No flies

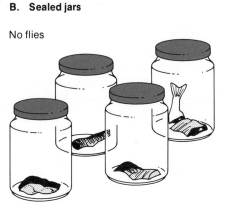

No maggots on meat

C. Cloth-covered jars

Flies present

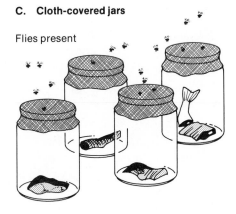

No maggots on meat

fine mesh cloth that would let air in but not flies (Fig. 2-10,C). Adult flies soon hovered over these jars. They even laid eggs on top of the cloth. But no maggots appeared in the decaying meat.

Redi went one step further and isolated some maggots in a separate container where he observed their development. He showed that maggots were a part of the life cycle of the fly (Fig. 2-9). In spite of all this evidence, supporters of the spontaneous generation theory remained largely unconvinced. Most agreed that Redi had proven that flies came only from flies. But they were unwilling to assume that the theory of biogenesis applied to all other organisms as well. Would you have felt the same way if you had been present then?

Needham's Investigation

After Hooke, van Leeuwenhoek, and others invented the microscope in the 17th century, the world of microscopic organisms, or **microorganisms**, was opened up to scientists. In addition to being one of the most important things to ever happen in biology, the microscope also helped supporters of spontaneous generation with their research. Under the microscope they observed that various mixtures such as meat broth and sugar solution contained numerous microorganisms such as bacteria, yeasts, and moulds. Even mixtures that were free of microorganisms at one time would have them later. Where did they come from?

Many biologists, including an Irish scientist, John Needham, hypothesized that spontaneous generation was occurring in these mixtures. Needham performed an experiment to prove that this hypothesis was correct. He began the experiment with an infusion of mutton gravy and various seeds. (An **infusion** is simply any mixture rich in nutrients.) He boiled the infusion for several minutes in flasks in order to kill any microorganisms that might be present. He then sealed the flasks with corks. Several days later he examined the infusion with a microscope and observed countless microorganisms. He repeated the experiment using different infusions. Each time microorganisms were observed. He therefore concluded that since the flasks were sealed, the microorganisms came from the material of the infusion by spontaneous generation.

Spallanzani's Investigations

The result of Needham's experiments were very convincing. In fact, about 25 years passed before Lazzaro Spallanzani, an Italian scientist, challenged Needham's work. Spallanzani did not believe in spontaneous generation. He felt that Needham had either not sealed the flasks tightly enough, thereby allowing microorganisms to enter, or that he had not boiled the flasks long enough to kill all the microorganisms originally present in the infusion. To prove this point, Spallanzani did a series of experiments similar to Needham's. He made an infusion by boiling some seeds in water. He then

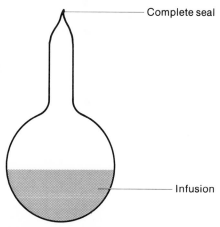

Complete seal

Infusion

Fig. 2-11 Spallanzani sealed his flasks completely by melting the glass together at the top.

sealed this infusion in flasks by melting the tops (as shown in Fig. 2-11). He then placed the flasks in boiling water for about 1 h to kill any microorganisms present. Spallanzani opened the flasks several days later and found no microorganisms. He reasoned that the complete seal had kept out any microorganisms and that the heating had killed those originally present. He concluded that microorganisms in infusions were not formed by spontaneous generation.

Supporters of spontaneous generation were not satisfied. Some claimed that the boiling destroyed an active substance in the infusion that was needed for spontaneous generation. Others claimed that air was needed for spontaneous generation. Therefore, Spallanzani proceeded to prove that boiling does not destroy an active substance. He boiled one set of loosely sealed flasks for 0.5 h, another for 1 h, a third for 1.5 h, and a fourth for 2 h. If the heat did destroy an active substance, fewer microorganisms would be present in the flask that was boiled the longest. However, microorganisms appeared in all four flasks (because they were loosely sealed) and in fact, more were present in the flask that was boiled longest. Clearly the heat did not destroy an active substance. On the contrary, it made the infusions even better suited for the growth of microorganisms. In spite of this excellent work, spontaneous generation was still accepted by many because Spallanzani had not allowed air into his flasks. About 50 years passed before Louis Pasteur dealt the theory of spontaneous generation its final blow.

Review

1. **a)** State the theory of biogenesis.
   b) What is meant by spontaneous generation?
2. **a)** Explain why Redi's first experiment did not convince his opponents.
   b) Which variable in Redi's first experiment was defective? In modern scientific terms, how would you describe this error?
   c) How did Redi modify his first experiment? Did his modified experiment convince his opponents that spontaneous generation did not occur?
3. **a)** What led Needham to assume that microorganisms were spontaneously generated from nonliving material?
   b) What is an infusion? Why did Needham use an infusion?
   c) Did Needham's results support his hypothesis? Explain.
4. **a)** Why did Spallanzani perform experiments similar to Needham's?
   b) Why was Spallanzani unable to change the opinions of supporters of spontaneous generation?

2.3 Pasteur's Investigations

Lazzaro Spallanzani showed that an infusion does not contain a substance necessary for spontaneous generation which is destroyed by boiling. Yet he was unable to convince supporters of spontaneous generation because he had kept air out of the flasks. Many of these supporters felt that air was needed for spontaneous generation. However, 50 years after Spallanzani did his work, Louis Pasteur, a famous 19th century French scientist, performed experiments that finally destroyed the theory of spontaneous generation.

In this section we describe Pasteur's experiments. You will then be asked to prepare a scientific paper to convince people that spontaneous generation does not occur. Study the information carefully and then follow the instructions in the "Review" section.

Background Information

Pasteur was studying the fermentation of sugars to produce alcohol. While doing so, he discovered that yeasts, bacteria, and other microorganisms were present in fermenting mixtures such as grape juice. He suspected that these microorganisms came from the air. He reasoned that they were present in the air in an inactive state. That stage is now called **spores**. Pasteur thought that when these spores got into a nutrient-rich medium (environment) such as an infusion or fruit juice, they became active and began growth, development, and reproduction.

To disprove spontaneous generation, Pasteur would have to keep these tiny spores out of his infusions. However, at the same time, he would have to let air in.

Pasteur's First Experiment

Pasteur reasoned that, if there were spores in the air, air at higher altitudes would contain fewer spores than air at lower altitudes. To test this idea, he prepared several infusions that would support the growth of yeasts, bacteria, and other microorganisms. He sealed each infusion in a flask and boiled it to kill all microorganisms. He then opened some flasks on a mountaintop, others by a dusty road, and still others in a variety of locations. When he examined the infusions a few days later, he discovered that those that were opened on the mountaintop contained much less life than the others. Some, in fact, contained no microorganisms at all. He had let air in but, in some cases, no microorganisms developed. He was now confident that the supporters of spontaneous generation were wrong and that his hypothesis was correct. Microorganisms did not come from the infusion with the help of air. Instead, they entered the infusion from the air as spores. He now made preparations to do a more controlled experiment in his laboratory.

Pasteur's Controlled Experiment

Follow the sequence of diagrams in Figure 2-12 as you read the following description.

Pasteur prepared a mixture of sugar solution and yeast and poured it

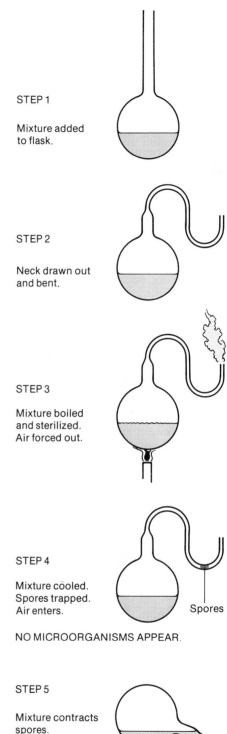

STEP 1

Mixture added to flask.

STEP 2

Neck drawn out and bent.

STEP 3

Mixture boiled and sterilized. Air forced out.

STEP 4

Mixture cooled. Spores trapped. Air enters.

Spores

NO MICROORGANISMS APPEAR.

STEP 5

Mixture contracts spores.

MICROORGANISMS APPEAR.

Fig. 2-12 The steps in Pasteur's experiment which proved that the theory of spontaneous generation was incorrect.

into several long-necked flasks. (Yeast is a microorganism that reproduces rapidly in sugar solution.) Next, he heated the neck of each flask and bent it into an S-shape. He then boiled each mixture for several minutes. The boiling killed any microorganisms in the mixture, and the steam killed the microorganisms on the walls of the flasks. The steam also drove the air out of the flasks. When the flasks cooled, air came back into them. However, dust and spores remained in the trap that was formed by the S-shape. Pasteur reasoned that, with this arrangement, air could always enter but the mixture would remain **sterile**, or free from microorganisms, indefinitely. And that is exactly what happened!

Pasteur went one step further. He tipped each flask over so that some of the liquid went up into the S-tube where the spores were trapped. A few days later all the mixtures contained countless microorganisms. Clearly boiling had not destroyed a substance in the mixture that was needed for spontaneous generation of microorganisms. Air had been allowed into the flasks so it, apparently, did not promote spontaneous generation. This step finally proved that the microorganisms came from the air. After two thousand years, the theory of spontaneous generation had been put to rest.

Review

Imagine that you are Louis Pasteur in the 19th century. You are an opponent of the spontaneous generation theory and you have completed the controlled experiment just described. Prepare a scientific paper that will convince its readers that there is no such thing as spontaneous generation. The following guidelines should help you with your report.

Problem

Begin your paper with a clear description of the problem that started you on this experiment.

Background information

Make a list of the related information that was known at that time.

Hypothesis

Using the results of Pasteur's first experiment and the background information, make a hypothesis concerning spontaneous generation. Your hypothesis should be a prediction of the results of the experiment. It should give you a goal to work toward.

Experiment

Describe in an organized way the procedure used in the experiment.

Observations

Record the results in a systematic manner.

Conclusion

Draw a conclusion based on the observations. Compare it to the hypothesis. Make a final statement regarding spontaneous generation.

Highlights

Living things, or organisms, have many common characteristics:

1. The need for energy;
2. Movement;
3. Cellular structure and organization;
4. Growth and development;
5. Maintenance and repair;
6. Reproduction;
7. Response to stimuli, or irritability;
8. Variation and adaptation;
9. Metabolism—ingestion, digestion, assimilation, respiration, and excretion.

The theory of spontaneous generation states that living things can be formed directly from nonliving things. Redi proved that this theory did not apply to the formation of maggots. Needham attempted to show that it did apply to the formation of microorganisms in infusions. Spallanzani proved, in part, that Needham was wrong. However, it was not until the 19th century that Pasteur finally proved that spontaneous generation did not occur.

Biologists today accept the theory of biogenesis, which states that, under the present conditions on earth and at the present time, only life can produce life.

Classification Of Living Things

Biologists believe that there may be over 2 000 000 different kinds of organisms (Fig. 3-1). Already over 1 500 000 different kinds have been identified and new ones are still being discovered. One biologist estimates that for each kind of organism now alive, another 400 kinds once lived but have since become extinct. Therefore, as many as 1 000 000 000 (one billion) different kinds of living things may have existed on the earth at one time or another!

How can we keep track of such a bewildering number of organisms? How can we even name the organisms now alive when no known language has 2 000 000 words in it? Biologists have answers to these questions. This chapter describes how they name living things and how they keep track of them through a classification system.

3.1 Introduction to Biological Classification

What is Classification?

Whenever we work with a large number and variety of things, we usually sort them into groups. Each group contains those things that are similar to

one another. We may then separate each of those groups into smaller groups that are even more alike. For example, if you have ever collected hockey cards, you may have placed all the Montreal Canadiens together, all the Vancouver Canucks together, all the Toronto Maple Leafs together, and so on. You may have gone even one step further and placed the forwards of each team together and the defencemen of each team together. You probably did all this without anyone telling you to do so. Why do humans behave this way? Scientists who study human behaviour say that we group similar things because we are bothered by confusing assortments. We seem to instinctively want to group similar things.

The grouping of similar things for a specific purpose is called **classification**. Although it may be instinctive for humans to classify things, there are also practical reasons for doing this. For example, a supermarket manager classifies the foods in her store by storing all the cereals together, all the meats together, all the cookies together, and so on. Stamp collectors classify their stamps. They may place all the Canadian stamps in one book and all the American stamps in another. The words in a dictionary are classified by alphabetical listings. Clearly, we classify things to make it easier to keep track of what we have, and to find particular items.

Early Biological Classification

Biologists have long recognized the need to classify living things. In fact, humans have been classifying living things for thousands of years. The earliest humans probably classified organisms as plant or animal. They may have further classified the plants as edible or poisonous and the animals as harmful or harmless. However, it was 300 B.C. before the first serious attempt was made to classify all the organisms known. This attempt was made by the Greek philosopher and scientist, Aristotle, and his students.

Fig. 3-1 How many different kinds of organisms can you see in this illustration? Name some kinds of organisms that are probably present but cannot be seen.

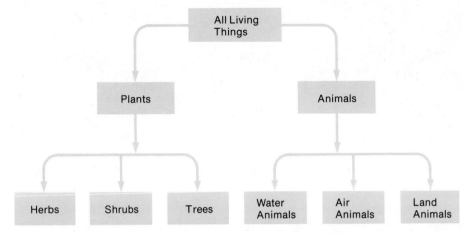

Fig. 3-2 Aristotle classified animals according to where they lived and plants according to the structure of their stems.

Since only about 1 000 kinds of organisms were known at that time, a very simple classification scheme could be used. Aristotle and his students first classified the organisms as plant or animal. They then classified the animals according to where they lived. This resulted in three groupings: air animals, water animals, and land animals. They classified the plants according to the structure of their stems. Those with soft stems were called herbs; those with a single woody stem were called trees; and those with many small woody stems were called shrubs (Fig. 3-2).

Aristotle's classification system survived for almost two thousand years. However, by the beginning of the 18th century, over 10 000 kinds of organisms were known and Aristotle's system was unable to classify them all. Many of the newly discovered organisms would not fit into any category of Aristotle's simple system. A new system was obviously needed.

Review

1. a) What is classification?
   b) Give three reasons why humans classify things.
2. a) How do librarians classify books? Why do they do this?
   b) How might coin collectors classify their coins? Why might they do this?
3. a) Make a list of all the examples of classification that you can find in your kitchen.
   b) List five other examples of classification found in your home.
4. a) Describe Aristotle's classification system for living things.
   b) Where in Aristotle's classification system would you place a seedling tree? Why?
   c) Where in Aristotle's classification system would you place each of the following animals: a frog, a penguin, a diving duck, a beaver? Be prepared to defend your answer.
   d) What faults do you see in Aristotle's method of classifying plants?
   e) What faults do you see in Aristotle's method of classifying animals?

3.2 Modern Biological Classification

Taxonomy *is the science that deals with the classification of organisms.* In this section you will see how this science began with the work of Linnaeus and developed into the system we use today.

The Contribution of Linnaeus

Carolus Linnaeus (1707–1778), a Swedish botanist, developed a simple classification system that forms the basis of our modern method of classification. During Linnaeus' lifetime, scientific discoveries were taking place at a rapid rate. At the start of the 18th century about 10 000 kinds of organisms were known. By the end of that century over 70 000 kinds were known. Linnaeus tried to develop a classification system for this large number of organisms. By 1753 his system was well developed and modern taxonomy began.

The basis for classification

When we think about the living things around us, we are first of all impressed by the diversity that exists. (Biologists use the word **diversity** to mean *differences, or the number of kinds of living things.*) There seem to be so many kinds of living things and they seem to be so different from one another. Yet, if we study them closely, we can see many likenesses. For example, at first glance lions, horses, humans, and mice seem to have little in common. A closer look however, shows that all have hair, a distinct head, four limbs, two eyes, two ears, and warm blood (Fig. 3-3). That is, they have similar **structural features**.

Linnaeus decided to use structural features as the basis for his classification system. Therefore, he grouped organisms according to their structural similarities. Those organisms with very similar structural features were considered to be the same **species**. Thus all modern-day humans belong to one species, all house cats belong to one species, and all sugar maple trees belong to one species.

Binomial Nomenclature

Once Linnaeus had decided on a basis for classifying organisms, he then developed a system for naming them. His system is quite simple. He gave each species a name that consisted of two words. This system is called **binomial nomenclature**. He used Latin words for these names because all scientists wrote in Latin in the time of Linnaeus. Thus, the human is *Homo sapiens,* the domestic (house) cat is *Felis domesticus,* and the white pine is *Pinus strobus.* The first word of each name is called the **genus** and the second word is called the **species**. Notice that the genus begins with a capital letter and the species does not. Notice also, that the genus and species are either

Fig. 3-3 These organisms have many similar structural features. Try to list five structural features that are not common to all four animals.

Fig. 3-4 Some species of maple. Note that all maples are members of the same genus, *Acer*.

Sugar maple
Acer saccharum

Silver maple
Acer saccharinum

Manitoba maple
Acer negundo

Red maple
Acer rubrum

Mountain maple
Acer spicatum

Striped maple
Acer pensylvanicum

printed in *italics* or <u>underlined.</u> Thus *Homo sapiens* and <u>Homo sapiens</u> are the correct forms for writing the scientific name of humans.

The genus concept

A **genus** (plural **genera**) groups species that are similar. For example, all maple trees belong to the genus *Acer.* You can see from Figure 3-4 that the leaves of the species in this genus are similar but not identical. Figure 3-5 shows three animals that are in the same genus, *Canis.* These animals are similar, but different enough that they are not the same species.

In the scientific naming of organisms, a generic name can be used only once. Thus *Acer* is used for nothing else except the maples and *Canis* is used for nothing else except a group of dog-like animals.

Every genus has characteristics that make it stand out clearly from other living things. You are not nearly as likely to confuse one genus with another, as one species with another. For example, you may have trouble telling the species *Acer rubrum* (red maple) from the species *Acer saccharum* (sugar maple). However it is unlikely that you would confuse any member of the maple genus, *Acer,* with a tree in another genus such as the oak genus, *Quercus,* or the walnut genus, *Juglans* (see Fig. 3-6). In the same way, you

German shepherd
Canis familiaris

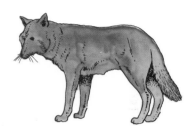

Coyote
Canis latrans

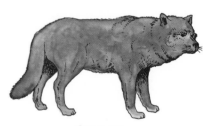

Gray wolf
Canis lupus

Fig. 3-5   These species have enough in common that they are classified in the same genus, *Canis*.

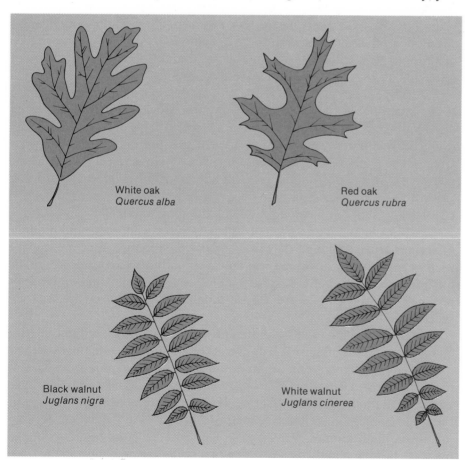

White oak
Quercus alba

Red oak
Quercus rubra

Black walnut
Juglans nigra

White walnut
Juglans cinerea

Fig. 3-6   Species that are in the same genus have many things in common.

may have trouble telling the species in the genus *Canis* apart, but you would not confuse them with members of the genus *Felis,* which includes the house cat, lion, and cougar.

The species concept

Linnaeus grouped as a species those organisms which he felt were very similar in structural features. But just how similar must the organisms be in order to be classified as the same species? Look at Figure 3-7. All the dogs in Group A have the same species name *familiaris.* Yet they are different in many ways. The thoroughbred and Shetland pony in Group B are in the same species but the donkey and zebra are not. Does not the donkey look as much like the thoroughbred as the Shetland pony does? The bears in Group C look similar in structure, yet they are not the same species either! The cottontail rabbit and jack rabbit of Group D also look alike, but they are even in different genera! Appearances can be deceiving. Clearly something more than similar structural features is needed to classify organisms into species.

Today a **species** (plural also **species**) is defined as *a group of individuals that are alike in many ways and interbreed under natural conditions to produce fertile offspring.* **Fertile** means that the animals are capable of producing offspring. This means that members of the same species can, if left alone in their natural environment, mate with one another to produce offspring which, in turn, are also able to produce offspring. Members of different species cannot fulfill all these conditions.

All dogs are in the same species because they can interbreed to produce fertile offspring. We have all seen the mongrel dogs that are formed when different breeds mate. These mongrels are still capable of having offspring. (Sometimes differences in size can prevent interbreeding. For example, a Great Dane and a Pekingese cannot interbreed. However, if a Great Dane interbreeds with a dog half its size and a Pekingese interbreeds with a dog twice its size, the resulting mongrels will be able to interbreed. Thus the Great Dane and Pekingese are listed as the same species.)

A thoroughbred and a zebra will not, in the wild, mate and produce offspring. Thus, although they look similar in structure, they are not the same species. A thoroughbred will however interbreed with a donkey. The offspring is called a mule. A mule is almost always **sterile**; that is, it cannot produce offspring. Therefore, the thoroughbred and donkey are separate species.

Sometimes individuals of different species will interbreed in captivity and produce fertile offspring. However, this seldom, if ever, happens in their natural environment. As an example, the polar bear, *Ursus maritimus,* and the brown bear, *Ursus arctos,* are different species. Yet, in a zoo, they have interbred to produce fertile offspring. However, they appear not to do so in the wild. The reason is simple. The brown bear lives in the woods, feeding mainly on small animals, fruits, and stream fish. The polar bear lives on the polar ice feeding mainly on seals. The two species seldom meet in the wild.

GROUP A	GROUP B	GROUP C

GROUP A

Collie
Canis familiaris

Irish terrier
Canis familiaris

Whippet
Canis familiaris

Boston bull terrier
Canis familiaris

GROUP B

Thoroughbred
Equus caballus

Shetland pony
Equus caballus

Donkey
Equus asinus

Zebra
Equus burchelli

GROUP C

Black bear
Ursus americanus

Grizzly bear
Ursus horribilis

Polar bear
Ursus maritimus

GROUP D

Cottontail rabbit
Sylvilagus floridanus

Jack rabbit
Lepus californicus

Fig. 3-7 Can organisms be classified as the same species just because they have similar structural features?

Cougar
Felis concolor

Jaguar
Felis onca

Tiger
Felis tigris

Fig. 3-8 Three members of the genus *Felis*.

You learned earlier that a generic name, or genus, can be used only once. It always refers to just one group of closely related species. However, a species name can be used often. Table 3-1 lists several organisms that have the same species name, *canadensis*. As you can see, some of these have little in common. Clearly then, it is necessary to include both the genus and the species to properly name an organism. The generic name may be abbreviated to a single capital letter if there is no danger of confusion.

COMMON NAME	SCIENTIFIC NAME	ABBREVIATED SCIENTIFIC NAME
Elk	*Cervus canadensis*	*C. canadensis*
Lynx	*Lynx canadensis*	*L. canadensis*
Beaver	*Castor canadensis*	*C. canadensis*
Bighorn sheep	*Ovis canadensis*	*O. canadensis*
Canada goose	*Branta canadensis*	*B. canadensis*
Canada waterweed	*Anacharis canadensis*	*A. canadensis*
Wild columbine	*Aquilegia canadensis*	*A. canadensis*

TABLE 3-1 Some Organisms with the Species Name Canadensis

Why Use Scientific Names?

You may wonder why biologists use Latin scientific names instead of common names. One reason is that common names can be confusing. Even in the same country, people often call an animal by different names. For example, *Felis concolor* (Fig. 3-8, A) is called a cougar, mountain lion, puma, panther, painter, and many other names. The jaguar (Fig. 3-8, B) is called "el tigre" in Spanish. This sounds very much like the English word "tiger", although it refers to a different animal (Fig. 3-8, C). There is no confusion if we call the jaguar *Felis onca* and the Asian tiger *Felis tigris*. The common name for a domestic cow is "la vache" in French, "die Kuh" in German, and "la vaca" in Spanish. However in all languages the scientific name is the same, *Bos taurus*. A common sparrow is called the House sparrow in Britain, and the English sparrow in the United States. In France it is "Moineau domestique", in Spain it is "Gorrion", in Portugal it is "Pardal", in Holland it is "Musch", and in Sweden it is "Hussparf". In Canada, where there are two official languages it is called either the House sparrow, the English sparrow, or "le Moineau domestique". To avoid confusion, biologists everywhere call it *Passer domesticus*.

Even books written in Chinese, Japanese, and Russian use the Latin alphabet for scientific names. Binomial nomenclature is the universal scientific language.

Scientific names need not be used all the time. However, they must be used when it is important to let others know exactly what species you mean.

The Main Classification Groups (Taxa)

Species and genus

You have seen that similar species are grouped to form a genus. For example, the three dog-like animals in Figure 3-5 are in the same genus, *Canis*. The three cat-like animals in Figure 3-8 are in the same genus, *Felis*. **Species** and **genus** are, however, only two of the main classification groups or **taxa** (singular: **taxon**).

Red fox
Vulpes fulva

Family

Just as similar species are grouped to form a genus, similar genera are grouped to form a taxon called **family**. The red fox, *Vulpes fulva,* and the Arctic fox, *Alopex lagopus,* look like dogs (Fig. 3-9). However, they are less like dogs than the coyotes and wolves. Thus, they are not in the genus *Canis*. However, they are enough like dogs to be placed in the same family, Canidae. In a similar manner, the Canada lynx, *Lynx canadensis,* is in the same family as the animals of the genus *Felis* (see Fig. 3-8). That family is called Felidae. Various bears make up the family Ursidae.

Arctic fox
Alopex lagopus

Fig. 3-9   The red fox, *Vulpes fulva*, and the Arctic fox, *Alopex lagopus*, are in the same family, Canidae, as are the members of the genus *Canis* (see Fig. 3-5).

Order

Similar families are grouped to form a taxon called **order**. You can probably think of several ways in which the bear family, Ursidae, the dog family, Canidae, and the cat family, Felidae, are similar. They are grouped in the order Carnivora (flesh-eaters). Rodents (squirrels, rats, mice, porcupines, beavers, and other gnawing mammals) are in the order Rodentia. Humans, apes, and monkeys, though in different families, are enough alike to be in the same order, Primates.

Class

Similar orders are grouped to form a taxon called **class**. Rodents (order Rodentia), carnivores (order Carnivora), and primates (order Primates), at first glance, seem to have little in common. However, they have enough in common to be grouped in the same class, Mammalia. One similarity is that females of all three orders produce milk for their young. Also, members of all three orders have skin with hair. Other similarities will be mentioned later in this text.

Phylum or division

Similar classes are grouped to form a taxon called **phylum** or **division**. (Zoologists favour *phylum* and botanists favour *division*.) Members of the class Mammalia, to which we belong, have a few things in common with members of the class Aves (birds), class Osteichthyes (fish), class Reptilia (reptiles), and class Amphibia (amphibians). Thus, these five classes are grouped in the phylum Chordata. The name of the phylum suggests one thing that these classes have in common. What is it?

Kingdom

Finally, all the phyla that contain animals are grouped in the **kingdom** Animalia. Members of these phyla have little in common. We, as members of the phylum Chordata, have little in common with grasshoppers of the phylum Arthropoda or clams of the phylum Mollusca. Yet chordates, arthropods, and molluscs are all in the kingdom Animalia because they have more in common with one another than with maple trees of the division Anthophyta or seaweed of the division Phaeophyta. That is why the latter two divisions are placed in a different kingdom, Plantae.

Summary

There are seven main taxa or classification groups:

Kingdom
Phylum or Division
Class
Order
Family
Genus
Species

As you follow the sequence from kingdom to species, you find organisms with more and more in common. Study Table 3-2 closely to make sure that you understand the relationships of the taxa.

Remember that the genus and species names are in italics and that the genus begins with a capital letter and the species with a small letter. All other taxa begin with a capital letter and are not in italics.

This system of classification can be compared to a tree. Many leaves (species) are on a tiny twig (genus). Several tiny twigs (genera) are on a larger

TAXON	HUMAN	GORILLA	DOG	GOLDEN EAGLE	HOUSEFLY	RED OAK	AMOEBA
Kingdom	Animalia	Animalia	Animalia	Animalia	Animalia	Plantae	Protista
Phylum or Division	Chordata	Chordata	Chordata	Chordata	Arthropoda	Anthophyta	Sarcodina
Class	Mammalia	Mammalia	Mammalia	Aves	Insecta	Angiospermae	Rhizopoda
Order	Primates	Primates	Carnivora	Falconiformes	Diptera	Fagales	Amoebida
Family	Hominidae	Pongidae	Canidae	Accipitridae	Muscidae	Fagaceae	Amoebidae
Genus	*Homo*	*Gorilla*	*Canis*	*Aquila*	*Musca*	*Quercus*	*Amoeba*
Species	*sapiens*	*gorilla*	*familiaris*	*chrysaetos*	*domestica*	*rubra*	*proteus*

Generally becoming less like humans >

TABLE 3-2 Classification Of Some Organisms

twig (family). Several larger twigs (families) are on a little branch (order). Some little branches (orders) are on a larger branch (class). Some larger branches (classes) are on a main limb of the tree (phylum). The few main limbs (phyla) make up the whole tree (kingdom).

More Taxa

As more and more information is collected about organisms, biologists need more than the seven main taxa. For example, there are more than 100 different breeds of dogs—collie, German shepherd, terrier, and hound, just to name a few. All can interbreed freely. Thus all are the same species, *Canis familiaris*. Clearly we need a taxon below species in order to tell all these breeds apart. This taxon is called **subspecies**. Table 3-3 gives the subspecies for four breeds of dogs.

COMMON NAME	SCIENTIFIC NAME
Terrier	*Canis familiaris polustris*
German shepherd	*Canis familiaris inostranzewi*
Hound	*Canis familiaris intermedius*
Collie	*Canis familiaris matris-optimae*

TABLE 3-3 Subspecies of Dogs

When required, biologists use a taxon called **superclass** which comes just above class and one called **subclass** found just below class. The same prefixes, **super** and **sub**, can be used in front of the other main taxa as well. Occasionally the prefix **infra** is used for taxa just below sub. Thus an infraclass occurs below a subclass. Other taxa commonly used are **group**, **tribe**, **variety**, and **form**. Table 3-4 shows three organisms and their detailed classifications. When still more information is discovered about these organisms further taxa may be added.

Note: For most of our work the seven main taxa are sufficient. Memorize their names in order. You do not have to memorize the taxa given in Table 3-4. That table was included simply to show you just how detailed taxonomy can become. Neither are you expected to memorize the scientific names or the classification of any organism, unless your teacher asks you to do so.

Some Problems in Classification

Taxa are human concepts; they are not real. They are simply a human attempt to put some order into a confusing array of facts about organisms. Biologists classify organisms for two main reasons: to show relationships among organisms and to make it easier to study them. However, biologists often differ on just how certain organisms should be classified. It is up to the taxonomist to use the facts at hand to convince others that the taxon he or she selects is the best one. The more the taxonomist knows, the more he or she will be respected. However, many respected taxonomists still differ

Taxon	Human	White-tailed deer	Pasture Rose
Kingdom	Animalia	Animalia	Plantae
Subkingdom	Metazoa	Metazoa	Embryophyta
Phylum (Division)	Chordata	Chordata	Anthophyta
Subphylum	Vertebrata	Vertebrata	Pteropsida
Superclass	Tetrapoda	Tetrapoda	Spermatophyta
Class	Mammalia	Mammalia	Angiospermae
Subclass	Theria	Theria	Dicotyledonae
Infraclass	Eutheria	Eutheria	—
Group	—	—	Calyciflorae
Order	Primates	Artiodactyla	Rosales
Suborder	Anthropoidea	Ruminantia	Rosineae
Infraorder	—	Pecora	—
Superfamily	Huminoidea	—	—
Family	Hominidae	Cervidae	Rosaceae
Subfamily	—	—	Rosoideae
Tribe	—	—	Roseae
Genus	*Homo*	*Odocoileus*	*Rosa*
Species	*H. sapiens*	*O. virginianus*	*R. carolina*
Subspecies	—	—	*R.c.carolina*
Variety	—	—	*R.c.c.carolina*
Form	—	—	*R.c.c.c.glandulosa*

TABLE 3-4 **Detailed Classification of Three Organisms**

widely on how they classify certain organisms. Even a common animal such as the African lion is called *Leo leo* by some, *Felis leo* by others, and *Panthera leo* by still others. The names *Leo leo* and *Panthera leo* place the lion in a different genus from the domestic cat, *Felis domesticus*. The name *Felis leo* places it in the same genus as the domestic cat. Similarly, the polar bear is called *Ursus maritimus* by some and *Thalarctos maritimus* by others.

Biologists do not even agree on the names for all the taxa. For example, most zoologists prefer the name "phylum" for the second main taxon. Most botanists, however, prefer the name "division."

Plant, animal, or neither?

The greatest problem that has faced taxonomists is deciding whether certain organisms should be in the kingdom Plantae, the kingdom Animalia, or some other kingdom. For some time bacteria were in the kingdom Plantae. Then, they were moved by some taxonomists to a new kingdom, Protista. Recently some taxonomists have moved them to still another kingdom, Monera. Yet there are still some taxonomists who prefer to leave them in the kingdom Plantae. Where should they be placed? How bacteria are classified depends on how taxonomists view their characteristics. This "kingdom problem" is dealt with in more detail in Section 3.3.

Modern Basis for Classification

Homologous structures

Linnaeus used structural features as the basis for his classification system. He grouped organisms according to their structural similarities. The more alike organisms were, the more taxa they shared. Today, taxonomists still use structural similarities as a basis for classification. They look for homologous structures just as Linnaeus did. **Homologous structures** *are structures that show the same basic pattern, the same general relationship to other parts, and the same pattern of development.* However, they need not have the same function. Study the human arm, the whale flipper, and the bat's wing in Figure 3-10. All these appendages are homologous structures. In all three animals, the bones of the structures show the same basic pattern. Also, all three appendages are found in the same part of the body. Finally, the bones in all three appendages develop in similar ways. Although their functions are different, they are homologous structures.

You may think that it makes more sense to group those organisms whose parts have similar functions. If that was the case, all animals that fly would be in one group. Thus, birds, bats, and insects would be closely grouped. However, these animals differ in so many other ways that the fact that they fly is not sufficient reason to group them closely.

According to Darwin's Theory of Evolution, animals with homologous organs may have a common ancestor. Thus humans, bats, and whales may have a common ancestor that had the same basic forelimb structure as these animals. Biologists who accept the Theory of Evolution feel that animals which have homologous structures are probably closely related and therefore, should be closely grouped.

Similar biochemistry

Biochemistry is the study of the chemical compounds formed by living things. Many biologists believe that closely related organisms form similar chemical compounds in their bodies. They use this belief to help classify organisms. For example, the horseshoe crab was, at one time, classified as a close relative of the true crabs (Fig. 3-11). However, chemical analysis showed that its blood was more like spider's blood than crab's blood. Thus the horseshoe crab is now classified as a close relative of spiders.

Genetic similarity

Most biologists agree that genetic similarity is the best evidence that organisms are closely related. You will have to wait until Unit 7 to find out what genetics is. For now, all you need to know is that every organism makes a special compound called DNA. This compound plays a key role in ensuring that offspring have many of the parents' characteristics. Thus, it seems reasonable to assume that the greater the similarity of DNA among organisms, the more closely they may be related.

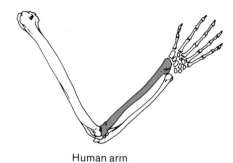

Human arm

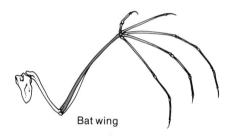

Bat wing

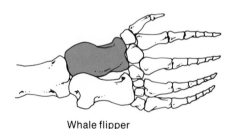

Whale flipper

Fig. 3-10 These three forelimbs are homologous structures.

Horseshoe crab

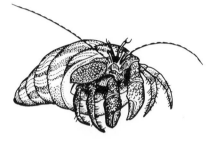

Hermit crab (a "true" crab)

Spider

Fig. 3-11 The horseshoe crab looks more like a true crab than a spider. However, biochemical studies show that it may be more closely related to spiders than crabs.

Review

1. **a)** What is taxonomy?
   b) What did Linnaeus use as the basis for his classification system?
   c) Why was Linnaeus' choice a better basis than something like colour?
2. **a)** What is binomial nomenclature?
   b) What is a genus?
   c) What is a species?
   d) What do you think would eventually happen if members of different species could freely interbreed in the wild to produce fertile offspring?
3. **a)** Why is it important that biologists in all countries use scientific names (binomial nomenclature) for organisms?
   b) How is it possible to name 2 000 000 organisms when no human language has that many words?
4. **a)** List, in order, the seven main taxa (classification groups), starting at kingdom.
   b) What taxa are used in the scientific name of an organism?
5. **a)** Table 3-5 gives the taxonomy of the dog and the domestic cat. Note that these animals are similarly classified from kingdom to order. They must, therefore, have some common structural features. List as many of these as you can.
   b) The dog and cat are in different families. Thus they must have some different structural features. List as many of these as you can.
   c) The wolf, coyote, and domestic dog are species within the same genus, *Canis.* The collie, German shepherd, and terrier are different subspecies of domestic dog. (Subspecies are sometimes called varieties or breeds.) Distinguish between a species and a subspecies by referring to the above animals.

TAXON	DOG	CAT
Kingdom	Animalia	Animalia
Phylum	Chordata	Chordata
Class	Mammalia	Mammalia
Order	Carnivora	Carnivora
Family	Canidae	Felidae
Genus	*Canis*	*Felis*
Species	*familiaris*	*domesticus*

TABLE 3-5 The Taxonomy of a Dog and a Cat

6. **a)** What taxa do you share with dogs and cats?
   b) What order are you in? What order are dogs and cats in? Why do you think you are in a different order from dogs and cats?
7. **a)** Name the three modern bases for classification.
   b) What are homologous structures?
   c) Why do most taxonomists feel that the presence of homologous structures may be good evidence that organisms are closely related?

3.3 Selecting a Classification System

In the last section we briefly discussed the "kingdom problem". This problem is that biologists do not agree on how many kingdoms are required to classify all organisms. Even those who agree on the number of kingdoms often do not agree on which organisms belong to those kingdoms.

Some biologists feel that two kingdoms, Plantae and Animalia, are enough to classify all living things. Others prefer three kingdoms, still others use four, and some use five kingdoms. This text uses a **5-kingdom system of classification**. We feel that it does best the two things that a classification system should do. First, it clearly shows the relationships among organisms. Second, it organizes them in a way that makes it easier to study biology.

You should realize however, that no classification system is perfect. Further, any system will likely change as new information is discovered. We can see how this change occurs by looking at the evolution of the 2-kingdom system into the 5-kingdom system. Although this book uses the 5-kingdom system, you should still be aware of the structure of other systems. This will prevent confusion when you read other biology books that do not use five kingdoms.

A 2-Kingdom System of Classification

This is the oldest and perhaps still the most widely used system of classification. It groups all organisms into two kingdoms, **Plantae** and **Animalia**. That is, all living things are assumed to be either plants or animals. Figure 3-12 summarizes this system. Organisms below the horizontal line are one-celled. Those above the line are many-celled. The

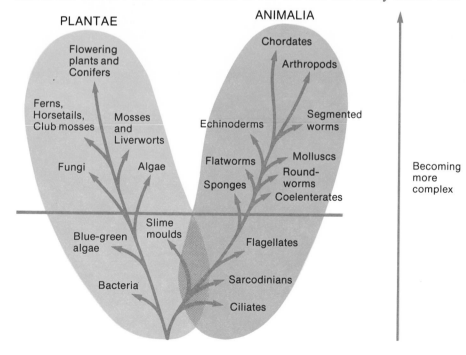

Fig. 3-12   A 2-kingdom system of classification. Only the main phyla and divisions are shown.

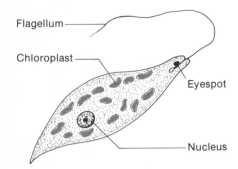

Fig. 3-13 The flagellate, *Euglena*: plant or animal?

complexity of the organisms increases from the bottom to the top of the diagram. The simplest plants are bacteria and blue-green algae. The simplest animals are ciliates such as the *Paramecium,* and Sarcodinians such as the *Amoeba.* The most complex plants are seed plants such as flowering plants and conifers. The most complex animals are chordates such as human beings.

This system works well for most organisms. Everyone agrees that trees, dandelions, tulips, ferns, geraniums, and mosses are plants. Everyone also agrees that humans, dogs, cows, chickens, earthworms, grasshoppers, and crayfish are animals. Problems arise however with some of the one-celled organisms, particularly the flagellates. These organisms have flagella (singular: flagellum) or whip-like "tails" that they use to propel themselves (Fig. 3-13). Because of this locomotion (movement from one place to another) they seem to be animals. However, they also contain chlorophyll, a green pigment, in their chloroplasts. They can, therefore, make their own food by photosynthesis. In this respect they are like plants. For decades botanists have claimed that the flagellates were plants and zoologists have claimed that they were animals. What are they?

Fig. 3-14 A 3-kingdom system of classification, showing the main phyla and divisions.

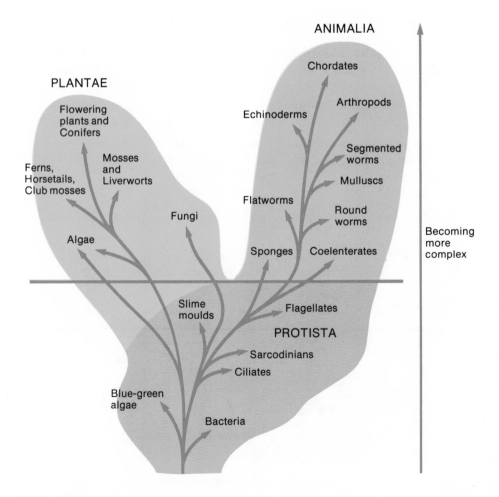

Three, Four, or Five Kingdoms?

The problem of the flagellates is handled by a 3-kingdom system of classification. This system still has the kingdoms Plantae and Animalia. However, it has a third kingdom, **Protista**. This kingdom contains most of the one-celled organisms, including the flagellates (Fig. 3-14). This system was recognized by many biologists during the first half of the 20th century and is still very popular. Yet it also has some problems. For example, the blue-green algae and bacteria are quite similar to one another, yet very different from the other protists. Their cells lack true nuclei unlike all other protists which have true nuclei. This fact, along with other evidence, led some biologists to propose a fourth kingdom, **Monera**, which contains only the bacteria and blue-green algae.

Still more recently, some biologists took a close look at the fungi. They decided that fungi were so different from the other organisms in kingdom Plantae that they should also be moved to a kingdom of their own. Fungi do not have chlorophyll and therefore cannot carry out photosynthesis. Thus a fifth kingdom, called **Fungi**, was created. You will probably agree that fungi such as bread mould and mushrooms are quite different from the usual organisms we call plants. Figure 3-15 shows a 5-kingdom classification system.

Even the 5-kingdom system is not without problems. Organisms that have traditionally been called algae are now spread over three kingdoms. The blue-green algae are in kingdom Monera, one-celled algae such as diatoms are in kingdom Protista, and many-celled algae such as kelp, a brown seaweed, are in kingdom Plantae. Needless to say, many biologists who specialize in the study of algae find this quite disturbing. As more information is gathered about algae, these problems may disappear. Already some biologists think that blue-green algae are not algae at all, but are simply bacteria.

Remember, no classification system is perfect. No classification system will remain unchanged. New kingdoms may be added. Phyla and divisions may be moved to other kingdoms. All this is done to make it easier to see and remember relationships among organisms.

There are many names in Figure 3-15 that you will not understand. Do not worry about it. You will learn about them all later in this book. Unit 3 deals with the kingdoms Monera, Protista, and Fungi. Unit 4 deals with the kingdom Plantae, and Unit 5 deals with the kingdom Animalia. You should refer to Figure 3-15 many times during this course in order to help organize the new information that you learn about organisms.

Review

1. a) What is the "kingdom problem"?
   b) Describe the two main things a classification system should do.
2. a) What kingdoms are used in a 2-kingdom system of classification?
   b) What advantage is there in this system?
   c) What major disadvantage does this system have?

Fig. 3-15 A 5-kingdom system of classification, showing the main phyla and divisions.

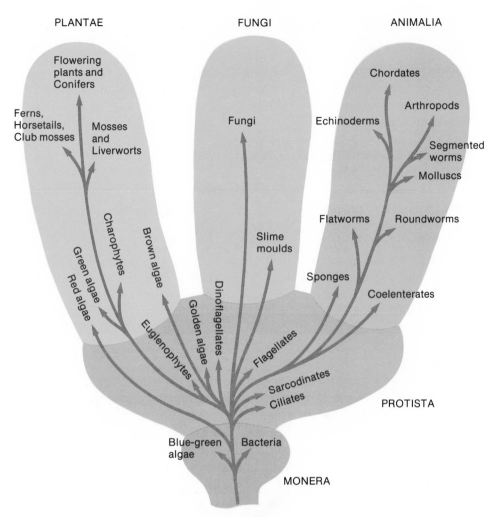

3. **a)** Name the kingdoms used in a 3-kingdom system of classification.
   b) What problem of the 2-kingdom system is solved by adding a third kingdom?
   c) What problem of the 3-kingdom system is solved by adding a fourth kingdom, Monera?
   d) Name the kingdoms used in a 5-kingdom system of classification.
4. Using a 5-kingdom system, state the kingdom in which each of the following organisms would be found: human, oak tree, bread mould, chicken, cat, cholera organism, earthworm, giant kelp (a seaweed), moss, mushroom, pond scum (a green alga), fern, clam, amoeba, flagellate, cow, grasshopper, housefly, diatom (golden alga), spider.

3.4 An Overview of the Five Kingdoms

The purpose of this section is to give you an overview of the major phyla and divisions in each of the five kingdoms. Diagrams have been included to illustrate some of the organisms in each phylum and division. Whenever possible, your teacher will show you specimens of these organisms.

Refer to Figure 3-15 as you study this section.

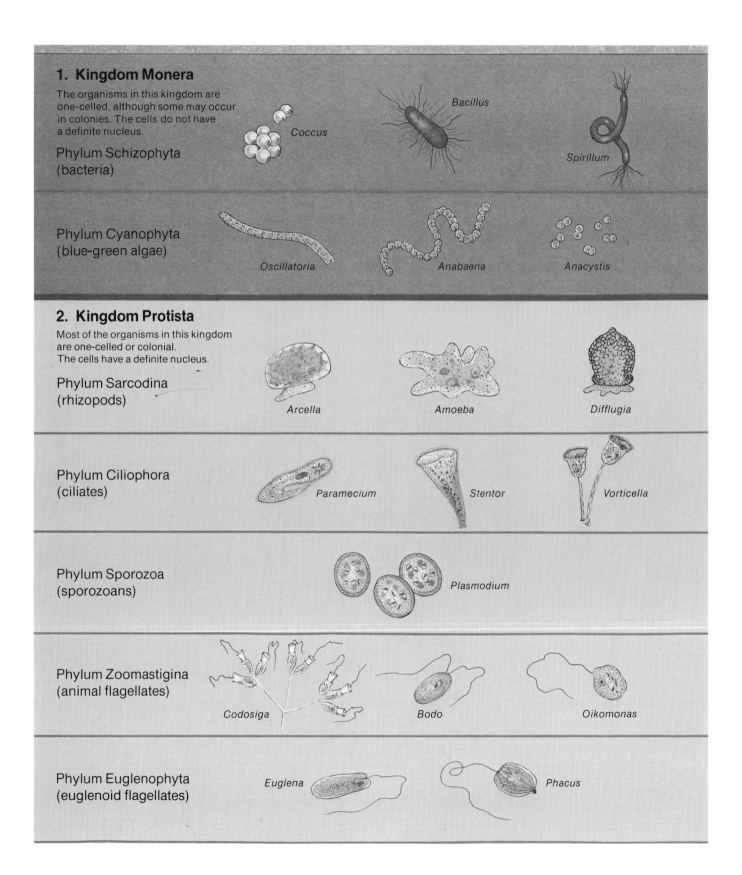

1. Kingdom Monera

The organisms in this kingdom are one-celled, although some may occur in colonies. The cells do not have a definite nucleus.

Phylum Schizophyta (bacteria)

Coccus

Bacillus

Spirillum

Phylum Cyanophyta (blue-green algae)

Oscillatoria

Anabaena

Anacystis

2. Kingdom Protista

Most of the organisms in this kingdom are one-celled or colonial.
The cells have a definite nucleus.

Phylum Sarcodina (rhizopods)

Arcella

Amoeba

Difflugia

Phylum Ciliophora (ciliates)

Paramecium

Stentor

Vorticella

Phylum Sporozoa (sporozoans)

Plasmodium

Phylum Zoomastigina (animal flagellates)

Codosiga

Bodo

Oikomonas

Phylum Euglenophyta (euglenoid flagellates)

Euglena

Phacus

Phylum Chrysophyta (golden algae—includes diatoms)	*Dinobryon* *Diatoma* *Tabellaria*	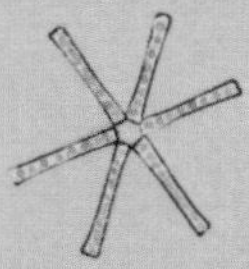 *Asterionella*

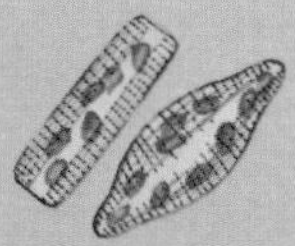

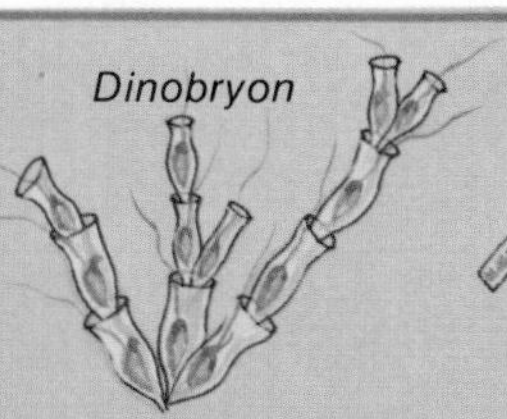

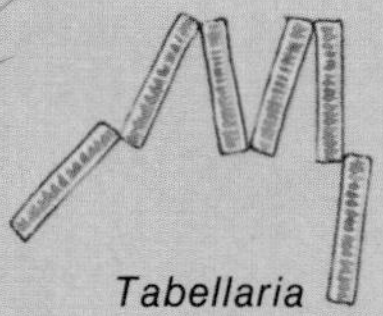

Phylum Pyrrhophyta (includes dino-flagellates)

Ceratium 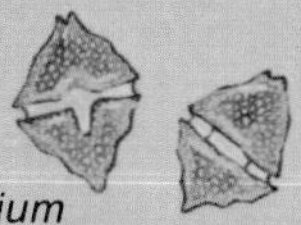 *Peridinium*

3. Kingdom Fungi

This kingdom includes fungi and slime moulds.

Division Myxomycota (slime moulds)

Physarum  *Lycogala*

Division Zygomycota (conjugation fungi)

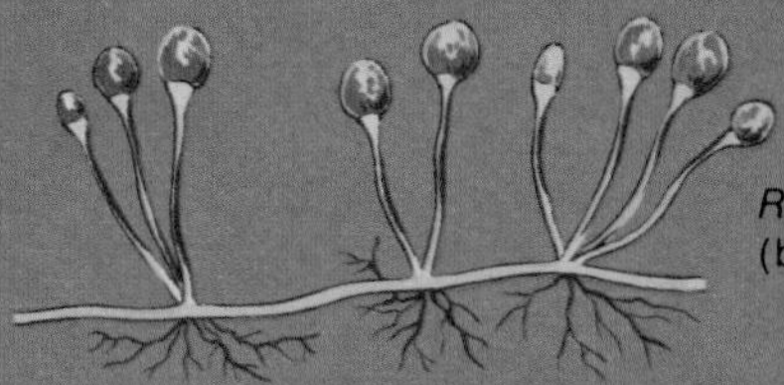

Rhizopus (bread mould)

Division Ascomycota (sac fungi)

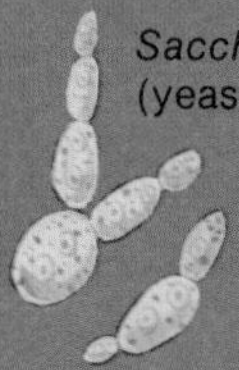

Saccharomyces (yeast)

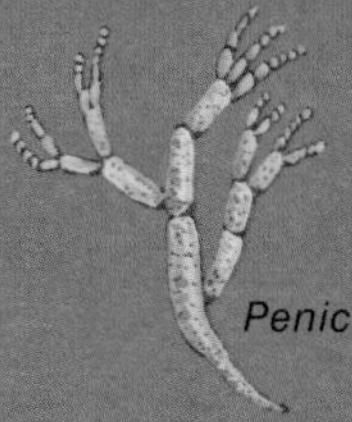

Penicillium

Morchella (morel)

Division Basidiomycota (club fungi)

Amanita  *Agaricus*

Division Deuteromycota (imperfect fungi)

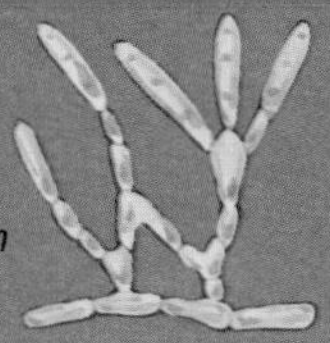

Fusarium

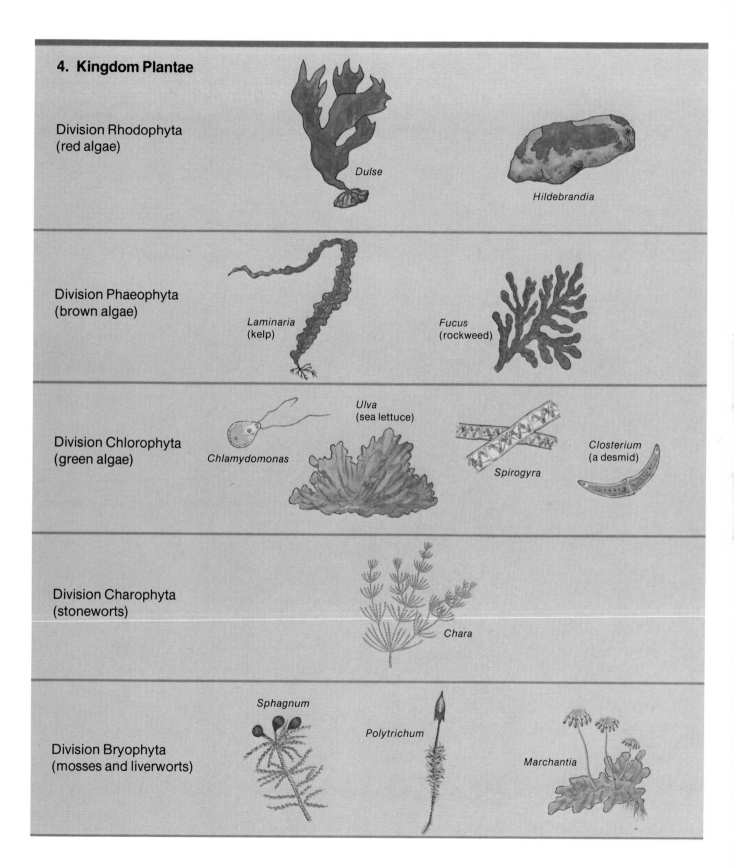

4. Kingdom Plantae

Division Rhodophyta
(red algae)

Dulse

Hildebrandia

Division Phaeophyta
(brown algae)

Laminaria
(kelp)

Fucus
(rockweed)

Division Chlorophyta
(green algae)

Chlamydomonas

Ulva
(sea lettuce)

Spirogyra

Closterium
(a desmid)

Division Charophyta
(stoneworts)

Chara

Division Bryophyta
(mosses and liverworts)

Sphagnum

Polytrichum

Marchantia

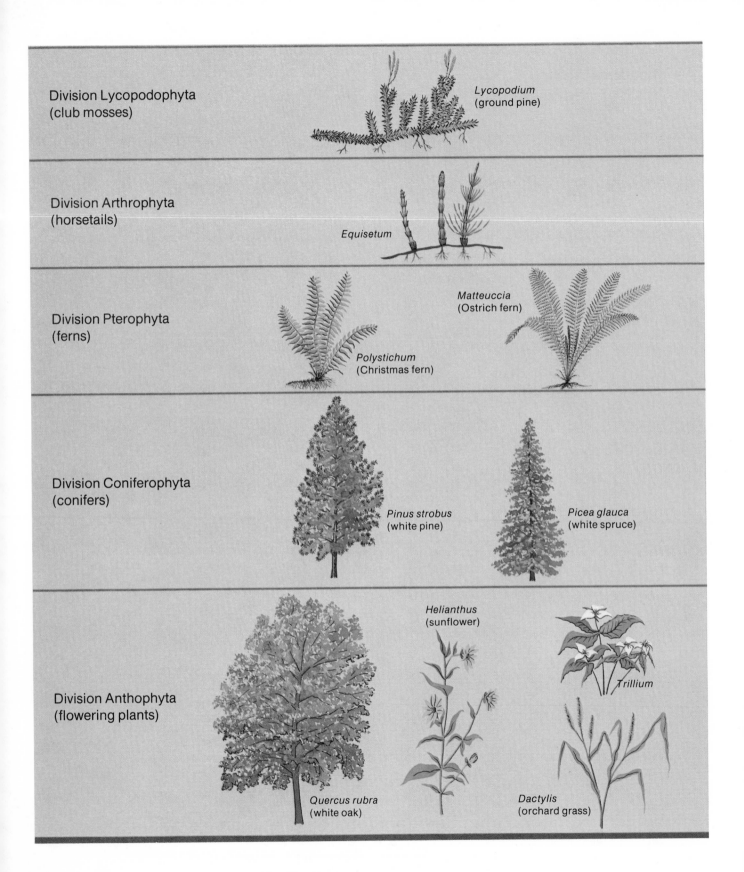

Division Lycopodophyta (club mosses) — *Lycopodium* (ground pine)

Division Arthrophyta (horsetails) — *Equisetum*

Division Pterophyta (ferns) — *Matteuccia* (Ostrich fern), *Polystichum* (Christmas fern)

Division Coniferophyta (conifers) — *Pinus strobus* (white pine), *Picea glauca* (white spruce)

Division Anthophyta (flowering plants) — *Helianthus* (sunflower), *Trillium*, *Quercus rubra* (white oak), *Dactylis* (orchard grass)

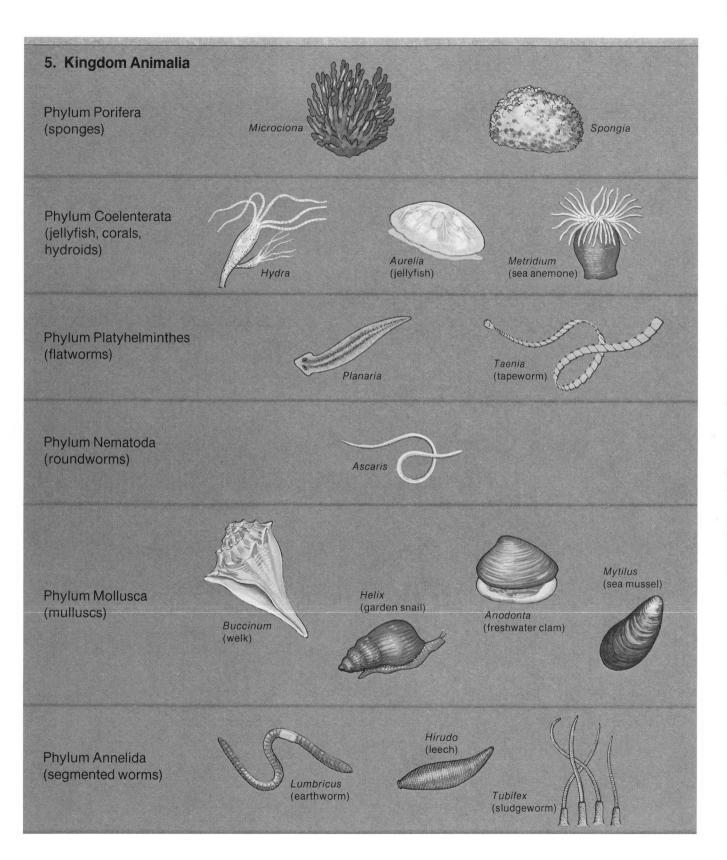

5. Kingdom Animalia

Phylum Porifera
(sponges)

Microciona

Spongia

Phylum Coelenterata
(jellyfish, corals,
hydroids)

Hydra

Aurelia
(jellyfish)

Metridium
(sea anemone)

Phylum Platyhelminthes
(flatworms)

Planaria

Taenia
(tapeworm)

Phylum Nematoda
(roundworms)

Ascaris

Phylum Mollusca
(mulluscs)

Buccinum
(welk)

Helix
(garden snail)

Anodonta
(freshwater clam)

Mytilus
(sea mussel)

Phylum Annelida
(segmented worms)

Lumbricus
(earthworm)

Hirudo
(leech)

Tubifex
(sludgeworm)

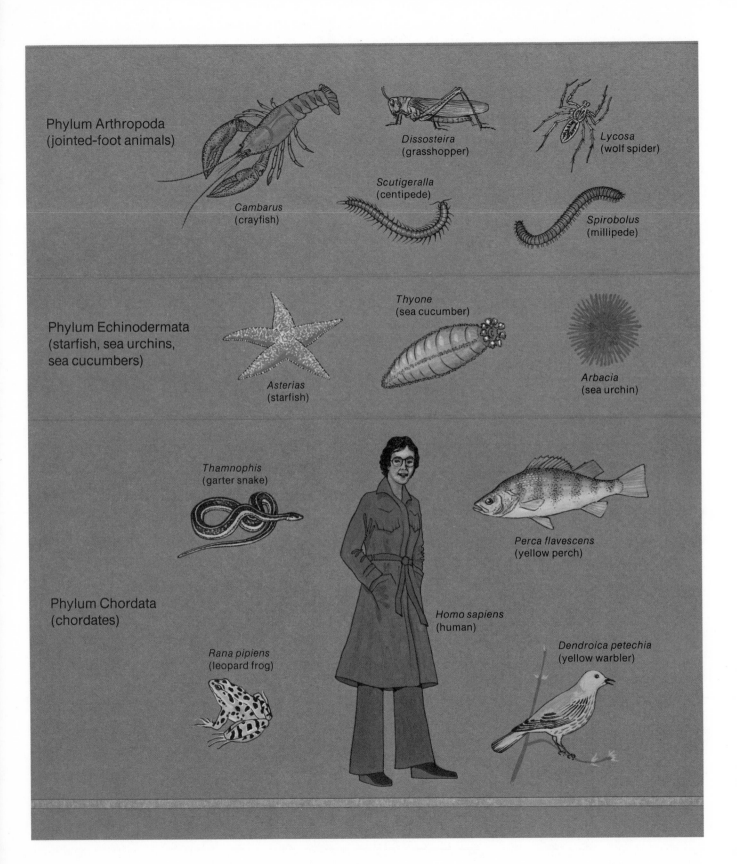

Phylum Arthropoda
(jointed-foot animals)

Cambarus
(crayfish)

Dissosteira
(grasshopper)

Lycosa
(wolf spider)

Scutigeralla
(centipede)

Spirobolus
(millipede)

Phylum Echinodermata
(starfish, sea urchins,
sea cucumbers)

Thyone
(sea cucumber)

Asterias
(starfish)

Arbacia
(sea urchin)

Thamnophis
(garter snake)

Perca flavescens
(yellow perch)

Phylum Chordata
(chordates)

Homo sapiens
(human)

Dendroica petechia
(yellow warbler)

Rana pipiens
(leopard frog)

Keep in mind that this section is only intended to provide an overview. The illustrations are included to show you the diversity of organisms. You are not expected to know much about the individual organisms in this section. Nor are you required to memorize the names of the organisms. Much more information will be given about each kingdom in later units. There you will have a chance to study individual organisms in detail. This section is only meant to give you a general understanding of the way in which all these organisms fit together to form five kingdoms.

Note: The 5-kingdom system of classification that is used in this text is an adaptation of that proposed by R.H. Whittaker in 1969. The names for the various taxa come from his writings and also from the writings of H.C. Bold.

3.5 Investigation

Classification Systems

You have learned so far in this chapter that classification is an important part of biology. It is also an important part of your work as a student of biology. As you know, classification is the grouping of similar things for specific purposes. The main purpose of classification is to make it easier to keep track of and find particular items. In this investigation you classify, or sort, pieces of paper. The pieces have different sizes, shapes, and colours. The purpose in classifying the pieces of paper is to find out exactly what is there. After this simple introduction to classification you will move on to the following sections on biological classification.

Materials

assorted pieces of paper of various colours, shapes, and sizes (30)

Procedure

a. Spread the pieces of paper on the top of your desk.
b. Examine them closely and decide on the main characteristics that distinguish the pieces from one another.
c. Begin your classification by sorting the pieces of paper into two piles, based on one of the main characteristics that you picked in step b. For example, if you were to choose size as a main characteristic, you would put all of the small pieces in one pile and all of the large pieces in another pile.
d. Now pick a second main characteristic and on that basis sort each of the two piles into two more piles.

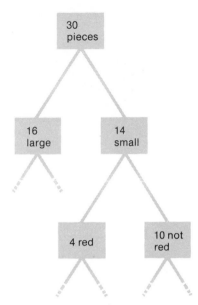

Fig. 3-16   A classification scheme for pieces of paper.

e. Continue separating each pile into two piles until the piles contain pieces that are very similar. Pick a different characteristic each time. Your classification is now complete.

Note: We want you to develop what is called a **dichotomous classification system.** You will see later that biologists use this type of system. *Di* means *"two"*. Therefore, every time you separate a pile, you must separate it into only two piles. Suppose for example, that your pieces of paper are three colours—red, green, and blue. Your first two piles could be "red" and "not red". Then you could split the "not red" pile into "green" and "blue".

f. Prepare a "tree diagram" that summarizes your classification of the pieces of paper. Figure 3-16 shows the start of such a diagram.

g. Repeat this entire procedure (steps a to f) starting with each of the other main characteristics that you picked in step b.

Discussion

1. a) What is a dichotomous classification system?

   b) What was accomplished by your classification of the pieces of paper?

   c) If you were in the paper-selling business, what two advantages would there be in classifying your paper as you did in this investigation?

2. Prepare a dichotomous classification system for about ten of your classmates. If you wish, you may prepare it for about ten teachers in your school. Just remember that biologists are nice people. They do not use nasty words in their classification systems!

3. Prepare a dichotomous classification system for ten cars in the school parking lot.

3.6 Investigation

Making a Biological Classification Key

Biological classification puts similar organisms in groups. This makes it easier to keep track of organisms. It also makes it easier to "find", or identify, a certain organism. For example, suppose you come across an unfamiliar animal and you want to know whether or not it is an insect. There are close to one million species of insect. To look through books for a picture of this animal to see whether or not it is an insect would be a hopeless task. Fortunately, biologists have classified, or grouped, all animals with three pairs of legs into the class Insecta of the phylum Arthropoda. Therefore, you only need to count the animal's legs to see if it is an insect.

In this investigation you prepare a dichotomous classification system for the leaves from several species of trees. If you set up your classification system properly, it can be used as a **classification key** which someone else could use to identify an unknown leaf. To set up a **dichotomous classification key**, you write down pairs of opposing characteristics in couplets. Study the following example to see how your key should look. Then proceed to make a classification system for the leaves provided. This system could be used as a classification key.

A

Example

The following is a dichotomous classification key for the five animals in Figure 3-17:

1a. Has wings . . . . butterfly
1b. Has no wings . . . . go to 2

2a. Has 8 or fewer legs . . . . spider
2b. Has more than 8 legs . . . . go to 3

3a. Has 10 or fewer legs . . . . crayfish
3b. Has more than 10 legs . . . . go to 4

4a. Body flattened; 1 pair of legs per body segment . . . . centipede
4b. Body rounded; 2 pairs of legs per body segment . . . . millipede

B

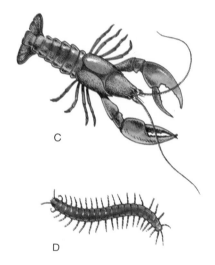

C

Materials

2 sets of leaves arranged as follows:

SET A: 10 leaves, each from a different species of tree, mounted on cards, numbered from 1 to 10, with the names of the leaves attached.
SET B: Identical to Set A, but without the names.

D

Procedure

a. Spread the leaves of Set A on the top of your desk.
b. Examine them closely and decide on the main characteristics that distinguish one leaf from another. In what ways is one leaf the same as the others? In what ways is it different? Use the terms shown in Figure 3-18.

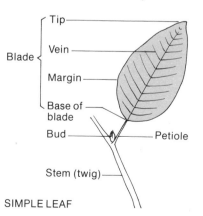

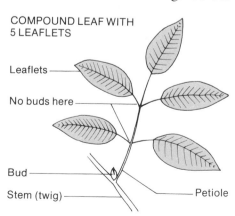

Fig. 3-17   Identify these anthropods by using the classification key.

SIMPLE LEAF

Blade
- Tip
- Vein
- Margin
- Base of blade

Bud — Petiole

Stem (twig)

COMPOUND LEAF WITH 5 LEAFLETS

Leaflets
No buds here
Bud
Stem (twig)
Petiole

Fig. 3-18   Some parts of a typical leaf.

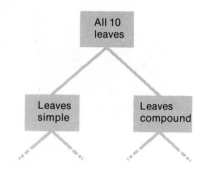

Fig. 3-19 A beginning of a classification scheme for leaves.

Ask yourself questions such as these:

1. Is the leaf simple (all one piece) or compound (divided into many separate leaflets)?
2. Is the leaf margin smooth, finely notched, coarsely notched, or wavy?
3. What shape is the blade—long and narrow, oval, heart-shaped, spear-shaped, circular?
4. Does the leaf have both a petiole and a blade?
5. Are the leaves smooth or hairy? On which side?
6. What is the length of the blade?
7. Do the veins fan out from the base of the blade or are they parallel to one another from the mid-rib?
8. Is the petiole flattened or rounded?
9. What colour is the upper surface of the leaf? the lower surface?

c. Now pick 2 contrasting characteristics, one which about half of the leaves have and the other which the rest of the leaves have. Separate the leaves into 2 groups based on these characteristics.

d. Repeat step c for each of the 2 groups. Continue to do this until you have only 1 leaf in each group.

e. Prepare a "tree diagram" for your classification system. Use the same procedure that you used in the last investigation. Figure 3-19 will give you a start. Of course, you may choose a different starting characteristic than we did.

f. Now take the information from your "tree diagram" and make a true dichotomous classification key out of it. Use the number-letter format that we used in the example. Your key might have the following as its first couplet:

1a. Leaf simple . . . . go to 2
1b. Leaf compound . . . . go to 3

g. Put Set A to one side. Without looking at Set A, try to identify each of the leaves in Set B by using your classification key.

h. Check your results by looking at Set A. If you made any mistakes, something is wrong with your key. Correct any errors in your key.

i. Here is the final test of your key: Give your key and Set B to another student. Ask that student to identify the leaves using your key. If the student has any problems, find out what they are. Then, if necessary, change your key to eliminate the problems.

Discussion

1. Take your key home. Collect two or three leaves that can be identified using your key. Have someone in your family use your key to try to identify the leaves. Write a brief report on the success (or failure) of your key when used by another person.

3.7 Investigation

Using a Biological Classification Key

You have made your own classification key and used it to identify leaves. Now you will use two dichotomous keys that were made by professional biologists.

Classification Key A

You should be able to identify all the organisms in Figure 3-20 using Key A. Your teacher may provide real specimens instead of the diagrams. This key

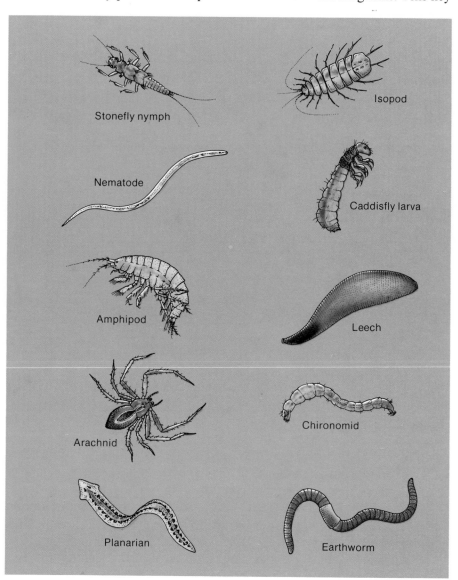

Fig. 3-20   Some common aquatic invertebrates.

Stonefly nymph

Isopod

Nematode

Caddisfly larva

Amphipod

Leech

Arachnid

Chironomid

Planarian

Earthworm

is for common invertebrates (animals without backbones) that live in ponds, lakes, and streams. If you are using the diagrams, begin with animal A and follow the key through until you have identified the animal. Check with your teacher if you have any problems. Then proceed to identify the remaining animals. If you are using real specimens, handle them carefully. They are expensive and easily damaged.

This key identifies some animals only as far as phylum. Most of them are identified to class and some as far as order. To find out the species, you would have to use a much more complex key.

Your teacher will give you the meanings of some terms in the key such as "lateral" and "ventral".

KEY A   KEY TO AQUATIC INVERTEBRATES

1a. Legs present . . . . go to 2
1b. Legs absent . . . . go to 5

2a. 3 pairs of legs . . . . class Insecta (except Diptera)
2b. More than 3 pairs of legs . . . . go to 3

3a. 4 pairs of legs . . . . class Arachnida
3b. More than 4 pairs of legs . . . . class Crustacea: go to 4

4a. Body flattened laterally . . . . order Amphipoda
4b. Body flattened dorso-ventrally . . . . . . order Isopoda

5a. With distinct head capsule . . . . order Diptera (true flies)
5b. Without distinct head capsule . . . . . . go to 6

6a. Unsegmented . . . . go to 7
6b. Segmented . . . . go to 8

7a. Body round . . . . phylum Nematoda
(roundworms)
7b. Body flat . . . . phylum Platyhelminthes
(flatworms)

8a. Suckers on both ends of
ventral surface . . . . class Hirudinea
8b. No suckers . . . . class Oligochaeta

Classification Key B

Key B is more challenging. It is to be used with real specimens. Your teacher will give you two or three animals to identify with the key. Again, remember to handle these fragile specimens carefully. They are expensive.

This key is for aquatic insects. These insects live on or near the bottom of ponds, lakes, and streams. The key helps you to identify the insects as far as order.

Your teacher will give you the meanings of terms such as "proleg", "tarsal", and "anal".

KEY B   KEY TO AQUATIC INSECTS (TO ORDER)

1a. Legs absent .... Diptera (true flies)
1b. Legs present .... go to 2

2a. Wings present .... go to 3
2b. Wings absent.................... go to 4

3a. Sucking mouthparts
    (look like a beak) .... Hemiptera (true bugs)
3b. Chewing mouthparts .... Coleoptera adult (beetles)

4a. Mouthparts can be extended .... Odonata: go to 5
4b. Mouthparts cannot be extended .... go to 6

5a. 3 plate-like gills at tip of abdomen .. damselflies
5b. Without plate-like gills .... dragonflies

6a. 2 or 3 "tails" (cerci) present;
    wing pads present.............. go to 7
6b. "Tails" absent; wing pads absent .... go to 8

7a. 2 tails; 2 tarsal claws; gills on thorax Plecoptera (stoneflies)
7b. 2 or 3 tails; 1 tarsal claw;
    gills on abdomen .... Ephemeroptera (mayflies)

8a. Anal prolegs usually absent;
    if present, no hooks............ Coleoptera larva (beetles)
8b. Anal prolegs present; with hooks ... go to 9

9a. 2 anal prolegs; 1 hook per proleg ... order Trichoptera (caddisflies)
9b. 2 anal prolegs; 2 hooks per proleg
    (or above may be replaced by a
    central tail) .... order Megaloptera (dobson
                                      flies, fish flies, alder flies)

Highlights

Classification is the grouping of similar things for a specific purpose.
Classification makes it easier to keep track of things and to find particular
items. Biological classification groups similar organisms. Such grouping
makes it easier to remember the characteristics of organisms. It also makes it
easier to identify them. Biologists use dichotomous classification keys for
classifying and identifying organisms.

A species is a group of individuals that are alike in many ways and
interbreed to produce fertile offspring under natural conditions. Species are
named using binomial nomenclature. Both the genus and the species must
be given to properly identify an organism. Scientific names are printed in
italics in books *(Homo sapiens)* or underlined when written by hand (Homo
sapiens).

The main classification groups, or taxa, are kingdom, phylum, class, order, family, genus, and species. The further one goes in this sequence, the more the organisms have in common.

Organisms can be classified in five kingdoms—Monera, Protista, Fungi, Plantae, Animalia. However, no classification system is perfect. As new evidence is obtained, kingdoms may be added or removed, organisms may be moved from one kingdom to another, and names may be changed. But all this is done to make it easier to see and remember relationships among organisms.

UNIT TWO

Cells And Their Processes

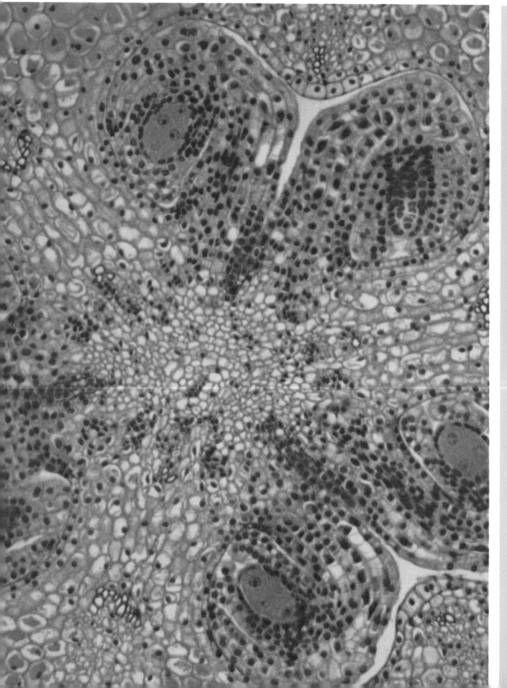

Unit One introduced you to the wide diversity of organisms that live on this earth. They range in size from tiny single-celled bacteria to giant many-celled trees and whales. But, whether they are one-celled or many-celled, all organisms must carry out most of the same basic processes. In this unit you study those processes.

The Chemical Basis of Life

Fig. 4-1 Some of the food that this sheep eats will be changed into living body material. It will become part of the animal. How does this change occur?

Until well into this century, biologists spent most of their time dealing with taxonomy and anatomy. That is, they identified organisms and studied their parts. After a time, some biologists became curious about how these parts worked. This curiosity opened up a new field of biology called **physiology**. Biologists now began to ask questions such as: How does water enter the roots of a tree? How does water get to the top of a tree? How do green plants use the sun's energy? Why does sap not leak out of the roots of trees? What makes a heart beat? What causes a muscle to move? How do animals get energy from food? How does an animal change food into living body material? (Fig. 4-1).

While seeking answers to such questions, biologists discovered that they needed a knowledge of two other sciences, physics and chemistry. In particular, they needed to know about the structure and behaviour of matter. You have reached much the same point in your study of biology. In the last unit, you classified organisms and looked at their parts. In this unit, you will find out what happens inside those organisms. However, to understand this, you need to know some simple chemistry and physics. This chapter provides that information. Although this section may seem very unrelated to biology, it provides important background information that you will use throughout the book.

You may have studied much of this material in an earlier course in science. If so, treat this as a review. Your teacher may choose to omit this chapter and instead, refer to it when necessary.

4.1 Properties of Matter

Matter and its Properties

Matter *is anything that has mass.* This means, of course, that matter takes up space, or has volume. In fact, everything, except energy, is matter.

States of Matter

Matter can occur in three states—solid, liquid, and gaseous (Fig. 4-2).

A **solid** has a definite shape and volume. It is rigid and not easily pressed together or compressed. Most solids slightly increase in volume (expand) when heated.

A **liquid** does not have a definite shape. In fact, the most obvious property of a liquid is that it can be poured. It is said to be **fluid**, because it pours. The volume of a liquid remains the same, regardless of its shape. Like solids, liquids are almost non-compressible. Liquids increase in volume when heated. This property is used in thermometers that contain alcohol or mercury.

A **gas** or **vapour** has neither a definite shape nor volume. It usually expands to fill its container. For example, if you spray some air freshener into a corner, the freshener does not stay there. It gradually spreads, or **diffuses**, throughout the room. Thus, a gas has both the volume and shape of its container. Unlike solids and liquids, gases are easily compressed. You have probably seen "cylinders" of compressed oxygen and other gases. A gas expands considerably when heated. Gases, like liquids, are fluids since they can be poured.

Particle theory

Scientists explain many of the properties of solids, liquids, and gases by assuming that matter is made up of particles. These particles are very small and invisible. In fact, one scientist estimated that the particles in the head of a pin are so small and numerous that all the people in the world would have to spend their entire lives counting to find out how many there are. The particle theory says that the particles have spaces between them which are much larger than the particles themselves. These particles attract one another. These attractive forces get stronger as the particles move closer together. Finally, the theory also says that the particles in matter are in constant motion and they move faster as the temperature is increased.

The particle theory explains the differences among solids, liquids, and gases (Fig. 4-3). For example, gases are the easiest to compress because they

Solid state (ice)

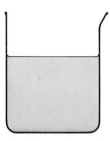
Liquid state (water)

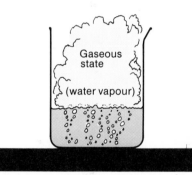
Hot-plate

Fig. 4-2   Water exists in all three states — solid, liquid, and gaseous.

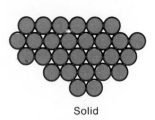

Solid

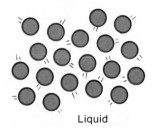

Liquid

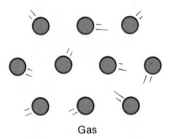

Gas

Fig. 4-3 The particle nature of solids, liquids, and gases. The spaces between the particles are *many* times greater than shown here.

Fig. 4-4 Changes of state.

have the largest spaces between their particles. Solids are the most rigid because they have the strongest forces between their particles. Liquids evaporate faster when they are heated because the heat gives the particles the energy needed to overcome the attractive forces between them.

Changes in the Properties of Matter

Physical changes

Many substances, such as water, can exist in all three states. Such substances can change from one state to another. Figure 4-4 shows that six **changes of state** can occur. **Melting evaporation**, and **sublimation** (solid changing to vapour) all require heat. The heat supplies the energy needed to overcome the attractive forces so the particles can move apart. The opposite changes of state—**freezing**, **condensation**, and **sublimation** (vapour changing to solid)—require removal of heat. Why?

These changes of state are called physical changes. *A physical change is any change in matter that does not form a new substance.* When ice changes to liquid water, no new substance is formed. The particles are still water particles. All that has changed are the spaces between the particles and the strength of the attractive forces. Similarly, when liquid water changes to vapour, no new substance is formed. Therefore this change is also a physical change.

Many physical changes are not changes of state. Changing wood into sawdust is one example. Crushing a piece of chalk is another.

Matter has **physical properties**. These are properties that describe matter. Colour and taste are physical properties. Physical properties do not describe matter that is changing to a new substance. The physical properties of a substance that are different from those of most other substances, are called **characteristic physical properties**. Density, melting point, and freezing point are characteristic physical properties. For example, liquid water has a density of 1 000 kg/m³, a melting point of 0°C, and a normal boiling point of 100°C. Few, if any, other substances have exactly those same values. Therefore, characteristic physical properties are often used to identify substances.

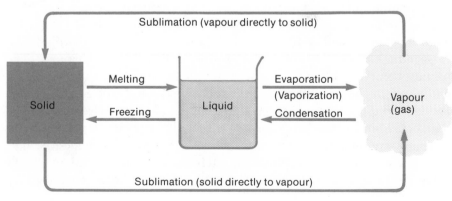

Chemical changes

A **chemical change** *is a change which forms one or more new substances.* Rusting of iron is a chemical change. The iron combines with oxygen from the air to form a new substance, iron oxide or rust. When wood burns it changes into several new substances. Two of them are carbon dioxide and water. Therefore burning of wood is a chemical change. The fact that iron can rust is a **chemical property** of iron. The fact that wood can burn is a chemical property of wood.

Matter and Energy

You learned in Chapter 2 that all living things require energy to support life processes such as movement, growth, and reproduction. You learned also, that almost all this energy originally comes from the sun. Green plants, through photosynthesis, store some of the sun's energy in compounds such as glucose. Then the plants, through respiration, "burn" or break down the glucose, releasing the energy needed to support their life processes. Animals eat plants or other animals that have eaten plants. In this way, they obtain glucose and other compounds which they too, break down through respiration to release energy to support their life processes. What is energy? How is it stored? How can it be changed from one form to another?

Energy *is the ability to do work.* It cannot be seen or felt, since it is not matter. However, you can see what it does. It boils water, moves cars, and makes it possible for you to move. Two main types of energy are of particular interest in the study of biology: kinetic energy and potential energy.

Kinetic energy

Kinetic energy is energy of motion. It is energy that is actually doing work. Its presence can be recognized by the changes that it causes in the position of matter. Whenever a mass is moving, kinetic energy is present. Thus a moving car, a running deer and even a moving molecule in a tiny living cell, have kinetic energy.

Kinetic energy often takes the form of **radiant energy. Heat energy** and **light energy** are two of the many forms of radiant energy. Both are important in many biological processes.

Potential energy

As the name implies, potential energy is a form of energy that has the potential to do work. However, it is not actually doing work, as kinetic energy is. A pile driver just about to begin its descent has potential energy. It has the potential to do work because its potential energy can be changed to kinetic energy as it drops. Water at the top of a waterfall has potential energy. It has the potential to do work because its potential energy can be changed to kinetic energy as it drops over the falls. That kinetic energy, in turn, can do work by turning a turbine to generate electricity.

The form of potential energy that most interests biologists is **chemical potential energy**. Substances such as coal and wood have chemical potential energy. This is stored in the bonds that hold the particles of these substances together. (You will learn about the nature of those bonds in Section 4.3.) When coal or wood is burned, its particles combine with oxygen particles. The bonds in the coal or wood must be broken for this to happen. When the bonds are broken the chemical potential energy that they contain is converted to kinetic energy. Generally, it appears as heat and/or light energy.

All chemical changes involve chemical potential energy. The reason is that all chemical changes involve the breaking and/or forming of bonds between particles. (Recall that a chemical change forms one or more new substances. Bonds between particles must be broken and/or formed for this to happen.)

Plants convert some of the sun's radiant energy into chemical potential energy. This energy is stored in the bonds of carbohydrates that are formed during photosynthesis. The plants then convert some of that chemical potential energy into kinetic energy which is used to support life processes. Animals that eat plants also convert that chemical potential energy into kinetic energy needed for their life processes.

Law of Conservation of Energy

You have seen that energy can be changed from one form into another. However, it cannot be destroyed. For example, when wood is burned, chemical potential energy in the wood is converted into heat and light energy. Chemical potential energy is also formed. It is found in the carbon dioxide and water that are by-products of the burning. The total amount of energy does not change. Only the form changes. This *conservation* of energy applies to all physical and chemical changes. Therefore, scientists made up this generalization called the **Law of Conservation of Energy**: *Energy can be neither created nor destroyed. It can only be converted from one form into another.*

Activation energy

You know that some of the chemical potential energy of wood can be changed to heat and light energy. However, you also know that it is necessary to set fire to the wood before this conversion will begin. In other words, you have to *activate,* or start, the chemical change. The energy required to do this is called **activation energy**. You will discover that many chemical changes within organisms require activation energy.

Figure 4-5 illustrates the relationships among activation energy, potential energy, and kinetic energy.

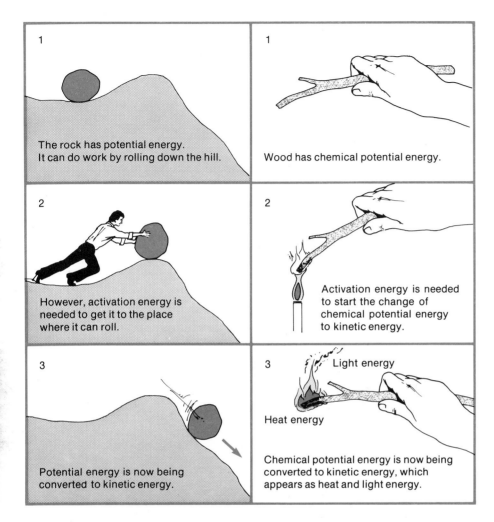

Fig. 4-5 The role of activation energy in starting the conversion of potential energy into kinetic energy.

1 The rock has potential energy. It can do work by rolling down the hill.	**1** Wood has chemical potential energy.
2 However, activation energy is needed to get it to the place where it can roll.	**2** Activation energy is needed to start the change of chemical potential energy to kinetic energy.
3 Potential energy is now being converted to kinetic energy.	**3** Light energy. Heat energy. Chemical potential energy is now being converted to kinetic energy, which appears as heat and light energy.

Review

1. Explain why biologists need to know some chemistry and physics.
2. **a)** What is matter?
   b) Name and describe the three states of matter.
   c) What is all matter made of?
3. **a)** What is a physical change?
   b) What is a physical property of matter?
   c) What is a characteristic physical property?
4. **a)** What is a chemical change?
   b) What is the basic difference between a physical change and a chemical change?
   c) Describe a chemical change other than the examples in this text.
   d) State a chemical property of gasoline.
5. **a)** Define energy.
   b) Explain the difference between kinetic energy and potential energy.
6. **a)** What is chemical potential energy?
   b) Explain why chemical potential energy is important to all living things.

7. **a)** State the Law of Conservation of Energy.
   b) Explain how the Law of Conservation of Energy will apply in the third steps of Figure 4-5.
8. What is activation energy?

4.2 Classification of Matter

In Unit One you learned that classification, or sorting into groups, makes it much easier to study and remember information. We classified organisms into taxa for that reason. Since there are many kinds of matter, it also makes sense to classify matter into groups.

Figure 4-6 shows a classification system for matter. Study the diagram and the following explanation. You need to understand these terms because you will use them many times in this course. Your teacher will show you examples of the various forms of matter. Examine them closely as you read the following information.

Fig. 4-6 A classification system for matter.

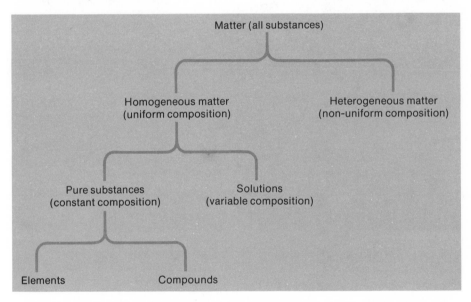

Homogeneous and Heterogeneous Matter

All matter is either homogeneous or heterogeneous. **Homogeneous matter** *is any substance that has a uniform composition and only one phase.* A phase is a visible portion of matter. The composition of a substance is its makeup. Therefore, you can see only one substance in homogeneous matter. However, there may be more than one substance present. All samples of a homogeneous substance have the same composition. Sugar, salt, iron, and after-shave lotion are examples of homogeneous matter. Although there may be many substances in these forms of matter, they appear to have only one.

Heterogeneous matter *is any substance that has a non-uniform composition and more than one phase.* The granite rock in Figure 4-7 is heterogeneous matter. You can see more than one substance in the rock; therefore it has more than one phase. Also, all portions of the rock are not the same; therefore it has a non-uniform composition. In contrast, the piece of steel, also in Figure 4-7, is homogeneous.

Types of Homogeneous Matter

Figure 4-6 shows that there are two types of homogeneous matter—pure substances and solutions. Both, of course, have a uniform composition throughout and only one phase. How do they differ?

A **pure substance** *is a homogeneous form of matter that consists of only one substance and has a fixed composition.* Sugar is a pure substance. A bag of sugar contains nothing but sucrose. Also, all samples of sugar are the same—100% sucrose. Similarly, diamond is also 100% carbon and table salt is 100% sodium chloride. Both, therefore are pure substances.

A **solution** *is a homogeneous mixture which has a variable composition and contains two or more pure substances.* Vinegar is a solution. It looks like only one substance so it is called homogeneous. However, it is actually a mixture of two pure substances, water and acetic acid. Also, its composition is not fixed. You can add water to vinegar and change its composition, but it still stays homogeneous. After-shave lotion is also a solution. It looks like one pure substance but it is really a homogeneous mixture of several pure substances—alcohol, water, perfume, oils, and colour pigments.

Elements and Compounds

There are two types of pure substances: elements and compounds.

An **element** *is a pure substance that consists of only one type of atom.* There are now 106 elements known to scientists. Among them are oxygen, nitrogen, iron, tin, aluminum, chlorine, and uranium. Elements are the building blocks of all matter. All forms of matter are made from them.

A **compound** *is a pure substance that consists of more than one type of atom.* Thus compounds are formed by the union of elements. About 2 000 000 different compounds are known and new ones are being discovered every day. All are made from the 106 elements. Table 4-1 shows some compounds and their elements.

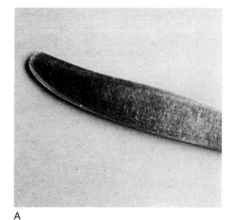

A

B

Fig. 4-7 Homogeneous and heterogeneous matter.

COMMON NAME	CHEMICAL NAME	ELEMENTS USED TO BUILD COMPOUND
Common salt	Sodium chloride	Sodium, chlorine
Table sugar	Sucrose	Carbon, hydrogen, oxygen
Lye	Sodium hydroxide	Sodium, hydrogen, oxygen
Bluestone	Copper sulfate	Copper, sulfur, oxygen

TABLE 4-1 Some Compounds and Their Composition

Fig. 4-8   The molecules of some pure substances.

Molecule	Composition	Element	Compound
Nitrogen 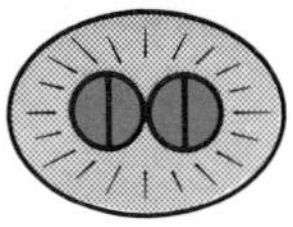	2 atoms of nitrogen	✓	
Magnesium Oxide 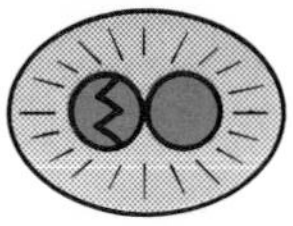	1 atom of magnesium 1 atom of oxygen		✓
Hydrogen 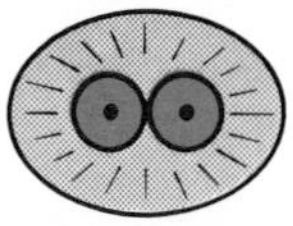	2 atoms of hydrogen	✓	
Carbon dioxide 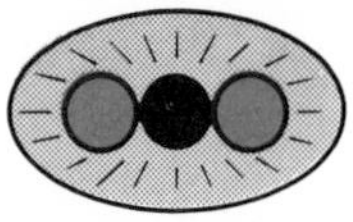	1 atom of carbon 2 atoms of oxygen		✓
Ammonia 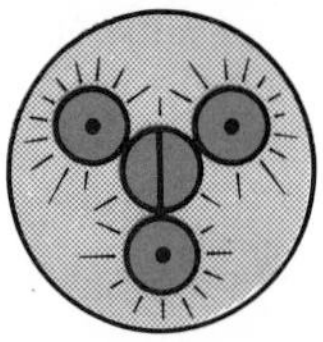	1 atom of nitrogen 3 atoms of hydrogen		✓

A single particle of an element is called an **atom**. An atom of iron is a single particle of the element iron. *A single particle of a compound is called a* **molecule**. A molecule of iron sulfide is made up of one atom of iron and one atom of sulfur. A molecule of table sugar is made up of 12 atoms of carbon, 22 atoms of hydrogen, and 11 atoms of oxygen. *A* **molecule** *is a particle composed of two or more atoms.* The two or more atoms can be the same or different. If the atoms are different, the substance is a compound. Figure 4-8 shows the molecules of some pure substances. As you will find out in Section 4.3, these diagrams are just models. No one knows what molecules actually look like.

Types of Heterogeneous Matter

Heterogeneous matter includes all kinds of mixtures except solutions. Therefore, before we deal with heterogeneous matter, let us review the properties of solutions.

Recall that a solution *is a homogeneous mixture of two or more pure substances.* It is clear, although it may be coloured. It consists of two portions, a solute and a solvent. The solute *is the substance that dissolves.* The solvent *is the substance that does the dissolving.* When a solute is added to a solvent, the particles of solute separate from one another and go among the molecules of the solvent. In some cases, the particles of solute are molecules and in other cases they are charged atoms, called ions. These small particles cannot be seen and will not settle out of the solvent on standing. A solution is said to be dilute when it contains little solute and concentrated when it contains much solute. When no more solute will dissolve in the solvent, the solution is called a saturated solution.

A suspension is one of the most common heterogeneous mixtures. It is a mixture in which some particles are large enough to be seen with the unaided eye and certainly with an ordinary microscope. It is cloudy and non-transparent. Some of the particles will settle out of the mixture on standing. Clay mixed with water is a suspension. Kerosene shaken with water also forms a suspension. This special suspension of liquid droplets in another liquid is called an emulsion.

Between a solution and a suspension is a type of mixture called a colloidal dispersion. It is slightly cloudy, semi-transparent and almost, but not quite, homogeneous. The particles cannot be seen with the unaided eye or even with an ordinary microscope. The particles will not settle out on standing. However, they are still much larger than particles of solute in a solution. Liquid laundry starch is a colloidal dispersion. Colloidal dispersions are quite common in living cells.

Review

1. Distinguish between homogeneous matter and heterogeneous matter.
2. a) What is a pure substance?
   b) How does a solution differ from a pure substance?
3. a) Name and define the two types of pure substances.
   b) Distinguish between an atom and a molecule.
   c) Can a molecule be an element? Explain.
4. a) Distinguish between a solution and a suspension.
   b) What is an emulsion?
   c) What is a colloidal dispersion?
5. a) Name and describe the parts of a solution.
   b) Distinguish between a dilute and a concentrated solution.
   c) What is a saturated solution?
6. Copy the names of the following substances into your notebook. After each name, indicate whether the substance is an element, compound, solution, or heterogeneous mixture. Iron, sodium chloride, vinegar, shaving lotion, orange juice, a soft drink, aluminum, oxygen, ketchup, tin, nitrogen, copper sulfate, homogenized milk.

MODEL A
An electron cloud model. The electrons
exist as a cloud about the nucleus.

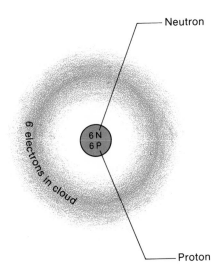

MODEL B
An energy level model. The electrons are
shown at their most probable distances
from the nucleus. One should *not* imagine
that they move around the nucleus just as
the planets move around the sun.

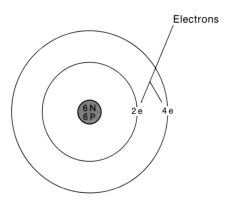

Fig. 4-9   Two models of a carbon atom.

4.3 Atoms and Molecules

Structure of Atoms

Of the 106 known elements, 92 occur naturally. The rest have been made by scientists. All matter is made from these 92 elements. In fact, only 15 of them form over 99% of all matter on earth. As you know, atoms are the particles of which elements are made. Thus they are the building blocks of all matter. During chemical changes, substances are broken down into the atoms of which they are made. The atoms themselves, however, never break down. They simply make new combinations which form new substances. To understand the chemical changes that take place in living things, you must know something about the structure of atoms.

Atoms are too small to be seen, even with the most powerful microscope. Therefore, the descriptions that follow are only **models**. They are mental pictures of what scientists think atoms must be like in order to have the properties that they do. The models discussed here are very simple ones. Much more complex models are now being used by scientists.

All atoms are made of the same kinds of particles. These are called **subatomic** or **elementary particles**. Scientists do not agree on how many kinds of elementary particles exist. However, they do agree that three kinds are basic to all matter. They are: protons, electrons, and neutrons.

All atoms, with the exception of hydrogen, have a **nucleus** that consists of densely packed **protons** and **neutrons**. Each proton carries a single positive charge. Neutrons have no charge. Moving rapidly around the nucleus are the **electrons**. Each electron carries a single negative charge. An atom has no charge on it because the number of electrons equals the number of protons. In other words, the number of negative charges equals the number of positive charges. Scientists believe that the electrons travel around the nucleus as bees swarm around a hive. They do not follow definite paths. Scientists call the resulting "picture" of the electrons an **electron cloud**. The electron cloud represents the space where the electrons will most likely be found. For each electron, there is a most probable location. This location, which is a certain distance from the nucleus, is called an **energy level**. Figure 4-9 shows two ways of representing a carbon atom using these ideas. The energy level model states that only two electrons can occupy the first energy level (nearest the nucleus). A maximum of eight electrons can occupy the second level. A maximum of 18 electrons can occupy the third level. Electrons generally fill the energy levels closest to the nucleus first.

Remember, diagrams like these are drawn to make it easier to understand how atoms behave. Atoms do not really look like this. From now on, we will use the energy level model since it is both simple and useful. However, you must not think that electrons are particles that travel around the nucleus just as planets travel around the sun.

The number of protons in the nucleus of an atom is called the **atomic number**. *The number of protons plus neutrons in the nucleus is called the*

mass number. Carbon (see Fig. 4-9) has an atomic number of six and a mass number of twelve.

Chemical symbols

An atom is represented by a chemical symbol. *A* **chemical symbol** *is a capital letter or a capital letter followed by a small letter.* In most cases these are the first or the first and second letter of the element's name. In many cases however, the symbol has come from the name in another language such as Latin. Remember that a symbol represents *one* atom of the element. Thus O represents one atom of oxygen. C represents one atom of carbon. Al represents one atom of aluminum. Table 4-2 shows the symbols of a few common elements.

ELEMENT	SYMBOL	ELEMENT	SYMBOL
Aluminum	Al	Iron (ferrum)	Fe
Barium	Ba	Magnesium	Mg
Bromine	Br	Mercury (hydrargyrum)	Hg
Calcium	Ca	Nitrogen	N
Carbon	C	Oxygen	O
Chlorine	Cl	Potassium (kalium)	K
Copper	Cu	Sodium (natrium)	Na
Hydrogen	H	Sulfur	S
Iodine	I	Zinc	Zn

TABLE 4-2 Symbols of Common Elements

Often, scientists write the symbol, atomic number, and mass number together as shown in Figure 4-10. Figure 4-11 gives a few examples of this way of representing atoms. Energy level diagrams are also included.

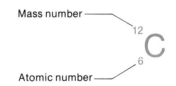

Fig. 4-10 One method of representing an atom.

Isotopes

Atoms of the same element can have different numbers of neutrons. Such atoms are called **isotopes** of one another. For example, hydrogen has three isotopes: "ordinary hydrogen", $_1^1H$; deuterium (heavy hydrogen), $_1^2H$; and tritium, $_1^3H$. Draw the energy level diagrams for these atoms.

Structure of Molecules

As you know, a molecule is a particle composed of two or more atoms. The force that holds the atoms together in a molecule is called a **chemical bond**. Atoms usually form bonds using the electrons in their outermost energy

Fig. 4-11 Energy level diagrams for some common atoms.

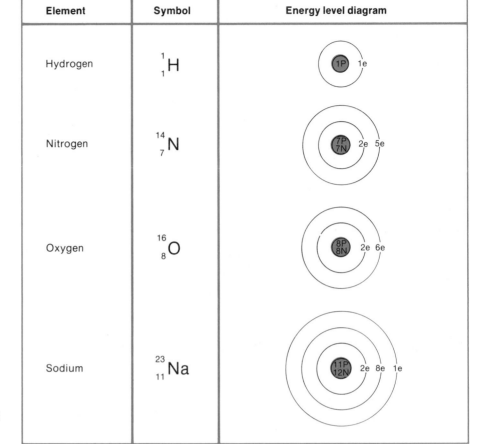

Element	Symbol	Energy level diagram
Hydrogen	$_1^1\text{H}$	1P 1e
Nitrogen	$_7^{14}\text{N}$	7P 7N 2e 5e
Oxygen	$_8^{16}\text{O}$	8P 8N 2e 6e
Sodium	$_{11}^{23}\text{Na}$	11P 12N 2e 8e 1e

HYDROGEN MOLECULE

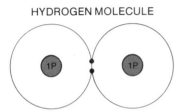

Two hydrogen atoms share a pair of electrons

WATER MOLECULE

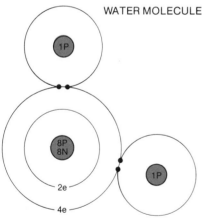

Fig. 4-12 Covalent bonds in the hydrogen molecule and the water molecule. The shared electrons are shown as black dots.

levels. In some cases, an atom *shares* one or more pairs of electrons from this level with another atom. The shared electrons go into new energy levels, or electron clouds, that belong to the molecule, instead of the individual atoms. *The sharing of pairs of electrons is called* **covalent bonding**. *One shared pair of electrons is called a* **single covalent bond**. Figure 4-12 shows one way of representing covalent bonding. Compounds that are formed by covalent bonding are called **covalent compounds**. As you will see later, biologists have a particular interest in the energy that is stored in covalent bonds. This energy is called **bond energy**.

Sometimes atoms form bonds by *transferring* electrons. Sodium and chlorine atoms unite this way (Fig. 4-13). Many atoms have a tendency to attract eight electrons into their outermost energy levels. Sodium and chlorine atoms have this tendency. This desired state can be achieved if a sodium atom transfers one electron to a chlorine atom. The transfer leaves the sodium atom with a positive charge. (It now has one more proton than electrons.) The transfer also gives the chlorine atom a negative charge. (It now has one more electron than protons.) *Atoms that have a charge, whether positive or negative, are called* **ions**. Since opposite charges attract one another, the sodium ions and chloride ions are attracted to each other, forming a chemical bond. *A bond that is formed by the transfer of electrons is called an* **ionic bond**.

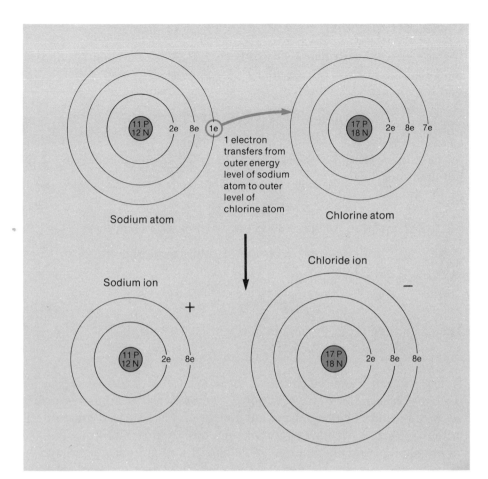

Compounds that are formed by ionic bonding are called **ionic compounds**. A crystal of sodium chloride (table salt) is an ionic compound because it consists of a network of sodium ions and chloride ions, held together by ionic bonds (Fig. 4-14).

Chemical formulas

A molecule of a compound is represented by a chemical formula. *A chemical formula consists of a combination of chemical symbols that represents the kind and number of each atom in the molecule.* A molecule of water consists of two hydrogen atoms and one oxygen atom. Its chemical formula is therefore H_2O. Note that the subscript $_2$ follows the hydrogen symbol. It tells us that there are two hydrogen atoms in the molecule. If there is no subscript, we assume that there is only one atom. One sodium atom unites with one chlorine atom. Thus the chemical formula of sodium chloride is NaCl. One hydrogen atom unites with one hydrogen atom. Therefore the formula for a hydrogen molecule is H_2. Formulas written in this manner are called **molecular formulas**.

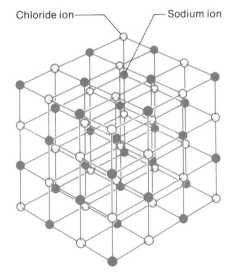

Chloride ion — Sodium ion

Fig. 4-14 A model that represents a crystal of sodium chloride. True molecules of sodium chloride really do not exist. Each ion is bonded to many neighbours, not just to one other ion.

Chemical Equations

In this course we will often write word equations. These summarize in words the chemical reaction that occurs between substances. For example, the burning of carbon is represented by the word equation:

$$\text{carbon} + \text{oxygen} \rightarrow \text{carbon dioxide}$$

Occasionally we will also use chemical equations. These use symbols and formulas to represent a chemical reaction. The chemical equation for the burning of carbon is:

$$C + O_2 \rightarrow CO_2$$

The word equation for the reaction of sodium with chlorine is:

$$\text{sodium} + \text{chlorine} \rightarrow \text{sodium chloride}$$

The unbalanced chemical equation is:

$$Na + Cl_2 \rightarrow NaCl$$

The balanced chemical equation is:

$$2Na + Cl_2 \rightarrow 2NaCl$$

Note that, in the balanced equation, the number of atoms of each element is the same on each side of the equation. Common sense says that this must be so. Atoms can be neither created nor destroyed. The balanced equation tells us that two atoms of sodium unite with one molecule of chlorine to form two "molecules" of sodium chloride. (The particles of an ionic compound are properly called **ion pairs**, not molecules.)

Note: You will not have to learn to write many formulas and equations in this course. You need to know much more chemistry before you can do that properly and easily. However, it is important to your progress in this course to understand what symbols, formulas, and equations are. Answer the following Review questions to make sure that you do understand. Check with your teacher if you have any problems.

Review

1. **a)** What is meant by the phrase "a model of an atom"?
   b) What are the three main elementary particles that make up all atoms?
2. **a)** Describe the nucleus of an atom.
   b) Explain the terms electron cloud and energy level.
   c) Define the terms atomic number and mass number.
3. **a)** What are isotopes?
   b) In what ways are the isotopes of hydrogen the same? How do they differ?
4. **a)** Distinguish between a chemical symbol and a chemical formula.
   b) Draw energy level diagrams for these atoms: $^{7}_{3}\text{Li}$, $^{24}_{12}\text{Mg}$, $^{32}_{16}\text{S}$.
5. **a)** What is a chemical bond?
   b) What is bond energy?

c) Distinguish between a covalent and an ionic bond.

d) Distinguish between an atom and an ion of the same element.

6. The molecular formula of a sugar called glucose is $C_6H_{12}O_6$.

   a) What does this formula represent?

   b) What information does this formula provide?

7. The balanced chemical equation for the burning of hydrogen gas is:

$$2H_2 + O_2 \rightarrow 2H_2O$$

   a) What does $2H_2O$ mean?

   b) How many *molecules* of hydrogen and oxygen reacted to form the $2H_2O$?

   c) How many *atoms* of hydrogen and oxygen reacted to form the $2H_2O$?

4.4 Some Basic Biochemistry

In Section 4.3 you learned some basic chemistry terms and ideas. Now you are ready to study that part of chemistry that deals with living things. It is called **biochemistry**.

Organic Compounds

All organisms can produce compounds, many of which are very complex. Such compounds were called *organic* compounds by the early chemists. They believed that only organisms could make such compounds. However, over the last one hundred years or so, chemists have made thousands of organic compounds in their laboratories. Some are the same as those made by organisms. Others are compounds that no organism ever makes. In spite of this, all such compounds are still called organic. The reason is that they all contain the element **carbon**. Chemists now define **organic compounds** *as those compounds that contain the element carbon.* A compound no longer needs to be associated with a living thing to be called organic. However, our interest here is in those organic compounds that are associated with living things.

Why carbon?

You may be asking: What is so special about carbon? Well, most important, the carbon atom has a special structure and behaviour (see Fig. 4-9). It has four electrons in its outermost energy level. The carbon atom therefore needs four more electrons to achieve the desired eight. It commonly gets these by forming four covalent bonds. Recall that many atoms try to get eight electrons in their outer energy level.

The carbon atom may form these bonds by joining with other atoms. An example of this is the compound methane (Fig. 4-15, A). It may also form some of those bonds by joining with other carbon atoms. It is this latter property that makes carbon so important to living organisms. The carbon

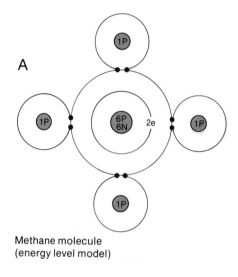

A

Methane molecule
(energy level model)

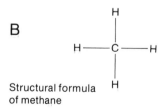

B

Structural formula
of methane

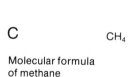

C

Molecular formula
of methane

Fig. 4-15   Methane, an organic compound. Each of the four hydrogen atoms shares its electron with one electron from the carbon atom. The result is a covalent bond between the carbon atom and each hydrogen atom.

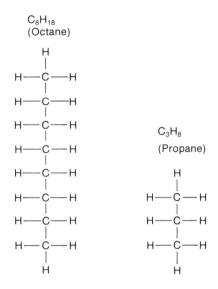

C_8H_{18}
(Octane)

C_3H_8
(Propane)

Fig. 4-16 Molecular and structural formulas for two common organic compounds. Note how the carbon atoms link together.

atoms often must form long chains and complex networks to build the large molecules required by living things. They can do this because they can bond to one another.

Structural formulas

Look at Part A of Figure 4-15. Clearly, it would take a long time and a lot of paper to draw a similar diagram for a molecule that has several carbon atoms in it. For that reason, chemists use **structural formulas**. *In a structural formula, the chemical symbol represents the atom and a line represents a single covalent bond.* Part B of Figure 4-15 is the structural formula for methane. Compare it closely with Part A. Part C shows the **molecular formula**, which you have already met.

Figure 4-16 shows the structural and molecular formulas of propane and octane. The former is the "bottled gas" that is used by many people for heating and cooking. The latter is a major ingredient of gasoline. Now let us look at some of the organic compounds that are important to organisms.

Carbohydrates

Carbohydrates are organic compounds that are composed of carbon, hydrogen, and oxygen. They contain hydrogen and oxygen in the same proportion as in water, namely two to one. Hence the name "carbo ... hydrate". There are three main classes of carbohydrates—sugars, starches, and cellulose. All these carbohydrates are vitally important to organisms. Among other things, most carbohydrates are energy sources.

There are many types of sugars. Among these are the **simple sugars** or **monosaccharides**. Most of their molecules contain six carbon atoms. Some simple sugars that are important to living things are glucose, fructose, galactose, and mannose. All have the same molecular formula, $C_6H_{12}O_6$. However, as Figure 4-17 shows, they have different structural formulas. Glucose is the main "fuel" for plant and animal cells.

Monosaccharides are the building blocks of more complex carbohydrates. For example, two monosaccharides can bond to form a **disaccharide** (Fig. 4-18). Some disaccharides are sucrose (one molecule of glucose linked to one molecule of fructose), lactose (one molecule of glucose linked to one molecule of galactose), and maltose (two molecules of glucose linked together). Before disaccharides can be used by organisms, they must be broken down into their monosaccharide units. The disaccharides have the molecular formula $C_{12}H_{22}O_{11}$. The following equation summarizes how two monosaccharide molecules form a disaccharide molecule by the loss of a water molecule.

$$C_6H_{12}O_6 + C_6H_{12}O_6 \rightarrow C_{12}H_{22}O_{11} + H_2O$$

Polysaccharides such as starch and cellulose, consist of many monosaccharide units bonded together. Some starches are made of thousands of

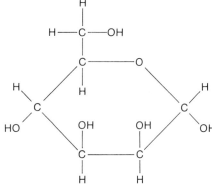

Glucose (grape sugar)
The only simple sugar known to exist
in a free state in the fasting human body.

Mannose
Found in manna.

Galactose
Found in milk sugar.

Fructose (fruit sugar)
Found in fruits and honey.

monosaccharide units. Plants often store their excess sugars in the form of starch. Potatoes, beans, and grains such as rice, corn, and wheat are examples of plants that store large quantities of starch. The starch molecules must be broken down by **hydrolysis** (the breaking down of molecules by water) into monosaccharides before they can be used as energy sources by organisms.

Humans and some animals produce a starch called **glycogen** in the liver. It is stored in the liver and in muscles. When extra energy is needed, the glycogen is broken down into glucose.

Cellulose is a polysaccharide that is still more complex than the starches. A cellulose molecule has glucose molecules bonded side by side and in long chains. The cell walls of plant cells are made of cellulose. Thus many natural products you use are cellulose-based. These include paper, cotton, and wood products.

Lipids

Lipids are another class of organic compounds important to organisms. They include **fats, oils,** and **waxes.** Lipids are similar to carbohydrates in that they contain only carbon, hydrogen, and oxygen. Many also serve as

Sucrose (cane or beet sugar)
The familiar sugar in domestic use.

Lactose
The principal sugar in milk.

Maltose (malt sugar)
Formed from barley starch.

sources of energy. In fact, a gram of fat can produce over twice as much energy as a gram of carbohydrate. Lipids also store energy. They differ from carbohydrates in one important way. The proportion of hydrogen to oxygen in carbohydrates is two to one. In lipids it is much greater.

A fat molecule consists of one molecule of glycerol and three molecules

Fig. 4-19 During digestion, a molecule of fat is broken down into its component parts: one molecule of glycerol and three molecules of fatty acid. Note that three molecules of water are used up in the process.

Fat + Three molecules of water → Glycerol + Three fatty acid molecules

of fatty acid (Fig. 4-19). During digestion, the fat is broken down into these simple molecules. Water molecules are required in this chemical reaction. Generally, fats are produced by animals. They are found in animal tissues and in dairy products such as milk and butter. Oils are usually produced by plants. They are liquid at ordinary temperatures. Some common vegetable oils are peanut, soyabean, and corn oil. Both plants and animals produce waxes. You have probably seen the waxy coating on some plant leaves. Beeswax is an example of a wax produced by an animal.

Proteins and Amino Acids

Like carbohydrates and lipids, **proteins** contain carbon, hydrogen, and oxygen. However, they also contain nitrogen. In addition, some proteins contain sulfur, and a few contain phosphorus and iron.

Proteins are large, complex molecules that consist of **amino acid** units linked together. There are about 20 different amino acids. All have the same general structure (Fig. 4-20). These amino acids are the building blocks of all proteins, both plant and animal. They link up with one another to build the wide variety of proteins that exist. For example, you have hundreds of kinds of proteins in your body and all are made out of those 20 amino acids.

The amino acid molecules join together by forming a bond called a **peptide bond**. When two amino acid molecules join, the resulting compound is called a **dipeptide** (Fig. 4-21). Dipeptides may also join together to form compounds called **polypeptides**. The polypeptides may join together to form proteins. Sometimes they join together in one long chain. In other cases parallel chains of polypeptides are joined together by cross links.

The number of amino acid molecules in a protein can vary from as low as 50 to as many as several thousand. The 20 different amino acid molecules combine in different numbers and sequences. As you might imagine, there is almost no limit to the number of different kinds of proteins that can be made from the 20 kinds of amino acids. Protein molecules are comparatively

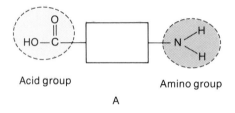

Acid group   Amino group

A

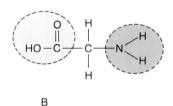

B

Fig. 4-20 Proteins consist of amino acid units. A:a generalized formula for an amino acid. B:a specific amino acid called glycine.

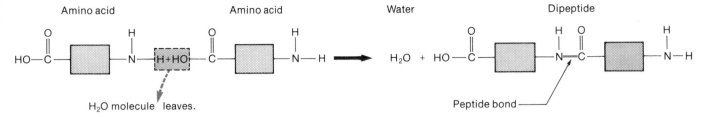

Fig. 4-21 Formation of a peptide bond. This is the way amino acid molecules join to one another in the formation of proteins.

large. For example, the hormone insulin is a protein. Its molecular formula is $C_{254}H_{377}O_{75}N_{65}S_6$.

Organisms of the same species have many proteins that are similar. In fact, you will see later that this is what makes them the same species. At the same time however, every individual of a species has proteins that are unlike those of other individuals of the same species. This is what makes each individual different.

Proteins have many uses in organisms. Some proteins are **structural proteins**. They form structural parts of cells such as the plasma membrane. **Enzymes** are proteins that aid processes such as digestion. Proteins can also serve as an energy source. Once carbohydrate and fat reserves are used up, proteins in muscles can be broken down to release energy. You will learn more about the uses of proteins later.

During digestion your body breaks down the plant and animal proteins that you ate into amino acids. Your body then puts the amino acids back together to make the different proteins that you require.

Nucleic Acids

Nucleic acids are large, complex organic molecules. They can consist of hundreds of thousands of atoms. Later, we will study two important nucleic acids. They are ribonucleic acid (RNA) and deoxyribonucleic acid (DNA). These molecules direct the making of proteins. They put the different amino acids together in the proper sequence to make the different proteins in an organism. As you will see in Unit 7, they also control heredity. DNA carries the genetic code. RNA assists the DNA in carrying out the message of the genetic code.

Review

1. a) How did the term "organic compound" originate?
   b) How is an organic compound defined today?
2. Describe the property of carbon that makes it so important to organisms.
3. a) What is a structural formula?
   b) Use an example to show that structural formulas give more information than molecular formulas.
4. a) What are carbohydrates?
   b) What are the three main classes of carbohydrates?
   c) Distinguish among monosaccharides, disaccharides, and polysaccharides.

d) Describe the importance of carbohydrates.

5. **a)** What are lipids?

   b) What are the three main classes of lipids?

   c) Describe the role of hydrolysis in the digestion of fats.

   d) Name one way organisms can use fats.

6. **a)** Distinguish among amino acids, dipeptides, polypeptides, and proteins.

   b) State three ways in which proteins are important to organisms.

7. **a)** Name two nucleic acids.

   b) Briefly state their functions.

Highlights

Matter is anything that has mass. It can exist in three states—solid, liquid, and gaseous. The particle theory explains the differences among these three states. It also accounts for many other properties of matter. This theory states that all matter is composed of tiny particles. These particles have large spaces between them. They are in constant motion, and they attract one another.

Matter can undergo changes. During physical changes, no new substance is formed. During chemical changes, one or more new substances are formed.

Energy is the ability to do work. Of special interest to biologists is chemical potential energy. This is the energy that is stored in the bonds of compounds.

Matter can be homogeneous or heterogeneous. Homogeneous matter includes solutions and pure substances. Pure substances are elements and compounds. Heterogeneous matter includes all mixtures except solutions.

All matter is made up of 92 naturally-occurring elements. The atoms of these elements join in countless combinations to form all the compounds on earth. All atoms are composed of elementary particles. The main elementary particles are protons, electrons, and neutrons. The protons and neutrons form the atomic nucleus and the electrons form an electron cloud around that nucleus. Energy level diagrams are often used to represent the atoms. Atoms are also represented by chemical symbols. Molecules are represented by chemical formulas and reactions are represented by chemical equations.

Atoms join to form either molecules or ion pairs. If electrons are shared during bonding, the bond is called covalent. If electrons are transferred, the bond is called ionic.

Organic chemistry deals with the study of the compounds of carbon. Four main groups of organic compounds are of interest to biologists. Carbohydrates are the main energy source for organisms. Lipids provide and store energy. Proteins build and maintain tissues. Nucleic acids control heredity.

Cell Structure And Function

In the last chapter you learned that organisms can make very complex organic molecules, such as carbohydrates, lipids, and proteins. These are all made by putting together small, simple molecules. How and where are these simple molecules put together? The answer is they are put together in cells. Here, complex molecules are made and organized into cellular material. In cells, non-living matter is converted into living matter.

Some organisms consist of only one cell. Others have trillions of cells. For example, an amoeba has only one cell. You have about sixty trillion cells (60 000 000 000 000). All organisms, regardless of size, are made of cells (Fig. 5-1). *Thus the* cell *is the basic structural unit of life.*

In this chapter you will use a compound microscope to examine several types of cells. Then you will study information that was gathered using high magnification electron microscopes. At the end of the chapter you should have a sound knowledge of cell structure and function.

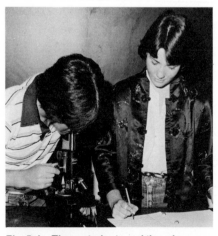

Fig. 5-1 **These students and the microscopic organisms that they are examining are all made of cells. The cells are about the same size in both the students and the microscopic organisms. The students are larger mainly because they have more cells.**

5.1 Development of the Cell Theory

Interest in the microscopic world began in earnest in the 1660s. It was then that Antony van Leeuwenhoek, a Dutchman, became interested in what he could see using a microscope. He built about 250 different microscopes in his lifetime. One was especially built for looking at the small organisms in pond water. Van Leeuwenhoek called them "little animals". Although he did not introduce the idea of cells, his work did a great deal to interest people in the study of microscopic life.

Also in the 1660s, an English scientist, Robert Hooke, discovered an interesting structure while looking at cork with a microscope. He saw regular rows of thick-walled units that he said looked like a honeycomb. Beekeepers call the units in honeycombs "cells". Hooke decided to use the same term. This appears to be the earliest time in history that the term "cell" was used in biological studies.

Many other scientists soon reported seeing cells in both plants and animals. However, no one seemed to realize the significance of this. Finally, in 1824, a French scientist, Henri Dutrochet, discovered that plants grew by increasing the number of their cells. After examining many plant and animal cells, he made up a generalization stating that, "the cell is truly the fundamental part of the living organism". From this point on, scientific developments came quickly.

In 1831, a Scottish scientist, Robert Brown, saw and described a centrally located body in the cell. He called it the **nucleus**. In 1835, another French scientist, Félix Dujardin, proposed the idea that all cells contained a "life-substance". He called it **protoplasm**. Then in 1838, Matthias Schleiden, a German botanist, concluded that all plants were made of cells. He reasoned further that those cells were alive and responsible for the operation of the organisms of which they were a part. In 1839, Theodor Schwann, a German zoologist, came to the same conclusions about animals. Scientists now knew that plants and animals had much more in common than anyone had ever imagined. As Schwann said, "We have thrown down a great barrier of separation between the animal and vegetable kingdoms." Finally, in 1858, Rudolf Virchow, also from Germany, pointed out that all cells came from other living cells. The foundation of the cell theory was now complete.

The Cell Theory

Today the work of all these scientists is summarized by the **cell theory**:

1. *Cells are the structural units of all living things.*
2. *Cells are the functional units of all living things.*
3. *All cells come from living cells.*

Review

1. Prepare a table on a page in your notebook, using the following column headings: Scientist; Contribution. In your table, list the names of the scientists mentioned in this section. Opposite the name, summarize his contribution to the development of the cell theory.
2. State the cell theory.

3. Select one scientist from this section whose work particularly interests you. Visit the library and collect further information on his life and work. Write a report of about 150 words on your findings.

5.2 Investigation

Care and Use of the Compound Microscope

Many of the investigations in this text use the compound microscope. Therefore, before you can do them, you need to know how to use it properly. Pay close attention to the details given in this investigation. You should develop the habit of using the microscope as we have outlined here. If you do, your findings will be more rewarding. Also, you will not damage this expensive instrument.

There are many kinds of microscopes (Fig. 5-2). Most school biology laboratories use the **monocular compound microscope**. Two styles are shown in Figure 5-2 (A and B). Most research laboratories use the more expensive **binocular compound microscope** (C). The advantage of the latter is that the observer uses both eyes which relieves eye strain. Also, it gives a three-dimensional appearance to the image. The **dissecting microscope** (D) is a low magnification microscope used for studying the features of whole objects.

This investigation introduces the monocular compound microscope. Usually the object to be examined is so thin that light can pass through it. Some parts of the object absorb more light than others and appear dark. Thus they stand out in contrast to the other parts. In this method of examining things, *transmitted* light is used. As Figure 5-2, D implies, a dissecting microscope often uses *reflected* light. The light is reflected or

Fig. 5-2 Some of the many types of microscopes.

A

B

C

D
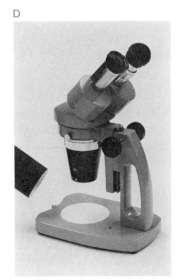

bounced off the object. Thus surface detail is about all that is seen with such a microscope.

Materials

monocular compound microscope
microscope slides (2)
cover slips (2)
paper towel
lens paper
piece of newspaper
piece of coloured picture from a magazine
medicine dropper
forceps (tweezers)
clear plastic ruler

Procedure A   Parts of the Microscope

Refer to Figure 5-3 while following these instructions.

a. Remove your microscope from its case or storage space in the room. Always grasp the **arm** firmly with one hand. Place the other hand under the **base**. This is the *only* way to lift and carry a microscope.
b. Place the microscope *gently* on the desk. The arm should point toward you and the stage away from you.
c. Follow Figure 5-3 closely as your teacher describes the parts of the microscope and their functions. If your microscope is not exactly the same as the one in the diagram, your teacher will point out any differences.

Procedure B   Preparation of the Microscope for Use

Every time you use the microscope, you should follow these preparatory steps:

a. Use the **coarse adjustment knob** to raise the **body tube** (or lower the **stage**) so that the **objectives** will not hit the stage when the **nosepiece** is turned.
b. Turn the nosepiece so that the **low power objective** (the shorter one) is above the hole in the centre of the stage. You will hear a click when it is in exactly the right position.
c. Close the **diaphragm** to its smallest opening.
d. Adjust the angle of the **mirror** so that light is reflected upward through the opening in the stage. Do not use direct sunlight. Look into the eyepiece (**ocular**). Continue to adjust the mirror until the **field of view** is illuminated evenly and brightly.
Note: If your microscope has a **lamp**, the mirror is not used.

Fig. 5-3   Parts of a compound microscope.

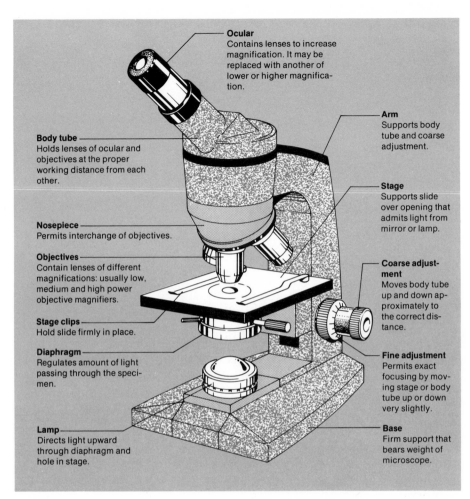

Ocular
Contains lenses to increase magnification. It may be replaced with another of lower or higher magnification.

Arm
Supports body tube and coarse adjustment.

Body tube
Holds lenses of ocular and objectives at the proper working distance from each other.

Stage
Supports slide over opening that admits light from mirror or lamp.

Nosepiece
Permits interchange of objectives.

Objectives
Contain lenses of different magnifications: usually low, medium and high power objective magnifiers.

Coarse adjustment
Moves body tube up and down approximately to the correct distance.

Stage clips
Hold slide firmly in place.

Fine adjustment
Permits exact focusing by moving stage or body tube up or down very slightly.

Diaphragm
Regulates amount of light passing through the specimen.

Lamp
Directs light upward through diaphragm and hole in stage.

Base
Firm support that bears weight of microscope.

e. Adjust the diaphragm until the field of view is of the proper intensity. Generally this means that it is bright but not glaring.

f. Clean the ocular and objective with a piece of lens paper if necessary. Use a gentle circular motion. Use a piece of lens paper only once. *Never* use any other type of paper or cloth. You may scratch the lenses.

Procedure C   Preparation of a Wet Mount

Follow this procedure every time you wish to prepare a wet mount of a specimen:

a. Hold a **microscope slide** by the edges and wet it with water. Wipe both sides dry with a paper towel. Always hold the slide by the edges or you will leave fingerprints on it.

b. Clean the **cover slip** in a similar manner. Keep in mind that it is very fragile. You can avoid breaking it by wiping both sides at once as shown in Figure 5-4.

c. Using the **medicine dropper**, place one drop of water near the centre of

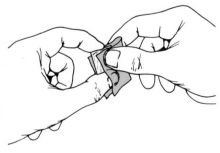

Fig. 5-4   **Proper technique for cleaning a cover slip.**

the microscope slide. This water is called a **mounting medium**. It helps produce a clear image.

d. Use the **forceps** to place a small piece of newspaper containing a small letter "a", "b", "d", "e", "g", or "h" in the drop of water. The smaller the letter, the better.

e. Hold a cover slip by the edges and lower one edge so that it touches one side of the drop of water at an angle of about 45° (Fig. 5-5). Now slowly lower the cover slip by supporting the upper edge with a pencil or dissecting needle. This will prevent the trapping of air bubbles under the cover slip. The bubbles will interfere with your viewing. A few small ones, however, will not cause serious problems. Just be sure that you do not confuse them with your specimen. Air bubbles appear as circular objects with thick, dark outlines.

Note: If the mount begins to dry out at any time, you can add water to it by placing a drop of water at one edge of the cover slip.

f. Place the prepared slide on the stage so that the letter is in a reading position and in the centre of the opening in the stage. Use the **stage clips** to hold the slide in place. The slide is now ready for viewing.

Procedure D   Focusing the Microscope

a. Using the **coarse adjustment**, turn the **low power objective** down (or raise the stage) as far as it will go. Watch what you are doing from one side, *not* through the ocular.

b. Look into the **ocular**. Slowly raise the objective (or lower the stage) by turning the coarse adjustment. Keep doing this until the letter comes into focus. You may have to move the slide slightly in order to find the letter. You can avoid eyestrain and headaches by keeping your other eye open while looking into the microscope. This takes a little practice.

c. Bring the image into the sharpest possible focus by using the **fine adjustment**. Try changing the **diaphragm** setting as well. Proper illumination increases the clearness of the image.

d. Move the slide slightly in various directions. Note carefully the corresponding movements of the image.

e. Now, proceed as follows to view the letter under medium power. Return the letter to the centre position on the stage. Make sure that the letter is still in sharp focus. Turn the nosepiece *carefully* so that the medium power objective is above the slide. Make sure that the objective *never* touches the slide or cover slip. It may be damaged if it does. Your microscope is likely *parfocal*. This means that, if one objective is focused, the other will also be in focus. Therefore you do not need to touch the coarse adjustment. A slight turning of the fine adjustment is sometimes required however. Now adjust the diaphragm for proper lighting.

Note: To be safe, *never use the coarse adjustment when you are on medium power*. Also, do not let the objective touch the cover slip or slide. This could scratch the objective.

Clean microscope slide

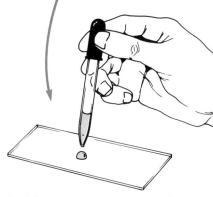

A   Add drop of pond water.

Cover slip

B   Add clean cover slip.

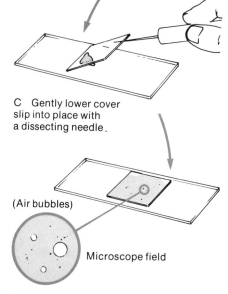

C   Gently lower cover slip into place with a dissecting needle.

(Air bubbles)

Microscope field

D   Ready for observation.

Fig. 5-5   Preparation of a wet mount.

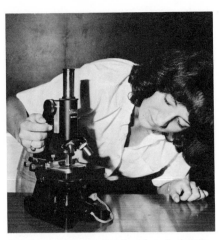

Fig. 5-6 Correct procedure for placing the high power objective above the slide. View the stage from the side to make sure the objective will not touch the cover slip.

f. Now, proceed as follows to view the letter under high power. Make sure that the letter is still in sharp focus. Then, while looking directly at the slide and cover slip from the side, turn the nosepiece *carefully* to bring the high power objective above the slide (Fig. 5-6).

g. Raise the body tube (or lower the stage) using the coarse adjustment. Remove the slide. (Keep it for Procedure G.) Then turn the nosepiece until the low power objective is in place. The microscope is now ready to return to its case or storage space. (Keep it, however, for the next procedure.)

Procedure E Magnification

Magnifications are expressed in **diameters**. The following example will illustrate the meaning of this term. If an object is magnified ten diameters (written 10 X), the image will be 10 times longer and 10 times wider than it would appear if the object were viewed from a distance of 25 cm with the unaided eye. The magnification is printed on each objective and ocular. Examine your microscope and note these magnifications. *The overall magnification of a microscope is the product of the magnifications of the objective and ocular being used.* Thus if a 10 X ocular is combined with a 40 X objective, the overall magnification is 10 x 40, or 400 diameters (400 X). Calculate the minimum and maximum magnifications that your microscope can produce.

Procedure F Resolving Power

The **resolving power** of a microscope is its ability to clearly separate details. If two objects are less than 0.1 mm apart, most people cannot resolve them with the unaided eye. That is, they cannot see them as separate objects. A microscope has a higher resolving power than your eye. Therefore it will let you see space between objects that are much closer than 0.1 mm. Study the resolving power of your microscope as follows:

a. Prepare a wet mount using the small piece of coloured picture from a magazine.

b. Examine the mount using low power. Describe what you see.

Procedure G Measuring with a Microscope

Objects that you examine under a microscope are usually so small that you cannot measure them with the smallest unit on your ruler, the millimetre (mm). Therefore you need a smaller unit. The one commonly used is the **micrometre** (μm). 1μm=0.001 mm; or, $1\,000\mu$m=1 mm.

Low power

Proceed as follows to estimate the size of a microscopic object under the low power objective of the microscope.

a. Place a clear plastic ruler on the stage. Focus on it using the low power objective and a 10 X ocular.

b. Move the ruler around until you have it in a position that measures the diameter of the field of view. Remember that one division is 1 mm. Note that the marks on the ruler appear very wide. Therefore consider 1 mm as being the distance from the *centre* of one mark to the *centre* of the next mark.

c. Record the diameter of the field of view in mm and in μm.

d. The size of an object can now be estimated by comparing it with the size of the field of view. Put your slide from Procedure D back on the stage. Estimate the height of the letter in mm and μm.

Medium power

The same procedure cannot be used for the medium power objective. The marks simply appear too far apart to measure accurately. Proceed as follows to estimate the size of an object under medium power:

a. Divide the magnification of the medium power objective by the magnification of the low power objective.

b. Now divide the diameter of the low power field by the number you obtained in step a. The result is the diameter of the medium power field of view. Suppose the medium power objective is 10 X and the low power objective is 5 X. Step a gives $\frac{10X}{5X}=2$. Suppose that the diameter of the low power field of view is 1 000μm. Then the diameter of the medium power field of view is $\frac{1\,000\mu m}{2}=500\mu m$.

High Power

The diameter of the field of view is calculated in the same way as was medium power. Simply substitute "high" instead of "medium" in steps a and b above.

Discussion

1. a) A hand lens (magnifying glass) is called a simple microscope. Why do you think your microscope is called a compound microscope?

　　b) What is the difference between a monocular and a binocular microscope?

2. a) Compare the orientation of the image of the letter with the orientation of the letter when you looked at it without the microscope.

　　b) Describe what happened to the image when you moved the letter in various directions.

3. a) What happened to the size of the letter when you switched to medium power?

b) What happened to the field of view when you switched to medium power?

c) What happened to the brightness of the field of view when you switched from low, to medium, to high power?

4. a) What is meant by the term "resolving power"?

b) Explain how Procedure F demonstrated the fact that your microscope has a greater resolving power than your unaided eye does.

5. a) State the diameter of the low power field of view of your microscope in mm and in μm.

b) What is the estimated height of the letter that you examined?

c) State the diameter of the medium and high power fields of view of your microscope in μm.

Note: Record the diameters of the low, medium, and high power fields of view where you can easily find them in your notebook. You will need these values in later investigations.

5.3 Investigation

Cell Structure—Onion Cells

Cells show wide diversity in structure. However, they also have many features in common. For example, all plant cells have three major parts: a **cell wall**, a **nucleus**, and **cytoplasm** (Fig. 5-7). Many plant cells also have **chloroplasts** and **vacuoles**. A number of other parts are common to most cells. Most of these, however, cannot be seen with your microscope.

The purpose of this investigation is to help you become familiar with some of the basic parts of a cell. The next two sections give you further experience in locating and identifying these parts. Then, in Section 5.6, you

Fig. 5-7 Most plant cells have many of the parts that are shown in this diagram.

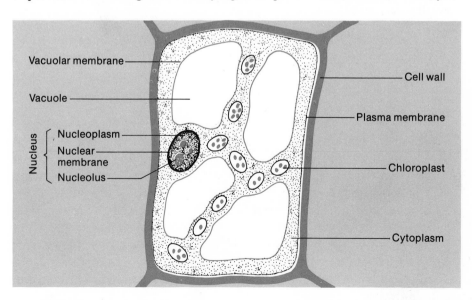

Vacuolar membrane

Vacuole

Nucleus { Nucleoplasm
Nuclear membrane
Nucleolus

Cell wall

Plasma membrane

Chloroplast

Cytoplasm

will study cell structure and function in considerable detail. To do so, you will use information that was gained using microscopes with much higher magnifications and resolving powers than those which you are using.

Materials

onion bulb scale	lens paper
compound microscope	medicine dropper
microscope slides (2)	forceps
cover slips (2)	iodine solution
paper towel	methylene blue solution

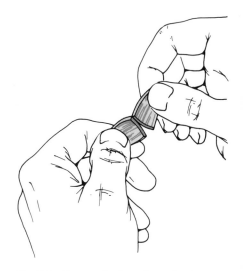

Fig. 5-8   How to obtain a piece of onion epidermis.

Procedure

a. Hold a piece of onion bulb scale so that the concave (inner) surface faces you. Then snap it backwards as shown in Figure 5-8. You should now be able to see the thin, transparent "skin", or **epidermis**, on the concave surface.

b. Use the forceps to pull off a small piece of this epidermis.

c. Prepare a wet mount of this sample as described in Procedure C of Section 5.2, page 90. Avoid wrinkling the epidermis.

d. Examine the sample under low power. Follow the steps outlined in Procedure D of Section 5.2, page 91. Continue your examination until you have completed as many as you can of the items in the Discussion section.

e. Now select one cell in which you can see the contents clearly. Move it to the centre of the field of view. Switch to medium power. Examine the contents closely. Focus up and down slightly with the fine adjustment as you do so. Continue to complete the Discussion items.

f. Now, switch to high power. BE CAREFUL. Refocus with the fine adjustment. Continue to complete the Discussion items.

g. Prepare a second wet mount of the onion epidermis. However, this time use iodine solution as the mounting medium instead of water. This solution is called a **stain**. Its purpose is to make some parts stand out more clearly.

h. Examine the stained sample under low, medium, and high power as you did in steps d to f. Continue to complete the Discussion items.

i. If time permits, repeat steps g and h using another stain, methylene blue.

Discussion

1. a) Describe the shape of a single cell of an onion epidermis.

  b) Describe the arrangement of the cells with respect to one another.

  c) Make a drawing of a group of five to ten cells, showing the features that you described in a) and b). The "lines" that run between the individual cells are the **cell walls**. They are non-living and made of cellulose.

2. **a)** Describe the **cytoplasm** of a cell. Include colour, clearness, and evidence of any motion. The outer edge of the cytoplasm is called the **plasma membrane** or **cell membrane**. It is difficult to see since it is normally pushed tightly against the cell wall.

   b) Describe the **nucleus** of a cell. If you saw them, include descriptions of the **nuclear membrane, nucleoplasm,** and **nucleoli** (there may be several). Are the nuclei always in the same position in the cell?

   c) Describe how the stain or stains that you used helped you to see cellular detail.

3. The "empty spaces" that you saw in the cytoplasm are called **vacuoles** They contain mainly water and dissolved substances. Each vacuole is surrounded by part of the cytoplasm called a **vacuolar membrane.** You probably noticed that some cells had only one vacuole that filled most of the cell. Explain why the nucleus in those cells was close to the cell wall.

4. The droplets in the cytoplasm are the oil that gives onions their smell and makes your eyes run. Describe an **oil droplet**.

5. Estimate the length of a single cell in micrometres (μm) using the method described in Procedure G of Section 5.2, page 92. You should already know the diameter of the field of view of your microscope, when using a 10 X ocular. Check the ocular and objective being used before making your calculations.

6. Make a drawing of a single cell as seen under high power. Make its length about 10 cm. Label all the following parts that you see: cell wall, nucleus, cytoplasm, plasma membrane, nuclear membrane, nucleolus, nucleoplasm, vacuole, vacuolar membrane, oil droplet.

5.4 Investigation

Cell Structure—*Elodea* Cells

Elodea, also called *Anacharis,* is a common plant of ponds and slow streams (Fig. 5-9). In this investigation you will examine the cells from a leaf of this plant. As you can see, *Elodea* is a green plant and therefore contains **chlorophyll**. You probably know that chlorophyll plays a role in the process of **photosynthesis** that is so important to life on earth. You will study that role in Chapter 7. At this time you will find out the location of the chlorophyll in these cells. In addition, you should see proof that cytoplasm is alive. You will also see an indication of the existence of the plasma membrane that you could not see in the last investigation.

Materials

Elodea leaf	paper towel	iodine solution
compound microscope	lens paper	5% salt solution
microscope slides (2)	medicine dropper	
cover slips (2)	forceps	

Fig. 5-9 *Elodea*, or *Anacharis*, is commonly called Canada waterweed. In ponds and slow streams it forms an excellent habitat for numerous small animals.

Procedure

a. Remove a young leaf from the tip of an *Elodea* plant. (Your teacher may have already removed the leaves. In that case, obtain a leaf from the container on the teacher's bench.)

b. Prepare a wet mount of the leaf as described in Procedure C of Section 5.2, page 90.

c. Examine the leaf under low power. Follow the steps outlined in Procedure D of Section 5.2, page 91. Continue your examination until you have completed as much as you can of the Discussion section.

d. Select one cell in which you can see the contents clearly. Move it to the centre of the field of view. Switch to medium power. Examine the contents of the cell closely. Focus up and down slightly as you do so. Continue to complete the Discussion items.

e. Now switch to high power. BE CAREFUL. Refocus with the fine adjustment. Continue to complete the Discussion items.

f. Stain the leaf by placing one or two drops of iodine solution at one edge of the cover slip. Then touch the water on the other side of the cover slip with a piece of paper towelling. The towelling will absorb the water and the iodine solution will be drawn under the cover slip to replace it. Note any changes in the appearance of the cell contents.

g. Obtain a fresh leaf from your teacher. Prepare a wet mount with it.

h. Examine the leaf under low power. When you have found a group of typical cells, switch to medium power.

i. Add some 5% salt solution to the leaf. Use the same procedure as in step f. Observe the result closely. It may take a short time before any change occurs.

Discussion

1. a) Describe the shape of a single *Elodea* cell.
   b) Describe the arrangement of the cells with respect to one another.
   c) Make a drawing of a group of five to ten cells. Show the features that you described in a) and b). Label the cell walls of two or three cells.

2. Describe any differences that you observe between this cell and the onion epidermis cell.

3. The green bodies in the cytoplasm are called **chloroplasts**. They contain the pigment chlorophyll. Iodine solution reacts with starch to form a dark blue or purple colour. What does step f of the Procedure indicate? Does iodine solution kill the cell? Explain.

4. You may have seen the green chloroplasts moving about the cell. They are unable to move by themselves. Therefore, the cytoplasm must be moving. This motion is called **cytoplasmic streaming**, or **cyclosis**. What does cyclosis tell you about the cell?

5. Make a drawing of a single stained *Elodea* cell as seen under high power. Make its length about 10 cm. Label all the following parts that you see: cell wall, nucleus, cytoplasm, plasma membrane, nuclear membrane, nucleolus, nucleoplasm, vacuole, vacuolar membrane, chloroplasts.

6. When the cell is placed in a salt solution that is more concentrated than the solution in the cell, a phenomenon called **plasmolysis** occurs. The concentrated salt solution causes water to move out of the cell. You will find out how this happens in Chapter 6. Describe and draw a plasmolyzed cell. Explain how these results indicate the presence of a plasma membrane at the outer edge of the cytoplasm. Does plasmolysis kill the cell? Explain.

5.5 Investigation

Cell Structure—Human Epidermal Cells

The plant cells that you examined in Sections 5.3 and 5.4 both had a cell wall. In addition, one of the cells contained chloroplasts. If you had the time to examine more types of plant cells, you would find that they all have a non-living cell wall that is usually made of cellulose. You would also see that many contain chloroplasts. How do animal cells compare to this?

In this investigation you examine human epidermal ("skin") cells that you obtain from the inside of your cheek. Your objective is to note the similarities and differences between animal and plant cells.

Materials

compound microscope	medicine dropper
microscope slide	toothpick
cover slip	iodine solution
paper towel	methylene blue solution
lens paper	

Procedure

a. Gently scrape the *inside* of your cheek with the blunt end of a toothpick. The purpose is to obtain a sample of the epidermal cells that are always flaking off in that region. You may have the best success if you scrape in the region of your molars (Fig. 5-10).

b. Prepare a wet mount of the scrapings. Thoroughly spread the scrapings through the drop of water before putting on the cover slip. Place the toothpick in the waste container.

c. Examine the scrapings under low power. Continue your examination until you have completed as many as you can of the Discussion items. *Note 1:* It is often difficult to distinguish cells from food particles. The cell mass should look much like irregular flagstones in a patio. *Note 2:* You will likely have to close the diaphragm further than you did for the previous investigations. These cells are quite transparent and cannot be seen if the light is too bright.

Fig. 5-10   The method for obtaining a scraping of cheek epidermal cells.

d. Now switch to medium power. Examine an individual cell closely. Look for the type of parts that you found in the plant cells. Continue to complete the Discussion items.

e. Now switch to high power. BE CAREFUL. Continue to answer the Discussion questions.

f. Stain the cells with methylene blue. Use the method outlined in step f of the Procedure for Section 5.4, page 97. This should darken the cell contents and make it easier for you to complete the Discussion items.

g. If time permits, make a fresh wet mount. Stain this one with iodine solution. Examine the cell contents closely under low, medium, and high power.

Discussion

1. a) Describe the shape of a single cheek epidermal cell.
  b) Describe the arrangement of the cells with respect to one another.
  c) Make a drawing of a group of five to ten cells showing the features that you described in a) and b).
2. a) List any differences that you observed between this cell and the plant cells.
  b) List any similarities that you observed between this cell and the plant cells.
  c) Are there more similarities than differences? What do you conclude from this fact?
3. Make a diagram of a human cheek epidermal cell as seen under high power. Make it 5-10 cm across. Label all the following parts that you see: nucleus, cytoplasm, plasma membrane, nuclear membrane, nucleolus, nucleoplasm, vacuole, vacuolar membrane.

5.6 Cell Structure and Function

Protoplasm and its Activities

You learned in Chapter 2 that living things have several common characteristics. They are called **life processes.** You learned in this chapter that all living things are made of cells (Fig. 5-11). Since cells are alive, it seems reasonable to assume that the life processes occur in the cells. We can further assume that they occur in the protoplasm of the cells.

Protoplasm is a granular, jelly-like substance that makes up most of the cell. It varies in composition from organism to organism and even from cell to cell in the same organism. It is about 70% water. The remaining 30% is carbohydrates, lipids, and proteins, with lesser amounts of many elements and compounds. Protoplasm is often called the **living material** of a cell. Yet, biologists do not agree as to whether or not it is actually alive. Certainly no one portion of a cell's protoplasm is alive. None of its components—water,

Fig. 5-11 There are countless millions of cells in this flower. Every living cell contains protoplasm. It creates the "living condition" of this plant.

Fig. 5-12 The electron microscope. The image can be viewed on a television screen. However, it is usually photographed so that it can be studied closely over a longer period of time. The world's first electron microscope to magnify more than a compound microscope was made at the University of Toronto in 1938. It was built by James Hillier and Albert Prebus, graduate students in the physics department.

carbohydrates, lipids, and proteins—is alive. Yet, somehow the protoplasm can organize all these substances into a "living condition".

Life processes occur in the protoplasm of cells. If an organism is unicellular (one-celled), then that one cell must perform all the life processes. However, in multicellular (many-celled) organisms, cells tend to specialize. They depend on one another. Different cells carry out different life processes. Yet there are many basic life processes that the protoplasm of all cells must be able to perform. Most important among these is metabolism.

You may recall that metabolism is the exchange of matter and energy between an organism and its environment, and the changes that occur in this matter and energy when they are within the organism. All cells must metabolize. That is, they must be able to ingest certain materials. They must also be able to digest, or break down, complex substances such as polysaccharides, fats, and proteins into simpler substances such as monosaccharides, fatty acids, glycerol, and amino acids. They must be able to assimilate, or put together, these simple substances into complex substances that are required to repair and maintain cell parts. They must be able to respire to release energy to maintain cellular activity. Finally, they must be able to excrete the waste by-products of metabolic activity. Metabolism is the total of all these processes. It is an activity of the protoplasm of living cells.

How does the protoplasm do all these things? The answer is that the protoplasm is organized. Certain parts carry out certain jobs. Many of these parts are called organelles ("little organs"). They are specialized parts that are found in the cytoplasm. Each organelle carries out certain functions. Yet it depends on the other organelles, too. The operation of the organelles in a cell can be compared to the operation of the organs in your body. Your stomach, lungs, brain, liver, and kidneys are organs. Each carries out certain functions. Yet they are all dependent on each other and must function together if your body is to survive. The organization of different structures for specific jobs is called division of labour. The term applies not only to cells but also to the relationships among the tissues and organs of multicellular organisms such as humans.

The electron microscope

Much of the information that we have about cell organelles and other cell parts has come from the electron microscope (Fig. 5-12). This type of microscope does not use light as the compound microscope does. Instead, it uses a beam of electrons. Magnifications as high as 200 000 X can be attained. In fact, if the resulting photograph is enlarged, the total magnification can be over 1 000 000 X. As well as high magnification, the electron microscope has a high resolving power. These two factors allow biologists to study fine cellular detail.

The remainder of this section describes the parts of a cell as they have

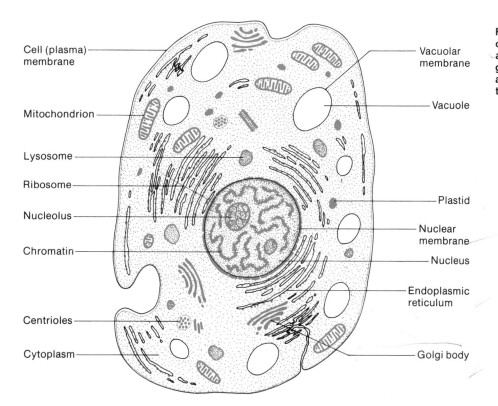

Cell (plasma) membrane

Mitochondrion

Lysosome

Ribosome

Nucleolus

Chromatin

Centrioles

Cytoplasm

Vacuolar membrane

Vacuole

Plastid

Nuclear membrane

Nucleus

Endoplasmic reticulum

Golgi body

Fig. 5-13 This diagram of a generalized cell was made from photographs taken with an electron microscope (electron micrographs). This cell is neither plant nor animal. No known cell looks exactly like this.

been seen with the electron microscope. In addition, it discusses the functions of these parts. Refer back to Figure 5-13 when you begin to read about a new part.

The Nucleus

Almost all types of cells contain a **nucleus** (Fig. 5-14). This spherical or oval body often lies near the centre of the cell, although sometimes it is far to one side. It is the control centre for all cell processes.

Nuclear membrane

The nucleus is surrounded by a **nuclear membrane**. Although very thin, this membrane actually consists of two membranes that are very close together. In various places the two membranes come together to form a **pore**. These are visible in an electron micrograph (Fig. 5-15). Biologists believe that these pores make it possible for certain substances to be exchanged between the nucleus and the cytoplasm.

Nucleolus

Inside the nuclear membrane is a thick colloidal dispersion called **nucleoplasm**. In most cells, one or more **nucleoli** (singular: **nucleolus**) are suspended in this nucleoplasm. Nucleoli are largely made of nucleic acids

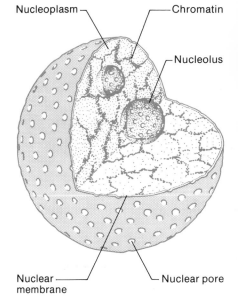

Nucleoplasm

Chromatin

Nucleolus

Nuclear membrane

Nuclear pore

Fig. 5-14 The nucleus as it would appear if a segment was removed from it.

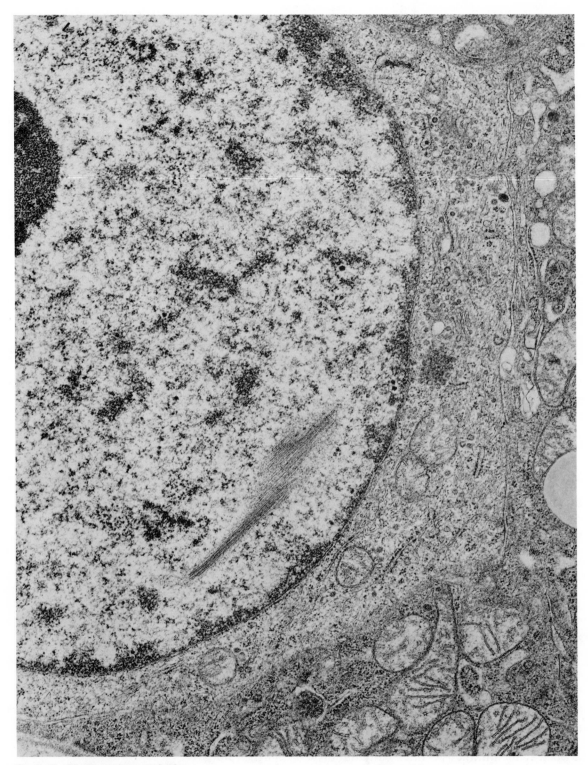

Fig. 5-15   Electron micrograph of a nucleus (6 000 X). Note the pores in the nuclear membrane. The dark area is a nucleolus.

(DNA and RNA) and proteins. They produce proteins and RNA that are used by organelles called **ribosomes** in the cytoplasm.

Chromatin

It has been known for some time that the nucleus contains **chromosomes**. However, these are normally present in an almost invisible state called **chromatin**. The chromatin strands thicken to form chromosomes which become visible at the time of cell division (see Chapter 8). Chromosomes are made of DNA and proteins. Recall that DNA is the hereditary material that carries the genetic code from generation to generation. This genetic code controls all the cell processes.

The Cytoplasm

The cytoplasm is the protoplasm that is outside the nucleus. You have seen it in three types of cells. It appears somewhat clear but granular, and is often moving, as you saw in the *Elodea* cell. The movement is called **cyclosis**, or **cytoplasmic streaming**.

Plasma membrane

The cytoplasm forms a membrane at its outer margin. It is called a **plasma membrane** or **cell membrane**. Its main function is to separate the cell from its environment or, in other words, to hold the contents of the cell together. It also controls the passage of certain materials in and out of the cell in order to keep the cell alive. How does it do this? Through your microscope, the plasma membrane appeared to be a single line that was the outer edge of the cytoplasm. However, the electron microscope shows that it, like the nuclear membrane, actually consists of two layers (Fig. 5-16). Much research has

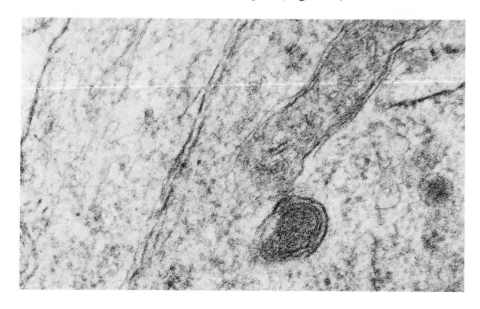

Fig. 5-16 Electron micrograph of a plasma membrane (86 500X). As you can see, the membrane consists of two layers.

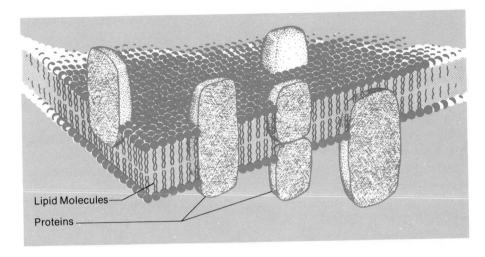

Fig. 5-17 Studies suggest that the plasma membrane may be composed of two layers of lipid molecules with protein molecules scattered throughout the lipid molecules.

Lipid Molecules

Proteins

been done on the makeup of the layers. Still no definite answer is available. However, biologists know that the membrane is made of lipids and proteins. Many of them think that the membrane consists of two layers of lipid molecules with protein molecules scattered through the lipids (Fig. 5-17).

All the molecules in the plasma membrane can move. Therefore this membrane is a very active part of the cell. Since it is not solid, molecules can pass through it. However, the membrane can control, to a considerable degree, the types of molecules that enter and leave the cell. You will study how this is done in Chapter 6.

Endoplasmic reticulum

Although cytoplasm appears almost clear under an ordinary microscope, it appears much more complex when viewed with an electron microscope (Fig. 5-18). In most cells it appears to contain a network of tube-like structures. Together, these structures are called the endoplasmic reticulum. The endoplasmic reticulum appears to consist of double membranes that are parallel to one another. As Figure 5-13 shows, the endoplasmic reticulum meets both the nuclear membrane and the plasma membrane.

Most biologists believe that this tube-like network transports materials within the cell. It would appear that substances could easily move through the network from the plasma membrane to the nuclear membrane. The endoplasmic reticulum also has ribosomes attached to it in various places.

The Cell Organelles

Organelles ("little organs") are specialized structures that are located in the cytoplasm. The chloroplasts that you saw in the *Elodea* cell are organelles. A few other types can be seen with the compound microscope, but many cannot. Their structural details were not known until the electron microscope came into use. Today, we know a great deal about their structure and function. It appears that each type of organelle directs a specific chemical

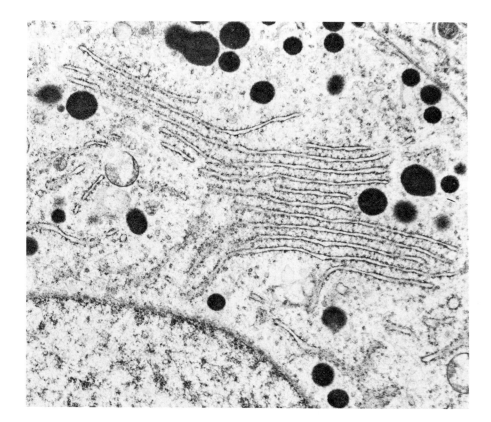

Fig. 5-18 Electron micrograph of the endoplasmic reticulum (13 000 X). Note how this tube-like structure is lined with ribosomes (the small dark dots). The endoplasmic reticulum transports substances within the cell. The ribosomes make proteins.

function in the cell. You will study this in more detail in the remaining chapters of Unit 2. Here we will describe the structure and location of the organelles, with only a brief treatment of how they perform their functions.

Ribosomes

These tiny, grain-like organelles are located either on the surface of the endoplasmic reticulum or floating freely in the cytoplasm. You can barely see them in Figure 5-18. They are composed mainly of the nucleic acid, RNA. They also contain enzymes that are involved in protein synthesis. Biologists believe that the ribosomes are the sites of protein synthesis in the cell. Here amino acids are put together to form proteins. (See Section 4.4, page 83). Perhaps the proteins that are made in the ribosomes move into the endoplasmic reticulum to be carried to all parts of the cell.

Golgi bodies

The electron microscope clearly reveals these organelles. They are often near the nucleus but can be scattered throughout the cytoplasm. They usually appear as a set of several flattened membranous tubes. The ends of the tubes have tiny sacs (Fig. 5-19). The exact function of the Golgi bodies is not yet known. However, these organelles appear to be involved in the processing of materials that are to be secreted from the cell or transported elsewhere in the

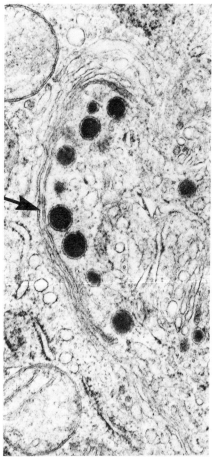

Fig. 5-19 Electron micrograph of Golgi bodies, also called Golgi apparatus (14 000 X).

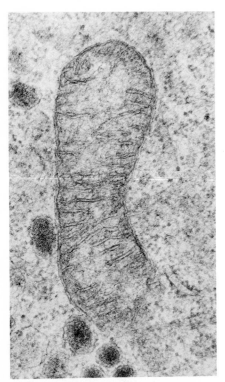

Fig. 5-20 Electron micrograph of a mito-chondrion (42 000 X). Note the inner and outer membranes and the infoldings that provide greater surface area.

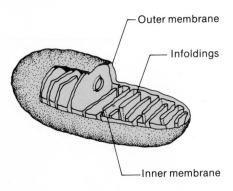

Outer membrane

Infoldings

Inner membrane

Fig. 5-21 A 3-dimensional drawing shows the folds in the inner membrane of a mito-chondrion better than an electron micro-graph does.

cell. They may also serve as a storage place for such materials. Some biologists believe that proteins travel from the ribosomes, through the endoplasmic reticulum to the Golgi bodies. There they unite with carbo-hydrates to produce some of the materials stored in or transported from the Golgi bodies. Golgi bodies are also thought to be possible sites for the synthesis of new endoplasmic reticulum.

Mitochondria (singular: mitochondrion)

You can see mitochondria in some cells with a compound microscope. They look like tiny specks, dots, or rods. However, under the electron micro-scope, their shape and inner structure become visible (Fig. 5-20). They vary from spherical to sausage-shaped and are filled with a fluid. A mitochon-drion is surrounded by two membranes. Each consists of a layer of lipid molecules and a layer of protein molecules, much like the nuclear membrane and plasma membrane. The inner membrane folds inward at many places (Fig. 5-21). These infoldings provide a larger surface area on which chemical reactions can occur.

Biologists discovered that cells which required great amounts of energy had more mitochondria than other cells. They therefore assumed that these organelles were involved in energy release. The assumption has been shown to be true. The mitochondria are the centres of respiration in the cell. They contain enzymes that help break down energy-rich organic compounds. This releases energy that can be transferred to other compounds and, eventually, to the cell as a whole. The mitochondria release the energy that powers all cell activities.

Lysosomes

Most cells seem to contain spherical organelles called lysosomes. They are surrounded by a single membrane and are usually slightly smaller than mitochondria (Fig. 5-22). Lysosomes appear to function as storage vessels for many powerful digestive enzymes. The membrane of the lysosome is able to resist the digestive action of the enzymes. If it did not, the enzymes would escape into the cell and digest its contents. In fact, if the lysosome membrane is broken, the enzymes do immediately begin to break down the surrounding cytoplasm. It is believed that these enzymes are used to break down, or digest, foreign materials such as bacteria. They also speed up the digestion of worn-out cell parts. The digestion of carbohydrates, lipids, and proteins within the cell also occurs inside the lysosomes.

The digestive enzymes in the lysosomes are produced by ribosomes. They pass through the endoplasmic reticulum into the Golgi bodies. They then collect in the tiny sacs at the ends of the Golgi bodies. These sacs enlarge to become lysosomes. Material is digested inside the lysosomes. Here, large molecules are broken down into smaller molecules. These smaller molecules then move from the lysosomes into the cytoplasm.

Plastids

These organelles are found mainly in the cells of green plants and some protists. Some plastids manufacture food; others store food. You have already seen the most common plastid, the chloroplast. It contains the chlorophyll that is needed for the process of photosynthesis. Its main function is to produce carbohydrates.

A chloroplast contains bodies called grana (Fig. 5-23). The grana consist of layers of lipids and proteins. The chlorophyll molecules are found between these layers. The light energy that is required for photosynthesis is trapped by the grana.

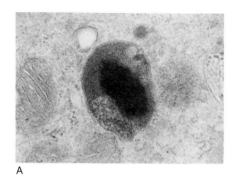
A

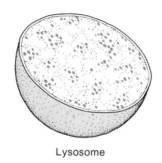
B

Lysosome

Fig. 5-22 Structure of a lysosome. A: as seen through the electron microscope. B: a 3-dimensional drawing.

Some chloroplasts contain other pigments as well as chlorophyll. Carotenes are orange and yellow pigments that are commonly found in chloroplasts. The xanthophylls, which are pale yellow pigments, are also sometimes found there.

Two other types of plastids are common in plant cells. Chromoplasts contain carotenes and xanthophylls as well as red or blue pigments, such as anthocyanin. They are found in the cells of many coloured fruits and flowers such as tomatoes and red roses. Leucoplasts serve mainly as a storage place for starch. They contain the enzymes that are required for the conversion of glucose to starch. Although this process does occur in some chloroplasts, most of it occurs in leucoplasts. Since starch is not soluble in water, it is not found anywhere else in the protoplasm except in the plastids. If energy is required elsewhere in the cell, the starch must be converted back to glucose. The glucose can dissolve in the water of the protoplasm.

You have probably seen leaves on trees change from green to many beautiful colours in the fall. Most people say this is caused by frost. However, this is not true. During the summer the leaves appear green because the green of the chlorophyll in the leaves hides the paler yellow and orange of the carotenes that are also present. But in the fall, the shorter period of daylight and the lower temperatures cause the chlorophyll to break down. The carotenes, however, do not break down and the leaves appear yellow or orange. Other chemical changes also occur. New pigments are often formed such as anthocyanin, a red pigment. The best environment for anthocyanin formation is found on warm autumn days followed by cool nights. In the daytime, sugars are produced in the leaves. The cool nights

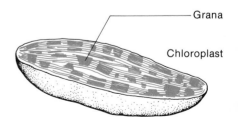

Grana

Chloroplast

Fig. 5-23 Structure of a chloroplast. Note the structure of the grana.

prevent the sugars from being transported from the leaves to other parts of the tree. At the same time, the abscission layer starts to grow. (This is the layer that eventually separates the dead leaf from the tree.) The abscission layer finally traps all the sugars in the leaves. There they are converted, along with other compounds, to the red pigment, anthocyanin. The replacement of chlorophyll by anthocyanin also occurs when many fruits, such as the tomato, ripen. The green colour changes to red as chloroplasts are converted into chromoplasts.

Centrioles

Most animal cells contain a pair of organelles called **centrioles**. These are two bundles of rod-like structures located at right angles to one another, usually near the nucleus. (See Fig. 5-13 for location and arrangement.) Each centriole consists of nine rod-like structures that are arranged to form a cylindrical shape (Fig. 5-24). Each structure, in turn, consists of three smaller rod-like structures. The centrioles play a key role in cell division (see Chapter 8).

The Cell Wall

Most plant cells have **cell walls** for support and protection. They consist largely of cellulose. The cell walls of soft plant structures such as leaves are relatively thin. However, in structures such as stems that must be hard and firm, the cell walls are thicker. A stem turns woody when more and more cellulose is deposited. These thick walls last long after the cell has died. We can often see them in a piece of wood. A piece of wood is made of dead cell walls.

Vacuoles

Both the onion epidermal cell and the *Elodea* cell that you examined contain **vacuoles**. Though they appear empty, each vacuole contains a fluid. The fluid is a colloidal dispersion. It consists of protein and sugar, molecules, other organic molecules, and many inorganic ions, all suspended in water. This mixture is called **cell sap**. Each vacuole is surrounded by a **vacuolar membrane**. This membrane is produced by the cytoplasm and is thought to be somewhat similar to the plasma membrane. Thus it controls the movement of materials in and out of the vacuole.

Some of the cells you examined had just one large vacuole. Such a vacuole is called a **central vacuole**. A central vacuole is common in older plant cells. As the plant cells grow and age, small vacuoles join together to form the one large vacuole. The central vacuole often crowds all the cellular contents against the cell wall.

Many aquatic protists such as the amoeba and paramecium contain vacuoles. In such organisms, some vacuoles serve mainly to "pump" excess water from the organism. The vacuoles, when full, move to the plasma membrane of the organism. There they contract, expelling their contents

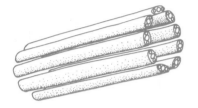

Fig. 5-24 Structure of a centriole. Note that the centriole consists of nine rod-like structures. Note, too, that each of the structures consists of three smaller rods.

from the organism. Because of this behaviour, they are called **contractile vacuoles**.

Review

1. What is protoplasm? What is it made of? Why is it better to say that protoplasm creates a "living condition" than to say protoplasm is alive?
2. **a)** What is metabolism?
   b) Name and describe the main metabolic processes that all cells must be able to perform.
3. **a)** What are organelles?
   b) What is meant by division of labour?
4. What advantages does the electron microscope have over the compound microscope?
5. **a)** What is the main function of the nucleus?
   b) Name and describe the four main parts of the nucleus.
   c) The structure of the nuclear membrane keeps certain substances in the nucleus but allows other substances to move freely in and out. Why is this important?
6. **a)** Describe the structure of the plasma membrane.
   b) Explain the importance of this structure.
   c) Describe the structure of the endoplasmic reticulum. Of what value is this structure to a cell?
7. Make a table with the column headings "Organelle" and "Function". Complete the table for all the organelles discussed in this section. In each case, summarize the function in just a few words.
8. Would you expect to find many mitochondria in a muscle cell? Explain.
9. **a)** In many respects a unicellular organism is more versatile than a multicellular organism. Explain why this is so.
   b) What disadvantages might there be in having all the activities of an organism concentrated in one cell?
10. Suppose that a scientist mixed in a flask all the substances that make up protoplasm. Would the mixture now be a "living condition"? Explain your answer.

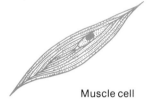
Muscle cell

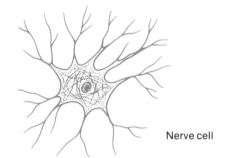
Nerve cell

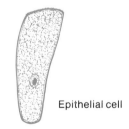
Epithelial cell

Fig. 5-25   Cell specialization. All the cells in your body do not perform the same functions. For what functions are each of these cells specialized?

5.7 Levels of Biological Organization

Organization is a characteristic of living things. That is, all the structures in a living thing must be organized in a way that produces and maintains a "living condition". Many levels of organization exist. Let us see what they are, beginning at the simplest level, the cellular level.

The Cellular Level

As you know, the cell is the basic structural unit of all living things. Nothing smaller than a single cell can be called "living". Thus one can say that the cellular level is the simplest level at which structures can be organized into a "living condition".

Clearly, unicellular organisms (monerans and protists) can be organ-

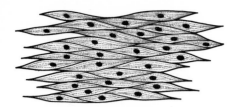

Muscle tissue

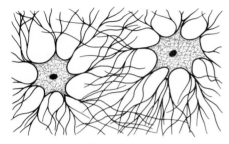

Nerve tissue

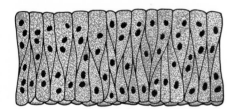

Epithelial tissue

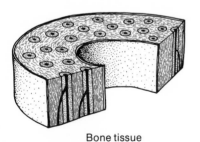

Bone tissue

Fig. 5-26   Some examples of animal tissues. Note that the cells in a particular tissue are alike.

ized only at the cellular level. Within each cell the tasks of carrying out life processes are spread among the organelles in an organized way. Each organelle performs its task for the organism as a whole. This is a simple example of division of labour. All the division of labour takes place inside just one cell.

Multicellular organisms (plants and animals) are much more complex. In order to function efficiently, they must have a more complex organization. However, their simplest level of organization is still the cellular level. They differ from unicellular organisms in that their cells show **specialization**. This means that each type of cell is specialized to perform a certain function (Fig. 5-25). For example, your body contains nerve cells, muscle cells, bone cells and many other types of cells, each of which has a special function.

The Tissue Level

A **tissue** *is a group of cells that have the same structure and function.* It is the second level at which structures can be organized in an organism. Alone in a multicellular organism, a single muscle cell could accomplish very little. Many muscle cells however, working together in a tissue, can move an arm. Most species of animals have many types of tissues. Some of the main types found in vertebrates are muscle tissue, bone tissue, nerve tissue, and skin tissue (Fig. 5-26). Most species of plants have tissues, too. Some of the main types found in flowering plants are conducting tissue, photosynthetic tissue, storage tissue, and epidermal tissue ("skin").

The Organ Level

An **organ** *consists of several tissues working together as a unit to perform a specific function.* It is the third level of organization of biological structures. Your heart is an organ. It consists of muscle tissue, nerve tissue, elastic connective tissue, fibrous tissue, and many other types. All these tissues work together to pump blood. If any one tissue failed to do its job, the organ would stop operating.

Your arm is also an organ. It is made of bone tissue, muscle tissue, skin tissue, nerve tissue, blood tissue, and several other types of tissues. Your stomach, brain, and kidneys are also organs.

Plants have organs, too. A plant leaf is an organ. It consists of conducting tissue, support tissue, epidermal tissue, photosynthetic tissue, and storage tissue. A plant stem is also an organ (Fig. 5-27).

The Organ System Level

An **organ system** *is a group of organs that work together to perform a specific function.* This fourth level of organization is found only in animals. A single organ may not be enough to permit a complex animal to carry out one of the major functions of its body. Two or more organs, working together, may be needed.

The human digestive system is an example of an organ system. In order to perform digestion properly in such a complex organism as the human, several organs are needed. Among these are the jaws, salivary glands, esophagus, stomach, intestine, and liver. Other organ systems in the human body are the breathing system, the circulatory system, the nervous system, and the skeletal system. In complex animals, there is an organ system to perform almost every life process.

Other Levels of Biological Organization

Biological organization does not end with organ systems. Organ systems function together to make a fifth level, the organism itself. This level is often called the individual by biologists.

The individuals of a certain species function together in what is called a population. Examples of populations are the herring gulls on a lake, the maple trees in a woodlot, and the Mallard ducks on a pond (Fig. 5-28).

Several different populations living together make up a community. Thus all the organisms in a woodlot make up a woodlot community. Some of the populations in a woodlot community are the maple tree population, the pine tree population, the oak tree population, the deer population, the snail population, and the trillium population. Similarly all the populations of organisms that live in and around a pond make up the pond community (Fig. 5-29).

Fig. 5-27   This leaf is a plant organ. It consists of several tissues, each of which has a certain function. What are some of those tissues?

Fig. 5-28   These Mallard ducks make up a population. At what level of organization is a single duck?

Fig. 5-29 How many different populations are in this pond community? See if you can name at least five different ones.

Fig. 5-30 What communities make up this alpine biome? Try to name at least five different ones.

All the different communities in a certain geographical region, with a characteristic climate, make up a **biome**. A desert is a biome. A desert biome consists of a sand dune community, an oasis community, a sand plain community, and several other types. An Arctic biome consists of a bog community, a coastal community, a pond community, a lake community, and others (Fig. 5-30).

Finally, all the world's biomes function together to make the highest level of biological organization, the **biosphere**. This level includes all life on earth.

You will learn more about the levels from population to biosphere in Unit 6, Ecology.

Levels Below the Cell

There are many levels of organization smaller than the cell. While these are not biological in nature, it is still interesting to think about them.

Below the cellular level is the **organelle** level. The organelles function together to make the cell. Below the organelle level is the **macromolecule** (giant molecule) level. The macromolecules such as lipids and proteins go together to make organelles. Below the macromolecule level is a level made up of **molecules** and **ions**. Can you name any levels below this?

Review

1. **a)** Make a list of the levels of biological organization. Begin with the simplest level and end with the most complex level.
   b) Make a list of the levels of organization of materials in which the cell is the most complex.

2. a) Explain the terms tissue, organ, and organ system. Give two examples of each.
  b) Describe the relationships between the following levels of biological organization: individual, population, community, biome, biosphere.

Highlights

The cell is the structural and functional unit of all life. Cells can exist singly as unicellular organisms. They can also be part of a multicellular organism. All cells contain protoplasm. It organizes cellular substances into a "living condition".

Almost all types of cells have a nucleus. The nucleus is the control centre for all cell processes. It contains chromatin which thickens into chromosomes during cell division. Chromosomes contain DNA, the hereditary material that carries the genetic code.

Organelles within a cell provide a simple form of division of labour. There is a specific type of organelle for almost every cellular process. For example, mitochondria are the centres of respiration for the cell. Ribosomes are the sites of protein synthesis.

Many levels of organization exist among the structures of living things. The simplest level is the cellular level. The most complex level is the world biosphere.

Cell Physiology: Diffusion And Osmosis

In Chapter 5 you studied the structure of cells. In other words, you learned the parts of a cell. You also looked briefly at the functions of those parts. That is, you learned what each part does for the cell as a whole. However, you did not find out *how* the parts function (Fig. 6-1). How does a water molecule get into a cell? How does a particle to be digested get into a lysosome? How does a chloroplast make carbohydrates? How does a mitochondrion break down carbohydrates to release energy?

The study of biological function is called physiology. The study of how the parts of a cell function is called cell physiology.

In this chapter we will centre our attention on the functioning of one particular part of the cell, the plasma membrane, or cell membrane. All cells have a plasma membrane. We know that, for the cell to remain alive, this membrane must allow certain materials to pass in and out of the cell. How does it do this? How can the plasma membrane allow only certain substances in? How can it let certain substances out but retain those that the cell needs?

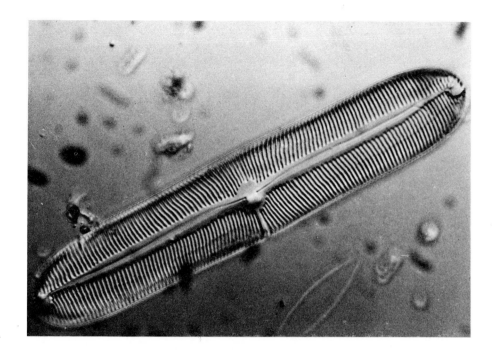

To answer such questions, we must first understand a process called diffusion. Once we do, we can understand a process called osmosis. Then we can explain how materials get in and out of cells through the plasma membrane.

6.1 The Process of Diffusion

If you release some air freshener into one corner of a room, the odour does not stay there. It gradually spreads, or **diffuses**, throughout the room. This happens even if the room is tightly sealed to prevent air currents. The odour seems to move by itself. Such movement is called **spontaneous**. What causes this spontaneous movement, or diffusion?

An Analogy of Diffusion

Suppose that all the students in your class were blindfolded and then crowded into one corner of the gymnasium. They were then told to begin walking in straight lines, if possible. Their movement would be random. At a certain time, a student would be walking in one direction. A few seconds later, he or she would be walking in another direction. A student would walk in a straight line until a collision with another student caused him or her to change direction. Clearly, the longest walks without collisions would occur in the direction where there were the fewest students—away from the corner. However, some students would, no doubt, end up back in the corner where they started. Still, in the early stages of this experiment, the *net* (overall) movement would be away from the corner. The reason is simple. There are more people in the corner to move out than there are people nearby to move

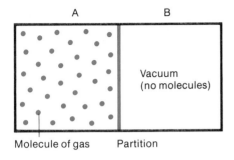

Molecule of gas   Partition

STEP 2
Partition has just been removed.

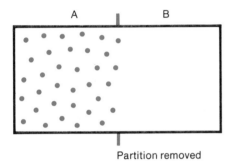

Partition removed

STEP 3
Some time after partition was removed.

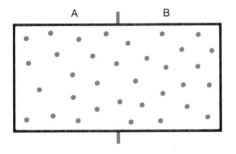

Fig. 6-2 Diffusion of a gas into a vacuum.

in. The time will come when as many students will be returning as leaving. At this time, your classmates will be spread more or less evenly throughout the gymnasium. This situation is called a **dynamic equilibrium**: *dynamic* because there is movement, and *equilibrium* because there is balance.

Suppose an identical experiment were performed except that this time, 100 other students, also blindfolded, were moving around the gymnasium floor before your class began to move. Would their presence stop your class from reaching a "dynamic equilibrium"? The answer is no. Each student in your class would simply have more collisions in a given time. The movement would still be random. More of your class would still be available for moving out of the corner than in. Therefore the same type of spreading would occur. It will simply take longer because of the increased collisions.

Let us now try to apply this analogy to particles.

The Particle Theory and Diffusion

The particle theory states that all matter is made of particles. Each particle is in constant motion. It moves randomly from place to place. It moves in a straight line until it collides with another particle. Then it will probably move off in a new direction, but still in a straight line. The particle is able to move because it has kinetic energy. Does this random motion of the particles explain why air freshener spreads evenly throughout a room? Do particles, like people, gradually spread out from an area where they are concentrated?

Look at Step 1 of Figure 6-2. The molecules of a gas are trapped by a partition in Side A of a box. Side B contains nothing. At this time, it is quite obvious that more molecules are available to move from A to B than from B to A. The same situation applies at the moment when the partition is removed (Step 2). Thus, there will be a net movement of molecules from A to B just after the partition is removed. Remember, "net" means "overall". Some molecules will move from B to A. But more will move from A to B, simply because there are more in A to do the moving. However, as more and more molecules move to Side B, more become available to move back to Side A. Eventually the rate at which they move from B back to A will equal the rate at which they move from A to B. A dynamic equilibrium has been set up.

Figure 6-3 shows a situation where Side B also contains a gas. The gas in Side B is different from the gas in Side A. The diagrams show that, after the partition is removed, both gases eventually reach dynamic equilibrium. There is a net movement of the A molecules to Side B and a net movement of the B molecules to Side A until the equilibrium is reached. The time required to reach equilibrium will be longer than in Figure 6-2. After the partition is removed, there are twice as many molecules in the same volume compared to the example in Figure 6-2. Thus more collisions will occur, slowing down the diffusion.

Definition of Diffusion

Using the particle theory, an understanding of diffusion, and the results of many experiments, scientists have produced a definition of diffusion:

Diffusion *is the process by which a substance spreads spontaneously in all directions from a region where there is a high concentration of the substance to a region where there is a lower concentration (perhaps none) of the substance.* Diffusion continues until a dynamic equilibrium exists between the regions that once had the high and low concentrations. During diffusion there is a *net* movement of particles from a region of high concentration of particles to a region of lower concentration of particles. At the point of dynamic equilibrium, the net movement is zero. Particles are still moving. But they are moving at the same rate in all directions.

Factors Affecting the Rate of Diffusion

Three factors affect the rate at which a substance will diffuse. These are concentration, temperature, and pressure.

Our previous discussions should have made it clear that a higher **concentration** of particles causes a more rapid rate of diffusion. An increase **temperature** also increases the rate of diffusion. At higher temperatures the particles move more rapidly. Thus they can move more rapidly from a region of high concentration to a region of lower concentration. An increase in **pressure** also increases the rate of diffusion. This is particularly true with a gas. An increase in pressure moves the particles closer together. This increases the concentration of the gas. Thus more particles in the region are available to move out.

Note: The explanations and definition of diffusion were all based on the particle theory. The particle theory has stood the test of time. Therefore we can be reasonably sure that our explanations and definition are correct. However, the scientific method suggests that we should consider these as tentative explanations until we have performed experiments to support them. The next section provides those experiments.

Review

1. **a)** What is physiology?
   b) What is cell physiology?
2. **a)** What does the term "spontaneous" mean?
   b) What is a dynamic equilibrium?
   c) What does "net movement" mean?
3. **a)** Define diffusion.
   b) Explain the definition of diffusion using the particle theory.
   c) State three factors that affect the rate of diffusion. In each case describe the effect that the factor has on the rate of diffusion.
   d) Will diffusion occur in a container in which the substances are in dynamic equilibrium? Explain your answer.
4. Explain why cooking odours from the kitchen can eventually be smelled throughout the entire house.

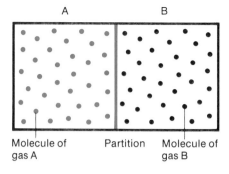

STEP 1
Partition in place.

Molecule of gas A — Partition — Molecule of gas B

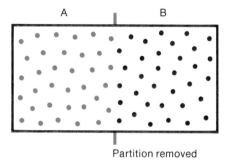

STEP 2
Partition has just been removed.

Partition removed

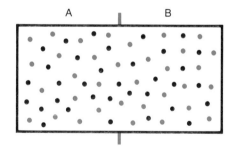

STEP 3
Some time after partition was removed.

Fig. 6-3   Diffusion of a gas into a gas.

6.2 Investigation

Diffusion

This investigation consists of two parts. The first part is a study of the diffusion of a gas. The second part is a study of diffusion in a liquid. If our explanation of diffusion in Section 6.1 is correct, it should explain the results of these experiments.

Part A Diffusion of Gases (Teacher Demonstration)

Materials

tall glass container
glass plate
sodium bromide
manganese dioxide
sulfuric acid (concentrated)
medicine dropper

CAUTION: 1. Wear safety goggles during this investigation.
          2. Bromine is poisonous. Do not inhale it. A fume-hood should be used. Perform the demonstration outside if a fume-hood is not available.

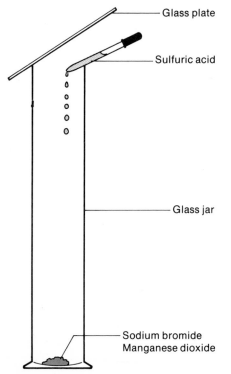

Fig. 6-4 **Studying the diffusion of a gas. Bromine gas is formed by a chemical reaction among sulfuric acid, manganese dioxide, and sodium bromide.**

— Glass plate
— Sulfuric acid
— Glass jar
— Sodium bromide
   Manganese dioxide

Procedure

a. Use the definition of diffusion and the particle theory to predict what will happen if some dense brown gas is released at the bottom of a tall container. The container is to be covered and left standing untouched for several minutes. Assume that the gas is more dense than air.

b. Bromine is a brown gas. It can be released at the bottom of a tall container using the method shown in Figure 6-4. Sulfuric acid, manganese dioxide, and sodium bromide react to form bromine gas. Bromine is more dense than air. Proceed as follows to make bromine gas in the bottom of the container. Mix about 1 cm³ of manganese dioxide with about 1 cm³ of sodium bromide. Place the mixture in the bottom of the container. Add two or three drops of sulfuric acid to the mixture. Immediately cover the container with a glass plate.

Discussion

1. Where was the gas most concentrated during most of this demonstration? Where was it least concentrated?
2. Use the definition of diffusion and the particle theory to explain the results of this experiment.

3. **a)** What do you think would have happened if the glass container had been twice as tall? Five times as tall?

b) Do you think that there is any limit to the distance through which diffusion will occur? Explain your answer using the definition of diffusion and the particle theory.

Part B   Diffusion in Liquids

Would you expect a substance to diffuse more rapidly through a liquid or through a gas? The particle theory and our discussions in Section 6.1 should help you answer this question.

Materials

petri dishes (2)
water
ethyl alcohol
iodine crystals
potassium permanganate crystals
overhead projector (optional)

CAUTION: Both potassium permanganate and iodine can cause damage to skin and clothes. Do not touch them.

Procedure

a. Fill one petri dish (A) with water and the other (B) with ethyl alcohol.
b. Place a few crystals of potassium permanganate in the centre of dish A. Place approximately the same volume of iodine crystals in the centre of dish B. Do not disturb the dishes after this point (Fig. 6.5).
c. Observe each dish carefully for 10 min. Make notes and sketches of the changes that occur. (If your classroom has an overhead projector, you could place the dishes on the projector and observe the change on the projection screen. The enlargement will make it easier for you to see what is happening. Step b must, of course, be performed with the dishes on the projector.)
d. Clean out both petri dishes. Fill one with cold water and the other with hot water. Place a few crystals of potassium permanganate in the centre of each dish. What are the differences between the observations of the two dishes?

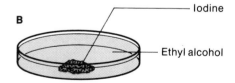

Fig. 6-5   Diffusion in liquids.

Discussion

1. Describe the appearance of the mixture as each of the solids diffuses through the liquid.
2. Use the definition of diffusion and the particle theory to explain the results of this experiment.

3. How does an increase in temperature affect the rate of diffusion? Why?
4. Does diffusion appear to be faster through a liquid or through a gas? Why?

6.3 Diffusion Through Membranes

Types of Membranes

Look back to Fig. 6-3 (page 117) for a moment. In that example, each type of particle diffused from a region of high concentration of that particle to a region of low concentration of that particle. Eventually the concentrations of both types of particles were uniform throughout the box. Dynamic equilibrium was reached.

Now, let us repeat that "experiment". However, this time we will replace the partition by a membrane. The membrane will be left in place. We will do the "experiment" three times. Each time we will use a different type of membrane. The types are: impermeable, permeable, and selectively permeable. In all three cases, the "experiment" begins with 10 particles of one substance on Side A and 10 particles of a different substance on Side B

Fig. 6-6   Three types of membranes. How do they affect diffusion?

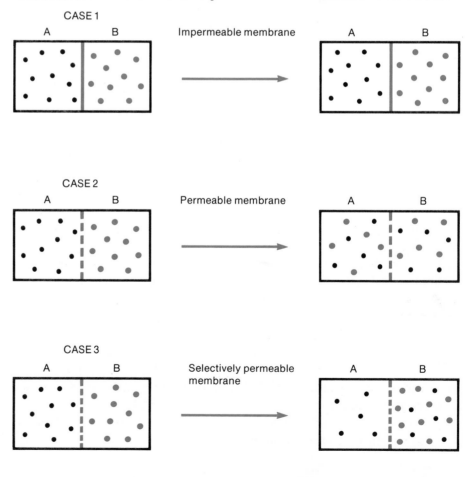

(Fig. 6-6). The purpose of this "thought experiment" is to acquaint you with the properties of these three types of membranes.

Case 1: Impermeable membrane. As the name implies, this type of membrane allows neither substance to pass through. Therefore no change occurs in the overall distribution of the particles. Rubber is an impermeable membrane for most substances.

Case 2: Permeable membrane. This type of membrane allows both substances to pass through. In other words, diffusion occurs almost as though the membrane was not there. (The diffusion may be slower however.) A net diffusion of A particles occurs toward Side B. A net diffusion of B particles occurs toward Side A. Eventually a dynamic equilibrium is reached when the concentrations of both particles are uniform throughout the box. Filter paper is a membrane permeable to water and any solute in the water.

Case 3: Selectively permeable membrane. Case 1 and Case 2 are extremes. The first allows no particles to pass through. The second allows both types of particles to pass through. A selectively permeable membrane is in between these extremes. It allows one substance to pass through but not the other. That is, it *selects* the type of particle that can pass through. You will learn later how the selection is made. As Figure 6-6 implies, the size of the pores in the membrane may be one method. Some particles may simply be too large to pass through the pores.

This type of membrane is also called a **differentially permeable** *membrane.* The membrane can *differentiate,* or tell the difference, between types of particles.

Clearly the plasma membrane of a cell must be a selectively permeable membrane. For example, it must allow water molecules to enter the cell. However, it must not allow carbohydrate, lipid, and protein molecules to leave. Our main objective in this chapter is to find out how materials move across the plasma membrane. Therefore we should study further the behaviour of selectively permeable membranes.

Osmosis

Water makes up over 70% of the mass of most organisms. Thus the movement of water in and out of cells is very common in living things. The water enters and leaves cells by moving through the plasma membrane. Since this process is so common, it is given a special name—osmosis. **Osmosis** *is the diffusion of water through a selectively permeable membrane.* Recall that diffusion is the *net* movement of particles from a region of high concentration of those particles to a region of lower concentration. Thus *osmosis is the net movement of water molecules through a selectively permeable membrane from a region of high concentration of water to a region of lower concentration of water.*

Study Figure 6-7 to help you understand the definition of osmosis. The

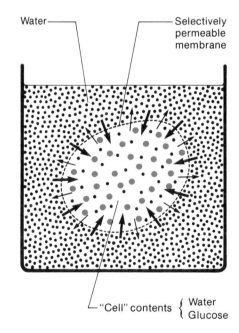

= Water molecule

= Glucose molecule

Fig. 6-7 Osmosis: the diffusion of water through a selectively permeable membrane.

diagram shows a "cell" that has a selectively permeable membrane. It is permeable to water but impermeable to glucose. In other words, water can pass through it but glucose cannot. The cell contains a solution of glucose in water. Therefore, it contains glucose molecules and water molecules. The concentration of water is higher outside the cell than inside. Therefore, if the membrane were not there, water would diffuse to where the cell was. What difference does the membrane make to the diffusion of water? Very little. It may slow down the diffusion somewhat, but that is all. The membrane is permeable to water. It lets water through. Therefore water will diffuse into the cell. The water has diffused through a selectively permeable membrane from a region of high concentration of water to a region of lower concentration of water. This is osmosis.

There is another way of looking at this process. The concentration of water is higher on the outside of the membrane. Therefore, in a given time, more water molecules will strike the membrane from the outside than from the inside. Therefore more water molecules will enter than leave.

Always remember that the water is moving both ways through the membrane. Only when the concentration of water is higher on the outside, will osmosis move water into the cell. If the concentration of water is higher inside the cell, then osmosis moves water out of the cell. What happens if the concentration of water is the same on both sides of the membrane? Water molecules still move through the membrane. However, they move in and out at the same rate. Since there is no *net* flow in one direction or the other, osmosis is not occurring.

Review

1. a) Distinguish among impermeable, permeable, and selectively permeable membranes.
   b) What type of membrane is the plasma membrane of a cell? How do you know?
2. a) Define osmosis.
   b) Distinguish between osmosis and diffusion.
   c) Why is osmosis so important to living things?
3. In Section 5.4 you studied the plasmolysis of an *Elodea* cell. Use your knowledge of osmosis to explain the phenomenon of plasmolysis.
4. a) In most parts of Canada, salt is put on the roads to melt snow. This causes the formation of a concentrated salt solution on the road. Heavy traffic sends that solution into the air as a fine spray. The spray drifts over nearby vegetation and often kills it. Explain why the salt spray kills vegetation.
   b) What do you think could be done about the problem of salt spray?

6.4 Investigation

Osmosis

This investigation demonstrates osmosis. As you know, osmosis is the diffusion of water through a selectively permeable membrane. Study the set-up of the apparatus as shown in Figure 6-8. Then predict what will happen in this investigation. Test your prediction by trying the investigation.

Materials

large beaker (600 mL or 1 000 mL)
thistle tubes (2)
selectively permeable membrane
    (dialysis tubing)
adjustable clamps (2)
ring stand

string or rubber bands
marking pen
sugar solution (50% sugar and
    50% water)
piece of thread
elastic band

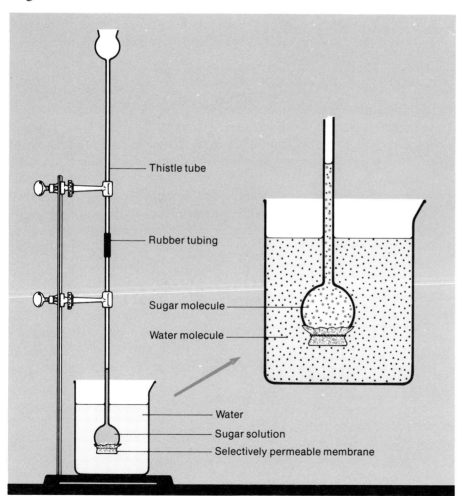

Fig. 6-8 Osmosis apparatus.

Procedure

a. Cut a piece of membrane that is large enough to cover the mouth of a thistle tube. Soak the membrane in water for several minutes to soften it.

b. Hold a thistle tube with the bulb end up. Close off the lower end by covering it with your finger. Now have your partner fill the bulb with sugar solution.

c. Place the membrane over the opening of the bulb. Fold down the edges of the membrane. Then hold the membrane in position with an elastic band. The elastic band will not hold the membrane tightly enough to prevent leakage. Therefore, tie the membrane tightly to the rim of the bulb with the thread.

d. Rinse off any sugar solution that is on the outside of the thistle tube and membrane. Check for leaks. You must have a perfect seal.

e. Attach a second thistle tube as shown in Figure 6-8.

f. Invert the sugar-filled thistle tube in a beaker of water. Clamp the thistle tubes in place.

g. Use the marking pen to mark the level of the sugar solution in the thistle tube.

h. Mark the level again at the end of the period, at the end of the school day, and three or four times during the next day or two, if possible.

Discussion

1. Describe and account for any change in the level of the sugar solution.

2. The membrane used here is called selectively permeable. It is supposed to be permeable to water and impermeable to sugar molecules. Describe any evidence that you saw which indicates that the membrane is not 100% selective.

3. Suppose that the membrane were 100% selective and only water could pass through. If osmosis were the only factor involved, how long would diffusion into the thistle tube continue? Explain your answer.

4. Osmosis of water through a membrane exerts a force. This force is called **osmotic pressure.** So long as this force is not opposed, the level of the sugar solution will continue to rise. What force opposes the osmotic pressure in the thistle tube? In other words, what force tends to work against the rise of the sugar solution? What will happen if this force equals the osmotic pressure? Did this happen in your experiment?

5. Suppose, after 2 d, you wanted to reverse the flow of water and bring the level of the sugar solution down again. How could you do this without taking apart the apparatus?

6.5 Investigation

Diffusion Through Membranes—Further Studies

Substances other than water diffuse through membranes. Such diffusion is not called osmosis. In Part B of this investigation you will study this type of diffusion. However, first you will look at another example of osmosis in Procedure A.

Materials

600 mL, or larger, beakers (4)   20 mL of glucose solution
pieces of dialysis tubing, 20 cm long (4)   20 mL of starch solution
pieces of string (4)   iodine solution
20 mL of molasses solution   Clinitest tablet

Procedure A   A Further Example of Osmosis

a. Make a tube for this investigation as follows. Open up a piece of the dialysis tubing by wetting it and rolling the end between your thumb and first finger. Then tie a knot near one end of the tubing.
b. Pour molasses solution into the tube until it is about 5 cm from the top.
c. Close the top of the tube by tying it tightly with a piece of string. You should leave a gap of about 2 cm between the top of the liquid and the string.
d. Rinse the outside of the tube with water to remove all molasses.
e. Place the tube in a beaker of water. Label this beaker "A" (Fig. 6-9).
f. Prepare a similar set-up except, this time, put water in the tube and molasses solution in the beaker. Because of the cost of molasses solution, your teacher may ask several of you to share the same beaker. Label this beaker "B".
g. Examine beakers "A" and "B" in your next class. Record your observations.

Procedure B   Diffusion of Other Substances Through a Membrane

a. Prepare another tube as in step a of Procedure A.
b. Pour starch solution into the tube until it is about 5 cm from the top.
c. Tie the tube and rinse it as in Procedure A.
d. Place the tube in a beaker of water. Label this beaker "C".
e. Add iodine solution to the water until the water is a yellow colour. Wait 20 min, then record any changes that occurred.
f. Prepare a similar set-up except, this time, put glucose solution in the tube. Also, do not add iodine solution. Be sure to rinse the tube before putting it in the beaker of water. Label this beaker "D".

Fig. 6-9 Diffusion through a selectively permeable membrane.

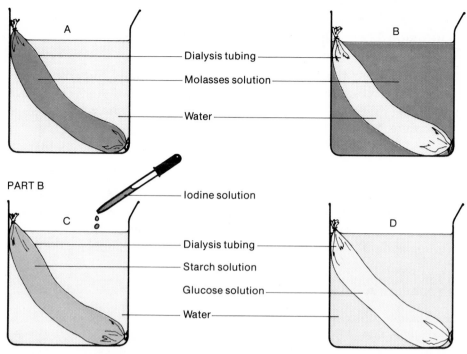

PART B

g. After about 20 min, pour about 1 mL of the water from beaker "D" into a test tube. Add a Clinitest tablet. Do not shake the test tube. Record the colour of the mixture about 15 s after the tablet stops bubbling. Changes that occur after that need not be recorded.

h. Let beaker "D" stand overnight. Record any changes that occurred.

Discussion A

1. Describe and account for your observations.
2. Is the membrane 100% selective? Explain your answer.
3. What did you learn from this investigation that you did not learn from Section 6.4?
4. Suppose that a tube filled with a 5% molasses solution was placed in a beaker filled with a 10% molasses solution. What would happen? Why?

Discussion B

In order to answer the following questions, you need the following information:

a) Starch molecules react with iodine molecules to form a dark blue or purplish compound.

b) Clinitest tablets are used to detect the presence of glucose. If no glucose is present, a blue colour forms. If glucose is present, the mixture will turn a series of greens. It may go orange or brownish if a great deal of glucose is present.

c) Scientists have established that water and iodine molecules are very small. Glucose molecules are much larger. Starch molecules are many times larger again.

1. Describe any changes that occurred in beaker "C". Use the information given above to account for the changes.
2. Describe the results of the Clinitest test on the water in beaker "D". Use the information given above to account for the change.
3. What happened to the tube in beaker "D" when it was allowed to stand overnight? What does this prove?
4. a) What molecules pass through the membrane with ease?
   b) What molecules pass through the membrane slowly?
   c) What molecules do not pass through the membrane?
5. Use your answers to question 4 and your knowledge of the sizes of these molecules to make up a hypothesis that accounts for the observations that you made in this investigation.

6.6 Diffusion and Living Cells

Factors Affecting Permeability

Molecules and ions of many types are always coming into contact with the plasma membrane of a cell. Some particles pass through the membrane with ease. Some pass through, but with difficulty. Others do not pass through at all. Biologists have learned that plasma membranes are generally permeable to water, some monosaccharide sugars, minerals, amino acids, and lipid-soluble substances. They have also learned that plasma membranes tend to be impermeable to proteins, polysaccharides, and other giant molecules. Why is this so? Why do some particles pass through and not others?

In Chapter 5 you learned that the plasma membrane of a cell may consist of two layers of lipid molecules with protein molecules scattered among them. You learned, also, that all these molecules can change positions somewhat. Thus, the plasma membrane is not solid. Molecules can move and open up holes in the membrane. If we combine this knowledge with your results from Section 6.5, we can make a list of the factors that determine whether or not a membrane will be permeable to a certain type of particle.

1. Size of particle

Many types of large molecules cannot pass through the plasma membranes of cells. Most types of small molecules, however, can do so easily. Thus the size of the particles seems to be one factor that affects permeability.

2. Size of pore

With an electron microscope, one can see pores in the surface of the plasma membrane of a cell. Biologists believe that certain molecules enter through these pores. Obviously, the larger the pore, the easier it is for large molecules to pass through.

3. Composition of the plasma membrane

The plasma membrane contains a layer of lipid molecules. These lipid molecules repel water-soluble substances, but allow lipid-soluble substances to pass through. Thus, it is easy to see how the membrane can be quite selective. Lipid-soluble substances probably move across the membrane by first dissolving in it.

4. Electrical charge

The plasma membrane probably has a positive electrical charge. If this is so, it is easy to understand why neutral particles can generally cross the membrane more easily than charged particles. Positive particles will be repelled from the surface of the membrane. Negative particles will be attracted to it. In either case, the particles will not be able to move through the membrane as easily as if they had no charge. However, negative particles generally move through more readily than positive particles. Why is this so?

5. Solubility of particles in water

Much of a cell's contents is water. Also, most cells have a water environment outside their plasma membranes. Therefore, a substance generally has to dissolve in water before it can even get to the plasma membrane. Thus the solubility of a substance in water affects the ease with which the substance can pass through the plasma membrane.

Osmosis and Living Cells

You learned in Section 6.4 that osmosis produces an osmotic pressure. This pressure is caused by the diffusion of water through the membrane. The osmotic pressure caused the sugar solution to rise in the thistle tube, in Investigation 6.4. Clearly, if the osmotic pressure gets too high, the membrane will burst. If this membrane is a plasma membrane, the cell will die. Fortunately, under normal conditions, the cell comes into **osmotic balance** with its environment. Let us see what this means by discussing three types of solution in which cells could be found.

1. Hypotonic solutions

*A **hypotonic solution** is one that has a greater concentration of water molecules than the cell does.* Therefore, if a cell is placed in a hypotonic solution, a net diffusion of water *into* the cell will occur. This osmosis builds up a pressure in the cell. This pressure is called **turgor pressure**. Turgor pressure may burst an animal cell. Plant cells, however, have rigid cell walls.

Thus, in plants, the turgor pressure simply pushes the cell contents against the cell wall. As the turgor pressure increases, the plant cells become more and more stiff. In this condition, the plant is said to be **turgid**. If the turgor pressure is low, the plant wilts. In this condition, the plant is said to be **flaccid**. Plants that are not watered sufficiently lose some of their turgor pressure and wilt.

During osmosis, a point is usually reached where the turgor pressure equals the osmotic pressure. This is called a state of dynamic equilibrium. The rate at which water enters the cell equals the rate at which it leaves the cell. The cell is in osmotic balance with its environment.

2. Hypertonic solutions

A **hypertonic solution** *is one that has a lower concentration of water molecules than the cell does.* Therefore, if a cell is placed in a hypertonic solution, a net diffusion of water *out of* the cell will occur. This outward diffusion causes a drop in turgor pressure. The cell contents shrink. In animal cells, this means that the entire cell will get smaller. In plant cells this means that the cell contents will shrink away from the cell wall. This process is called **plasmolysis**. You observed this in Section 5.4. Prolonged plasmolysis can cause the death of a cell.

3. Isotonic solutions

An **isotonic solution** *is one that has the same concentration of water molecules as the cell does.* If a cell is placed in an isotonic solution, it will immediately be in osmotic balance with the solution. Water will enter and leave the cell at the same rate. The cell will have no net gain or loss of water.

Animal Cells and Turgor Pressure

Since plant cells have rigid cell walls, they can withstand high turgor pressures without bursting. Eventually the turgor pressure equals the osmotic pressure and osmosis stops. However, this is not so with animal cells. Lacking a cell wall, they will keep swelling until they burst, if they are placed in a hypotonic solution. Bursting due to turgor pressure is called **cytolysis**. Animal cells clearly need some means for pumping water out to keep the turgor pressure down.

Many aquatic protists such as the amoeba have one or more **contractile vacuoles** that get rid of excess water. These vacuoles fill with water and move to the cell membrane. There they burst, expelling their watery contents from the cell (Fig. 6-10). Aquatic animals such as crayfish and fish also take in excess water. They get rid of this water by excreting it as urine. You will find out later how this is done. You probably know that humans excrete excess water through use of the kidneys. Sweat glands also aid in getting rid of excess water. Even breathing expels some water.

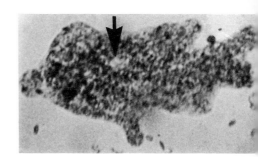

Fig. 6-10 The circle is the contractile vacuole of this amoeba.

Review

1. List and explain five factors that can determine whether or not substances can pass through a plasma membrane.
2. **a)** What is osmotic pressure?
   b) What is meant by the term "osmotic balance"?
3. **a)** What is a hypotonic solution?
   b) What is turgor pressure?
   c) Describe how osmotic balance is reached in a cell.
   d) A saltwater fish may die if placed in fresh water. Why?
4. **a)** What is a hypertonic solution?
   b) What is plasmolysis?
   c) Look back and review your results for the experiment in which you plasmolyzed an *Elodea* cell (Section 5.4, page 96). Was the salt solution hypotonic or hypertonic to the *Elodea* cell contents? Why did plasmolysis occur?
   d) What would be the best thing to do to help a plasmolyzed aquatic plant recover? Why?
5. **a)** A freshwater fish may die if placed in salt water. Why?
   b) Explain why a salt solution is a good antiseptic (kills bacteria).
   c) A small amount of fertilizer will make the grass of a lawn grow. Too much fertilizer will "burn" or kill the grass. Why?
6. **a)** What is an isotonic solution?
   b) The red blood cells in your blood are suspended in a solution called blood plasma. Is the blood plasma isotonic, hypotonic, or hypertonic? Explain your answer.
   c) How can a fish live in salt water without becoming dehydrated?
7. **a)** Describe how some aquatic protists make use of contractile vacuoles to prevent cytolysis.
   b) What would happen to an amoeba if it were placed in a hypertonic solution? Why?

6.7 Investigation

Osmosis and Living Cells

This investigation has two parts. Each part deals with osmosis through a living membrane. If you understand Section 6.6, you should be able to easily explain the results of these studies.
Note: Your teacher may suggest that you try this investigation at home.

Materials

eggs (2)
dilute hydrochloric acid or vinegar
250 mL beakers or glasses (4)
spoon

distilled water
concentrated salt solution or molasses solution
potato

Procedure A   Osmosis Through an Egg Membrane

The membrane just under the shell of an egg is selectively permeable. It is permeable to oxygen and carbon dioxide. Oxygen must enter and carbon dioxide must leave in order for the egg cell inside to stay alive. Is this membrane permeable to water? To find out, you will place one egg in a hypotonic solution and another in a hypertonic solution. The shells must be removed first.

a. Remove the shells from the 2 eggs by covering them with dilute hydrochloric acid in a beaker. (If you are doing this at home, you can use vinegar). The shells may dissolve in 30-40 min or it may take several hours. Turn the eggs from time to time with a spoon. Tap them gently with the spoon to see if the shells are gone. Do not leave the eggs in the acid for very long after the shells are gone. Some of the contents will be destroyed and the experiment will not work. When the shells are gone, remove the eggs with a spoon. Gently rinse them with water. The membrane is delicate. Handle the eggs carefully.

b. Place one egg in a beaker and fill the beaker with distilled water (Fig. 6-11).

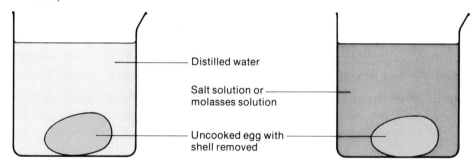

Distilled water

Salt solution or molasses solution

Uncooked egg with shell removed

Fig. 6-11   Studying osmosis through an egg membrane.

c. Place the other egg in a beaker and fill the beaker with concentrated salt solution or molasses solution.

d. Observe the eggs at the end of the period, at the end of the day, and in the next science period. Record your observations.

Procedure B   Osmosis in Potato Cells

a. Cut 6 slices of potato that are all about 0.5 cm thick.

b. Place 2 slices in a beaker and fill the beaker with distilled water (Fig. 6-12).

c. Place 2 slices in a beaker and fill the beaker with concentrated salt solution or molasses solution.

d. Examine the remaining 2 slices closely. Make notes on their turgidity (stiffness) and texture. This is your control experiment.

e. After 20-30 min, remove the slices from the water. Compare their turgidity and texture to the 2 slices used for the control.

f. Repeat step e for the slices that were in the salt solution or molasses.

Fig. 6-12 Studying osmosis in potato cells.

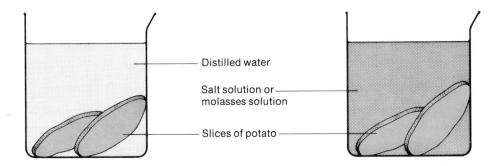

Distilled water

Salt solution or molasses solution

Slices of potato

g. Return the slices to their proper beakers. Leave them overnight. Repeat your examination of their turgidity and texture the following period.

Discussion A

1. Was the water hypotonic or hypertonic to the egg contents? How do you know?
2. Was the salt or molasses solution hypotonic or hypertonic to the egg contents? How do you know?
3. Use the results of this experiment to explain these terms: turgid, flaccid, turgor pressure, plasmolysis, and osmotic balance.

Discussion B

1. **a)** Did the cells of the potato in the water undergo an increase or a decrease in turgor pressure? How do you know?
   b) Is water hypotonic or hypertonic to the contents of potato cells? How do you know?
2. **a)** Did the cells of the potato in the salt solution or molasses undergo an increase or a decrease in turgor pressure? How do you know?
   b) Is the salt solution or molasses solution hypotonic or hypertonic to the contents of potato cells? How do you know?
3. Peter was preparing some potatoes for cooking. He put some water in a pot and added some salt. He cut up the potatoes and put them in the salt solution. Then the telephone rang and he spent 30 min talking to George. When he returned to put the potatoes on the stove, they were flaccid (soft). Explain what happened to the potatoes.

6.8 Passive and Active Transport

Passive Transport

Diffusion is a physical process. Particles move as a result of their kinetic energy. When particles pass through a plasma membrane solely as a result of diffusion, the process is called **passive transport**. The cell plays little or no

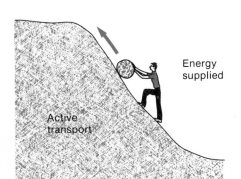

No energy supplied

Passive transport

Energy supplied

Active transport

Fig. 6-13 Passive and active transport of a rock.

role in the movement of the particles. It provides little or no energy to the particles. Hence the term "passive". Passive transport was involved in all the experiments you have done in this chapter. Diffusion is a passive process.

During passive transport, the particles move in the direction of the diffusion pressure. That is, they move through the plasma membrane from a region of *high* concentration of that particle to a region of *low* concentration.

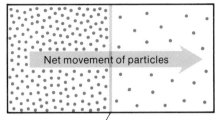

Passive transport

Net movement of particles

Plasma membrane

Direction of osmotic pressure

Active Transport

Biologists have known for many years that some particles move through the plasma membrane *against* the diffusion pressure. That is, they move from a region of *low* concentration of that particle to a region of *high* concentration of that particle (Figs. 6-13 and 6-14). Such movement cannot be explained by anything that you have learned so far. However, you do know that if the particles were moving solely as a result of their own energy (kinetic energy), they would move *with* the diffusion pressure, not against it. Therefore they must be receiving outside energy. This energy is supplied by the cell. Since the cell is actively involved in the movement of the particles, this process is called **active transport**.

Biologists do not fully understand how the cell provides particles with the energy needed to cross the plasma membrane against the diffusion pressure. One theory suggests that a "carrier molecule" uses energy to carry the particles through the membrane.

The cells in the root hairs of plants use active transport to bring in needed mineral ions from the soil. Some ions like manganese may be present in the soil in only trace (very small) amounts. The concentration of manganese in the soil may be much less than that in the root hairs. However, by active transport, the root hairs can still move those needed ions against the diffusion pressure.

Many marine (saltwater) algae, or seaweeds, have concentrations of iodine that are many times greater than that of the water. Yet they still bring in more iodine by using active transport.

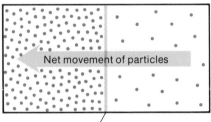

Active transport

Net movement of particles

Plasma membrane

Fig. 6-14   A comparison of passive and active transport of particles.

Review

1. Distinguish between passive and active transport.

6.9 Entry of Large Particles into Cells

Small molecules such as water can pass through pores or spaces in the plasma membrane. However, large molecules like lipids and proteins often cannot do this. Yet they still get into cells. How?

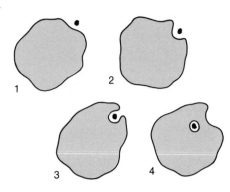

Phagocytosis

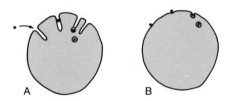

Pinocytosis

Fig. 6-15 Two methods by which large particles can be taken in by a cell.

Phagocytosis

Large solid particles are often "engulfed" by a cell. During this process, portions of the cell flow around a particle and enclose it in a chamber. Figure 6-15 shows how this occurs in an amoeba. The chamber then breaks away from the plasma membrane. The resulting body is called a **food vacuole**. The process is called **phagocytosis**.

The white blood cells in your body engulf and destroy bacteria using phagocytosis. After the vacuole is formed, it combines with a lysosome. The powerful digestive enzymes of the lysosome destroy the bacteria.

Pinocytosis

When the particles to be taken in are liquid droplets or smaller solid particles, a slightly different process occurs. It is called **pinocytosis**. The basic difference is that the particle is not surrounded by large "arms". Rather, the particle seems to become attached to the plasma membrane. Then the membrane may flow inward in a deep channel as in A. A food vacuole forms at the end of the channel. Sometimes, however, the food vacuole may form right at the surface, as shown in B.

Review

1. Distinguish between phagocytosis and pinocytosis.
2. Why are these processes sometimes necessary?

6.10 Homeostasis

What is Homeostasis?

The environment of any organism is always changing. Some of the factors that can change are the temperature, the food supply, the availability of water, and the light intensity. If an organism is to survive, it must adjust to the changing conditions. It must maintain a balance in its internal life processes even though changes occur in the external environment. This balance is called homeostatis.

Homeostasis *is the maintenance of a balanced state.* Homeostasis occurs at all levels of biological organization from the cell level to the organism level. In fact, the principle of self-regulated balance even applies at levels above the organism—population, community, biome, and even biosphere.

You can probably better understand the meaning of homeostasis by first considering this analogy. In the winter your home is kept at a fairly constant temperature by **feedback control**. The thermostat feeds signals back to the furnace controlling whether it comes on or goes off. When the temperature drops, the thermostat turns the furnace on. When the temperature gets up to a certain point, the thermostat turns the furnace off. This system is self-regulated. It maintains constant or nearly constant conditions in the home without any outside help. You do not have to keep turning the furnace off and on.

A similar type of feedback control regulates homeostasis in your body. For example, your body temperature must remain close to 37.5° C for you to stay healthy. If your body gets too warm, circulation of blood to the skin increases. Your face flushes. This allows some of the excess heat to radiate from your body. Also, you begin to sweat more. The evaporation of sweat uses up heat. If your body gets too cold, blood withdraws from the skin to conserve heat. You begin to shiver. Shivering produces heat that warms the body. All these processes are self-regulating. You do not have to say to yourself, "I am cold; I should start shivering," or "I am hot; I should start sweating." A homeostatic feedback mechanism in your body regulates these processes.

You will meet other examples of homeostasis in animals and plants in Units 4 and 5. You will see in Unit 6 that balance at levels above the organism is also important for the survival of life on this earth. For example, if one population in a community is not kept in balance, it could destroy the entire community.

Homeostasis and the Cell

Every cell, whether in a one-celled organism or a multicellular organism, must be in a state of balance with its environment. Otherwise it will not survive for long. For example, an amoeba is normally in a state of balance with the water in which it lives. But, if we add a little salt to the water, the balance is upset. Water leaves the organism and plasmolysis begins. However, feedback control in the amoeba immediately tries to restore a balance. For example, the contractile vacuole stops "pumping" as much water from the organism.

Figure 6-16 shows some factors that must remain in balance if an amoeba is to survive. Since the amoeba uses up oxygen during respiration, more oxygen must diffuse in than out in order to maintain a balance. Since the amoeba produces carbon dioxide during respiration, more carbon dioxide must diffuse out than in to maintain a balance. Since more water diffuses in than out, the contractile vacuole operates to maintain a balance. All these self-regulating processes help to maintain constant or nearly constant conditions in the amoeba. That is, they help to maintain homeostasis.

Homeostasis obviously depends to a large extent on movement of materials in and out of the cell. Thus the plasma membrane plays an important role in homeostasis. You have already studied how it controls the entry and exit of materials from the cell.

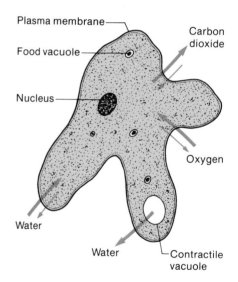

Fig. 6-16 Homeostasis is necessary for the survival of this one-celled organism. A balance must be maintained between materials entering and leaving the cell.

1. a) Define homeostasis.
   b) Hunger determines, to a large extent, when and how much you eat. It is involved in a type of feedback control that contributes to homeostasis in your digestive system. Try to figure out how this occurs. Consider these questions to help get you started: How do you know when to eat? Why do most people eat just about the right amount—not too much and not too little?
2. a) Describe how an amoeba maintains a homeostatic balance in its water content.
   b) Why is it important that the amoeba maintain such a balance?
3. Discuss briefly the role of the plasma membrane in cell homeostasis.

Highlights

Cell physiology is the study of how the parts of a cell function. This chapter dealt mainly with osmosis, an important cell function that involves the plasma membrane of the cell.

Diffusion is the process by which a substance spreads spontaneously from a region where there is a high concentration of the substance to a region where there is a lower concentration of the substance. Osmosis is a special type of diffusion. It is the diffusion of water through a selectively permeable membrane. Thus, during osmosis, there is a net movement of water molecules through a selectively permeable membrane, from a region of high concentration of water to a region of lower concentration of water. A selectively permeable membrane is one which allows certain substances to pass through with ease. But it prevents, or slows down considerably, the passage of other substances.

Five main factors regulate the passage of a substance through the plasma membrane of a cell—the size of the particles of the substance, the size of the pores in the plasma membrane, the composition of the plasma membrane, electric charges of the membrane, and the solubility in water of the substance.

If a plant cell is placed in a hypotonic solution, a net flow of water into the cell usually occurs. A pressure called turgor pressure is set up. The cell becomes stiff, or turgid. If an animal cell, or any other cell without a cell wall, is placed in a hypotonic solution, it may burst. If a cell is placed in a hypertonic solution, a net flow of water out of the cell usually occurs. The cell becomes soft, or flaccid. It is said to be plasmolyzed. In an isotonic solution, the cell is in osmotic balance with its environment.

Passive transport is the passage of particles through a plasma membrane solely as a result of diffusion. Active transport requires energy since the particles are moving against diffusion pressure. That is, they are moving from a region of low concentration of that particle to a region of high concentration.

Particles too large to enter a cell by either passive or active transport do so by phagocytosis or pinocytosis.

Homeostasis is the maintenance of a balanced state. This is achieved by self-regulating feedback control. Homeostasis is as vital to a single living cell as it is to a complete organism or to whole populations of organisms. At the cellular level, the plasma membrane plays a key role in maintaining homeostatic conditions in a living cell.

Cell Physiology: Photosynthesis And Respiration

This chapter continues the investigation of cell physiology that was started in Chapter 6. Recall that cell physiology is the study of how the parts of a cell function. In Chapter 6 we concentrated mainly on the plasma membrane and the process of osmosis. In this chapter we will concentrate on two of the most important processes that occur at the cellular level—photosynthesis and respiration. We will seek answers to questions such as: Where do cells get the energy required for life processes? In what organelles do photosynthesis and respiration occur? What happens during these processes? How are these processes related? Of what value are photosynthesis and respiration at the cellular level, the organ level, and the biosphere level?

7.1 Photosynthesis and Respiration: An Overview

You have probably studied photosynthesis and respiration before. This chapter was written with the assumption that you have. If you have not done so, or if you feel you have forgotten what you learned, do not worry. This section is a review of the basic things you should know in order to proceed with this chapter. The rest of the chapter consists of investigations and narratives that should extend your knowledge and appreciation of these important processes.

The Need for Energy

All living things need a continuous supply of energy (Fig. 7-1). This energy is needed to support life processes such as movement, growth, and reproduction. As you know, almost all the energy used by living things originally comes from the sun (see Fig. 2-1, page 14). Green plants and many other organisms contain chlorophyll. These living things, through photosynthesis, store some of the sun's energy in the bonds of glucose molecules. They do this by converting light energy into chemical potential energy. The glucose that is made during photosynthesis is the basic energy source for almost all organisms. Organisms break down glucose during the process of respiration. This releases some of the energy that was stored in the bonds of the glucose molecules.

Organisms that produce their own food are called autotrophs. Therefore organisms that contain chlorophyll are autotrophs. They use the glucose that they produce during photosynthesis as an energy source during respiration. Some of the glucose may be temporarily stored as starch. (You may recall from Chapter 4 that monosaccharides like glucose can join to form polysaccharides like starch.) However, when glucose is needed, the stored starch is digested (broken up) into glucose.

Organisms that depend on other organisms for food are called heterotrophs. All heterotrophs depend directly or indirectly on autotrophs

Fig. 7-1 **The main source of energy for this marsh hawk is small rodents such as mice. What is the main source of energy for the mice?**

for food. The dependence is direct in the case of plant eaters, or **herbivores**, such as rabbits, deer, and cows. They eat plants and convert stored carbohydrates in the plants into glucose. The glucose is then "burned" during respiration to release needed energy. You, too, show a direct dependence when you eat lettuce, spinach, potatoes, and cereals. The dependence is indirect in the case of flesh eaters, or **carnivores**, such as cougars, wolves, and bobcats. These animals feed on herbivores such as rabbits. The rabbits eat autotrophic organisms such as grass. Therefore, indirectly, the carnivores depend on autotrophs for the glucose that is needed for energy.

Summary

Remember these two definitions:

1. **Photosynthesis** *is the process by which light energy is changed to chemical potential energy and stored in the bonds of glucose molecules.*
2. **Respiration** *is the process by which living cells break down glucose molecules and release the stored chemical potential energy.*

Photosynthesis takes place *only* in cells that contain chlorophyll. Photosynthesis also requires light energy. In contrast, respiration takes place in *all* cells, *all* the time.

The Composition of Glucose

As you know, glucose belongs to a family of organic compounds called carbohydrates. Its molecular formula is $C_6H_{12}O_6$. If you write its formula as $C_6(H_2O)_6$, you can see why it is called "carbo . . . hydrate". It contains carbon. It also contains hydrogen and oxygen in the same proportion as water.

Where does a green plant get the carbon, hydrogen, and oxygen atoms needed to make glucose? (Fig. 7-2). Atoms cannot be created or destroyed. They can only be rearranged to make new substances. Therefore green plants and other photosynthetic organisms must get carbon, hydrogen, and oxygen atoms from their environment.

A person who has not studied biology usually thinks that plants are built entirely of materials that were once part of the soil. However, an interesting experiment performed many years ago casts doubts on that belief.

The van Helmont Experiment

About three hundred years ago, Johann van Helmont, a Belgian physician and chemist, began wondering where the substances for plant growth came from. Therefore he conducted an experiment. He placed 90.90 kg of perfectly dry soil in a large pot. Then he planted in the soil a willow tree that

Fig. 7-2 This Norfolk Island pine has been growing for several years in the same pot. The tree has grown from a 10cm seedling to its present height of 120cm. Yet the quantity of soil appears to be unchanged. Where did the tree get the atoms of which it is made?

had a mass of 2.30 kg. For five years he did nothing to the tree but water it with distilled water or rain water. After the five year period the willow tree had a mass of 76.80 kg. He then carefully collected all the soil from the pot and dried it. To his surprise, the dry soil had a mass very close to what it had when the experiment was started. Its mass was 90.84 kg. It had lost only 0.06 kg or 60 g of its mass! Van Helmont concluded that 74.50 kg of "wood, bark, and root had arisen from water alone".

We know today that a tree is not made only of water. A tree is a living organism. All living organisms contain carbon. However, van Helmont's experiment does suggest that water *could be* the source of the hydrogen and oxygen that are in the glucose and other molecules of the tree. But where do the carbon atoms come from? Carbon is black. Therefore you might first think of soil as a source. However, van Helmont's experiment makes us wonder if the slight change in mass of the soil could amount to enough carbon to make the tree. There is another source of carbon in the tree's environment—the carbon dioxide in the air. Perhaps this is the source of the tree's carbon atoms. In the next section you will do experiments to find out just what is involved in making a plant. You will study the process of photosynthesis.

Review

1. **a)** Define photosynthesis.
   b) Define respiration.
   c) What connection do you see between photosynthesis and respiration?
   d) Why is glucose such an important molecule to cells?
2. **a)** Distinguish between an autotroph and a heterotroph.
   b) Describe the importance of photosynthesis to all living things. Make specific reference to autotrophs, herbivores, and carnivores.
3. **a)** Describe and explain what you think would happen to living things on earth if the sun's light no longer reached the earth. (Assume that the sun's heat rays still reach the earth.) Make specific references to autotrophs, herbivores, and carnivores. Also consider organisms such as fungi that feed on dead organic matter.
   b) Describe and explain what you think would happen to living things on earth if the sun continued to shine but all autotrophs somehow disappeared. Make specific reference to herbivores, carnivores, and organisms such as fungi that feed on dead organic matter.

7.2 Investigation

Photosynthesis

The purpose of this investigation is to study the process of photosynthesis under controlled conditions. By doing so, you should be able to determine the substances required for photosynthesis, the substances produced, and any other requirements of the process.

You should not begin any experiment without having a hypothesis. In this case, on the basis of what you already know, you should predict the answers to the following:

a) What substances are required for photosynthesis?
b) What special conditions are required for photosynthesis?
c) What products are produced by photosynthesis?

After you have answered these questions, proceed with the investigation. You will be able to design some experiments in the latter parts of the investigation.

Part A Composition of Glucose

Section 7.1 stated that glucose is formed during photosynthesis. In this part of the investigation, you decompose, or break down, glucose by heating it. Then you examine the products. Perhaps they will suggest what substances went together to make the glucose in the first place.

Since the test tube is hard to clean after this investigation, your teacher may ask you to work in groups of five or six.

Materials

test tubes (2)	glucose (about 1 cm³)
Bunsen burner	starch (about 1 cm³)
adjustable clamp	Cobalt chloride test paper

CAUTION: Wear safety goggles during this investigation.

Procedure

a. Place about 1 cm³ of glucose in the bottom of a clean dry test tube.
b. Heat the glucose gently for a minute or two. Keep the test tube on an angle as shown in Figure 7-3. Record any changes that occur.
c. Test the liquid that forms to see whether or not it is water. Cobalt chloride test paper turns red in the presence of water.
d. Heat the glucose strongly until no further changes occur. Record a description of the substance that remains.
e. Repeat steps a to d using 1 cm³ of starch.

Discussion

1. Based on your observations, what two substances do you think may have been present in the glucose?
2. Perhaps a plant made the glucose during photosynthesis. Where do you think it might have obtained the substances of which the glucose was made?

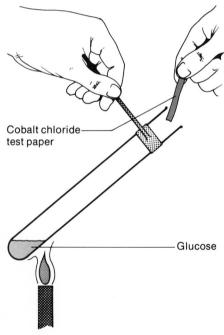

Cobalt chloride test paper

Glucose

Fig. 7-3 Decomposition of glucose by heat.

3. What evidence do you have that glucose and starch are carbohydrates?
4. Glucose is a monosaccharide. Starch is a polysaccharide. What do these terms mean? How is starch related to glucose? (If you have forgotten, look back to Section 4.4, page 80.)

Part B A Test For Starch

In several parts of this investigation, you will need a test which proves that photosynthesis occurred. In other words, you will need to prove that glucose was formed. There is no simple way that you can test for glucose in cells and tissues. However, biologists have shown that, during photosynthesis, much of the glucose produced is converted to starch. Thus starch is an indirect product of photosynthesis. Therefore you can use its presence as indirect evidence that photosynthesis occurred.

You may have done the iodine test for starch earlier in this course. If you have not done it before, try it now.

Materials

glass plate
stirring rod
starch (about 1 cm³)
iodine solution

Procedure

a. Make a paste of starch and water on a glass plate.
b. Add a few drops of iodine solution. Record any changes that occur.

Discussion

1. Describe a test for the presence of starch.

Part C Testing for Starch in Green Plants

In some parts of this investigation, you need to know how to test for starch in the leaves of plants. This part of the investigation describes that test.

Materials

150 mL beaker
600 mL beaker
hot plate (or Bunsen burner and
  ring stand assembly)
crucible tongs

glass plate
ethyl or isopropyl alcohol
geranium plant that has been under
  bright light for 48 h
iodine solution

CAUTION: Wear safety goggles during this investigation.

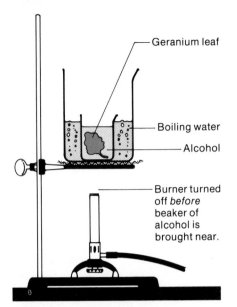

- Geranium leaf
- Boiling water
- Alcohol
- Burner turned off *before* beaker of alcohol is brought near.

Fig. 7-4 The boiling point of alcohol is lower than that of water. Therefore, water near its boiling point can make alcohol boil with no flame present.

Procedure

a. Remove a leaf from the geranium plant.

b. Immerse the leaf in boiling water for a few seconds. Remove it as soon as it becomes limp.

c. Transfer the leaf to a beaker containing boiling alcohol. If possible, use a hot plate to heat the alcohol. If one is not available, place a small beaker of alcohol in a larger beaker that is partly full of boiling water (Fig. 7-4). Turn off the burner *before* you bring the alcohol near the boiling water.

CAUTION: Do not heat the alcohol with an open flame. It will ignite.

d. When the leaf is white, remove it from the alcohol. Soften it by dipping it in boiling water for 2-3 s.

e. Spread the leaf on a glass plate. Cover the leaf with iodine solution. Record your observations.

Discussion

1. What do you think was the purpose of dipping the leaf in boiling water at the start of this experiment? (The appearance of the leaf should tell you what happened at the cellular level.)

2. Is chlorophyll soluble in water? How do you know?

3. Is chlorophyll soluble in alcohol? How do you know?

4. Was starch present in the leaf? How do you know?

5. Does this experiment prove that green plants make starch? Explain.

Part D Light and Photosynthesis

Is light required for photosynthesis? Based on your reading and your knowledge that plants will not live long without light, you will probably answer "yes". However, you have not performed any experiments. Therefore your "yes" is really just an educated guess or a simple type of hypothesis.

In this part of the investigation you are to perform a controlled experiment to see whether or not light is required for photosynthesis. You are required to design part of the experiment on your own.

You must have a hypothesis before you start the experiment design. We will give you some help to get you started. Your hypothesis could be something like this: "Light is required for photosynthesis. Therefore, if I do . . . to the plant, . . . will happen." Read the procedure that follows. Make up your hypothesis. Then design and conduct the experiment.

Materials

The same as for Part C except that the geranium plant has been in the dark for 48 h. You may also require aluminum foil and paper clips.

CAUTION: Wear safety goggles during this investigation.

Procedure

a. Remove a leaf from a geranium plant that has been in the dark for 48 h.
b. Perform the test for starch as described in Part C. Record your results.
c. You may think that the results you obtained here, combined with what you obtained in Part C, answer the question "Is light required for photosynthesis?". However, you used two different plants in these experiments. A scientist would say that you did not have very good **controls** in the experiment. The different results could be due to the fact that you used different plants. They could also be due to the fact that the experiments were done on different days. To eliminate such possibilities, you should use only one plant or, preferably, one leaf of that plant. See if you can design such an experiment. Have your teacher check the design of your experiment before you perform it.

Discussion

1. Why was it important to begin this experiment by testing a leaf from a plant that had been in the dark for 48 h?
2. Describe the experiment you designed to study the relationship between light and photosynthesis.
3. Is light necessary for photosynthesis? How do your results support your answer to this question?

Part E Chlorophyll and Photosynthesis

Is chlorophyll necessary for photosynthesis? Many plants have leaves that contain chlorophyll in some regions and not in others. Examples are silver-leaf geraniums and some *Coleus* plants (Fig. 7-5). Could one of these leaves be used to prove whether or not chlorophyll is needed for photosynthesis?

Materials

The same as for Part C.
Your teacher will provide a silver-leaf geranium or suitable *Coleus* plant that has been under bright light for 48 h.

CAUTION: Wear safety goggles during this investigation.

Procedure

Design an experiment to determine whether or not chlorophyll is required for photosynthesis. Write out the complete procedure. Have your teacher check it before you begin. Do the experiment. In your writeup, include labelled "before and after" drawings that illustrate your experimental results.

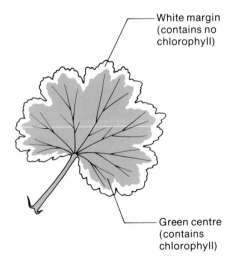

White margin (contains no chlorophyll)

Green centre (contains chlorophyll)

Fig. 7-5 A leaf from a silver-leaf geranium. Note the distribution of chlorophyll.

Discussion

1. Why was it important to begin this experiment with a plant that had been in bright light for 48 h? Suppose you began with a plant that had been in the dark for 48 h. What would your results have been?
2. Explain carefully how your procedure and results answer the question, "Is chlorophyll necessary for photosynthesis?"
3. *Iresine* is a plant that has deep red stems and leaves. If an *Iresine* plant is placed in bright light for 48 h, a positive starch test is obtained. If it is placed in darkness for 48 h, a negative starch test is obtained. How are these results possible?

Part F Carbon Dioxide and Photosynthesis

Does a green plant use carbon dioxide during photosynthesis? Our discussions in Section 7.1 suggested that the carbon dioxide gas in the atmosphere might be the source of the carbon that is used to make glucose during photosynthesis. In this part of the investigation you find out whether the suggestion is true or not.

Materials

The same as for Part C. Your teacher will provide a geranium plant that has been in the dark for 48 h. You will also need about 3-4 cm³ of sodium hydroxide or potassium hydroxide.

CAUTION: Wear safety goggles during this investigation. Also, sodium and potassium hydroxide are very corrosive. Do not touch them with your hands.

Procedure

a. Study Figure 7-6 closely to get an idea of how to proceed with this

Fig. 7-6 Control and experimental set-ups used to find out if carbon dioxide is required for photosynthesis.

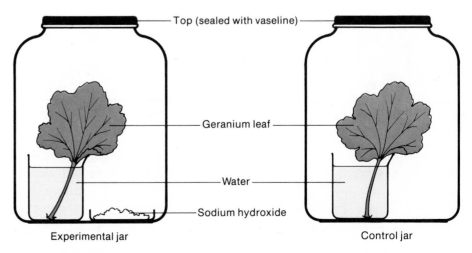

investigation. Sodium or potassium hydroxide reacts chemically with carbon dioxide. This removes the carbon dioxide from the air.

b. Write out the complete procedure that you plan to use. Have your teacher check it before you begin.

Discussion

1. Why was it important to begin this experiment with a plant that had been in the dark for 48 h?
2. Explain carefully how your procedure and results answer the question, "Does a green plant use carbon dioxide during photosynthesis?" Be sure to discuss the role of the sodium or potassium hydroxide. Also, explain why the control was necessary.
3. Bromthymol blue is an acid-base indicator. That is, it tells us whether a solution is acidic or basic. Bromthymol blue is yellow in an acidic solution, green in a neutral solution, and blue in a basic solution. Carbon dioxide forms an acid when it reacts with water. Therefore carbon dioxide will turn a bromthymol blue solution yellow. The same solution will turn green or blue if the carbon dioxide is removed.
    a) Use this information and the materials shown in Figure 7-7 to design an experiment to show that carbon dioxide is used in photosynthesis. If time permits, your teacher may let you try your method. Do not forget to include a control in your experiment design.
    b) Can you say beyond any doubt that this experiment proves that carbon dioxide is used in photosynthesis? Why?
    c) The procedure outlined in parts a) and b) of this question is called an

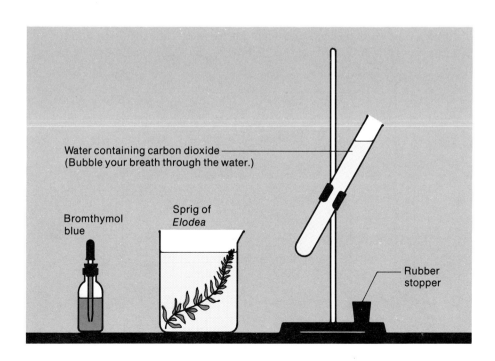

Fig. 7-7   An indirect method for showing that carbon dioxide is used during photosynthesis.

Water containing carbon dioxide
(Bubble your breath through the water.)

Bromthymol blue

Sprig of
Elodea

Rubber stopper

indirect method. The procedure that used sodium hydroxide is called a **direct method**. Explain why the terms direct and indirect are used. Which procedure do you think is better? Why?

Part G Oxygen and Photosynthesis

You may know that, like almost all living things, green plants require oxygen for respiration. You may also have heard that green plants produce oxygen gas during photosynthesis. The purpose of this experiment is to find out whether or not photosynthesis does produce oxygen gas.

Materials

test tubes (2)
1 000 mL beakers (2)
funnels (2)
rubber stopper
Elodea
sodium bicarbonate
wooden splint

Procedure

a. Set up the experimental beaker and control beaker as shown in Figure 7-8. The water must contain carbon dioxide during the entire experiment. Otherwise photosynthesis cannot occur. If you add 2 or 3 cm³ of sodium bicarbonate to the water it will supply the necessary carbon dioxide. Insert the cut ends of the *Elodea* sprigs into the stem of the funnel. The funnel should be deep in the water. The test tube and funnel must be full of water at the beginning of the experiment. You can

Fig. 7-8 Does photosynthesis produce oxygen gas? What is the function of the control?

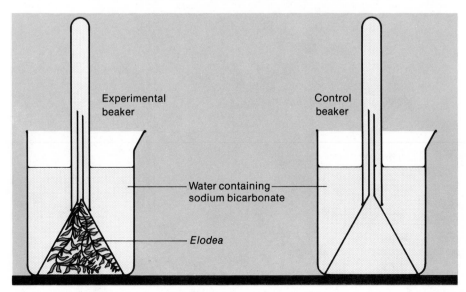

Experimental beaker

Control beaker

Water containing sodium bicarbonate

Elodea

accomplish this by submerging the entire apparatus (beaker, funnel, and test tube) in a sink full of water. The control beaker must be identical to the experimental beaker, except that it does not contain *Elodea*.

b. Shine a bright light on the entire setup for at least 2 d. If no noticeable change has occurred, extend the time another day or two.

c. Test the gas that collects in the test tube to see if it is oxygen. Use the glowing splint test. Oxygen is one of very few gases that will cause a glowing splint to burst into flame. Proceed as follows to perform the test. Fill the beaker full of water. Lift the test tube off the funnel but do not let any air in. Put a stopper in the mouth of the test tube. Lift the test tube out of the beaker. Turn it upright. You should now have gas on top of water as shown in Figure 7-9. Remove the stopper and immediately try the glowing splint test. (Have the glowing splint ready before you remove the stopper.)

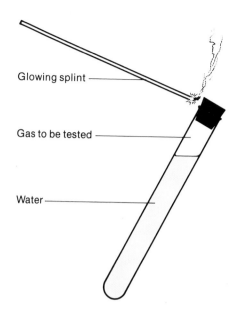

Glowing splint

Gas to be tested

Water

Fig. 7-9 You must not allow air to dilute the gas before you test it. Keep the stopper in place until the glowing splint is ready.

Discussion

1. Describe the gas that collected in the experimental beaker.
2. Describe the effect of that gas on the glowing splint. What do you think the gas is? Have you proven beyond any doubt that it is that substance?
3. Only two other gases seem at all likely to be given off in this situation. What are they? Where might they come from? What would they do in the presence of a glowing splint?
4. What was the function of the control in this experiment?

7.3 Investigation

Properties and Composition of Chlorophyll

You established in Section 7.2, Part E that chlorophyll is necessary before photosynthesis can take place. In this investigation you study chlorophyll more closely. First you extract it from some leaves. Then you study its composition and some of its properties.

Materials

green leaves (preferably 5-10 geranium leaves and several spinach leaves)
alcohol
600 mL beaker
1 000 mL beaker
hot plate (or Bunsen burner and ring stand assembly)
crucible tongs
projector (slide or filmstrip)
small flat-sided bottle

large flat-sided glass container
triangular glass prism
green food colouring
large test tube (20 x 150 mm)
filter paper (or chromatography paper, if available)
fine-tipped dropper
piece of wooden splint
aluminum foil
mortar and pestle
clean quartz sand

CAUTION: Wear safety goggles during the extraction portion of this investigation.

Procedure A   Extraction of Chlorophyll from Leaves

a. Put about 600 mL of water in the 1 000 mL beaker. Bring the water to the boiling point.

b. Drop 5-10 geranium leaves into the boiling water. Remove them as soon as they become limp.

c. Transfer the leaves to a 600 mL beaker containing about 300 mL of boiling alcohol. If possible, use a hot plate to heat the alcohol. If one is not available, place the beaker of alcohol in the 1 000 mL beaker containing about 600 mL of boiling water (see Fig. 7-4, page 144). Turn off the burner *before* you bring the alcohol near the boiling water.

CAUTION: Do not heat the alcohol with an open flame. It will ignite.

d. Boil the leaves in the alcohol for about 5 min. If the extraction does not appear complete, continue the boiling for a few more minutes.

e. Discard the leaves. Retain the green solution for Procedure B.

Procedure B   Some Properties of Chlorophyll

a. Pour chlorophyll solution into the small flat-sided bottle until it is almost full.

b. Shine a bright light on one side of the bottle. (Use the projector.) View the solution from directly above by looking through the open neck of the bottle. Record your observations.

c. Your teacher has a similar flat-sided bottle that contains a solution of green food colouring of about the same green colour as the chlorophyll solution. Obtain this bottle from your teacher. Repeat step b with it. Record your observations.

d. Pour all your chlorophyll solution into the large flat-sided glass container on the teacher's bench. After several members of the class have done this, your teacher will have enough solution to demonstrate light absorption by chlorophyll as described in the next step.

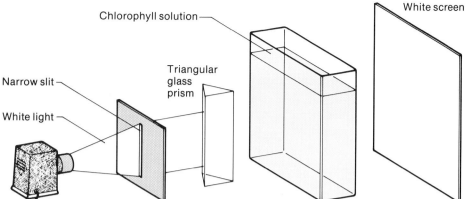

Chlorophyll solution

Triangular glass prism

Narrow slit

White light

White screen

Fig. 7-10 Absorption and transmission of light by chlorophyll solution.

e. Light absorption. You can study the absorption and transmission of light by chlorophyll as follows: Set up the apparatus as shown in Fig. 7-10, but *without* the chlorophyll solution in place. Adjust the positions of the prism, projector, and screen until a complete colour spectrum appears on the screen. Note carefully the colours observed. Then insert the flat-sided container full of chlorophyll solution in the path of the beam of light as shown in Figure 7-10. Note any changes that occur in the spectrum.

Procedure C Separation of the Pigments in Chlorophyll Extract

In this procedure you use a method called **paper chromatography** to separate the pigments that are in chlorophyll extract. The way this works is easily understood. You have seen water soaked up by a paper towel. You have probably seen alcohol move up the wick of an alcohol burner or kerosene move up the wick of a kerosene lamp. Liquids appear to move well through some materials like paper and cloth.

Suppose a substance is dissolved in the kerosene of a kerosene lamp. This substance will be carried up the wick by the kerosene. How far it is carried up depends on several factors. One factor is the mass of the substance's particles. Another factor is the solubility of the substance in kerosene.

Paper chromatography works much the same way. Paper replaces the wick of the kerosene lamp and a solvent that will dissolve the substances to be separated replaces the kerosene. In this investigation the substances to be separated are the pigments in the chlorophyll extract. The solvent is alcohol.

a. Collect a few green leaves. Spinach leaves are especially good for this procedure. Prepare a concentrated chlorophyll solution as follows.

b. Tear the leaves into pieces. Put the pieces in a mortar. Add about 5 mL of alcohol and a pinch of quartz sand. Grind the mixture until a deep green (almost black) solution is obtained (Fig. 7-11,A). Add more leaves if the solution is not dark enough. This deep green solution is the chlorophyll extract.

A Grind leaves in alcohol with a mortar and pestle. Some quartz sand will assist the grinding.

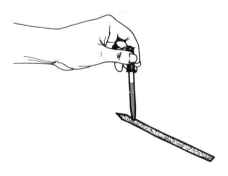

B Use a fine-tipped dropper to spread a line of extract across the paper.

Fig. 7-11 Preparation for separation of pigments in chlorophyll extract.

Section 7.3 **151**

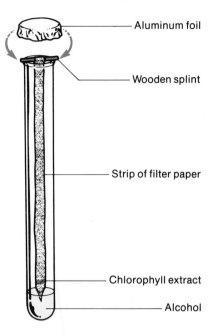

Aluminum foil

Wooden splint

Strip of filter paper

Chlorophyll extract

Alcohol

Fig. 7-12 Making a chromatograph of chlorophyll extract.

c. Prepare a strip of filter paper (or chromatography paper) about 1 cm wide and 20 cm long. Tape one end as shown in Figure 7-11,B.

d. Using a pencil (*not* pen), draw a line across the filter paper about 2 cm above the tip of the paper. Use the dropper to spread one drop of the pigment extract along the line. Allow the paper to dry completely. (Gentle fanning will help.) Now spread a second drop of the pigment extract along the line. Allow it to dry. Keep repeating this until a dark green strip is present across the filter paper.

e. Place 3-4 mL of alcohol in the test tube.

f. Hang the filter paper over the wooden splint as shown in Figure 7-12. Only the tip of the filter paper should be in the alcohol. The strip of chlorophyll extract should be about 1 cm above the surface of the alcohol. Make sure that the filter paper hangs in the centre. It must not touch the wall of the test tube.

g. Cover the top of the test tube with aluminum foil to prevent escape of alcohol vapour.

h. Let the system stand until the solvent has almost reached the top of the paper. Record your observations immediately. (Some of the colours change if they stand too long.) In particular, note the number of bands of colour, the colour of each band, and the relative widths of the bands. You may wish to remove the paper from the test tube, dry it, and mark this information on the paper.

Discussion A

1. The chlorophyll extract that you obtained is green. The presence of a green pigment could create this colour. Is it possible that pigments with colours other than green are present? Explain.

Discussion B

1. What did you observe when you looked down on chlorophyll solution that was being illuminated from the side? Try to explain what you observed.

2. a) What effect did the chlorophyll extract have on the spectrum?
   b) What colours of the spectrum does chlorophyll absorb?
   c) What colours of the spectrum does chlorophyll transmit?

3. a) What colours of the spectrum do you think are most involved in photosynthesis? Why?
   b) What colours of the spectrum do you think are least involved in photosynthesis? Why?
   c) Why does a geranium leaf appear green in daylight?
   d) What colour do you think a geranium leaf would appear if it were illuminated only with red light? Why? (Try it!)

Discussion C

1. How many different bands of colour were formed?
2. The yellow or yellow-orange pigments that travel highest on the paper are **carotenes**. The other yellow or yellow-green band (sometimes two bands) consists of **xanthophylls**. The dark green band is **chlorophyll** *a*. The lighter green (sometimes blue-green or even yellow-green) band is **chlorophyll** *b*. List the components of your chlorophyll extract in the order in which they appeared on the paper, from the top down. Beside their names place the colour and the relative width of the bands. Draw any conclusions you can from the relative widths and the distance they rose.
3. Why were you unable to see all these colours in the leaf?

7.4 Photosynthesis

You had a brief introduction to photosynthesis in Section 7.1. You learned more about it in the investigations that you did in Sections 7.2 and 7.3. In this section you will put all that together with some new knowledge. This will give you a better picture of what happens during photosynthesis.

The Summation Equation for Photosynthesis

Photosynthesis *is the process by which light energy is changed to chemical potential energy and stored in the bonds of glucose molecules.* You are familiar with this definition. But you should understand it much better now than when you first met it.

As the name implies, light energy (*photo*) is used in building complex substances from simple substances (*synthesis*). You know now that this process occurs only in organisms that contain chlorophyll. You know, too, that the simple substances used in photosynthesis are water and carbon dioxide. You also know that oxygen is produced as well as glucose. During photosynthesis light energy is changed to and stored as chemical potential energy in the bonds of glucose molecules. This energy can be released from those bonds by the oxidation, or "burning", of glucose during respiration by living things. Biologists summarize the description of photosynthesis in this paragraph by using a word equation or a chemical equation:

$$\text{Carbon dioxide + Water + Light Energy} \xrightarrow{\text{Chlorophyll}} \text{Glucose + Oxygen}$$

$$6\ CO_2 + H_2O + \text{Light Energy} \xrightarrow{\text{Chlorophyll}} C_6H_{12}O_6 + 6\ O_2$$

You must remember that these equations are just **summation equations**. They simply "sum up" *what* happens during photosynthesis (Fig. 7-13). They tell us *what* photosynthesis starts with and *what* it ends with. They indicate that chlorophyll must be present. However, these equations do not tell us *how* the process occurs.

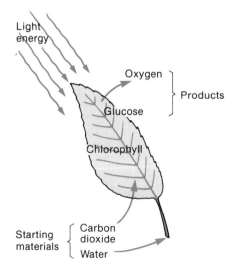

Fig. 7-13 Like a summation equation, this diagram sums up what happens during photosynthesis. Water and carbon dioxide enter the leaf. Glucose and oxygen are produced.

Light

Light is the energy source for photosynthesis. Therefore you must know something about the nature of light in order to understand how it contributes to the process of photosynthesis.

Light *is that form of energy which the eye detects.* It makes up what is called the **visible band** of the **electromagnetic spectrum** (Fig. 7-14). The visible band is made up of those wave lengths from about 390 nm (violet) to 770 nm (red). One nanometre= 1 nm = 0.000 000 001 m. When the wave lengths in this region are mixed in the proper proportions, white light is produced. In Part B of Section 7.3 you saw that a prism will break white light into its component colours—red, orange, yellow, green, blue, and violet.

Fig. 7-14 As you can see, the visible band is only a small portion of the electromagnetic spectrum.

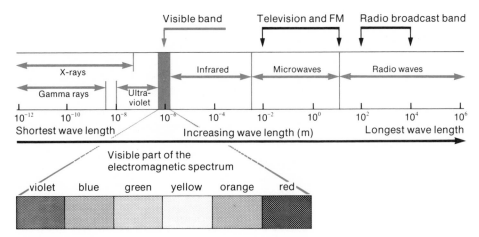

As you can see from Figure 7-14, electromagnetic radiation comes in a wide range of wave lengths. But of all these possible wave lengths, only certain wave lengths in the visible band participate in photosynthesis. Some simple colour theory will tell you which wave lengths do so.

Figure 7-15,A shows why a white object appears white when you shine white light on it. The object absorbs none of the white light. It is all reflected to your eye. Therefore the object appears white.

Figure 7-15,B shows why a black object appears black when you shine white light on it. White light is made up of the spectral colours red, orange, yellow, green, blue, and violet. Apparently there are pigments in the object that absorb all these spectral colours. Nothing remains to be reflected. Thus the object has no colour and appears black. If you have ever worn black clothes on a sunny day, you will know that the absorbed light energy is changed into heat energy. This makes you very warm.

Figure 7-15,C shows why a green object appears green when you shine white light on it. Pigments in the object absorb wave lengths at the red and blue ends of the spectrum. They do not absorb green and may not completely absorb the colours on either side of green. Therefore green (and perhaps some blue and yellow) is reflected and the object appears green.

Similarly, a green transparent object allows only green (and perhaps some blue and yellow) to pass through it (Fig. 7-15,D).

The Role of Chlorophyll

A green leaf appears green in white light. Therefore it must contain pigments that absorb the red and blue wave lengths of the spectrum but reflect green wave lengths. Of course, these pigments are chlorophyll. In Part B of Section 7.3 you saw that a solution of chlorophyll transmits mainly green and yellow light. Thus the solution must be absorbing the red and blue ends of the spectrum. Figure 7-16 shows the **absorption curve** for chlorophyll. The curve was obtained by using sensitive scientific equipment. Study the graph carefully. Note that much of the violet and nearly all the blue light is absorbed by the chlorophyll. Also, much of the orange and red light is absorbed. However, little green and yellow light is absorbed. Therefore, plants that contain chlorophyll as their main pigment appear green because chlorophyll reflects mainly green and yellow light. (The green dominates the yellow.)

There are at least five types of chlorophyll molecules. All are somewhat alike in structure and properties. The two most common types are chlorophyll *a* and chlorophyll *b*. You probably saw them in Part C of Section 7.3. The molecular formula of chlorophyll *a* is $C_{55}H_{72}O_5N_4Mg$. The molecular formula of chlorophyll *b* is $C_{55}H_{70}O_6N_4Mg$. As you might imagine, these structural formulas are quite complex.

You learned in Chapter 5 that chlorophyll is found in cell organelles called **chloroplasts**. The chloroplasts contain bodies called **grana**. The grana consist of layers of lipid and protein molecules. The chlorophyll molecules are between these layers. Apparently the light energy required for photosynthesis is trapped by the grana. Most of the reactions of photosynthesis occur on the grana. The chlorophyll functions as a catalyst in the process of photosynthesis. *A catalyst is a substance that affects the rate of a chemical reaction without being permanently changed itself.* Thus chlorophyll speeds up the process of photosynthesis, but is not used up in the process.

Biologists are not certain how living things make chlorophyll. They do know that it is made in chloroplasts (except in some monerans). They also know that little, if any, chlorophyll is made in the absence of light. You may have seen the sprouts that bean seeds and potatoes form if they are allowed to grow in the absence of light. These sprouts are white. If you give them light for a few days however, they turn green. This indicates that these plants cannot make chlorophyll without light. Also, you have probably seen how grass turns white under a board or other opaque object that is left on a lawn for a few days. The grass cannot replace worn-out chlorophyll unless it has light.

Apparently other pigments, the carotenes and the xanthophylls, also assist in photosynthesis. It is believed that they aid chlorophyll in absorbing the light energy needed for photosynthesis. In fact, in some monerans and algae, carotenes seem to take the place of chlorophyll.

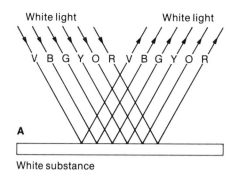

White substance

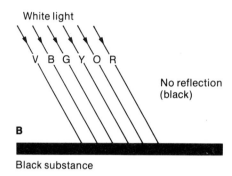

Black substance

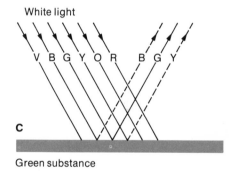

Green substance

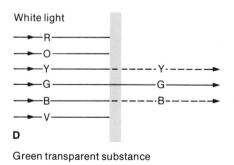

Green transparent substance

Fig. 7-15 The pigments present in objects determine their colour.

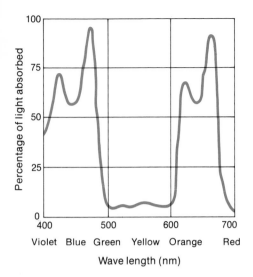

Fig. 7-16 The absorption curve for chlorophyll.

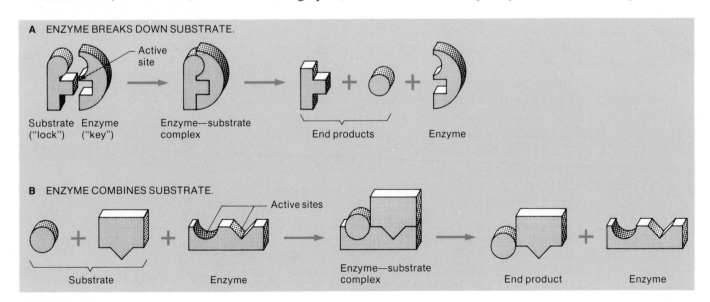

Fig. 7-17 Action of an enzyme. A substance and its enzyme are often referred to as a lock and key. When a lock and a key fit exactly, the lock will open. When the substrate and enzyme fit exactly, the conversion of substrate to end products is catalyzed.

Enzymes

Many substances called enzymes are involved in photosynthesis. *An enzyme is a protein that acts as a catalyst.* Enzymes help reactions occur during photosynthesis, but they are not used up during the process. In other words, they do not become part of the end product.

The main function of an enzyme is to *lower* the activation energy of the reaction in which it participates. You may recall from Chapter 4 that many reactions require an activation energy to get them started. For example, wood has to be given activation energy by heating it before it will burn. Clearly a cell cannot use heat as a means of overcoming a large activation energy. The heat would destroy the cell. The only alternative is to lower the activation energy required. That is what an enzyme does.

Biologists do not know exactly how enzymes work. However, they do know that enzymes are proteins. They know that each enzyme is **specific**. This means, it will catalyze only one of the many reactions that occur in cells. Biologists also know that enzymes are not used up or chemically changed during a reaction. On the basis of this and other knowledge, many models have been proposed to explain how enzymes work. One of these models is illustrated in Figure 7-17. In this model, the enzyme is shown to have one or more **active sites**. An active site is part of the enzyme molecule that fits with a molecule of the **substrate**. (The substrate is the substance that the enzyme causes to react.) The idea of an active site explains why enzymes are specific for a certain substrate. The enzyme and the substrate unite to form what is called an **enzyme-substrate complex**. This complex breaks down quickly. However, instead of forming enzyme and substrate again, it forms enzyme and product. Thus the enzyme has assisted in the conversion of substrate to product. But the enzyme itself, is unchanged.

As Figure 7-17, A shows, an enzyme can assist in the *decomposition*, or breaking apart, of a substrate. The hydrolysis of fats into fatty acids and glycerol is an example of such a reaction (see Section 4.4, page 82). The hydrolysis of starch into glucose is another example.

Figure 7-17,B shows that an enzyme can assist in the *combination* of substances. The combination of amino acids into dipeptides is an example of such a reaction (see Section 4.4, page 83). The combination of glucose molecules to form starch is another example.

In summary, an enzyme lowers the activation energy required for a certain chemical reaction. Thus an enzyme allows a reaction to proceed without the need for excessive heat that would destroy the cell. An enzyme also speeds up a reaction. Without enzymes, most cellular reactions would not proceed fast enough to maintain a living condition.

Coenzymes

A coenzyme is a non-protein molecule that works with an enzyme to catalyze a reaction. Many enzymes need such molecules to assist them in lowering the activation energy of a reaction. A coenzyme often acts as a **transfer agent**. For example, in certain reactions, an enzyme removes a hydrogen atom from one substance in the substrate. Then the coenzyme transfers the hydrogen atom to another substance in the substrate. The enzyme could not complete its catalysis of the reaction without the coenzyme to assist in the hydrogen transfer.

A coenzyme is not as specific as an enzyme. For example, the coenzyme that accepts hydrogen atoms from one enzyme can probably accept hydrogen atoms from other enzymes as well.

If coenzymes are not proteins, what are they? Most are molecules of vitamins or parts of such molecules. Suppose you lack vitamin C in your diet. The absence of this vitamin means that you will be lacking one coenzyme in your cells. As a result, one or more enzymes may not be able to perform their tasks. This causes serious disruptions in the metabolic activities of the cells. The result is a **deficiency disease**. In the case of vitamin C deficiency, the disease is scurvy.

Review

1. **a)** Define photosynthesis.
   b) Write a summation equation for photosynthesis. Include both a word equation and a chemical equation.
2. **a)** Define light.
   b) Why does a geranium appear green when it is illuminated with white light?
   c) What colour do you think a geranium would appear if it were illuminated only with red light? Why?
3. **a)** What does the absorption curve of chlorophyll tell you about chlorophyll?
   b) What wave lengths (colours) of light do you think participate most in photosynthesis? Why?

c) What wave lengths (colours) of light do you think participate least in photosynthesis? Why?

4. a) How does chlorophyll *a* differ from chlorophyll *b*?

b) Describe the structure of a chloroplast.

c) What is the function of the chlorophyll in the chloroplast?

5. a) What is an enzyme?

b) Why is it necessary to have enzymes helping with most cellular reactions?

c) Explain briefly how an enzyme does its job.

d) What is a coenzyme?

6. a) Explain what the process of photosynthesis does for the plant in which it takes place.

b) What does photosynthesis do for heterotrophs?

c) What does photosynthesis do for the biosphere (the whole of life on earth)?

7. Design and conduct a controlled experiment to show whether or not the production of chlorophyll in a plant requires light. Prepare a report including your procedure, results, and conclusions for submission to your teacher. Pay particular attention to the use of controls. Your teacher may ask you to do this experiment at home.

7.5 Cellular Respiration

What is Cellular Respiration?

As you know, **respiration** *is the process by which living cells break down glucose molecules and release stored chemical potential energy.* In order to emphasize that this process occurs in cells, it is usually called **cellular respiration**. It occurs in the mitochondria. You must not confuse respiration with breathing. Breathing is simply the exchange of gases between an organism and its environment. Cellular respiration is the "burning" of glucose in cells to release the energy required to support life processes. The word "burning" is placed in quotation marks because the glucose really does not burn. A better term to use is oxidation. **Oxidation** *is the addition of oxygen to or the removal of hydrogen from a substance.* In cellular respiration, that substance is usually glucose.

In some respects, oxidation in cells, or cellular respiration, is similar to the oxidation, or burning, of fuels such as coal, wood, and oil. When a fuel is burned, chemical potential energy in the molecules of the fuel is released as heat and light energy. In a similar manner, when glucose is oxidized in cells, chemical potential energy in the glucose molecules is released partly as heat energy. However, there is one important difference between the burning of fuel and the oxidation of glucose in living cells. The burning of fuel produces very high temperatures. Clearly such high temperatures cannot occur in cells. The cells would be destroyed. Thus, during cellular respiration, the rate of oxidation must be controlled. As a result, it occurs in several small steps. Each step is assisted by an enzyme. The enzyme permits that step to take place at the normal temperature of the organism.

The Summation Equation for Cellular Respiration

You know that cellular respiration begins with two substances, glucose and oxygen. You know also, that it produces energy. Oxidation of most organic compounds produces carbon dioxide and water. Therefore we will assume that they are formed in cellular respiration. (You likely know that you breathe out carbon dioxide and water vapour produced in your body by cellular respiration.) You will get a chance to verify that these products are formed in Section 7.6. Based on what you know at this stage, the following are the summation equations for cellular respiration:

$$\text{Glucose} + \text{Oxygen} \xrightarrow{\text{Enzymes}} \text{Carbon Dioxide} + \text{Water} + \text{Energy}$$

$$C_6H_{12}O_6 + 6\ O_2 \xrightarrow{\text{Enzymes}} 6\ CO_2 + 6\ H_2O + \text{Energy}$$

Remember that this summation equation only "sums up" *what* happens during cellular respiration. It tells us *what* cellular respiration starts with and *what* it ends with. However, it does not tell us *how* this process occurs.

Recall that the glucose was formed, in the first place, by photosynthesis. During that process, chemical potential energy was stored in the bonds of the glucose molecules. During cellular respiration, this glucose is broken down by the action of oxygen, with the help of enzymes. Much of the energy stored in the bonds of the glucose is released.

Energy Transfer by ATP

In most cellular processes, enzymes alone are not enough to start a reaction. The reaction needs an "energy boost" as well. For this purpose, cells have a substance that can quickly store or release energy in any part of the cell. The substance is called **adenosine triphosphate**, or **ATP**. ATP has a complex structural formula. We will write it simply as $A - $ (P)~(P)~(P) . The A stands for adenosine and each (P) stands for one phosphate group. The wavy lines (~) represent **high-energy bonds**.

ATP is made in the mitochondria of a cell. It is then distributed to all parts of the cell. The bond between the last two phosphate groups has very high energy. This bond plays the key role in the storage, transfer, and release of energy in a cell. The last phosphate group can be quickly removed. When this happens, the energy stored in the bond is *released*. This leaves behind **adenosine diphosphate**, **ADP**. ADP can quickly pick up a phosphate group and form ATP again. This process *requires* energy. It is *stored* in the high energy bond. The process can be summarized in this equation:

$$A - \text{(P)}\sim\text{(P)}\sim\text{(P)} \rightleftarrows A - \text{(P)}\sim\text{(P)} + \text{(P)} + \text{energy}$$

or

$$\text{ATP} \rightleftarrows \text{ADP} + \text{(P)} + \text{energy}$$

In words the equation says: ATP produces ADP, a phosphate group, and energy. Also, ADP combines with a phosphate group and energy to produce ATP.

The change from ATP to ADP and from ADP to ATP occurs over and over again throughout the cell. The energy is stored in ATP and transported to where it is needed. There the ATP converts to ADP, releasing the needed energy.

The Role of ATP in Cellular Respiration

The energy stored in ATP molecules is the only energy that is *directly usable* by cells. As a result, the energy released by glucose during cellular respiration must be stored in the bonds of ATP. Figure 7-18 shows how the ATP-ADP cycle connects with the process of cellular respiration.

Enzymes permit the energy of glucose to be released in a slow, controlled manner. As it is released, it combines with a phosphate group, $\textcircled{P}$, and ADP to form high energy bonds in ATP. The ATP travels to places in the cell where energy is needed. It then loses a phosphate group and becomes ADP. The energy stored in the high energy bond is released at this time. That energy powers all life processes. Of course, cellular respiration is one of these life processes. Therefore some of this ATP energy is used to start some of the reactions in cellular respiration. However, much more energy is produced by cellular respiration than is used to keep it going.

Biologists have shown that, for every molecule of glucose that is oxidized, 38 molecules of ATP are formed. Therefore, the summation equation for cellular respiration showing the involvement of the ATP-ADP cycle could be written as:

$$C_6H_{12}O_6 + 6\ O_2 + \boxed{38\ P + 38\ ADP} \xrightarrow{\text{Enzymes}} 6\ CO_2 + 6\ H_2O + \boxed{38\ ATP}$$

Fig. 7-18 The energy released during cellular respiration changes ADP to ATP. That is, the energy is stored in high energy phosphate bonds. Now it is available for the maintenance of life processes when ATP changes to ADP.

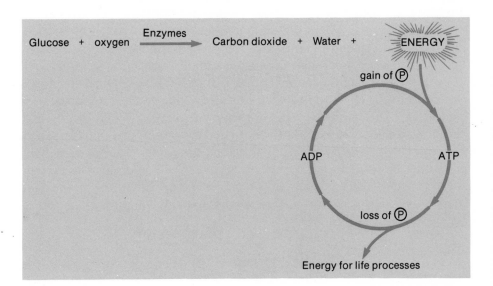

Cellular respiration consists of many complex reactions. Some occur in the cytoplasm of the cell. Others occur in the mitochondria. Many enzymes and coenzymes assist in these reactions.

Anaerobic Respiration

The process of cellular respiration just discussed requires free oxygen, or oxygen gas. As a result, it is called aerobic respiration. ("Aerobic" means "with air".) However, many small organisms such as yeasts and some bacteria can respire without free oxygen. This type of respiration is called anaerobic respiration. ("Anaerobic" means "without air".) During anaerobic respiration, glucose is broken down to release the chemical potential energy in its bonds. In that respect, anaerobic respiration is like aerobic respiration. However, in anaerobic respiration, oxygen is not involved. Enzymes do the job alone. Of course, some energy must be put into the process from ATP to keep it going. But, again, more energy is released by the process than is put into it.

Bacteria and yeasts that live in areas where there is no oxygen must respire anaerobically. Many of these areas exist, for instance, deep in the soil, deep in a pile of garbage, in the muck of a bog or swamp, in the bottom of a brewer's vat. The summation equation for this type of anaerobic respiration is:

$$\text{Glucose} \xrightarrow{\text{Enzymes}} \text{Ethyl Alcohol} + \text{Carbon Dioxide} + \text{Energy}$$

$$C_6H_{12}O_6 \xrightarrow{\text{Enzymes}} 2\ C_2H_5OH + 2\ CO_2 + \text{Energy}$$

Since one of the end products is alcohol, this process is often called alcoholic fermentation. Notice that no free oxygen is shown in this equation.

Anaerobic respiration does occur in some of your body cells from time to time. In fact, it often occurs in the muscle cells of many animals. During periods of heavy physical activity, your muscle cells may not be able to get oxygen fast enough to carry out the usual aerobic cellular respiration. When this occurs, your muscle cells begin to respire anaerobically. Glycogen, or animal starch, is stored in your muscles (see Section 4.4, page 81). During anaerobic respiration, this glycogen is changed to glucose. The glucose is then broken down anaerobically. No free oxygen is involved. The summation equation for this type of anaerobic respiration is:

$$\text{Glucose} \xrightarrow{\text{Enzymes}} \text{Lactic Acid} + \text{Energy}$$

$$C_6H_{12}O_6 \xrightarrow{\text{Enzymes}} 2\ C_3H_6O_3 + \text{Energy}$$

Since the end product is lactic acid, this process is often called lactic acid fermentation. Even without oxygen, muscle cells can still break down glucose to get needed energy. The accumulation of lactic acid in muscles causes the soreness that often follows exertion.

Examine the formulas of lactic acid and ethyl alcohol. You can see that these products of anaerobic respiration have more than one carbon atom in each molecule. Therefore, not all the bonds in the glucose were completely

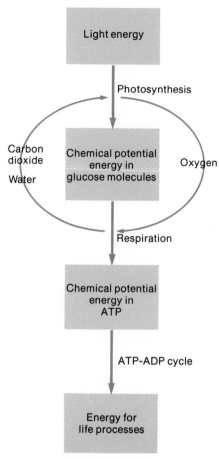

Light energy

↓ Photosynthesis

Carbon dioxide

Water

Chemical potential energy in glucose molecules

Oxygen

↓ Respiration

Chemical potential energy in ATP

↓ ATP-ADP cycle

Energy for life processes

Fig. 7-19   The flow of energy from the sun to life processes. Note the relationship between photosynthesis and respiration.

broken during anaerobic respiration. This is not true in aerobic respiration. Thus, the same amount of glucose will release more energy during aerobic respiration than during anaerobic respiration.

Relationship between Photosynthesis and Respiration

You have likely noticed a relationship between photosynthesis and aerobic respiration. If you have not, look closely at the summation equations for photosynthesis (page 153) and cellular respiration (page 159). Figure 7-19 illustrates this relationship. Light energy is converted to chemical potential energy in glucose molecules during photosynthesis. This process produces oxygen as a by-product. This oxygen can participate in the oxidation of glucose during cellular respiration. That process releases energy that is stored in the energy-rich bonds of ATP. The by-products of respiration, carbon dioxide and water, may now be photosynthesized into more glucose. The ATP energy is used to power cell activities.

Review

1. **a)** Distinguish between breathing and cellular respiration.
   b) In what way is cellular respiration similar to the oxidation (burning) of a fuel?
   c) In what major way is cellular respiration different from the oxidation of a fuel? Why must this be so?
2. **a)** Write a summation equation for cellular respiration in words and in symbols.
   b) Explain how the conversion of ATP to ADP and the conversion of ADP to ATP is involved in energy transfer in cells.
   c) Outline the role of the ATP-ADP cycle in cellular respiration.
   d) Write a summation equation for cellular respiration that shows the involvement of ATP and ADP.
   e) In what parts of the cell does cellular respiration occur?
   f) Why is cellular respiration necessary?
3. **a)** Distinguish between aerobic and anaerobic respiration.
   b) Give a description of alcoholic fermentation. Include the summation equation.
   c) Give a description of lactic acid fermentation. Include the summation equation.
4. Describe how photosynthesis and respiration participate in the flow of energy from the sun to the life processes of organisms.
5. **a)** Aerobic cellular respiration and both types of fermentation begin with glucose. Also, all three release energy from glucose. How do the three processes differ?
   b) One gram of glucose will release more energy during aerobic than during anaerobic respiration. Why?

7.6 Investigation

Cellular Respiration

In Section 7.5 you studied the process of cellular respiration. In this investigation you use the knowledge that you gained to explain the results of some experiments that deal with respiration.

Note: Vacuum flasks are required for parts of this investigation. Because they are expensive your teacher may demonstrate those parts or have you work on them in larger groups.

Materials

250 mL Erlenmeyer flasks (2)
thistle tubes (2)
test tubes (2)
limewater (50 mL)
glass tubing (bent as shown in Fig. 7-20)
absorbent cotton
germinating pea or bean seeds (soaked in water overnight)
vacuum flasks (2)
one-hole rubber stoppers (2)
–10°C to 110°C thermometers (2)
10% formalin solution
25% molasses solution or fresh apple juice (300 mL)
dry yeast (¼ package)

Procedure A   Respiration of Germinating Seeds

Seeds are alive. Therefore they must respire. While they are germinating, cellular activity is at a peak. This means that cellular respiration must also be at a peak, since it provides the energy to power the cellular activities.

In this procedure you test germinating seeds to see if carbon dioxide is produced as a result of cellular respiration.

a. Obtain about 100 mL of germinating pea or bean seeds from your teacher. Germination was started by allowing the seeds to stand in water overnight.
b. Add the seeds to the Erlenmeyer flask assembly that is shown in Figure 7-20. If you are assembling this apparatus yourself, follow your teacher's instructions carefully. Otherwise you could cut your hands. Note that the bottom of the thistle tube is covered with water. The seeds must not be covered with water. The limewater is not in place at this stage.
c. Set up a similar apparatus as a control. However, this time, use dead seeds instead of living ones. Your teacher will give you 100 mL of seeds that were killed by soaking them for 1-2 h in 10% formalin solution.

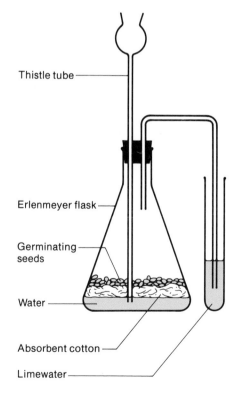

Thistle tube

Erlenmeyer flask

Germinating seeds

Water

Absorbent cotton

Limewater

Fig. 7-20   Do germinating seeds respire? If they do, what products will be formed?

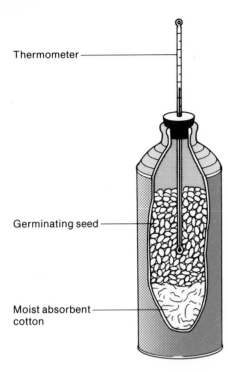

Thermometer

Germinating seed

Moist absorbent cotton

Fig. 7-21   Do germinating seeds give off heat energy?

d. After 24 h, test each flask as follows to see whether or not carbon dioxide was produced. Carbon dioxide turns limewater milky. Therefore, place a test tube containing about 10 mL of limewater over the end of the glass tubing. Force the gas in the Erlenmeyer flask to bubble through the limewater by pouring 100 mL of water down the thistle tube. Do this for the control as well. Record your results.

Procedure B   Energy Production by Germinating Seeds

Respiration converts the chemical potential energy in glucose molecules into other forms of energy. One of these is heat energy. Therefore you should be able to detect heat production by germinating seeds.

a. Place a wad of moist absorbent cotton in the bottom of a small vacuum flask.

b. Add germinating pea or bean seeds to this flask until it is nearly full (Fig. 7-21).

c. Insert the thermometer far enough through the one-holed stopper that the bulb is in the middle of the seeds. Follow your teacher's instructions in order to avoid breaking the thermometer.

d. Prepare a control that uses dead seeds instead of living ones. Your teacher will provide seeds that were killed by soaking them in 10% formalin solution for 1-2 h.

e. Observe the temperatures immediately and several times during the remainder of the period. If possible, note the temperatures 2 or 3 times during the remainder of the day. Take final readings the next day.

Procedure C   Alcoholic Fermentation

Alcoholic fermentation is a form of respiration. Therefore it should convert the chemical potential energy of glucose into other forms of energy. One of these is heat energy. In this procedure you will check to see if heat is formed during alcoholic fermentation. You will also test for the two products of this process.

a. Fill a small vacuum flask nearly full of 25% molasses solution or fresh apple juice. Then add about ¼ package of dry yeast. Swirl the flask to mix the contents.

b. Assemble the apparatus as shown in Figure 7-22. Follow your teacher's instructions in order to avoid breakage and possible cuts.

c. Set up a similar apparatus as a control. However, this time do not add the yeast.

d. Observe any changes that occur in the limewater during the remainder of the period.

e. Record the temperatures immediately. Then record them as often as possible during the next 72 h.

f. Smell the contents of both flasks at the end of the 72 h period.

Discussion A

1. Compare the appearance of the limewater in the two test tubes. What gas was produced?
2. Explain why the control was necessary.
3. What conclusion can you draw from Procedure A?

Discussion B

1. Describe any temperature changes that occurred during Procedure B.
2. Explain why the control was necessary.
3. What conclusion can you draw from Procedure B?

Discussion C

1. What evidence do you have that some form of respiration occurred in Procedure C? Did it occur in both flasks? If not, in which one did it occur? If it occurred in both flasks, was there a difference in rate?
2. Are the conditions within the molasses solution (or apple juice) aerobic or anaerobic? Explain.
3. Describe the smell of the contents of each flask at the end of the experiment. What do you conclude from this observation?
4. Describe any temperature changes that occurred.
5. What conclusion can you draw by putting all your observations together?

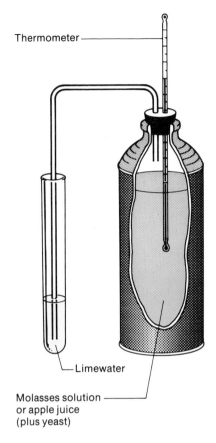

Thermometer

Limewater

Molasses solution or apple juice (plus yeast)

Fig. 7-22 What are the end-products of alcoholic fermentation?

Highlights

All living things need a continuous supply of energy. Almost all organisms get most of their energy for life processes by breaking down glucose molecules in their cells. This releases the chemical potential energy that is stored in the bonds of the glucose molecules. This process is called cellular respiration.

Almost all the energy used by living things comes, in the first place, from the sun. Autotrophic organisms make glucose through photosynthesis. Heterotrophic organisms obtain their glucose either directly or indirectly from autotrophs.

The following are the summation equations for the main processes discussed in this chapter:

1. Photosynthesis

$$\text{Carbon Dioxide} + \text{Water} + \text{Light Energy} \xrightarrow[\text{Enzymes}]{\text{Chlorophyll}} \text{Glucose} + \text{Oxygen}$$

2. Cellular Respiration (Aerobic)

$$\text{Glucose} + \text{Oxygen} \xrightarrow{\text{Enzymes}} \text{Carbon Dioxide} + \text{Water} + \text{Energy}$$

3. Alcoholic Fermentation

$$\text{Glucose} \xrightarrow{\text{Enzymes}} \text{Ethyl Alcohol} + \text{Carbon Dioxide} + \text{Energy}$$

4. Lactic Acid Fermentation

$$\text{Glucose} \xrightarrow{\text{Enzymes}} \text{Lactic Acid} + \text{Energy}$$

The first two equations show that photosynthesis and aerobic cellular respiration are complementary. The substances that are reactants for the one process are the products of the other. The last three equations show that all forms of respiration release energy.

    The energy stored in ATP molecules is the only energy directly usable by cells. ATP converts to ADP as its stored energy is released. ADP can be "recharged" with a phosphate group and energy to form ATP again. Both photosynthesis and respiration use the ATP-ADP cycle for energy transfer within cells.

Cell Reproduction and Development

The cell theory was introduced in Chapter 5. It consisted of three points:

1. Cells are the structural units of all living things.
2. Cells are the functional units of all living things.
3. All cells come from living cells.

In Chapter 5 you saw that the first point seems to be true. You examined several living things. All were made of cells. In Chapters 6 and 7 you studied some processes that are important to living things—osmosis, photosynthesis, and respiration. All these occurred at the cellular level. Therefore the second point seems to be true. The cell appears to be the functional unit of living things. In this chapter you will look at the third point in the cell theory. How do cells come from other living cells? Why does this occur?

    Cells are always making new materials. Much of this material is used for the repair and replacement of worn-out parts. However, cells usually make more material than they need for those purposes. How can they store this material? One way is to grow larger. But a cell cannot increase in size forever. In fact, there seems to be a maximum size that a given type of cell can maintain. Once the cell goes beyond that size, the plasma membrane simply has too small a surface area to serve the needs of the cell contents. It cannot move materials in and out fast enough to maintain a living condition. The alternative is for the cell to divide when it reaches its maximum size.

    Therefore, in addition to cell enlargement, it seems that growth results in division of cells. However, it is probably not the direct cause of division. In fact, biologists admit that they still do not know why a cell divides. Whatever the cause, cells do divide. In the early part of the 19th century, many scientists observed dividing cells. The process seemed quite simple. A single cell grows to a certain size and then divides into two cells. Each of these grows to its full size, then divides. This process continues, resulting in the growth of the organism.

We know today that the process is not this simple. Let us see how it works.

8.1 Mitosis

Many years ago, biologists discovered that the nucleus plays a key role in cell division. They noted that, every time a cell divides, the nucleus **replicates** (**duplicates**) itself. After a series of predictable changes, a nucleus forms two identical copies of itself. *The process of nuclear replication is called* **mitosis**.

Importance of Mitosis

The main task of mitosis is quite clear. It must ensure **genetic continuity**. The genetic code is stored in the nucleus and controls cell activities. The **daughter cells** formed by cell division must be genetically identical to the **mother cell**. (During cell division, the cell that divides is called the mother cell. The cells that are formed are called daughter cells.)

The nucleus contains the genetic material of a cell. Therefore, the nucleus must divide in a manner that gives each daughter nucleus an exact replicate (duplicate) of its genetic material. If exact replication did not occur, the daughter cells would be unlike the mother cell. If this happened in the root hair cells of a plant, for example, the daughter cells would be unable to perform the functions required of root hair cells. Most likely, they would die. Then the plant would die.

Interphase

When a nucleus is not actively dividing, it is said to be in the **interphase** stage. During this stage all cell activities, except reproduction, continue. The cell is growing, repairing itself, metabolizing, and so on. Most of the cell's time is spent in this stage.

Interphase is not a stage of mitosis. Rather it is the stage just before and just after mitosis. However, you must know what the nucleus is like in this stage in order to understand mitosis.

The cell nucleus was described in Section 5.6, page 101. The description was of the interphase stage. You learned then that the nucleus consists of three main parts. It has a **nuclear membrane** that controls the exchange of materials between the nucleus and the cytoplasm. It has one or more nucleoli. These consist largely of DNA, RNA, and proteins. It has chromosomes. During interphase the chromosomes are spread throughout the nucleus in a delicate network called chromatin. The chromosomes are made of DNA and protein. DNA is the hereditary material that carries the genetic code. Therefore, if we want to understand how a nucleus replicates itself, we must centre our attention on the chromosomes.

Phases in Mitosis

Mitosis is a continuous process. However, four fairly distinct phases can be recognized. It is useful to base a discussion of mitosis on and around these four phases. But keep in mind that there is no sharp division between them. The phases are called prophase, metaphase, anaphase, and telophase. Let us see how an animal cell goes through these phases. Follow the drawings in Figure 8-1 as you read this description.

Mitosis begins near the end of interphase. At this point the chromosomes replicate themselves. (You will find out how this occurs in Unit 7.) They remain in the form of long, thin strands. The rest of mitosis is

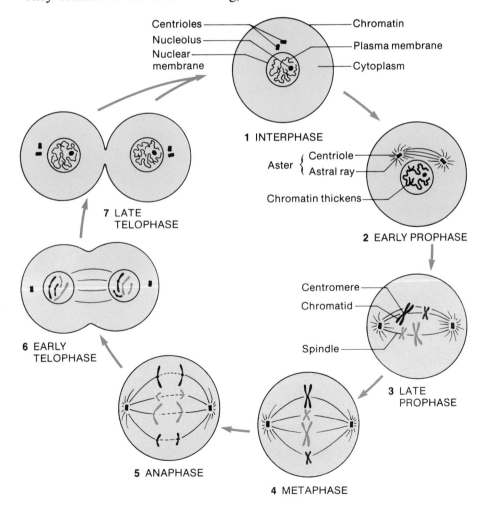

Fig. 8-1 Mitosis in an animal cell.

1 INTERPHASE

Centrioles — Chromatin
Nucleolus — Plasma membrane
Nuclear membrane — Cytoplasm

Aster { Centriole — Astral ray
Chromatin thickens —

2 EARLY PROPHASE

Centromere —
Chromatid —
Spindle —

3 LATE PROPHASE

4 METAPHASE

5 ANAPHASE

6 EARLY TELOPHASE

7 LATE TELOPHASE

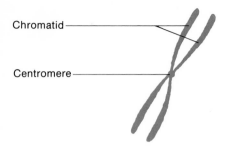

Chromatid ⎯⎯⎯⎯⎯⎯

Centromere ⎯⎯⎯⎯⎯⎯

Fig. 8-2 A chromosome that has replicated to form a pair of chromatids.

concerned with separating the replicated chromosomes into two identical sets—one for each daughter nucleus.

Prophase

In an animal cell, mitosis begins with the movement of the **centrioles** to opposite sides of the nucleus. At the same time, tiny fibres of protein, called **astral rays**, form around each centriole. A centriole with its astral rays is called an **aster**. The **chromosomes** appear to thicken and become shorter. Even at this early stage in prophase, you can easily see the chromosomes with an ordinary compound microscope.

By mid-prophase, the **nucleoli**, if present, have usually disappeared. Also, the **nuclear membrane** begins to break down. At this stage it is easy to see that the chromosomes have replicated. They appear as a double strand along their entire length. They are attached to one another at a single point called a **centromere** (Fig. 8-2). Each of the strands is called a **chromatid**. The two chromatids are identical.

In the late prophase, the nuclear membrane completely disappears. More protein fibres form between the centrioles. Some stretch from one centriole to the other. Others stretch only from a centriole to the centre ("equator") of the cell. An oval structure results that is called the **spindle**.

At the end of prophase, the paired chromatids are moving toward the equator of the cell.

Metaphase

During this stage, the paired chromatids line up at the equator of the cell. Each chromatid pair then separates into two identical chromosomes. The centromere of each chromosome becomes attached to a protein fibre. Each chromosome of a pair is attached to a protein fibre that goes to a different centriole.

Anaphase

Each matching pair of chromosomes now separates. The two chromosomes move to opposite poles of the cell. Each of the chromosomes seems to move along one of the protein fibres of the spindle. Biologists are not sure what causes the movement. However, it appears that the protein fibres shorten.

By the end of the anaphase, the chromosomes have reached opposite poles of the cell.

Telophase

During telophase, everything that occurs is the opposite of prophase. The nuclear membrane reappears. The nucleoli reappear. The chromosomes lose their thick, short appearance and become a thread of chromatin again.

The spindle and asters disappear. Two nuclei have been formed. Finally, the centrioles replicate. Each daughter cell now has a pair of centrioles.

While all this is happening, the cytoplasm also divides. The plasma membrane pinches inward at the equator. Eventually two new daughter cells are formed and they separate from one another. They now enter interphase. They will grow and develop. Eventually they, too, will undergo mitosis and cell division.

The end result

The identical pairs of chromosomes lined up at the equator during metaphase. Then, during anaphase, one member of each pair went to one pole of the cell and the other member went to the other pole. Thus one of each kind of chromosome is present in the nucleus of each daughter cell. Therefore, each daughter cell has the same number and kinds of chromosomes as its mother cell. In this way, genetic continuity is ensured. The daughter cells will have the same genetic code as the mother cell and, therefore, will function in the same way.

Mitosis and Cell Division in Plant Cells

Basically, mitosis and cell division in plant cells is very much like that in animal cells. However, there are two main differences:

1. Plant cells do not have centrioles. As a result, they do not form asters. They do, however, develop a spindle.
2. In late anaphase, a structure forms in the middle of the spindle. It gradually grows across the cell to form the two daughter cells. It is called the **cell plate**. By the end of telophase, cellulose has been added to the cell plate to form a new **cell wall**.

Review

1. Why does cell division appear to be necessary?
2. **a)** Define mitosis.
   b) What is genetic continuity?
   c) How does mitosis contribute to genetic continuity?
3. Give a brief description of a nucleus while it is in interphase.
4. Summarize the events that occur in each of the four stages of mitosis.
5. **a)** Distinguish among chromatin, chromosomes, and chromatids.
   b) Distinguish between a spindle and an aster.
6. How does mitosis in plant cells differ from mitosis in animal cells?

8.2 Investigation

Mitosis and Cell Division

The tip of a root is a place of active growth. Therefore you can expect to see mitosis and cell division in that region. In this investigation an onion is placed in the dark with its base in water. After a few days new white roots appear. You will examine a wet mount of these roots to look for mitosis and cell division in plant cells.

Before you look for mitosis and cell division in these living cells, you will look at prepared slides of both plant and animal cells that are dividing. After a "warm-up" using the prepared slides, you should have little trouble recognizing the stages of mitosis in your own wet mount.

Materials

compound microscope	fresh onion
microscope slide	Bunsen burner
cover slip	acetocarmine stain
paper towel	beaker
lens paper	toothpicks
medicine dropper	watch glass
forceps	prepared slide of onion root tip
scalpel or razor blade	prepared slide of whitefish cells

Procedure A   Mitosis and Cell Division in Animal Cells

a. Examine the prepared slide of whitefish cells under low power. When you have found a region that shows mitosis switch to medium power, and, finally, to high power.
b. Find a cell that appears to be in interphase. Sketch it. Label your diagram.
c. Now seek out a cell that is in prophase. Sketch it. Label your diagram. Include labels that indicate any important changes that have occurred.
d. Repeat step c for cells that are in metaphase, anaphase, and telophase. You may wish to include early and late stages for some of these, if you can find them.

Procedure B   Mitosis and Cell Division in Plant Cells

a. Examine the prepared slide of the onion root tip under low power. Scan the root tip until you find a region where mitosis was underway when the cells were killed. The nuclei in this slide have been stained. Therefore they appear much darker than the rest of the cell. The section of the root tip with the greatest number of nuclei will likely be the section that you want.
b. Switch to medium power and then to high power. Examine the section closely.

c. Find a cell that appears to be in interphase. Sketch it. Label your diagram.

d. Now proceed as you did in steps c and d of Procedure A. Seek out cells that are in prophase, metaphase, anaphase, and telophase. Sketch these. Label your diagrams carefully. Include labels that indicate important changes that have occurred.

Procedure C   Mitosis and Cell Division in Living Onion Root Tips

Having examined prepared slides in Procedures A and B, you should know what the stages of mitosis look like. Now see if you can prepare your own wet mount that shows these stages.

a. Obtain an onion that is as firm and fresh as possible. Using a scalpel or razor blade, cut off the dead roots quite close to the base of the onion bulb.

b. Add water to a beaker until it is about ⅔ full.

c. Use 3 or 4 toothpicks to support the onion in the water as shown in Figure 8-3. You may have to add more water.

d. Place the beaker and its contents in a cool dark place until new white roots appear that are about 1-2 cm long. You should check the onion every day. However, usually 2-4 d are required for the roots to become long enough.

e. Half fill the watch glass with acetocarmine stain.

f. Remove the onion from the water. Use a sharp scalpel or razor blade to cut each root about 0.5 cm up from the tip. Let the tips drop into the acetocarmine stain. Do not handle the tips with forceps.

g. Heat the stain plus tips *gently* for 3-5 min. *You must not boil the stain.*

h. Place a fresh drop of acetocarmine stain on a microscope slide.

i. Select 2 tips that appear well stained. Use the forceps to add them to the drop of stain on the microscope slide. Grasp the tips at the cut end, not the tip end.

j. Use a sharp scalpel or razor blade to cut off the last 2 mm of each tip. Leave those pieces in the drop of stain. Discard the long pieces.

k. Use the scalpel or razor blade to cut the two tips into pieces that are smaller than a pinhead.

l. Carefully place a coverslip over the pieces of root tip. Tap the coverslip lightly with a pencil to separate and spread the cells of the root tips.

m. Use your thumb to apply an even pressure on the coverslip. Use the procedure shown in Figure 8-4. *Do not let the coverslip slide over the sample.* The intent is to squash the small pieces of root tip so that the material is only one cell thick. If you slide the coverslip, you may damage the cells. You have just made what is called a "squash" preparation.

n. Examine your wet mount under low power, medium power, then high power. Add any new information that you gather to the drawings you made in Procedure B. Examine the slides of other classmates if they find particularly good stages that you cannot find.

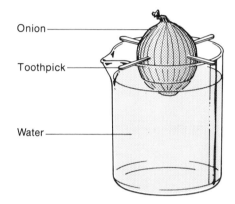

Fig. 8-3   A method for growing onion root tips that will show the stages of mitosis.

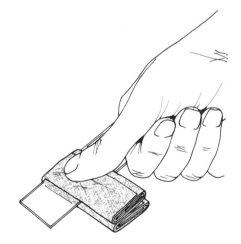

Fig. 8-4   Making a squash preparation of onion root tip cells. Fold a facial tissue several times until it is about as wide as a microscope slide. Lay it on the table. Carefully, place your prepared slide on it. Then fold the tissue over to make a "sandwich". Press evenly downward with your thumb. Do not make the coverslip slide or twist. Carefully remove the tissue from the slide.

Discussions A and B

1. Your completed diagrams (fully labelled) are the main part of your writeup for this investigation.
2. Compare the methods of separation of the daughter cells from one another in plant and animal cell division.
3. What structures did you find in mitotic animal cells that you did not find in mitotic plant cells?

Discussion C

1. Why were just the tips selected for this investigation?
2. What was the main function of the acetocarmine stain?
3. What advantages do you feel your wet mount offered over the prepared slide?
4. What advantage do you feel the prepared slide offered over your wet mount?

8.3 Cell Development

What is Cell Development?

After mitosis, the daughter cells must undergo **development**. You probably know in a general way what is meant by development. It generally means that the cell or organism will increase in size and mass. It may also involve new features and functions. Normally it slows down when a certain mature size is reached.

You have developed through many stages: fertilized egg to embryo to unborn baby to child to adolescent. Your development has not stopped yet. You will continue to develop physically. You will move through young adulthood to middle age to old age and, finally, to death.

We usually associate development with the early life of an organism. However, development continues into old age and, in fact, right up to death. For example, a human body is always forming new blood cells and tissues to heal wounds. Even processes that we associate with aging are a form of development. An example is the calcification of joints, or arthritis.

After mitosis, the daughter cells do one of two things, depending on the type of organism involved. They can separate or they can remain together. If they separate, then one unicellular organism has formed two unicellular organisms. Development of these new organisms involves only **cell growth**. That is, the daughter cells must do nothing more than grow to become mature organisms. An amoeba reproduces in this way. If the cells remain together, a series of cell divisions will form a mass of connected cells. In some organisms these cells are all alike. Such organisms are said to be colonial. However, in many organisms the cells are not alike. You learned that mitosis ensures that the daughter cells will have the same genetic code as the mother cell. Yet cells that came from the same mother cell can develop in different ways. You began life as one cell but now you contain many different kinds of cells. Clearly, more than cell growth has occurred. Cells have become specialized to perform different functions. Biologists say that **cell differentiation** has occurred.

Development of unicellular and colonial organisms involves only growth. Development of multicellular organisms involves both growth and cell specialization, or differentiation. Let us look more closely at growth and differentiation.

Growth

Growth *is an increase in size and mass.* Growth can be accomplished in two ways. First, a cell can simply increase in size and mass. It takes in substances from its environment and assimilates them, or puts them together, to form more cellular material. However, such growth seems to cease when a cell reaches a certain size. Then growth can only be accomplished by the second way, an increase in cell numbers by cell division.

A human egg has a mass of about 1 μg (one microgram). That is 10^{-6} g. A human sperm has a mass of about 1 ng (one nanogram). That is 5×10^{-9} g. The two combine to start the development of a human. At birth, an average child has a mass of about 3 kg or 3 000 g. This mass is about one billion times greater than the mass of the egg and sperm. Growth has certainly occurred. However, it is easy to see that such growth involves more than just an increase in size and mass. As an organism develops, it takes on form, or shape. This type of growth is called **developmental growth**. The shape changes constantly during development. Some parts grow faster or slower than others. Thus some features appear sooner than others. Compare the proportions of head to body and head to legs in Figure 8-5. Clearly, some parts of the human body develop faster or slower than other parts. Some features appear sooner than other features.

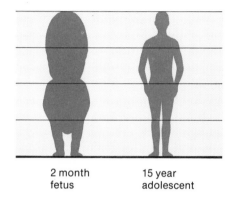

2 month fetus   15 year adolescent

Fig. 8-5 A comparison of proportions in the human body for a 2 month fetus and a 15-year-old adolescent. What percentage of each is head? What percentage is body? What percentage is legs? (Use length to get a rough idea of these percentages.)

Summary

Growth is an increase in size and mass. It can be accomplished by enlargement of cells and by division of cells. Developmental growth is much more than this. Different centres of growth are active at different times. Further, they may have different rates of development. The work of all these centres of growth is coordinated to produce a certain form. It is this form and the way it functions that makes a horse different from you and makes you different from an oak tree.

Differentiation

In addition to growth, development also involves differentiation. **Differentiation** *is the gaining of specific structural features and functions by a cell.*

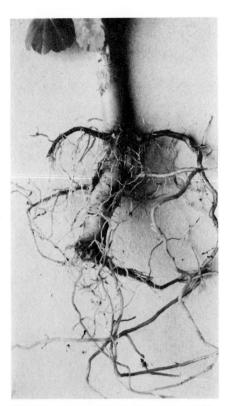

Fig. 8-6 This geranium stem was placed in a "root environment". The genetic information on roots that was stored in the stem cells expressed itself. Differentiation and growth produced root cells.

After differentiation, cells become specialized to carry out certain activities in an organism. As you grew from one cell to your present sixty trillion, certain cells became specialized to perform certain functions. Some became cells of the nervous system. Others became cells of the skeletal system. Still others became cells of the excretory, breathing, circulatory, and digestive systems. All these diverse cells, with their various forms and functions, began with the same single cell.

It is clear that differentiation is a very orderly process. Cells do not differentiate in a haphazard way. For example, in order for a "dog cell" to end up as a dog, the growth and differentiation of one cell must be coordinated with the growth and development of all other cells. Otherwise, one would have a strange looking dog, indeed! Such coordination requires some sort of communication between the different cells in the developing organism. Development, therefore, depends on coordinated growth and differentiation.

Biologists do not know exactly *how* differentiation occurs. But they do know that it involves three things:

1. There must be communication among the cells of the developing organism.
2. Cells only use part of the genetic information in their chromosomes, depending on the type of cell to be formed. (Remember that mitosis ensures that all these cells have the same genetic information.)
3. The environment in which the cell is located must play a role in determining how a cell expresses its genetic information.

We have already discussed the need for the first point. Let us now look at the other two.

At one time biologists thought that differentiation occurred because certain cells lost some genetic information. They reasoned that some plant cells became stem cells because they somehow lost the genetic information that might direct them to become root cells. However, a simple experiment shows that this is not so. If you place a stem cutting from a geranium in water, it will form roots (Fig. 8-6). Apparently the stem cells did not lose the genetic information necessary for them to function as root cells.

Many other experiments have been done to show that differentiation is not the result of changes in genetic information. Biologists now believe that only certain parts of the genetic code are expressed in a given cell type. They also believe that the environment in which the cell is located determines when and how the genetic information is used. For example, in plants, stem cells contain genetic information relating to root function. But they do not express that information until they are in a "root environment". As another example, you know that a plant grown in the dark does not produce chlorophyll. Undeveloped plastids are present but they do not perform photosynthesis. However, when light is added to the plant's environment, the chloroplasts develop. Photosynthesis soon begins. Therefore, the genetic information for chlorophyll-making was always in the cell nucleus. But it did not express itself until a certain environment—light—was present.

Summary

Mitosis ensures that daughter cells contain the same genetic information as the mother cell. However, the daughter cells may differentiate, or take on different forms and functions, by making use of different parts of this genetic information. Which information they use depends on the environment in which the cell is located.

In Section 5.7, page 109, you studied levels of biological organization—cell, tissue, organ, and organ system. You should now understand the relationship between these levels and cell differentiation. The number of cells increases by mitosis. The resulting cells differentiate into specialized cells. The specialized cells of one type are grouped to form a tissue. Different tissues are grouped to form organs. Different organs are grouped to form organ systems.

Aging

Different species of organisms have different average life-spans (Fig. 8-7). The life-span may vary from a few days for some insects to a few thousand years for redwood trees (Fig. 8-7). However, sooner or later, an organism grows old and dies. What causes aging? Why does an organism grow old and die? Why does a person become less active, get grey hair, suffer hearing and eyesight loss, and get wrinkled, sagging skin? Why is death a normal part of development?

Cells, like organisms, also have a life-span. No single cell lives forever. On the other hand, no unicellular organism like an amoeba ever grows old. It just redistributes its protoplasm when it divides into two new organisms. Just think: no amoeba has ever lost its "grandfather" through death! (Where is "grandpa amoeba"?)

The situation is much more complex in multicellular organisms. Some cells just stop dividing. They gradually lose their vigour. They become "old", and die. Why? Many biologists suspect that environmental conditions around the cell cause aging. One environmental factor that may cause aging is the sun's radiation. Other biologists believe that changes occur during differentiation that upset some metabolic processes. This would lead to a gradual slowing down, or aging, of the cell.

Cell replacement

When an organism dies, its cells die. However, cells within an organism can die without the organism dying. Some cells simply wear out. Fortunately, they are replaced for most of the life-span of an organism. In one sense, you get a new body about every seven years. It takes about that long for all your worn-out cells to be replaced by new ones. However, not all cells can be replaced. For example, a nerve cell, if destroyed, cannot be replaced. Once differentiation into a nerve cell begins, the cell loses its ability to reproduce. Further, no nerve cell replacement centre exists in your body.

Fig. 8-7 Maximum life-spans of some organisms. The redwood bar goes beyond the scale. The page would have to be 4-5m long to include it.

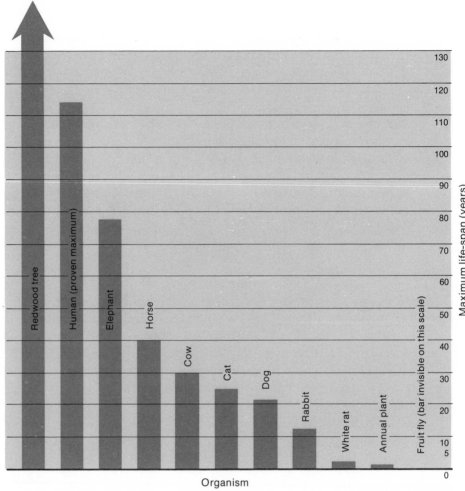

Maximum life-span (years)

Organism

Redwood tree

Human (proven maximum)

Elephant

Horse

Cow

Cat

Dog

Rabbit

White rat

Annual plant

Fruit fly (bar invisible on this scale)

Some biologists believe that up to 2% of your cells die each day. If no cells died, your mass would double in about 100 d, provided that cell division continued normally and you maintained your proper diet. But cells do die, billions every day. Your mass stays fairly constant because the rate of cell replacement equals the rate of cell death.

The epidermis, or outer surface of your body, is an area of rapid cell division and replacement. The underlying cells of your skin are constantly dividing and pushing outward. The outer cells harden, die, and peel away (Fig. 8-8). Several days are required for the cells to be pushed from the dividing layer to the outer layer. The friction caused by clothing and by washing constantly removes the dead outer layers of cells.

Cell replacement is especially high in the blood of your body and in your intestine. About two billion red blood cells must be replaced every day. About seventy billion cells in the lining of your intestine must be replaced every day. It is easy to see that, for an organism to survive, the loss and replacement of cells must stay in balance. If loss exceeds replacement, tissues will go into a state of disrepair.

The problem of aging and death is complex. As with differentiation, no one knows exactly why these processes occur. We do know that *to stay alive, a cell must divide. If it differentiates, it will eventually die.* This fact explains why most trees live much longer than mammals, including humans. The mammalian body tries to preserve most of its differentiated cells in a living, functioning state. For example, you keep your muscle cells year after year. In contrast, a tree replaces most of its differentiated cells every year. For example, it drops its leaves and fruit. It converts other differentiated cells into dead conducting and support tissue, or wood. Then it replaces those cells. Cell division and differentiation go on almost indefinitely in a tree. Since cells are dividing, the tree maintains its vigour.

Much research is now being conducted on the causes of differentiation and aging. Because many diseases have been conquered, humans are living longer and longer. However, aging cells do cause some medical problems. If the cause of aging could be determined, perhaps these medical problems could be eliminated or at least reduced.

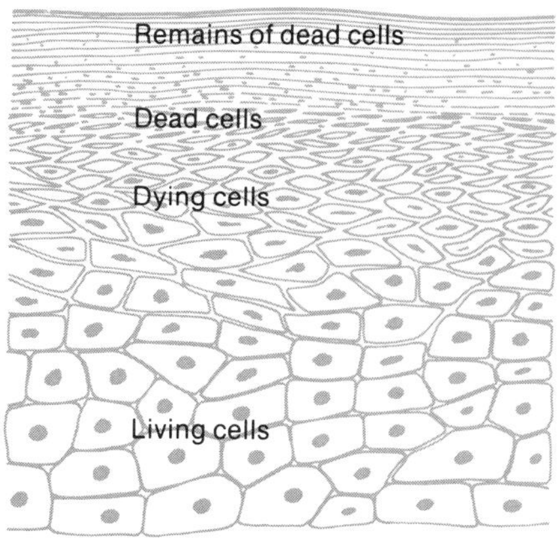

Fig. 8-8   Cells that die at the surface of the skin are constantly replaced by living cells.

Review

1. a) What is meant by cell development?
   b) What is cell growth? How does it occur?
   c) What is developmental growth?
2. a) What is cell differentiation?
   b) Why must cell differentiation be coordinated throughout an organism?
   c) What three things must be involved in cell differentiation?
3. You learned earlier that mitosis ensures that daughter cells contain the same genetic information as the mother cell. If this is so, how can a multicellular organism have many types of cells that are quite different in form and function?
4. a) Explain why no amoeba has ever lost its "grandfather" through death.
   b) Why is the same thing not true for multicellular animals like humans?
5. What two factors do biologists think may cause aging?
6. Describe the importance of cell replacement to organisms.
7. If a cell differentiates, it signs its own death warrant. Explain this statement.
8. Why is much of the money donated for cancer research given to biologists who are studying cell differentiation?

8.4 Meiosis

You have learned that mitosis ensures genetic continuity. That is, it ensures that the daughter cells carry the same genetic information as the mother cell. It does this by making sure that each daughter cell has the same number and kind of chromosomes as the mother cell. During mitosis, the chromosomes of the mother cell duplicate. They then arrange themselves in identical pairs at the equator of the cell. Then one member of each pair moves to a pole of the cell to become part of a daughter cell nucleus. In this way each daughter cell ends up with the same number and kind of chromosomes as the mother cell. Genetic continuity is ensured.

Every body cell in an organism has the same number of chromosomes. In fact, the body cells of all normal individuals of the same species have the same number of chromosomes. For example, all normal human cells have 46 chromosomes, all onion plant cells have 16, and all crayfish cells have 200.

This sounds reasonable and straightforward. However, there is one problem. You began life when a **sperm cell** from your father united with an **egg cell** from your mother. A cell called a **zygote** was formed by this union. If the sperm and egg each had 46 chromosomes, the zygote would have 92. You developed from the zygote by mitosis. Therefore every cell in your body would have 92 chromosomes. But this is not so. If it were, you would not be a human. Clearly something has happened along the way to reduce the number of chromosomes from a possible 92 to 46. A process called **meiosis** ensures that the chromosome number stays constant from generation to generation. In this section you will find out what meiosis is, how it occurs, and why it occurs. However, before you can understand meiosis, you need some background information.

Types of Reproduction

Sexual reproduction

This is the type of reproduction just discussed. In sexual reproduction, two special **sex cells** unite. These reproductive cells are called **gametes**. In some species, the gametes are alike. However, in many species the two gametes are different. In this case, one is called a **male gamete**, or **sperm**. The other is called a **female gamete**, or **egg**. Usually a male parent produces the sperm and a female parent produces the egg. However, some species can produce both sperm and eggs in one individual. Such individuals are called **hermaphrodites**. The earthworm is a hermaphrodite as are many species of snails. Most flowering plants are also hermaphrodites. They produce male and female gametes in each flower.

When a male gamete unites with a female gamete, a cell results that is called a **zygote**. The process by which gametes unite is called **fertilization**.

Asexual reproduction

This type of reproduction does not involve the union of sex cells. Two types of asexual reproduction are binary fission and budding.

Binary fission. Mitosis followed by cell division is a form of asexual reproduction when it occurs in unicellular organisms. When the same process occurs in a multicellular organism, it is not reproduction. Cells may be reproduced, but no new individuals are formed. Cells may increase in

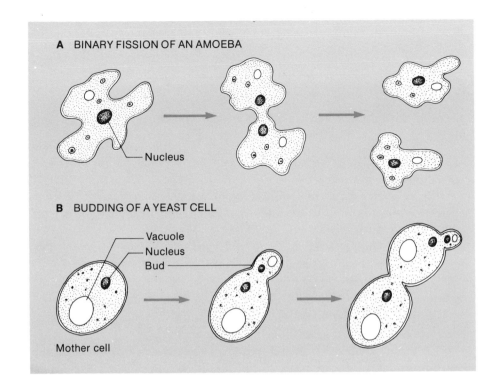

Fig. 8-9   Two examples of asexual reproduction.

A   BINARY FISSION OF AN AMOEBA

Nucleus

B   BUDDING OF A YEAST CELL

Vacuole
Nucleus
Bud

Mother cell

number in various parts of the organism. This results in growth. Or cells may simply replace worn-out cells. In either case, the reproduction of individuals has not occurred. However, when a unicellular organism such as an amoeba undergoes mitosis and cell division, two new individuals are formed (Fig. 8-9,A). Reproduction of individuals has occurred.

This is called **binary fission**. Mitosis ensures genetic continuity during binary fission.

Budding. Yeast cells reproduce by an interesting process called **budding** (Fig. 8-9,B). This process is similar to binary fission because it begins with mitosis. Two nuclei are formed at opposite poles of the cell. However, in binary fission the cytoplasm is split more or less evenly between the daughter cells during cell division. This does not happen in budding. Most of the cytoplasm stays with one cell. As a result it keeps the name mother cell. Only a small part of the cytoplasm goes to the bud (daughter cell). The bud then makes more cytoplasm, grows to full size, and also produces one or more buds. The buds may stay together to form long chains of cells. They may also break apart to form separate cells. Here too, mitosis ensures genetic continuity.

Summary

Genetic continuity is ensured in asexual reproduction because mitosis is all that is involved. However, in sexual reproduction, a union of cells occurs. To prevent a doubling of the chromosome number, something more than

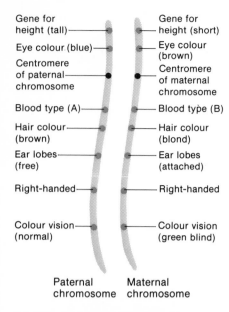

Gene for
height (tall)

Eye colour (blue)

Centromere
of paternal
chromosome

Blood type (A)

Hair colour
(brown)

Ear lobes
(free)

Right-handed

Colour vision
(normal)

Gene for
height (short)

Eye colour
(brown)

Centromere
of maternal
chromosome

Blood type (B)

Hair colour
(blond)

Ear lobes
(attached)

Right-handed

Colour vision
(green blind)

Paternal   Maternal
chromosome   chromosome

**Fig. 8-10 Chromosome structure as
shown by one imaginary homologous pair
of chromosomes. The two chromosomes
are similar in shape but not in the genes
they carry. A human cell contains 23 pairs
of chromosomes carrying thousands of
genes.**

mitosis is needed. To ensure genetic continuity in sexual reproduction, the process of **meiosis** occurs.

Chromosome Number and Structure

You need to know a few facts about chromosomes before you can understand meiosis. You have already met the first two points.

1. *Every body cell in an organism has the same number of chromosomes* (**chromosome number**).
2. *The body cells of all normal individuals of the same species have the same chromosome number.* A normal human cell has 46, a crayfish 200, a fruit fly 8, and a garden pea 14.
3. *The chromosomes in each body cell normally occur in pairs.* The chromosomes in each pair are similar in form and in the arrangement of the genetic information they carry. Such a pair of chromosomes is called a **homologous pair** (Fig. 8-10). Thus the 46 chromosomes in a human cell make up 23 homologous pairs. The two chromosomes in a homologous pair are called **homologues**. Each homologue carries genetic information for the same hereditary trait as its partner. The genetic information of the chromosomes is located at specific points called **genes**. The two chromosomes of a homologous pair carry genes for the same traits at the same positions. However, the two chromosomes are not genetically identical. For example, both homologues may carry information relating to hair colour. But the one homologue may be coded for brown hair. The other homologue may be coded for blond hair (see Fig. 8-10). The final hair colour of the offspring is determined by the combination of these two pieces of genetic information. Each homologous pair can carry genetic information for thousands of hereditary traits.

The number of chromosomes in a complete set of homologous chromosomes is called the **diploid** number. This number is represented by **2n**. Thus the diploid number (2n) for humans is 46. Every body cell that you have contains the diploid number of chromosomes. That is, 2n=46. This diploid number of chromosomes is arranged in 23 homologous pairs.

4. *A gamete contains one homologue from each homologous pair of chromosomes.* In other words, it has half the diploid number of chromosomes. This number is called the **haploid** number. It is represented by **n**. The haploid number for humans is 23. That is, n=23.

A human sperm (male gamete) contains 23 chromosomes (one of each homologous pair). A human egg (female gamete) also contains 23 chromosomes (the homologues of the 23 in the sperm). During fertilization, the sperm and egg unite to form a zygote. The zygote will have 46 chromosomes, in 23 homologous pairs. Thus fertilization has restored the diploid number of chromosomes. All your body cells came from that zygote by mitosis and cell division. Therefore all your body cells also have the diploid number of chromosomes, which is 46. This

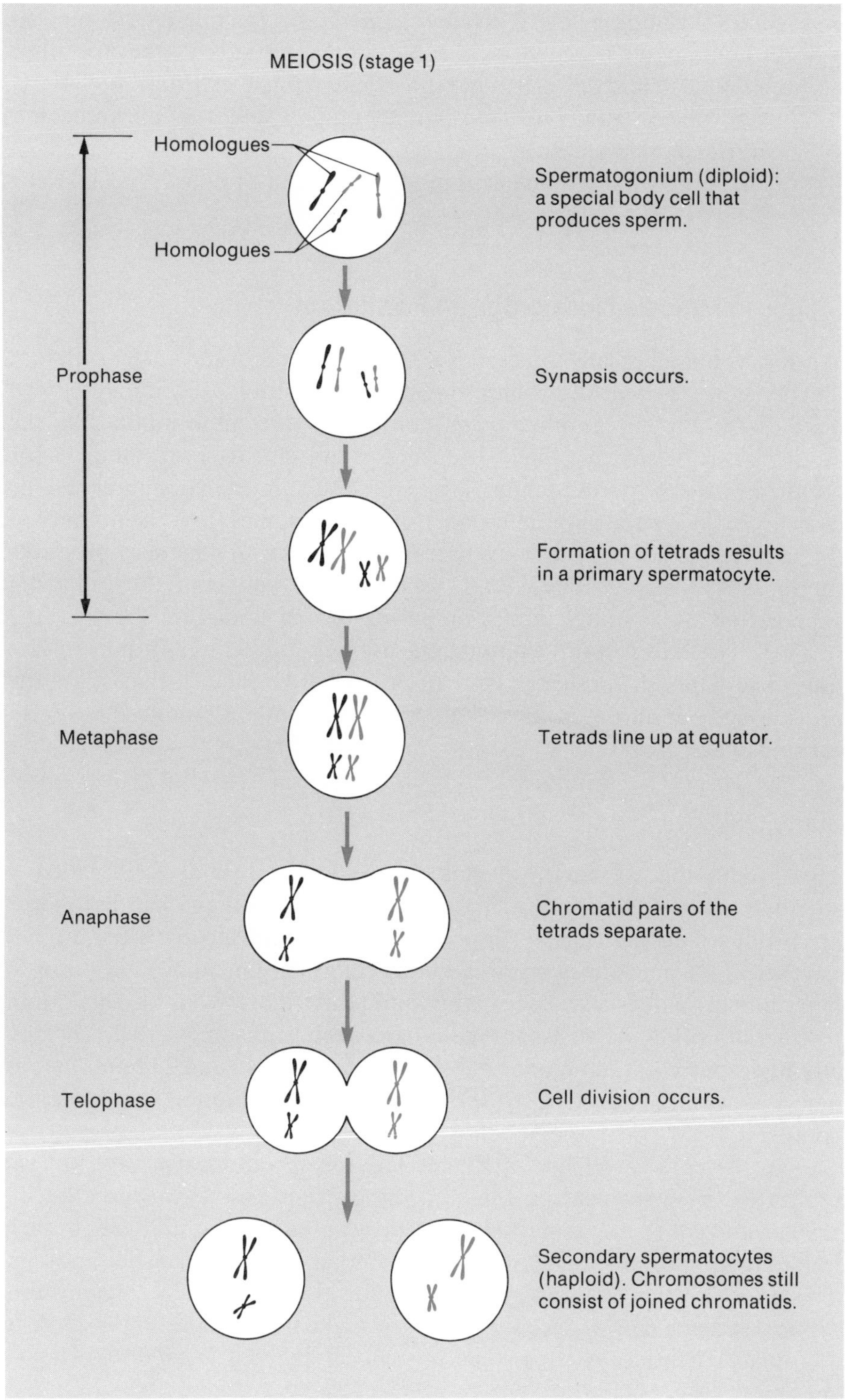

MEIOSIS (stage 1)

Homologues

Homologues

Prophase

Spermatogonium (diploid):
a special body cell that
produces sperm.

Synapsis occurs.

Formation of tetrads results
in a primary spermatocyte.

Metaphase

Tetrads line up at equator.

Anaphase

Chromatid pairs of the
tetrads separate.

Telophase

Cell division occurs.

Secondary spermatocytes
(haploid). Chromosomes still
consist of joined chromatids.

Fig. 8-11 The first stage of meiosis for
sperm production. Only the chromosomes
are shown.

includes the body cells that produce gametes. As you might have guessed, one homologue of each pair came from one parent and the other homologue from the other parent. Each human cell has 23 **paternal** chromosomes (from the male parent) and 23 **maternal** chromosomes (from the female parent).

How do diploid body cells produce haploid gametes? The answer is by **meiosis**.

Stages in Meiosis During Sperm Formation

Mitosis, you may recall, consists of a single division of the nucleus. Before the division, the chromosomes replicate. Therefore each daughter cell receives the diploid number of chromosomes (the same number as the mother cell). In contrast, meiosis occurs in two stages. Two cell divisions occur to produce four daughter cells. This results in each daughter cell, or gamete, receiving the haploid number of chromosomes (half the number of the mother cell). The following narrative explains how meiosis produces sperm cells. Consult Figures 8-11 and 8-12 as you read the following explanation. To simplify things, only four chromosomes are shown in the first cell. Therefore, in this example, the diploid number of chromosomes is four. The haploid number is two. In other words, the cell that starts the process has four chromosomes. The sperm cells that are produced have two chromosomes.

Stage One

The production of sperm begins with a special body cell called a **spermatogonium** (plural: **spermatogonia**). In the human, spermatogonia are produced in the testes. These spermatogonium cells have the diploid number of chromosomes, as do all body cells. The chromosomes occur in homologous pairs. The spermatogonium in Figure 8-11 has a diploid number of four, and two homologous pairs of chromosomes. (In actual fact, a human spermatogonium has a diploid number of 46 and 23 homologous pairs. However, if we tried to show all these, the diagrams would be too confusing.)

In the early prophase of Stage One, the chromosomes have not yet replicated, as they have by the same step in mitosis. However, they do thicken and shorten considerably. In mid-prophase the homologues of each homologous pair move side by side, with similar parts opposite one another. This process is called **synapsis**. In the late prophase, each chromosome replicates. Each pair of chromosomes is called a **tetrad** since it now consists of four chromatids. At this stage of meiosis, the cell is called a **primary spermatocyte**.

As prophase proceeds, the nuclear membrane and nucleoli disappear. Also, the centrioles take up positions at the poles. An aster and a spindle form. All this is similar to what happens in the prophase of mitosis. The

MEIOSIS (Stage 2)

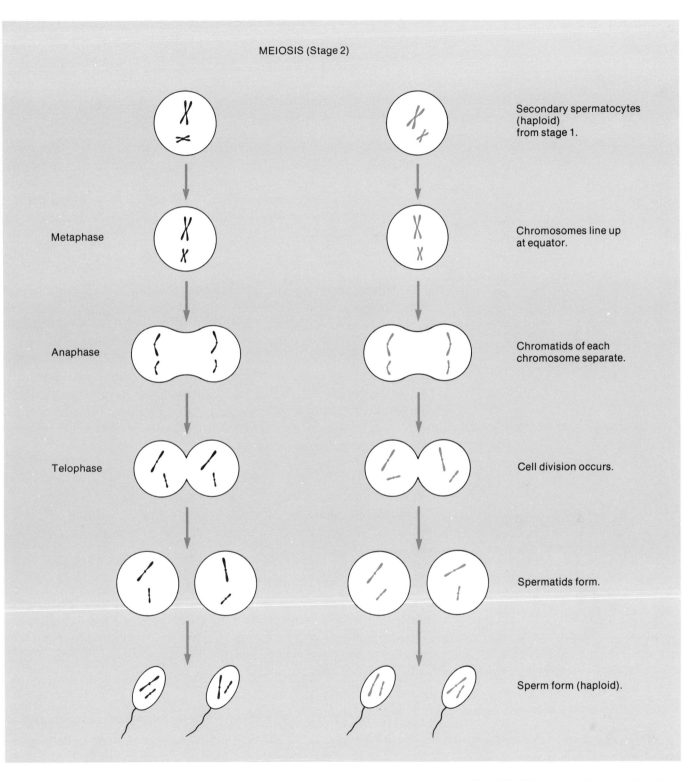

Secondary spermatocytes (haploid) from stage 1.

Metaphase — Chromosomes line up at equator.

Anaphase — Chromatids of each chromosome separate.

Telophase — Cell division occurs.

Spermatids form.

Sperm form (haploid).

Fig. 8-12 The second stage of meiosis for sperm production. Only the chromosomes are shown.

main difference is that the homologues of each homologous pair of chromosomes pair up. By the end of prophase, the tetrads are moving toward the equator of the cell.

In metaphase, the tetrads line up at the equator. Then each tetrad becomes attached to a protein fibre of the spindle. In anaphase, one pair of chromatids from each tetrad moves toward each pole of the cell. In telophase, cell division occurs. Each daughter cell has one homologue of each homologous pair. Each daughter cell now has the haploid number of chromosomes (half of the number that were in the spermatogonium that started the process). However, each of these chromosomes still consists of two chromatids that are joined together. When cell division is complete, the haploid daughter cells are called **secondary spermatocytes**.

Stage Two

The secondary spermatocytes go immediately into the second stage of meiosis. In this stage, normal mitosis occurs except for one thing. The chromosomes do not replicate. (They did so when tetrads were formed.) Thus the nuclei of the secondary spermatocytes go directly into metaphase. The chromatid pairs line up at the equator of the cell. Each chromatid pair then separates into two single chromosomes. The centromere of each chromosome becomes attached to a protein fibre. Remember that each chromosome of a pair is attached to a fibre that goes to a different centriole. Therefore, in anaphase, the two single chromosomes of each homologous pair move to opposite poles of the cell.

As you can see, the daughter cells, or **spermatids**, that form during telophase have the haploid number of chromosomes. These chromosomes are no longer double as they were after the telophase of Stage One.

During interphase, the spermatids mature and grow a "tail". They are now called **sperm**.

Egg formation

Female gametes, or eggs, are also produced by meiosis. The steps are similar to those outlined for sperm production. Of course, the names are different. **Oögonium** replaces spermatogonium. **Primary oöcyte** replaces primary spermatocyte and **secondary oöcyte** replaces secondary spermatocyte. Finally, **oötid** replaces spermatid and **egg** replaces sperm.

It is interesting to note that, when the primary oöcyte divides, it forms two daughter cells that are unequal in size. The larger one is the secondary oöcyte. The smaller one is called the **first polar body**. These two cells enter the second stage of meiosis. But cells produced by the first polar body do not survive. Meiosis stops at this point until a sperm cell triggers the process again. Then the secondary oöcyte divides unequally to produce a large cell called an **oötid** and a small cell called a **second polar body**. The polar body dies, but the oötid develops into an egg.

Summary

Meiosis ensures genetic continuity by reducing the chromosome number of the male and female gametes to the haploid number. Then, when these gametes unite to form a zygote during fertilization, the diploid number is restored. The zygote develops through mitosis and cell division into an organism with body cells that have the diploid number of chromosomes (Fig. 8-13).

Genetic Variation

You should be aware that meiosis does more than just ensure genetic continuity. It mixes up the genetic information in an interesting way. Figure 8-14,A shows the possible chromosome combinations in the sperm if the tetrads are lined up as shown during the metaphase of Stage One. Look back to Figures 8-11 and 8-12 to see this. However, the tetrads can line up another way. It is shown in Figure 8-14,B. Note that the chromosome combinations in the sperm are different.

The chromosomes labelled 1 and 2 are homologues. They carry genetic information for the same traits. However, their specific information is usually different. In a similar manner, chromosomes 3 and 4 are homologues. In the sperm there are four different combinations of chromosomes. Each sperm carries in its chromosomes genetic information for the same traits. However, since the homologues have been shuffled around, the specific information that each carries will likely be different.

What does this mean? It means that genetic continuity will be ensured. The species will be maintained. The same traits are passed on to the next generation through the sperm. Therefore, if these are human cells, it means that humans will be produced. However, it also means that the individuals produced will show variation. They will not be exactly the same.

A species in which the individuals show variation will have a better chance of survival if environmental conditions change. Certain individuals may not be able to live through an adverse change. But others, who are a little different, may live through it. They live to perpetuate the species. Biologists say that the species has undergone adaptation.

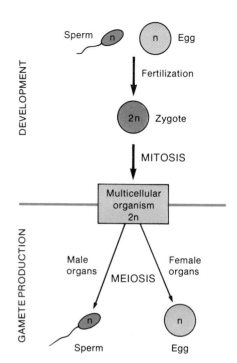

Fig. 8-13 Gamete production occurs by meiosis. Growth and development occurs by mitosis.

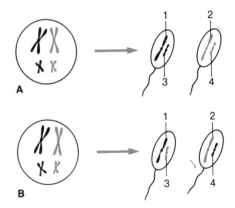

Fig. 8-14 The gametes formed during meiosis have a chromosome combination that was determined by how the tetrads lined up during the metaphase of stage 1.

Review

1. **a)** What is genetic continuity?
   b) How does mitosis contribute to genetic continuity?
2. Distinguish between sexual and asexual reproduction.
3. **a)** What is binary fission?
   b) What is budding?
   c) How is genetic continuity ensured during asexual reproduction?
4. **a)** What is meant by "chromosome number"?
   b) What are homologues?
   c) Distinguish between the diploid and the haploid number of chromosomes.

Fig. 8-15 How many combinations of chromosomes are possible in the zygotes formed from these sperm and eggs? What significance does this result have? (Chromosomes in the gametes are represented by numbers. Chromosomes 1 and 2 are one homologous pair. Chromosomes 3 and 4 are a second homologous pair.)

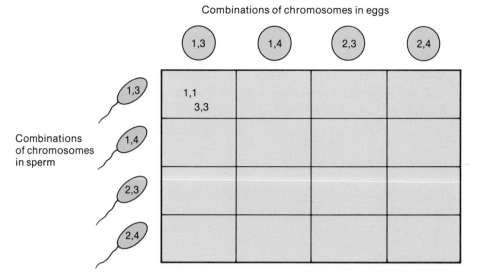

Combinations of chromosomes in eggs

Combinations of chromosomes in sperm

5. **a)** Why must gametes contain only the haploid number of chromosomes?

 b) Why is it important that a gamete contain one homologue from each homologous pair?

 c) What is the origin of each homologue of a homologous pair?

6. Outline, step by step, what occurs during meiosis in which a spermatogonium produces sperm.

7. Female gametes, or eggs, are formed by meiosis. The process begins with an oögonium and ends with an egg. Make a series of diagrams to illustrate this process. Pattern your diagrams on Figures 8-11 and 8-12. Consult the description of egg formation on page 186. Label your diagrams fully. Assume that 2n=4 for the oögonium.

8. **a)** How does the first stage of meiosis differ from mitosis?

 b) Compare the second stage of meiosis with mitosis.

9. **a)** How does meiosis contribute to genetic continuity? Explain your answer by discussing the process of fertilization.

 b) Describe how meiosis produces variations.

 c) Explain how variations produce adaptations that may increase the survival chances of a species.

10. Figure 8-14 shows that four combinations of chromosomes are possible in the sperm from a spermatogonium with 2n=4. Suppose that the same four combinations of chromosomes show up in the eggs from an oögonium of the same species. How many different combinations of chromosomes are possible in the zygotes formed by the union of these sperm and eggs? If you complete a chart such as the one in Figure 8-15, it will help you determine the number.

Highlights

The process of nuclear replication is called mitosis. Mitosis ensures genetic continuity by making sure that the daughter cells are genetically identical to the mother cell.

Four fairly distinct phases can be recognized in mitosis. They are prophase, metaphase, anaphase, and telophase. As a nucleus passes through these phases, its chromosomes, which had previously replicated, separate into two identical sets. Each set forms a new nucleus identical to the mother nucleus.

Cell development can involve both cell growth and cell differentiation. Growth is an increase in size and mass. Cell growth can be accomplished two ways. A cell can simply increase in size and mass. Or it can increase its numbers by cell division.

Cell differentiation is the gaining of specific structural features and functions by a cell. During differentiation, cells become specialized to carry out certain activities in an organism. Differentiated cells of one type may group together to form a tissue.

A

Fig. 8-16 All of these organisms began their existence as a result of meiosis. They arrived at their present size and shape through mitosis, differentiation, and development.

B

C

Aging is a biological puzzle that has not yet been solved. It may be due to environmental factors or changes that occur during cell differentiation. To stay alive, a cell must divide. If it differentiates, it will eventually die.

Asexual reproduction may be of several types. Two of these are binary fission and budding. In all these, genetic continuity is ensured since mitosis is all that is involved.

Sexual reproduction involves the union of gametes during fertilization. The chromosome number of these cells must be half that of the body cells before they unite. This reduction occurs during meiosis.

During meiosis, the homologues of each homologous pair of chromosomes pair up, undergo tetrad formation, and separate. Eventually the diploid number of chromosomes is reduced to the haploid number in the sperm and eggs.

Meiosis ensures genetic continuity by reducing the diploid number to the haploid number before fertilization. In addition, it promotes variation in a species. This, in turn, means that the species is more likely to be able to adapt to environmental changes.

UNIT THREE

Monerans, Protists, And Fungi

In Chapter 3 you learned that all living things can be classified into five kingdoms. In this unit, you study the form and function of some organisms from each of the three simplest kingdoms: kingdom Monera, kingdom Protista, and kingdom Fungi.

Monerans And Viruses

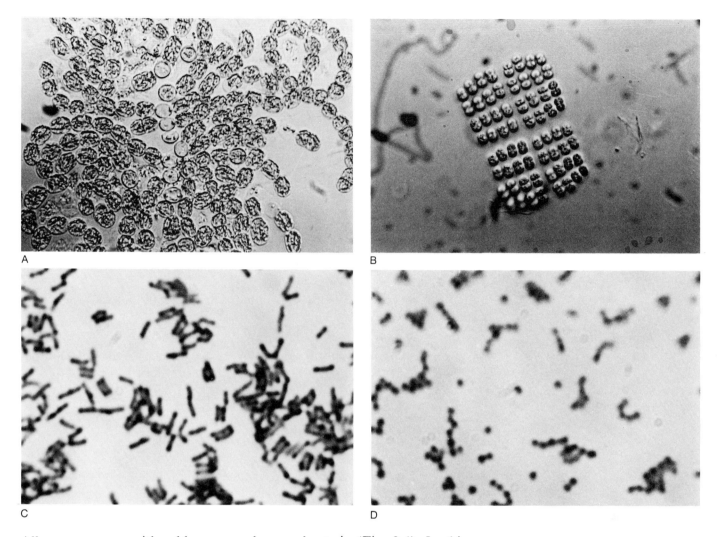

A

B

C

D

All monerans are either blue-green algae or bacteria (Fig. 9-1). In this chapter you will study the form and function of representative organisms from each of these groups.

In the 5-kingdom system of classification, about 3 100 species of organisms are placed in the kingdom Monera. This is only 0.2% of the 1 500 000 or so species of organisms currently known. In other words, two species out of every 1 000 are monerans.

The number of species is small. The organisms themselves are certainly small. However, you will see as you study this chapter that their importance is very great. Also, the number of individuals existing today is enormous. Monerans are the most numerous living things on earth.

You will also study viruses in this chapter. You may have noticed in Chapter 3 that viruses were not placed in the 5-kingdom classification system. That is because debate is still going on as to whether they are living or non-living. However, it is known that they are very simple in structure. Therefore, they are discussed in this chapter with the simplest living things, the monerans.

Fig. 9-1 Some representative monerans: *Anabaena* **(blue-green alga) (A);** *Merismopedia* **(blue-green alga) (B); a bacillus bacterium (C); a coccus bacterium (D). Some species are single cells. Others are colonial. Still others are filamentous.**

9.1 Characteristics of Monerans

The kingdom Monera consists of two phyla:

a) Phylum Cyanophyta—the blue-green algae (about 1 500 species);
b) Phylum Schizophyta—the bacteria (about 1 600 species).

No one can describe with absolute certainty what primitive life on earth was like. However, fossils resembling modern monerans have been found that are over three billion years old. This discovery is an indication that monerans may be the earliest organisms to have lived on earth. Their simple structure suggests that they can survive in very primitive conditions. In fact, many monerans live in such conditions today. They are found almost everywhere—from the highest mountain to the deep sea sediment. Some species thrive at extremely high temperatures. Others live in sub-zero conditions. They occur in salt water and fresh water. They live in and on the soil. Some species even grow on rocks.

Monerans have many features in common. Let us look at them before we begin a detailed study of blue-green algae, and bacteria.

1. *All monerans are unicellular (one-celled).* The cells may occur singly or in groups. When cells group together, they may form a **colony**. Such forms are said to be **colonial**. They may also form a chain of cells called a **filament**. Such forms are called **filamentous**. Figure 9-1 shows all three types.
2. *The cells of monerans lack a true nucleus.* They do not have a nuclear membrane. Therefore nuclear material such as DNA is spread throughout the cell.
3. *The cells of monerans lack internal membranes.* In addition to lacking a nuclear membrane, monerans also lack other internal membranes. They have no membranes to enclose cell organelles. As a result, they lack membrane-bound organelles such as mitochondria, plastids, Golgi apparatus, and endoplasmic reticulum.
4. *The cells of monerans have rigid cell walls with a unique composition.* The cell walls of both bacteria and blue-green algae contain substances that are not found in the cell walls of any other organisms.
5. *Moneran cells often secrete a slime coating.* This jelly-like coating protects the cells. It also helps to hold colonies and filaments together.
6. *Most monerans reproduce only by asexual means.* Sexual reproduction occurs in only a few species. Most species reproduce by binary fission.
7. *Many monerans can form spores.* Under adverse conditions such as extreme temperature or dryness, many monerans form spores. These have very thick walls to protect the contents from the adverse conditions.

Review

1. Summarize the characteristics shared by most monerans.
2. How would a simple structure permit an organism to live under primitive conditions on earth?

3. What disadvantage could a species experience if it reproduced only by asexual means? Does this appear to have affected monerans?

9.2 Blue-Green Algae (Phylum Cyanophyta)

What Are Algae?

Fig. 9-2 This "pond scum" is only one of thousands of species of algae.

Most people know that oceans, lakes, ponds, and streams contain **algae** (singular: **alga**). However, few people know what algae are or how important they are. To the ordinary person, they are just "pond scum", "seaweed", or "water moss" (Fig. 9-2). But, to the biologist, they are much more. They are chlorophyll-bearing organisms in which sexual reproduction, when it occurs, is quite different from that in higher plants. The gametes unite to form a zygote. However, the zygote does not develop into an embryo as it does in higher plants.

Algae are often inconspicuous. Yet they are the most common chlorophyll-bearing organisms on earth. They are the dominant vegetation of the oceans and most freshwater habitats. Most species are aquatic. However, several are terrestrial (land dwellers). These algae live in and on the soil, on moist rocks, and on wood.

Algae vary in size from microscopic forms to giant "seaweeds" as long as 100 m. They may be free-floating, or they may be attached to something. They may occur as single cells, in colonies, and in long filaments. Some look very much like "ordinary" plants.

Classification of algae

As you can see, algae are very diverse in form and function. Because of the differences among them, algae are classified into three kingdoms which are divided into seven different phyla and divisions.

1. **Kingdom Monera**

   Phylum Cyanophyta (blue-green algae)

2. **Kingdom Protista**

   Phylum Euglenophyta (euglenoid flagellates)
   Phylum Chrysophyta (golden algae, including diatoms)
   Phylum Pyrrhophyta (includes dinoflagellates)

3. **Kingdom Plantae**

   Division Chlorophyta (green algae)
   Division Rhodophyta (red algae)
   Division Phaeophyta (brown algae)

Look back to Section 3.4, page 50, to see what some organisms in these phyla and divisions look like. This chapter deals with the algae in the kingdom Monera. Chapter 10 deals with those in the kingdom Protista.

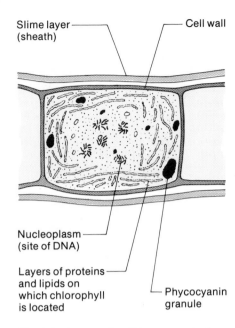

Slime layer (sheath)

Cell wall

Nucleoplasm (site of DNA)

Layers of proteins and lipids on which chlorophyll is located

Phycocyanin granule

Fig. 9-3   A generalized blue-green algal cell. This cell is part of a filament (chain of cells).

Chapter 13 deals with those in the kingdom Plantae.

Many factors are used to classify algae. Cell structure is one factor. Methods of reproduction is another. The nature of pigments is a third factor. Finally, the nature of products used to store food is also used. For example, both green and golden algae make simple sugars during photosynthesis. However, green algae store surplus sugar as starch and golden algae store it as oils.

Characteristics of Blue-green Algae

The most primitive algae are in the phylum Cyanophyta of the kingdom Monera. These are the blue-green algae. About 1 500 species of algae are placed in this phylum. These algae are closely related to bacteria. In fact, some botanists claim that they are bacteria. However, they do differ from bacteria in several ways. For example, they contain some different pigments. Also, they have several different structural features.

Pigments

Blue-green algae contain **chlorophyll** *a* as do all green plants and algae. They also contain **carotenes** and **xanthophylls**, which are yellow pigments. In addition, they contain the blue pigment **phycocyanin** and the red pigment **phycoerythrin**. The combination of these pigments generally produces the blue-green colour that gives these algae their name. However, blue-green algae can be almost any colour. Some contain large amounts of the blue and red pigments. Therefore they appear purple or even black.

In typical plant cells, the chlorophyll is located on layers of proteins and lipids which are in bodies called grana within the chloroplasts. Blue-green algae lack chloroplasts. Thus the layers of proteins and lipids lie free in the cytoplasm. Often they are near the edge of the cell (Fig. 9-3).

Reproduction

Blue-green algae are among the most primitive living things. They have no nucleus. However, as Figure 9-3 shows, they often have a granular region near the centre of the cell. This region is called the **nucleoplasm**, even though it lacks a definite boundary. The genetic material, DNA, is often in this region.

Chromosomes are either lacking or organized differently from plant cells. Therefore true mitosis does not occur in blue-green algae. No stages of mitosis have ever been observed. Yet cell division still occurs by **binary fission**. How the genetic information is divided between the daughter cells is not yet known.

Most blue-green algae reproduce only asexually. In addition to binary fission, some species ensure reproduction by forming **spores**. These spores are thick-walled. As a result, they can remain dormant during unfavourable conditions. When favourable conditions return, the spores produce active cells again. Many colonial and filamentous species reproduce by **fragmentation**. The colony or filament breaks into pieces. Each piece starts a new colony or filament.

Recent studies have shown that there may be a type of sexual reproduction in some species. However, it is uncommon and the process is not well understood.

Occurrence of Blue-green Algae

Blue-green algae are among the most widely distributed living things. Whether the environment is hot or cold, wet or dry, some species of blue-green alga will likely live there.

Some species grow in hot, dry deserts. They live beneath stones where there is some moisture. Of course, they can only live beneath translucent stones because they must have light for photosynthesis. They also live in the sand and soil of the desert. They are particularly common around any water holes that occur in the desert.

Most species live in or near water. A few species are marine (live in salt water). But the majority live in fresh water. Blue-green algae are also common on damp surfaces. These include moist rocks, moist soil, flower pots in a greenhouse, and tree trunks in a damp forest.

Some of the aquatic species are **planktonic** (free floating). They are called **plankton** or, more accurately, **phytoplankton**. (*Phyto* means plant. Small free floating animals are called **zooplankton**.) Other aquatic species are attached to rocks, logs, plants, and organic debris in the water.

Some species of blue-green algae have been discovered in hot springs and streams. They thrive at temperatures over 75°C. Much of the attractive colour of hot springs such as those in Yellowstone National Park is due to blue-green algae. Other species live in sub-zero conditions. They inhabit permanent ice-fields, where they often colour the ice and snow green or blue-green.

Importance of Blue-green Algae

Blue-green algae are important for several reasons. Four of the main reasons are discussed here.

Producers

Blue-green algae are **producers** This means they make simple sugars by photosynthesis. All other algae and green plants are also producers. These organisms are also called **autotrophs**, or "self-feeders". However, they also feed other organisms. Producers may be eaten by **first-order consumers** (**herbivores**). These organisms use the food stored in the producers. Then the herbivores may be eaten by **second-order consumers first-order carnivores**). The second-order consumers may be eaten by **third-order consumers** (**second-order carnivores**). In this way a **food chain** is set up. Here is an example of a food chain in which blue-green algae are involved:

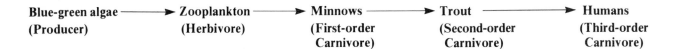

Blue-green algae ⟶ Zooplankton ⟶ Minnows ⟶ Trout ⟶ Humans
(Producer)   (Herbivore)   (First-order   (Second-order   (Third-order
                                             Carnivore)   Carnivore)   Carnivore)

As you can see, blue-green algae are at the base of a food chain in which humans are involved.

Almost all food chains begin with green plants or algae. Humans are at the end of many food chains that begin with blue-green algae.

Oxygen production

A by-product of photosynthesis is oxygen. The oxygen that blue-green algae produce helps to replace atmospheric oxygen used by most living things for respiration.

Blue-green algae that live in water play an important role in putting oxygen into the water. Oxygen that they produce dissolves in the water. It can then be used by fish and other aquatic animals. The roots of aquatic plants also require oxygen. Some scientists believe that blue-green algae supply a large portion of the oxygen required by the roots of rice plants in rice paddies.

Nitrogen fixation

All organisms require nitrogen for growth and development. Yet very few are able to use the abundant nitrogen in the atmosphere. Producers (algae and green plants) must get it from the soil or water in which they live. Consumers obtain it from the producers. However, the nitrogen must normally be in the form of a nitrate before most producers can absorb it. The molecules of nitrogen in the atmosphere, therefore, must first be **fixed**. This means they must be converted to nitrogen compounds. These compounds, in turn, must be converted into nitrates before the producers can use the nitrogen.

Several species of blue-green algae can fix nitrogen. They can absorb molecules of atmospheric nitrogen and use them in their metabolic processes. When these algae die and decay, this fixed nitrogen converts to nitrates. Plants absorb the nitrates to use in their metabolic processes.

Nitrogen fixation by blue-green algae is very important in deserts, since little other nitrogen is available there. Also, nitrogen fixation is responsible for the main supply of nitrogen for rice crops. Many countries depend on rice as their main food. Yet some countries cannot afford to buy fertilizer to make the rice grow. Fortunately, some nitrogen fixing blue-green algae live in the water and soil of rice paddies. The algae fix atmospheric nitrogen. Then, when they die and decay, they provide free nitrogen fertilizer to nourish the rice plants.

Water pollution

Blue-green algae are normal and important organisms in most bodies of water. However, under certain conditions, they may multiply rapidly. The result is called an **algal bloom**. Algal blooms often ruin swimming beaches. They also clog the filters in water purification plants. Some species of blue-green algae give water an unpleasant taste and odour. Thus towns and cities must be careful not to let blue-green algal blooms occur in their water reservoirs.

Some species of blue-green algae secrete **toxins** (poisonous substances) that can harm aquatic life. In fact, massive fish kills have been traced to such toxins. Some toxins will make cattle and other terrestrial animals sick if they drink the water. In some cases, the animals die.

As algae from a bloom die, bacteria feed on them. The bacteria increase in numbers. They use oxygen from the water for cellular respiration. During the day, the living blue-green algae can replace the oxygen by photosynthesis. However, at night, when photosynthesis stops, the bacteria often rob the water of much of its oxygen. The death of fish and other aquatic organisms usually follows.

You will study the causes and controls of algal blooms in Chapter 24.

Review

1. **a)** What does the word "algae" mean?
   b) Why are algae classified in three different kingdoms?
   c) List the seven phyla and divisions of algae. Indicate the kingdom in which each phylum occurs.
2. **a)** Describe the pigments found in blue-green algae.
   b) Describe the reproduction of blue-green algae.
3. **a)** Distinguish between phytoplankton and zooplankton.
   b) Explain why blue-green algae are considered to be among the most widely distributed living things.
4. **a)** What is an autotroph? Are blue-green algae autotrophic? Explain.
   b) Describe how blue-green algae form the base of some food chains in which humans participate.
   c) Describe the role of blue-green algae as oxygen suppliers.
5. Explain the role of blue-green algae in making nitrogen available to other organisms.
6. **a)** What is an algal bloom?
   b) Summarize the harmful effects of algal blooms.

9.3 Some Representative Blue-Green Algae

Blue-green algae may be unicellular, colonial, or filamentous. They occur in a wide variety of shapes and arrangements. A few of the important genera are discussed in this section. As you read, keep in mind the common features of all blue-green algae.

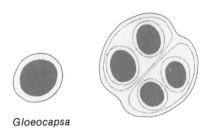

Gloeocapsa

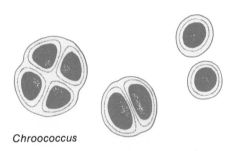

Chroococcus

Fig. 9-4   Some representative unicellular or small colonial blue-green algae. What do they have in common? How do they differ?

Unicellular and Small Colonial Forms (Fig. 9-4)

The unicellular blue-green algae are probably the most primitive forms. At some time in the past, they likely gave rise to colonial and filamentous forms.

Gloeocapsa

Gloeocapsa is an example of a blue-green alga that can occur as a single cell. It can also occur as small groups of cells, or colonies. Each cell secretes a **slime layer**, or **gelatinous sheath**, around itself. When fission occurs, each new cell secretes a sheath within the old one. Further fission produces still more sheaths. Therefore, if the cells do not separate, a colony surrounded by a common sheath results.

Some species of *Gloeocapsa* live in water in a planktonic form. Others live attached to logs and rocks in the water. However, most species live in moist terrestrial habitats. You may have seen this alga as a thin olive-green film on moist rocks and soil. It is also commonly found on clay flowerpots in moist habitats such as greenhouses.

Chroococcus

This blue-green alga has spherical cells much like *Gloeocapsa*. It is unicellular. However, it often occurs in groups of two and four. This is because the cells fail to separate immediately after fission.

Most species of *Chroococcus* are planktonic. However, some species form films on moist rocks and soil, and on aquatic plants.

Colonial Forms (Fig. 9-5)

Anacystis

Most species of *Anacystis* consist of tiny oval cells. Each cell is covered with a thin gelatinous sheath. The cells are usually loosely packed but evenly spaced, covered with a common gelatinous sheath.

Anacystis is one of the blue-green algae that secretes a toxin. It is dangerous to humans and other animals that drink or live in the water.

Microcystis

This alga is planktonic. It often forms large, irregularly shaped colonies. The individual cells are spherical. They are held together by a mucilage-like substance. Some species contain air pockets. These pockets cause the alga to float high in the water. There the alga appears as surface scum. This scum causes many problems in lakes and reservoirs.

Microcystis blooms easily. Also, like *Anacystis,* it secretes a toxin. Thus a bloom may poison fish with the toxin. It may also suffocate them by robbing the water of its oxygen at night. Cattle and aquatic birds are often killed by the *Microcystis* toxin.

Merismopedia

Merismopedia forms small rectangular colonies only one cell thick. The cells may occur singly or in pairs. Usually the cells are in regular rows. A colony may contain from four to hundreds of cells. *Merismopedia* is a planktonic alga of lakes and ponds. It reproduces by fragmentation. The colony breaks into pieces and each piece starts a new colony.

Eucapsis

This genus of blue-green alga forms colonies with the cells arranged in cubes. Each colony may have many cells. A separate gelatinous sheath binds together each group of eight to sixteen cells.

Note that *Eucapsis* looks much like *Merismopedia*. Its cells are in regular rows. However, cell division in *Merismopedia* occurs in just two directions. Thus *Merismopedia* forms flat colonies one cell thick. Cell division in *Eucapsis* occurs in three directions. Thus *Eucapsis* forms cubic colonies.

Only one species of *Eucapsis* occurs in North America. However, it is widely spread throughout the lakes of this continent.

Filamentous Blue-green Algae (Fig. 9-6)

Oscillatoria

This genus is one of the most common and widespread blue-green algae. Most marine and freshwater habitats contain *Oscillatoria*. It may be planktonic or it may grow attached to rocks and plants. Some species grow on and in the soil. Others inhabit ditches and ponds. Almost any sample of water collected from a ditch or pond will contain *Oscillatoria*.

Oscillatoria grows in hair-like strands or filaments. These generally form dense mats. Each filament may be covered with a thin gelatinous sheath. Sometimes this sheath is absent. Reproduction is by fragmentation. If conditions are right, *Oscillatoria* grows and fragments very rapidly in the summer. Only a week or so is required for this blue-green alga to turn a lake or pond olive green.

If you watch *Oscillatoria* under a microscope, you will see how it got its name. Most species oscillate, or move from side to side. Why and how this motion occurs is not clearly understood at this time.

Nostoc

Nostoc is a filamentous blue-green alga. However, each filament forms a large gelatinous sheath. The sheaths of several filaments join together to

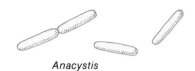
Anacystis

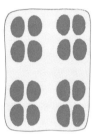
Microcystis

Merismopedia

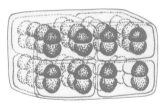
Eucapsis

Fig. 9-5 Some representative colonial blue-green algae. What do they have in common? How do they differ?

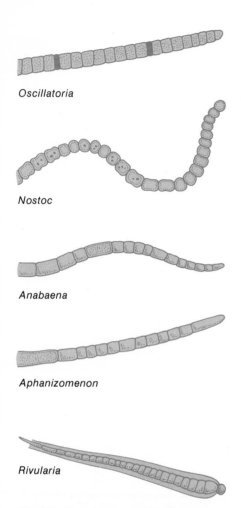

Oscillatoria

Nostoc

Anabaena

Aphanizomenon

Rivularia

Fig. 9-6 Some representative filamentous blue-green algae. What do they have in common? How do they differ?

form a large spherical "blob". This gelatinous mass can often be seen floating on the water of ponds and lakes. In fact, the mass is sometimes so large that it is called a *Nostoc* apple or *Nostoc* ball. Some people eat these as a delicacy.

Several species of *Nostoc* grow on moist soil and rocks. A shady environment is preferred. Here, too, gelatinous masses are formed. They are commonly called mares' eggs, witches' butter, and star jelly. The masses often reach the size of a baseball when they grow in wet meadows.

The individual cells are usually spherical. In fact, a single filament resembles a string of beads. Certain cells in a filament are sometimes larger than others. These cells serve two functions. They perform nitrogen fixation and they cause fragmentation.

Nostoc has been collected at depths up to 1 m (metre) in the soil.

Anabaena

Most species of this genus are aquatic. Several are planktonic. *Anabaena* is a frequent companion of *Microcystis* in algal blooms. Like *Microcystis,* it secretes a toxin that is harmful to humans and other animals. Cattle frequently die after drinking water from ponds containing an *Anabaena* bloom. *Anabaena* even blooms in relatively unpolluted water. This can occur since *Anabaena* produces its own fertilizer through nitrogen fixation.

The filaments of *Anabaena* resemble those of *Nostoc*. However, *Anabaena* filaments often occur singly. When they do form colonies, the colonies are not regular in shape like those of *Nostoc*.

Aphanizomenon

This genus, along with *Microcystis* and *Anabaena,* is among the dominant phytoplankton of Lake Erie. Like its companions, *Aphanizomenon* can release enough toxin during blooms to kill fish, cattle, and other animals.

The filaments of this genus lie parallel to one another in bundles. The bundles are large enough to see with the unaided eye. In fact, during a bloom, the water appears to contain finely chopped grass.

Rivularia

All species of this genus are attached to submerged objects such as rocks and logs. The filaments are usually parallel to one another. They are held in that position by a very firm gelatinous sheath. The sheath surrounds large clusters of the filaments.

A thick mat of *Rivularia* often feels hard. This is because a crust of limestone is often deposited on the mats. When *Rivularia* is abundant, the water takes on a musty smell. This alga often clogs the filters in water purification plants.

Review

1. How do unicellular, colonial, and filamentous blue-green algae differ?
2. **a)** What is the main method of reproduction of unicellular species?
   b) Explain how fission can produce a rectangular colony of *Merismopedia* that is only one cell thick.
   c) Explain how fission can produce a colony of *Eucapsis* that is cube-shaped.
   d) What is fragmentation? How does it benefit a blue-green alga?
3. **a)** Name four genera of blue-green algae in which there are species that secrete harmful toxins.
   b) Name two genera of blue-green algae that can perform nitrogen fixation.
4. Some species of blue-green algae are said to be indicators of water pollution. How might they indicate that water is polluted?

9.4 Investigation

Examination of Some Blue-Green Algae

You should have sufficient knowledge from Sections 9.2 and 9.3 to recognize blue-green algae when you see them. In this investigation you will first use prepared slides to practice identification. You can also study many characteristics of blue-green algae using those slides. Then you will prepare wet mounts of living algae. You should try to identify these algae, as well. Be sure to note any characteristics of blue-green algae that were not noted with the prepared slides.

Materials

compound microscope
prepared stained slides of several genera of blue-green algae
microscope slides (2)
cover slips (2)
paper towel
lens paper
medicine dropper
forceps
samples of living blue-green algae
immersion oil
xylol
identification guides for blue-green algae

Procedure

a. Your teacher will provide prepared stained slides of several types of blue-green algae. Some of those that you studied in Section 9.3 are included.

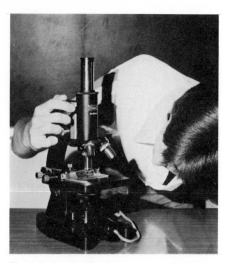

Fig. 9-7 Lower the oil immersion objective, or any objective, slowly and carefully while watching from the side.

Examine as many different types as time permits. Make a sketch of each type. Note features that all the specimens have in common. Note whether a specimen is unicellular, colonial, or filamentous. Look for evidence that reproduction was occurring when the cells were killed.

Note: Blue-green algae are very small. You will be able to see their features best if you use an **oil immersion objective** of 100 X. If one is available, proceed as follows. Place a small drop of **immersion oil** on the cover slip, over the place where the specimen is located. Lower the lens into the drop of oil (Fig. 9-7). As usual, watch carefully from the side to prevent grinding the objective into the cover slip. Now focus on the specimen using the fine focus adjustment.

b. Prepare a wet mount of a living blue-green alga. The specimen can be obtained from: the scum (brown or blue-green) on the inner walls of an aquarium; the green slime on rocks, tree trunks, and soil; the green film on clay flowerpots; blue-green or brown "blobs" floating in ponds.

c. Examine your wet mount using an oil immersion lens, if possible. Sketch the specimen. Note whether it is unicellular, colonial, or filamentous. Determine the genus of the specimen. (Your teacher has identification manuals if your specimen is not one you studied in Section 9.3 or in step a of this procedure.) Note any special features of your specimen.

d. If time permits, prepare and study wet mounts of specimens from some of the other habitats listed in step b.

e. Moisten a piece of lens paper with xylol. Use this to wipe the oil from the oil immersion objective.

CAUTION: Do not inhale the xylol fumes.

Discussion

Your laboratory notes and sketches make up most of the writeup for this investigation. Be sure that they are complete. Then answer these questions:

1. What evidence do you have that blue-green algae may be very primitive living things?
2. Most blue-green algae grow best in habitats that have a rich supply of organic matter. What evidence do you have to support this statement?
3. What were the advantages of prepared stained slides over your wet mount? What were the disadvantages?

9.5 Introduction to Bacteria (Phylum Schizophyta)

About 1 600 species of bacteria (phylum Schizophyta) share the kingdom Monera with the blue-green algae (phylum Cyanophyta). These organisms have so much in common that some biologists say that the two phyla should be one. In fact, blue-green algae are called blue-green bacteria by these biologists.

Do you remember what bacteria and blue-green algae have in common? If you do not, read Section 9.1 again before you begin this study of bacteria.

Discovery of Bacteria

You may recall Anton van Leeuwenhoek's discovery of "little animals". Van Leeuwenhoek sent several reports on his work to the Royal Society in London. These reports included drawings and descriptions of small animals and many protists. A report that he submitted in 1864 included drawings and descriptions of what we know today as bacteria. Van Leeuwenhoek seems to have been the first person to observe bacteria (Fig. 9-8).

The science of bacteriology remained undeveloped until the middle of the nineteenth century. Then, in 1854, Louis Pasteur was appointed professor of chemistry at the University of Lille, in France. This university was located in the heart of an area that produced alcoholic beverages by fermentation. Therefore it is not surprising that Pasteur got involved in studies of fermentation. You have already seen how these studies disproved the theory of spontaneous generation. The same studies began the science of bacteriology.

Before Pasteur's time, yeast cells had been observed in fermenting mixtures. However, no one realized that they were the cause of fermentation. Then, one day, several vats of fermenting mixture were found to be sour. Pasteur was asked to find out what caused this souring. He studied the fermenting mixture using his microscope. He noticed that as the mixture aged, the number of yeast cells increased. At the same time, the alcohol content increased. Therefore Pasteur reasoned that the yeast played a role in the production of alcohol. (You know from Section 7.5, page 161, that this is so.)

While trying to prove his hypothesis, Pasteur examined some of the sour mixture. He found that it did not contain alcohol. Instead, it contained lactic acid. Also, instead of yeast cells, it now contained much smaller cells. These cells were alive. They moved and they increased in numbers as the amount of lactic acid increased.

After many years of study, Pasteur proved that microorganisms were responsible for both fermentation and souring. Yeast produced the alcohol. What we today call bacteria produced the lactic acid.

Pasteur continued his studies of bacteria. From these studies came the conclusion that bacteria cause diseases. Because of this important work, Pasteur is known as the father of bacteriology. Modern bacteriology and microbiology are based on his work. His bacterial theory of disease is considered to be a major turning point in the field of medicine. However, the theory was not firmly established until the work of Robert Koch, a German physician, in 1877.

In addition to these achievements, Pasteur developed the process of pasteurization and prepared the first vaccine for rabies. He also developed

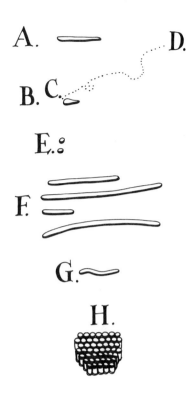

Fig. 9-8 These drawings of bacteria by van Leeuwenhoek were the first ever published.

an immunization process for the deadly anthrax disease that was killing whole flocks of sheep. He is truly one of the greatest scientists of all time.

What Are Bacteria?

Compared to most other cells, bacteria are extremely small. You have seen that blue-green algae are small. The largest bacteria are about the size of the smallest blue-green algae. They are so small that they are measured in **nanometres**. (One nanometre = one billionth of a metre, or 1 nm = 0.000 000 001 m.) Some bacterial cells are spherical in shape. Most have diameters between 500 nm and 1 000 nm. Other bacterial cells are rod-shaped. These range in length from 1 000 nm to 10 000 nm. These numbers are probably hard for you to visualize. However, many thousands of average-sized bacteria could fit on the smallest dot you can make with your pencil.

You can see many species of bacteria with your microscope. Of course, you must use high power (400 X to 600 X). However, you will not see much more than their general shape. To see most bacteria well you must use an oil immersion objective at 1 000 X to 1 500 X. No compound microscope has sufficient magnification to let you see cellular details in bacteria. For this purpose, biologists use electron microscopes with magnifications of 100 000 X or more (see Fig. 5-12, page 100).

You should not call bacteria "germs". To most people, "germs" are microorganisms that cause diseases. Yet, the majority of bacterial species do not cause diseases. Further, many species of organisms that are not bacteria do cause diseases. Among these are some protists, fungi, and viruses.

Occurrence of Bacteria

Most biologists believe that bacteria were the first form of life on earth. These early bacteria probably obtained the energy needed for life processes from inorganic compounds. They probably used the chemical potential energy in compounds of nitrogen, iron, and sulfur. Most present-day organisms use the chemical potential energy in glucose and other organic compounds.

As green plants developed, new species of bacteria also developed that could use organic compounds for energy. Then, as animals developed, species of bacteria developed that could live in and on the animals.

No one knows for sure that bacteria evolved in this manner. But we do know that a wide variety of forms exist today. We know, too, that bacteria live in almost any environment. They are present in tremendous numbers on and in soils, plants, animals, and rotting organic matter such as wood and animal feces. Some species live in salt water; others live in fresh water. Some require oxygen; others die if oxygen is present. Some need organic matter; others need only inorganic matter. Some can live for many years at sub-zero temperatures; others have been found living at temperatures as high as

90° C. Bacteria have been found on the highest mountain icefields and in the hottest deserts. Almost every environment, natural and artificial, is host to some species of bacteria.

Review

1. **a)** Describe van Leeuwenhoek's contribution to the development of bacteriology.
   b) Describe Pasteur's contribution to the development of bacteriology.
   c) What is the difference between bacteriology and microbiology?
2. **a)** Why are blue-green algae sometimes called blue-green bacteria?
   b) How do blue-green algae and bacteria compare in size?
   c) Why are blue-green algae and bacteria placed in the same kingdom?
3. **a)** What method might bacteria have used to obtain energy for life processes before the earth had any organic compounds?
   b) Why do most species of bacteria use different sources of energy today?

9.6 Structure of Bacteria

Classification of Bacteria

The classification of bacteria has traditionally been based on the shapes of individual cells. Three basic shapes exist—spherical, rod-shaped, and spirally twisted. The spherical bacteria are called **cocci** (singular: **coccus**), the rod-shaped ones **bacilli** (singular: **bacillus**), and the spirally twisted ones **spirilla** (singular: **spirillum**). Almost all unicellular bacteria and bacterial colonies are based on these three shapes. Table 9-1 shows how colonies are named. Examples of these forms are shown in Figure 9-9.

COCCI

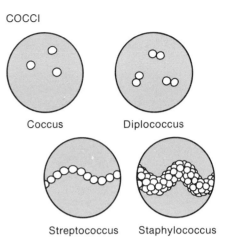

BACILLI

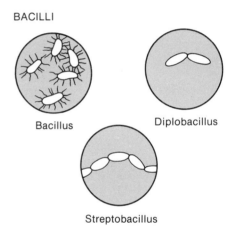

SPIRILLA

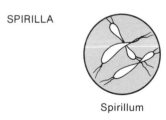

Spirillum

Fig. 9-9 Some basic shapes of bacterial cells and colonies.

NAME	DESCRIPTION
Coccus (plural: **Cocci**)	Spherical cells occurring singly
Diplococcus	Spherical cells in pairs or short filaments
Streptococcus	Spherical cells in long filaments
Staphylococcus	Spherical cells in clusters
Bacillus (plural: **Bacilli**)	Rod-shaped cells occurring singly
Diplobacillus	Rod-shaped cells in pairs
Streptobacillus	Many rod-shaped cells forming a filament
Spirillum (plural: **Spirilla**)	Spirally twisted cells occurring singly

TABLE 9-1 Classification of Bacteria Using Shape

Table 9-2 lists some bacterial diseases and the shapes of their cells.

DISEASE	BACTERIUM	DISEASE	BACTERIUM
Boils	*Staphylococcus*	Whooping cough	Bacillus
Strept throat	*Streptococcus*	Typhoid fever	*Diplobacillus*
Pneumonia		Tetanus	*Diplobacillus*
(bacterial)	*Diplococcus*	Diphtheria	Bacillus
Scarlet fever	*Streptococcus*	Tuberculosis	Bacillus
Gonorrhoea	*Diplococcus*	Leprosy	Bacillus
Syphilis	Spirillum	Bubonic plague	Bacillus
Cholera (Asiatic)	Spirillum	Botulism	Bacillus

TABLE 9-2 **Bacterial Diseases and Cell Shape**

Present day bacteriology requires a better basis for classification than just shape. Bacteria are now identified and classified using many other factors. These include colour, manner of growth, and the effect of temperature and food on their rate of growth. The nature of the metabolic products formed and the effects of the bacteria on various stains, dyes, and culture media are also used in classification.

Structure of a Bacterial Cell

The **cell wall** gives a bacterial cell its characteristic shape (Fig. 9-10). There are two patterns of cell walls. One type is made of tightly connected molecules that make the wall rigid and strong. This particular type of wall allows a special stain to react with the cell. The stain was developed in 1884 by a Danish physician, Hans Christian Gram. Thus such bacteria are called **gram-positive bacteria**. The other type of cell wall is more fragile. It contains an extra protein layer that is not present in the first type. These walls do not allow the special stain to react with the cell. Thus such bacteria are called **gram-negative bacteria**.

Scientists are spending a great deal of time studying the structure of the cell walls of bacteria. They have discovered that many antibiotics fight infections by preventing cell wall formation. For example, penicillin prevents bacterial cells from making a certain substance needed to build cell walls. That substance is much more common in gram-positive than in gram-negative bacteria. Thus scientists know that penicillin works best against gram-positive bacteria. Another antibiotic may work best against gram-negative bacteria.

The cell wall is commonly surrounded by a **slime layer**, or **gelatinous sheath**. This sheath may protect the cell against sudden environmental changes and infection by viruses.

The gelatinous sheath varies in thickness from species to species. The thickness of the sheath is thought to be an indication of the virulence (strength) of the species. The thicker the sheath, the more virulent the species will be. Serious infections are normally produced by species with very thick

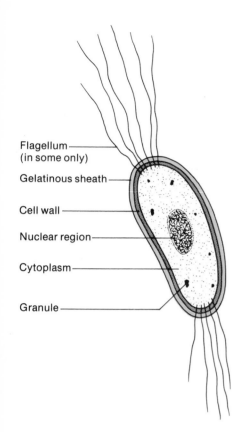

Flagellum
(in some only)

Gelatinous sheath

Cell wall

Nuclear region

Cytoplasm

Granule

Fig. 9-10 Structure of a generalized bacterial cell. In what ways is it similar to the blue-green algal cell in Fig. 9-3? How is it different from that cell?

sheaths, or **capsules**. The capsule may protect the bacterium from the host's natural defense mechanisms. For example, *Diplococcus pneumoniae,* the pneumonia bacterium, has a capsule. This bacterium produces a serious infection. However, a strain of this bacterium exists that has no capsule. Otherwise it is identical to the one with the capsule. The strain without a capsule is almost non-infectious.

Like all cells, the cytoplasm of a bacterial cell is surrounded by the **plasma membrane**. It performs the same role in bacterial cells that it does in all other cells.

The **cytoplasm** appears granular under the electron microscope (Fig. 9-11). This granular nature is due to the presence of numerous **ribosomes**. In most cells, these are found on the endoplasmic reticulum. However, bacterial cells lack membrane-bound organelles such as the endoplasmic reticulum. Therefore the ribosomes are suspended in the cytoplasm. Ribosomes are responsible for protein synthesis. Their large numbers allow rapid protein production. This, in turn, allows rapid cell division in bacteria.

Bacteria lack a nuclear membrane. However, a fibrous **nuclear material** is present in the cells. It often occurs near the centre of the cell. It is made of DNA. In many species it has been shown to be one long **chromosome**. This chromosome replicates and divides when the cell splits. Each daughter cell gets the same genetic information that was stored in the mother cell. Mitosis, however, has not been observed in bacterial cells.

The cytoplasm often contains **granules** of stored food such as starch. It may also contain small **vacuoles** of water. Some bacteria are photosynthetic. The cytoplasm of these forms contains chlorophyll on layers of lipids and proteins as in blue-green algae. No chloroplasts are present.

Some species of bacilli and spirilla have **flagella** (singular: **flagellum**). These are hair-like projections that originate in the plasma membrane. Usually they are longer than the cell (Fig. 9-12). They propel the bacteria by a whip-like action.

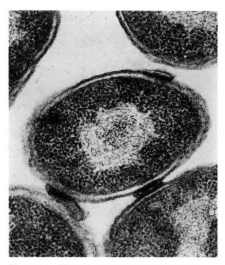

Fig. 9-11 This electron micrograph of a bacterial cell shows the granular cytoplasm, nuclear material, and cell wall.

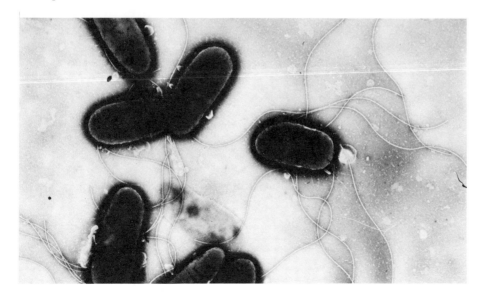

Fig. 9-12 This electron micrograph shows the numerous flagella of the bacterial species *E. Coli* (12 000 X).

Review

1. **a)** Name and describe the three shapes used to classify bacteria.
   b) What do the prefixes *diplo, strepto,* and *staphylo* mean?
2. **a)** Distinguish between gram-positive and gram-negative bacteria.
   b) Why might doctors want to know if infections in their patients consist of gram-positive or gram-negative bacteria?
3. **a)** What is a capsule?
   b) Describe the relationship between the presence of a capsule and the virulence (strength) of an infectious bacterium.
4. **a)** Of what importance is the fact that bacterial cells usually contain numerous ribosomes?
   b) What structures may be found in the cytoplasm of a bacterial cell? What are their functions?
   c) Describe the structure and function of a flagellum.
5. Describe the nuclear region of a bacterial cell.

9.7 Investigation

Examination of Some Forms of Bacteria

In this investigation you will study the shape and arrangement of representative bacteria. You will use prepared slides. If your teacher has training in microbiology, you may be allowed to examine living bacteria.

Materials

compound microscope
prepared stained slides of several types of bacteria
lens paper
immersion oil
xylol

Procedure

a. Your teacher will provide prepared slides of several types of bacteria. Examine as many different types as time permits. Make a sketch of each type. Make notes on each sketch of any unique features.
Note: You should use an oil immersion objective if one is available. See Section 9.5 for information on how to use it.
b. Identify each type, if possible. Refer to Figure 9-9 for help.
c. Moisten a piece of lens paper with xylol. Use this to wipe the oil from the oil immersion objective.

CAUTION: Do not inhale the xylol fumes.

Discussion

1. Your laboratory notes and sketches are the writeup for this investigation.

9.8 Some Life Processes of Bacteria

Reproduction and Spore Formation

Binary fission

When conditions are suitable, bacteria reproduce by **binary fission** (Fig. 9-13). The chromatin divides without going through the usual stages of mitosis. As the nuclear material is dividing, the plasma membrane bends inwards from both sides near the centre of the cell. Eventually it divides the cell in half. Cell wall material is then deposited between these membranes. The cells separate to complete the process.

Under ideal conditions a bacterium grows to full size and undergoes binary fission in about 20 min. If conditions remained ideal there would be four bacteria after 40 min, eight bacteria after 60 min, and sixteen bacteria after 80 min. If this process continued unchecked for 48 h, about 2×10^{43} bacteria would be produced. That is 20 000 000 000 000 000 000 000 000 000 000 000 000 000 000 bacteria! The mass of this number of bacteria would be about 2×10^{25} t (tonnes). That is 20 000 000 000 000 000 000 000 000 t, many times the mass of the entire earth! Fortunately, ideal conditions are seldom encountered for long. Competition for food by the bacteria and pollution of their environment by their own metabolic wastes create less-than-ideal conditions. However, is it surprising that fresh milk sours quickly when exposed to warmth and open air? Or that potato salad may become

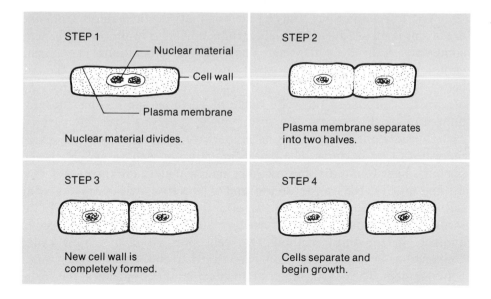

Fig. 9-13 Binary fission of a bacterial cell.

STEP 1

Nuclear material

Cell wall

Plasma membrane

Nuclear material divides.

STEP 2

Plasma membrane separates into two halves.

STEP 3

New cell wall is completely formed.

STEP 4

Cells separate and begin growth.

dangerously contaminated in just a few minutes on a summer afternoon picnic? Or that a human bacterial infection can quickly cause illness or even death?

Spore formation

Many types of bacteria can form a highly resistant spore called an **endospore**. They do this when they are exposed to unfavourable conditions such as extreme dryness and temperatures. The cell wall thickens and the cell goes into a resting stage. For most species this is not a method of reproduction. It is simply a means of survival until favourable conditions return. When favourable conditions do return, the endospore absorbs water and swells. The cell wall breaks open and one bacterial cell emerges and resumes activity.

Endospores of many species can survive several hours in boiling water. Endospores of some species have survived temperatures as low as –250° C. Scientists have collected endospores from deep in the Antarctic ice cap. Some were estimated to be millions of years old. Yet, when they were provided with favourable conditions, the cells resumed activity.

Fortunately, most disease-producing bacteria do not form spores. If they could, it would be very difficult to keep diseases under control. The tetanus bacillus that causes lockjaw is one that does form spores. This species can live as an endospore for long periods in soil. Then, if you step on a rusty nail or piece of glass that is in contact with that soil, the endospores enter your body and return to activity.

Sexual reproduction

Although most species of bacteria rely only on binary fission for reproduction, a few do have a form of sexual reproduction.

You learned in your study of meiosis that genetic material from two organisms is mixed during sexual reproduction. This results in new genetic combinations in the offspring. These new combinations may create variations in the species. Some variations may produce organisms that are better suited to their environment than the parents were. Perhaps this kind of variation produced the wide range of bacterial species that are now present on earth.

Considerable research has been conducted since the 1940s on sexual reproduction in bacteria. The details are still not well known. However, scientists have found that bacteria mix genetic material in three ways. One method involves the transfer of a strand of DNA from a dead cell to a living cell. The living cell incorporates that strand into its genetic makeup. In the second method, viruses transfer genetic material from one bacterial cell to another. In the third method, two cells pair up in a process called **conjugation**. A bridge of cytoplasm, called a **conjugation tube**, forms between them. The genetic material flows across the conjugation tube from one cell to the other.

Sexual reproduction occurs in only a few species and even in those species, it occurs rarely.

Nutrition

Like all other organisms, bacteria have nutritional requirements. They must obtain certain substances from their environment to remain alive. These substances must be converted into food materials such as glucose, lipids, and proteins. In turn, these substances are used to make cellular components such as cytoplasm and organelles.

Autotrophic bacteria

A few species of bacteria can make their own food from carbon dioxide and water through **photosynthesis**. As you know, energy is required for this process. This energy enters the cells in the form of light energy. Bacterial photosynthesis is quite different from that of other organisms. For one thing, no oxygen is produced as a by-product.

A few other species of bacteria make their own food by a process called **chemosynthesis**. They use no light energy. Instead, they get the energy required to make food by breaking down inorganic compounds. Usually these are compounds of iron, sulfur, and nitrogen. No organisms except these bacteria can carry out chemosynthesis.

Organisms that can make their own food from inorganic substances are called **autotrophs**.

Heterotrophic bacteria

Most species of bacteria cannot make their own food. Therefore, these species depend on other organisms for food. *Organisms that depend on other living things for their food are called* **heterotrophs**. There are two kinds of heterotrophs—saprophytes and parasites.

Saprophytes *are organisms that obtain their food from non-living organic matter*. Saprophytic bacteria feed on dead plants, dead animals, fecal matter, and other non-living organic matter. They are often called **decomposers**. Obviously they feed on many of the same things that you eat. Think about the last meal you ate. How much of it was dead plant or animal matter? How much of it came from plants or animals? You will see later that bacteria often spoil food as they feed on it.

Parasites *are organisms that obtain their food by living attached to or inside living organisms*. Parasitic bacteria invade the tissues of another organism. There they feed directly on organic matter in the tissues. *An organism that supports a parasite is called a* **host**. *Parasitic bacteria that cause a disease in the host are called* **pathogens**.

Saprophytic bacteria feed by secreting enzymes on the intended food. A series of chemical reactions converts the food into particles small enough to pass through the plasma membrane. Some bacteria can break down cellulose products such as wood. Few other organisms can do that.

Parasitic bacteria usually do not have as many enzymes as saprophytes. As a result, they must use the enzymes of the host. This explains why they are found only in living organisms.

Respiration

Aerobic and anaerobic respiration

In Chapter 7 you learned that cellular respiration that requires oxygen is called **aerobic respiration**. You also learned that cellular respiration that occurs in the absence of oxygen is called **anaerobic respiration**. Some bacteria are aerobic; some are anaerobic. Others are in between these two extremes.

Bacteria that must have oxygen for cellular respiration are called **obligate aerobes**. The bacteria responsible for tuberculosis, pneumonia, diphtheria, and cholera are in this group.

Bacteria that must live in the absence of atmospheric oxygen are called **obligate anaerobes**. The bacteria responsible for botulism and tetanus are in this group. These organisms not only cannot use oxygen but are actually harmed by it. Oxygen either kills them or inhibits their growth.

Some species of bacteria grow best in the presence of oxygen. However, if necessary, they can function in its absence. Such organisms are called **facultative anaerobes**. The bacteria responsible for scarlet fever and typhoid fever are in this group. The *E. coli* bacteria of the human intestine are also facultative anaerobes.

Though uncommon, some species of bacteria function mainly in the absence of oxygen. However, if necessary, they can function in a limited way in the presence of oxygen. These organisms are called **facultative aerobes**.

Products of anaerobic respiration

You know that some bacteria produce ethyl alcohol when they respire anaerobically. This process is called **alcoholic fermentation**.

$$\text{Glucose} \xrightarrow{\text{Enzymes}} \text{Ethyl alcohol} + \text{Carbon dioxide} + \text{Energy}$$

$$C_6H_{12}O_6 \xrightarrow{\text{Enzymes}} 2\ C_2H_5OH + 2\ CO_2 + \text{Energy}$$

Other species of bacteria produce lactic acid when they respire anaerobically. This process is called **lactic acid fermentation**.

$$\text{Glucose} \xrightarrow{\text{Enzymes}} \text{Lactic acid} + \text{Energy}$$

$$C_6H_{12}O_6 \xrightarrow{\text{Enzymes}} 2\ C_3H_6O_3 + \text{Energy}$$

Still other species produce methane gas when they respire anaerobically. These bacteria obtain their energy for life processes by combining hydrogen gas and carbon dioxide gas.

$$\text{Hydrogen} + \text{Carbon dioxide} \xrightarrow{\text{Enzymes}} \text{Methane} + \text{Water} + \text{Energy}$$

$$4\,H_2 + CO_2 \xrightarrow{\text{Enzymes}} CH_4 + 2\,H_2O + \text{Energy}$$

If you have ever pushed a stick into the muck at the edge of a pond, swamp, or bog, you probably smelled methane gas. Anaerobic respiration takes place deep in the organic sediment of such areas. The stick simply releases some of the gas from the sediment. Methane is often called marsh gas. It is the main component of natural gas.

Methane gas is often released from sewers and landfill sites (garbage dumps). In these areas, bacteria act on organic matter under anaerobic conditions.

Summary of Characteristics of Bacteria

Let us now summarize some of the main things we have learned about bacterial form and function:

1. All bacteria are unicellular. Some species may occur in colonies or filaments. The cells may be spherical (cocci), rod-shaped (bacilli), or spiral-shaped (spirilla).
2. Bacterial cells lack nuclei and internal membranes.
3. Most bacterial species lack chlorophyll.
4. Bacteria reproduce mainly by binary fission. A few species have a form of sexual reproduction. Some species can form protective spores.
5. A few species of bacteria are autotrophic. Some of these are photosynthetic. Others are chemosynthetic. Most species of bacteria are heterotrophic. Some of these are saprophytic. Others are parasitic.
6. Some species of bacteria respire aerobically. Others respire anaerobically. Still others are in between these extremes.

Review

1. How does binary fission in a bacterium differ from that in an amoeba?
2. a) What is an endospore?
   b) What is the main function of spore formation by most species of bacteria?
3. a) Of what value might sexual reproduction be to a bacterial species?
   b) What is conjugation?
   c) Describe two other methods of sexual reproduction in bacteria.
4. Distinguish between photosynthesis and chemosynthesis.
5. Distinguish between autotrophic bacteria and heterotrophic bacteria.
6. a) Distinguish between a saprophyte and a parasite.
   b) Explain the terms host and pathogen.
   c) Why can parasitic bacteria function only in a living host?

7. a) Distinguish between obligate aerobes and obligate anaerobes.

   b) Distinguish between facultative aerobes and facultative anaerobes.

8. Briefly describe three forms of anaerobic respiration that bacteria perform.

9.9 Importance of Bacteria: Useful Activities

The earth as we know it today could not exist without bacteria. These organisms are involved in thousands of processes. Some processes are harmful to humans. However, most are quite harmless. In fact, some are actually helpful to humans and essential to our existence.

Decay of Organic Matter

Bacteria play an important role in keeping the soil fertile. They break down non-living organic matter, releasing nutrients.

All living organisms require certain nutrients. These include elements such as nitrogen, phosphorus, and potassium. Animals get these nutrients by eating other animals or plants. Most plants get them from the soil. However, if it were not for bacterial action, the soil's supply of nutrient elements would soon be exhausted. Bacteria (and fungi) break down the tissues of dead organisms. This releases the nutrient elements and makes them available to living plants. When those plants (or the animals that eat them) die, bacteria once again release the nutrient elements. In this way, elements are recycled over and over again. You will study nutrient cycles in Unit 6.

There are many interesting things involved in nutrient cycles. For example, because of the cycling of nutrient elements by bacteria, you could have atoms in your body that were once part of a wolf. The wolf died. Bacteria broke down its tissues. Nutrient elements were released into the soil. Some grass plants absorbed the nutrient elements. A cow ate the grass. You ate a piece of the cow for dinner. Your body's metabolic processes used the nutrient elements to build tissues. Part of the wolf is now part of you.

Bacterial decay of organic matter does more than just increase soil fertility. It gets rid of dead organisms. If it were not for bacterial action, dead animals, dead trees, and other dead organisms would litter the earth. (As you will see in Chapter 11, fungi assist bacteria in decomposition.) Also, bacteria assist greatly in the breakdown of wastes in septic systems and in sewage disposal plants.

Nitrogen Fixation

A few species of bacteria are capable of **nitrogen fixation**. Such bacteria are very important in keeping soil fertile. All plants require nitrogen. However, they are unable to use atmospheric nitrogen. They must absorb nitrogen in the form of nitrates (occasionally ammonia) from the soil.

Nitrogen-fixing bacteria use atmospheric nitrogen to make their protein molecules. When these bacteria die and decompose, this nitrogen is released as nitrates that plants can absorb. This is similar to what happens in some species of blue-green algae.

Some nitrogen-fixing bacteria occur free in the soil. Others are associated with certain plants. Plants of the legume family (peas, beans, clover, alfalfa, soybeans) develop lumps or **nodules** on their roots (Fig. 9-14). These nodules contain nitrogen-fixing bacteria of the genus *Rhizobium*. The nitrogen fixed by these bacteria is made available to the legume plant. However, some of it is returned to the soil where it assists the growth of other plants. Farmers often grow legumes in a field the year before they grow a crop such as wheat which requires a great deal of nitrogen.

In return for providing the legume with nitrogen, the *Rhizobium* bacteria receive carbohydrates that the plant made through photosynthesis. They get this food from the roots of the plant.

A relationship in which two organisms live together in close association is called **symbiosis**. In the relationship of *Rhizobium* with legumes, both organisms benefit. The plant grows better because it has nitrogen. The bacteria obtain needed carbohydrates. *The condition in which both organisms benefit from their relationship is called* **mutualism**. In Section 9.8 you met a form of symbiosis in which only one organism benefited. The other was harmed. What is the name of that form of symbiosis?

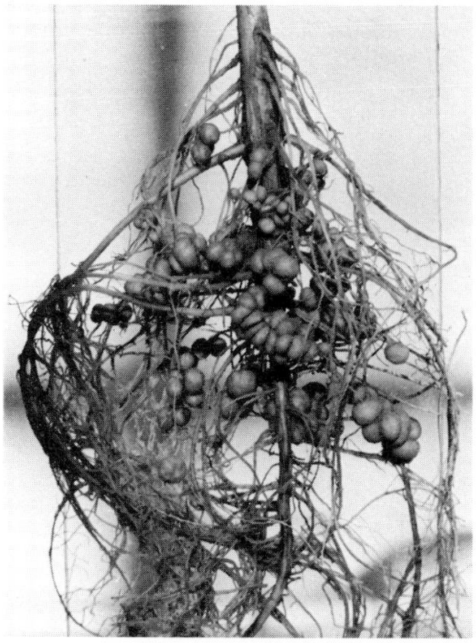

Fig. 9-14 The nodules on the roots of this soybean plant contain the nitrogen-fixing bacterium *Rhizobium*. Both the soybean and the bacteria benefit from this relationship called mutualism.

Industrial Uses of Bacteria

Making dairy products

Most people know that bacteria can harm dairy products. For example, if milk is not pasteurized, bacteria will cause it to go sour. However, many processes in the dairy industry actually need bacteria.

During cheese-making, bacteria change lactose (milk sugar) into lactic acid. This acid causes the solids in the milk to coagulate (come together). The resulting mass is called curds. The curds are then separated from the whey, or liquid portion. The curds are inoculated with certain bacteria that act on the casein (a compound in milk) in the curds. The result is a cheese with a certain flavour and texture. Different species of bacteria produce different flavours and textures of cheese.

The making of butter also requires bacteria. Butter is made from cream. Pasteurized cream is inoculated with a species of bacterium that produces lactic acid. The cream becomes sour. The sour cream is then churned. This separates the fat particles, or butter, from the rest of the mixture which is called buttermilk.

The making of cottage cheese and yogurt also require bacteria.

Making silage

You have probably seen large cement or metal silos beside barns (Fig. 9-15). These contain a food material called silage. Silage consists of chopped corn, grass, alfalfa, or other green crops. It is blown into the silo and packed

down. The packing creates anaerobic conditions. Certain anaerobic bacteria act on the silage. They change sugars in the silage into lactic acid. Lactic acid is of great value in milk production in dairy animals.

Making vinegar

In Section 9.5, page 205, we pointed out that bacteria can sour wines. They act on the ethyl alcohol formed during fermentation, and produce acetic acid. Dilute acetic acid is called vinegar.

Vinegar is normally made by allowing yeast cells to convert sugars in fruit juice into ethyl alcohol. Apple juice is often used. Bacteria then convert the alcohol into acetic acid. Most vinegar is about 5% acetic acid.

Making sauerkraut

To make sauerkraut, cabbage leaves are placed in a container under anaerobic conditions. Anaerobic bacteria act on the sugars in the leaves. Both alcoholic fermentation and lactic acid fermentation occur. Ethyl alcohol and lactic acid are produced. These give sauerkraut its characteristic flavour.

Tanning leather

Special strains of bacteria are used to tan leather. First, all the flesh and hair are removed from the animal hide. Then the hide is placed in a tank that contains, among other things, special kinds of bacteria. These bacteria begin feeding on the hide. The enzymes they secrete soften the hide and make it pliable (flexible). In this condition, the hide is said to be tanned.

Review

1. **a)** Outline the role of bacteria in the maintenance of soil fertility through the decomposition of organic matter.
   b) Your body could contain atoms that were once part of a dinosaur. Explain how this could happen.
2. **a)** What is nitrogen fixation?
   b) What other organisms besides certain bacteria can fix atmospheric nitrogen?
   c) Of what value is nitrogen fixation to plants?
3. **a)** What is symbiosis?
   b) What is mutualism?
   c) Describe the symbiotic relationship that exists between the bacterium *Rhizobium* and legumes. What is this relationship called?
   d) Name and describe another form of symbiosis.
4. Select one industrial use of bacteria described in this section. The descriptions are quite brief. Prepare a more detailed description. Research the use that you chose. Visit the school resource centre and the community library. Write to government agencies and other sources, if necessary. Then prepare a written report of about 500 words. Include diagrams if they help to explain the written work.

   If you select a use such as making butter or sauerkraut, you may wish to demonstrate the process to your class.

9.10 Importance of Bacteria: Food Spoilage and Its Control

Like all living cells, bacteria carry out nutrition and cellular respiration. While doing so, they often harm things that are of economic value to humans. For example, plant crops and animals may be harmed. Some bacteria also harm humans directly by causing disease. This section describes the major harmful effects of bacteria.

Food Spoilage

You learned in the last section that bacteria are considered helpful because they cause the decay of organic matter. They break down dead organisms, fecal matter, and other organic waste. The products formed enrich the soil.

However, bacteria do not restrict their action to these organic wastes. They may also attack foods. Sour milk, rotten fruits, and spoiled meat are caused by bacterial action.

Usually such spoilage of food is harmless to human health. You could normally drink a glass of sour milk or eat a rotten banana without suffering any ill effects. However, serious diseases can result from eating foods that have been acted upon by certain types of bacteria. These diseases are commonly called **food poisoning**. Three types are discussed here.

Staphylococcus food poisoning

This type of food poisoning is sometimes called ptomaine poisoning. It is the most common type of food poisoning. It is caused by the bacterium *Staphylococcus aureus*. This bacterium is normally present on the skin and in the nose and throat of humans. It multiplies rapidly in many foods. Dairy products and cream-filled bakery products are excellent environments for the growth of this bacterium. It gets into foods when they are handled in an unsanitary manner. Then, if the food is not properly refrigerated, the bacteria multiply rapidly.

This type of food poisoning is caused by **toxins**, or poisons, secreted by the bacteria while they are digesting the food.

The symptoms of this disease are nausea, diarrhea, and abdominal cramps. Symptoms usually appear from 1 to 5 h after the infected food is eaten. Complete recovery usually occurs in a few days. Death seldom results.

Most people have probably had a mild case of this disease. Also, most of these people probably got the disease the same way. They ate chicken or egg salad sandwiches, potato salad, or any food with salad dressing in it. They may also have eaten cold meats or cream puffs. These foods were likely eaten at a picnic on a hot day. The foods were probably not kept refrigerated from the moment they were prepared to the moment they were eaten.

Do you remember how fast bacteria multiply under ideal conditions? Well, dairy products in a warm place are almost ideal conditions for *Staphylococcus aureus*.

Botulism

This type of food poisoning is caused by the bacterium *Clostridium botulinum*. This bacterium is common in many soils. In an aerobic environment it seldom causes trouble. However, under anaerobic conditions it produces toxins that are among the most deadly known. They are more lethal than rattlesnake venom.

Most cases of botulism arise from eating home-canned vegetables that have not been thoroughly cooked. String beans seem to be an ideal environment for this bacterium. *Clostridium botulinum* produces spores. Some spores may get on the vegetables before they are canned. If the vegetables are not cooked thoroughly, the spores may not be killed. Once the jar is sealed, anaerobic conditions are created. The spores become active bacteria. They secrete toxins as they feed on the vegetables.

The toxins are destroyed by high temperatures. Therefore, even though some canned food may be contaminated, botulism may be prevented by heating the food before eating it. When purchasing canned food, never buy a can that has a bulge in it. Cans of infected food often bulge because of gases given off during bacterial respiration.

The symptoms of botulism are double vision, weakness, and paralysis. The paralysis begins in the neck and spreads to other parts of the body. The symptoms usually begin in 12 to 36 h after the infected food has been eaten. The disease is fatal in about 70% of cases. Therefore it is important that the bacteria be killed during food processing.

Salmonella poisoning

This disease is also commonly called food poisoning. However, it would be more accurate to call it a food infection. In both *Staphylococcus* poisoning and botulism, the bacteria release toxins, or poisons, into the food. Thus the food is truly poisoned. However, *Salmonella* does not do this. The food simply transfers the *Salmonella* bacteria from an infected host to a new host. No symptoms arise until after the *Salmonella* begin to grow in the intestines of the host.

Almost all species of *Salmonella* cause diseases in humans. One, *Salmonella typhi,* causes deadly typhoid fever. A number of other species cause infections of the digestive system. These food-borne species originate

Fig. 9-16 *Salmonella* **often spreads from infected food animals to humans.**

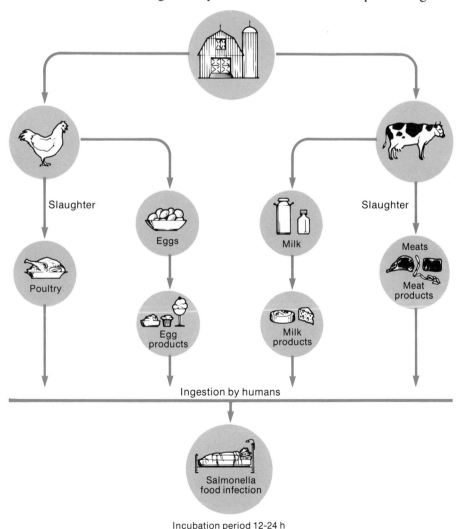

Slaughter

Eggs

Milk

Slaughter

Poultry

Meats

Egg products

Milk products

Meat products

Ingestion by humans

Salmonella food infection

Incubation period 12-24 h

in infected humans or other warm-blooded animals. An infected food handler may introduce *Salmonella* into food. Also an infected animal such as a cow or chicken may be the source of the disease (Fig. 9-16). The products usually involved are meat pies, cured meats, sausage, poultry, eggs, milk, and milk products.

Proper cooking kills the bacteria. However, many of these products are often eaten uncooked or partly cooked. For example, many cases of *Salmonella* poisoning have been traced to products made from uncooked eggs or partly cooked eggs. These include eggnog, meringues, cream cakes, and custards. Many other cases have been traced to cooked foods that have remained unrefrigerated for a time. Still other cases have been traced to canned foods. These foods stood in a warm place after being opened and handled by an infected person.

The food must contain a fairly high concentration of *Salmonella* for infection to occur. Therefore infected food normally has to stand in a warm place for a while to let the bacteria multiply.

Symptoms usually occur 12 to 24 h after infected food is eaten. They usually begin with a headache or chills. Then nausea, diarrhea, and abdominal pain follow. Complete recovery usually occurs, even without treatment, in 2 to 3 d. This disease is rarely fatal.

Control of Food Spoilage

As you have seen, bacteria in food can destroy food and cause diseases. Many methods are used to prevent these harmful effects. In general, two things can be done. In some cases, the bacteria in the food can be killed. However, if they cannot be killed, they can be kept inactive, or nearly so, by exposing them to unfavourable environmental conditions. Note how each of the following methods uses one or both of these general methods.

Pasteurization

This process, named after Louis Pasteur, is used to preserve milk, fruit juices, and wines from bacterial action. It involves two main steps. The first step uses high temperatures to kill bacteria. The second step uses low temperatures to keep inactive any bacteria that survived the first step.

The temperature of the first step must be carefully controlled. Too high a temperature changes the taste of most beverages. It can also destroy certain nutrients. However, the higher the temperature, the shorter the time required to kill bacteria. For example, some dairies pasteurize milk by heating it at 62°C for 30 min. Other dairies heat the milk at 71°C for just 15 s. Either process kills about 90% of all bacteria in the milk. This includes almost all pathogens (disease-causing bacteria). Rapid and immediate cooling to 10°C greatly reduces the activity of any survivors.

Refrigeration

Foods are usually refrigerated at temperatures around 10 to 15°C. A low temperature does not kill bacteria. It just reduces their activity. Therefore foods cannot be kept indefinitely in a refrigerator. The bacteria will slowly multiply. Eventually they will destroy the food.

Freezing

Most freezers use temperatures around −10 to −20°C. At all temperatures below 0°C, bacterial activity is reduced to almost zero. Bacteria require water for metabolic activities. Yet all water in food is frozen at a few degrees below 0°C. As a result, foods will keep almost indefinitely at sub-zero temperatures.

Modern day Antarctic explorers have reported the discovery of food caches (storage areas) that were set up decades earlier by explorers. These caches often contain edible meat. Russian scientists have discovered wooly mammoths buried in Siberian icefields. They had been there 25 000 years, and most of their flesh is still present. These examples show that freezing greatly reduces bacterial action.

Dehydration

Dehydration is the removal of most of the water from food. As we just pointed out, bacteria require water for metabolic processes. If most of the water is removed, foods will keep indefinitely. Most species of bacteria die. Others produce inactive spores. Fruits and vegetables can be kept indefinitely if 80% of the water is removed. Eggs and milk must have 90% of the water removed.

Dehydration has other advantages. Dehydrated foods do not have to be kept cool. Also, they are easier to package. Finally, they are less expensive to ship, since one is not paying to ship water.

Canning

Several species of bacteria can survive the temperature of boiling water (100°C). Therefore, during the canning of fruits and vegetables, the temperature is raised to 120°C for about 30 min. A pressure cooker is often used to reach such temperatures.

The sterilized food must be sealed in air-tight containers that have also been sterilized.

Preservatives

Salt is a common preservative. When added to foods, it plasmolyzes the bacterial cells. This usually kills them. Pickles are often preserved in salt solution (brine). Fish and pork are also often preserved with salt.

Fruits are usually preserved with a concentrated sugar solution. This also plasmolyzes the bacteria.

Vinegar is another common preservative. It is used to "pickle" cucumbers, corn, and other vegetables. Bacteria cannot survive in the acidic solution of vinegar.

A few years ago, chemical preservatives were commonly used. The government has been withdrawing some of these from use. They have been shown to have harmful effects on humans.

Radiation

The advantage of this method is that the food can be packaged before the bacteria are killed. The food is sealed in its containers. Then the containers are exposed to high energy radiation such as X-rays and gamma rays. These rays kill all bacteria.

The government carefully controls the foods that are preserved in this manner.

Smoking

Smoke-curing is used mainly to preserve bacon and ham. Chemicals in the smoke kill many bacteria. Usually salt and sugar are used along with smoking to kill the remaining bacteria. All three treatments give the pork products their characteristic flavour.

Chlorination and ozonation

Water is not a food. However, its sterilization is discussed here because water is usually ingested with most foods.

Boiling for 30 min will kill most bacteria present in water. But this process obviously could not be used on a large scale. Instead, the bacteria are oxidized, or "burned", by a chemical. Chlorine and ozone are two chemicals commonly used. One of these is added to the water at the purification plant.

Such treatment is particularly important when a community's water supply comes from the same lake into which its sewage flows. Certain pathogenic bacteria, such as those that cause typhoid fever and cholera, are transmitted in water. They leave the intestines of a carrier and enter the lake via the sewage system. They will return to the community through the water system if they are not destroyed by chlorination or ozonation.

Review

1. a) Describe the causes and symptoms of *Staphylococcus* food poisoning.
   b) How can this type of poisoning be prevented?
2. a) Describe the causes and symptoms of botulism.
   b) How can botulism be prevented?

3. **a)** Why should *Salmonella* poisoning not be called food poisoning?
   b) Describe the causes and symptoms of this type of poisoning.
   c) How can *Salmonella* poisoning be prevented?
4. Pasteurization involves two main steps. What are they? What is the function of each step?
5. **a)** What are the advantages of freezing, in comparison to refrigeration? What disadvantages might there be?
   b) What are the advantages of dehydration, in comparison to freezing? What disadvantages might there be?
6. How does canning preserve foods?
7. Name and describe the effects of three common preservatives.
8. What advantages does radiation have in preserving foods? What disadvantages might it have?
9. What is involved in the preparation of a smoke-cured ham?
10. Why is chlorination or ozonation of drinking water necessary?

9.11 Viruses

What Are Viruses?

Viruses are not monerans. In fact, if you look at the classification of living things in Chapter 3, you will not find viruses listed in any kingdom. Why are they included in this chapter?

A virus, alone, seems to be a non-living particle. It shows no signs of life. But in a living cell, a virus seems very much alive. Are viruses living or non-living? No one knows. The answer depends on how one defines life. However, if viruses are living, they certainly are the simplest of all organisms. Therefore, the study of viruses is included in this chapter because it deals with the simplest known living things, the monerans.

You have likely heard a sick person say, "I've got a virus." Just what does this person have? Biologists have been seeking an answer to that question for years.

Before viruses were discovered, biologists thought they knew what life was. They had drawn up a list of the characteristics of living things. You saw this list in Chapter 2. Then viruses were discovered. A virus has no metabolism of its own. It cannot grow or respire. Also, it has no cellular organization. This means it has no nucleus, no cell organelles, no cytoplasm, and no surrounding membrane. It is many times smaller than a cell. Yet it is larger than a molecule. Viruses are so simply organized that they can be crystallized from a solution. The process is similar to the crystallization of salt from a solution. The crystals can be stored in a bottle just like the crystals of salt or sugar. All this suggests that viruses are non-living. However, there is one important difference between viruses and non-living matter. Within a cell, a virus behaves quite unlike any non-living matter. It can undergo reproduction, one of the most basic characteristics of living things. The virus takes over chemical control of the cell. It then "directs" the cell to make new virus particles. Since they can reproduce, viruses appear to be living. Yet this method of reproduction is quite unlike that of any living organism.

If a virus is non-living, it certainly differs from other non-living matter. No other matter affects cells as viruses do. If a virus is living, it certainly differs from other living things. It must be life at a simple molecular level. Perhaps a virus is non-living at times and living at other times. Or, perhaps the definition of life needs to be changed to include viruses. Perhaps, in viruses, biologists have found the link between the living and the non-living worlds.

What are viruses? Biologists cannot tell us whether they are living or non-living. Nor can they tell us how viruses direct cells to make more viruses. However, they do know a great deal about them. After a brief look at the history of the discovery of viruses, we will see what biologists know.

Discovery of Viruses

Not all viruses cause diseases. However, over 300 kinds of viruses do. Diseases caused by viruses include the common cold, influenza, polio, rabies, and smallpox. Therefore it is not surprising that scientists first learned about viruses while trying to find cures for diseases.

In 1796, Edward Jenner vaccinated a boy against the smallpox virus. Yet he did not even know that a virus caused smallpox. In 1885, Louis Pasteur vaccinated a child who had been bitten by a rabid dog. Yet he, too, knew nothing of viruses.

In the late 1800s, early bacteriologists like Louis Pasteur and Robert Koch learned that bacteria caused several diseases. However, one thing puzzled them. In some cases, they could not see the organisms that caused the disease, even with the best microscopes. Then, in 1892, a chain of experiments began that finally led to the discovery of viruses.

In that year, Dmitri Iwanowski, a Russian biologist, was investigating a disease of tobacco plants. It is called the tobacco mosaic disease since it produces a *mosaic,* or pattern, of yellow and light green areas on the leaves.

During his experiments, Iwanowski squeezed juice from an infected leaf. He rubbed this juice on healthy leaves. Soon these leaves were also infected. Then Iwanowski repeated the experiment, but with one change. Before he rubbed the juice on healthy leaves, he filtered it through a special filter. This filter was known to stop the passage of bacteria. Examination of the filtrate with a microscope revealed no bacteria. However, this filtered juice still caused the disease when rubbed on healthy plants.

Iwanowski knew nothing of viruses. Therefore he assumed that bacteria caused the disease and that they had given off a toxin (poison) that was able to go through the filter paper.

In 1898 a Dutch botanist, Martinus Beijerinck, carried out experiments similar to those of Iwanowski. He got the same results. However, he discovered one additional thing. Whatever caused the disease was able to multiply within the plants. Toxins cannot multiply. Therefore Beijerinck

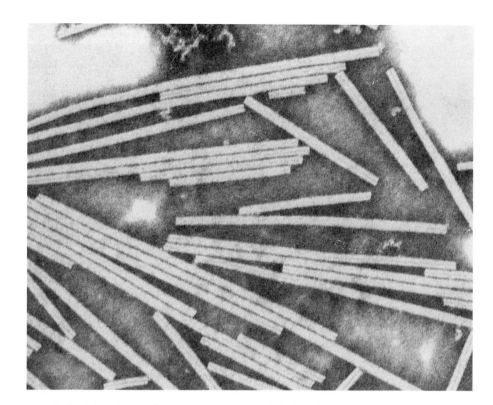

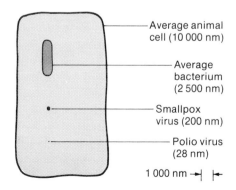

Fig. 9-17 An electron micrograph of crystals of the tobacco mosiac virus (77 000 X).

concluded that bacteria were not responsible for the disease. Instead, he said that the disease must be caused by a "living fluid" (*living* because it can reproduce and *fluid* because it can pass through very fine filters). This living fluid is now called a **virus**.

Finally, in 1935, Wendell Stanley, an American microbiologist, isolated the tobacco mosaic virus. He removed the fluid from over a ton of infected tobacco leaves. He then extracted the virus from that fluid. He obtained about 2-3 cm³ of virus crystals. In a bottle, these crystals were lifeless. However, when they were added to water and rubbed on tobacco leaves, they multiplied greatly and produced the tobacco mosaic disease. The first virus had been isolated.

The electron microscope was invented in the same year that Stanley made his discovery. Individual virus particles cannot be seen with an ordinary compound microscope. However, under the electron microscope, they are clearly visible (Fig. 9-17). The electron microscope has made a major contribution to the discovery of the many viruses that infect humans, other animals, plants, and even bacteria.

Characteristics of Viruses

Size and shape

Viruses are many times smaller than bacteria (Fig. 9-18). They are usually measured in **nanometres (nm)**. A nanometre is one billionth of a metre

Average animal cell (10 000 nm)

Average bacterium (2 500 nm)

Smallpox virus (200 nm)

Polio virus (28 nm)

1 000 nm →| |←

Fig. 9-18 A comparison of the sizes of two viruses with bacterial and animal cells.

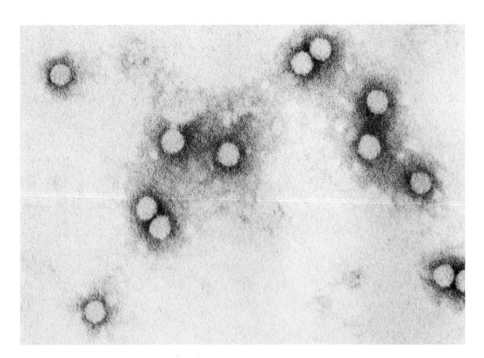

Fig. 9-19 An electron micrograph of the polio virus (200 000 X).

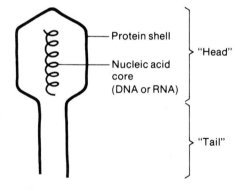

Protein shell

Nucleic acid
core
(DNA or RNA)

} "Head"

} "Tail"

Fig. 9-20 A simplified diagram showing the basic parts of a virus.

(0.000 000 001 m). You have seen bacteria under a microscope. A small bacterium is about 500 nm in length. It is just a tiny speck under the compound microscope at high power. Viruses range in size from about 30 nm to 300 nm. Thus the largest viruses are just visible under a high quality compound microscope.

Viruses have many shapes. Some, like the tobacco mosaic virus, are rod-shaped (see Fig. 9-18). Others, like the polio virus, are spherical (Fig. 9-19). Still others are cubic and rectangular. Some are shaped like a tadpole, with a "head" and a "tail" region.

Structure

A virus has two basic parts, a core and a shell (Fig. 9-20). The **core** is made of nucleic acid. In most viruses, the core is ribonucleic acid, RNA. In some it is deoxyribonucleic acid, DNA. The belief that viruses are living is supported by the composition of this core.

The **shell** is made of protein. It makes up at least 95% of the virus. A few viruses contain other substances such as lipids and carbohydrates.

Reproduction

On its own, a virus shows no signs of life. Before it does show signs of life, it must come in contact with the contents of a certain type of cell. A cell that supports the reproduction of a virus is called a **host cell**. The following is a description of how a virus uses an animal host cell to reproduce itself.

The virus must first come in contact with the host cell. Usually the virus is then taken into the cell by phagocytosis (see Section 6.9, page 133). It then releases its RNA or DNA into the cell. Biologists think that this nucleic acid now takes over protein synthesis in the cell. It uses the cell's materials, ATP, enzymes, and cell parts to replicate its nucleic acid and to make new protein coats. In effect, it directs the cell to build virus nucleic acids and proteins rather than cell nucleic acids and proteins. Thus new nucleic acid cores and protein shells are built. These join to make new viruses which are then released from the cell.

Effects of viruses

Viruses vary greatly in the effects that they have on host cells. In some cases no effect is observed. These are called **latent viruses** (Fig. 9-21,A). In other cases the viruses are very **virulent** (powerful). Cells may die soon after infection (Fig. 9-21,B). In still other cases, scientists have found more and more evidence that some viruses can turn normal host cells into cancer cells (Fig. 9-21,C). Apparently some viruses do not kill cells but, instead, cause enough changes in them to produce cancer cells. (Cancer cells are cells with an uncontrolled growth rate.)

Each type of virus normally attacks only a certain type of host cell. For example, many viruses invade humans. But a certain virus may only be able to become active if it enters a certain type of tissue cell such as nerve tissue. The polio virus is an example. It attacks only one kind of nerve cell in the spinal cord and brain of the human host. A plant virus may attack only leaf epidermal cells, and a bacterial virus may invade only a certain species of bacterium.

Fig. 9-21   Three ways in which viruses can affect host cells.

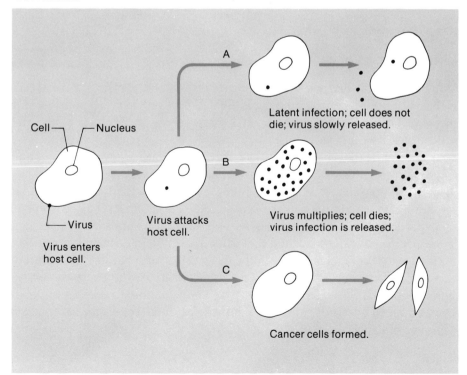

Cell — Nucleus

Virus

Virus enters host cell.

Virus attacks host cell.

A   Latent infection; cell does not die; virus slowly released.

B   Virus multiplies; cell dies; virus infection is released.

C   Cancer cells formed.

STEP A

Phage attaches itself to bacterium.

STEP B

DNA of phage enters bacterium.

STEP C

DNA takes over chemical control of the cell and replicates itself.

STEP D

Protein shells are formed.

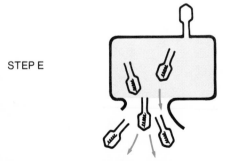

STEP E

Cell breaks open, releasing phages.

Fig. 9-22   The cycle of a virulent phage.

Bacteriophages

Bacteriophages, or **phages** for short, are viruses that attack bacteria. The diagram in Figure 9-20 shows the general shape of a phage. The "head" region may have many sides as in our diagram. However, in some cases, it is circular. The "tail" is equipped with protein fibres that attach the virus to the bacterium. The core of a phage is usually DNA.

Figure 9-22 shows the cycle of a phage attacking a bacterium and replicating itself. The phage attaches itself by the tail end to the bacterium. An enzyme is secreted by the tail which dissolves a hole in the cell wall of the bacterium. The DNA of the phage is then injected into the bacterium (Fig. 9-23). A few minutes later, the DNA of the phage destroys the DNA of the cell. It then takes over control of protein synthesis in the cell.

The DNA directs the cell to make phage DNA instead of cell DNA. It then directs the cell to make protein shells for this DNA. In effect, the cell has become a virus factory, with the virus DNA in charge of operations.

In the end, a single bacterial cell may contain as many as 200 to 300 virus particles. At this point, the cell breaks open. The new phages are released to attack other bacterial cells. The entire cycle requires about 45 min.

You may be wondering why doctors do not use phages to destroy pathogenic bacteria in humans. At one time, scientists thought that this was a real possibility. Now they are less hopeful. One problem is bringing the viruses into contact with the bacteria. For example, if you had a bacterial disease deep in some tissue of your body, how could the viruses be brought into contact with the bacteria?

Virus Diseases of Humans

You have already learned that viruses are responsible for the common cold, influenza, polio, rabies, and smallpox. They are also responsible for measles, warts, virus pneumonia, cold sores, yellow fever, chicken pox, and a host of less common diseases.

A great deal of cancer research is directed toward a study of viruses and their characteristics. Scientists know that DNA controls cell growth. They also know that cancer is abnormal cell growth. Further, they know that viruses can take over the work of DNA in cells. Therefore it seems reasonable to hypothesize that some types of viruses may be responsible for cancer.

Viruses have been isolated that cause leukemia in mice and some other mammals. (Leukemia is cancer in the tissues that form blood.) As a result of this and other research, viruses are suspected to be the cause of several human cancers. Some day scientists may know for sure that a specific virus causes a specific type of cancer. It is hoped that then a cure can be found for that type of cancer.

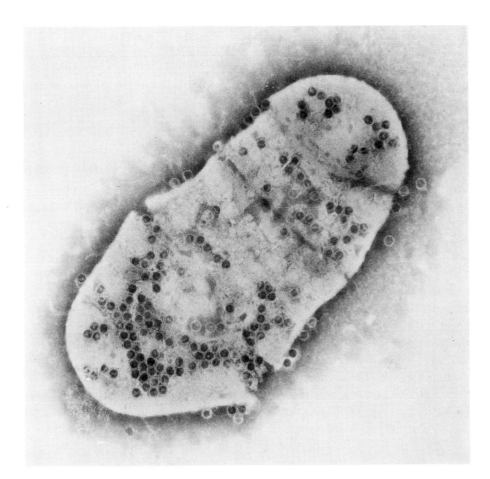

Fig. 9-23   Electron micrograph of phage viruses attacking a bacterium. Some of the phages have already injected their DNA into the bacterium. As you can see, the bacterium has become a factory for the making of phage viruses (49 000 X).

Review

1. **a)** What evidence suggests that viruses may be living?
   b) What evidence suggests that viruses may be non-living?
2. Outline the history of the discovery of viruses.
3. **a)** Compare the size of an average virus to that of an average bacterium.
   b) What shapes may viruses have?
4. Describe the general structure of a virus.
5. Describe how a virus reproduces itself.
6. Viruses may affect host cells in three ways. What are these ways?
7. **a)** What is a bacteriophage?
   b) Describe how a bacteriophage attacks a bacterium and destroys it while reproducing itself.
8. **a)** Name ten virus diseases of humans. Indicate which viruses can be prevented by some means.
   b) Why do scientists think that viruses may be responsible for some human cancers?

9.12 Human Diseases and Their Control

Many kinds of organisms cause diseases in humans. Among these are certain bacteria, viruses, protists, and fungi. This section deals mainly with pathogenic bacteria and viruses.

How Pathogens Harm Humans

In Section 9.10, page 217, you learned about food poisoning. The bacteria that cause food poisoning are examples of **pathogenic bacteria**, or bacteria that cause diseases. These **pathogens** harm humans by producing a **toxin**, or poison. In fact, most pathogenic bacteria affect humans by producing toxins.

Often the toxin travels through the circulatory system and damages tissues far from the originally infected area. For example, rheumatic fever is caused by a species of streptococcus bacterium that thrives in the human throat. As a by-product of their metabolism, these bacteria release a toxin. It travels throughout the body by means of the circulatory system. It often damages the heart valves and joints of the infected person. In a similar manner, a toxin is released by diphtheria bacteria in the throat. It, too, damages tissues far from the throat, such as the limb muscles and heart.

Tetanus bacteria can enter the body through a cut made by a dirty nail or piece of glass. The toxin that the bacteria release travels throughout the body. It affects all body muscles. Since this disease often causes paralysis of the jaw muscles, it is sometimes called lockjaw.

Some species of bacteria do not release their toxins until they die and break down. Then the toxins are released with devastating results. For example, the toxins of typhoid fever bacteria destroy cells of the intestinal wall. The toxins of tuberculosis normally destroy cells in the lungs. Cholera toxins spread from the intestines to the muscles and other organs, including the kidneys. Agonizing cramps occur and the urine stops flowing.

The Spread of Diseases

Airborne diseases

Many diseases are spread through the air. Among these are colds, influenza, whooping cough, diphtheria, measles, mumps, tuberculosis, pneumonia, and scarlet fever. They are commonly spread by coughing and sneezing. Since these diseases enter the host by means of the breathing system, the first sites of the infection are generally the nose, throat, and lungs.

Foodborne and waterborne diseases

A wide range of diseases are spread through food and water that are handled in an unsanitary manner. Many diseases in this group are diseases of the digestive system.

Infected persons have the pathogens in their feces. In our society, feces are commonly released into water that is later used for drinking purposes. Therefore, other persons will be infected if the wastewater is not properly treated. Of course, an infected person can also spread the disease by handling food and drinks that are ingested by others.

Food poisoning, cholera, typhoid fever, and dysentery are common diseases spread in this manner.

Contact diseases

Many diseases are most commonly spread by direct contact with an infected person or with materials that have been handled by the infected person. Many airborne, foodborne, and waterborne diseases can be transmitted in this manner also.

A number of diseases of the mucous membranes and the skin are spread by contact. For example, smallpox produces sores on the skin. These sores give off a **pus** that contains the smallpox virus. These viruses spread to a new host if that person contacts the sores directly or handles any object that has the pus on it.

Chickenpox is spread in a similar manner. The venereal diseases, gonorrhea and syphilis, are also contact diseases.

Wound diseases

Many diseases enter the body through breaks in the skin. Thus all wounds should be immediately cleansed and properly treated to prevent infection. Some diseases simply cause local infection. However others, such as tetanus, can be deadly.

You may recall that tetanus bacteria are obligate anaerobes. They must have an oxygen-free atmosphere if they are to survive and multiply. A nail puncture provides an ideal anaerobic environment. Suppose there are tetanus spores on a nail in a board. You step on the nail. The spores enter your body through the wound. Because of the small size of the puncture and the nature of the flesh on the foot, the wound quickly closes over. Anaerobic conditions are created. The spores become active bacteria, and you get the deadly disease, tetanus.

Carriers

You have seen that infected people and objects can be carriers, or transmitters, of diseases. Two special types of carriers are discussed here.

Many people are **immune carriers**. These people carry the disease in their bodies yet show no symptoms of the disease. People who have recovered from a disease often carry it for many months afterwards. Polio, typhoid fever, and diphtheria are often spread by immune carriers. Some people think that a lake may be safe for swimming and drinking because no one near the lake has any symptoms of disease. However, a carrier of typhoid fever could be contaminating the water regularly.

Animal carriers are another serious problem. Some diseases are spread

Anopheles mosquito
(carrier of malaria,
a protozoan)

Aedes mosquito
(carrier of yellow
fever, a virus)

Culex mosquito
(carrier of
encephalitis,
a virus)

Human body louse
(carrier of typhus
fever, a bacterium)

Housefly (carrier of
many diseases)

Tsetse fly
(carrier of African
sleeping sickness,
a protozoan)

Fig. 9-24   Some arthropods that are carriers of human diseases.

by direct or indirect contact with infected mammals. For example, tuberculosis can be spread from cattle to humans through unpasteurized milk. Fortunately, most diseases of mammals, including those of pet dogs and cats, cannot affect humans.

One group of animals, the arthropods, spreads many diseases. The arthropods include insects, spiders, mites, and other animals with jointed legs (Fig. 9-24). Houseflies are major carriers of disease. They walk on infected people or objects and then carry the disease on their feet to food and humans.

You likely know about the role that mosquitoes play in the spread of diseases. Three serious diseases, malaria, encephalitis, and yellow fever, are spread by mosquitoes.

The tsetse fly spreads African sleeping sickness to countless thousands of people every year. Typhus fever (not typhoid fever) is spread by the human body louse. It is believed that this disease was a major factor in the downfall of Napoleon's army in 1812. The louse became established because of unsanitary conditions. It then carried the disease from soldier to soldier.

Defenses Against Diseases

The human body has three general methods of fighting diseases. First, it has certain structures that prevent the entry of bacteria. Second, it contains certain cells that attack bacteria. Finally, it produces substances called antibodies that act against diseases.

Structural defenses

The **skin** is one of the body's main defenses against bacteria. It is almost impossible for most bacteria to get through the skin. Some microorganisms do get through the pores and hair follicles. However, human perspiration (sweat) contains salt and acids that kill many organisms that attempt to enter the body in this manner.

Many bacteria enter the body through the breathing, digestive, and reproductive systems. All these systems are lined with mucous membrane cells. These cells secrete a slimy material called **mucus** that traps bacteria. Mucus contains enzymes that dissolve cell walls of many types of bacteria.

The breathing system is also lined with hair-like projections. These projections sweep mucus-covered bacteria, dust, and other foreign material up the throat. Eventually these substances are expelled by coughing or are swallowed. In the stomach, bacteria are killed by the acid in the gastric juice.

Further structural defenses include tears. These wash bacteria from the eyes into the tear ducts. The tear ducts empty into the back of the nasal passage. Tears contain enzymes that dissolve some bacterial walls. Many species of bacteria that survive the effect of tears are attacked by the enzymes in the mucus of the nasal passage.

Cellular activity

Some bacteria get past the structural defenses of the body. However, once they get into the body, they are normally attacked by special cells called **phagocytes**. These cells engulf the bacteria by phagocytosis (See Section 6.9, page 133). Then enzymes in the phagocytes dissolve the cell walls of the bacteria.

Some phagocytes are blood cells called **leucocytes**. These small cells are able to leave the circulatory system by passing through the thin walls of the capillaries. They travel through the fluid that bathes body tissues to the site of an infection. There they begin to attack the infectious organisms. The resulting "battle" leaves a mass of destroyed pathogens, dead leucocytes, and blood serum. This mixture is called **pus**. You have likely seen pus form at cuts and other breaks in your skin.

You may also have noticed that infected areas often turn red and swollen. This is caused by the extra blood that flows to the infected area to combat the pathogens. Often pus does not form as the leucocytes fight the pathogens. Destroyed bacteria and leucocytes do not build up. A fluid in the blood, called **lymph**, picks up the bacteria and leucocytes. It then carries them to **lymph ducts**. At certain points along the lymph ducts are **lymph nodes** where bacteria and leucocytes are removed before the lymph returns to the blood system. Serious infections often cause swelling of the lymph nodes near the infection.

You may have had swelling of the lymph nodes in your armpit due to a cut in your hand. More likely, you may have had swelling of the lymph nodes in your neck region as a result of infections of the nose and throat. The tonsils and adenoids are a mass of lymph nodes. They often become so swollen that they must be removed.

Throughout the body are other cells that help the leucocytes combat pathogens. These cells are found in many tissues such as the lungs, spleen, bone marrow, and liver. Like leucocytes, these cells also destroy pathogens by phagocytosis.

A common symptom of infection is a **fever**. The fever may be caused by substances released by the pathogen. It may also be caused by the body's reaction to the pathogen. In either case, the higher temperature stops or slows down the growth of many species of bacteria. Also, it speeds up other body reactions, such as antibody production, that fight infections. A low fever is often helpful in fighting infections. However, body cells can be destroyed along with pathogens if the fever is too high or lasts too long.

Antibody production

Foreign proteins are not readily accepted by the human body. Bacteria, their toxins, and other pathogenic microorganisms contain proteins. Therefore, when they enter the body, the body cells react defensively. Foreign proteins that cause such reactions are called **antigens**. Pollen grains and red blood cells from a different person are two common antigens that affect many people.

When an antigen gets into the body, it causes the spleen and lymph nodes of the host to produce a special protein called an **antibody**. The antibody eventually reacts with the antigen and destroys it. For example, diphtheria bacteria release a toxin (the antigen) when they get into the human body. This toxin causes the production of an antibody, called an **antitoxin**, that destroys the toxin. A host shows the symptoms of the disease when antibody production does not keep up with antigen production.

Immunity

Immunity *is the ability of an organism to oppose the pathogens that can cause an infectious disease.* Immunity may be inherited or it may be acquired. **Inherited immunity** *is that which is present in the individual from birth.* Inherited immunity is also called **species immunity**, since all members of a species will normally be immune to the same diseases of other species. Humans have inherited immunity to most diseases of plants and animals other than humans. For example, the pathogens that cause running eyes in cats simply cannot live in the human body. However, we do not have immunity to all diseases that infect animals. Two serious diseases, rabies and tuberculosis, are examples of these.

Acquired immunity *is that which is developed during the individual's life.* Acquired immunity is classified as active or passive, depending on how it was obtained.

Active immunity

In this type of immunity, the individual makes the required antibodies to fight a disease. This type of immunity may be naturally acquired or artificially acquired.

Naturally acquired active immunity may be present in a host that has recovered from a certain disease. The disease (an antigen) entered the host's body. The body cells produced an antibody to fight the antigen. Often, some of this antibody remains with the host. The cells often continue to produce it, even after recovery from the disease. In such cases permanent, or lifetime, immunity from that disease often results. Diphtheria, mumps, scarlet fever, and red measles produce permanent naturally acquired active immunity in humans.

Artificially acquired active immunity may be given to an individual with a vaccine. *A* **vaccine** *consists of dead or attenuated (weakened) microorganisms or their toxins.* The vaccine (an antigen) is introduced into the individual to be immunized. The vaccine is too weak to cause the disease. However, it does stimulate the production of antibodies.

The Salk polio vaccine consists of polio viruses that have been killed. The vaccines for whooping cough and typhoid fever are made from killed bacteria. The Sabin oral polio vaccine consists of attenuated (weakened) polio viruses. If the vaccine consists of weakened toxins instead of dead or weakened microorganisms, it is usually called a **toxoid**. Toxoids are used to create artificially acquired active immunity against diphtheria, tetanus, and scarlet fever.

After an antigen enters a person, whether it was naturally or artificially introduced, several weeks must pass before immunity results.

Passive immunity

In this type of immunity, the antibodies themselves are introduced directly into the person. This type of immunity may also be naturally or artificially acquired.

Naturally acquired passive immunity against some diseases is often present in a new-born child. While the child is still within its mother, it receives antibodies from the mother's blood. A mother's milk may also give certain antibodies to a child for the first few days. This type of immunity usually lasts for less than a year.

Artificially acquired passive immunity is given to an individual with an injection of antibodies that have been produced by an infected host. Since the person does not have to produce the antibodies, immunity is immediate. However, it usually lasts for just a few weeks. This type of immunization is used for people who have come in contact with infected people and require immediate protection.

The substance used to create artificially acquired passive immunity is called a **serum**. This is the part of the infected host's blood that contains the antibody. Diphtheria serum can be prepared as follows. Diphtheria bacteria (an antigen) are injected into a horse. The horse produces an antibody, in this case an antitoxin, that can destroy the diphtheria toxin. Some of the horse's blood is removed. It is then filtered to remove bacteria, corpuscles, and other particles. Only the blood serum and antitoxin remain. This serum is then injected into the person requiring immediate immunity.

Antibiotics

An **antibiotic** *is a substance formed by a living organism that prevents or slows the growth of microorganisms.*

You read in Chapter 1 about how Sir Alexander Fleming discovered penicillin, the first antibiotic, in 1929. It is produced by a mould called *Penicillium notatum.* Although Fleming discovered that it destroyed many types of bacteria, it was not used to treat diseases until World War II, 1939-45.

Streptomycin is another widely used antibiotic. It is formed by a soil organism called *Streptomyces griseus.* This antibiotic is quite effective against tuberculosis, dysentery, some types of pneumonia, and many other diseases. Recently, more effective antibiotics have replaced it for most diseases except tuberculosis. Terramycin, aureomycin, neomycin, erythromycin, and chloromycetin are among the many antibiotics in use today.

Terramycin is called a **broad spectrum** antibiotic. It affects a wide range of microorganisms with few side-effects on the human body.

Fig. 9-25 A bioassay to determine the best antibiotic to use in the treatment of a certain infection. In this case, chloromycetin (C) proved most effective. The other antibiotics tested were neomycin (N), terramycin (T), aureomycin (A), erythromycin (E), penicillin (P), and streptomycin (S).

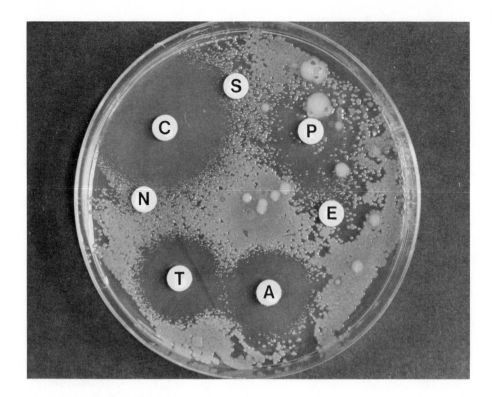

Care must always be taken in the use of antibiotics. Many people have taken overdoses of antibiotics that killed the normal bacteria in their digestive systems. Severe disorders result, some of which end in death. Because of this problem, non-prescription throat lozenges and skin ointments that contain antibiotics have been banned.

Extended use of a particular antibiotic tends to result in strains of bacteria that are resistant to that antibiotic. Penicillin was over-used when it was first available to the public. For many years it effectively controlled staphylococcus bacteria. Then some strains of these bacteria developed that were resistant to penicillin. Erythromycin will control most of those. But, sooner, or later, strains will develop that are resistant to erythromycin.

Antibiotics must only be taken as prescribed by a doctor! Dangerous side-effects can result from improper use of antibiotics.

Doctors often perform a **bioassay** to decide what antibiotic to prescribe for an infection (Fig. 9-25). A sterilized culture medium is streaked with a swab of the infected area. Then antibiotic discs, each one containing a different antibiotic, are placed on the culture. If the bacterium fails to grow near a certain disc, the doctor knows that the antibiotic on that disc is effective against that bacterium.

Sulfa Drugs

A family of chemicals called **sulfa drugs** is used to destroy pathogens. Two common sulfa drugs are sulfadiazine and sulfanilamide. The sulfa drugs were introduced in the 1930s. They saved countless lives before most of them were replaced by the more effective antibiotics. However, sulfa drugs are still the best treatment for most diseases of the excretory system.

Disinfectants and Antiseptics

A **disinfectant** is a chemical that kills microorganisms on contact. Disinfectants are very strong and cannot be used on human flesh. Lye, carbolic acid (phenol), and formaldehyde are good disinfectants. Solutions of these substances are often used to scrub floors of dairy barns and hospitals.

An **antiseptic** is a chemical that slows down the growth of microorganisms. It may not kill them. Solutions of ethyl alcohol and boric acid are antiseptics. Very dilute solutions of phenol may also be used as an antiseptic. Antiseptics can be used on human flesh.

Review

1. **a)** What is a pathogen?
   b) What is a toxin?
2. **a)** Describe two examples of diseases in which the toxins produce symptoms far from the site of the infection.
   b) Give two examples of diseases in which the toxins do their damage at the site of the infection.
3. **a)** Name several diseases that are caused by airborne bacteria and viruses.
   b) Describe how a waterborne disease such as typhoid fever may travel from an infected person to a new host.
4. **a)** Name two skin diseases that are transmitted by contact.
   b) Name two venereal diseases that are transmitted by contact.
5. Describe how a wound disease such as tetanus can become established in a human.
6. **a)** What is an immune carrier?
   b) Describe the role of arthropods in the spreading of human diseases.
7. Describe how the skin and mucus help to prevent the entry of pathogens into the body.
8. **a)** Describe how leucocytes combat microorganisms.
   b) What is pus?
   c) What causes swellings of the lymph nodes?
9. **a)** What is an antigen?
   b) What is an antibody?
   c) What is an antitoxin?
10. **a)** What is immunity?
    b) What is inherited, or species, immunity?
    c) What is acquired immunity?
11. **a)** Distinguish between active immunity and passive immunity.
    b) Distinguish between naturally acquired and artificially acquired active immunity.
    c) Distinguish between naturally acquired and artificially acquired passive immunity.

12. **a)** What is a vaccine?
    b) What is a toxoid?
    c) What is a serum?
13. **a)** What is an antibiotic?
    b) Name five commonly used antibiotics.
    c) Describe two dangerous consequences of prolonged use of an antibiotic.
14. **a)** What is a bioassay?
    b) How is a bioassay performed?
15. **a)** Organ transplants are regularly performed today. For example, people often receive kidneys from other people. However, unless the donor (giver) is an identical twin of the receiver, the receiver's body may reject the organ. Why?
    b) After an organ transplant, the receiver is often given drugs that lower the body's natural immunity to antigens. Why is this done? What dangerous consequences can this have?
16. Polio and smallpox are caused by viruses. Viral diseases are not helped by either sulfa drugs or antibiotics. Can anything be done to fight either disease once symptoms appear? What is the best protection against these diseases?

9.13 Investigation

Bacteria and Other Microorganisms in Your Environment

In this investigation your class will culture (grow) microorganisms from several different sources that you are exposed to on most days.

Materials

petri dish containing sterilized medium (environment)
incubator

CAUTION: Treat all the petri dishes as though they contained pathogens. Never remove the tops after the medium has been inoculated. Follow these and your teacher's directions closely.

Procedure A   Preparation of the Medium

This procedure will have been carried out by your teacher or a small group in your class prior to this day.

a. *Sterilization of equipment.*   All equipment to be used in this procedure must be sterile, or free from microorganisms. Wash the tongs, flasks, and petri dishes using a detergent. Rinse everything several times with running water. Autoclave the equipment at 170°C to 200°C for at least 15 min. If you do not have an autoclave or oven for this purpose, boil the equipment in water for about 30 min.

b. *Preparation of the medium.* Use sterilized nutrient agar, if possible. The bottles of this mixture need only be heated to liquefy the mixture. After heating the bottles, pour enough of the mixture into each sterilized petri dish to half fill it. Cover each dish with a sterilized lid. Powdered nutrient agar is cheaper to use. If you prefer to use it, follow the directions on the package.

c. *Sterilization of the medium.* Place the petri dishes containing the nutrient agar in an oven at a temperature of 110°C to 120°C for about 15 min. Let the nutrient agar cool. The culture medium is now ready for use. It should be a clear, amber jelly.

Procedure B   Inoculation and Incubation

a. Obtain one of the petri dishes containing sterilized nutrient agar from your teacher.

b. Inoculate the medium by doing one of the following:
   1. Use a sterile swab to transfer some dust from the window ledge to the medium.
   2. Let a living insect walk across the medium.
   3. Expose the medium to room air for 5 min. Different students should try different rooms—the classroom, the gymnasium, the locker room, the lunchroom, the corridor of the school.
   4. Drag a coin over the medium.
   5. Touch the medium with the tips of your fingers. Do the same with a second medium after washing your hands thoroughly.
   6. Use a sterile swab to transfer some water from an aquarium to the medium.

CAUTION: As soon as you have inoculated the medium, put the top back on. Seal it in place with transparent tape. Never remove the top after this time.

c. Incubate the culture as follows. Place the inoculated medium in an incubator at 25°C to 35°C. Place it upside down. This keeps the moisture in the medium. Look at your culture every day for the next 3-4 d. Make careful notes on the shape, size, structure, texture and colour of any colonies that appear.

Discussion

1. How large do you think each colony was when it first started?
2. By what process did the colonies grow?
3. Are conditions in the dish ideal for the growth of microorganisms such as bacteria? How do you know?
4. What procedure produced the largest population of microorganisms? Why is this so?
5. What procedure produced the largest number of different kinds of microorganisms? Why is this so?
6. Mould colonies will generally be fuzzy and larger than bacterial colonies. Which procedures produced mould cultures?

Highlights

The kingdom Monera contains two phyla of organisms. The blue-green algae make up the phylum Cyanophyta and the bacteria make up the phylum Schizophyta.

All monerans are unicellular. Their cells lack nuclei and internal membranes. Monerans reproduce mainly by asexual means.

Blue-green algae contain chlorophyll. They are producers in many food chains. They are also important producers of oxygen and fixers of nitrogen.

Pathogenic bacteria harm humans by releasing toxins, or poisons.

Diseases may be spread by means of air, water and food. Diseases may also be spread by contact with infected people and contaminated objects. Immune carriers and animal carriers also spread many diseases.

The body combats diseases in three ways. It has structural defenses such as mucus and skin that prevent the entry of many microorganisms into the body. It also has cellular defenses in the form of leucocytes that ingest microorganisms. Finally, the body produces antibodies that destroy antigens (foreign proteins such as bacteria and their toxins). Antibodies that destroy toxins are called antitoxins.

Immunity is the ability of an organism to oppose an infectious disease. Many types of immunity exist. They are summarized in the following chart. Review the appropriate sections if you cannot remember the descriptions of each type of immunity.

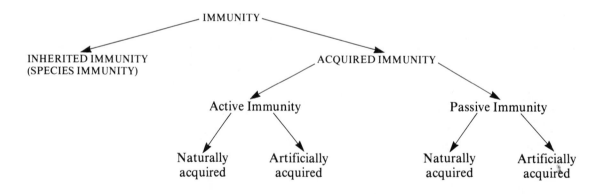

Antibiotics are substances that are formed by living organisms that prevent or slow down the growth of microorganisms. Along with sulfa drugs, these "wonder drugs" have saved millions of lives in the past three to four decades. Disinfectants and antiseptics also play an important role in the control of diseases.

Protists: Protozoans And Algae

10

The kingdom Protista is made up of a diverse group of organisms that are relatively simple in form and function (Fig. 10-1). However, they are more complex in many ways than the organisms of the kingdom Monera that you studied in Chapter 9. Which of the following characteristics make protists more complex, or advanced, than monerans?

Characteristics of Protists

1. Protists are aquatic. They either live in water or in very damp terrestrial (land) habitats.
2. Protists are unicellular or colonial. The colonial species show no differentiation of tissue.

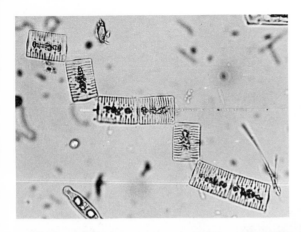
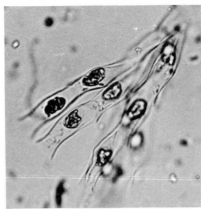
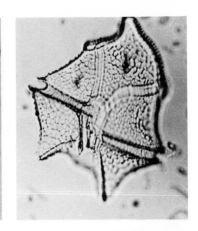

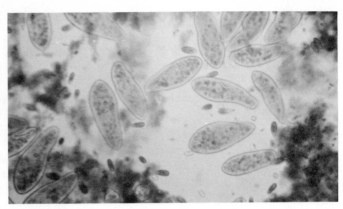

Fig. 10-1 Some representative organisms of the kingdom Protista: *Dinobryon* (A); *Tabellaria* (B); *Peridinium* (C); *Amoeba* (D); *Paramecium* (E).

3. The cells of protists have nuclei.
4. The cells of most protists have internal membranes. In other words, they have membrane-bound organelles such as mitochondria.
5. Many protists can reproduce sexually.

The kingdom Protista consists of two sub-kingdoms:

Sub-kingdom Protozoa (about 20 000 species)
Sub-kingdom Algae (about 11 000 species)

This chapter consists of two main parts. The first part deals with the sub-kingdom Protozoa, or the **Protozoan Protists**. The second deals with the sub-kingdom Algae, or the **Algal Protists**.

PROTOZOAN PROTISTS

Protozoans are important to humans in many ways. Many protozoans serve as food for small animals. In this way, they help to support food chains of which we may be a part. Other protozoans are important in the decomposition of sewage sludge in sewage treatment plants. Many species break down and recycle organic debris in the bottom of ponds and lakes. A few species cause deadly diseases. For example, protozoans are responsible for malaria, African sleeping sickness, and amoebic dysentery.

Protozoans are among the most widely studied organisms. Although they consist of only one cell, that one cell must perform for a protozoan all the functions that your complex body performs. Much has been learned about the functioning of many-celled organisms by first studying protozoans.

10.1 Characteristics of Protozoans

In the two-kingdom system of classification, protozoans are in the kingdom Animalia. In that system they are called the simplest animals. In fact, the word *protozoan* means "first animal".

In the five-kingdom system of classification, the 20 000 species of protozoans are grouped in the sub-kingdom Protozoa of the kingdom Protista. They are put together because they have the following common characteristics:

1. Most protozoans are unicellular. A few are colonial, but each cell in the colony performs all the required life processes. In other words, there is no differentiation of tissue.
2. A few species can change shape. However, most have a relatively fixed shape such as oval, spherical, and many others.
3. Protozoan cells have a distinct nucleus. Some species have more than one nucleus per cell. Organelles are present in the cells of many protozoans. However, the cells are never organized into tissues.
4. Locomotion of protozoans is by means of **pseudopods** (false feet), **cilia** (hair-like projections) or **flagella** (whip-like tails). Most species are motile (able to move). A few are sessile (fixed in position).
5. Some species have protective shells. Many can form protective **cysts** (spores) during unfavourable conditions.
6. Most species are free-living. However, many live in a symbiotic relationship with other organisms.
7. Protozoans undergo asexual reproduction by binary fission. Many species can undergo sexual reproduction.
8. Protozoans may feed on microorganisms such as bacteria, yeasts, algae, and other protozoans. They may also feed on dead organic matter.
9. Most protozoans are microscopic. They are usually measured in **micrometres** (μm). 1μm$=0.000\,001$ m. Some protozoans are only 2 or 3 μm long. Most species are less than 250 μm long. *Spirostomum* (see Fig. 10-3) can be 3 mm long, and *Porospora gigantea* grows to 16 mm in length.

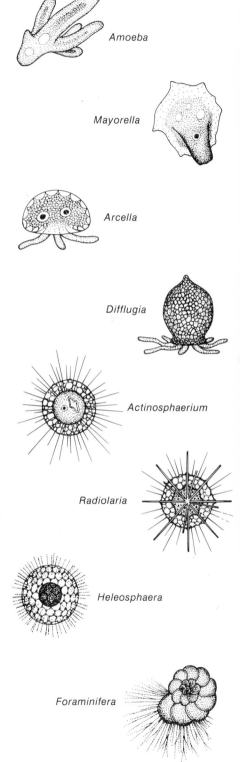

Amoeba

Mayorella

Arcella

Difflugia

Actinosphaerium

Radiolaria

Heleosphaera

Foraminifera

Fig. 10-2 Some representative sarcodinans. What do they have in common?

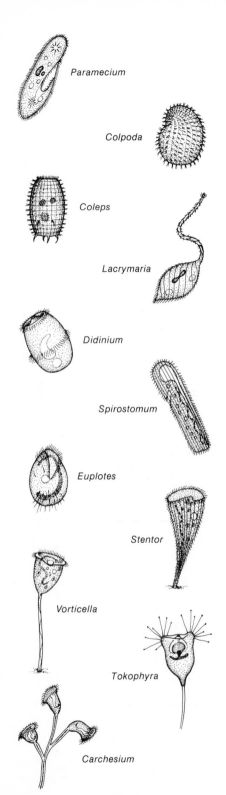

Paramecium

Colpoda

Coleps

Lacrymaria

Didinium

Spirostomum

Euplotes

Stentor

Vorticella

Tokophyra

Carchesium

Fig. 10-3 Some representative ciliates. What do they have in common? How do they differ from the sarcodinans? Which ones appear to be sessile?

Phyla of Protozoa

Phylum Sarcodina (Fig. 10-2)

There are about 8 000 species of **sarcodinans**. Among them is the famous *Amoeba*. This particular sarcodinan has no definite shape. Instead, it continually flows into extensions of its cytoplasm called **pseudopods** (false feet). *The presence of pseudopods is the characteristic feature of all members of this phylum.* Some sarcodinans have pseudopods that are more or less fixed in position.

Sarcodinans move by using their pseudopods. This motion is called **amoeboid movement**.

Some sarcodinans, like *Radiolaria,* have shells made of silica, a component of sand. These have numerous, stiff, long pseudopods sticking out from the shell. Other sarcodinans, like *Foraminifera,* have shells made of calcium carbonate, or limestone. This is the main substance in the shells of clams and oysters.

Many sarcodinans live in ponds, lakes, puddles, and wet terrestrial sites. Most of the types that form shells live in the oceans. A wide variety of amoeba-like sarcodinans are parasites in humans and other animals. One of these is responsible for amoeboid dysentry. This disease is common in areas where sanitation is poor. The disease is caught by eating food that was handled by an infected person.

Phylum Ciliophora (Fig. 10-3)

There are about 5 000 species of **ciliates**. They are among the largest of the protozoa. In fact, several genera, like *Spirostomum,* can be seen with the unaided eye when they are well-illuminated.

The ciliates are covered with hair-like projections called **cilia**. *The presence of cilia is the characteristic feature of all members of this phylum.* The cilia are used to propel the organism through the water in which it lives. They do so by beating in a rhythmic manner.

All ciliates have a definite shape. Many genera, like *Paramecium,* move actively through the water. These have a definite anterior (front) and posterior (rear) end. Some genera, like *Vorticella,* are sessile. They live attached to aquatic plants, wood, stones, leaves, and other submerged objects.

Phylum Sporozoa

There are about 2 000 species of **sporozoans**. As the name implies, *the formation of spores is a characteristic feature of all members of this phylum.* A mature cell reproduces by forming numerous spores. The spores are not true resting spores like those formed by algae, bacteria, and fungi. However, they are similar in many ways.

Members of this phylum have other common characteristics. First, the adults have no method of locomotion. Second, sporozoans do not ingest food particles by phagocytosis as do sarcodinans and ciliates. Instead, they absorb the food in soluble form as do bacteria. Finally, all sporozoans are parasites. This means they depend on a host for food. Some sporozoans have little effect on the host. Others may cause the host to become weak and sick. Still others can cause the host to die. The most dreaded sporozoan is *Plasmodium,* the cause of malaria (Fig. 10-4).

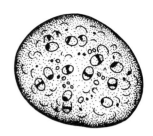

Fig. 10-4 *Plasmodium vivax*, one of four species of the genus *Plasmodium*, that are responsible for malaria. The organisms are shown here in a human blood cell.

Phylum Zoomastigina (Fig. 10-5)

There are about 1 200 species in this phylum. These organisms are commonly called **animal flagellates** or **colourless flagellates**. These names distinguish them from the algal flagellates (euglenoid flagellates and dinoflagellates), both of which contain chlorophyll (see Section 10.7).

These organisms move by lashing their **flagella**. *The presence of flagella and the absence of chlorophyll are characteristic features of all members of this phylum.*

Many animal flagellates are found in fresh water, salt water, and soil. Many species feed by ingesting bacteria and other microorganisms smaller than themselves. Some species feed on dead organic matter.

Many species of animal flagellates live in symbiotic relationships with other organisms. Some species are parasitic. For example, *Trypanosoma* parasitizes humans, causing African sleeping sickness. Other species are mutualistic. (**Mutualism** *is a relationship between two organisms in which both organisms benefit.*) For example, one species of animal flagellate lives in the digestive tract of termites. They digest the wood that the termites eat. Termites cannot digest wood. Therefore they benefit from the digestive action of the flagellates. In return, the flagellates benefit from the protection, moisture, and food provided by the termites.

Review

1. **a)** What two sub-kingdoms make up the kingdom Protista?
   b) What characteristics do all protists have in common?
2. **a)** In what ways are protozoans important to humans?
   b) Summarize the common characteristics of protozoans.
3. There are four main phyla of protozoans. For each of these phyla:
   a) State the name of the phylum.
   b) Give the number of species in the phylum.
   c) State the characteristic feature(s) of the species in that phylum.
   d) State the habitat(s) in which most of the species live.
4. **a)** What are pseudopods?
   b) What are cilia?
   c) What are flagella?
5. Parasitism and mutualism are two types of symbiosis. How are they alike? How are they different?

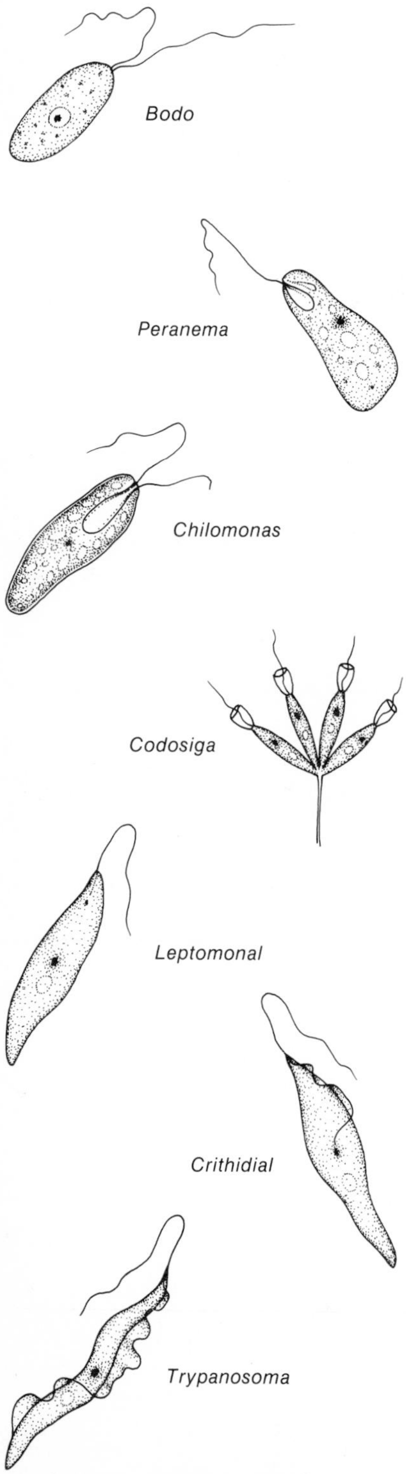

10.2 Investigation

Protozoan Diversity

Wherever there is water you can find protozoans. They live in oceans, lakes, ponds, and streams. They even live in extreme environments such as stagnant puddles, sewage lagoons, hot springs, and icefields.

Protozoans from these varied environments have much in common. Yet they also differ in many ways. In this investigation you will examine many species of protozoans. Note how they are alike and how they are different.

Materials

compound microscope	prepared stained slides of several
microscope slides (2)	species of protozoans
cover slips (2)	large beaker (over 500 mL)
paper towel	jar
lens paper	ONE OR MORE OF:
medicine dropper	mixed protozoan culture (living)
1.5% methyl cellulose solution	hay infusion
vital stain	pond water culture

Procedure A   Examination of Prepared Slides

a. Read carefully the characteristics of protozoans that are listed in Section 10.1, page 245.

b. Examine prepared slides of as many species of protozoans as possible. For each species, sketch the shape of the cell and draw in any prominent features. Label as many of the following parts as you can find: plasma membrane, cytoplasm, nucleus, pseudopods, cilia, flagella, protective shells.

Procedure B   Examination of a Mixed Protozoan Culture

Perform this procedure if your teacher provides you with a culture that contains several species of protozoans.

a. Use the medicine dropper to obtain a drop of the culture as directed by your teacher.

b. Prepare a wet mount of the culture, but do not add a cover slip.

c. Observe the culture under low power. If the protozoans are moving too quickly, add a drop of 1.5% methyl cellulose solution. It will slow them down without killing them.

d. Add a cover slip. Switch to medium power and then to high power and study each species closely.

Bodo

Peranema

Chilomonas

Codosiga

Leptomonal

Crithidial

Trypanosoma

Fig. 10-5   Some representative animal flagellates. How do they differ from other flagellates in the kingdom Protista? What do animal flagellates have in common? How do they differ from ciliates?

e. For each species, sketch the shape of the cell and draw in any prominent features.

f. For each species, describe carefully its method of locomotion.

g. Make careful notes on any interesting behaviour that you see. This could include feeding, interactions with other organisms, and response to contact with non-living material.

h. Add a drop of vital stain to one edge of the cover slip. Draw it under the cover slip by using the paper towel.

i. Some cell structures will be made more visible by this stain. Add these structures to your sketches.

Procedure C   Study of a Hay Infusion

A hay infusion is a rich source of protozoans. If time permits, your teacher may ask you to prepare and study a hay infusion as follows.

a. Half fill a large beaker with dry grass (hay) that has been cut into pieces that are 1 cm long or shorter.

b. Cover the grass with water.

c. Boil this mixture for about 30 min. Add more water as it is needed. This will form a hay "tea" that is called an **infusion**.

d. Allow the mixture to cool. Then add a few pinches of cut grass (Fig. 10-6).

e. Inoculate the infusion with a little sludge from the bottom of a jar of pond water.

f. Let the infusion stand for several days until it turns cloudy. The cloudiness is probably due to the presence of bacteria. Most species of protozoans either feed on bacteria or on protozoans that eat bacteria. Therefore you will soon have a plentiful supply of protozoans in the surface scum and the bottom debris. Your culture will develop most rapidly if both heat and light are moderate. The bacteria may not grow well at extreme temperatures or in bright light.

g. Use the medicine dropper to obtain a drop of the surface scum or bottom debris.

h. Prepare a wet mount and study the protozoans present as described in steps b to i of Procedure B.

Fig. 10-6   A hay infusion can be prepared by adding chopped dry grass (hay) to water, followed by boiling for 30 min.

Procedure D   Study of a Pond Water Culture

In addition to or instead of the hay infusion described in Procedure C, your teacher may ask you to prepare and study a pond water culture.

a. Fill a jar with pond water.

b. Add some filamentous algae (pond scum) and other aquatic plants. Also, include a little bottom debris such as dead leaves, twigs, and plants.

c. Let the jar stand in moderate light at room temperature for several days. The algae and other plants will likely die and settle to the bottom. The resulting sludge usually contains numerous protozoans.

d. Use the medicine dropper to obtain a drop of the sludge.

e. Prepare a wet mount and study the protozoans present as described in steps b to i of Procedure B.

Discussion

1. Your diagrams and laboratory notes are your writeup for this investigation. Make sure they are complete.

10.3 The Amoeba

A wide diversity of protozoans exists. Therefore we cannot possibly describe in detail representatives of the many different species. However, it is possible to obtain a fairly good idea of how protozoans function by closely studying two organisms. Those two organisms are the sarcodinan *Amoeba* and the ciliate *Paramecium. Amoeba* represents the simplest protozoans. *Paramecium* represents the most complex protozoans.

The common amoeba, *Amoeba proteus,* lives in clean freshwater that contains abundant plant life. It is about as simple as any organism could be. It is an independent cell with a nucleus and cytoplasm. However, it has no permanent organelles. It seems to be life near the most basic level. As a result, it has been the subject of much study by biologists who are seeking an answer to the question, "What is life?"

In spite of its simple structure, the amoeba can move, reproduce, capture and ingest food, digest food, egest waste solids, metabolize, excrete wastes, respire, and respond to stimuli. It can do almost everything you can do. Yet it does these things without all the specialized organs that you have. Let us see how this is possible.

Structure of an Amoeba

An amoeba is a mass of colourless jellylike protoplasm (Fig. 10-7). Some species are as long as 0.6 mm. However, most are much smaller than this. The amoeba does not have a fixed shape. It changes shape by sending out extensions of its protoplasm called **pseudopods** (false feet).

The **cytoplasm** consists of two layers. Just inside the **plasma membrane** is a clear watery layer called the **ectoplasm**. Inside the ectoplasm is a more dense, granular region called the **endoplasm**. It makes up most of the cell's cytoplasm.

The endoplasm itself has two layers. The outer layer, the **plasmagel**, is stiff and does not flow easily. The inner layer, the **plasmasol**, is more fluid. You can watch it flow with a microscope.

The endoplasm contains the **nucleus**. The nucleus is not easily seen in a living non-stained cell. The endoplasm also contains a spherical fluid-filled **contractile vacuole** and one or more **food vacuoles**. The food vacuoles

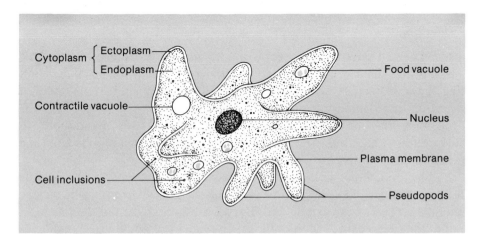

Fig. 10-7   Structure of *Amoeba*.

Cytoplasm { Ectoplasm—
Endoplasm—

Contractile vacuole—

Cell inclusions—

Food vacuole

Nucleus

Plasma membrane

Pseudopods

contain food in various stages of digestion. Finally, the endoplasm contains a wide range of **cell inclusions**, such as other vacuoles, oil droplets, and tiny solid particles.

Locomotion

An ameoba moves by extending temporary **pseudopods** (see Fig. 10-7). These can form at any place on the surface of the organism. An amoeba cannot swim. In order to move, it must be on a solid surface.

Although many theories have been developed to explain the movement of an amoeba, biologists admit that they are still not sure exactly how it occurs. However, they do know that plasmagel and plasmasol can change quickly from one to the other. It is this change that is responsible for the movement. Plasmasol (the fluid portion) flows in the direction in which a pseudopod is to form (Fig. 10-8). As it spreads out at the tip of the pseudopod, it changes to plasmagel (the stiff portion). This gives the anterior end shape and rigidity. At the same time, plasmagel at the posterior end changes to plasmasol and moves forward into the developing pseudopod. Whether it is pushed forward or pulled forward is a matter of debate.

Although this method of locomotion is difficult to explain, it is so common that it is given a name—**amoeboid movement**. Many protozoa and some types of vertebrate white blood cells show amoeboid movement.

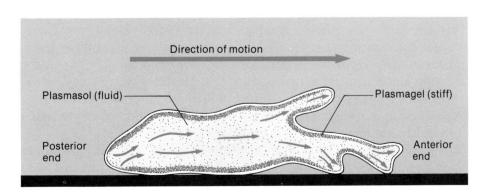

Fig. 10-8   A side view of amoeboid movement. Plasmagel changes to plasmasol at the posterior end. Plasmasol changes to plasmagel at the anterior end.

Direction of motion

Plasmasol (fluid)—

Plasmagel (stiff)

Posterior
end

Anterior
end

STEP 1
Amoeba senses food.

STEP 2
Pseudopods extend toward food.

STEP 3
Pseudopods surround food.

STEP 4
Food is engulfed by pseudopods; enters cell by phagocytosis.

STEP 5
Food vacuole formed; it circulates through cytoplasm as digestion occurs.

STEP 6
Indigestible material is left behind.

Fig. 10-9   Feeding by an amoeba. Less than 10 min is the average time for the steps shown here.

Feeding

The amoeba feeds on microscopic algae, other protozoans, microscopic animals and dead organic matter. Its preference seems to be ciliates and flagellates. In fact, several paramecia can often be seen in various stages of digestion in an amoeba.

Once an amoeba has sensed its food, it extends pseudopods to surround it (Fig. 10-9). The food is engulfed by the pseudopods and then taken in, or **ingested**, by phagocytosis.

The food, along with some water, becomes a **food vacuole**. The vacuole moves through the cell as the endoplasm moves. Enzymes from the endoplasm enter the vacuole and digestion begins. Digested food diffuses out of the vacuole into the endoplasm. The vacuole gradually gets smaller as **digestion** continues. Undigested substances are eliminated, or **egested**, anywhere on the plasma membrane.

Reproduction

The amoeba reproduces by **binary fission** (see Fig. 8-9, page 181). The organism must reach a certain size before fission can occur. Once it has reached this size, the cell becomes almost spherical, with many short pseudopods on its surface. Then the nucleus divides by mitosis. As the two daughter nuclei move apart, the plasma membrane forms a constriction across the centre of the cell. Eventually the constriction divides the protozoan into two daughter cells. Each daughter cell resumes activity and growth.

Clearly, an amoeba is almost immortal. It never dies of old age. Unless it dies by an accident such as the drying up of the pond, its protoplasm becomes the next generation.

Under unfavourable conditions, many species of amoeba can form **cysts**. The protoplasm of the cell withdraws into a spherical shape. Then a thick protective covering forms around the cell. When favourable conditions return, the cyst splits open and the amoeba resumes activity.

Gas Exchange and Excretion

Like most living things, the amoeba requires oxygen for cellular respiration. This gas enters the amoeba by diffusion across the plasma membrane. Oxygen diffuses into the amoeba when the concentration of oxygen is greater in the surrounding water than it is in the amoeba.

The amoeba produces carbon dioxide as a result of cellular respiration. This gas is excreted from the organism by diffusion. Carbon dioxide diffuses out of the amoeba when the concentration of carbon dioxide is greater in the amoeba than it is in the surrounding water.

The metabolism of nitrogen-containing compounds such as amino acids produces poisonous wastes such as urea. These are also excreted from the organism by diffusion.

The **contractile vacuole** plays a minor role in the excretion of carbon dioxide and other metabolic wastes. Its main function is to pump excess water from the amoeba. Of course, this water will contain some metabolic waste.

The vacuole begins as a tiny sphere. It gradually grows as it fills with water. When it reaches a certain maximum size, it moves to the plasma membrane. Then it bursts, expelling its contents outside the amoeba.

The water in which an amoeba lives is a hypotonic solution (see Section 6.6, page 128). Therefore, a net diffusion of water into the amoeba is always occurring. The actions of the contractile vacuole prevent this water from destroying the cell.

Behaviour

For such a simple organism, an amoeba has a surprisingly complex behaviour. Like all other living things, the amoeba responds to stimuli. It usually does so for one of two reasons: to protect itself from harm or to lead itself to a source of food. *Response to stimuli by an organism is called* **behaviour.** A response is called *positive* if the organism moves *toward* the stimulus. It is called *negative* if the organism moves *away from* the stimulus.

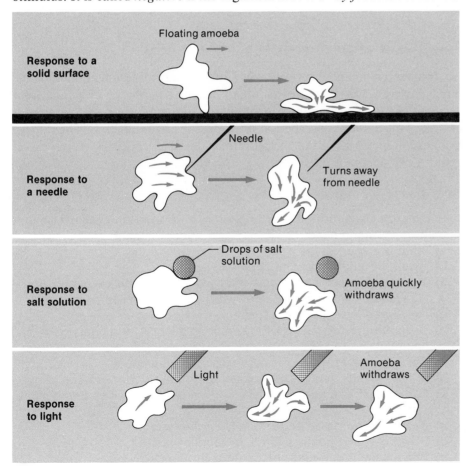

Fig. 10-10 Response to stimuli of an amoeba.

Response to touch

You already know that an amoeba responds positively to contact with food by engulfing it with pseudopods. An amoeba that is floating in water responds positively to a solid surface by attaching itself to it (Fig. 10-10). An amoeba on a solid surface responds negatively to a sharp object by withdrawing from it and moving away.

Response to chemicals

An amoeba responds positively to chemicals diffusing from foods. Once it senses these chemicals, the amoeba begins to move toward the food. In contrast, an amoeba responds negatively to harsh chemicals such as salt or acids. If a drop of salt solution is placed in contact with an amoeba, the amoeba will withdraw from it.

Response to light

An amoeba shows little or no response to a gradual change in light intensity. However, it responds negatively to a sudden bright light by moving away from it.

Response to temperature change

An amoeba responds negatively when its normal environmental temperature is lowered. For example, its rate of feeding and its rate of locomotion decrease. At 0°C, feeding and locomotion stop completely.

An amoeba functions best at a temperature around 20°C. The rate of locomotion increases as the temperature is raised above 20°C. However, at about 30°C, locomotion stops.

Review

1. Summarize the structure and function of an amoeba in a table. Use the following column headings: Part; Description; Function.
2. a) What is a pseudopod?
   b) Give a description of amoeboid movement.
   c) The plasmagel of the endoplasm of an amoeba is thick. It cannot flow. How, then, can an amoeba carry out amoeboid movement?
3. a) On what does an amoeba feed?
   b) Describe the processes of ingestion, digestion, and egestion in an amoeba.
4. a) Describe the process of binary fission in an amoeba.
   b) Is this process sexual or asexual reproduction? How do you know?
   c) Part of the first amoeba that ever lived is still living today. Explain this statement.
5. a) Describe how an amoeba obtains needed oxygen.
   b) How does an amoeba excrete metabolic wastes?
   c) What is the function of the contractile vacuole?
   d) Describe how the contractile vacuole operates.

6. Briefly summarize in a table the response of an amoeba to the following stimuli: contact with food, sensing food nearby, contact with a solid surface, contact with a sharp point, a bright light, a harsh chemical, a decrease in temperature below 20° C, an increase in temperature above 20° C.

7. If an amoeba is cut into two pieces, the plasma membrane quickly surrounds each piece. The part without the nucleus can still ingest food and move. However, it cannot digest or assimilate food. It soon dies. The part with the nucleus grows and reproduces. A nucleus by itself cannot survive. What does this information tell you about the role of the nucleus in the cell? Can the nucleus perform this role unaided? How do you know?

8. a) If an amoeba is placed in a hypertonic solution, its contractile vacuole disappears and does not reappear. Why?
   b) A marine (saltwater) amoeba has no contractile vacuole. Why?
   c) A marine amoeba develops a contractile vacuole if it is placed in fresh water. Why?

10.4 The Paramecium

The genus *Paramecium* is in the phylum Ciliophora. In other words, the paramecium is a **ciliate**. Like all ciliates, its surface is covered with many hairlike projections called **cilia**. These are used for locomotion and feeding.

The paramecium is one of the most common and widely distributed unicellular organisms. It prefers stagnant water such as a quiet pond. It often forms a scum on such water. You probably noticed this in Section 10.2.

The amoeba is one of the simplest protozoans. In contrast, the paramecium is one of the most complex protozoans. Like the amoeba, the paramecium is unicellular. Unlike the amoeba, the paramecium has a definite shape. It also uses division of labour within its cytoplasm. That is, certain sections of the cytoplasm carry out certain functions, such as feeding, on behalf of the whole organism.

The paramecium and other ciliates seem to be as complex as any unicellular organism could be. Yet, with all this complexity, can they perform any functions that the simple amoeba cannot perform? Does this complexity offer any advantage to the ciliates? Keep these two questions in mind as you study this section.

Structure of a Paramecium

The paramecium has a definite shape (Fig. 10-11). It is blunt, or rounded, at the anterior (front) end and pointed at the posterior (rear) end. The outer surface is covered with an elastic membrane called the **pellicle**. This membrane is stiff enough to give the organism a definite overall shape, yet flexible enough to allow small changes. The pellicle is covered with numerous hairlike projections called **cilia**.

Just inside the pellicle is a thin clear layer of cytoplasm called the

Fig. 10-11   Structure of *Paramecium*.

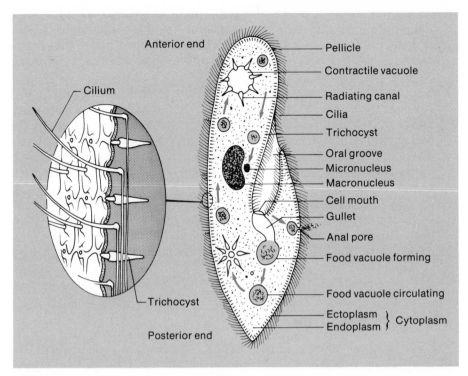

An indentation called the **oral groove** begins near the anterior end and runs about half way along the organism. It has a **cell mouth** at its posterior end. The cell mouth opens into a **gullet**. Near the gullet is the **anal pore**. This opening in the pellicle can be seen only when the animal is discharging waste.

The endoplasm contains a large nucleus called the **macronucleus** and a smaller nucleus called the **micronucleus**. It also contains two **contractile vacuoles**, one near each end of the cell. Each contractile vacuole has a central spherical region and several branches called **radiating canals**. The endoplasm also contains several **food vacuoles**.

Locomotion

The cilia beat backward to push the paramecium forward in the water. The cilia beat at an angle. Therefore the animal rotates on, or rolls around, its

Fig. 10-12   Locomotion of a paramecium. This organism follows a spiral path, rotating on its long axis as it does so.

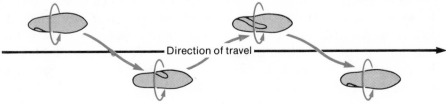

long axis (Fig. 10-12). In addition, the oral groove has more cilia than any other part of the pellicle, and they beat faster. This causes the anterior end to turn in a large circle. The combination of these two rotations causes the animal to move forward in a spiral pattern (Fig. 10-12).

The cilia beat forward to move the organism backward.

Feeding

A paramecium feeds on microorganisms such as bacteria, algae, yeasts, and small protozoans. The beating of the cilia in the oral groove sweeps the food and some water to the cell mouth. From there the food moves into the gullet. At the end of the gullet, a food vacuole is formed that contains food and water. When the vacuole reaches a certain size, it breaks away.

Cytoplasmic streaming (cyclosis) carries the vacuole through the cell on a definite path. It moves to the posterior end first. Then it moves to the anterior end, along the side of the cell that is opposite the oral groove.

As the vacuole moves along this path, enzymes from the cytoplasm enter it. They act on the food to digest it. Digested food is actively transported out of the vacuole into the cytoplasm. Thus the vacuole gets smaller and smaller as it moves along its path. Finally, the vacuole arrives at the anal pore near the oral groove. Undigested waste is expelled from the protozoan through this opening in the pellicle.

Reproduction

Asexual reproduction

Like the amoeba, the paramecium reproduces by binary fission. A paramecium has two nuclei. The larger one, the macronucleus, directs most cell functions. The smaller one, the micronucleus, directs reproduction. Some species of paramecia have two or more micronuclei.

During binary fission, the micronucleus undergoes mitotic division and the daughter nuclei move to opposite ends of the cell (Fig. 10-13). The macronucleus divides by pinching into two parts. Mitosis is not involved in the division. A second gullet forms and two more contractile vacuoles form. Then a constriction forms across the centre of the cell, splitting the cell into

Fig. 10-13 Binary fission of a paramecium. This process is directed by the micro-nucleus.

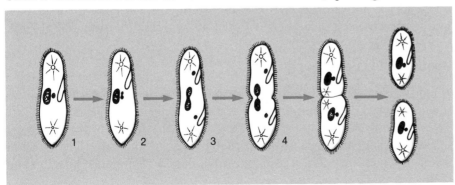

two daughter cells. The daughter cells feed and grow. Under ideal conditions, they will be ready to divide in about 6 to 12 h.

Sexual reproduction

From time to time, paramecia undergo a form of sexual reproduction called **conjugation**. Two individuals join together in the region of their oral grooves. Then they exchange material from their micronuclei. After this exchange, the organisms divide by binary fission.

The micronuclei contain genetic material of the cells. As you know, the mixing of genetic material creates variations in a species. Variations, in turn, tend to increase the chances that the species will survive if environmental changes occur.

Gas Exchange and Excretion

Gas exchange and the excretion of metabolic wastes in a paramecium is similar to an amoeba. Oxygen and metabolic wastes such as carbon dioxide and urea move through the pellicle by diffusion.

A **contractile vacuole** is located at each end of the cell. Although their main function is to pump out excess water, they also eliminate some metabolic waste.

Each contractile vacuole is equipped with several **radiating canals** that extend out into the cytoplasm. They collect water from the cytoplasm and direct it to the contractile vacuole. The contractile vacuoles contract alternately at 10-20 s intervals. Each vacuole expels its contents through a tiny pore in the pellicle. The radiating canals are seen most easily when the vacuole is small.

Behaviour

Response to touch

When a paramecium runs into a solid object, the cilia begin to beat forward. This causes the organism to move backward a short distance. The organism then turns slightly and moves forward again. If it runs into the object again, it repeats the above process. It keeps repeating this pattern until it can pass by the object. Such behaviour is called an **avoiding reaction** (Fig. 10-14).

Response to chemicals

A paramecium responds positively to a weakly acidic medium. It will move toward a weak acid solution such as dilute acetic acid (vinegar) or dilute carbonic acid (carbon dioxide in water). This behaviour is probably related to the fact that paramecia are adapted to living in acidic water. The stagnant water in which they live is usually acidic, due to bacterial respiration and their own respiration.

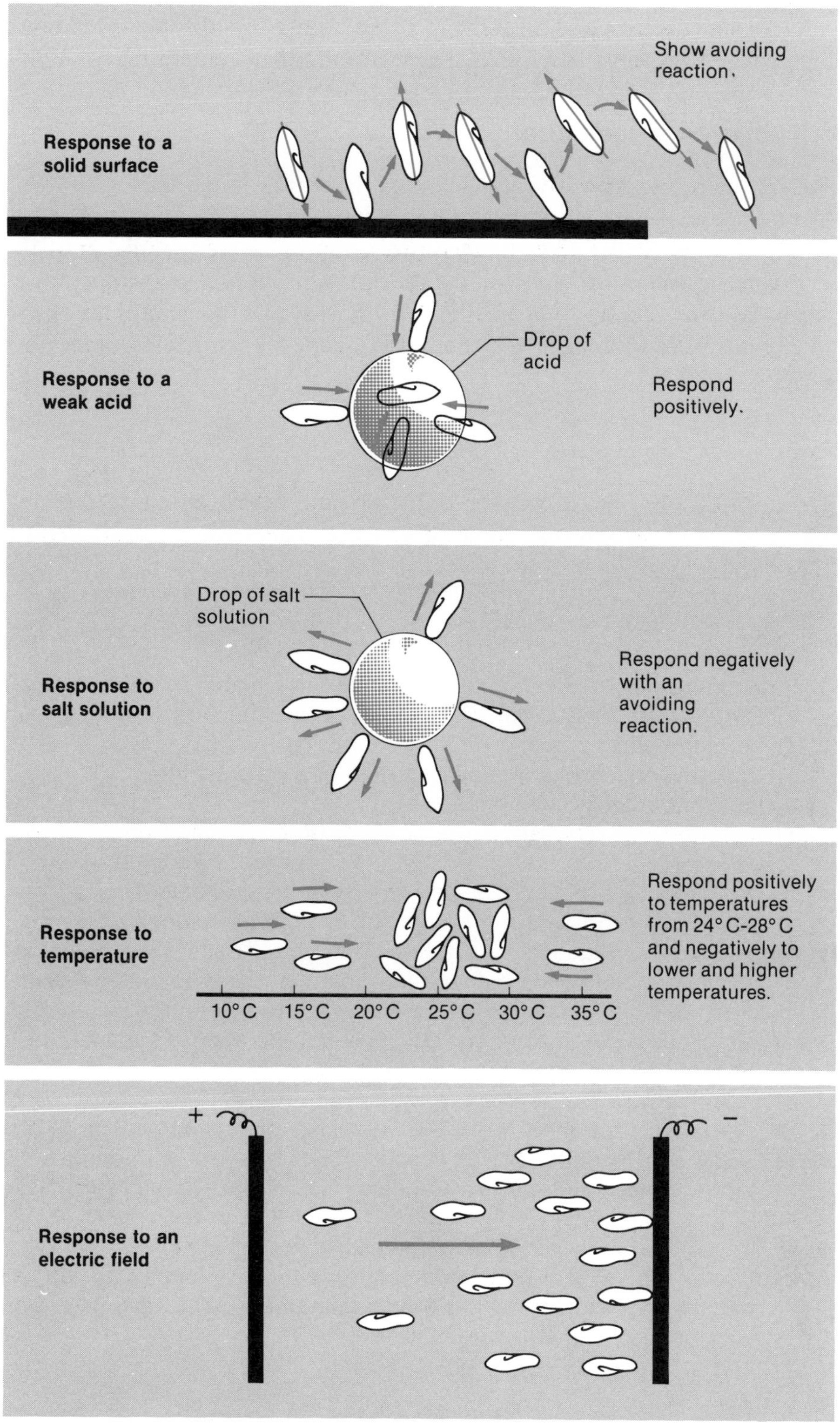

Fig. 10-14 Behaviour of a paramecium.

Response to a solid surface

Show avoiding reaction.

Response to a weak acid

Drop of acid

Respond positively.

Response to salt solution

Drop of salt solution

Respond negatively with an avoiding reaction.

Response to temperature

10° C 15° C 20° C 25° C 30° C 35° C

Respond positively to temperatures from 24° C-28° C and negatively to lower and higher temperatures.

Response to an electric field

+   −

Paramecia respond negatively to solutions of salt and most other chemicals. They show an avoiding reaction and will not enter such solutions.

Response to temperature

Paramecia respond positively to temperatures in the range of 24°C to 28°C. They respond negatively with avoiding reactions to temperatures above and below this range. At high temperatures, motion is very rapid until the organisms escape through avoiding reactions, or die. At low temperatures the organisms also try to escape. However, their metabolic rates may be so low that they cannot move fast enough to escape before they become numb and die.

Response to an electric field

If paramecia are placed between positively and negatively charged plates, they always move toward the negative plate. This indicates that, overall, they carry a positive charge.

Review

1. Summarize the structure and function of a paramecium in a table. Use the following column headings: Part; Description; Function.
2. a) What are cilia?
   b) Describe and explain the locomotion of a paramecium.
3. a) On what does a paramecium feed?
   b) Describe the processes of ingestion, digestion, and egestion in a paramecium.
4. a) Describe the process of binary fission in a paramecium.
   b) Describe the process of conjugation in a paramecium.
   c) Why can conjugation be considered a type of sexual reproduction?
   d) What advantage does conjugation offer compared to binary fission?
5. a) How does gas exchange take place in a paramecium?
   b) How does a paramecium get rid of metabolic wastes such as urea?
6. a) What is the function of the contractile vacuoles?
   b) Describe how the contractile vacuoles operate.
7. Briefly summarize in a table the response of a paramecium to the following stimuli: contact with a solid object, contact with dilute acid, contact with salt solution, exposure to a temperature range from 0°C to 40°C, exposure to an electric field.
8. a) In what ways is a paramecium similar to an amoeba?
   b) In what ways does a paramecium differ from an amoeba? Do you feel that any of these differences give the paramecium an advantage over the amoeba? Explain your answer.

10.5 Investigation

A Detailed Study of Two Protozoans

It is easy to read about the structure and function of an organism as you have done in Sections 10.3 and 10.4. However, it is a test of your skill with a microscope to apply what you read to the study of living specimens.

In this investigation you will study the structure and function of two protozoans, the amoeba and the paramecium. If you have not already done so, read Sections 10.3 and 10.4 carefully before you begin this investigation.

Materials

compound microscope	1.5% methyl cellulose solution
microscope slides (2)	0.5% salt solution
cover slips (2)	sharp needle
paper towel	amoeba culture
lens paper	paramecium culture
medicine dropper	carmine red solution

Procedure A   The Amoeba

a. Use the medicine dropper to suck up a drop or two of the cloudy sludge from the bottom of a jar containing amoeba culture.

b. Prepare a wet mount of the sample. Lower the cover slip slowly and carefully to avoid crushing the organisms.

c. Scan the sample slowly under low power until you find an amoeba. It may take you several minutes to find one among the debris.

d. Once you have found an amoeba, study its locomotion closely. Make sketches of its shape at 30 s intervals for 2 or 3 min. Put arrows on these sketches to show the direction of motion of the cytoplasm.

e. Switch to medium and then to high power. Observe cytoplasmic streaming closely. In what part of the cell does the cytoplasm move most rapidly? Note the appearance of the cytoplasm and its motion. Record this information in your notebook.

f. Observe the formation of a pseudopod. Describe this process carefully in your notebook.

g. If the contractile vacuole is close to the plasma membrane, watch it closely. It is about to burst. Continue to observe the organism for a few minutes after the vacuole bursts. You should see a new one form.

h. If you are fortunate enough to see your specimen feed, share your microscope with other students.

i. Study the response of the amoeba to these stimuli: contact with a sharp object and a salt solution. Design your procedure carefully before you begin. Describe your findings in your notebook. Include sketches to illustrate your notes.

Procedure B   The Paramecium

a. Use the medicine dropper to suck up a drop or two of the scum from a jar containing paramecium culture.

b. Prepare a wet mount of the sample. Before you put on the cover slip, add a drop of 1.5% methyl cellulose solution. This will slow down the organisms so you can study them. Lower the cover slip slowly and carefully to avoid crushing the paramecia.

c. Observe the sample under low power. Make notes on the locomotion of paramecia. You must watch one individual closely to get all the information on this process. You may wish to continue the study after you switch to high power.

d. Study one individual under high power. Locate the food vacuoles. Make notes on the path that a vacuole follows after it leaves the gullet.

e. Note and describe the motion of the cilia in the oral groove. You can see this best if you add a drop of carmine red solution to your wet mount. (Draw it under the cover slip using a paper towel.)

f. Observe for a few minutes the region where the anal pore is located. Watch for the discharge, or egestion, of solid waste.

g. Observe the pellicle closely to see if any trichocysts have discharged their hairs.

h. Observe and describe the contractile vacuoles and their operation.

i. Study the response of a paramecium to as many of these stimuli as time permits: contact with a solid object; dilute acid solution; and 0.5% salt solution. Design your procedure carefully before you begin. Describe your findings in your notebook. Include sketches to illustrate your notes.

Discussion

1. Your sketches and laboratory notes are your writeup for this investigation. Make sure they are complete.

10.6 Pathogenic Protozoans

Several species of protozoans can invade the bodies of humans and other animals. Some are harmless. Others simply live on food in the intestine and rob the host of some of its food. Still others enter the blood and cause serious diseases. This section deals with some of these parasitic species.

Parasitic Sarcodinans

About six or seven species of amoeba can live in the human body. Only one, *Entamoeba histolytica,* causes serious harm to humans. It is responsible for **amoebic dysentery.**

This disease is most common in the tropics. Often everyone in a village will have the disease. However, it is not restricted to the tropics. Some areas of the United States have an infection rate as high as 23%. Yet the disease rarely occurs in Canada.

This parasite is spread from person to person through contaminated water and food. Flies and other arthropods assist in the spreading of the disease.

The infection centres in the intestine of the host. The parasites feed on blood cells and cells from the wall of the intestine. Bleeding ulcers often result. Infected people release cysts of the parasite in their feces. If sanitation is poor, if drinking water is not sterilized, or if human feces are used for fertilizer, the cysts find their way to new hosts. Many people are carriers of the disease without showing any symptoms. If these people handle food or drink, they can transmit the disease to others.

Canadians who travel in tropical countries often get this disease if they are careless about the food they eat and the water they drink. Aureomycin and terramycin are the most effective antibiotics for this infection.

Parasitic Sporozoans

All sporozoans are parasites. The most feared is *Plasmodium,* the cause of malaria (see Fig. 10-4).

Malaria is spread by the female mosquito of the genus *Anopheles.* This mosquito must first bite a person who has malaria. When this happens, some *Plasmodium* organisms enter the mosquito's stomach. In the stomach they grow and reproduce. Eventually they get into the mosquito's blood and travel to the salivary glands. Then, when the mosquito bites another person, *Plasmodium* gets into that person's blood. The parasite travels to the host's liver where it develops for about two weeks. It then re-enters the blood stream where it attacks and destroys red corpuscles. The corpuscles burst at very regular intervals (48 h in the case of *Plasmodium vivax*), releasing spores of *Plasmodium.* When this happens, the characteristic symptoms of malaria appear—fever, chills, and sweating.

The life cycle of *Plasmodium* is much more complex than we have described here. It can be broken by keeping infected people away from the *Anopheles* mosquito and by controlling the mosquito itself. In wet tropical areas these steps are virtually impossible. However, in Canada, the United States, and Europe, malaria has been almost wiped out by destroying the *Anopheles* mosquito. Other parts of the world are not so fortunate. Today there are about 100 000 000 cases of malaria in the world. However, prior to 1945, there were over 300 000 000 cases. At that time, about 3 000 000 people died each year from malaria. That was about half the total deaths in the world per year! Malaria has earned its name as the world's number one killer.

Many drugs work well in the treatment and prevention of malaria. Quinine and several other drugs relieve the symptoms but do not completely cure the host. However, a complete cure can be attained by combining one of those drugs with another drug, primaquine.

Parasitic Animal Flagellates

Animal flagellates are very common in the digestive tract and blood of humans and many other vertebrates. Many of these parasites cause untold misery throughout the world. They affect humans directly as well as the livestock on which humans depend. The worst parasites are the **trypanosomes** (see *Trypanosoma* in Fig. 10-5). These cause African sleeping sickness.

Trypanosomes live in the blood of many mammals in Africa. They are transmitted to humans by the bite of the tsetse fly in the same manner that the malaria parasite is transmitted. In the human host, the organisms live and grow mainly in the blood. They multiply for several weeks before the first symptoms occur. These are fever, headache, and fatigue. The host becomes weak and very anemic as more and more blood cells are destroyed. Eventually the trypanosomes attack the central nervous system. Paralysis sets in, causing the characteristic drowsiness of the victim. Coma and death usually follow.

Local mammals carry the disease but are generally not harmed by it. Therefore a reservoir of trypanosomes always exists. As a result, the most effective way to wipe out this disease is to destroy the tsetse fly. DDT has been used for this purpose. However, its use causes many serious environmental problems. Recently, experiments have been done to wipe out the fly using biological methods. As yet, none have been completely effective.

Several drugs will kill trypanosomes in the human blood. However, there is little hope for a cure once the organisms get to the nervous system.

Symbiotic Ciliates

Only one parasitic ciliate affects humans. However, a wide variety of ciliates live in the digestive tracts of herbivorous mammals—the intestines of horses and the stomach of sheep, deer, cattle, and other cud-chewers. A mature cow may have from 10 000 000 000 to 50 000 000 000 ciliates in its stomach. These ciliates digest carbohydrates, lipids, and proteins in the host's food. When they die, they are digested by the host. They do not harm the host. Therefore they are not parasites. The ciliates are not necessary to the health of the host either. The host grows just as well without them. Therefore the relationship is not mutualism. The ciliates benefit from the relationship but the animal does not. *A relationship in which one organism benefits and the other organism receives neither benefit nor harm is called* **commensalism**.

Review

1. Our description of these diseases and their causes, symptoms, and cures has been brief. Select one protozoan disease discussed in this section or one approved by your teacher. Visit your resource centre or community library and prepare a more thorough description of the disease under these headings: Characteristics of the parasite; Life cycle of the disease;

Symptoms of the disease; Location of the disease in the world; Methods of controlling the disease; Methods of treating and curing the disease; Economic and social effects of the disease.

Prepare a paper of about 500-700 words. Include diagrams to illustrate your written work.

ALGAL PROTISTS

You may recall from Chapter 9 that algae are spread over three kingdoms. You have already studied the blue-green algae of the kingdom Monera. In the next unit you will study the algae of the kingdom Plantae. Here you look at the algae that are in the kingdom Protista, the algal protists.

A wide variety of organisms make up the algal protists. These organisms vary greatly in form and function. Yet they have much in common. All are unicellular or colonial. Practically all contain chlorophyll. All are aquatic or live in damp terrestrial environments. All have nuclei and all have membrane-bound organelles.

There are about 11 000 species of algal protists. These are spread over three phyla. Sections 10.7 to 10.9 describe the general features of the species in each of these phyla.

10.7 Phylum Euglenophyta: The Euglenoid Flagellates

About 500 species of algal protists are placed in this phylum (Fig. 10-15). They are called **euglenoid flagellates** since *Euglena* is a well-known and typical member of this group. They are also called **coloured flagellates** to distinguish them from the animal flagellates (colourless flagellates) of the phylum Zoomastigina (see page 247).

Characteristics

Most euglenoid flagellates are unicellular, but a few are colonial. Also, most are active swimmers, propelling themselves with one or two **flagella**. A few move by amoeboid movement.

Most species lack a cell wall when they are in the active state. Instead, the outer covering of the cell is a firm but elastic **pellicle** similar to that of the ciliates. Like the ciliates, many flagellates can change shape as they move about.

Most euglenoid flagellates can form a protective **cyst** if they are exposed to unfavourable conditions. The cyst is normally covered by a thick cellulose cell wall.

The cells of all euglenoid flagellates have a gullet. Food is ingested by many species through this gullet. The flagella are attached near the gullet.

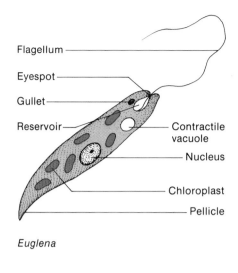

Euglena

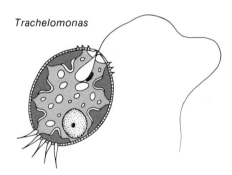

Trachelomonas

Fig. 10-15 Two representative euglenoid flagellates. What do they have in common?

The euglenoid flagellates contain chlorophyll *a*, chlorophyll *b*, carotene, and xanthophylls. The chlorophylls dominate, giving the cells a bright green colour. The chlorophyll is present in chloroplasts. Surplus food made during photosynthesis is stored in the form of a carbohydrate called **paramylum**. This compound is similar to starch, but it does not react to the iodine test for starch.

A few euglenoid flagellates appear to have lost their ability to produce chlorophyll as they evolved through the ages. Although many euglenoid flagellates are both autotrophic and heterotrophic, these particular species are only heterotrophic. They absorb dissolved food through the pellicle or ingest solid food through the gullet.

Most species have an **eyespot** that is sensitive to light. It is generally red. These organisms can respond either positively or negatively to light as a result of this eyespot.

Reproduction of euglenoid flagellates is asexual by binary fission. Sexual reproduction is thought to occur in some species.

Occurrence and Importance

Euglenoid flagellates occur in almost all freshwater environments. A few species also occur in saltwater environments. Some even live in moist soil. A number of species thrive in polluted water. Among these are some species of the genus *Euglena*. When suitable nutrients are present in the polluted water, the *Euglena* blooms. Asexual reproduction by fission occurs at a rapid rate and the water turns a bright green colour. Occasionally the water turns red because of the red eyespots of the *Euglena*.

In polluted water, *Euglena* and other euglenoid flagellates are often the main producers. They are the base of the food chain. They trap light energy and convert it to chemical potential energy. This energy supports most of the life in the water. Also, the flagellates add needed oxygen to the water.

The euglenoid flagellates have been widely studied by biologists who are seeking a link between the plant and animal worlds. Why do you think this is so?

Review

1. **a)** Distinguish between animal flagellates and euglenoid flagellates.
   b) Describe the structure of a representative euglenoid flagellate.
2. **a)** What pigments are found in euglenoid flagellates?
   b) Why are these organisms a bright green colour?
3. **a)** Where do euglenoid flagellates live?
   b) How do these organisms benefit other organisms that live in the same body of water?
4. A few euglenoid flagellates do not contain chlorophyll. In fact, if *Euglena* itself is grown at high temperatures with streptomycin in the water, it loses its chloroplasts and lives only as a heterotroph. Should these colourless euglenoid flagellates be classified as animal flagellates (phylum Zoomastigina)? Hint: Compare the drawings of euglenoid flagellates in Figure 10-15 to those of animal flagellates in Figure 10-5.

5. Why do you think *Euglena* and other euglenoid flagellates are widely studied by biologists who are seeking a link between the plant and animal kingdoms?

10.8 Phylum Pyrrhophyta: The Dinoflagellates

Approximately 1 000 species of algal protists are classified as dinoflagellates (Fig. 10-16).

Characteristics

Most dinoflagellates are unicellular; a few are colonial. Also, most are active swimmers, propelling themselves with *two* **flagella**. A few species do not move.

The cells of most species of dinoflagellates have a true **cellulose cell wall**. A few species have a pellicle similar to that of the euglenoid flagellates. In most species with cell walls, the walls are made of overlapping plates that resemble armour. You can see this in the drawing of *Peridinium*. The patterns that result place dinoflagellates among the most beautiful of all organisms to observe under a microscope.

Dinoflagellates contain chlorophyll *a,* chlorophyll *c,* carotenes, and xanthophylls. The carotenes and xanthophylls generally dominate. As a result, these organisms are greenish-brown or golden brown. Each cell contains from one to many chloroplasts.

Most dinoflagellates are autotrophic. The surplus food that they produce is stored in the form of starch. A few species store oils and fats in large quantities. This gives the cells a deep brown or reddish-brown colour. Some species are heterotrophic. These can ingest solid nutrients or absorb soluble nutrients directly from the water.

Dinoflagellates reproduce asexually by binary fission. Each daughter cell gets one flagellum from the mother cell. It develops the second flagellum as it matures. Only a few species have been observed to have sexual reproduction.

Some dinoflagellates such as *Noctiluca* are luminescent. This means they give off flashes of light. They do this most often when the water in which they live is disturbed. As a result, waves and the wake of ships often glow brightly when dinoflagellates are present. The phylum name **Pyrrhophyta** means "fire plants".

Occurrence and Importance

Dinoflagellates are quite common in both fresh water and the oceans. The salt water species are particularly abundant in warm water, where they live mainly in the upper metre of water.

Dinoflagellates are at the base of many ocean and fresh water food chains. In fact, next to the diatoms (see Section 10.9), they are the most important phytoplankton in the world. In these food chains, the dinoflagel-

Peridinium

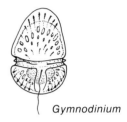
Gymnodinium

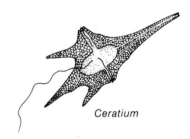
Ceratium

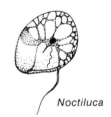
Noctiluca

Fig. 10-16   Some representative dinoflagellates. What do they have in common? How do they differ from euglenoid flagellates?

lates convert light energy to chemical potential energy. This energy is passed on to microscopic crustaceans and other animals when they eat the dinoflagellates. Fish eat small animals to obtain their energy for life processes. Humans eat some of these fish. Thus dinoflagellates are, indirectly, of great importance to humans as a source of food. They are also responsible for the production of a significant proportion of the earth's oxygen gas.

Some species of dinoflagellates bloom easily, and with dangerous results. When conditions are favourable, fission occurs at a rapid rate. At the peak of a bloom, the surface water may contain an average of 20 000 000 dinoflagellate cells in a litre of water! Such blooms are responsible for the famous "red tides" that occur throughout the world from time to time. They are common along all the shores of North America. During a bloom, the concentration of dinoflagellates becomes so high that the water is coloured red, reddish-brown, or yellow. Areas of several square kilometres may be affected.

Some dinoflagellate blooms produce deadly toxins. These toxins fall into three categories. First, some kill fish but very few other aquatic animals. Blooms of *Gymnodinium breve* are in this category. This dinoflagellate blooms from time to time on the west coast of Florida and in other places in the warm coastal waters of the United States. Countless billions of fish are killed each year by this toxin. Second, other species of dinoflagellates release toxins that kill small aquatic animals but few fish. Of course, these indirectly affect the fish by destroying part of their food supply. Third, some species of dinoflagellates release toxins that kill very few aquatic animals. However, these toxins become concentrated in shellfish such as clams, oysters, scallops, and mussels. When humans eat those animals, they get **paralytic shellfish poisoning (PSP)**. One toxin that causes PSP is 100 000 times more dangerous than cocaine. PSP causes death by paralyzing the muscles that control breathing and beating of the heart.

Review

1. **a)** Compare the structure of a dinoflagellate to that of a euglenoid flagellate. Use a table to record your comparisons.
   b) Give two characteristics, other than structural ones, in which dinoflagellates differ from euglenoid flagellates.
2. Describe the importance of dinoflagellates to humans. Include both positive and negative points.
3. Some authorities say that the Red Sea in the mid-East got its name because it actually was red from time to time in the past. What might be responsible for the red colour? Are conditions suitable in the Red Sea for such an event to happen? Explain.

10.9 Phylum Chrysophtya: Golden Algae, Yellow-Green Algae, and Diatoms

About 9 000 species of algal protists are classified as **chrysophytes**. They are spread throughout six classes. The organisms in four of these classes are golden in colour. Those in the other two classes are yellow-green. This section deals briefly with the three main classes:

Class Chrysophyceae (golden algae);
Class Xanthophyceae (yellow-green algae);
Class Bacillariophyceae (diatoms—also called golden algae).

The algae in these three classes differ considerably from one another (Fig. 10-17). However, they also share many common features. If they did not,

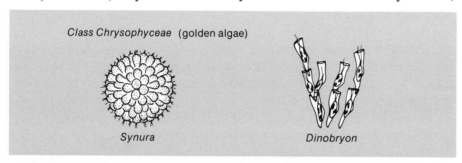

Class Chrysophyceae (golden algae)

Synura

Dinobryon

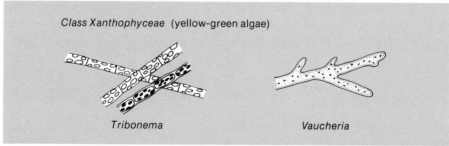

Class Xanthophyceae (yellow-green algae)

Tribonema

Vaucheria

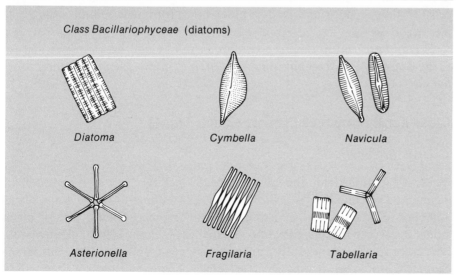

Class Bacillariophyceae (diatoms)

Diatoma

Cymbella

Navicula

Asterionella

Fragilaria

Tabellaria

Fig. 10-17 Some representative golden algae from three different classes. In what ways are the classes similar? How are they different?

they would not be in the same phylum. The common characteristics are:

1. They do not store food in the form of starch. Instead, it is stored in the form of a different type of carbohydrate.
2. All algae in this phylum contain chlorophyll *a* and chlorophyll *c*. They lack chlorophyll *b*. They also contain xanthophylls and carotenes. The golden algae and diatoms contain a brown xanthophyll. A certain mix of these pigments creates a yellow-green colour. When the xanthophylls and carotenes dominate, the algae have a golden colour.

Class Chrysophyceae (Golden Algae)

Characteristics

Most golden algae live in freshwater environments. A few species live in salt water. The golden algae prefer relatively cool and clean water.

Most species are autotrophic, making their food by photosynthesis. A few species are heterotrophic. These absorb soluble nutrients by osmosis or take in solid particles by phagocytosis. Some species are both autotrophic and heterotrophic.

Some golden algae are enclosed by just the plasma membrane. Others are covered by a cellulose cell wall or hard plates of silica.

Reproduction of golden algae is commonly by binary fission. However, some colonial forms also reproduce by fragmentation (breaking apart of the colonies). Sexual reproduction has been observed in a few species.

These algae appear golden because the chlorophylls are masked by the brown pigment, xanthophyll.

Two golden algae

Synura consists of free-swimming colonies of cells (see Fig. 10-17). A colony may consist of a few or many cells. A colony grows by binary fission of its cells. In time, groups of cells may pinch off to form new colonies. Each cell has two flagella.

Dinobryon is another free-swimming colony of cells. In some waters it is the dominant phytoplankton. When it is, it often gives the water a foul, fishy odour.

Class Xanthophyceae (Yellow-Green Algae)

Characteristics

These algae differ from the golden algae by lacking the brown xanthophyll pigment. The chlorophylls, carotenes, and xanthophylls that remain, blend together to give these algae a yellow-green colour. Looking at these algae under a microscope, it is easy to confuse them with the cells of green plants. The chloroplasts of these algae are yellow-green which is not much different from those of green plants which are green. There is a simple way of telling them apart, however. Green plants store food as starch. These algae do not. Therefore, a drop of iodine solution added to cells that have been exposed to light for a few hours will help identify them. If the cells belong to a green plant, the chloroplasts will turn dark blue. If they belong to a yellow-green alga, they will not be affected.

Yellow-green algae are many shapes and forms. Some are unicellular and cannot move. Others are unicellular and can move by the use of flagella or amoeboid movement. Some species are colonial. A few of these occur in the form of long filaments.

Most yellow-green algae have true cellulose cell walls. In many species, this cell wall contains silica as well. The nuclei of these algae are very small and cannot be seen in detail with even the best compound microscope.

Like the golden algae, most yellow-green algae are autotrophic, making their food by photosynthesis. However, a few species are heterotrophic, obtaining food both by diffusion and by phagocytosis.

Like all other algae, the yellow-green algae are important to humans for several reasons. They are producers in many food chains of which humans may be a part. Also, they give off oxygen gas as a by-product of photosynthesis. They bloom easily when the water in which they are growing gets rich in nutrients. These blooms are very dense and cause many problems in the purification of drinking water. They clog filters and produce undesirable odour and taste. Yellow-green algal blooms are also responsible for many fish kills.

Two yellow-green algae

Tribonema is common in many freshwater ponds. It is also found in some streams. As Figure 10-17 shows, it consists of an unbranched filament that is made up of many cells arranged end to end. Since it is filamentous and blooms easily, it causes many problems in water used by humans.

Vaucheria is one of the most widespread of all algae. It is common in both freshwater and saltwater environments. One often finds large felt-like mats of *Vaucheria* along the moist bank of a stream or at the edge of a lake. Clumps of this alga can even be found in damp regions of a greenhouse such as the space under a bench.

Like *Tribonema, Vaucheria* is filamentous. Unlike *Tribonema,* a filament of *Vaucheria* is one giant cell with many nuclei. In other words, there are no walls across the filament to separate it into cells. The nuclei are small and can seldom be seen because of the numerous chloroplasts.

Both asexual and sexual reproduction are common. Asexual reproduction may be by fragmentation. However, it often occurs by the formation of zoospores (Fig. 10-18). These are spores propelled by flagella. A zoospore commonly forms at the end of a filament. The end breaks open and the zoospore, covered with many flagella, swims out. When it comes to a suitable environment, it germinates to form a new filament. The filament grows to form a mature filament called a gametophyte. This is the sexual stage which produces sex cells, or gametes.

Fig. 10-18 Both asexual and sexual reproduction occur in *Vaucheria*. Is the mature gametophyte diploid or haploid? Is the zygote diploid or haploid? Compare your answers to what you learned about sexual reproduction when you studied meiosis in Section 8.4, page 179.

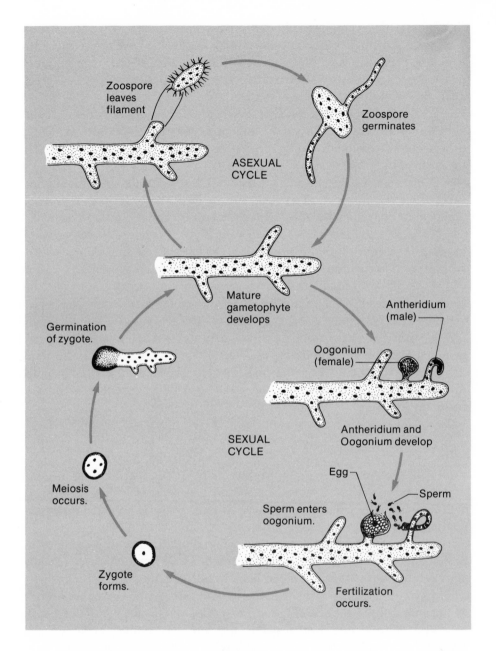

Zoospore leaves filament

Zoospore germinates

ASEXUAL CYCLE

Mature gametophyte develops

Antheridium (male)

Oogonium (female)

Germination of zygote.

Antheridium and Oogonium develop

Egg

Sperm

SEXUAL CYCLE

Meiosis occurs.

Sperm enters oogonium.

Zygote forms.

Fertilization occurs.

Sexual reproduction is also illustrated in Figure 10-18. An **oogonium** (egg-forming body) and an **antheridium** (sperm-forming body) develop on the filament. The tips of both open. Sperm swim out of the antheridium and some of these find their way to the oogonium. One sperm unites with the one egg produced in the oogonium. The resulting **zygote** forms a thick wall and enters a resting stage, normally over the winter months. When favourable conditions return, the zygote becomes active and undergoes meiosis. It then grows into a mature gametophyte.

Class Bacillariophyceae (The Diatoms)

Characteristics

The diatoms are the most important and the most widely distributed of all algae. They have also been studied longer than most other algae. This is because their attractive symmetry and beautiful shapes caught the attention of the early microscopists (see Fig. 10-17).

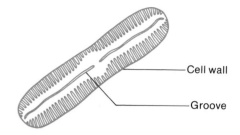

Fig. 10-19 *Pinnularia*, a diatom that can move.

Almost all diatoms are unicellular. Some can move and others cannot. The species that move do so in a very interesting way. Figure 10-19 shows one diatom that can move. Note that it has no cilia or flagella. Further, it has a stiff cell wall. Yet this diatom can slide along on a solid surface. How it does this is not completely understood by biologists. However, many biologists believe that the movement is caused by the secretion of a mucous material into the groove on the surface of the cell. Only diatoms with such a groove are able to do this.

Diatoms have cell walls. However, they contain little, if any, cellulose. The basic material in the cell wall is usually **pectin**. However, in many species, the cell wall may be made almost entirely of **silica**, the main substance in glass. Also, the cell wall of a diatom consists of two overlapping halves. The halves fit together just like the top and bottom of a petri dish.

Most diatoms are autotrophic, making their food by photosynthesis. These species are brown or greenish-brown, since the brown xanthophyll dominates the chlorophylls in the chloroplasts. Only a few species are heterotrophic. However, many of the autotrophic species can live heterotrophically if there is insufficient light to make their own food.

Diatoms reproduce asexually by binary fission. Most species can also reproduce sexually, although this method of reproduction does not occur frequently in natural populations.

Occurrence and importance

Diatoms occur in almost all freshwater, saltwater, and moist terrestrial habitats. In fact, next to bacteria, there are probably more diatoms on earth than any other type of organism. In lakes, ponds, and streams, many diatom species are **planktonic** (free-floating). Other species live attached to plants, logs, rocks, and bottom sediment. Diatoms often form a thick brown layer on the rocks in the bottom of a river or stream.

Marine (saltwater) diatoms occur in all the oceans on earth. They are the most abundant marine phytoplankton. They normally live in the upper 2 or 3 m of the ocean where the light intensity is the greatest. Many species live on plants and rocks along the shores of oceans.

Many species of diatoms live in or on the soil. Some species live on moist stones, plants, and soil. Others actually live as far down in the soil as 20 cm.

Without question, diatoms are among the most important organisms on earth. Since they are the most abundant producers in many bodies of water, they make up much of the base of many food chains. Humans are a part of many of these food chains. Because of their great numbers, diatoms are also responsible for producing a large percentage of the earth's oxygen.

When diatoms die, everything decays except the silica portion of the cell walls. This material settles to the bottom of the lakes or ocean. Over the years, this material forms deep layers of a fine solid called **diatomaceous earth**. This process has been going on for many millions of years. Some of the lakes and seas in which this occurred have dried up long ago. Today many of these deposits of diatomaceous earth are mined to recover this useful material. Diatomaceous earth is used as an abrasive in polishing and scouring preparations. It is also used as a filtering medium in the refining of sugar and gasoline. In addition, it is used in some paints and some kinds of insulation.

Review

1. **a)** Give the common names of the three main classes of chrysophytes.
   b) What do these three classes have in common?
   c) Account for the colours of the organisms in these three classes of algae.
2. **a)** Summarize the main characteristics of the golden algae (class Chrysophyceae).
   b) Summarize the main characteristics of the yellow-green algae (class Xanthophyceae).
   c) Summarize the main characteristics of the diatoms (class Bacillariophyceae).
3. **a)** What two main benefits do the chrysophytes provide for humans and other animals?
   b) In what negative way do some chrysophytes affect humans and other animals?
4. What is the difference between sexual and asexual reproduction? Refer to the life cycle of *Vaucheria* to illustrate your answer.
5. **a)** What is diatomaceous earth?
   b) In what ways is it used by humans?

10.10 Investigation

Algal Protists

Algal protists are divided into three phyla:

1. Phylum Euglenophyta—the euglenoid flagellates;
2. Phylum Pyrrhophyta—the dinoflagellates;
3. Phylum Chrysophyta—the chrysophytes (golden algae, yellow-green algae, and diatoms).

In this investigation you study representative organisms from all three phyla. If you have not done so, read Sections 10.7, 10.8, and 10.9 before you begin.

Materials

compound microscope
microscope slide
cover slip
paper towel
lens paper
medicine dropper
1.5% methyl cellulose solution
prepared slides of euglenoid flagellates, dinoflagellates, and chrysophytes
Euglena culture
Optional: cultures of living dinoflagellates and diatoms

Procedure A   Algal Protist Diversity

a. Obtain a set of prepared slides of algal protists from your teacher. The set contains euglenoid flagellates, dinoflagellates, and chrysophytes.
b. Examine each slide under low and high power. Do the following for each different species that you observe:
  1. State the phylum to which the organism belongs. Is it an euglenoid flagellate, a dinoflagellate, or a chrysophyte? If the organism is a chrysophyte, what class of chrysophyte is it? Is it a golden alga, yellow-green alga, or diatom?
  2. Make a sketch of the organism. Sketch the general shape of the cell and then draw in any prominent features. Label as many parts as you can.
  3. Write a brief paragraph explaining why you placed the organism in the phylum that you did in step 1.

Procedure B   *Euglena*

a. Use the medicine dropper to obtain a drop of the culture as directed by your teacher.
b. Prepare a wet mount of the culture, but do not add the cover slip.
c. Observe the *Euglena* under low power. Pay particular attention to the locomotion of the organisms. Make notes on their locomotion and on their behaviour.
d. Add a drop of 1.5% methyl cellulose solution. It will slow down the *Euglena* without killing them.
e. Add a cover slip. Then switch to high power and study one *Euglena* closely.
f. Make sketches of *Euglena* at intervals of about 30 s. Do this until you have 5 or 6 sketches. What happened to the shape of the cell? What kind of motion occurred? How does the flagellum propel *Euglena?*

g. Make a large sketch of *Euglena.* Draw in and label as many of these parts as you can find: pellicle, chloroplast, nucleus, contractile vacuole, reservoir, gullet, eyespot, flagellum, anterior end, posterior end.

Procedure C   Diatoms

If time permits, your teacher will allow you to study a culture of living diatoms.

a. Prepare a wet mount of the culture.
b. Observe the culture under low power. Scan the sample and note how many different species are present.
c. Study a cell of each species under medium then high power. Sketch the shape of the cell. Draw in and label any prominent features. Pay particular attention to those features that are characteristic of diatoms.

Discussion

1. Your laboratory notes and drawings are your writeup for this investigation. Make sure that they are complete.

Highlights

The following chart summarizes the classification of protists introduced in this section:

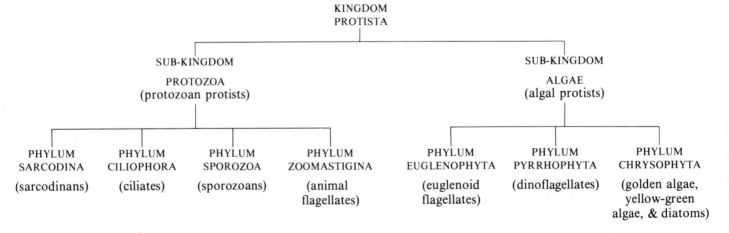

The protozoan protists are unicellular or colonial. Locomotion is by pseudopods, flagella, or cilia. A few species cannot move. Protozoans are heterotrophic. They feed on bacteria, yeasts, algae, and other protozoans.

The algal protists are also unicellular or colonial. Almost all species contain chlorophyll. As a result, most species are autotrophic. However, many species are also heterotrophic. They feed by taking in dissolved nutrients by diffusion or by ingesting solid nutrients by phagocytosis.

The Fungi

11

About 80 000 species of organisms are classified in the kingdom Fungi. These organisms vary widely in appearance. Some, like yeast, are microscopic. Others, like mushrooms, are quite large. As a group, fungi are among the most important living things on earth.

11.1 Introduction to Fungi

Characteristics

The body of a fungus is called a thallus (plural: thalli). Although the different groups of fungi have greatly varying thalli, they do have some important common characteristics.

1. *All fungi lack chlorophyll.* As a result, all fungi are heterotrophic. They cannot make their own food, but must get it from an outside source.

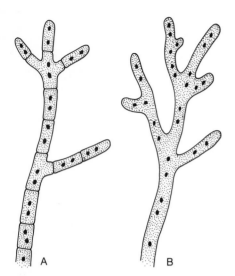

Fig. 11-1 The hyphae of some fungi are multicellular (A). The hyphae of other fungi are not divided into cells (B).

Fungi cannot move about. Therefore they must live in close contact with their food.

Fungi are classified in two basic groups, according to how they obtain their food. Some fungi are **parasites**. These species obtain their nutrients from a living host. Other fungi are **saprophytes**. These obtain their nutrients from non-living organic matter such as dead plants, dead animals, and animal waste products. A few species can obtain nutrients from both living and non-living organic matter.

Fungi feed by secreting enzymes into the food material. These enzymes break down the food into small particles. Some are absorbed by the fungi and used for nutrition. The "left-overs" may be used by other organisms such as green plants.

2. *The thallus of most fungi consists of microscopic threads or filaments.* Each filament is called a **hypha** (plural: **hyphae**). The hyphae may be multi-cellular. That is, they may be divided into cells by cross-walls. In some species the hyphae lack cross-walls. For these, the hyphae are just long tubes full of cytoplasm and many nuclei (Fig. 11-1). The hyphae often appear as a mass of separate tiny hairs. You may have seen such a mass in bread mould or in the moulds that grow on fruits. In mushrooms and some other fungi, the hyphae are tightly woven together to form a fleshy mass. A mass of interwoven hyphae is called a **mycelium**.

Some species of fungi, the yeasts, are unicellular.

3. *Most fungi reproduce both asexually and sexually.* Asexual reproduction is by fragmentation (the breaking away of pieces of the thallus) and by spore formation. Sexual reproduction is similar to some of the algal protists that you have studied.

Occurrence and Importance

Fungi live in almost all environments. However, they are most abundant in moist, warm, organic-rich environments. They are frequently found on dead trees and in the upper few centimetres of the soil. Some species live in water. Others live on and in living plants and animals, including humans. The one thing that every environment must have in order to support fungal growth is organic matter. That organic matter may be a dead tree, a slice of bread, an orange, or a human hand.

Fungi decompose, or break down organic matter very rapidly and efficiently. They share with bacteria the important role of **decomposers** of dead organic matter. Green plants require nutrients in order to carry out metabolic activities such as photosynthesis. These plants get the nutrients from water or soil. If it were not for fungi and bacteria, the supply of nutrients in the water and soil would soon be exhausted. Fungi and bacteria break down plant and animal remains into particles small enough to be absorbed by green plants. By doing so, they also prevent the earth from becoming buried in a deep layer of leaves, logs, dead animals, and animal waste.

Unfortunately, not all the decomposing action of fungi is beneficial. They also decompose foods, leather, cloth, wood, paper, some plastics, and rubber. Also, parasitic species destroy living plants, including farm crops. They also infect animals, including humans.

Some fungi, such as the common field mushroom, *Agaricus campestris,* are used for food. Other fungi, the yeasts, are used in baking and production of alcoholic beverages. Still others produce antibiotics, vitamins, and other important organic compounds.

Classification of Fungi

Fungi are classified into seven divisions, or phyla. (Most botanists prefer to use the term "divisions".) Six divisions are briefly described here. You will study them all more closely in the rest of this chapter.

1. Division Basidiomycota: Club Fungi

This division consists of the most advanced fungi. All species in this division produce spores on a special structure called a **basidium**. Included in this division are mushrooms, bracket fungi, puffballs, the rusts that infect white pine trees and wheat, and the smuts that infect corn, barley, and oats.

2. Division Zygomycota: Conjugation Fungi

This division consists of the simplest fungi. All species in this division are filamentous fungi in which the hyphae lack cross-walls. Sexual reproduction is by **conjugation**, and a **zygospore** is produced. Included in this division are moulds such as bread mould.

3. Division Ascomycota: Sac Fungi

This is the largest division of fungi. All species in this division produce spores in sacs called **asci** (singular: **ascus**). Included in this division are the yeasts, blue and green moulds such as *Penicillium,* powdery mildews, and the parasites that cause apple scab, Dutch elm disease, and black knot of cherry and plum trees.

4. Division Oomycota: Water Moulds

Like the zygomycetes, this division consists of filamentous fungi in which the hyphae lack cross-walls. In this division, sexual reproduction is by the fertilization of large eggs by nuclei from a male hypha. The hypha functions as an **antheridium** (sperm-producing body). Included in this division are moulds that parasitize fish and many saprophytic moulds that feed on damp, non-living organic matter. Many downy mildews and blights that parasitize important food crops are water moulds.

5. Division Deuteromycota: Imperfect Fungi

This division may not, in fact, be a true division. It contains fungi for which no sexual life cycle is known. If sexual reproduction is discovered in a species, that species is then moved to another division. The parasites that cause athlete's foot and ringworm are presently in this division.

6. Division Myxomycota: Slime Moulds

The thalli of these moulds are a mass of protoplasm that contain many nuclei and move by amoeboid movement. These coloured masses often appear on damp, rotten logs in dark locations.

Review

1. **a)** What is the thallus of a fungus?
   b) State the three common characteristics of most fungi.
2. **a)** What is the difference between a hypha of a fungus and the mycelium of a fungus?
   b) Distinguish between parasitic and saprophytic fungi.
3. **a)** Describe the role of fungi as decomposers. Mention both positive and negative aspects.
   b) State three important uses of fungi.
4. Summarize the classification of fungi in a table. Use the following column headings: Division; Special Characteristics; Examples.

11.2 Division Basidiomycota: Club Fungi

About 14 000 species of fungi are classified as basidiomycetes. You have probably seen several members of this division—mushrooms, puffballs, and bracket (shelf) fungi. However, you may not be familiar with the rusts and smuts that parasitize many plants (Fig. 11-2). This section describes the form and function of some important basidiomycetes.

Fig. 11-2 Some representative basidiomycetes (club fungi): *Amanita Muscaria* or fly agaric mushroom (A); puff ball (B); bracket or shelf fungus (C); corn smut (D).

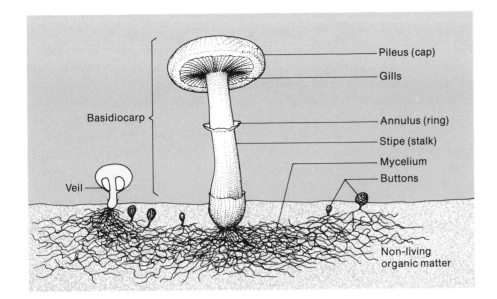

Fig. 11-3   Structure of a mushroom basidiocarp.

Labels in figure: Pileus (cap), Gills, Annulus (ring), Stipe (stalk), Mycelium, Buttons, Basidiocarp, Veil, Non-living organic matter

Structure of a Mushroom

You probably have not seen the largest part of a mushroom. This part is found in the organic matter in which the mushroom is growing. It consists of a tangled mass of **hyphae** called the **mycelium** (Fig. 11-3). For just one mushroom, the mycelium can extend for several metres beyond the part that you see above ground. The mycelium secretes enzymes into the organic matter to decompose, or digest it. The resulting nutrients are absorbed by the mycelium to be used for the nutrition of the mushroom.

When conditions are suitable, some of the mycelium forms a tight bundle called a **button**. The button quickly grows into a fruiting body, or **basidiocarp**. This is the part that people commonly call a mushroom. As you can see, it is only part of a mushroom.

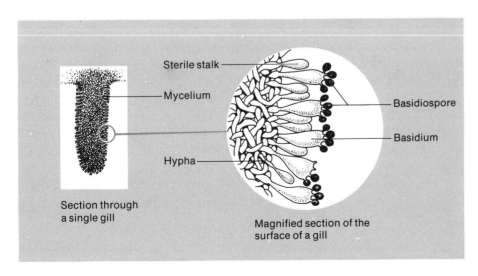

Fig. 11-4   Structure of the gills of a mushroom.

Labels in figure: Sterile stalk, Mycelium, Hypha, Section through a single gill, Basidiospore, Basidium, Magnified section of the surface of a gill

The basidiocarp consists of a **stipe** (stalk) with a **pileus** (cap) on top. When a button first bursts through the ground surface, a **veil** extends from the pileus to the stipe. This veil protects the delicate **gills** on the underside of the pileus. As the pileus grows, the veil stretches. Finally it breaks, leaving tattered remains called an **annulus** (ring) on the stipe.

The gills are thin delicate plates that hang down from the pileus. Sticking out perpendicular to the surface of each gill are countless club-shaped stalks called **basidia** (singular: **basidium**). Each basidium produces four spores called **basidiospores** (Fig. 11-4). When the spores are mature, they are ejected from the basidium. They then drift in the wind to new locations where, if conditions are suitable, germination occurs.

Reproduction of Mushrooms

Asexual reproduction is common among mushrooms and most other basidiomycetes. It usually occurs by fragmentation, or the breaking away of parts of the mycelium.

Sexual reproduction occurs by the joining of hyphae of opposite sexual strains. These strains are called **positive (+)** and **negative (−)**. Follow Figure 11-5 as you read this description of the life cycle in which such joining occurs.

Life cycle

Let us begin the cycle with a **basidiospore** of a mushroom landing in a suitable environment. The spore germinates to produce a mass of hyphae. These hyphae are called **primary hyphae.** Let us suppose that a spore of the opposite strain germinates nearby. It, too, produces a mass of primary hyphae. When the two opposite strains of hyphae meet, they join. The cells that join now have two nuclei. These cells divide rapidly, forming a mass of **secondary hyphae.** These hyphae have cells with a + strain nucleus and a − strain nucleus. When conditions are suitable, the secondary hyphae become quickly organized into a mass of **tertiary hyphae** called the **basidiocarp.**

Some of the hyphae that make up the basidiocarp end in the gills. The last cells of some of these hyphae develop into **basidia.** The basidia, in turn, produce the **basidiospores** that begin the cycle again. Note that the cell which forms a basidium begins with two nuclei, one from each strain. These fuse to form a zygote. It is diploid, or 2n. The zygote immediately divides by meiosis to form four haploid (n) basidiospores that move to the outside of the basidium. After a basidiospore is mature, it is shot away from the basidium to begin a new cycle.

A single mushroom can produce over two billion (2 000 000 000) spores. These are shot into the air from time to time at the rate of a few million per minute for several days.

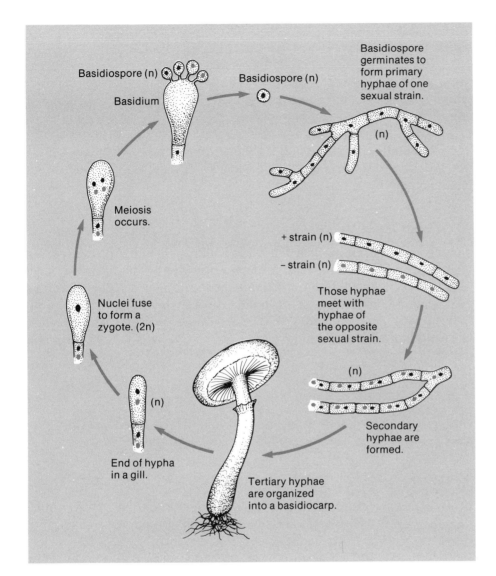

Basidiospore (n)

Basidium

Basidiospore (n)

Basidiospore germinates to form primary hyphae of one sexual strain.

(n)

Meiosis occurs.

+ strain (n)

− strain (n)

Those hyphae meet with hyphae of the opposite sexual strain.

Nuclei fuse to form a zygote. (2n)

(n)

(n)

Secondary hyphae are formed.

End of hypha in a gill.

Tertiary hyphae are organized into a basidiocarp.

Edible and Poisonous Mushrooms

CAUTION: Never eat any mushroom or other fungus unless you are absolutely sure it is edible.

About half of all known mushrooms are edible. Of course, this means that the remainder are poisonous or distasteful. The most poisonous mushrooms belong to the genus *Amanita*. Small amounts of these mushrooms cause hallucinations, drunken behaviour, and even death. In fact, the most poisonous of all fungi is *Amanita verna,* the destroying angel. It is so deadly that one basidiocarp contains enough poison to kill several people. Terrible pain, vomiting, diarrhea, cramps, and finally death occur. The unfortunate victim often suffers for 3 or 4 d.

Many primitive tribes have used these mushrooms in their religious rites. In fact, the amanitas have probably been a part of witchcraft and black magic for hundreds of years.

There are no simple tests or rules that you can use to tell poisonous from edible mushrooms. You must be able to identify the species. Therefore, unless you are a mushroom expert, you should eat only one kind of mushroom—the kind you buy in stores.

A large industry exists to produce *Agaricus bisporus,* a close relative of the common field mushroom. It is the one you buy in stores. It is also put in soups and added to pizzas. It is a good source of many minerals and vitamins.

Mushroom culture

Mushrooms are grown in windowless buildings. (Fungi do not require light.) A compost bed about 30 cm deep is prepared by mixing humus-laden soil with well-rotted manure. Chunks of mushroom mycelium are then planted throughout this bed. The humidity is kept high and the temperature is kept between 10°C and 15°C. Under these conditions, the mycelium develops quickly. Harvesting begins about five weeks after the initial planting.

Parasitic Basidiomycetes

Two groups of basidiomycetes, the **rusts** and the **smuts**, are parasitic fungi that are of great importance to humans.

Rusts

Rusts attack a wide range of plants that are important to humans. Among these plants are white pine, wheat, coffee, and many ornamental plants (Fig. 11-6).

Outbreaks of wheat rust have often destroyed entire crops. Rust-resistant strains of wheat have been developed. However, wheat rust still destroys a significant portion of the world's wheat crop each year. While the wheat rust is growing on the wheat, it harms the wheat by destroying photosynthetic tissue. Thus the kernels of wheat are much smaller on infected plants.

The life cycle of wheat rust is very complex. In fact, it is beyond the scope of this course. Five different types of spores are produced during the cycle. Also, two different hosts are involved. Part of the life cycle is on the wheat plant and part of it is on the barberry plant. The barberry is called an **alternate host**. The following is a simplified description of the life cycle.

In the spring, **Type 1 spores** of wheat rust drift in the air (Fig. 11-7). If one of these lands on the upper surface of a barberry leaf, it may germinate to start the infection. A small body develops at the site of the infection. This body produces **Type 2 spores**, which are haploid. There are two different strains of Type 2 spores. These spores are covered with a sticky liquid which

Fig. 11-6   White pine blister rust destroys countless millions of pine trees every year.

In the spring, Type 5 spores give rise to Type 1 spores.

Infection over-winters in the form of Type 5 spores produced on the wheat plant.

Type 1 spores drift in the air.

Type 4 spores spread infection to new wheat plants.

Type 1 spores land on upper surface of barberry leaf. They germinate and produce Type 2 spores.

Type 3 spores land on wheat. Type 4 spores are produced.

Type 2 spores of opposite sexual strains unite and produce Type 3 spores on lower surface.

attracts flies and other insects. They carry the spores from place to place on their bodies. Eventually these spores may come in contact with Type 2 spores of the opposite sexual strain. If this happens, fertilization occurs. The resulting hyphae grow through the barberry leaf. On the lower surface of the leaf, a new body is formed that produces **Type 3 spores** (Fig. 11-8). These are released into the air and are spread by air currents.

When a Type 3 spore lands on the stem or leaf of a wheat plant, it may germinate. The resulting hyphae penetrate into the tissue. Soon a body is formed that produces **Type 4 spores**. These are released into the air and eventually infect new wheat plants. These spores can spread for hundreds of kilometres. They continue to infect new plants throughout the entire growing season.

When the wheat matures, the fungus begins to form **Type 5 spores**. These are very resistant to low temperatures. Therefore the fungus overwinters in this stage. In the spring these spores divide by meiosis to form **Type 1 spores**. These are carried by air currents to barberry plants to start the cycle all over again. The Type 1 spores are actually **basidiospores**, which is why wheat rust is in the phylum Basidiomycota.

Biologists discovered in the 1920s that barberry was an alternate host

Fig. 11-8 The spore-producing bodies on the lower side of a barberry leaf.

in this life cycle. Therefore the government ordered all barberry plants destroyed, hoping that this would break the life cycle. However, many plants remained in remote areas to keep the disease alive. Also, the Type 4 spores spread the infection so well that few barberries are needed to keep the cycle going.

Biologists have developed many rust resistant strains of wheat. They feel that this is the best way to control the disease. However, the disease still causes serious economic problems and food shortages in many countries as well as in parts of Canada.

Smuts

Like rusts, smuts are important to humans because they damage many food crops. Among the crops attacked by smuts are grains such as oats, barley, rye, and corn. Smuts also attack many pasture grasses such as brome grass.

As you can see from Figure 11-2, smuts cause great damage to their hosts. They stunt the growth of the plants and often destroy the entire fruit.

Fig. 11-9   Life cycle of corn smut.

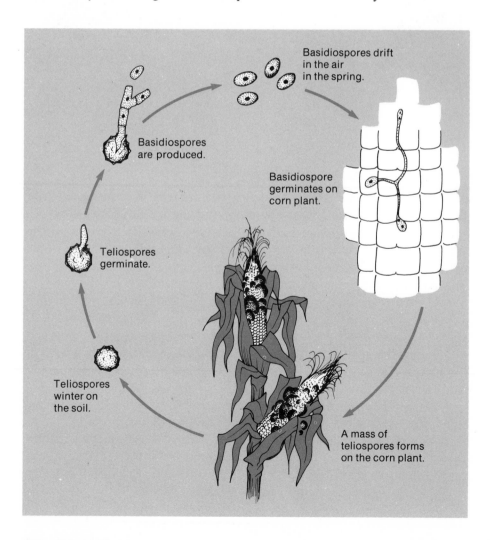

Basidiospores drift in the air in the spring.

Basidiospores are produced.

Basidiospore germinates on corn plant.

Teliospores germinate.

Teliospores winter on the soil.

A mass of teliospores forms on the corn plant.

The life cycle of a smut is more simple than that of a rust. Only one host is involved and only three types of spores are produced. Let us see how this life cycle develops.

In the spring, **basidiospores** of corn smut drift in the air (Fig. 11-9). If one of these lands on a corn plant, it may germinate and send hyphae into the plant. When two hyphae of opposite sexual strains meet, they join. Fertilization occurs and new hyphae are sent out that continue to infect the plant. Eventually large gray masses of hyphae are formed that may completely cover the ears of corn. Both the original hyphae and the new hyphae can produce summer spores that spread the infection during the growing season.

As winter approaches, thick-walled **teliospores** (winter spores) form. They look like a large powdery black mass. These spores live on the ground during the winter. In the spring, they germinate, sending out a short hypha called a **basidium**. The basidium produces four **basidiospores**. These are released into the air to begin the cycle again.

As you can see from the life cycle, this disease can be partly controlled by removing and burning infected plants. However, this method is impractical in large fields of corn. Plowing the soil in the fall also helps to control the disease by burying the teliospores. The most effective control of corn smut is the development of smut-resistant strains of corn. The disease has been greatly reduced since farmers began to plant these strains. However, corn smut is still a major problem in many parts of the world.

Review

1. What common feature do all members of the phylum Basidiomycota have? Why are they called club fungi?
2. **a)** Where is most of the mycelium of a mushroom located? What is its function?
   b) What is the basidiocarp of a mushroom? Describe its structure and function.
3. **a)** Summarize the life cycle of a mushroom.
   b) At what point in the life cycle does sexual reproduction occur?
4. **a)** Are most mushrooms parasites or saprophytes? How do you know?
   b) Describe the kinds of environments in which you have seen mushrooms growing.
   c) Mushroom growers provide ideal conditions for the mushrooms in order to get the largest possible crops. What are those conditions?
5. **a)** Is wheat rust a parasite or a saprophyte? How do you know?
   b) How does wheat rust harm wheat plants?
   c) Explain how the knowledge of the life cycle of wheat rust has helped humans control this disease.
   d) Strains of wheat that are resistant to wheat rust often lose that resistance in a few years. Why do you think this happens?
6. **a)** Is corn smut a parasite or a saprophyte? How do you know?
   b) How does corn smut harm corn plants?
   c) Explain how the knowledge of the life cycle of corn smut has helped humans control this disease.

11.3 Investigation

Structure of a Mushroom

In this investigation you use a hand lens and a microscope to study the structure of the common edible mushroom *Agaricus bisporus*. Refer to Section 11.2 if you have forgotten the meanings of any of the terms used here.

Materials

fresh basidiocarp of *Agaricus bisporus*
scalpel or razor blade
hand lens
compound microscope
microscope slides (2)
cover slips (2)

paper towel
lens paper
medicine dropper
forceps
optional: prepared slide
    of mushroom gills

Procedure

a. Hold the mushroom basidiocarp in your hand. Using the hand lens, note the appearance of each of the following parts: stipe, pileus, annulus, gills. Record your findings in a table under the column headings "Part" and "Description".

b. Break the stipe of the mushroom. Examine the broken area with a hand lens.

c. Using the forceps, remove a small piece of the stipe from the broken area. Prepare a wet mount of this piece. Examine it under low, medium, and high power with a compound microscope. Record a description of the structure of the stipe.

d. Make a longitudinal cut (top to bottom) through the pileus. Examine the cut surface with a hand lens.

e. Using the forceps, remove a small piece of the pileus from the cut surface. Prepare a wet mount of this piece. Examine it under low, medium, and high power with a compound microscope. Record a description of the structure of the pileus.

f. Using the forceps, carefully remove a piece of gill from the pileus. Prepare a wet mount of this piece. Examine it under low, medium, and high power. Locate and describe a basidium and a basidiospore.
Note: Your teacher may give you a prepared slide to study at this time.

g. Hold the pileus of a mature mushroom over a microscope slide. Tap it gently to dislodge some spores. Examine the spores under low and high power. Record a description of the spores and make a sketch of one.

h. (Optional)   If enough mushrooms are available, prepare a **spore print** of the mushroom as follows. Cut off the pileus of a fresh mature basidiocarp. Place it, gills down, on a piece of white paper. Cover it with a large beaker (Fig. 11-10). Let it sit untouched for a day or two. Then remove the beaker and carefully lift the pileus from the paper. Examine the spore print with a hand lens. Record your observations.

Discussion

1. **a)** What is the stipe of a mushroom made of?
   b) What is the pileus of a mushroom made of?
   c) What are the gills of a mushroom made of?
2. What evidence did you see that explains why mushrooms are classified as basidiomycetes, or club fungi?
3. Spore prints are often used to help identify mushrooms. The colour of the spores is generally different for most species. If you can find any other species of mushroom, prepare and examine their spore prints.

CAUTION: Wear gloves when handling unknown mushrooms. They may be poisonous.

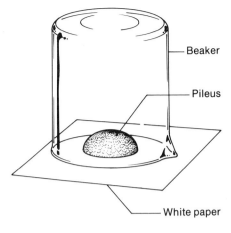

Fig. 11-10   Preparation of a spore print.

11.4 Division Zygomycota

Conjugation Fungi

About 1 000 species of fungi are classed as **zygomycetes**. They occur in most moist terrestrial environments, where they are among the most efficient decomposers. You have likely seen several zygomycetes. Bread mould is probably the most familiar to you. This fungus attacks fruits such as strawberries and grapes, as well as bread. This section describes the form and function of this important zygomycete.

Structure of Bread Mould

Rhizopus, the common bread mould, can be found in the upper centimetre of almost any kind of soil anywhere on the earth. Its spores are in the air at most times. Like most zygomycetes, it is saprophytic, and grows rapidly wherever moisture and nutrients are available. It is also parasitic at times, attacking fruits such as grapes, strawberries, and cantaloupes.

The mould first appears as a fluffy mass of fine white threads. The mass later turns black as the spores mature. As a result, zygomycetes are often called **black moulds**. Within a few days, a piece of bread or fruit can be completely covered with this fungus (Fig. 11-11).

The thallus (plant body) of bread mould consists of a branching mass of **hyphae** that lack cross-walls (Fig. 11-12). Together, these hyphae make up the **mycelium** of the bread mould. The mycelium consists of three types of hyphae. One type, called **stolons**, runs parallel to the surface of the medium on which the mould is growing. Another type, called **rhizoids**, develops at

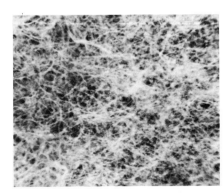

Fig. 11-11   *Rhizopus*, the common bread mould. This mass of fungus developed within hours after a spore lit on an over-ripe cantepole that was stored in a damp place.

Fig. 11-12 The structure of bread mould. Note the three types of hyphae: stolons (horizontal hyphae), sporangiophores (erect hyphae), and rhizoids (root-like hyphae).

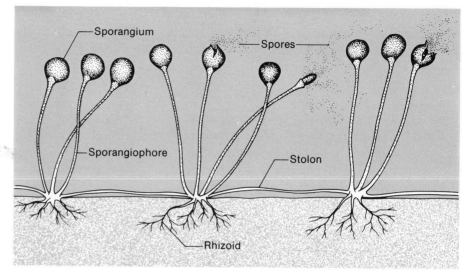

Sporangium — Spores — Sporangiophore — Stolon — Rhizoid

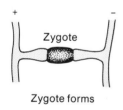

Hypha from a + mycelium / Hypha from a − mycelium

Branch

Branches meet

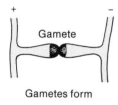

+ −

Gamete

Gametes form

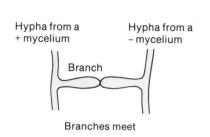

+ −

Zygote

Zygote forms

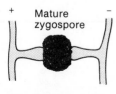

+ Mature zygospore −

Zygospore matures

Fig. 11-13 Sexual reproduction in bread mould. Hyphae of opposite sexual strains join to form a zygospore.

swellings along the stolons. The rhizoids penetrate the medium and secrete enzymes into it that digest it. They then absorb nutrients from the digested material. Erect hyphae, called **sporangiophores**, also develop at each swelling. Each sporangiophore bears a spore case, or **sporangium**, that can contain up to 70 000 **spores**. The sporangia are white at first. However, they turn black when the spores mature.

Reproduction of Bread Mould

Asexual reproduction is common in bread mould and other zygomycetes. It usually occurs by the formation of spores in a sporangium. As each sporangium matures, thousands of black spores are formed within it by rapid mitotic division. Since no union of gametes (sex cells) occurs, these spores are called **asexual spores**.

When the spores are fully developed, the wall of the sporangium splits open to release the spores into the air. These spores are very buoyant and therefore drift long distances from the source. If a spore lands in a suitable habitat, it germinates and forms a new mycelium.

Sexual reproduction occurs by **conjugation**, or the joining of hyphae. It generally occurs only when conditions become unfavourable for the growth of bread mould.

The process begins when one hypha from each of two mycelia of opposite sexual strains touch one another. These strains are called **positive** (+) and **negative** (−), instead of male and female, since there are no structural differences between them. At the point of contact, each hypha sends out a short branch (Fig. 11-13). The two branches join at the tip. Then a **gamete** (sex cell) forms at the tip of each branch. The two gametes join to form a **zygote**. The zygote develops a thick wall to become a **zygospore**.

A zygospore can withstand unfavourable environmental conditions for many months. When favourable conditions return, it germinates and

Fig. 11-14 Life cycle of bread mould.

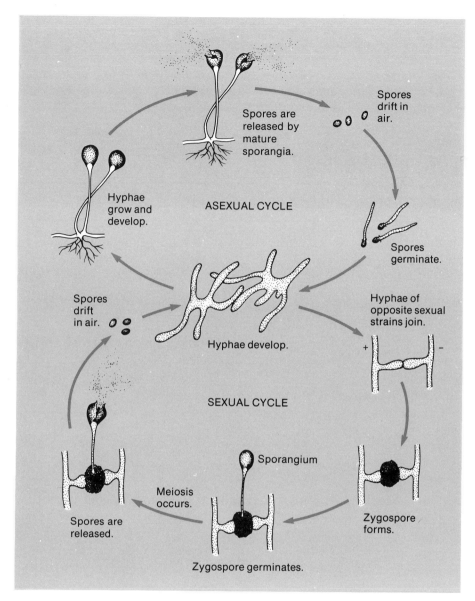

produces a **sporangium**. Meiosis occurs in the sporangium, resulting in numerous haploid spores. These are released into the air. When they land in a suitable habitat, they germinate to form new mycelia.

The life cycle of bread mould is summarized in Figure 11-14.

Review

1. **a)** Why do you think this division of living things was given the name Zygomycota?
   b) Why are the zygomycetes called black moulds?
   c) Why are the zygomycetes also called conjugation fungi?
2. **a)** Name the three types of hyphae that make up the mycelium of bread mould.

b) Describe each of the types of hyphae and state its function.
3. **a)** Describe asexual reproduction in bread mould.
   b) Describe sexual reproduction in bread mould.
   c) What advantage(s) does asexual reproduction offer to bread mould?
   d) What advantage(s) does sexual reproduction offer to bread mould?
4. Explain how bread mould destroys bread and fruits.

11.5 Investigation

Culture and Structure of Bread Mould

In this investigation you will culture (grow) bread mould and study its structure.

Materials

slice of bread	cover slip
petri dish	paper towel
paper towel	lens paper
hand lens	medicine dropper
dissecting microscope (optional)	forceps
compound microscope	dissecting needle
microscope slide	

Procedure A   The Culture of Bread Mould

a. Line the bottom of a petri dish with paper towelling. Wet the paper, then pour off any excess water.
b. Wipe a piece of bread over a table top, bench, or floor to collect bread mould spores. Place the bread on the paper towelling in the petri dish.
c. Add several drops of water to the bread. Do not soak the bread. Place the cover on the petri dish.
d. Store the petri dish in a cool dark place for several days. Examine the contents each day with a hand lens and, if possible, with a dissecting microscope. Record your observations. *Do not remove the cover at any time.* If you do, you will destroy much of the mycelium of the mould. Continue these daily examinations until the mould turns noticeably black.

Procedure B   Structure of Bread Mould

a. Carefully remove the cover from the petri dish.
b. Using the forceps, remove a few strands of the mycelium that contain the parts shown in Figure 11-12.
c. Prepare a wet mount of this sample. A dissecting needle should help you

to spread the mass of hyphae so that you can study them under the microscope.

d. Examine the mount under low and high power. Locate and describe each of the following parts: sporangiophore, sporangium, stolon, spores, and rhizoids. (In order to see rhizoids, you will have to include a tiny piece of bread in your sample.)

Discussion

1. Your teacher provided special bread for this study which did not contain a **mould inhibitor**. Most bakeries add a mould inhibitor to their bread. What is a mould inhibitor? Why do bakeries add it to their bread?
2. Bread mould can often be cultured by exposing the bread to the air for at least 1 h before the lid is placed on the petri dish. Why will this procedure work?
3. Describe the development of the mould culture from the day it began until it turned black.
4. **a)** You probably noticed an odour when you removed the lid of the petri dish. What do you think caused it?
   b) What evidence did you see that bread mould is a saprophyte?
5. **a)** What do sporangiophores, stolons, and rhizoids have in common?
   b) How do these three structures differ?
6. Design and carry out an experiment to show that bread mould is also parasitic.
7. When you performed this experiment, you assumed that bread mould develops best in a moist, cool, dark place. Design and carry out a controlled experiment that will prove whether or not these are the best conditions for the growth of bread mould.

11.6 Division Ascomycota: Sac Fungi

About 15 000 species of organisms are classified as **ascomycetes**. You have probably seen many of them. Some common ascomycetes are yeast, blue-green moulds, morels, black moulds, and powdery mildews. Apple scab and many other serious plant pathogens are also ascomycetes. This section describes the structure and function of some of the common ascomycetes.

The Yeasts

Yeasts are widely distributed in nature. They occur throughout the world and are especially common in substances that contain a high quantity of sugar. Thus decaying fruits, grains, and many other foods contain yeasts. Fruit orchards and vineyards are excellent habitats for these fungi. The soil and the air near your home probably contain them.

Most yeasts are saprophytes. However, a few species are parasites. These cause plant and animal diseases. Some species even attack humans.

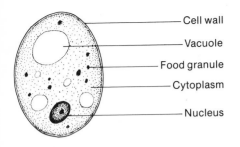

Cell wall

Vacuole

Food granule

Cytoplasm

Nucleus

MATURE YEAST CELL

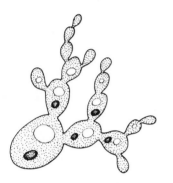

CHAIN OF YEAST CELLS

Fig. 11-15 Yeast is a single-celled organism. However, long chains often form if the rate of reproduction is rapid.

Structure

Yeast cells are very small. Just 1 g of the yeast you can buy in a grocery store contains over 6 000 000 000 yeast cells! Still, these cells are larger than most bacterial cells. However, you can see little cellular detail with an ordinary compound microscope.

Yeasts are one-celled organisms. However, chains of cells often form when the culture is growing rapidly (Fig. 11-15). The cells are usually egg-shaped. Each cell is surrounded by a thin cellulose cell wall. Within the cell wall are the nucleus and cytoplasm. The cytoplasm normally contains one large vacuole and often some smaller ones. It also usually contains several food granules.

Nutrition

Most species of yeast require sugars for their nutrition. You learned in Chapter 7 that yeast cells respire anaerobically according to this equation:

$$\text{Glucose} \xrightarrow{\text{Enzymes}} \text{Ethyl alcohol} + \text{Carbon dioxide} + \text{Energy}$$

Yeast cells secrete a mixture of enzymes called **zymase**. The zymase digests the glucose, and energy is released to power cell activities. Ethyl alcohol and carbon dioxide are produced as by-products.

Reproduction

Yeast cells reproduce asexually by **budding** (Fig. 11-16). When a yeast cell reaches a mature size, a bulge called a **bud** forms on its side. Mitosis occurs and one nucleus moves into the bud. The other nucleus stays in the mother cell. The bud grows and eventually breaks away from the mother cell. Under ideal conditions, budding occurs so rapidly that buds begin to form on buds before they break away from the mother cell. A chain of cells results that looks much like the hyphae of many other fungi (see Fig. 11-15).

Most yeasts can also reproduce sexually by **spore formation** (see Fig. 11-16). When conditions become unfavourable for budding and growth, cells of opposite sexual strains unite to form a diploid cell. This cell acts as a zygote and also becomes the **ascus**, or sac, that holds the spores. The zygote is diploid. It divides by meiosis to form four haploid nuclei. These, in turn, often divide by mitosis to form eight haploid nuclei. The cytoplasm divides more or less equally among the eight nuclei to form eight **ascospores**.

When favourable conditions return, the ascus splits open to release the ascospores. These spores germinate and produce mature yeast cells.

Importance

Yeasts are used by the baking, brewing, and wine-making industries, because they can change certain sugars into ethyl alcohol and carbon dioxide.

Baker's yeast multiplies and grows rapidly in dough. As it does so, it

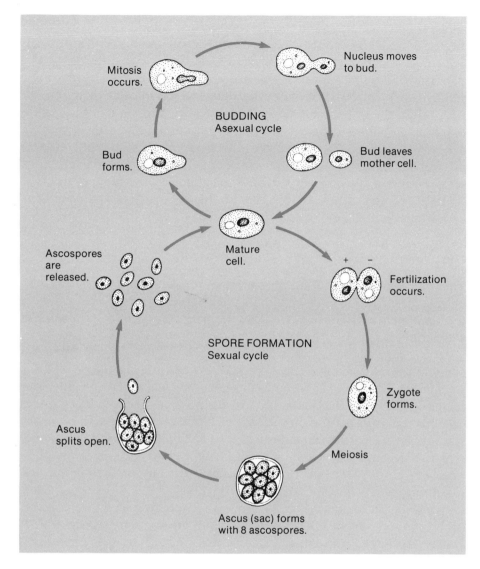

Fig. 11-16   Life cycle of yeast.

Mitosis occurs.

Nucleus moves to bud.

BUDDING
Asexual cycle

Bud forms.

Bud leaves mother cell.

Mature cell.

Ascospores are released.

+ − Fertilization occurs.

SPORE FORMATION
Sexual cycle

Zygote forms.

Ascus splits open.

Meiosis

Ascus (sac) forms with 8 ascospores.

gives off carbon dioxide. Bubbles of this gas cause the dough to rise. The ethyl alcohol that is produced is driven off by the heat.

Brewer's yeasts and wine-making yeasts are special species that are selected and cultured because of their ability to produce high alcohol concentrations and certain tastes. In wine-making, the yeasts act on sugars in ripe fruits such as grapes. In brewing and in making some spirits, the yeasts act on sugars that are formed by the digestion of starches in grains such as barley, corn, and rye.

Apple Scab

Importance

Apple scab is a parasitic fungus that causes great economic losses for apple growers in many regions of Canada. You have likely seen this disease on

Fig. 11-17   Apple scab. Minor infections simply change the appearance of the apple. Serious infections destroy the shape and stunt the growth of the apple.

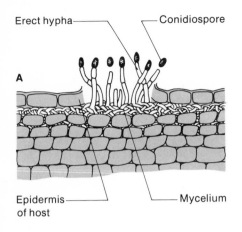

Erect hypha — Conidiospore

A

Epidermis of host — Mycelium

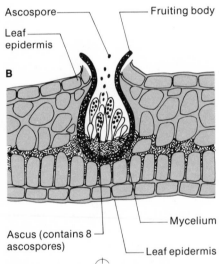

Ascospore — Fruiting body

Leaf epidermis

B

Ascus (contains 8 ascospores) — Mycelium

Leaf epidermis

Fig. 11-18 The parasitic (A) and saprophytic (B) stages of apple scab.

apples (Fig. 11-17). It appears as circular scabs on both the leaves and fruit. At first the infection is olive-green and soft. Later it becomes brown or black and rough. Often scabs run together to form large black masses.

The fungus first appears in the spring. It infects the young leaves, twigs, and blossoms. Infected leaves are often damaged so much that they drop off. As a result, the tree cannot produce as much food to store in the apples. Many young apples drop off before they reach maturity. Others grow to maturity but are small and deformed. Such apples are so unsightly that no one will buy them. The scabs often crack open. This lets bacterial infections enter the apples, causing them to rot. Apples that become infected late in the summer are not deformed or stunted. However, the scabs ruin their appearance and make them difficult to sell.

Life cycle

The cycle begins in the spring when an ascospore lands on a moist fruit, leaf, or twig. The fruit, leaf, or twig must remain moist for many hours before germination occurs. Since water aids germination, the disease is the worst in wet periods.

The germinating spore sends out **hyphae** just under the epidermis (skin) of the infected plant (Fig. 11-18,A). In about 10-15 d the **mycelium** is well-developed. It breaks through the epidermis and sends up many **erect hyphae**. These form the young scab that is olive-green and soft. Black spores called **conidiospores** form on the ends of the erect hyphae. At this time the scab turns brown or black and rough. The conidiospores are continuously sent into the air. If they land on a moist fruit, leaf, or twig, they start a new infection. In dry weather, they can do no harm. In wet weather, a new infection of scabs can appear every 10-15 d.

During the winter the fungus lives as a saprophyte on the apple leaves on the ground. In this stage, the mycelium grows deep into the leaf. A **fruiting body** is formed that contains several **asci**, or sacs. Each ascus contains eight **ascospores** (Fig. 11-18,B).

When the spring rains come, the fruiting body opens. The asci burst, shooting the ascospores into the air. These spores may land on moist fruit, leaves, and twigs to begin the cycle again.

Control

Like most fungus diseases, apple scab can be controlled by breaking the life cycle at some point. For example, an apple grower could rake up the dead leaves in the fall and burn them. This would prevent ascospore formation the following spring. The grower could also plough the leaves under in the fall. However, both these methods are impractical. They involve a great deal of time and expense. Also, they are of little value unless neighbouring growers do likewise.

In most orchards, apple scab is controlled by the use of **fungicides** Fungicides are chemicals that kill fungal hyphae and spores. The fruit,

leaves, and twigs are sprayed with a fungicide until they are completely wet (Fig. 11-19). Spraying must be done early in the spring to prevent ascospores from infecting the trees. It must be repeated several times during the growing season to prevent condiospores from infecting the trees. At least five sprayings are used. Many growers use up to fifteen sprayings, particularly in wet years.

Other Ascomycetes

Blue-green moulds

Penicillium is a well-known blue-green mould. It is best known for the production of antibiotics by certain species. Another species, *Penicillium roqueforte,* produces the special taste and colour of roquefort cheese. However, the same species often destroys the taste of other cheeses.

Many species of *Penicillium* and its closely related genus *Aspergillus* are destructive saprophytes. They attack most moist organic materials, including leather, cloth, bread, meat, oranges, lemons, and other fruits (Fig. 11-20). The hyphae of these species penetrate deep into the organic material. They secrete enzymes to digest the material and then absorb the digested material for nutrition. This enzyme action usually destroys the organic material.

A few species of *Penicillium* are pathogenic to humans. The conidiospores of these species germinate on the skin and even in the lungs. The disease is usually not dangerous on the skin. However, it can be troublesome and serious when it occurs in the lungs.

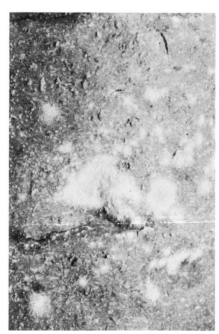

Fig. 11-20 This blue-green mould destroys many organic materials by digesting them.

Fig. 11-21 The gray patches on this leaf are a powdery mildew.

Fig. 11-22 The morel, *Morchella*, is highly prized for its flavour.

Powdery mildews

These fungi are serious plant pathogens. They attack a wide variety of food crops, including wheat, barley, and other grains. They also attack fruits such as grapes and apples. In addition, they attack many ornamental plants such as roses and lilac trees (Fig. 11-21).

When a spore from one of these ascomycetes lands on a leaf of the host, it produces a mass of hyphae on the surface of the leaf. Short hyphae penetrate the leaf and absorb food from it. The surface hyphae produce numerous gray spores. They become so numerous that the surface of the leaf appears to have been dusted with powder. By damaging the leaves, powdery mildews reduce photosynthesis by the plant. As a result, growth and development of the plant are affected.

Powdery mildews are especially infectious if the surfaces of the leaves are damp. Water assists the germination of the spores. Thus rose gardeners try to avoid wetting the leaves of the rose bushes when they water their gardens. Grape growers seldom irrigate their vineyards at night, because the leaves would remain damp for too long a period.

Edible ascomycetes

Although **morels** look like mushrooms, they are ascomycetes, not basidiomycetes (Fig. 11-22). Because of their delicate flavour, they are collected from the woods in the spring. However, they have not yet been cultivated like the field mushroom. Although they are one of the easiest fungi to identify, care must be taken when collecting them. They could be confused with some deadly mushrooms.

The **truffles** are highly prized as gourmet foods, particularly in Europe. These ascomycetes grow and complete their entire life cycle underground. Therefore they are very difficult to find and quite expensive. However, they do have a strange odour that pigs can easily smell. As a result, many truffle hunters train pigs as "truffle sniffers". Since pigs are not easily led, they are often transported to the truffle grounds in wheelbarrows. Then they are released to sniff out the truffles.

More plant pathogens

Three destructive plant pathogens are ascomycetes. The fungus that causes **Dutch elm disease** is one of these. This fungus has almost eliminated the elm tree from the parks, streets, and woods of most of Canada (Fig. 11-23).

The American chestnut used to be one of the largest, most abundant, and most valuable tree species in the forests of the eastern United States. Around 1925, the **chestnut blight** infected these trees. This ascomycete apparently came from the Orient. Today a few dead trees still stand to remind us of the tragedy for which there has been no equal in any forest on earth. Sprouts still grow from some old stumps. However, they soon are killed by the fungus.

Ergot of rye is one of the most dangerous fungi. This ascomycete parasitizes many grasses, including cultivated rye. It reduces the yield of rye in many parts of the world. It also produces a poisonous substance that can kill humans and domestic animals that eat the rye or products made from it.

Review

1. **a)** Why was this division of living things given the name Ascomycota?
   b) Name five common ascomycetes.
   c) Why are the ascomycetes placed in the kingdom Fungi?
2. **a)** Describe where yeast occurs.
   b) Describe the structure of a mature yeast cell.
   c) How do yeast cells obtain the energy that they require?
3. **a)** Describe asexual reproduction of yeasts.
   b) Describe sexual reproduction of yeasts.
   c) During sexual reproduction, yeast cells of opposite sexual strains join to form a zygote. These cells are called + and – instead of male and female. Why?
4. **a)** Describe the importance of the yeasts.
   b) Describe the importance of apple scab.

5. Summarize the life cycle of apple scab by drawing a diagram that is similar to those shown for other fungi in this chapter.
6. If possible, obtain an apple that is infected with apple scab. Wild trees are almost always infected. Examine the infection with a hand lens and answer these questions:
    a) How has the fungus affected the shape of the apple?
    b) How has the fungus affected the size of the apple?
    c) Describe any evidence that the fungus has made the apple more susceptible to bacterial infections.
7. Some species of *Penicillium* are beneficial; others are harmful. Explain this statement.
8. a) What is a powdery mildew?
    b) How can this disease be partly controlled without using a fungicide?
    c) Name two forms of edible ascomycetes.
9. Select one of Dutch elm disease, American chestnut blight, or ergot of rye. Research the disease in the library. The Ministry of Agriculture and the Ministry of Natural Resources also have information on these diseases. Write a report of about 300 words that describes the source of the disease, how it is spread, its effects, and how it can be controlled.

11.7 Investigation

Culture, Structure, and Function of Yeast

In this investigation you will culture (grow) yeast cells under almost ideal conditions. You will then study the structure of the cells, their nutrition, and reproduction.

Materials

test tube rack
test tubes (4)
10% aqueous solution of sugar
granular yeast
iodine solution
methylene blue stain
one-hole rubber stoppers (2)
glass elbows (2)
rubber tubing (2)
limewater

oil (e.g. cooking oil)
compound microscope
microscope slide
cover slip
medicine dropper
lens paper
paper towel

Procedure

a. Fill two test tubes about two-thirds full with 10% sugar solution. Label one test tube "Control" and the other "Experimental".
b. Add a pinch of granular yeast to the "Experimental" test tube. Shake the test tube to mix the yeast well with the sugar solution.

c. Examine a drop of the mixture from each test tube under low, medium, and high power. Note the structure of the cells.

d. Stain a fresh sample of the mixture from the "Experimental" test tube with iodine solution. Note any additional structures that you can see.

e. Repeat step d using methylene blue stain.

f. Connect each test tube to the assembly shown in Figure 11-24. The end of the rubber tubing should be covered by about 5 cm of limewater. Cover the limewater with a thin layer of oil to keep air away from the limewater.

g. Stand the test tubes in a test tube rack. Store them in a warm place, preferably at 25°C to 30°C. Note any changes that occur during the next two days. Record any changes in appearance and odour.

h. After two days, examine a drop of the mixture from the "Experimental" test tube under low, medium, and high power. Examine an unstained mount, a mount stained with iodine solution, and a mount stained with methylene blue. Note and sketch any changes that took place in the appearance of the yeast cells.

Discussion

1. What is the function of the "Control" test tube in this investigation?
2. a) Describe yeast cells as they appear in the unstained mount.
   b) What structures were made visible by Lugol's iodine stain?
   c) What further structures were made visible by methylene blue stain?
3. a) Limewater turns cloudy in the presence of carbon dioxide gas. Excess carbon dioxide may turn it clear again. What do you conclude from the change that occurred in the limewater in this investigation?
   b) What other evidence did you see to support your conclusion?
4. a) Describe the change in odour that occurred in this investigation. What product is responsible for this change? (Hint: Review "Anaerobic Respiration" in Section 7.5, page 161.)
   b) Write a summation equation for anaerobic respiration by the yeast cells in the mixture.
5. Describe and account for any changes that occurred in the appearance of the yeast cells after two days.

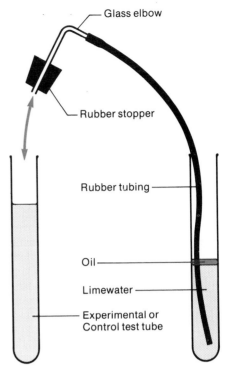

Fig. 11-24 **Studying the properties of yeast.**

11.8 Other Fungi

You have carried out a fairly detailed study of three divisions of fungi: Basidiomycota (club fungi), Zygomycota (conjugation fungi), and Ascomycota (sac fungi). This section briefly describes the form and function of fungi that belong to three other divisions: Oomycota (water moulds), Myxomycota (slime moulds), and Deuteromycota (imperfect fungi).

Division Oomycota: Water Moulds

About 500 species of fungi are classified as **oomycetes**. Like the zygomycetes, the oomycetes consist of filamentous fungi in which the hyphae lack cross-walls. However, in this division, sexual reproduction is by the fertilization of large eggs in an **oogonium** (egg-producing body) by nuclei from a male hypha. The hypha functions as an **antheridium** (sperm-producing body). Oomycetes are common in almost any aquatic or damp terrestrial habitat. Thus, they are called **water moulds**.

A few water moulds are parasitic. One of these, the **potato blight**, played an important role in world history. In 1845 and 1846 it destroyed the entire potato crop in Ireland, causing a disastrous famine. Hundreds of thousands of Irish people died. Large numbers immigrated to the United States and Canada.

The hyphae of this fungus penetrate the leaves of the potato plant. They destroy most of the leaves, thereby preventing them from making carbohydrates for storage in the potato tuber. As a result, the potatoes are very small. The fungus can also invade the soil and attack the tubers directly, causing them to rot. This fungus also attacks a close relative of the potato, the tomato.

Water moulds of the genus *Saprolegnia* often parasitize fish. You may have seen a grey fluffy mass of this fungus on an aquarium fish (Fig. 11-25). The hyphae of this fungus invade the body of the fish. They rob the host of food and, in addition, make it possible for bacterial infections to enter. In the end, the host usually dies. The fungus usually feeds as a saprophyte on the dead fish.

Saprolegnia and many other water moulds can be collected by placing a dead insect in some water, preferably aquarium or pond water. Within a few days, the insect will usually be completely covered by the fluffy mycelium of a water mould.

Fig. 11-25 This fish was killed by a severe infection of a water mould of the genus *Saprolegnia*. You can see the mould near the fin.

Division Myxomycota: Slime Moulds

About 400 species in the kingdom Fungi are classified as **myxomycetes** or *slime moulds*. These fascinating organisms are difficult to classify. In some ways they are much like giant amoebas. As a result, they could be placed in the kingdom Protista. However, they form sporangia and spores like fungi. Therefore, they could also be placed in the kingdom Fungi. Most biologists prefer not to call slime moulds true fungi. However, the five-kingdom system of classification places them in the kingdom Fungi because they are more like fungi than like any other kingdom.

The thallus (plant body) of a slime mould is called a **plasmodium**. It consists of a mass of protoplasm with many nuclei. However, the protoplasm is not divided into cells. The plasmodium is surrounded by a flexible plasma membrane that allows it to change shape.

The plasmodium usually looks like a slimy mass of living material (Fig. 11-26). It occurs on decaying organic matter such as rotting logs, dead leaves, and the soil of woodlots. It moves over the organic matter by amoeboid movement. As it moves, it ingests particles of the organic matter. It prefers bacteria for food. Digestion takes place in food vacuoles.

Asexual reproduction of a slime mould occurs by **fragmentation**. Pieces of the organism break away from the main thallus. If a piece has one or more nuclei in it, it can develop into a new thallus.

Asexual reproduction requires moisture and a moderate temperature. Therefore it usually stops during dry or cold periods. At this time, sexual reproduction often occurs. The plasmodium becomes inactive and forms **sporangia**. **Spores** are produced within the sporangia. When they are mature, the spores are released into the air. If a spore lands in a moist environment, it germinates to form an amoeboid cell or a flagellated cell. Two such cells, of opposite sexual strains, join to form a zygote. The zygote divides many times to form a mature plasmodium.

Fig. 11-26 The slime mould, *Physarum*. This organism moves over decaying organic matter like a giant amoeba.

Division Deuteromycota: Imperfect Fungi

Sexual reproduction has been observed in the five divisions of fungi that have been discussed so far in this chapter. However, many fungus-like organisms exist for which sexual reproduction has not been observed. These organisms reproduce asexually by spore formation. Such organisms are classified as **deuteromycetes**, or **imperfect fungi**. If, at any time, sexual reproduction is observed in one of these fungi, it is moved to one of the other divisions.

Many imperfect fungi are parasites of animals and plants. For example, athlete's foot and ringworm are imperfect fungi, as are several plant diseases of tomatoes, cabbages, lettuce, apples, and grains.

Review

1. **a)** Why are water moulds given the division name Oomycota?
   b) How are water moulds similar to zygomycetes like bread mould?

c) How do water moulds differ from zygomycetes?
2. a) Name an oomycete that is a plant parasite.
   b) Why is it called a parasite?
   c) Describe its effects on the host plant.
3. a) Name an oomycete that is an animal parasite.
   b) When might this parasite become a saprophyte?
   c) Describe its effects on the host animal.
4. a) Account for the classification of slime moulds in the kingdom Fungi.
   b) What is a plasmodium?
   c) Why are slime moulds usually found on decaying organic matter?
   d) Describe the asexual reproduction of slime moulds.
   e) Describe the sexual reproduction of slime moulds.
5. a) What are imperfect fungi?
   b) Name two parasitic fungi that attack humans. Find out how these fungi can be destroyed.

11.9 Lichens

About 15 000 species of organisms are classified as lichens. *A lichen is actually a close association of a fungus and an alga.* Therefore, some botanists classify lichens as fungi. Others classify them as algae. Still others classify the fungus and alga of a lichen separately.

Characteristics

Structure

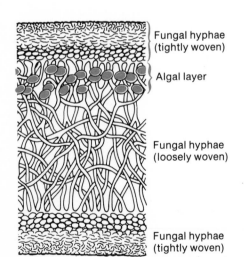

Fungal hyphae (tightly woven)

Algal layer

Fungal hyphae (loosely woven)

Fungal hyphae (tightly woven)

Fig. 11-27 Structure of a generalized lichen.

The thallus (plant body) of a lichen consists of an alga and a fungus. The alga is a green or blue-green alga. The fungus is usually an ascomycete. In a few genera, it is a basidiomycete.

The hyphae of the fungus wind through and around clusters of algal cells (Fig. 11-27). The association of the hyphae and algal cells is so close that the lichen appears to be a single organism.

Most lichens consist of an upper and lower layer made of tightly woven fungal hyphae. Just beneath the upper layer are the algal cells. Loosely woven fungal hyphae occur in the centre and among the algal cells.

Symbiosis

A symbiotic relationship exists between the alga and the fungus in a lichen. In many species, this relationship is **mutualism**. (You may recall that mutualism is a relationship between two organisms in which both organisms benefit.) The alga is autotrophic. It can manufacture foods through photosynthesis. It shares these foods with the fungus. The fungus, in turn, provides the alga with water, minerals, and support.

Recent studies have shown that the symbiotic relationship is more correctly called **parasitism** in some species. The fungus is parasitic on the

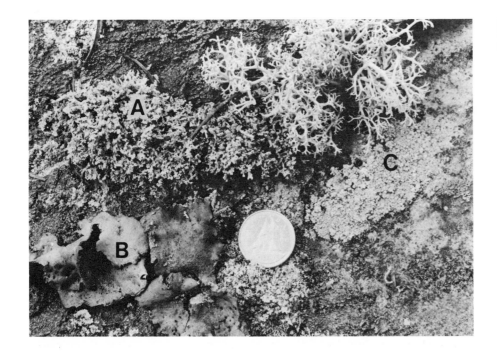

Fig. 11-28   The three general types of lichens: fructicose (A), foliose (B), and cructose (C).

alga. It actually grows an "algal garden" for its own benefit. Fungal hyphae have been observed penetrating algal cells, just as the hyphae of plant parasites such as apple scab penetrate the host plant.

Types

Three general types of lichens exist (Fig. 11-28). **Crustose** lichens appear as finely textured, coloured patches on rocks and trees. They send rhizoids (root-like hyphae) into the rock or tree. **Foliose** lichens are leaf-like in appearance. They are also attached to rocks and trees by hyphae. However, the hyphae are usually not as firmly attached as those of crustose lichens. **Fruticose** lichens are many shapes. However, most are lacy and branched like the reindeer moss, *Cladonia,* shown in Figure 11-28.

Occurrence and Importance

Occurrence

Lichens occur in almost all environments, from the Arctic tundra, to the tropical rain forest, to the desert.

In moist tropical and temperate regions, lichens often are epiphytes. *An **epiphyte** is an organism that lives as a commensal in the branches of trees.* (Recall that **commensalism** is a relationship in which one organism benefits and the other neither benefits nor suffers harm.) For example, "Spanish moss" is a common epiphyte on oak trees in the southern United States. This fruticose lichen uses the oak tree as a home but causes no harm to the tree.

Fig. 11-29 The vegetation of the Arctic tundra is a mixture of lichens, grasses, sedges, and other plants.

Crustose lichens are the most common type in cold climates, although the other two types are present as well. In the Arctic tundra, lichens are among the most common organisms (Fig. 11-29). They form a spongy mat many centimetres thick. Lichens are also common at the tops of mountains. Few other organisms can survive the harsh conditions of the alpine environment.

Lichens are also common in deserts. Most desert lichens are of the crustose type. They are hard to see except in the spring and after a rainfall. At those times, they often cover the desert with a brilliant display of colour.

Importance

Many Arctic herbivores graze on lichens such as reindeer moss. However, the nutritional value of lichens is generally quite low.

Lichens are most important for the role they play in plant succession. **Plant succession** *is the replacement of one community of plants by another.* It occurs when one species of plant invades a certain habitat and changes it so another species of plant can live in it.

Lichens are the **pioneer species** of plant succession on rocks. That is, they are the first plant-like organisms to colonize rocks. Generally crustose and foliose lichens invade the rock first. To obtain needed minerals, these lichens send rhizoids into the rocks. The rhizoids secrete acids that dissolve the rock so that its minerals can be absorbed. The weakened rock crumbles, forming parent soil material in which mosses and fruticose lichens such as reindeer moss can grow. These branching organisms trap wind-blown earth and organic matter. They also add further organic matter to the soil as they die and decay. Eventually the soil will support the growth of ferns, grasses, and other herbaceous plants. The build-up continues until even trees can grow on the rocks.

You will study more about succession in Chapter 23.

Review

1. **a)** What is a lichen?
   b) Name and describe the two kinds of symbiotic relationships that exist among the lichens.
   c) Name and describe the three general types of lichens.
2. **a)** What is an epiphyte?
   b) Explain why "Spanish moss" is called an epiphyte.
3. **a)** What is plant succession?
   b) What is a pioneer species?
   c) Why are lichens called pioneer species in rocky regions?
   d) Outline the process of plant succession, beginning with bare rock.

11.10 Investigation

Collection and Structure of Lichens

Lichens are many colours. Most of them are gray or grayish-green. However, others are brown, black, yellow, orange, and white. The crustose type is usually hard and finely textured. The foliose type has a flattened leafy appearance. The fruticose type is usually lacy and branched.

In this investigation you will collect lichens from many habitats. Then you will study their structure.

Materials

bags for collecting specimens	medicine dropper
lichen specimens	lens paper
hand lens	paper towel
compound microscope	dissecting needle
microscope slide	forceps
cover slip	

Procedure A   Collection of Lichens

Most species of lichen are killed by air pollution. Therefore lichens are not common in or near cities. As a result, you can do this part of the investigation only if you do not live in a city or if you make a trip to the country.

Examine rocks, tree bark, fence posts, and soils for lichens. When you find a lichen, carefully remove a small piece. Place it in a bag with a note that gives the date, the place of collection, and the type of lichen (crustose, foliose, or fruticose).

Procedure B   Structure of Lichens

For this part of the investigation, use the samples you collected or samples provided by your teacher.

a. Use a hand lens to study the overall structure of each of the three types of lichens. Study the shape of the thallus and the colours of the upper and lower surfaces for each type. Look for evidence of branching and for rhizoids or other holdfasts. Look for fruiting bodies. Make a careful sketch of each type.
b. Prepare a wet mount of a lichen as follows. Place a small piece of lichen on a microscope slide in a drop of water. Carefully break it apart with a dissecting needle to separate the hyphae and the algal cells. Cover the sample with a cover slip.
c. Study the mount under low and high power. Note the colour and shape of the algal cells. Note the colour and shape of the fungal hyphae. Do the

algae and hyphae touch one another? Do any hyphae penetrate algal cells?

d. If time permits, repeat steps b and c for other types of lichens.

Discussion

1. Prepare a table that has the following column headings: Place of Collection; Type of Lichen; Description of Lichen. Complete the first two columns using information you collected in the field. Complete the third column using information you collected in the laboratory using a hand lens.
2. In step c of Procedure B you studied the structure of a lichen. Record all the information you discovered. What evidence did you see that a symbiotic relationship existed between the algal and the fungus? Do you think the relationship was mutualism or parasitism? Why?

Highlights

Fungi lack chlorophyll. The thallus of most fungi is made of hyphae. Most fungi reproduce both asexually and sexually.

Fungi are efficient decomposers of organic matter. Some of this decomposing action is beneficial. Fungi help to break down organic matter and return nutrients to the soil. However, some saprophytic fungi destroy leather, paper, and other valuable goods. Parasitic fungi attack many plants and animals, including humans.

Fungi are classified into seven divisions. The division Basidiomycota includes mushrooms, wheat rust, and corn smut. All produce spores on a special structure called a basidium. The division Zygomycota includes black moulds such as bread mould. These moulds reproduce by conjugation, during which a zygospore is produced. The division Ascomycota includes yeasts, blue-green moulds, powdery mildews, and several serious plant pathogens. These fungi produce spores in sacs called asci. Water moulds, slime moulds, and imperfect fungi make up three more divisions of fungi.

Lichens are fascinating organisms that consist of an alga and a fungus growing in close association with one another. The relationship is mutualistic in some species and parasitic in others. Three general types of lichens exist: crustose, foliose, and fruticose.

UNIT FOUR

Plant Form And Function

In Unit Three you studied the form and function of organisms in the three simplest kingdoms of living things: kingdom Monera, kingdom Protista, and kingdom Fungi. In this unit you will investigate the form and function of organisms in the fourth kingdom, the kingdom Plantae.

12 Plant Diversity

Over 300 000 species of living things are classified in the kingdom Plantae. Some botanists estimate that only about half or two-thirds of the plants now living on earth have been discovered and classified. Therefore, over 500 000 species could be living at this time.

Close estimates of the number of species in each division cannot be given. However, to give you an idea of the diversity in the plant kingdom, Table 12-1 shows the ten main divisions of the kingdom Plantae and the approximate number of known species in each of these divisions.

DIVISION	COMMON NAME	NUMBER OF SPECIES
Rhodophyta	red algae	2 500
Phaeophyta	brown algae	1 500
Chlorophyta	green algae	6 000
Charophyta	stoneworts	100
Bryophyta	mosses and liverworts	24 000
Lycopodophyta	club mosses	1 100
Arthrophyta	horsetails	30
Pterophyta	ferns	10 000
Coniferophyta	conifers	550
Anthophyta	flowering plants	250 000

TABLE 12-1   **The Main Divisions of the Kingdom Plantae**

This chapter describes the common characteristics of the organisms in the kingdom Plantae. It also gives an overview of the ten main divisions. Then, the remaining chapters of Unit Four describe more fully the form and function of some representative organisms in each division.

12.1 Characteristics of the Kingdom Plantae

The organisms in the kingdom Plantae share most of the following characteristics:

1. Most plants cannot move from place to place. They are anchored in one place by roots, rhizoids, or holdfasts of some sort.
2. All plants are made of cells. A few of the simplest plants are one-celled. However, most plants are many-celled (Fig. 12-1).

   The simplest many-celled plants are made of cells that are all alike. For example, most of the algae that make up pond scum are filaments of identical cells. However, most many-celled plants are made of cells that have differentiated to form specialized tissues, organs, and systems. For example, a flowering plant is made of millions of cells that differ widely in both form and function.

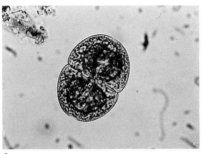
A

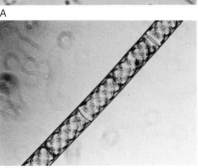
B

C

Fig. 12-1 Three levels of complexity in the cellular structure of plants. The desmid (A) is a one-celled plant. The green alga (B) consists of many cells. The flowering plant (C) consists of cells that are specialized to perform certain functions.

3. The cell walls of most plant cells are made largely of **cellulose**.
4. The cells of most plants have a nucleus and organelles such as mitochondria, Golgi bodies, lysosomes, and ribosomes.
5. Most plant cells contain **chlorophyll** in plastids called chloroplasts. As a result, most plants are **autotrophic**. This means they can make their own food through photosynthesis.

   Plastids often contain other pigments such as xanthophylls, which range from pale yellow to brown in colour, and carotenes, which range from yellow to orange in colour.

   The colour of a plant is usually determined by the balance of its pigments. If chlorophylls dominate, the plant is green. If brown xanthophylls dominate, the plant is brown.

6. Many plants can reproduce asexually by processes such as **fragmentation**. During fragmentation, a small piece of the plant breaks off. It then begins to grow by itself. Only mitosis is involved in asexual reproduction.

Most plants can also reproduce sexually. During sexual reproduction, gametes are produced. The gametes unite to form a zygote. The gametes have the haploid (n) number of chromosomes and the zygote has the diploid (2n) number of chromosomes.

All plants and, for that matter, all living things have a life cycle. A **life cycle** is the course of events occurring between the appearance of a mature organism of one generation and the development of a mature organism of the next generation. For example, Figure 12-2 shows the life cycle of a human. Mature male and female humans, which are diploid, produce gametes by meiosis. The gametes are haploid. These join through fertilization to form a zygote which is diploid. The zygote grows and develops into a mature human.

Fig. 12-2 Life cycle of a human.

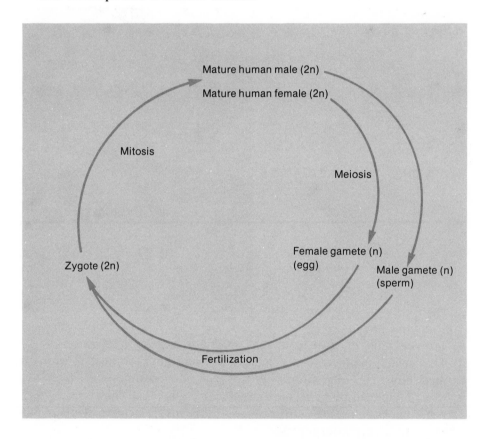

The life cycle of most plants is quite different from that of humans and most other animals. One basic difference is that the mature *diploid* plant of most species does not form gametes directly as does a mature diploid animal. Instead, it produces spores by meiosis (Fig. 12-3). Each spore germinates and grows into a *haploid* plant called a **gametophyte**. This plant produces the gametes. These gametes unite to produce a zygote. The zygote,

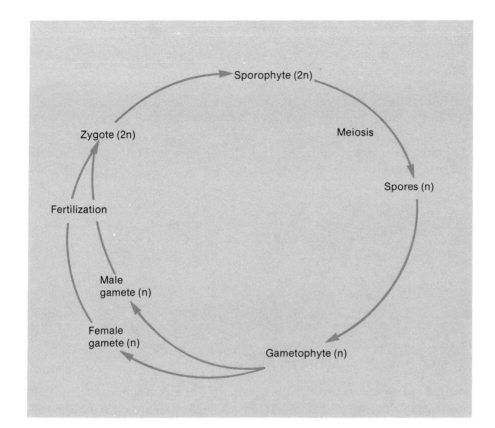

Fig. 12-3 The life cycles of most plants follow this pattern: an alternation of generations.

which is diploid, grows and develops into a *diploid* plant called a **sporophyte**. The sporophyte produces spores by meiosis to continue the cycle. Note that *gametophyte* means "gamete-producing plant" and *sporophyte* means "spore-producing plant".

With few exceptions, the life cycles of plants are similar to this general example. They alternate between a sporophyte generation of plants and a gametophyte generation of plants. Biologists call such a cycle an **alternation of generations**. Of course, the various species of plants differ in the form of the two generations and in the length of time spent in each stage.

Review

1. **a)** How do organisms in the kingdom Plantae differ from organisms in the kingdom Monera? (See Section 9.1, page 194.)
   b) How are plants and monerans alike?
2. **a)** How do organisms in the kingdom Plantae differ from organisms in the kingdom Protista? (See Section 10.1, page 245.)
   b) How are plants and protists alike?
3. **a)** How do organisms in the kingdom Plantae differ from organisms in the kingdom Fungi? (See Section 11.1, page 277.)
   b) How are plants and fungi alike?

12.2 A Survey of the Main Plant Divisions

Division Rhodophyta: Red Algae

The thallus (plant body) of most red algae is many-celled. Red algae contain red and brown pigments, as well as chlorophyll, xanthophylls, and carotenes.

Most red algae occur in marine (salt water) habitats. They prefer warm water. They grow either floating or attached to rocks in tidal zones.

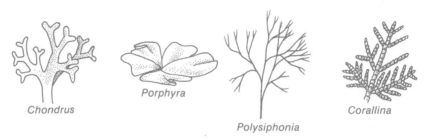

Chondrus   *Porphyra*   *Polysiphonia*   *Corallina*

Division Phaeophyta: Brown Algae

Like the red algae, the brown algae are mainly marine organisms. However, most brown algae prefer cold water. The thallus of all brown algae is many-celled. Brown algae contain brown pigments as well as chlorophyll.

Brown algae vary in size from microscopic filamentous species to huge kelps up to 30 m long. Most brown algae have holdfasts that attach them to rocks.

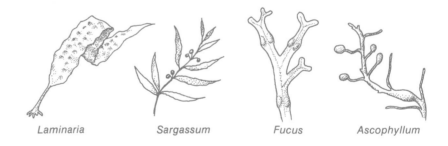

Laminaria   *Sargassum*   *Fucus*   *Ascophyllum*

Division Chlorophyta: Green Algae

Green algae occur in both freshwater and marine habitats. Chlorophyll usually dominates other pigments. As a result, these algae normally appear green.

The thallus of some green algae is one-celled. The thallus of others consists of a filament of a few identical cells. Many green algae however, have a thallus that consists of millions of cells.

Green algae store food in the form of starch.

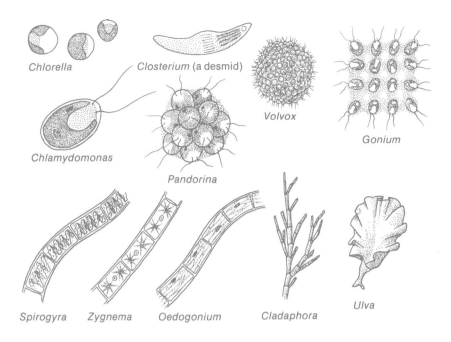

Chlorella

Closterium (a desmid)

Volvox

Gonium

Chlamydomonas

Pandorina

Spirogyra Zygnema Oedogonium Cladaphora

Ulva

Division Charophyta: Stoneworts

These strange plants are divided into nodes and internodes, with a whorl (circle) of branches at each node. They store food in the form of starch like the green algae do. However, in most respects they differ greatly from the green algae.

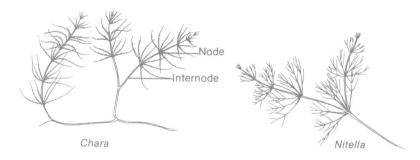

Node

Internode

Chara

Nitella

Division Bryophyta: Mosses and Liverworts

Bryophytes are small many-celled plants. Most are under 20 cm tall and all are less than 70 cm tall. Bryophytes lack true roots, stems, and leaves. They also lack conducting tissue. However, most of them have structures that resemble stems and leaves.

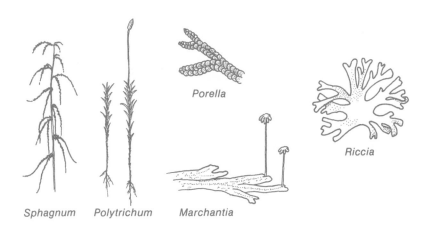

Porella

Riccia

Sphagnum   Polytrichum   Marchantia

Division Lycopodophyta: Club Mosses

These small evergreen plants are usually less than 40 cm tall. They have true roots, stems, and leaves. They spread by horizontal stems that run along the surface of the soil or just under the surface. Upright branches grow from these horizontal stems. Club mosses reproduce by spores that are produced on special leaves. In some species these leaves form a club-like cone at the tip of an upright branch.

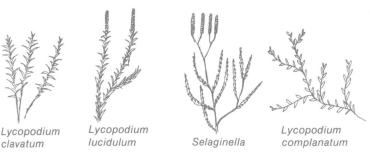

Lycopodium
clavatum

Lycopodium
lucidulum

Selaginella

Lycopodium
complanatum

Division Arthrophyta: Horsetails

Horsetails vary in height from a few centimetres to several metres. The tall species are found in the tropics. These plants have true roots and hollow stems that grow horizontally underground. Jointed, hollow branches grow up from the stems. The leaves are very small. The tips of some branches bear a cone-like body that produces spores.

The tissues of horsetails contain silica, one of the substances in sand. As a result, they are abrasive to the touch. *Equisetum hyemale* is called "scouring rush", because the pioneers used it to scour, or clean, their pots and pans.

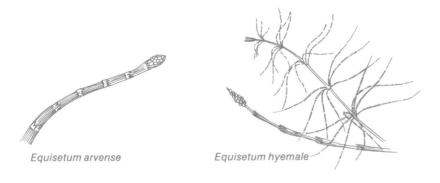

Equisetum arvense

Equisetum hyemale

Division Pterophyta: Ferns

Most ferns are less than 1 m tall. However, some tropical species are as high as trees.

Ferns have true roots, stems, and leaves. The stems are often underground and perennial. New leaves shoot up from them every spring. The leaves of ferns are generally quite large.

Ferns reproduce by spores. The spores often form on the undersides of the leaves. In some species the spores form on special leaves that are quite different from the other leaves.

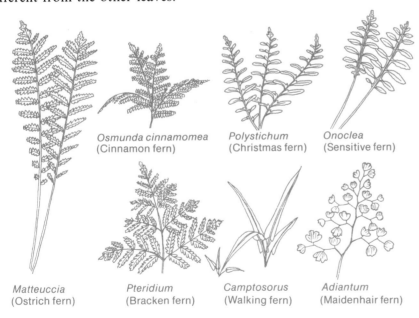

Osmunda cinnamomea
(Cinnamon fern)

Polystichum
(Christmas fern)

Onoclea
(Sensitive fern)

Matteuccia
(Ostrich fern)

Pteridium
(Bracken fern)

Camptosorus
(Walking fern)

Adiantum
(Maidenhair fern)

Division Coniferophyta: Conifers

The conifers are woody perennial trees or shrubs. They produce seeds that are found on the upper surface of scales. The scales are grouped into cones.

The leaves of conifers are usually very small and are evergreen in most species. The tamarack is a native species that is deciduous. That is, it drops its leaves as winter approaches.

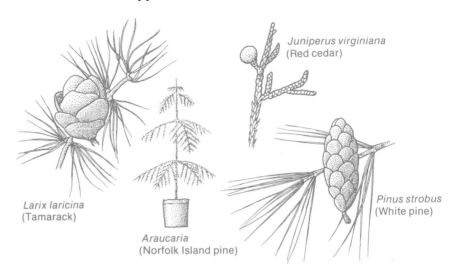

Juniperus virginiana
(Red cedar)

Larix laricina
(Tamarack)

Araucaria
(Norfolk Island pine)

Pinus strobus
(White pine)

Division Anthophyta: Flowering Plants

This division includes a wide variety of shrubs, trees, and herbaceous (soft-stemmed) plants. Among these are the flowers and vegetables that you see in gardens. The pasture grasses that nourish livestock are also flowering plants. Among the most important flowering plants are grains such as wheat, corn, rye, oats, and barley. Although you often do not see the flowers, all trees that are not conifers are flowering plants.

All these plants have flowers at some time in their life cycles. They reproduce by seeds which are formed within ovaries. The ovaries develop into a fruit.

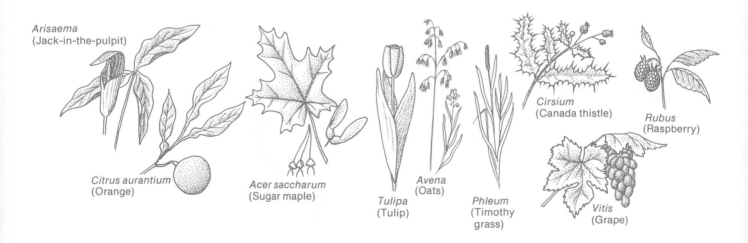

Arisaema
(Jack-in-the-pulpit)

Citrus aurantium
(Orange)

Acer saccharum
(Sugar maple)

Tulipa
(Tulip)

Avena
(Oats)

Phleum
(Timothy grass)

Cirsium
(Canada thistle)

Rubus
(Raspberry)

Vitis
(Grape)

1. Your teacher has prepared several stations around the room. At each station there is a plant that belongs to one of the ten divisions discussed in this section.

    Visit each station and study the plant closely. Decide which division it is in. Record your reasons for placing it in that division.

Highlights

Most plants are anchored in one place. Most plant cells have cell walls made of cellulose. These cells contain chlorophyll in plastids.

The life cycle of most plants show an alternation of generations. That is, they alternate between a sporophyte, or spore-producing plant, and a gametophyte, or gamete-producing plant. The sporophyte is diploid (2n) and the gametophyte is haploid (n).

About 300 000 species of plants have been discovered and classified. Over 250 000 of these are flowering plants. About 24 000 species are mosses and liverworts, while 10 000 species are ferns. Another 10 000 species are green, brown, and red algae. The remainder are largely stoneworts, club mosses, horsetails, and conifers.

13 Simple Green Plants

Most people think of a "plant" as an organism that has roots, stems, and leaves. However, there are many plants that do not have these structures. Such plants are organized very simply and are known as **primitive plants** (Fig. 13-1). Some examples are the algae, the stoneworts, the mosses, and the liverworts. On land, these primitive plants seldom grow very large. This is because they lack the specialized vascular (conducting) tissues to carry water and food throughout the plant body. The plants are small and lie close to the ground, where they can reach water easily. We often overlook these non-vascular plants, even though they are common in many places. In this chapter, you will become familiar with a few of them.

Other primitive plants do contain vascular tissues, but their structure is still simple compared to flowering plants. Examples of these are the ferns, the club mosses, and the horsetails. You may recognize some of these simple vascular plants as you read through the second half of this chapter.

13.1 The Algae

The word "algae" is confusing. It may cause you to think of seaweed or the green scum that forms on quiet ponds. Biologists use the term algae to

Fig. 13-1 Some primitive plants: the green alga, *Cladophora* (A); peat moss, *Sphagnum* (B); the liverwort, *Marchantia* (C); the sensitive fern, *Onoclea* (D); Shining club moss, *Lycopodium lucidulum* (E); the field horsetail, *Equisetum arvense* (F).

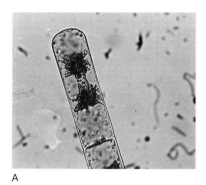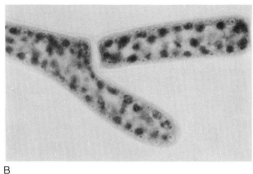

A   B

Fig. 13-2 Two types of filamentous algae. The pond scum, *Zygnema* (A), has a thallus that consists of a single strand of cells. *Cladophora* (B) has a branched thallus. It is found on rocks, in rivers and at the shores of lakes. Which alga do you think has holdfasts?

describe many different groups that are not related to each other. In Chapters 9 and 10, you were introduced to some of the simplest algae, such as blue-green algae and diatoms. Here you will look at a few of the more complex forms.

The plant body of an alga is known as a **thallus** (plural: **thalli**). A thallus contains no vascular tissue and is not separated into roots, stems, or leaves. It may be a single row of cells that forms a threadlike filament. In some filamentous algae, the filaments are branched (Fig. 13-2). In many species, the lowermost cell of the filament is specialized into a "holdfast". A holdfast anchors the filament to rocks or mud.

The thalli of other species of algae are composed of many closely packed filaments that form ribbons or sheets. The largest thalli are the kelps of the Pacific Ocean. Some kelps reach a length of over 30 m.

Although algae are simple in structure, they are very common and very successful. Most live in water, either attached to rocks or floating freely. Freshwater species can be found in fast-flowing streams, lakes, quiet ponds, or even drainage ditches. Saltwater species are most common near shores and in tidal zones. Some are found in the surface waters of the open ocean where sunlight penetrates. All algae use sunlight to produce food by photosynthesis. In fact, algae are the main food producers in the oceans. To produce food they require the green pigment chlorophyll, which is found in the cells of every species. One of the by-products of photosynthesis is oxygen. Because of their great numbers, the algae are, therefore, important suppliers of oxygen.

Review

1. What is meant by "primitive plant"?
2. Why is the word algae confusing?
3. **a)** What is a thallus?
   b) What is a holdfast?
4. List a few places where you might go to find algae.
5. In what ways are algae important to us?

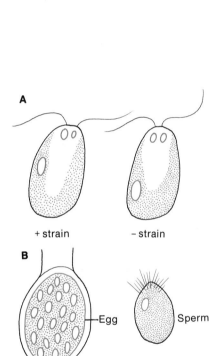

A

+ strain   – strain

B

—Egg   Sperm

Fig. 13-3 Sexual reproductive cells. In some species of algae, all gametes appear the same. These are called isogametes (A). In other species the male and female gametes are not alike. These are called heterogametes (B).

13.2 The Classification of Algae

For a long time, the classification of algae into groups was based on their colours. Although all algae contain the green pigment chlorophyll, the green colour is often masked by other pigments in the cells. Some algae are red, others are brown, some are green, and some are yellowish. However, since the colour of two closely related species may be different, features other than colour are also used for classification.

Biologists have found that different groups of algae store different kinds of food products in their cells. The type of storage product has therefore become a useful classification tool. The size and structure of the thallus and the method of reproduction are also important for separating the thousands of species of algae into groups.

Reproduction of Algae

Algae reproduce in many different ways. Most have both sexual and asexual methods of reproducing. In **sexual reproduction**, specialized cells called gametes are produced. A **gamete** is a sex cell that unites with another gamete to produce a new plant. Male gametes are usually called **sperm** and female gametes, **eggs**. If the sperm and eggs are alike in size and shape, they are called isogametes (Fig. 13-3). In this case, they are called positive strain and negative strain instead of sperm and egg. In many algae, reproduction is isogamous. This of course means that the sperm and egg involved are isogametes. If the gametes are two different kinds, they are called **heterogametes**. In heterogamous algae, the female gametes are usually large and unable to move. The sperm are smaller and may move by using whiplike threads that lash the water. When a sperm reaches an egg cell they unite to form a **zygote**. The zygote then grows into a new plant.

In **asexual reproduction**, no gametes are produced and no union of cells occurs. Instead, several things might happen. A unicellular alga may simply divide into two new cells by mitosis. A filament may break apart into several pieces, each of which can survive as a separate plant. This breaking apart is called **fragmentation**. Some algae reproduce asexually by producing zoospores. **Zoospores** are small reproductive cells with at least two threads called flagella that allow them to swim about (Fig. 13-4). Each zoospore is able to develop into a plant.

Divisions of Algae

Using colour, method of food storage, size and structure of thallus, and method of reproduction, biologists have classified most of the multicellular algae into three divisions:

The **Chlorophyta**, or green algae;
The **Phaeophyta**, or brown algae;
The **Rhodophyta**, or red algae.

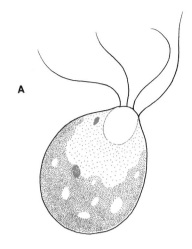

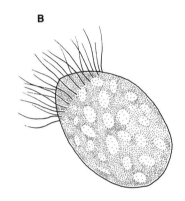

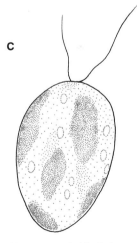

Fig. 13-4   Zoospores. A: *Ulothrix* zoospore with four flagella. B: *Oedogonium* zoospore with ring of many flagella. C: *Tribonema* zoospore with two flagella.

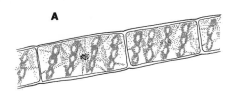

A

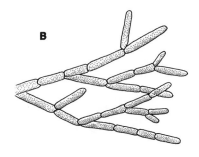

B

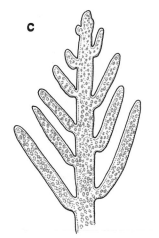

C

Fig. 13-5 Some representative green algae. *Spirogyra*, a simple filament (A); *Cladophora*, a branched filament (B); *Bryopsis*, a tubular type (C).

All these divisions except the Phaeophyta include unicellular as well as multicellular species.

Review

1. What four characteristics do biologists use to classify algae into groups?
2. What is the difference between isogametes and heterogametes?
3. What are three asexual methods of reproduction in algae?
4. What is a zoospore?

13.3 The Green Algae

The Chlorophyta, or green algae, are among the most abundant and most widely distributed of all algae. Both unicellular and multicellular forms are found almost everywhere. The multicellular types include simple filaments, branched filaments, flat sheets of cells, and tubular forms. A few of the more common ones are illustrated in Figure 13-5. The flat, thin blade of the green alga *Ulva* looks so much like a leaf of lettuce, that it is often called "Sea Lettuce" (Fig. 13-6). Its wavy green blades sometimes grow to one metre in length. They are found in quiet tidal pools where they attach to rocks by multicellular holdfasts.

Although the various green algae are not at all similar in appearance, they are classified into one group. About 6 000 species have been described. All contain chlorophylls and certain other pigments. They also all store starch in their cells. It is interesting to note that higher plants also store starch and contain the same pigments as the green algae. This has led biologists to hypothesize that the higher plants developed from certain types of green algae in ancient times.

The life cycles of the different green algae are just as varied as the structures of plants. Many reproduce asexually by producing zoospores. Most species also reproduce sexually by producing either isogametes or heterogametes. When two gametes unite, the resulting zygote cell contains all the genetic material from both gametes. In other words, it has double the number of chromosomes. Therefore, somewhere in the life cycle, the number of chromosomes must be reduced to the original number. This is achieved by a special type of cell division called **meiosis**. During meiosis, the cells with two sets of chromosomes, called **diploid** cells, divide to form cells with only one set of chromosomes. These cells are called **haploid**.

Because the timing of cell union and meiosis varies from species to species, the life cycles of different green algae also vary. In some species, the plant seen growing on a pond or in the ocean is in the haploid stage. Because this haploid stage produces male and female gametes, it is called a **gametophyte**. The gametes produced by this gametophyte join to form a zygote which is diploid. The diploid zygote immediately divides by meiosis to form haploid spores which grow into new haploid gametophyte plants. Because the diploid stage produces spores, it is called a **sporophyte**. It lasts only briefly in some life cycles. For example, the zygote is the sporophyte stage in the life cycle just described.

In other green algae, such as Sea Lettuce, both a diploid and a haploid plant body are prominent (quite noticeable) in the life cycle (Fig. 13-7). The diploid zygote does not immediately divide by meiosis but, instead, grows into a diploid plant. The plant may look identical to the haploid plant, as in Sea Lettuce, or it may be different. When the diploid plant, or sporophyte, is mature, it produces spores by meiosis. The spores germinate into the haploid gametophytic plants which will later produce gametes.

Although the length of the two phases varies, a haploid generation always alternates with a diploid one. This **alternation of generations**, as it is called, occurs in all higher plants as well as in algae.

Review

1. What types of plant bodies are found in green algae?
2. Why do biologists think that ancient green algae were the ancestors of higher plants?
3. a) How does the chromosome number of a zygote compare to that of a gamete?
   b) What name is given to each of these chromosome numbers?
4. Why is meiosis necessary during the life cycle of a green alga?

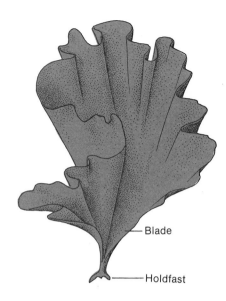

Fig. 13-6 *Ulva*, a green alga known as Sea Lettuce. It is common in tidal zones.

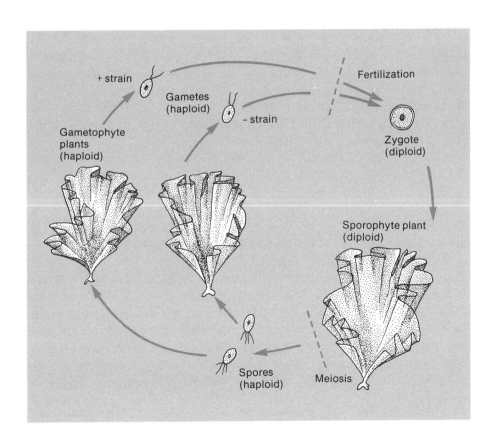

Fig. 13-7 The life cycle of Sea Lettuce (*Ulva*). The gametophyte looks identical to the sporophyte in this alga, but they produce different types of cells.

5. a) What is the function of the gametophyte stage of the life cycle?
   b) What is the function of the sporophyte stage?
6. What is an alternation of generations?

13.4 Investigation

Three Green Algae

In this investigation, you will study the structure and the life cycle of three common green algae: *Spirogyra, Ulothrix,* and *Oedogonium.*

Materials

compound microscope
cover slip
microscope slides
dissecting needles
water
eye dropper
fresh or preserved specimens of *Spirogyra, Ulothrix,* and *Oedogonium*

Spirogyra

Spirogyra is a filamentous alga that forms a bright green scum on quiet ponds and streams. Many people call it "pond scum". The filaments are unbranched and of various lengths. They are covered with a jelly-like substance that protects them from injury. Each cell in a filament contains one or two green ribbon-like **chloroplasts** that are spiral in shape. Can you guess why pond scum is called *Spirogyra?*

Located on the chloroplasts are small dark bodies called **pyrenoids**. They are used to store starch. The central part of a *Spirogyra* cell is a large **vacuole** containing watery fluid. Surrounding the vacuole are threads of **cytoplasm** in which a **nucleus** is located.

Sexual reproduction in *Spirogyra* involves the fusion (union) of two gametes in a process called **conjugation**. Two filaments line up side by side. Each cell in each filament develops a small projection on one side (Fig. 13-8). The projections meet and fuse. This forms a tube called a **conjugation tube** between each pair of cells. The contents of one cell then move through the conjugation tube to the other cell and fuse with it. Usually all the cells of a filament conjugate at the same time. Each conjugation produces a diploid zygote which later develops a thick spiny wall. The thick-walled **zygospore**, as it is now called, is able to survive the winter. In the spring it germinates and develops into a new filament. Just before germination, it divides by meiosis so that the new filament will be haploid like the parent cells.

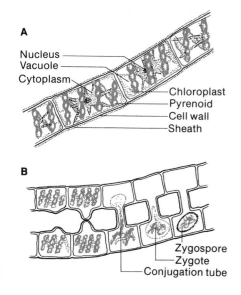

A
Nucleus
Vacuole
Cytoplasm
Chloroplast
Pyrenoid
Cell wall
Sheath

B
Zygospore
Zygote
Conjugation tube

Fig. 13-8 *Spirogyra.* **A: the vegetative cells of the filament. B: conjugation.**

Procedure A   A Study of *Spirogyra*

a. If the specimens of *Spirogyra* are freshly collected from a pond or stream, pick up some filaments and feel their texture.

b. With a dissecting needle, transfer a few filaments of *Spirogyra* into a drop of water on a slide. Cover the drop with a cover slip.

c. Observe a filament on low power. Try to determine where one cell ends and the next begins. Describe in words the shape and size of the cells.

d. Make a large labelled drawing of a few cells to show how they are arranged in the filament.

e. Turn to medium or high power and focus on one cell. Using the fine adjustment knob, focus on one chloroplast and study its spiral arrangement. Repeat this step for a few other cells.

f. Observe one cell in detail. Look for the nucleus, cytoplasm, pyrenoids, vacuole, and jelly-like sheath (covering). Draw what you see and label the various parts.

g. Search for filaments that are conjugating. Your teacher may give you prepared slides of this. Observe them on low and medium power. Make a drawing of several conjugating cells and zygospores, and label the following: conjugation tube; gamete; and zygospore.

Ulothrix

Ulothrix is also an unbranched filamentous green alga. The filaments are made of short cylindrical cells. Inside each cell is a band-like chloroplast. The cell at the base of a filament forms a holdfast that attaches to rocks. Figure 13-9 shows a typical filament. You would most likely find *Ulothrix* in fast flowing streams.

Sexual reproduction in *Ulothrix* is quite different from that in *Spirogyra*. Conjugation does not occur. Instead, a cell of a filament divides to produce 8 to 64 gametes. These are released into the water, where they meet with gametes from other cells. Since all the gametes look alike, reproduction is isogamous. A gamete from one filament (+ strain) fuses with a gamete from another filament (– strain) to form a diploid zygote. The zygote, in turn, becomes a thick-walled zygospore. The zygospore is the only diploid stage in the life cycle. After a resting period, it divides by meiosis to form four haploid zoospores. Each spore is able to germinate and develop into a new filament. The life cycle of *Ulothrix* is shown in Figure 13-10.

Procedure B   A Study of *Ulothrix*

a. Make a wet mount of a few filaments of *Ulothrix*.

b. Observe a filament on low power. Describe the shape of the cells and their arrangement. Make a sketch of one filament.

c. Use medium or high power to make a detailed drawing of one cell. Label the chloroplast, pyrenoids, nucleus, cytoplasm, and cell wall.

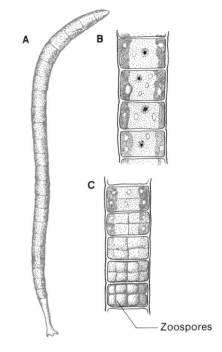

Zoospores

Fig. 13-9 *Ulothrix*. A: a holdfast cell anchors the unbranched filament to mud. B: each cell has a single chloroplast. C: cells in the filament are able to produce zoospores asexually.

Fig. 13-10   The life cycle of *Ulothrix*.

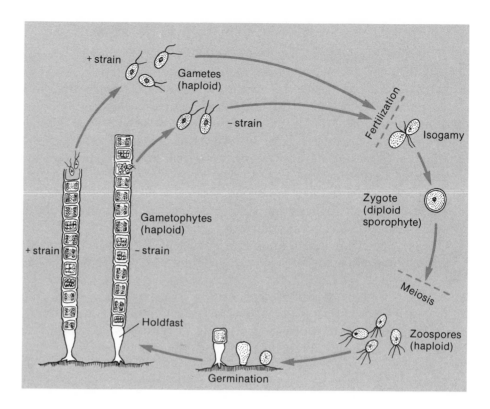

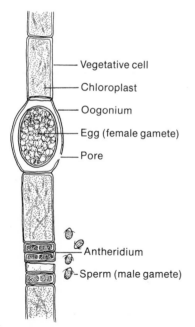

Fig. 13-11 *Oedogonium*. The large cells in the filaments are the oogonia, and the small ones are the antheridia. *Oedogonium* is heterogamous.

Oedogonium

Oedogonium is another filamentous green alga that is unbranched. It is commonly found in ponds. The young filaments are anchored by holdfasts, but the older ones float free. Each cell has one net-like chloroplast.

Sexual reproduction in *Oedogonium* is heterogamous. Only a few special cells in a filament develop into reproductive cells (Fig. 13-11). Large cells, called **oogonia** (singular: **oogonium**), each contain a female gamete (egg). Special short cells called **antheridia** (singular: **antheridium**) each produce two male gametes (sperm). The sperm are released from the antheridia and swim to the oogonia. A sperm enters through a pore in the oogonium and fertilizes the egg, forming a zygote. After a while the wall of the oogonium decays and the zygote is released. It forms a thick wall and remains dormant for a few months. Then it divides by meiosis into four zoospores that germinate and become new filaments.

Procedure C   A Study of *Oedogonium*

a. Mount a few strands of *Oedogonium* in a drop of water and observe them on low power. Look for cells of different sizes.

b. Sketch part of a filament. Show the different cell types, if you can find them.

c. Switch to a higher power and make a large labelled drawing of one vegetative (non-reproductive) cell. Label the cell wall, chloroplast, cytoplasm, nucleus, and vacuole.

   d. Draw and label a row of antheridia and one oogonium. If you cannot find any, ask your teacher for a prepared slide.

Discussion A

1. What characteristics of *Spirogyra* would help you identify it if you were looking for it in a pond?
2. What are pyrenoids?
3. During the life cycle of *Spirogyra,* what acts as a gamete?
4. Is *Spirogyra* isogamous or heterogamous? Explain.
5. Is the *Spirogyra* filament a gametophyte or a sporophyte? How do you know?
6. How does *Spirogyra* survive the winter?
7. Why do filaments of *Spirogyra* feel slimy?
8. Do all the chloroplasts in a filament of *Spirogyra* spiral in the same direction?
9. Describe a zygospore.
10. Do all the conjugating cells of two filaments of *Spirogyra* produce zygospores?
11. Does alternation of generations occur in *Spirogyra?* Explain.

Discussion B

1. Describe a vegetative filament of *Ulothrix.*
2. Which generation is the dominant one in the life cycle of *Ulothrix?*
3. List two differences between the life cycles of *Ulothrix* and *Spirogyra.*
4. List three ways in which the life cycles of *Ulothrix* and *Spirogyra* are similar.

Discussion C

1. How could you distinguish a filament of *Oedogonium* from one of *Spirogyra* or *Ulothrix?*
2. What is an oogonium?
3. What is an antheridium?
4. What is produced when a sperm cell fertilizes an egg?
5. What represents the sporophytic generation in *Oedogonium?*
6. How does the life cycle of *Oedogonium* differ from that of *Ulothrix?*
7. Figure 13-10 shows the life cycle of *Ulothrix.* Draw a similar diagram of the life cycle of *Oedogonium.* Label all the stages and include meiosis and fertilization in your diagram.

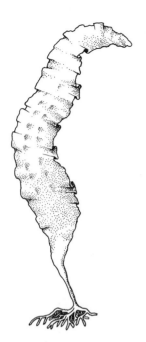

Fig. 13-12 The brown alga, *Laminaria*. Notice the branched holdfast, the stem-like stipe, and the thin leafy blade.

Fig. 13-13 Brown algae of the Pacific Ocean.

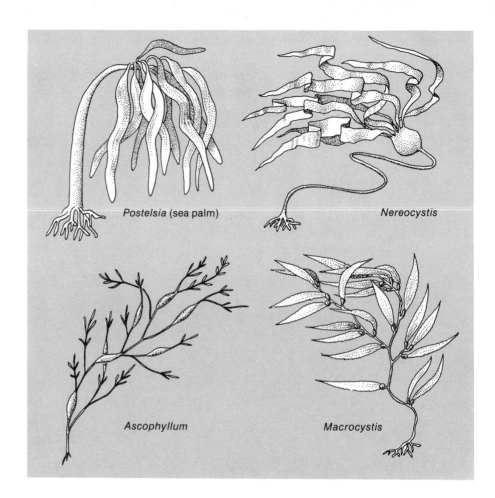

Postelsia (sea palm)

Nereocystis

Ascophyllum

Macrocystis

13.5 Brown and Red Algae

The Brown Algae

Almost all the Phaeophyta, or brown algae, are found in cold ocean waters. Most of the large seaweeds are brown algae. They are classified together because they have complex plant bodies, many-celled reproductive organs, and similar coloured pigments. They appear brownish because the dark pigments hide the green of the chloroplasts.

None of the brown algae are unicellular. Some are simple filaments with holdfasts that attach them to stones or pilings. Others form large flat ribbons that may grow to 30 m in length. The largest ones are the kelps, found in cold waters on rocky shores. The thallus of a kelp has a branched holdfast, a stem-like stipe (stalk), and a thin leafy blade (Fig. 13-12). Sometimes the blade is divided into many sections, as in the giant kelps of the Pacific (Fig. 13-13). Some kelps are harvested by humans to produce a substance called algin. Algin is added to ice cream to improve its texture.

The rockweeds are brown algae whose blades have air-filled sacs called bladders to help keep them afloat. The rockweed *Fucus* has a circular holdfast that attaches it to rocks near shorelines where the tide goes in and out. Its plant body is flattened and branched (Fig. 13-14). Air bladders along the sides of the thallus lift the forked blades toward the surface.

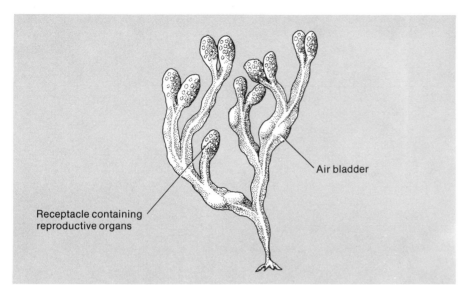

Receptacle containing reproductive organs

Air bladder

Fig. 13-14 The rockweed, *Fucus*. Rockweeds have air bladders to help keep them afloat. The gametes are produced in receptacles at the branch tips.

Another rockweed, *Sargassum* (Fig. 13-15), is found in the open ocean where it floats near the surface. A large part of the Atlantic Ocean, between the Bahamas and the Azores, is known as the Sargasso Sea, because *Sargassum* is so plentiful there.

The life cycles of most brown algae are very complicated. In *Fucus,* the reproductive cells are produced at the tips of the blades in swollen structures called **receptacles**. The egg and sperm cells produced there are released into the water where they unite to form a zygote. This zygote develops into a new plant. Other brown algae have somewhat different life cycles than *Fucus,* but all show an alternation of generations.

The Red Algae

The Rhodophyta, or red algae, contain red pigments that give them their colour. Sometimes several pigments blend together, producing a brownish or purplish plant body. Most red algae are found in warm ocean waters, either floating or attached to rocks in tidal zones. Some occur deep in clear tropical waters where there is only a small amount of light. A few species live in freshwater lakes and streams.

Many red algae are very delicate in structure and rather small in size. The largest ones are only about one metre long, and the smallest ones are unicellular. The blades may be feathery and branching, filamentous, or thin and leafy (Fig. 13-16). The life cycles are very complex and are still unknown for some species.

Fig. 13-15 *Sargassum filipendula*. Sargassum is the brown alga that gives the famous "Sargasso Sea" its name.

Fig. 13-16 Some red algae.

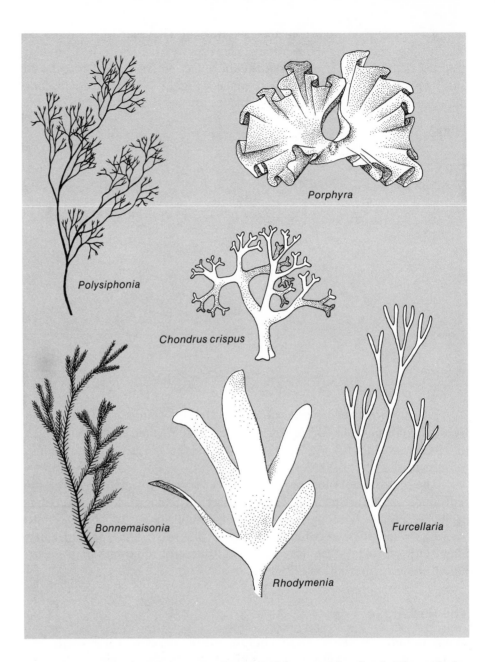

Porphyra

Polysiphonia

Chondrus crispus

Bonnemaisonia

Rhodymenia

Furcellaria

A number of red algae are harvested by man for food. The red alga *Porphyra* is regularly eaten in the Orient. Known as "laver", it is grown in the shallow waters near Japan. It is eaten in soups and stews, and is used as a covering around boiled rice. It is rich in vitamins B and C and in protein. Laver is also popular in Ireland and Britain, where its thin sheet-like thalli grow on the rocky coasts.

Dulse is another red alga that is harvested for food. Either fresh or dried, it has a very salty taste.

Carrageen, or Irish Moss *(Chondrus crispus),* is a branched red alga that is used as a base for some candies. Extracts of it are used in ice cream, chocolate milk, and salad dressings. The harvesting of this alga is an important industry in the Maritime provinces, New England, and Great Britain.

Red algae are also important as the source of a substance known as agar. Agar is used by microbiologists who study bacteria. It acts as a solidifying agent in the culture media used to grow bacteria for study.

Review

1. **a)** In what three ways are the golden algae similar to the green algae? (See Section 10.9, page 270.)
   b) In what three ways are they different?
2. Are brown and red algae likely to be found in the same habitat? Explain.
3. What is kelp?
4. **a)** Where is the Sargasso Sea?
   b) How did it gets its name?
5. Humans have made use of many kinds of brown and red algae.
   a) List three that are harvested by humans.
   b) What use is made of each of these?
6. What is agar and where does it come from?

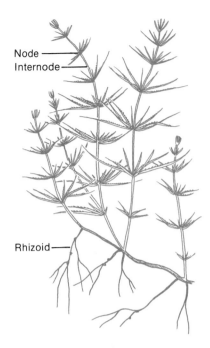

Fig. 13-17 **The stonewort, *Chara*. The rhizoids produce many erect shoots with whorls of branches.**

13.6 The Stoneworts

The Charophyta, or stoneworts, are unusual freshwater plants that seem to be related to both the green algae and the mosses. They form large colonies in clear lakes and in limestone streams and quarry basins. Some are able to remove lime from the water and cover themselves with a brittle layer of it. For this reason they are known as stoneworts.

A stonewort is an easy plant to recognize. The plant body is an erect filament up to one metre long. It has whorls of short branches on it (Fig. 13-17). The branches attach at points called **nodes**. The bare lengths of filament between the nodes are the **internodes**. It is anchored at the base by small branched filaments called **rhizoids**. As the rhizoids grow through the sand or mud, they send up new erect shoots. In this way, the plant spreads asexually.

Sexual reproduction in stoneworts is heterogamous as in some green algae. The male and female sex organs are different in structure (Fig. 13-18). They are found at the nodes on highly specialized branches. The female **oogonium** contains a single egg cell surrounded by five spirally twisted cells. The male **antheridium** is a spherical structure with some sterile cells surrounding the sperm. The antheridium is usually orange in colour. When a sperm is released into the water, it floats to an egg cell and fertilizes it. The resulting zygote develops a thick wall, falls off the plant, and remains dormant for some time. Eventually its cells begin to divide and a new plant develops.

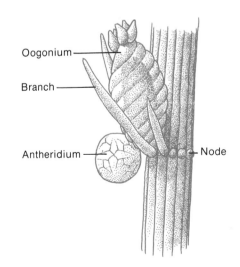

Fig. 13-18 *Chara*. **The sex organs of *Chara* are found at the nodes. The male antheridium is spherical; the female oogonium is oval and covered with spiral cells.**

1. How did stoneworts get their name?
2. What features would you look for if you were trying to identify a stonewort?
3. What type of sexual reproduction occurs in stoneworts?
4. Does alternation of generations occur in stoneworts? Explain your answer.

13.7 The Mosses and Liverworts

Both the mosses and liverworts are small greenish plants that grow in damp environments. They are classified together in the Division Bryophyta and are commonly called **bryophytes**. The mosses were probably the very first plants to survive on land. For this reason, they are of special interest to biologists. They are thought to have evolved from green algae about 350 000 000 years ago.

Like the first land plants to which they are related, modern bryophytes are primitive in structure. They have never adapted fully to the land environment. They are much less successful than flowering plants for several reasons. Since they depend on water to reproduce, they cannot survive in dry areas. They cannot grow very tall because they have no vascular tissue for support or for carrying water through the plant body. They have no true roots to reach deep into the soil for water. Nevertheless, they have survived on land for millions of years and are quite common wherever there is moisture.

Dark forest floors, banks of streams, and wet rock surfaces are the most likely places to find mosses and liverworts. Some grow on tree bark and others grow at the entrances of caves. A few species can survive short periods of dryness by slowing their growth until water returns. Many mosses grow in tiny cracks in rocks. As they grow, they break down the rocks into soil. Once enough soil has formed, other plants are able to take hold and crowd out the mosses.

The Mosses

Some plants that are called mosses are not mosses at all. For example, Reindeer moss is a lichen and club mosses are relatives of ferns. Most true mosses are short plants less than 20 cm high that grow in clumps. The individual plants in a clump consist of leafy stalks that are held down by rhizoids. There are no true roots, stems, or leaves. The shoots have whorls of small leaf-like blades that are only one cell thick. They contain chloroplasts that produce food by photosynthesis.

The life cycle of a moss shows the same pattern as algae. A gametophytic generation alternates with a sporophytic one (Fig. 13-19). In all mosses, the gametophyte is the dominant generation. The leafy green

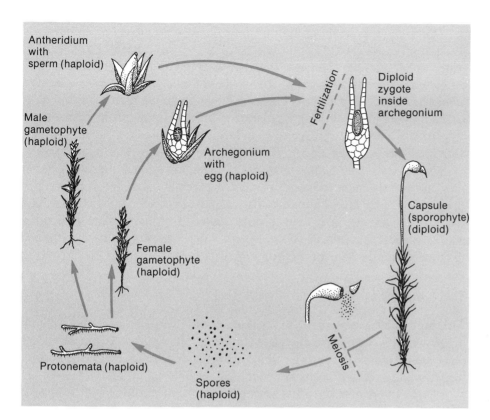

Fig. 13-19   The life cycle of a moss.

Antheridium
with
sperm (haploid)

Male
gametophyte
(haploid)

Archegonium
with
egg (haploid)

Fertilization

Diploid
zygote
inside
archegonium

Capsule
(sporophyte)
(diploid)

Female
gametophyte
(haploid)

Meiosis

Protonemata (haploid)

Spores
(haploid)

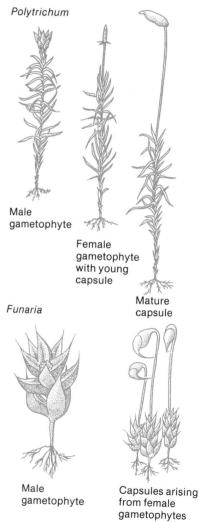

Polytrichum

Male
gametophyte

Female
gametophyte
with young
capsule

Mature
capsule

Funaria

Male
gametophyte

Capsules arising
from female
gametophytes

Fig. 13-20 *Polytrichum,* or Haircap Moss, and *Funaria,* or Cord Moss, are both quite common.

clumps of moss that you find in the woods are the gametophytes. Figure 13-20 shows both the gametophyte and sporophyte of two common mosses. When mature, the gametophytes produce sex organs at the tips of the shoots. The male antheridia are multicellular stalked structures that produce many sperm. The egg-producing structures are also multicellular and stalked. They are called **archegonia** (singular: **archegonium**). They produce only one egg each. (You may recall that single-celled egg-producing structures are called oogonia.) In some mosses, both sex organs are produced on one plant; in others the sex organs are separate.

For fertilization to occur in mosses, there must be a film of rainwater or dew between an antheridium and an archegonium. When an egg cell is mature, the archegonium around it opens and the sperm swim into it. Fertilization produces the first cell of the sporophyte, the zygote. The zygote does not fall off the plant, but begins to divide inside the archegonium. It develops into a stalked **capsule** in which many spores are produced. The sporophyte depends on the leafy gametophyte for water and food, and never becomes independent. Most mosses produce sporophytes and release their spores in the spring or fall.

When a spore reaches a moist environment, it germinates into a long green thread called a **protonema** (plural: **protonemata**). The protonema looks much like a filament of a green alga. It produces rhizoids to absorb water, and soon begins to send up the familiar leafy shoots of the gametophyte. In one season, the life cycle is completed.

One of the most common mosses is **peat moss**, or **sphagnum**. Sphagnum moss is used by gardeners to loosen clay soils and to help hold water in sandy soils. Some of the cells of sphagnum are empty and have pores in their walls which allow them to hold great amounts of water, even when they are dead. Dead peat moss is used as packing material around the roots of living plants to keep them moist during shipping from a nursery.

In the wild, peat moss grows in large floating mats in bogs, swamps, and small lakes. Each mat is made up of many individual plants packed closely together. The mats sometimes thicken and spread out so much that they completely cover the water surface. As the older plants die and settle to the bottom, thick layers build up. The pressure of the upper layers compacts the lower part into a brownish mass called **peat**. In Ireland and other countries, peat is used as a source of fuel. Large amounts of peat are mined in Quebec and the Maritimes for use by gardeners.

The Liverworts

Although you have probably often seen mosses, you may never have seen a liverwort. Liverworts are much less common. They grow with mosses in wet areas near streams and springs. Some grow flat on the ground and on rotting wood, while others float in the water.

Most liverworts have flat leathery thalli that are lobed (in sections). The lobes looked so much like the human liver to the early botanists that they called all members of this group liverworts. However, some liverworts are moss-like, with two rows of flat leafy structures along the shoots. The thalli of several kinds are shown in Figure 13-21.

Fig. 13-21   Some liverworts.

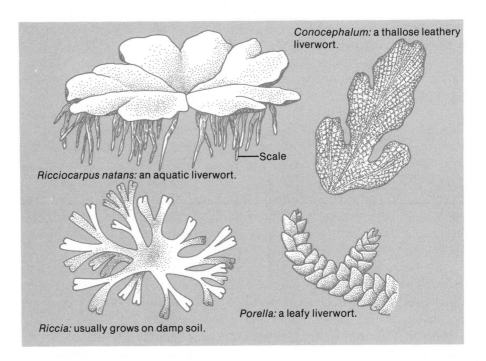

Conocephalum: a thallose leathery liverwort.

Ricciocarpus natans: an aquatic liverwort.

—Scale

Porella: a leafy liverwort.

Riccia: usually grows on damp soil.

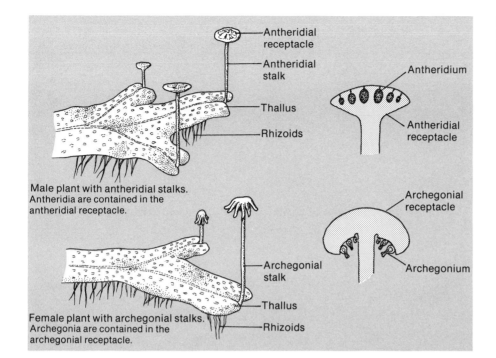

Antheridial receptacle
Antheridial stalk
Thallus
Rhizoids

Antheridium
Antheridial receptacle

Male plant with antheridial stalks. Antheridia are contained in the antheridial receptacle.

Archegonial receptacle
Archegonial stalk
Thallus
Rhizoids

Archegonium

Female plant with archegonial stalks. Archegonia are contained in the archegonial receptacle.

Fig. 13-22 The liverwort *Marchantia*. The male and female gametophytes separate in *Marchantia*. Each produce sex organs that are stalked.

One of the most common liverworts is *Marchantia*. The gametophyte of *Marchantia* is a branched leathery thallus that lies close to the ground. It is anchored by **rhizoids**. The gametophytes are either male or female. A male plant produces sperm cells in umbrella-like structures called **antheridial stalks** (Fig. 13-22). The eggs are produced in **archegonial stalks** on female plants. There must be water for fertilization to occur. The sperm swim to the eggs and fertilize them, forming zygotes which develop into sporophytes. As in mosses, the sporophyte is a capsule that releases spores into the air. These spores grow into new gametophytic plants when they land in a suitable, damp spot.

Review

1. **a)** What limits the success of bryophytes on land?
   b) In what sorts of environments are bryophytes found?
2. What adaptations have allowed mosses to survive on land?
3. Which generation is dominant in the life cycle of a moss?
4. Describe a typical moss gametophyte.
5. **a)** Why is a moss capsule called a sporophyte?
   b) What is found inside the capsule?
6. What structure is formed when a moss spore germinates? To which generation does it belong?
7. Describe how peat is formed.
8. How did liverworts get their name?
9. How do the sex organs of *Marchantia* differ from those of mosses?

13.8 Investigation

The Life Cycle of a Moss

In this investigation, you will examine both the gametophyte and sporophyte of a moss. Keep in mind the life cycle shown in Figure 13-19 as you observe the specimens.

Materials

fresh or dried specimens of moss gametophytes with sex organs
fresh or dried specimens of gametophytes with capsules
hand lens
compound microscope
slides
cover slips
water
eye dropper
dissecting needle

Procedure

a. Use the hand lens to observe a clump of moss plants. Separate one leafy shoot and identify the rhizoids, stalk, and leafy blades. Draw one plant and label it fully.

b. Describe the arrangement of the leaf-like structures on the shoot.

c. Search the leafy stem tips for the antheridia and archegonia. Notice that each shoot has either one, but not both. Describe any differences between the two types of shoot.

d. Carefully remove the leaves from the tip of a male shoot and squeeze out an antheridium into a drop of water on a slide. Use the microscope on low power and draw what you observe.

e. Repeat step d for a female archegonium.

f. Obtain a gametophyte that has a capsule on it. This is the sporophyte. Observe it with a hand lens and draw the various parts. Label the capsule, stalk, and gametophyte.

g. Determine how a capsule opens to release spores. Then carefully crush the capsule into a drop of water on a slide. Observe and draw a few of the spores.

Discussion

1. Are the shoots in a clump of moss attached to each other?

2. a) How are the leaf-like structures of the gametophyte arranged?
  b) Why can we not call them leaves?

3. a) Describe a moss antheridium. What does it produce?

b) Describe a moss archegonium. What develops inside it?

4. How do you know that the moss sporophyte depends on the gametophyte for food?

5. Does alternation of generations occur in mosses? Explain.

13.9 The Ferns

Ferns suggest lush forests and cool streams. Most ferns grow in shaded places, such as deep woodlands, ravines, cracks in rocks and even caves. They need shade and moisture to grow. A few have returned to the water environment as tiny floating plants. Some others grow up to 15 m tall in humid tropical jungles. The occasional species, such as the bracken fern, can survive in open sunny fields and roadsides (Fig. 13-23).

Although there are about 10 000 species of ferns living today, there were many more than that about 300 000 000 years ago. Long before dinosaurs roamed the earth, dense forests of tree ferns towered above the plant world. Smaller ferns, horsetails, club mosses, and true mosses thrived under the tree fern canopy. Because ferns were so abundant then, this period of time is called the Age of Ferns. The ferns grew quickly in the warm humid climate. Older dying plants accumulated in large swamps. As generation after generation died, the layers built up and were compressed into the coal beds that we use today for fuel. Since coal is made of carbon, the Age of Ferns is also called the Carboniferous Period.

Why are ferns not the dominant plants still? Probably the climate changed and the swamps dried up. Many ferns could not adapt to the drier conditions. Other plants that could cope better with dryness began to compete with the ferns and crowd them out. Today, most ferns are small and

Fig. 13-23 The bracken fern is one of the few ferns that can live in direct sunlight.

Fig. 13-24 A few common ferns.

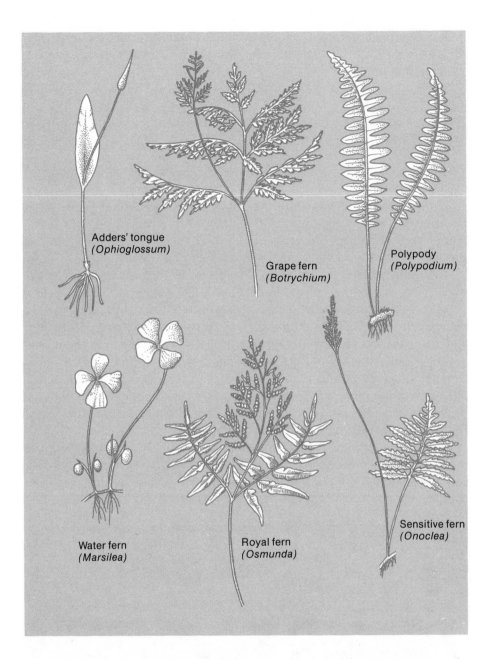

Adders' tongue
(Ophioglossum)

Grape fern
(Botrychium)

Polypody
(Polypodium)

Water fern
(Marsilea)

Royal fern
(Osmunda)

Sensitive fern
(Onoclea)

grow only in shady damp places. A few grow in the arctic, and some grow in temperate woodlands, but most are found in tropical rain forests.

The ferns belong to the division Pterophyta. These plants have vascular (conducting) tissue like seed plants. However, they have never become as highly evolved as the seed plants.

The Fern Life Cycle

In the life cycle of a fern, there is an alternation of generations similar to that of mosses. However, unlike mosses, the sporophyte is the dominant stage in

Fig. 13-25 The parts of a fern sporophyte.

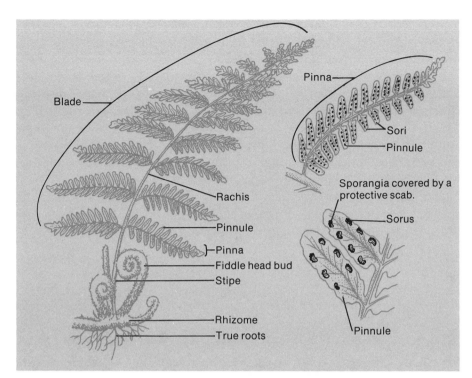

Fig. 13-25 The parts of a fern sporophyte.

the fern life cycle. The typical fern plant that you see is the sporophyte. Figure 13-24 shows a few common species.

Although the giant tree ferns of the tropics have sturdy, erect stems, most temperate ferns do not. Instead, a thick stem runs horizontally under the soil (Fig. 13-25). It is usually covered with scales. Underground stems of this type are called **rhizomes**. Inside the rhizomes are vascular tissues that carry water and food through the plant. Fern rhizomes live from year to year. They gradually grow outward through the soil and produce numerous roots and leaves each season. Because of the creeping rhizomes, ferns are often found in large clumps.

Unlike mosses, ferns do have true roots. Clusters of roots develop from the rhizome. They anchor the fern in place and absorb water from the soil.

The leaves of ferns are usually large and are known as **fronds**. In temperate climates, the fronds die in the winter. A new set develops from the rhizome each spring. A frond consists of three parts: a **blade**, a **rachis**, and a **stipe** (Fig. 13-25). The large expanded part of the frond is the blade. The midrib that supports the blade is the rachis. The stipe is the stalk that attaches the rachis to the rhizome.

The fronds of some species of ferns are simple blades, but most are divided into many leaflets called **pinnae** (sing. **pinna**). The pinnae are arranged in two rows along the sides of the rachis. In ferns that appear very feathery, the pinnae are again divided into **pinnules** (Fig. 13-25).

As new fronds develop in the spring, they unroll from tightly coiled buds produced by the rhizome. Because they look like the curved end of a violin, these young fronds are called **fiddleheads**. The fiddleheads of the

Fig. 13-26 Fiddleheads of the Ostrich Fern are edible. They arise among the old fertile fronds in the spring.

Fig. 13-27 Sorus types.

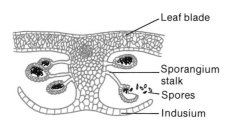

Fig. 13-28 A section of a fern sorus shows the arrangement of sporangia under the indusium.

ostrich fern (Fig. 13-26) are edible and taste somewhat like asparagus. However, the fiddleheads of other ferns are thought to cause cancer and other illnesses and should not be eaten.

For most species of ferns, as the fronds mature, raised brown dots appear on the lower surfaces of the pinnae. These spore dots are called **sori** (sing. **sorus**). Their shape and position vary with the species of fern (Fig. 13-27). Each sorus contains a cluster of stalked spore cases called **sporangia** (singular: **sporangium**). The sporangia produce thousands of dust-like spores. In most ferns the sporangia are covered with a scale-like membrane, the **indusium** (Fig. 13-28). The indusium protects the spores as they develop but withers away when they are ripe. In some ferns the indusia are umbrella-shaped. Other ferns, such as bracken, have no indusia, but the sporangia are protected by the leaf which is rolled down at the edges.

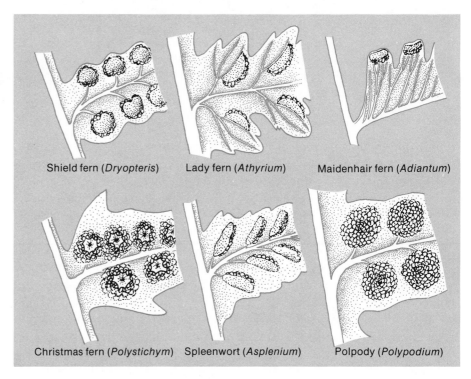

Shield fern (*Dryopteris*)   Lady fern (*Athyrium*)   Maidenhair fern (*Adiantum*)

Christmas fern (*Polystichym*)   Spleenwort (*Asplenium*)   Polpody (*Polypodium*)

Each sporangium in a sorus has a row of special cells around it that helps to spread the spores when they mature. This ring of cells is called the **annulus** (Fig. 13-29). The cells of the annulus have thickened walls that will straighten out in response to dryness. As they straighten, the sporangium tears open. The spores are thrown several centimetres into the air as the annulus snaps back into position. If a spore lands in a warm damp place, it germinates and develops into the first cell of the gametophyte generation.

A fern gametophyte starts as a short filament of cells with a few rhizoids to anchor it. It then widens into a flat sheet of cells that is called a **prothallus** (Fig. 13-30). The prothallus is heart-shaped, green, and less than 1 cm in diameter. It has rhizoids but no true roots, stems, or leaves. Most

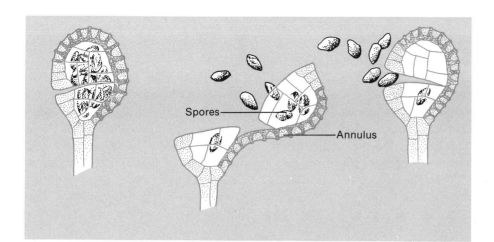

Spores

Annulus

people have never seen fern prothalli, because they are so tiny and short-lived. The function of the prothallus is to produce gametes.

Both male and female sex organs develop on the lower surface of the prothallus. The male antheridia are found near the base of the prothallus among the rhizoids. They are dome-shaped structures that release many spirally-coiled sperm cells. The sperm have many flagella to help them swim.

Fig. 13-30 A fern gametophyte. The archegonia are found near the notch of the prothallus, and the antheridia are found among the rhizoids.

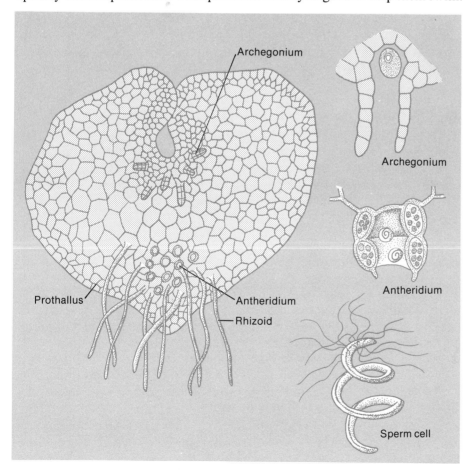

Archegonium

Archegonium

Antheridium

Prothallus

Antheridium

Rhizoid

Sperm cell

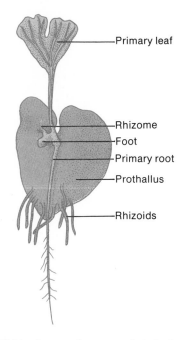

Primary leaf

Rhizome
Foot
Primary root
Prothallus

Rhizoids

Fig. 13-31 A young fern sporophyte is at first dependent on the gametophyte for food. Once the primary leaf of the sporophyte has formed, the prothallus begins to wither and die.

The female sex organs, or archegonia, are found near the notched end of the prothallus. They are flask-shaped with a short neck that faces the ground. One egg cell is produced in each archegonium.

On a single prothallus, the egg and sperm cells mature at different times. Therefore there must be other prothalli close by in order for fertilization to occur. When there is a constant film of water between the prothalli and the soil, the sperm swim to nearby prothalli and enter the archegonia. Only a single sperm fertilizes each egg. The resulting zygote, with two sets of chromosomes, is the first cell of the sporophyte generation.

While still inside the archegonium, the zygote begins to grow into a sporophyte. At first it is parasitic on the prothallus. It develops a "foot" that absorbs food and water from the prothallus. Then a primary leaf, root, and stem are produced (Fig. 13-31). The young sporophyte has chlorophyll and is able to produce its own food. As the sporophyte develops, the tiny prothallus, on which it was once dependent, withers and dies.

Although the heart-shaped prothalli are similar in appearance from species to species, the sporophytes of ferns are very different. The size and shape of the fronds vary a great deal. The walking fern is one species whose fronds do not look at all fern-like. They are spear-like, with heart-shaped bases and long tips (Fig. 13-32). The plant is called the walking fern because its long leaves bend over and produce new plants where the tips touch the soil. As each new plant takes root, the colony slowly "walks" across the ground. This unusual fern is found only on mossy limestone or sandstone rocks in shady forests.

Review

1. In what habitats could you find ferns?
2. a) Explain briefly how coal beds are formed.
   b) Why is the Age of Ferns called the Carboniferous Period?
3. Why are ferns not the dominant type of plant today?
4. What are the two stages in the life cycle of a fern?
5. a) What is a rhizome?
   b) Describe the structure of a fern frond.
   c) What are fiddleheads?
6. a) What are sori?
   b) Where are sori found?
   c) What is inside a sorus?
   d) What is an indusium?
7. How does the structure of a fern sporangium help to spread the spores?
8. Describe a fern gametophyte.
9. Why is water necessary for ferns to reproduce?
10. Which stage in the life cycle of a fern is parasitic? Which stage is parasitic in the life cycle of a moss?
11. How did the walking fern gets its name?

13.10 Investigation

Ferns

Although you may have a fern or two in your home or garden, perhaps you have never taken a good look at one. In this investigation you will be able to examine a fern sporophyte in detail. If the fern has sori, you will also be able to see the spores.

Materials

potted fern plant, a freshly
   collected specimen, or
   a dried specimen that has
   fertile fronds
hand lens
compound microscope

slide
cover slip
dissecting needle
water
eye dropper
glycerin

Fig. 13-32 The Walking Fern, *Camptosorus rhizophyllus*. The leaf tips touch down, root, and develop new plants. The whole colony "walks" across the ground.

Procedure

a. Remove the fern plant from the soil and carefully wash away any soil from the roots.

b. Examine the rhizome and the fibrous roots growing from it. Describe the shape and appearance of the rhizome.

c. Search for young fiddleheads coming from the rhizome and describe their appearance.

d. Examine a single frond and note how the pinnae and pinnules (if present) are arranged.

e. Make a large drawing of the sporophyte. Include the rhizome, roots, fiddleheads, and one frond. Label the pinnules, pinnae, blade, rachis, stipe, frond, rhizome, roots, and fiddleheads.

f. Compare the size of the pinnae at the bottom, middle, and tip of the blade.

g. Locate the sori on the lower surface of a mature frond and examine one sorus with a hand lens. Describe the shape, colour, and position of the sori.

h. Using a dissecting needle, carefully put part of a sorus into a small drop of water on a slide. Separate it with the needle and observe the sporangia without a cover slip on low power. Look for the annulus. Draw and label one sporangium.

i. Add a drop of glycerin to the sporangia and observe what happens.

Discussion

1. Are any scars left from old fronds visible on the rhizome?
2. Where are the youngest leaves located on the plant?

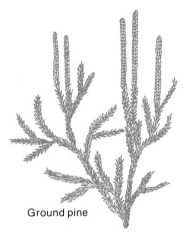

Ground pine

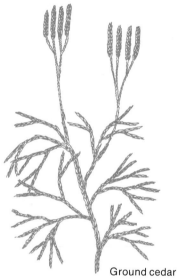

Ground cedar

Fig. 13-33 Club mosses are usually low-growing plants with small scale-like leaves. The sporangia are sometimes found in "clubs" on top of the plants.

3. Describe the shape of the young leaves as they develop.
4. How does the position and shape of the sori help disperse spores?
5. Describe the structure of a sporangium.
6. What happens when glycerin is added to a sporangium?

13.11 The Fern Allies: The Club Mosses and The Horsetails

Since the club mosses and horsetails are closely related to the ferns, they are often called the **fern allies**. They have similar life cycles and live in similar habitats. During the Carboniferous Period, they dominated the great swamp forests along with the ferns. Many have since become extinct, but their fossil remains are found all over the world.

The Club Mosses

The club mosses belong to the division Lycopodophyta. Today there are about 1 500 living species. Most belong to the two genera *Lycopodium* and *Selaginella*. Most species grow in the tropics, but a few thrive in the cool moist forests of temperate regions, including Canada.

Although they are called club mosses, these plants are not related to mosses at all. Unlike mosses, the sporophyte is dominant in the life cycle. The sporophytes are evergreen plants with small scale-like leaves that grow close to the ground. They spread by branched horizontal stems that grow along the top of the soil or underneath it. They also have erect branches that vary from 4 cm to 25 cm in height. Some species are rather bushy and look like seedling evergreen trees. They are known as ground pines and ground cedars (Fig. 13-33).

Like the ferns, club mosses develop sporangia that produce spores. In some species, the sporangia are scattered along the leafy shoot. In others, they are grouped at the tips of the branches in club-shaped cones (Fig. 13-33). This explains how club mosses got their name.

The spores of club mosses germinate into delicate prothalli like those of ferns. The prothalli are short-lived, but are essential for the completion of the life cycle.

The Horsetails

All horsetails belong to the genus *Equisetum,* which is classified in the division Arthrophyta. There are very few of these today since most became extinct millions of years ago. The horsetails grow in colonies with underground rhizomes. The erect shoots are hollow, jointed, and branched, and look somewhat like horses' tails. From the stem joints, the branches spread out in whorls (Fig. 13-34). The leaves are tiny scale-like structures that encircle the shoots at each joint.

The horsetails feel gritty to touch because of large amounts of silica in their cells. In pioneer days, they earned the name "scouring rush" because of this. The gritty branches were often used by the pioneers to scour (clean) out dirty pots and pans.

You may have walked through clumps of horsetails many times without even knowing it. They are very common in damp fields and roadside ditches, and near lakeshores and streams. Many horsetails look very feathery.

Early in the spring, horsetails produce sporangia in special reproductive cones. In some species, the cones are formed on special non-leafy shoots. In others, the cones develop right at the tips of the leafy shoots that are carrying on photosynthesis. In either case, the sporangia produce thousands of spores which are carried by air currents to new locations.

Review

1. What features of a club moss would help you identify it in the woods?
2. What is one major difference between the life cycles of club mosses and true mosses?
3. Why are some horsetails called scouring rushes?
4. In what kind of habitat could you find horsetails?
5. In what two ways do horsetails reproduce?

Highlights

Although the flowering plants dominate our vegetation today, they have not always done so. Many types of primitive plants have at one time or another dominated the flora. Groups such as the algae, mosses, liverworts, and ferns have evolved along different lines over time. Relics of these ancient groups are still thriving today in all parts of the world.

The development of vascular tissue used to transport materials contributed greatly to the success of plants on land. Plants lacking vascular tissues, such as the mosses and liverworts, have never become very large. Neither have they spread into dry environments, because they depend on water for reproduction. Only the seed plants, which you will meet in the next chapter, have overcome that barrier.

Despite differences in their sizes and structures, all primitive plants have two things in common. They all use chlorophyll, found in their cells, for photosynthesis. In all species that reproduce sexually, a haploid generation alternates with a diploid one during the life cycle. The alternation of a gametophyte with a sporophyte also occurs in the seed plants, as you will see in Chapter 14.

Fig. 13-34 Horsetails (*Equisetum*). Some horsetails have separate sterile and fertile stems, while others have only one stem.

14 The Seed Plants

By far the most complex and widely distributed plants on earth today are the seed plants. Like the ferns, seed plants contain vascular tissues that carry food and water. Unlike the ferns, they reproduce by seeds instead of spores. Probably most of the plants you can think of are seed plants. Pines, maples, grasses, roses, and dandelions are just a few examples. Although they look different, they all reproduce by seeds. The evolution of the seed has made seed plants the largest and most successful group of plants so far.

14.1 Success of the Seed Plants

Why are seed plants the most successful plants on earth today? Although a few are water plants, the majority live on land. What has allowed them to become so successful there? Certainly the vascular tissue that carries water, nutrients, and food to all parts of the plant has been a factor. Because the roots and stems conduct water upward from the soil, seed plants can live in dry environments. Vascular tissue also helps to support tall plants. It keeps them from falling over. Some of the tallest living plants are the redwood trees of California (Fig. 14-1). These are seed plants with very efficient and strong vascular systems.

What else makes seed plants successful on land? The specialization of the plant bodies into roots, stems, and leaves has made them very efficient. Each part is specialized to perform specific functions. Also adding to their success has been the evolution of a method of reproduction that does not require a water environment. You will see later how this occurs. By far the most important factor in the success story is the seed itself. A **seed** is a miniature sporophyte plant surrounded by stored food and a protective coat. The immature plant inside it is called an **embryo**. In the proper conditions of warmth and moisture, a seed is able to germinate and its embryo develops into a complete sporophyte plant. Seeds may remain in a resting state, or dormancy, for long periods of time, until conditions are good for germination. Dormant seeds are easily spread by wind, water, and animals to new places. Seed plants are now found all over the world because their dormant seeds have been carried everywhere. The more primitive plants, such as mosses, liverworts, and ferns, depend on microscopic one-celled spores for dispersal.

There are several distinct groups of seed plants. They differ primarily in their method of reproduction. The more primitive groups produce seeds in cone-like structures and are often called **conifers** or **gymnosperms**. The word gymnosperm comes from two Greek words that mean "naked seed". It is used because the seeds are uncovered. Of the four divisions of gymnosperms, only the Coniferophyta are discussed in this chapter.

The most advanced seed plants belong to the division Anthophyta, or flowering plants. This group includes all the plants that produce flowers. They are sometimes called **angiosperms** because their seeds are enclosed inside the parts of the flower. The Greek words that mean "covered seed" are "angeion" and "sperma". The tissues that surround the seeds develop into **fruits**. Fruits are unique to flowering plants.

Fig. 14-1 The giant redwood trees are the largest seed plants. They grow in California.

Review

1. In what two ways do seed plants differ from ferns?
2. Explain why vascular tissue is so important to survival of plants on land.
3. What are four reasons that seed plants have become so successful?
4. What is a seed?
5. What is the major difference between gymnosperms and angiosperms?

14.2 Division Coniferophyta: The Conifers

General Characteristics

The best known and largest group of gymnosperms is the Coniferophyta, or the conifer group. **Conifers** are cone-bearing trees and shrubs, many of which have thin, needle-like leaves. Pines, spruces, cedars, junipers, and

Fig. 14-2 Bristlecone pines are the oldest living plants.

yews are all conifers. Although most species of conifers are now extinct, there are several hundred living today. They grow in every part of the world except Antarctica.

The ancestors of the conifers evolved long ago during the Carboniferous period when ferns were the most plentiful type of vegetation. The ancestors were fern-like plants that developed seeds instead of just spores. As the conifers evolved, they became the dominant land plants because of their efficient dispersal methods and their ability to withstand dry conditions. They reached their peak of development during the Mesozoic, or Dinosaur Era. Since then they have declined in number of species. The better adapted flowering plants have since become dominant.

The conifers are important in the present-day landscape because of the large area of land they cover. Thick forests of pines, firs, and spruces grow throughout northern Canada, Europe, and Asia. In British Columbia and eastern Canada, coniferous forests are the source of large quantities of resins, pulp and paper, and lumber.

The several hundred species of conifers living today are usually classified into seven families. The largest is the pine family, which includes about 220 species. Pine, spruce, balsam, hemlock, and tamarack (larch) all belong to the pine family. All except the tamarack keep their leaves over the winter and are called **evergreens**. Tamarack is a **deciduous** conifer. Its leaves turn bright yellow and fall off in the autumn.

Two conifers that grow in the United States have special significance. The Sequoias, or redwoods, of California are the tallest land plants. They sometimes reach a height of more than 100 m. The bristlecone pines are the oldest known living plants (Fig. 14-2). Some are more than 4 000 years old, although they are less than 20 m high. Their dense, gnarled wood is full of substances which resist insect attack, and fungal and bacterial decay. These magnificent trees are found only at high elevations in California and neighbouring states.

Reproduction in Conifers

Conifers reproduce by seeds that are formed in cones. A cone is made of scales. Scales are modified leaves arranged spirally around a short branch. Cones are produced by the sporophyte, usually in the spring. The cones are of two kinds, male and female.

The life cycle of conifers involves **alternation of generations** just like the life cycles of algae, mosses, and ferns. As in ferns, the sporophyte is the dominant generation. A pine tree, for example, is a sporophyte, or spore-producing plant. It has roots, a trunk, branches, and leaves. Unlike ferns, the seed plants produce separate male and female spores, which develop into separate male and female gametophytes. The pine tree produces soft **male cones** in clusters at the base of new spring shoots (Fig. 14-3). These cones last only one or two weeks. Each of their scales produces haploid male spores by meiosis. These spores are called **pollen grains**. Before a pollen grain is shed

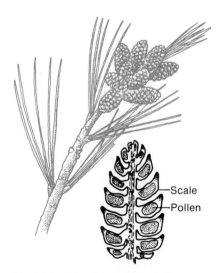

—Scale

—Pollen

Fig. 14-3 Pine: the maple or pollen cones are produced in clusters at the base of new spring shoots. Each cone is made up of many scales that bear pollen.

from a male cone, the cell inside it divides to form the **male gametophyte**. This gametophyte is protected by a thick wall around the pollen grain. In pine, part of the wall bulges out to form two wings (Fig. 14-4). When released from the cone, the pollen grains are carried about by the wind. Of the millions of pollen grains released each spring, only a few ever reach the female cones.

The **female cones**, or seed cones, of conifers are much larger and harder than the male cones. They become quite woody in many species as they mature. The typical "pine cone" that you might collect on a forest floor is a woody female cone. Cones of various conifers are many shapes and sizes, and are used to help identify conifer trees (Fig. 14-5). Some are less than 1 cm long, but the sugar pine, which grows in the northwestern United States, has cones as long as 0.5 m.

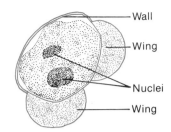

Fig. 14-4 The pollen grains of pine are winged and thick-walled.

Fig. 14-5 The female cones of different conifer species are useful in identification because they differ in colour, shape and size.

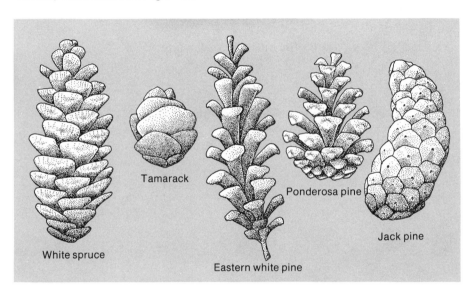

White spruce

Tamarack

Eastern white pine

Ponderosa pine

Jack pine

On the upper surface of each scale of a female cone are two raised masses called **sporangia** or **ovules**. Each ovule produces four female spores by meiosis (Fig. 14-6). Only one spore out of the four survives to form a **female gametophyte**, or **embryo sac**. In conifers, this gametophyte is neither independent nor photosynthetic like that of ferns. It is simply a small group of cells that is parasitic on the sporophyte cone scale. The scale protects it as it develops its **archegonia** containing **female gametes**, or **eggs**. A female cone that is ready to be fertilized usually sits upright on a branch with the scales open to receive pollen.

The transfer of pollen (male spores) from a male to a female cone is called **pollination**. Wind is the main agent of transfer. Once some pollen reaches the female cones, the female cone scales close up. The pollen grains germinate inside the female cone. Each produces a **pollen tube** which digests its way to an egg in the archegonium of the female gametophyte. Two male gametes, the **sperm nuclei**, are produced in the pollen tube. These gametes do not have cytoplasm and a plasma membrane around them. This is not

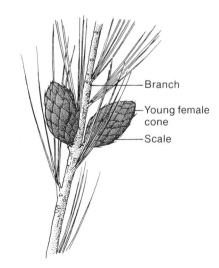

Fig. 14-6 Female or seed cone of pine. On each scale are two sporangia called ovules that produce the female spores.

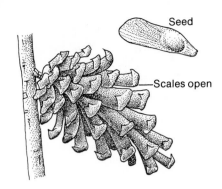

Fig. 14-7   A mature female pine cone sheds its seeds to the wind.

necessary since they travel directly to the eggs through the pollen tube. Unlike primitive plants, a film of water is not required for fertilization to occur. This has allowed conifers to reproduce and thrive in dry environments where algae, mosses, and ferns cannot.

Fertilization of the egg produces the first diploid cell of the sporophyte generation, the **zygote**. The zygote develops into an **embryo**, or embryonic sporophytic plant, using food that was stored in the egg cell. The embryo, the stored food, and a hard coat that develops around them, make up the **seed**. Two seeds develop on each scale. From the time of pollination, it takes two years for pine seeds to mature, and one year for many other conifers. At maturity, the scales of the female cone open and the seeds are released to the wind (Fig. 14-7). Pine seeds have a wing-like structure that helps with dispersal in the wind. Many seeds are dispersed over great distances. They often remain dormant for a long time until conditions suitable for germination occur.

Figure 14-8 illustrates the life cycle of the pine.

Review

1. Of what importance are conifers in Canada?
2. What is the difference between a conifer and an evergreen?
3. What does deciduous mean? Give an example.
4. What is a cone?

Fig. 14-8   The life cycle of the pine.

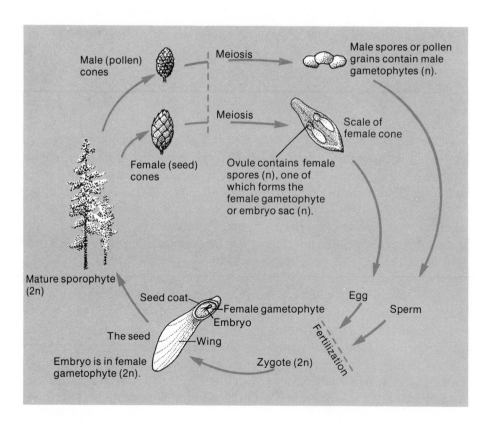

5. Compare male and female pine cones in terms of their size, duration, and location on the tree.
6. Why is it necessary for pines to produce thousands of pollen grains?
7. How are pine pollen grains and seeds adapted for wind dispersal?
8. In what three ways are conifers more advanced than the ferns?

14.3 Investigation

Coniferous Trees

Because there are relatively few species of coniferous trees in Canada, it is fairly easy to learn to recognize them. Your teacher will supply you with some tree specimens to identify, and a key to determine the names of the unknown specimens.

Remember that a key is simply an identification tool. At each step of the key you must choose between two alternatives, and then follow that branch of the key that leads from your first choice. You will have to determine if the leaves of your specimens are needle-like or scale-like and whether they are in clusters or single (Fig. 14-9). For some characteristics, such as whether the leaves are stalked or the twigs hairy, you will need a hand lens. For some steps of the key you will need cones. By working carefully, you should be able to arrive at the correct names for all your specimens.

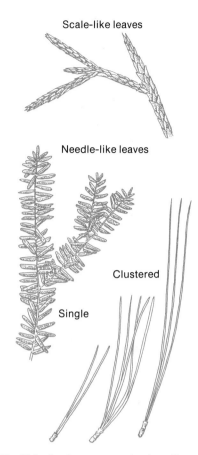

Fig. 14-9 Leaf arrangements of conifers are used in identification.

Materials

fresh or pressed specimens of various conifers, with cones
hand lens
key to the conifers

Procedure

a. Carefully examine the twigs, leaves, and cones of one specimen. In your notebook, describe the specimen in words. Include the size, shape, and arrangement of the leaves and cones.
b. Use the key provided to identify the specimen. Record the name with your description.
c. Repeat steps a and b for as many specimens as you can in the time allowed.
d. Choose one species of pine and one other conifer species to draw. Make a large clear drawing of the twig, leaves, and cone of each species. Be sure to label and title each drawing.

Discussion

1. **a)** What characteristics would a forester look for first, when trying to identify a coniferous tree?
   b) Why is it important to be able to distinguish one species from another?
2. What characteristic distinguishes all pines from other types of conifers?
3. How could you distinguish a cedar from a spruce?

14.4 Division Anthophyta: The Flowering Plants

General Characteristics

The tremendous success of the flowering plants becomes obvious when you look around you. Over 70% of all living plant species are flowering plants. They began in the Jurassic Period, which started 180 000 000 years ago. Dinosaurs and gymnosperms were most abundant during this period. They increased greatly in numbers during the Cretaceous Period, which began 135 000 000 years ago, and have continued to do so ever since. No one is sure if the flowering plants have yet reached the peak of their development.

What has contributed to their success? The evolution of the flower has made all the difference. The flower has one major function: to reproduce. It protects the developing embryo with tissue called the **ovary**. Ovaries are usually pear-shaped and found in the centres of flowers. Flowers usually have additional parts that help in some way or another in getting the eggs fertilized. Many flowers are brightly coloured to attract insects that pollinate them. After a flower is fertilized, its ovary enlarges to form a **fruit** around the seeds. Fruits protect the seeds from damage and help with their dispersal.

Over the years since the Cretaceous Period, flowering plants have evolved into many different shapes and sizes. Some are trees, others are shrubs, herbs, or vines. Flowers of every imaginable colour and form exist everywhere. Some flowering plants live only one season and are called **annuals**. Petunias, marigolds, and zinnias are colourful annuals that are planted in gardens each spring. Other plants, such as delphiniums and roses, can survive the winters and live year after year. These are called **perennials**. A few plants live for only two years, producing leaves the first year and flowers the second. Carrots and foxglove are two examples of two-year, or **biennial**, plants. The great diversity in flower type, growth cycle, and life form has allowed flowering plants to live in every environment on earth, from the tropics to the arctic.

Flowering plants have adapted to many different climate and soil conditions. Some have developed structures that reduce water loss, so that they are able to survive in dry deserts. Others have air passages between their cells that allow them to float in water (Fig. 14-10). Some are specially adapted to survive in salt marshes. A few have left the ground completely and sit perched as epiphytes on tall trees in tropical jungles. Some have lost

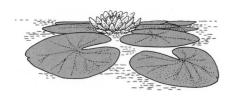

Fig. 14-10 Water lilies are adapted to float on the surface of the water in lakes and ponds.

their chlorophyll and depend either on dead organic material for their food or on a living photosynthetic host (Fig. 14-11). Some plants like mistletoe, which have chlorophyll, have become partially parasitic on other plants. Altogether there are over 250 000 species of flowering plants, with more being discovered every day by botanists all over the world.

Monocots and Dicots

Early in their evolution, the flowering plants began to develop along two separate and distinct lines. The two groups are classified today as the Subclass Monocotyledonae, or **monocots**, and the Subclass Dicotyledonae, or **dicots**. These words refer to the number of seed leaves, or **cotyledons**, found inside the seeds (Fig. 14-12). *Monocots have only one cotyledon in each seed and dicots have two.*

Some seeds are too tiny to see the number of cotyledons without a microscope. However, there are other characteristics that help us to distinguish monocots and dicots at a glance. Monocots have flowers that have three of each part. In dicots the parts are arranged in fours or fives. In monocots the leaves have many *parallel* veins. In dicots the leaves are usually *net-veined*. The vascular tissue inside the stems is also arranged differently in the two groups. Figure 14-12 shows the four major characteristics that separate monocots from dicots. Using these characteristics, you should be able to tell if a plant is a monocot or a dicot just by looking at it.

Because there are so many different kinds of monocots and dicots, botanists have classified them into much smaller groups called *families*. Each member of a family has a certain combination of characteristics that is different from plants in other families. Some families are very large and others contain only one or two species. Two of the largest families are the grass family and the aster family. The *grass family* includes pasture grasses,

Fig. 14-11 Indian Pipe is a parasite that depends on other plants for food. It is all white, with no chlorophyll to produce its own food. You find Indian Pipe in shady woodlands. Its roots obtain food by joining on to the roots of trees, usually spruce or pine.

Fig. 14-12 Monocots and dicots differ in four significant ways.

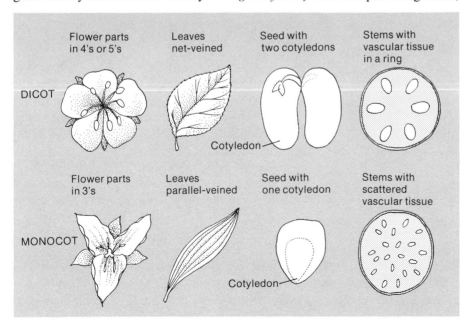

lawn grasses, the cereal grains (corn, wheat, oats, barley, and rice), sugar cane, and bamboo. The *aster family* includes goldenrods, daisies, chrysanthemums, and dandelions. Other large families are the legume family, which includes beans, peas, and peanuts, and the orchid family, which is largely tropical.

Importance

Flowering plants are important to humans in many ways. They produce the fruits, grains, and vegetables that we eat. Oils, natural rubber, cotton, spices, coffee, and tea are all products of flowering plants. Trees such as black cherry, oak, and maple provide us with valuable lumber. The beauty of flowering plants is important too. Gardens, parks, greenhouses, and flower shops are full of flowers. Botanical gardens in many Canadian cities display labelled specimens of many kinds of beautiful trees, shrubs, and flowers for the public to enjoy. Like all organisms, flowering plants are important links in many food chains, on which survival of the biosphere depends.

Review

1. Why are flowering plants more successful than gymnosperms?
2. **a)** What part of a flower forms a fruit?
   b) What are two functions of fruits?
3. **a)** What is the difference between an annual and a perennial plant?
   b) What is a biennial?
4. Explain why flowering plants have spread to every habitat on earth.
5. What characteristics would help you to distinguish a monocot from a dicot?
6. Why are flowering plants of more direct benefit to humans than other plant groups?

14.5 Flowers

You now know that the widespread success of flowering plants is largely due to the development of the flower. What exactly is a flower? It is a highly specialized reproductive structure that produces gametes. All its parts are modified leaves on a short branch. Most of the parts are so specialized that they do not even look like leaves at all.

Parts of a Flower

Calyx

The most conspicuous parts of a flower are usually the calyx and corolla (Fig. 14-13). The **calyx** is the outer whorl of green leaf-like structures. The parts of the calyx are called **sepals**. These may be separate or joined to each other. Sepals protect the other flower parts in the bud stage and support

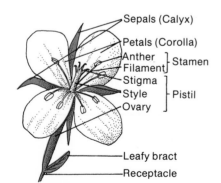

Fig. 14-13 This flower of Fireweed has all the parts of a typical flower. Note that the stigma is separated into four sticky lobes. Is Fireweed a monocot or a dicot?

them when the flower opens. They sometimes remain green and leafy as the flower develops into a fruit, as in the strawberry (Fig. 14-14).

Corolla

The **corolla** is composed of one or more whorls of **petals** that are often brightly coloured. Petals attract pollinating insects and serve as landing platforms for them. The petals are often fragrant. At the base of some petals are **nectaries** that secrete a sweet liquid called **nectar** that also attracts pollinators. A few flowers, such as columbines and delphiniums, have spurred corollas with nectar in the spurs (Fig. 14-15).

Stamens

Inside the calyx and corolla are the reproductive parts of the flower, the stamens and the pistil. The **stamens** produce the male spores, or **pollen grains**, in sac-like **anthers** at their tips. The anthers are supported by thin stalk-like **filaments** (see Fig. 14-13). Most flowers have several stamens, often twice as many as the number of petals.

Pistil

The female reproductive structure, or **pistil**, is located in the centre of the flower. A pistil is usually pear-shaped and made of an ovary, a style, and a stigma (see Fig. 14-13). The **ovary** is the large bulb-shaped part at the base of the pistil that contains the ovules which may become seeds. Above the ovary is the **style**, a stalk that supports the enlarged tip called the **stigma**. The stigma is often sticky or feathery in order to hold pollen. The pistil of a flower may be **simple**, as just described, or compound. A **compound** pistil is made of several simple pistils that are fused together. Inside the ovary of a pistil, whether it is simple or compound, **ovules** are attached to the ovary wall by short stalks. There may be hundreds of ovules, a few, or only one per flower. Figure 14-16 shows some of the ways the ovules may be attached to the ovary wall.

Fig. 14-14 The calyx of a strawberry flower remains on the fruit.

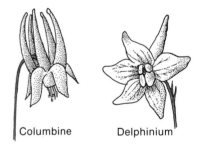

Fig. 14-15 Columbine flowers have five spurs and delphinium flowers have none. The spurs contain nectar that attracts pollinators.

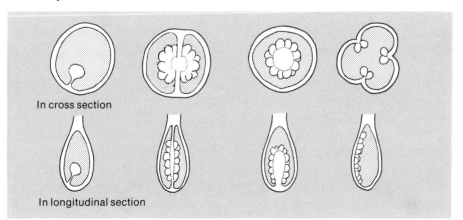

In cross section

In longitudinal section

Fig. 14-16 Attachment of ovules. There are many ways in which ovules can be attached to the ovary wall.

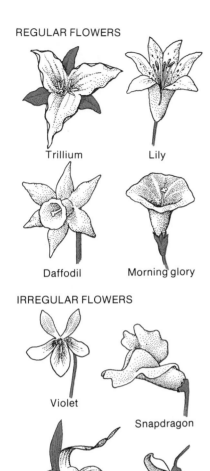

REGULAR FLOWERS

Trillium

Lily

Daffodil

Morning glory

IRREGULAR FLOWERS

Violet

Snapdragon

Lady slipper orchid

Touch-me-not

Fig. 14-17 Flower symmetry. What difference is there between regular and irregular flowers?

Remember that, in conifers, the ovules developed on the upper surface of the scales in the female cone. The conifer ovules were "naked" and not enclosed by an ovary as in the flowering plants.

Receptacle

The calyx, corolla, stamens, and pistil are all attached to the enlarged tip of a flower stalk, called the **receptacle**. A receptacle sometimes forms part of a fruit, as in an apple. However, its most important function is to support the flower and fruit as they develop.

Types of Flowers

There are over 250 000 variations in flower structure, and yet all flowers have the same function of producing seeds. The flowering plants are classified into families on the basis of their flower structure because flowers are less variable than leaves and stems within a species. The number and arrangement of the floral parts on the axis and the fusion of parts with each other are characteristics used in classification. The number of flower parts classifies plants as monocots or dicots. The symmetry of the flower parts helps in identification, too. Flowers whose petals are of equal size and are evenly spaced around the axis are called **regular** flowers. They could be sliced in half in any direction and the two halves would be mirror images of each other (Fig. 14-17). **Irregular** flowers, like the pea, have petals of different sizes and shapes. There is only one plane in which you could cut an irregular flower so that the two halves would be mirror images.

The fusion of floral parts is also important in classification. In some flowers the sepals are fused together in a cup. However, in others they are separate or even missing. Petals also are often fused in tubes or cups, or may be lacking altogether. Sometimes the stamens are attached to the petals instead of to the receptacle.

The position of the ovary with respect to the other flower parts also varies. If the ovary is **superior**, as in May apples, it sits on top of the receptacle with the other floral parts attached below it (Fig. 14-18,A). In flowers with **inferior** ovaries, such as cucumbers and sunflowers, the calyx, corolla, and stamens are attached to the top of the ovary (see Fig. 14-18,B).

Although most plants have both a pistil and stamens in each flower, some do not. Flowers that have both male and female parts are called **perfect**. Those that lack one or the other, or both, are called **imperfect**. Imperfect flowers are of three types. **Staminate** flowers have stamens but no pistils. **Pistillate** flowers have pistils but no stamens. **Sterile** flowers have neither. Corn is an example of a plant with imperfect flowers. The staminate flowers (tassels) are separate from the pistillate flowers (ears) found below them. Since both types are found on each corn plant, corn is called **monoecious**. (*Monoecious* means "one home".) Manitoba maples and willows produce pistillate and staminate flowers on completely separate plants. Such plants are called **dioecious**. (*Dioecious* means "two homes".)

The pollen from the male plants has to be carried to the female plants for reproduction to occur. The male and female flowers of willow are shown in Figure 14-19, along with the sterile flowers of a High-bush cranberry shrub.

How did so much variation in flower structure evolve? It took a long, long time. Much of the diversity is related to how the flowers are pollinated. Insect-pollinated flowers are generally large and colourful. Wind-pollinated flowers are smaller and often lacking in parts. The co-evolution of flowers and their pollinators is an interesting example of the complex interrelationships of the natural world.

Review

1. **a)** In what ways is a flower similar to a cone?
   b) How does it differ from a cone?
2. What would you consider to be the essential parts of a flower? Why?
3. List three functions of the corolla.
4. What part of a flower serves the same function as the pollen cone of a conifer?
5. Why would it be an advantage for a flower to have a compound pistil?
6. **a)** Explain the difference between regular and irregular flowers.
   b) Give an example of each.
7. List five characteristics of a flower that would enable you to identify it with a key.
8. **a)** What is the difference between perfect and imperfect flowers?
   b) Name and describe the three types of imperfect flowers.
   c) Distinguish between superior and inferior ovaries.
9. **a)** What is the difference between a monoecious and a dioecious species?
   b) Could a species be dioecious and have perfect flowers? Explain.

14.6 Investigation

Structure of a Simple Flower

The best way to learn the parts of a flower is to examine and compare the structures of several different types. Although the variations from species to species seem infinite, all flowers have some or all of the typical flower parts shown in Figure 14-13. Since flowering plants are grouped into families according to their flower structure, a knowledge of one member gives you a general picture of the flower type of all the members in that family.

Materials

fresh flowers of petunia, snapdragon, lily, gladiolus, or other simple flower	hand lens	eye dropper
	forceps	compound microscope
razor blade or scalpel	microscope slide	lens paper
	cover slip	paper towel

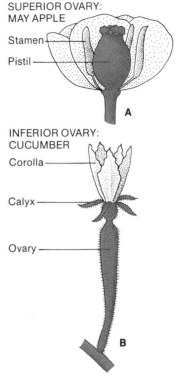

Fig. 14-18 Superior and inferior ovary positions.

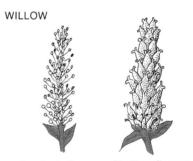

WILLOW

Staminate flowers   Pistillate flowers

HIGH-BUSH CRANBERRY   Fertile flowers

Sterile flowers

Fig. 14-19 Imperfect flowers.

Procedure

a. Carefully examine one of the specimens available. Locate the four floral parts (calyx, corolla, stamens, and pistil). Do not tear the flower apart.

b. Make a large sketch of the whole flower to show its shape and the normal positions and relative sizes of the flower parts with respect to each other.

c. Observe the calyx. Note the number of sepals and whether or not they are fused. Describe the shape, colour, and texture of the sepals.

d. Repeat step c for the petals of the corolla.

e. Remove some of the petals so you can see the stamens. Are the stamens attached to the petals? Observe the stamens with a hand lens. Count them. Note where they attach to the rest of the flower. Find the anthers and the filaments.

f. Remove one anther. Crush it gently in a drop of water on a slide to shake off the pollen. Cover the preparation with a cover slip. Observe the pollen under low, medium, and high power. Make a large labelled drawing of a few pollen grains.

g. Locate the pistil in the centre of the flower. Describe its shape. Note whether the ovary is superior or inferior with respect to the other flower parts. Observe the stigma with the hand lens. Note its texture and shape.

h. Use the razor blade to cut carefully down through the centre of the pistil from top to bottom. With the hand lens, observe the ovules inside the ovary. Describe their appearance and how they are attached to the ovary wall. Count or estimate the number of ovules inside the whole ovary.

i. Make a final drawing of all the parts of the flower as they appear when some of the petals have been removed and the ovary has been cut lengthwise. Label all the parts.

j. If time is available, repeat this lab for a second type of flower. Compare its parts with the first species as you go along.

Discussion

1. Does this flower contain all the four floral parts? If not, what is missing?

2. What is the relationship between the numbers of sepals, petals, and stamens?

3. Is the flower a dicot or a monocot? How do you know?

4. Complete Table 14-1 in your notebook.

FLOWER_____:					
STRUCTURE	NUMBER OF PARTS	FUSION OF PARTS	WHERE PARTS ARE ATTACHED	DESCRIPTION	FUNCTION
Calyx					
Corolla					
Stamens					
Pistil					

TABLE 14-1 **Parts of a Flower**

5. How is the stigma of the flower adapted to trap pollen?
6. How are the pollen grains adapted for transfer from the anthers to a pistil?
7. How do you think the flower gets pollinated? Give reasons for your answer.
8. Estimate how many seeds the flower could possibly produce.

14.7 Investigation

Structure of a Composite Flower

One of the largest plant families is the Asteraceae, or aster family, with over 20 000 species. Daisies, thistles, dandelions, chicory, ragweed, dahlias, sunflowers, zinnias, mums, asters, and goldenrods all belong to this group. It is one of the most successful plant families, and has spread to every part of the world. Its flowers are so different in structure from all other plant families, that it deserves special mention.

Although you would probably look at a daisy or a dandelion and call it a flower, it is actually a whole cluster of tiny flowers on a single stalk. This flower cluster is called a **flower head** and is unique to the aster family (Fig. 14-20). The base of the head is called the **receptacle**. It is usually covered with several rows of leaf-like **bracts** called an **involucre**.

Fig. 14-20 The composite head of ox-eye daisy.

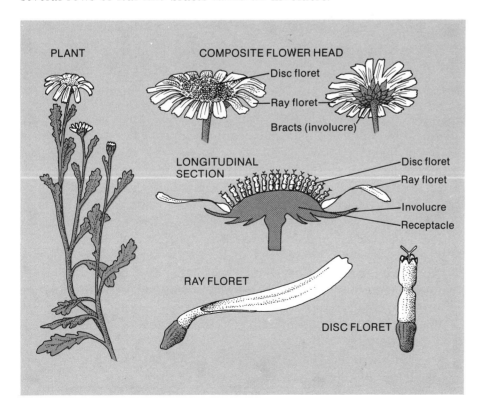

Each tiny flower of a composite head is called a **floret**. The florets of a composite flower like a daisy are of two types (see Fig. 14-20). The outer whorl of florets are called **ray florets**. Their petals are fused together to form a strap-shaped ray that looks like a single petal. Often the stamens and pistil of ray florets are missing or sterile. The calyx usually consists of a row of hairs, teeth, or spines and is called a **pappus**. It is not at all like the green leafy sepals of a simple flower like a rose.

The inner florets of a daisy are called **disc florets**. Their petals are small and fused together in the shape of a tube. The disc florets usually contain fertile stamens that are attached by their anthers in a ring around the pistil. The pistil is composed of an inferior ovary, a slender style, and a feathery divided stigma. Each fertile floret contains only one ovule and can produce only one seed. However, because there are usually many florets on a single head, composite flowers probably produce more seeds per flower than any other plant.

Most composites have both ray and disc florets on the same head. However, some, such as dandelion and chicory, have only ray florets. Others, like thistles, have only disc florets. In many cultivated species, such as chrysanthemums, the heads are very crowded and many of the florets are sterile. For this reason, composite flowers from the home garden are sometimes difficult to study.

Materials

fresh specimen of chicory, daisy, dandelion, or other composite flower head	hand lens forceps razor blade or scalpel

Procedure

a. Observe the entire head with a hand lens. Find out if it has ray florets, disc florets, or both. Count and record the approximate number of each kind of floret.

b. Locate the small leaf-like bracts of the involucre at the base of the head. Count the number of rows of bracts. Describe their appearance.

c. Use a razor blade to split the head in half by cutting vertically down through the receptacle. Make a sketch of the head as it appears from the side. Label the ray and disc florets, receptacle, and involucre.

d. If any ray florets are present, remove one of them with the forceps. Examine it with a hand lens. Find out if all four flower parts are present. Locate the pappus (calyx) at the top of the ovary and describe it. Find the corolla and describe its shape and colour. Note where it is attached to the floret. Count the number of teeth at its tip to determine how many petals are fused together.

e. Note that the anthers of the stamens are attached to each other in a ring. Count the number of stamens. Determine where their filaments are attached. Find the pistil (if present) and describe its stigma.

f. Make a large drawing of one ray floret. Label all the parts you observed.

g. If disc florets are present, examine them in the same way that you examined the ray florets. Make a labelled drawing of one disc floret.

Discussion

1. Why is it wrong to call a daisy a flower?
2. a) What is an involucre? What do you suppose is its function?
   b) Which structures perform this same function in a simple flower?
3. What are two differences between a typical ray floret and disc floret?
4. What is a pappus?
5. Approximately how many seeds could be produced by a single head of the species you observed?
6. a) How many petals make up the corolla of a floret?
   b) How does this number compare to the number of stamens?
   c) Is the corolla of a ray floret regular or irregular? Explain.
7. What is the probable function of each type of floret in the species you observed?
8. To what other flower part are the stamen filaments attached in florets of the aster family?
9. How are the stigmas of your specimen adapted to hold pollen?
10. Why would it be an advantage for so many tiny flowers to be crowded together in one head?

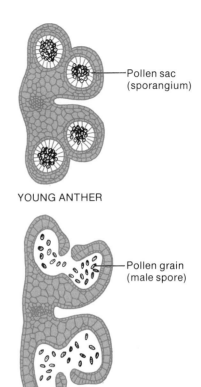

YOUNG ANTHER

Pollen sac (sporangium)

Pollen grain (male spore)

MATURE ANTHER

Fig. 14-21 Cross section of an anther at two stages of development.

14.8 Reproduction in Flowering Plants

The stamens and pistil of a flower are the reproductive parts of the sporophyte. They produce the male and female spores of a flowering plant. The male and female gametophytes are microscopic as in conifers. The reduction of the gametophyte to only a few cells is an evolutionary trend that can be seen throughout the plant kingdom. An independent gametophyte dominates the life cycle of many algae and all mosses. In ferns the gametophyte is independent, but reduced to a flattened sheet of cells. However, in conifers and flowering plants, the gametophyte is reduced to a few cells that are completely dependent on the sporophyte generation.

The Male Spores

The male spores, or pollen grains, are produced by the stamens of the sporophyte. The anther of a stamen contains four sporangia or **pollen sacs**. These are clearly visible in cross section (Fig. 14-21). Meiotic division of the cells inside produces the **pollen grains** that are shed from the anther at maturity. A mature pollen grain is a single cell that contains two haploid nuclei. It is surrounded by a thick wall that helps to prevent it from drying out. The wall is often textured with grooves, warts, or other surface features to help in dispersal (Fig. 14-22). The two nuclei inside the pollen grain are

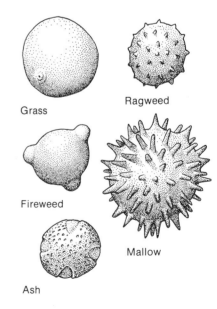

Grass

Ragweed

Fireweed

Mallow

Ash

Fig. 14-22 Pollen grains vary in their ornamentation. Grains carried by wind are often smooth, and those carried by insects are deeply warted or grooved.

Fig. 14-23 Development of pollen grains (male spores). A mature grain contains two haploid nuclei.

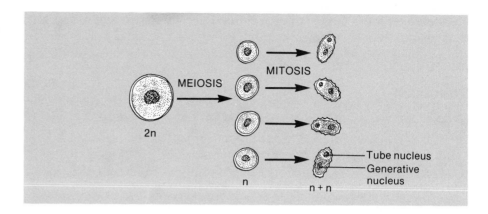

called the **tube nucleus** and the **generative nucleus** (Fig. 14-23). These two nuclei represent the entire male gametophyte of a flowering plant. One of them, the generative nucleus, divides to form two sperm nuclei or male gametes.

The Female Spores

The female spores are produced deep inside the tissues of the ovary in a structure called an **ovule**. One ovary may contain many ovules or only one.

Fig. 14-24 Development of the ovule and female gametophyte in a flower.

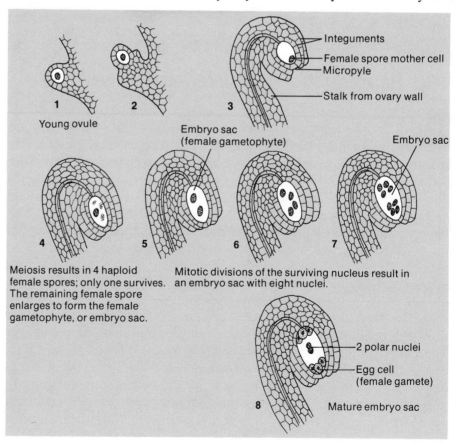

Each ovule grows out from the ovary wall on a short stalk, and is surrounded by protective layers of tissue. The protective layers, called **integuments**, meet at a small pore called the **micropyle** (Fig. 14-24).

Early in the development of an ovule, one nucleus inside it divides by meiosis into four haploid nuclei. These are the female spores. Three of them die. The fourth one becomes the female gametophyte, or **embryo sac**. The nucleus of this cell divides three times to form eight nuclei, one of which forms the female gamete, or **egg** (see Fig. 14-24). Two of the eight nuclei are called **polar nuclei**. These are located at the centre of the embryo sac when it is mature. The female gametophyte, or embryo sac, does not produce archegonia. In this respect flowering plants differ from all other groups of plants.

Pollination and Fertilization

For the mature egg inside the ovary to become a diploid zygote, it must be fertilized by a male gamete, or sperm nucleus. The transfer of pollen from an anther to a pistil is called **pollination**. When a pollen grain (male spore) is deposited on the stigma of a mature pistil, it is stimulated to grow by chemical substances in the tissues. The pollen grain absorbs moisture. It then produces a **pollen tube** that grows through the stigma and style to the ovary (Fig. 14-25). The **tube nucleus** of the pollen grain enters the tube and

Fig. 14-25   Longitudinal section of a flower ready for fertilization.

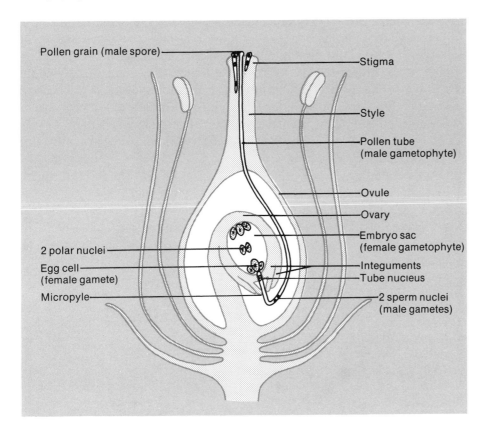

Pollen grain (male spore)

Stigma

Style

Pollen tube (male gametophyte)

Ovule

Ovary

Embryo sac (female gametophyte)

2 polar nuclei

Egg cell (female gamete)

Integuments

Tube nucleus

Micropyle

2 sperm nuclei (male gametes)

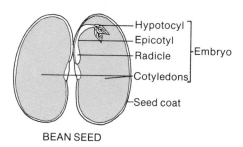

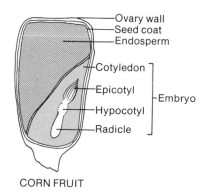

Fig. 14-26 The embryos of a bean seed and a corn fruit.

remains near the tip. The **generative nucleus** follows it and divides by mitosis into two male gametes called **sperm nuclei**, before fertilization occurs. (In some plants the generative nucleus divides before the pollen tube forms.) When the pollen tube reaches the ovule, it enters it through the pore called the **micropyle**.

The distance from the stigma to the ovule varies from a few millimetres in some flowers to several hundred in others (for example, the silk of a corn kernel). The pollen tube secretes enzymes at its tip that digest the tissues of the style ahead of the tube. The pollen tube acts as a vehicle of sperm transport. In that sense, it carries out the same function as water does in the life cycle of a moss or fern.

The pollen tube, as it penetrates the ovule, ruptures and releases the two sperm nuclei into the embryo sac. One sperm nucleus fertilizes the egg and forms a diploid **zygote**. The other unites with the two polar nuclei to form a triploid nucleus called the **endosperm nucleus**. This nucleus divides many times to form the **endosperm tissue**. This is a mass of cells that becomes a food supply for the developing embryo.

Because there are two sperm nuclei involved in the fertilization of an ovule, the whole process is called **double fertilization**. Double fertilization occurs only in flowering plants. Without it, no seeds are produced.

Development of the Embryo

Following fertilization, the zygote begins to divide by mitosis to produce an **embryo**. At first the embryo is just a mass of cells that pushes into the endosperm. It soon develops into a recognizable, although miniature, plant with four parts: the epicotyl, the hypocotyl, the radicle, and the cotyledons (Fig. 14-26). The **epicotyl** will become the leaves and the **radicle** will become the root of the young sporophyte plant. The **hypocotyl** forms the central axis of the embryo and, later, becomes the stem. The fleshy **cotyledons**, or seed leaves, store food materials. Remember that monocots have only one cotyledon and dicots have two.

As the embryo develops, it partly or totally absorbs the endosperm tissue, which it uses as a source of energy. Meanwhile, the integuments around the ovule harden and become the **seed coats**. The embryo, together with the seed coats and the endosperm, make up the **seed**.

Review

1. What structure produces the pollen grains of a flowering plant?
2. Describe the structure of a pollen grain.
3. Does alternation of generations occur in flowering plants? Explain.
4. Explain briefly how an embryo sac with eight nuclei develops inside an ovule.
5. Define pollination.
6. How do the sperm nuclei get from the top of a stigma to the micropyle of an ovule?

7. **a)** What is endosperm tissue?
   b) How does it arise?
8. Why is reproduction in flowering plants called double fertiization?
9. Describe the structure of a mature embryo.
10. What is a seed? Into what will it grow?

14.9 Pollination

Adaptations for Pollination

The methods by which pollen grains get from the anthers to the stigmas of flowers are quite varied. Many flowers are pollinated by insects such as bees, moths, and butterflies. Others are pollinated by birds and bats. The wind is an important pollinating agent, too. Flowers have many special adaptations that are related to pollination.

Insect-pollinated flowers usually have large brightly coloured corollas that attract insects visually and act as landing platforms (Fig. 14-27). They are usually scented and often contain nectar. Flowers pollinated by butterflies are frequently red. Those pollinated by moths are often white. Some bee-pollinated flowers have markings called nectar guides on them that are visible to bees. These markings cannot be seen by humans unless the flowers are put under ultra-violet light.

Flowers that attract birds and bats often need to have large petals for landing. Bat-pollinated flowers open mostly at night. Flowers like delphiniums, that are pollinated by hummingbirds, have long spurs into which the bird can reach for nectar.

Wind-pollinated flowers are much different in design from insect-pollinated ones. They are often missing the calyx and corolla, and have no nectar. Their stigmas are frequently large and feathery. Their stamens produce large amounts of smooth, light-weight pollen. Figure 14-28 shows the wind-pollinated flowers of a maple tree. Maples produce their flowers in clusters at the tips of branches early in the spring before there are any leaves to block the wind.

Types of Pollination

Pollination can be of two types. **Self-pollination** is the transfer of pollen from the anther to the stigma of the same flower, or to another flower on the same plant. Tomatoes, peas, and many cacti are self-pollinated.

Cross-pollination is the transfer of pollen from one plant to another. The seeds that result from cross-pollination contain genetic information from both parents. Because the seeds germinate into plants that have some characteristics of each parent, species that are cross-pollinated have more variations than those that are self-pollinated. You may recall that variability is helpful in a changing environment. It increases the chances of survival for

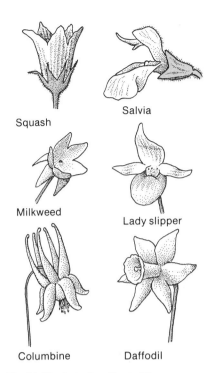

Squash

Salvia

Milkweed

Lady slipper

Columbine

Daffodil

Fig. 14-27 Insect-pollinated flowers are usually very showy, with large colourful corollas.

Fig. 14-28 Maple flowers are wind-pollinated.

at least some of the population. Plants that are cross-pollinated can often compete better in nature and adapt more easily to changing conditions.

Because cross-pollinated plants have such an advantage, many different mechanisms have evolved to prevent self-pollination. Some species produce imperfect flowers, with staminate and pistillate flowers on separate plants. Others produce a chemical that inhibits the germination of their own pollen. Many species produce perfect flowers in which the anthers and stigmas mature at different times. Often the anthers ripen and shed their pollen before the stigma is matured. Because different individuals of the same species flower at different times, there is always pollen available from other plants to fertilize a stigma when it is ready.

Although self-pollinated plants adapt poorly over the long term, they do have some advantages. They can produce seeds when the weather is rainy and cold. At such times, insect-pollinated plants often remain unfertilized.

Review

1. What are three floral adaptations for insect-pollination?
2. What is a nectar guide?
3. **a)** List three characteristics of flowers that are adapted for wind pollination.
   b) Explain how each of these characteristics assists pollinations.
4. Use a diagram to explain the difference between self- and cross-pollination.
5. Compare the advantages of cross- and self-pollination.
6. List three adaptations that prevent self-pollination in flowering plants.

14.10 The Development of Fruits

What are Fruits?

As a seed begins to develop after an egg is fertilized, the ovary surrounding the ovule begins to grow into a fruit. Sometimes the calyx or the receptacle also develop into parts of the fruit. But more frequently, these flower parts wither and die since they are no longer needed. The **fruit** of any flowering plant consists of the ripened ovary with one or more seeds inside, with or without other floral parts. A fruit always has two scars on the outside. One scar is where the fruit was attached to the parent plant. A stalk may still be joined to the fruit at this point. The other scar is left by the withered style and stigma. Where the ovary is inferior, the remains of the calyx and corolla may also be present at the style end of the fruit. All seeds have only one scar. It occurs at the point where they were attached to the inside of the ovary wall.

Most fruits are adapted to protect the seeds and to aid in their dispersal from the parent plant. The adaptations are nearly as varied as the structures of the flowers from which they develop. Many fruits have hooks or spines that catch on animals. Some, like the coconut, will float on water. Maple

Raspberry

Receptacle—

Strawberry

Receptacle—

Fig. 14-29 Aggregate fruits: raspberry and strawberry.

seeds have wings, and dandelion fruits have a parachute of hairs that help their dispersal in the wind. Fruits like that of the touch-me-not snap open and fling the seeds outward. Others, like cherries, are fleshy and get eaten by birds and animals. Cherry seeds pass through the digestive system and are dropped many kilometres from the parent plant.

When you see the word fruit, you probably think of apples, oranges, cherries, and similar foods. They are definitely fruits, but so are corn kernels, maple keys, bean pods, tomatoes, cucumbers, pumpkins, and poppy capsules. *All fruits are ripened ovaries containing seeds.*

Classification of Fruits

Because all 250 000 or so flowering plants in the world produce fruits, a classification system for the most common types has been created. Most fruits are either simple, aggregate, or multiple. A **simple fruit** develops from a single ripened ovary. Apples, cherries, acorns, and bean pods are all simple fruits. Aggregate fruits, such as strawberries and raspberries (Fig. 14-29), and multiple fruits, such as figs and pineapples (Fig. 14-30), are both made of many small ovaries fused together. The difference is that **aggregate fruits** develop from many separate ovaries in one flower, while **multiple fruits** develop from a cluster of separate flowers that fuse together later on.

Since most flowers contain only one pistil, simple fruits are the most common type. The group can be separated into fleshy and dry fruits, depending on how the ovaries develop. In **fleshy fruits**, like peaches and

Fig. 14-30   Multiple fruit: pineapple.

Fig. 14-31   A classification of simple fruits.

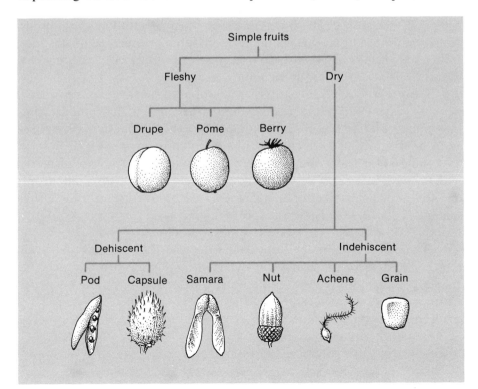

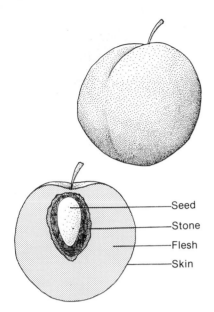

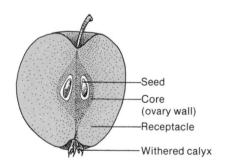

Fig. 14-32   Drupe: peach. The skin, flesh, and stone are all produced from the ovary wall.

Fig. 14-33   Pome: apple (longitudinal section).

tomatoes, the mature ovary wall, or **pericarp**, is soft and succulent. In **dry fruits**, such as acorns and milkweed pods, the pericarp is hard and often very tough. Further separations of fleshy and dry fruits results in a classification like that in Figure 14-31.

Simple Fleshy Fruits

The three types of fleshy fruits shown in Figure 14-31 are the drupe, pome, and berry. A **drupe**, such as a peach, cherry, olive, or plum, has a hard pit inside it. The pit is formed by the inner layer of the pericarp (Fig. 14-32). The other layers of the pericarp form the fleshy edible part and the outer skin.

A **pome** is a fleshy fruit in which the flower receptacle has become greatly enlarged around the ovary (Fig. 14-33). Apples and pears belong to this group. The core of an apple or pear is the ovary wall that has developed around the seeds. The fleshy edible part is the receptacle.

A **berry** (Fig. 14-34) is a fleshy fruit that has a thick skin, a thick flesh, and many seeds. All the layers come from the ovary wall. Grapes, tomatoes, and currants are classified as berries. Aggregate fruits such as strawberries are not. Some berries such as oranges, cucumbers, and watermelons have thick outer rinds surrounding the pulp. Others, such as bananas, have no seeds.

Simple Dry Fruits

In Figure 14-31, dry fruits are separated into two groups according to their manner of dehiscence. The word *dehiscence* means "splitting open". **Pods**, such as milkweed pods, that split open at maturity are **dehiscent fruits** (Fig. 14-35). Other pods that dehisce to release their seeds are peas, peanuts, and beans. Pods are produced from simple pistils only.

A **capsule** (Fig. 14-35) is another type of dry fruit that dehisces when it matures. Capsules differ from pods in that they have many compartments that come from the sections of a compound pistil. Poppies, irises, violets, and snapdragons all form capsules that contain many seeds.

Indehiscent fruits do not split open when mature. They are dispersed with the seeds and break down when the seeds germinate. Most contain only one seed. The winged keys of maple are a type of indehiscent fruit called a **samara** (Fig. 14-36). They are spread about easily by the wind. Other examples of samaras are the fruits of ash and elm.

An **achene** is an indehiscent fruit with a hard ovary wall that is not attached to the seed. A **grain** is similar, but its pericarp is firmly attached to the seed inside it. Buttercups, sunflowers, and dandelions produce achenes. Corn and all other grasses produce grains (Fig. 14-36).

The final type of indehiscent fruit is the **nut** (Fig. 14-36). Acorns, pecans, and chestnuts are good examples. A nut contains a single seed that is surrounded by a very hard ovary wall.

Review

1. **a)** What parts of a flower develop into a fruit?
   b) Explain how you could distinguish between a fruit and a seed.
2. What are the two main functions of fruits?
3. **a)** Why is it important for seeds to be dispersed away from the parent plant?
   b) Give four examples of adaptations that ensure seed dispersal.
4. **a)** Define and give an example of an aggregate fruit.
   b) Explain how a pineapple forms.
5. **a)** What is the difference between a drupe and a berry?
   b) What flower part are you eating when you bite into an apple?
6. **a)** What does dehiscent mean?
   b) Give two examples of dehiscent fruit types. Explain how they are different.
7. What is a samara?
8. Explain how the fruit of a sunflower differs from the fruit of a grass.
9. Why is a peanut not classified as a nut?

Fig. 14-34   Berry: tomato.

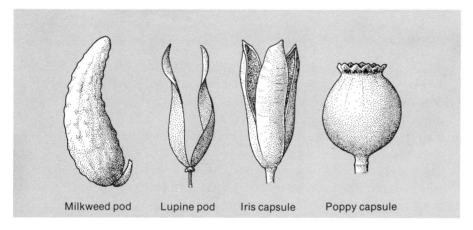

Milkweed pod   Lupine pod   Iris capsule   Poppy capsule

Fig. 14-35   Dry dehiscent fruits: pods and capsules.

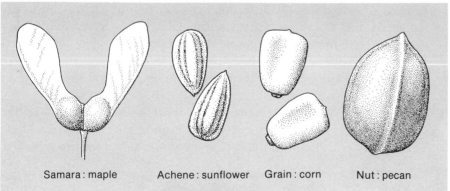

Samara : maple   Achene : sunflower   Grain : corn   Nut : pecan

Fig. 14-36   Dry indehiscent fruits: samara, achenes, grains and nuts.

14.11 Investigation

Fruits

If you are a bit confused about the classification of fruit types, some practical experience should help you become more familiar with the groups. Your teacher will have a selection of fruits for you to observe and compare as you work through this investigation.

Materials

apple
tomato, orange, or grapefruit
selection of other fruits (bean, maple, cucumber, milkweed, iris, poppy, peanut, acorn, cherry, sunflower, plum, corn ... or others)
hand lens
razor blade or scalpel

Procedure

a. Begin your investigation with an apple. Observe the end of the apple opposite to the stalk to determine what flower parts remain there. Make a drawing of the entire apple and label all the parts. Is the ovary in apple superior or inferior? How can you tell?

b. Cut the apple in half crosswise. Examine the pulp, core, and seeds. The core should appear as a star-shaped compartment in the centre. Count the number of divisions in the core. Make a drawing of your sectioned apple. Label the receptacle, ovary wall, and seeds. Title your drawing with the name of the fruit type for which apple is an example.

c. If a tomato, orange, or grapefruit is available, observe it and make a labelled drawing. Cut the fruit open crosswise. Make a drawing of the interior. Label all the parts that you can. Classify the fruit and title your drawings. Use the text as a guide, if you need it.

d. Examine one dry fruit from those available. Classify it, if you can. Make a labelled drawing. Then split or crack the fruit open. Determine how many seeds it contains.

e. Practise classifying fruit types by comparing the structures of the remaining specimens with the descriptions and figures in Section 14.10. Try to identify as many of the specimens as you can in the time allowed.

Discussion

1. a) Describe any evidence of flower parts at the end of an apple opposite to the stalk.
   b) Which flower parts are there?
   c) Why do the flower parts dry up as the apple develops?
   d) Is the ovary superior or inferior? How can you tell?

2. **a)** What part of the flower produces the fleshy part of the apple?
   b) What part produces the core?
3. **a)** How many seeds do most apples contain?
   b) How are apple seeds dispersed?
4. From your observations, what differences are there between a pome and a berry?
5. What advantage do dehiscent fruits have over indehiscent fruits?
6. How could you be sure that cucumbers, peanuts, corn grains, or tomatoes are fruits?

Highlights

The seed plants are the most advanced and most successful group of terrestrial plants on earth today. Seeds are easily dispersed and are able to remain dormant for long periods of time. Therefore, the seed plants have become widely distributed throughout the world. Coniferous forests cover large land areas in temperate climates. Flowering plants have adapted to almost every environment available.

The two largest groups of seed plants are the conifers and the flowering plants. The conifers include trees and shrubs whose seeds are produced naked on the scales of cones. Most conifers have needle-like evergreen leaves or overlapping scale-like leaves. Many are important economically as sources of pulp and paper and lumber.

Flowering plants differ from conifers in two major respects. Their reproductive parts are enclosed in whorls of specialized leaves that make up the flower. The male reproductive parts, or stamens, produce pollen grains. These are transferred to the female reproductive structure, the pistil, inside which fertilization occurs. Unlike mosses and ferns, water is not required for fertilization to occur. The male gametophyte produces a pollen tube which carries the sperm nuclei directly to the egg in the ovule. After fertilization, the pistil grows into a fruit. The fruit protects the seeds from damage and helps to disperse them away from the parent plant.

There is much variability in the structure of flowering plants. Leaves, stems, roots, flowers, and fruits all exist in many shapes, colours, and sizes. This great diversity has allowed flowering plants to adapt to many different climates and habitats. Their diversity of form and their highly advanced method of reproduction are the main reasons the flowering plants have become so dominant in the landscape today.

15

Root, Stem, And Leaf: Form And Function

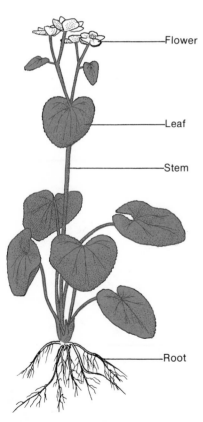

Fig. 15-1 Marsh marigold (*Caltha palustris*) shows the parts of a typical flowering plant.

Most seed plants are composed of roots, stems, and leaves. These organs are specialized to perform certain functions that add to the survival of the plant. The parts of a typical plant are shown in Figure 15-1.

Materials enter and leave a plant only at certain points. They must be transported to where they are needed. Complex terrestrial (land) plants have efficient vascular, or transport, systems that link all parts of the plant. Water and nutrients are absorbed from the soil into the plant by the roots. Roots also store food in their cells and hold the plant in place. Stems transport materials, and give support to the leaves and flowers. Stems that are green also produce food by photosynthesis. Leaves are specially designed to trap sunlight and convert its energy into chemical potential energy in food. They also allow carbon dioxide, oxygen, and water to enter and leave the plant.

15.1 Cell Types in Vascular Plants

Each organ of a plant is made up of many types of cells. All these cells originate by mitosis from one kind of cell called a **meristematic cell**. **Meristems** are regions in a plant that contain these embryonic cells. The buds on a stem and the tips of roots and shoots are meristems. The cells there are small, thin-walled, and usually cube-shaped. They are able to divide again and again to produce new cells. Because they keep dividing throughout the life of a plant, growth is unlimited. The new cells produced by a meristem, mature into many different types of cells. This process is called **differentiation**. As cells differentiate in structure, they become specialized for different functions such as transport, support, protection, photosynthesis, and storage. Just how cells are able to differentiate is still one of the mysteries of biology. (See Section 8.3, page 175.) When very similar cells are organized to perform a certain function, they form a **tissue**.

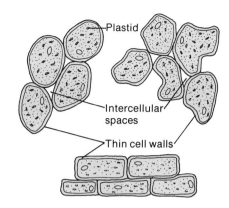

Fig. 15-2 Parenchyma cells.

Parenchyma Tissue

One of the simplest kinds of cells that develop from a meristem is parenchyma. **Parenchyma** cells are large, thin-walled cells. Many roots, stems, and leaves are made mostly of these cells. Figure 15-2 shows a few parenchyma cells. In roots, parenchyma tissue is used for storage of starches, sugar, and water. In leaves, parenchyma tissue produces food by photosynthesis. There, the cells contain the green pigment chlorophyll.

Sclerenchyma Tissue

A second type of cell that develops from meristematic cells is sclerenchyma. **Sclerenchyma cells** vary in shape, but have very thick walls. Their main function is to strengthen and support other plant tissues. Most lose their protoplasm at maturity. Only the thickened, dead cell walls remain. Long narrow sclerenchyma cells are called **fibers**. These are present in many stems (Fig. 15-3).

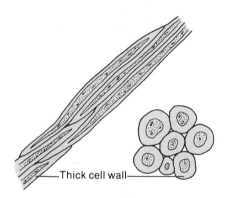

Fig. 15-3 Sclerenchyma fibers viewed in longitudinal and cross-sections.

Vascular Tissue

Seed plants are also given strength and support by **vascular tissues**. Vascular tissues are chiefly specialized for transporting water, foods, minerals and wastes. There are two types of vascular tissue: xylem and phloem.

Xylem

The xylem cells carry water and dissolved minerals from the soil to all parts of the plant. They are found in roots, stems, and the veins of leaves. Their thick walls contain lignin, a substance that gives them strength. A common name for xylem tissue is **wood**.

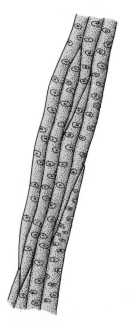

Fig. 15-4   Vessel elements. The end walls of vessel elements are perforated with many small holes or one large hole. The cell walls are thickened with lignin in ring-like or pitted patterns.

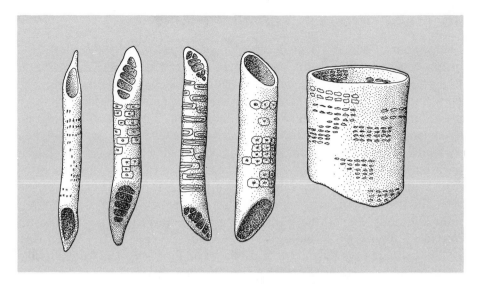

There are two kinds of xylem cells: tracheids and vessel elements. **Vessel elements** are wide tubular cells whose protoplasm has died (Fig. 15-4). Their end walls often have large holes, or even disappear completely. Columns of vessel elements form continuous tubes called **vessels**. These are very efficient at carrying water. The side walls of vessels are thickened with lignin and usually have tiny holes called pits in them. These pits allow materials to pass through to neighbouring cells.

Tracheids are much narrower xylem cells which, like vessel elements, die at maturity. Their side walls are frequently pitted. Their end walls are angular and do not disappear as in most vessels (Fig. 15-5). They also carry water and dissolved nutrients through the plant.

Fig. 15-5   Tracheids. Tracheids are long narrow conducting cells with thick pitted walls and pointed ends.

Phloem

Phloem is a tissue whose main function is to transport food produced in the leaves to other parts of the plant. It is made mainly of sieve tube cells and companion cells. **Sieve tube cells** are tube-like cells that lose their nuclei at maturity. They have plate-like end walls with many holes that allow materials to pass from cell to cell (Fig. 15-6). Columns of sieve tube cells, called **sieve tubes**, conduct food materials throughout the plant, under the control of the companion cells.

Companion cells are small cells with nuclei. They are located next to the sieve tube cells. When a companion cell dies, its associated sieve tube cell stops functioning.

Vascular tissues (xylem and phloem) from different plants provide a wide variety of commercial products. These include wood pulp for paper and lumber, and fibers such as jute, hemp, and the flax used for linen.

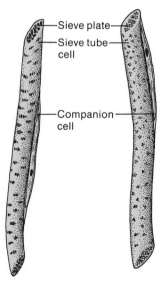

- Sieve plate
- Sieve tube cell
- Companion cell

Fig. 15-6   Phloem. Phloem tissue includes sieve tube cells that carry food products and companion cells that exert some control over the sieve tube cells.

Epidermal Tissue

All the inner tissues of plant organs are protected by a covering tissue called an **epidermis**. Epidermal tissue is usually made of a single layer of cells that have flat outer surfaces (Fig. 15-7). On leaves and stems, the epidermis has a waxy coating called a **cuticle** that prevents too much water from leaving the plant. In young roots, the epidermal cells near the root tip have short projections called **root hairs** that absorb water from the soil. In woody tree trunks, the epidermal tissue is replaced by another protective layer called **cork**. The bark of a woody stem consists of the cork layer together with some of the underlying tissues.

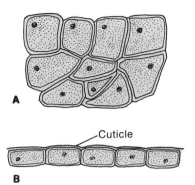

Fig. 15-7 Epidermal cells. In surface view, epidermal cells often have irregular shapes (A). In cross section, they appear as a single layer with a thick outer layer of cutin (B).

Review

1. Why are specialized plant parts more necessary in terrestrial (land) plants than in aquatic (water) ones?
2. **a)** What is a meristem?
   b) Where in a plant would you find a meristem?
3. Explain why plants keep growing continually until they die.
4. What is cell differentiation?
5. Compare in a table the cells of parenchyma and sclerenchyma tissue in terms of their walls, contents, and functions.
6. What is the major function of xylem?
7. Name two types of cells that make up xylem. List two differences and two similarities between them.
8. **a)** What are three similarities between the sieve tube elements of phloem and the vessel elements of xylem?
   b) What is one important difference?
9. What are two functions of epidermal cells?
10. What commercial products are obtained from vascular tissues?

Fig. 15-8 Aerial roots of the spider plant, *Chlorophytum*.

15.2 Roots

An **organ** is a group of different tissues organized to perform a major function or functions. Plants have four main types of organs—roots, stems, leaves, and flowers. This section describes the form and function of roots. The rest of the chapter describes the form and function of stems and leaves.

Kinds of Roots

If someone asked you to define a root, you might say it is the part of a plant that grows underground. For most plants this is true. However, in some plants, roots grow above the ground. Orchids that grow as epiphytes in the tops of tropical trees have long **aerial roots** that extend out into the air. Many house plants also have aerial roots (Fig. 15-8). The stems of ivy often develop clusters of short aerial roots that are called **adventitious roots** (Fig. 15-9). Adventitious refers to any root that does not come from another root.

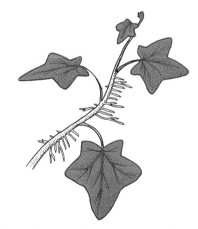

Fig. 15-9 Adventitious roots. English ivy often develops adventitious roots along its stem. The roots cling to walls, trellises or other supports.

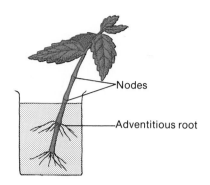

Fig. 15-10 Adventitious roots. Stem cuttings of *Coleus* will produce roots in a few days if put into water. The roots always appear from the joints, or nodes, of the stem.

Fig. 15-11 Prop roots of the Banyan Tree. The prop roots grow down from the branches of these huge trees.

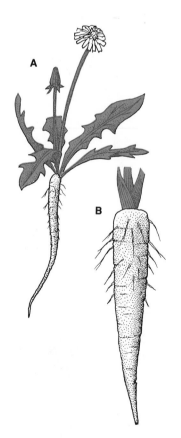

Fig. 15-12 Taproots. Dandelions (A) and carrots (B) are taprooted plants.

The adventitious roots of ivies often cling to walls of buildings. You may have grown adventitious roots yourself by putting stem cuttings of geraniums or coleus into a glass of water (Fig. 15-10). The huge banyan trees of the tropics also have adventitious roots that grow down from the branches to the soil, helping to support the trunks (Fig. 15-11).

Sometimes the underground parts of a plant are not roots at all. Remember that ferns have underground stems, called rhizomes, that look like roots. Potato plants have underground stems that grow into fleshy edible tubers that we call potatoes.

True roots develop from the radicle of the embryo in a seed. When a seed germinates, the first root, or **primary root**, pushes out of the seed coat. In some plants, this primary root grows into the main root, producing a large **taproot** that reaches deep into the soil (Fig. 15-12). Dandelions, carrots, radishes, and oaks are taprooted plants. Because of their long taproots, these plants can stay green during dry periods when more shallow-rooted plants wilt.

The primary roots of many other plants begin to branch into **secondary roots** almost as soon as they leave the seed. The secondary roots branch again and again, forming a mass called a **fibrous root system**. The fibrous root system of a weed called plantain is shown in Figure 15-13. Grasses have fibrous roots that are quite shallow and spread outward just beneath the soil. They help hold the soil together and are able to absorb surface water quickly. Most woody plants have fibrous root systems with many secondary roots that reach far from the main trunk. Some cacti have very extensive fibrous roots that reach far into dry desert soils for water (Fig. 15-14).

The Root Tip

The tip of a root is the region that absorbs water from the soil. Like other plant parts, it is made of several kinds of cells (Fig. 15-15). At the end of the root tip is a rounded cap of cells that protects the growing point as it pushes through the soil. This **root cap** is made of loosely attached cells that tear off easily. They are quickly replaced by new ones that develop from the meristem.

The **meristem**, or growing point, is located just behind the root cap. It is only one or two millimetres long. Its cells keep dividing by mitosis to produce all the root tissue above it. It is in the meristem that the epidermis, parenchyma, and vascular tissues originate.

Just above the meristem is the **region of elongation**. The cells in this area increase greatly in length, but not in number. As they expand, they push the root tip through the soil. The whole region of elongation is only a few millimetres long.

Behind the region of elongation is the **region of differentiation**. The first fully differentiated cells are found here. Because they are the first permanent root tissues, they are called the **primary tissues**. There are three primary tissues; the vascular cylinder, the cortex, and the epidermis. The innermost layers of cells in the root become the xylem and phloem of the **vascular cylinder**. The middle layers of cells form the **cortex**, a region of parenchyma cells. The innermost layer of the cortex forms the **endodermis**.

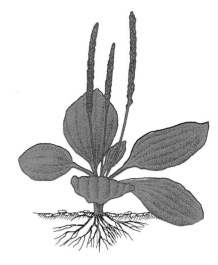

Fig. 15-13 Fibrous roots. The roots of plantain branch repeatedly, forming a fibrous root system.

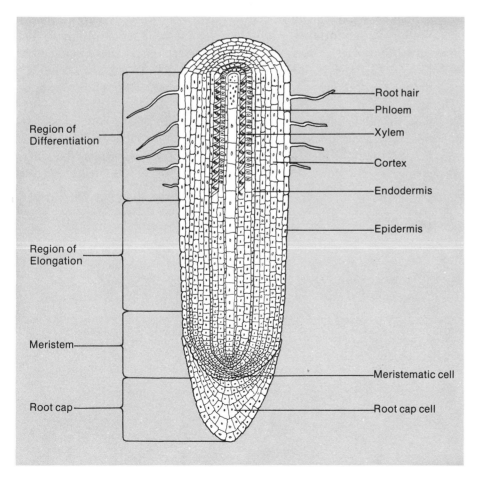

Region of Differentiation

Region of Elongation

Meristem

Root cap

Root hair
Phloem
Xylem
Cortex
Endodermis
Epidermis
Meristematic cell
Root cap cell

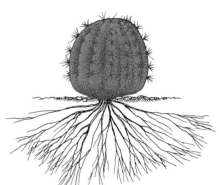

Fig. 15-14 Fibrous roots. The roots of many cacti spread outward over great distances.

Fig. 15-15 The root tip — Longitudinal section. The root tip is differentiated into four regions that are visible in longitudinal section.

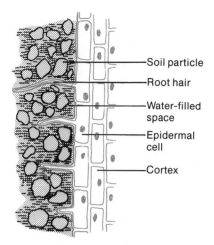

Soil particle
Root hair
Water-filled space
Epidermal cell
Cortex

Fig. 15-16 Root hairs. Root hairs greatly increase the surface area of the root for absorption of water and minerals.

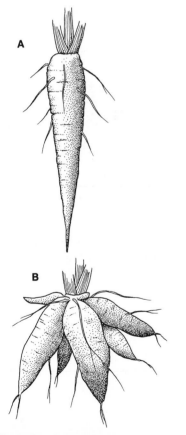

A

B

Fig. 15-17 Storage roots. Carrots (A) and dahlias (B) are storage roots with many layers of cells in the cortex.

Fig. 15-18 Cross section of a dicot root.

The outermost layer of cells develops into the **epidermal tissue**, or **epidermis**.

Root hairs grow out from the epidermal cells and absorb water from the soil. Each root hair is a thread-like extension of a single epidermal cell (Fig. 15-16). The root hair zone of a root measures only two to five millimetres long. Some floating plants lack root hairs completely because water can easily enter any part of the epidermis. In terrestrial plants, root hairs last only a few days. They are constantly being replaced by new ones closer to the tip. Their only function is to increase the surface area used for absorbing water and minerals. Water passively moves into a root hair by osmosis whenever the concentration of water in the soil is greater than in the cell. When the epidermal cell absorbs water from the soil, its water content becomes greater than in the adjacent cells. The resulting osmotic pressure causes water to move from cell to cell across the root to the xylem, which transports it up the plant.

Above the root hair zone, the epidermal cells form a single layer of smooth-walled cells all around the outer surface. Here their main function is to protect the inner tissues of the root. Immediately inside the epidermis is a thick region of loosely packed parenchyma cells called the **cortex**. Inside the cells are large vacuoles that store starch. In some roots, such as carrots and dahlias, the cortex forms a thickened storage area (Fig. 15-17).

Water passes from cell to cell through the selectively permeable cell membranes of the cortex toward the vascular cylinder by osmosis. The innermost layer of the cortex, next to the vascular tissue, is called the **endodermis** (Fig. 15-18). It is a single layer of thick-walled waxy cells. The endodermis controls how much water is able to pass through the vascular cylinder.

The vascular cylinder is the conducting core of the root. It usually contains four kinds of tissue, whose arrangements vary depending on whether the root is a dicot or a monocot. In most dicot roots, the xylem forms an "X"-shaped central mass of cells (Fig. 15-18). Phloem cells

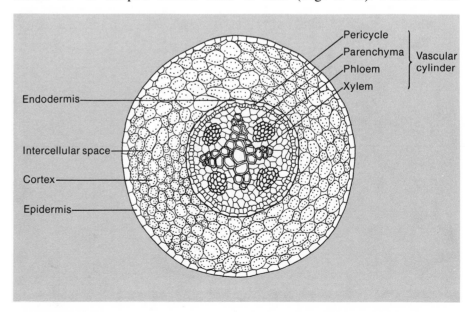

Endodermis

Pericycle
Parenchyma
Phloem
Xylem
} Vascular cylinder

Intercellular space

Cortex

Epidermis

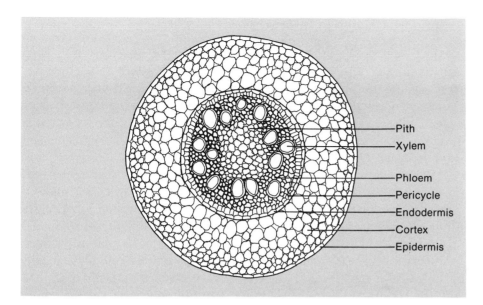

alternate with the "arms" of the xylem. Between and surrounding the xylem and phloem are parenchyma cells. In many monocots, there is no central xylem. Instead, the centre of the root is a mass of parenchyma cells called a pith. The pith may be surrounded by a ring of xylem and phloem strands (Fig. 15-19), or may have xylem and phloem scattered through it.

In both dicot and monocot roots, the outermost layer of the vascular cylinder is the pericycle (Fig. 15-18). The pericycle is a single layer of parenchyma cells immediately inside the endodermis. Its cells are able to divide like meristems and produce branch roots. Figure 15-20 shows a young branch root pushing out from the pericycle. As the branch root develops, its cells differentiate like those of the main root. The xylem and phloem of the branch root connect with the xylem and phloem of the main root, maintaining a continuous flow of materials.

Fig. 15-20 Branch roots. This dicot root is producing a branch root from the pericycle layer.

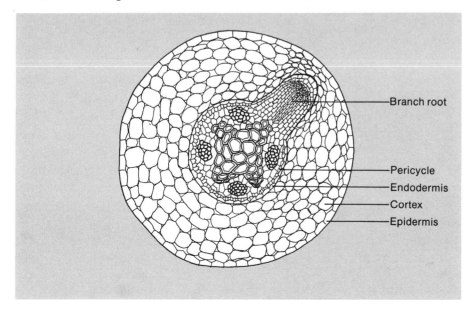

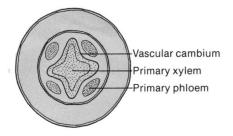

Fig. 15-21 Vascular cambium of a dicot root.

Secondary Growth in Roots

All the tissues shown in Figure 15-18 are **primary tissues** of the root. These primary tissues all develop from the **primary meristem** at the root tip. Although the root increases in length in the zone of elongation, the primary tissues are not able to increase in diameter. For growth in diameter, a new meristem, called **secondary meristem**, is necessary. The secondary meristem develops from immature parenchyma cells in the roots of dicots and gymnosperms. They are not present in monocots.

Additional vascular tissues for conduction and support develop from a secondary meristem called the **vascular cambium**. It is a single cylinder of cells that develops from the parenchyma between the primary xylem and primary phloem. It forms a wavy layer outside the "X"-shaped xylem in a dicot (Fig. 15-21). By mitotic division, the vascular cambium produces secondary xylem cells to the inside and secondary phloem cells to the outside of the vascular cambium. These new cells increase the diameter of the root (Fig. 15-22). As more and more cells are produced, the vascular cambium smoothes out into a cylinder of dividing cells around a core of xylem. In older roots, the xylem builds up over time into layers of wood. You will learn more about wood later when you study stems.

The cylinder of vascular cambium surrounds the primary xylem and phloem. New cylinders of secondary xylem and phloem are added each year. The increase in diameter of this central core stretches the endodermis, cortex, and epidermis on the outside of the core. Eventually these tissues break up and fall away. Before this happens, another secondary meristem called the **cork cambium** arises from the parenchyma cells of the pericycle. It produces a tissue called **cork** that performs the function of the epidermis which has peeled away with the cortex and endodermis (Fig. 15-22). The cork may be many layers thick.

Fig. 15-22 Cross section of a woody dicot root. The vascular cambium and cork cambium produce the cells that result in an increase in root diameter.

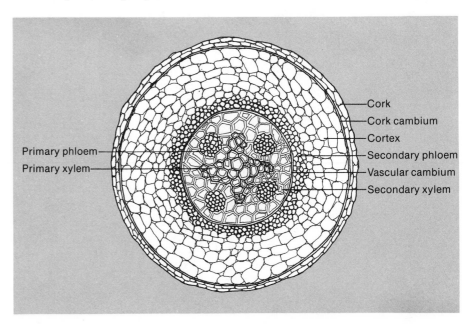

Review

1. Why is it difficult to define the word "root"?
2. **a)** What are adventitious roots?
   b) Give two examples of plants that produce adventitious roots. Give the functions of the roots in each case.
3. What is the difference between a taproot and a fibrous root?
4. List the three main functions of roots.
5. What protects the tissue of a root from wear and tear as it pushes through the soil?
6. Why is the root cap not worn off completely by soil particles?
7. **a)** What part of the root tip results in the root pushing deeper into the soil?
   b) How is this growth accomplished?
8. What are the three primary tissues that develop from a root meristem?
9. **a)** What is a root hair?
   b) On what part of a root are root hairs found?
   c) Why do root hairs not get scraped off as the root tip pushed through the soil?
10. Explain how water is absorbed from the soil into the xylem of the root.
11. Why is it important to keep a ball of soil around the roots of a plant when you transplant it?
12. Describe the cells and the function of the cortex.
13. What is an endodermis and what is its function?
14. What are the four types of cells found in the vascular cylinder of a root?
15. What is a pith?
16. Explain how branch roots develop.
17. **a)** Distinguish between primary tissues and secondary tissues.
    b) Why does a plant need to produce secondary tissues?
18. In what two important ways do the roots of dicots and monocots differ?
19. What is the difference between a vascular cambium and a cork cambium?
20. Why is it necessary for a cork layer to form in roots?

15.3 Investigation

Root Structure

By doing this investigation you will become familiar with all the root tissues outlined in Section 15.2. You will also be able to observe first-hand some examples of tap and fibrous root systems.

Materials

fresh or dried specimens of a fibrous rooted plant (grass, plantain, or others)
fresh or dried specimens of a taprooted plant (dandelion, carrot, or others)
hand lens and compound microscope
prepared slide of the longitudinal section of a root tip (corn or other)
prepared slide of the cross section of a dicot root (buttercup or other)

Procedure

a. By observing the root systems of the specimens provided by your teacher, decide which plants are taprooted and which are fibrous.

b. Choose one specimen of each type and examine its roots carefully. Make a large labelled diagram of each specimen. Be sure to label the primary and secondary roots.

c. If the specimens are fresh, take the hand lens and try to locate any root hairs that may still be attached to the roots. Make sure you are looking in the correct place. Record whether you found any. Then show their locations on your diagrams.

d. Examine a prepared slide of the longitudinal section of a root tip under low power on the microscope. Remember that a longitudinal section is made by cutting vertically through the root in very thin slices. Move the slide around until you are familiar with the outline of the whole root tip. Determine where the root cap, the meristem, the region of elongation, and the region of cell differentiation are located.

e. Make a large outline drawing of the root tip without drawing any of the individual cells. Using large brackets, mark the four regions you located in step d.

f. Copy the following table into your notebook:

ZONE OF THE ROOT TIP	DESCRIPTION OF THE CELLS	FUNCTION
1. Root cap		
2. Meristem		
3. Region of elongation		
4. Region of cell differentiation: a) epidermis b) cortex c) vascular cylinder		

g. Using medium or high power, carefully observe the cells in each zone of the root tip and fill in the table. Start with the root cap and move slowly up the root.

h. Go back to your outline drawing and draw in a few cells of each region as they appear through the microscope. Label everything you observe.

i. Return the slide to your teacher and obtain a cross-sectional slide of a dicot root. This type of section is circular. You will be looking at the same layers of tissue as you observed in the region of cell differentiation in the previous slide.

j. Using low power, familiarize yourself with the outline of the tissues by moving the slide slowly back and forth. Starting at the outside edge, locate each of the following layers: epidermis, cortex, endodermis, pericycle, phloem, and xylem. If you have trouble, refer to the diagrams in Section 15.2 or ask your teacher.

k. Without drawing the cells, make a large sketch of the various layers of the cross-section.

l. Beginning with the epidermis, observe each cell type on medium and high power. Add a few cells of each layer to your drawing. Be accurate with the relative size and shape of the cells. Finally, label all the layers.

m. The roots you have been observing are both made of primary tissues only. Your teacher may have other slides of woody roots that show secondary growth from the vascular and cork cambium. If not, you will be able to see them when you study woody stems.

Discussion

1. To what type of environment is a taproot system especially well adapted?
2. **a)** Where on the root are root hairs located?
   b) If you could not find any on your specimens, what could be a reason for their absence?
3. What are the two major functions of epidermal tissue?
4. How does the structure of the cells of the endodermis compare to the cells of the pericycle?
5. **a)** Describe the pattern of xylem in this dicot root.
   b) What is the main function of the xylem?
   c) How does the structure of the xylem cells help them perform this function?
6. **a)** How is the phloem arranged?
   b) List all the differences you observed between the cells of the xylem and those of the phloem.
7. Make a list of all the root tissues that contain simple parenchyma cells.
8. **a)** Why are the differentiated tissues of a dicot root considered an evolutionary advance over the simple tissues of primitive plants?
   b) Why is tissue differentiation an advantage?

15.4 Investigation

Root Functions

In this investigation, water absorption and water transport are studied in two separate experiments. Both parts require several days to complete.

Materials

radish seeds	compound microscope
paper towel	microscope slide
masking tape	cover slip
petri dishes	razor blade or scalpel
scissors	large carrot
water	small beaker
hand lens	green or blue food colouring

Fig. 15-23 Cucumber vines. Cucumber stems usually lie horizontally on the surface of the ground.

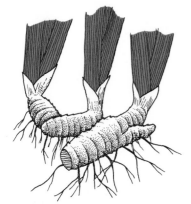

Fig. 15-24 Rhizomes of iris. Rhizomes are underground stems.

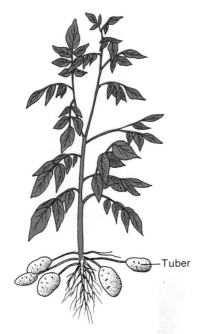

Fig. 15-25 Tubers of potato. Tubers are fleshy stems that store food.

Procedure

a. Cut three pieces of paper towel into a circular shape that will fit into a petri dish. Insert the papers into the dish and soak them with water. Pour off the extra water, leaving the papers wet but not floating.

b. Put 4 or 5 radish seeds on the paper. Then cover the petri dish with a lid. Seal the dish with masking tape to keep the moisture in. Keep the seeds in a warm place for 3 or 4 d.

c. When the roots are about 2 cm long, examine them carefully with a hand lens. Describe their general appearance in your notebook. Try to find the four zones that you observed in a longitudinal section of a root tip in Investigation 15.3.

d. A short distance from the tip are the root hairs. Describe their colour, length, and location in your notes. Make a large drawing of the whole seedling, showing where the root hairs are. Label the root cap, root hairs, seed, and shoot.

e. Take one seedling and carefully cut out a thin cross-section of tissue in the root hair region. Put it in a drop of water on a slide and cover it with a cover slip.

f. On low power, locate and examine one epidermal cell and a root hair. This is sometimes difficult to get clearly focused. Make a labelled drawing of what you observe.

g. For the second part of this investigation, take a carrot and cut off 2 cm from the lower end and 1 cm from the top. Put the lower end into a beaker that contains 2 cm of water that has been strongly coloured with green or blue food colouring. Set the beaker and carrot aside for 24 h.

h. Observe the carrot and record any changes that have occurred. Cut off 1 cm from the lower end and observe the cross section. Repeat at the upper end of the carrot. Compare the two cross sections and record your observations.

i. Choose a deeply stained region and cut a very thin cross section with a razor blade. Mount the section on a slide. Observe the stained cells in the microscope and describe their structure. Make a drawing of the cross section, showing all areas that took up the food colouring.

j. Use the razor blade to cut the rest of the carrot lengthwise through its core. Note the location of the stained cells.

Discussion

1. **a)** How far from the tip of the radish root are the root hairs produced?
   b) Which one of the four regions of a root tip produces root hairs?
2. **a)** What is the function of root hairs?
   b) How is this function accomplished?
3. From your observations, are root hairs cells? Explain.
4. Why is it an advantage for a radish seedling to have so many root hairs?
5. What use is made of the water absorbed by roots?
6. **a)** What region of the carrot absorbed the coloured water?

b) What name is given to the tissue found there? Describe its cells.

c) When you cut the carrot lengthwise, how was the stain distributed?

7. In which direction does the xylem of flowering plants conduct materials?

15.5 Stems

Kinds of Stems

Stems are similar to roots in many ways. They are varied in size and shape, and may grow above or below ground. They are made of the same types of tissues as roots. They also share two important functions with roots: they transport food, water, and nutrients, and they store food materials in their cells.

Although most stems grow upright, some such as strawberries **trail** along the ground. Vines such as cucumbers can also grow horizontally (Fig. 15-23). Some horizontal stems grow underground and are known as **rhizomes**. Figure 15-24 shows the rhizomes of iris. Other underground stems like those of the potato grow into fleshy storage organs called **tubers** (Fig. 15-25).

Some stems are **annual**, lasting only one season. Others are **perennial** and increase in size from year to year. Perennial stems become **woody** with age and develop into shrubs and tall trunks and branches of trees. Trees like pines and spruces have one main trunk that grows much faster than the lateral branches (Fig. 15-26). Others such as maples and willows have several large branches that spread outward. Most botanists can identify tree specimens from a distance just by their shapes.

The Shoot Tip

The stem of any plant originates from the part of the embryo called the **epicotyl**. In a seed such as a bean, the epicotyl can be seen as a tiny shoot between the two cotyledons (Fig. 15-27). At germination, this shoot emerges from the seed coats after the radicle. Then it quickly grows by mitotic cell divisions at its tip. The tip continues to produce new cells as long as the plant lives. Its organization is somewhat more complex than that of the root tip because it produces leaves as well as the stem.

Just like that of the root, the tip of a stem is a meristematic region that keeps producing new cells. However, it is not protected by any covering similar to a root cap. Why might this be so? As in the root, the three primary tissues (epidermis, cortex, and vascular tissue) all develop in a region a short distance behind the tip. Arising from the tip meristem are the leaves which first appear as small bumps (Fig. 15-28). In each young leaf there is a group of cells that remains meristematic until all the tissues of the leaf are produced. The young leaves may grow singly, in pairs, or in whorls. They always develop in an orderly pattern. No one is sure why this is so. However,

White spruce

Sugar maple

Fig. 15-26 Branching patterns of trees. The shape of trees varies with the dominance of the central trunk.

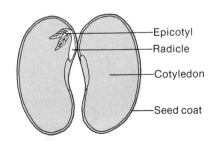

Epicotyl

Radicle

Cotyledon

Seed coat

Fig. 15-27 Shoot of a bean seedling. The two cotyledons have been separated.

Fig. 15-28 Shoot tip: longitudinal section.

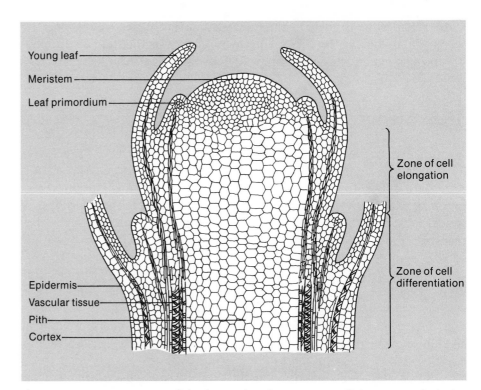

Young leaf
Meristem
Leaf primordium

Zone of cell elongation

Zone of cell differentiation

Epidermis
Vascular tissue
Pith
Cortex

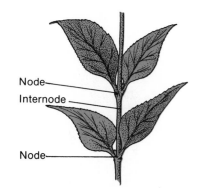

Node
Internode

Node

Fig. 15-29 Nodes and internodes.

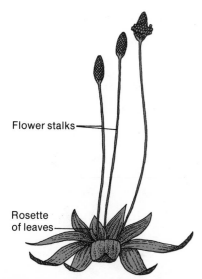

Flower stalks

Rosette of leaves

Fig. 15-30 Rosette plant. If the stem nodes are all close together, the leaves form a rosette at the base, as in plantain.

the pattern arrangement of the leaves has become a useful characteristic for identifying flowering plants.

At the tip, the new leaves are formed very close together. In the elongation zone, the young leaves become spaced out along the stem. The point at which a leaf attaches to a stem is called a node. The length of stem between two nodes is called an internode (Fig. 15-29). The internodes lengthen both by cell elongation and by cell divisions for some distance from the tip. This increases the length of the stem. In different plant species, the amount of elongation varies. Some stems remain condensed, with all their leaves crowded in a whorl at the base called a rosette (Fig. 15-30). Others become very elongated with extended internodes. In grasses, the bases of the internodes remain meristematic for a long time. This allows the plants to continue to grow after you cut their tops off with a mower.

While increasing in length, the stem also increases in thickness. The cells of the first primary tissues expand as they begin to develop below the tip. The parenchyma cells of the cortex and the central pith become especially large.

The angle formed between a leaf and the stem is called a leaf axil. Almost immediately after a young leaf is produced, a bud begins to form in its axil (Fig. 15-31). A bud is a small group of cells that remains meristematic and later produces a lateral branch. The cells inside the bud become organized in the same pattern as those in the tip meristem. They often remain dormant for long periods because hormones from the growing tip inhibit their development. If the tip is damaged or cut off, the flow of hormones is reduced, and the lateral buds begin to produce branches.

Gardeners make use of this phenomenon when they prune shrubs and trees into attractive shapes.

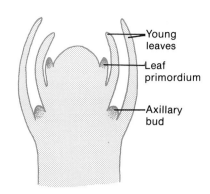

Fig. 15-31 Shoot apex in longitudinal section. Buds begin to form in the axils of the leaves early in their development.

Review

1. In what ways are roots and stems similar?
2. The text shows how cucumbers, irises, and pines have different growth habits. List five more plants whose stems grow in different patterns.
3. In what three ways is the shoot tip different from the root tip?
4. Explain how the meristematic zone of a shoot is organized.
5. Which region of a stem contains parenchyma tissue? How does this compare with a root?
6. What is the difference between nodes and internodes?
7. Explain how stems grow.
8. **a)** What is a bud?
   b) Where are buds found on a stem?
9. How would you prune a rose bush to prevent new growth from criss-crossing on itself in the middle of the bush?

15.6 Cell Structure of Herbaceous and Woody Stems

Herbaceous Stems

Any stem that remains green and flexible and does not become woody is called a **herbaceous stem**. Herbaceous stems die back to ground level each winter. The stems of many common plants, such as corn, bean, buttercup,

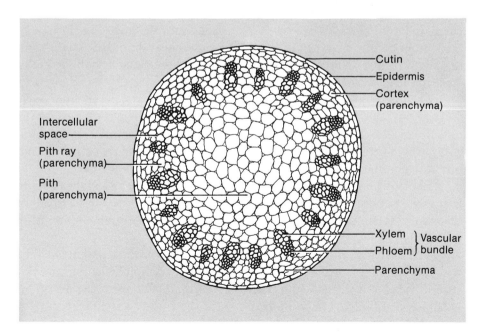

Fig. 15-32 Cross section of a herbaceous dicot stem.

clover, and potato are herbaceous. As a group, herbaceous stems have similar cell structures, except for a different arrangement of vascular tissues between dicots and monocots.

The outer layer of a mature herbaceous stem is the **epidermis** (Fig. 15-32). It is a single layer of living cells that keeps increasing in size as the stem grows in diameter. The cells have thickened outer walls, with a thin waxy layer called the **cuticle** on the outside. The thick walls and the cuticle both help to reduce water loss to the air. Scattered throughout the epidermis are small holes called **stomata** (singular: **stoma**) that allow gases in and out of the stem. You will learn more about stomata when you study leaves.

The **cortex** of the stem consists mainly of parenchyma tissue with spaces throughout it. The cortex cells usually have chloroplasts that allow them to photosynthesize. Some cells close to the epidermis have extra thick walls containing lignin that gives extra support to the stem. In most angiosperm stems, there is no endodermis similar to that of roots.

The central part of a stem is made of the vascular tissue and the pith. The whole central core is called a **stele**, which means "a column". It forms the main axis of the stem. The stele of a stem does not have a pericycle separating it from the cortex as does the stele of a root. There is often no sharp differentiation between the parenchyma of the cortex and the parenchyma of the pith. The function of the pith parenchyma is primarily food storage.

The vascular tissue in a herbaceous stem is composed of individual bundles of xylem and phloem cells that are known as **vascular bundles**. There is no central core of xylem as in a root. In dicot stems, the vascular bundles are generally arranged in a circle around the pith. They may be in an unbroken cylinder or separated by parenchyma that extend from the pith to the cortex. This parenchyma forms channels called **pith rays**. These rays transport water and food between the vascular bundles. In monocots, the vascular bundles are usually scattered throughout the pith (Fig. 15-33). In most monocots, there are more vascular bundles near the outside of the stem than in the centre. The cortex is very thin because the vascular bundles are so close to the epidermis.

Each vascular bundle is made of xylem, phloem, and parenchyma cells (Fig. 15-34). In most flowering plants the primary phloem is located on the outside of the xylem. However, in some plants, the primary phloem completely surrounds the primary xylem. Outside of the phloem there are often strong fiber cells that give the stem extra support.

In dicots, a single layer of meristematic cells called a **vascular cambium** is present between the xylem and phloem. This layer is able to produce secondary xylem and phloem tissue by mitosis. It is completely absent in monocots.

The vascular bundles of the stem form an unbroken system with the vascular system of the root and the leaves. Near each node, some vascular strands bend outward to join with the vascular bundles in the leaf. This produces a small gap, called a **leaf gap**, in the ring of vascular tissue just above the node (Fig. 15-35).

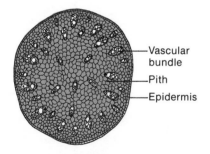

Vascular bundle

Pith

Epidermis

Fig. 15-33 Cross section of a monocot stem — corn. The vascular bundles are scattered through the pith although most of them are near the outside.

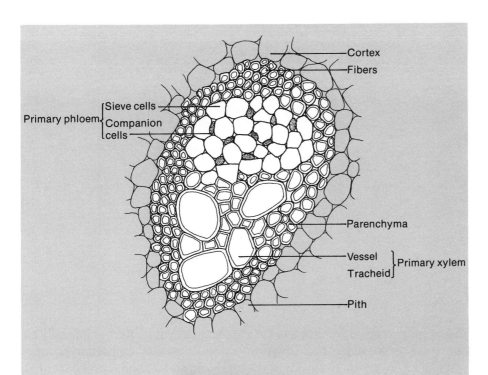

Fig. 15-34 A typical vascular bundle.

Cortex
Fibers

Primary phloem { Sieve cells
Companion cells
cells

Parenchyma

Vessel
Tracheid } Primary xylem

Pith

Woody Stems

After their first season of growth, the stems of many dicots develop into woody branches and trunks. This secondary growth is possible because of the vascular cambium between the xylem and phloem. The vascular cambium is a meristematic layer that is responsible for increases in the diameter of the stem. The bulk of new stem tissue produced by the vascular cambium is **wood**, which is simply secondary xylem.

Most monocots never develop woody stems because they lack the vascular cambium. No secondary xylem or phloem are ever produced in such monocot stems. As a result, most monocots remain relatively small in size, with long thin stems.

Woody dicot stems develop tissues very similar to those of herbaceous dicots during their first year. In temperate climates, the tissues become dormant after the first season and remain so over the winter. They become active and resume growth again the following spring. In tropical climates, growth continues throughout the year. However, it often slows during dry seasons.

The ring of vascular bundles in a stem gradually becomes a continuous cylinder as a new vascular cambium is produced in the rays between the original vascular bundles (Fig. 15-36). The completed ring of vascular cambium produces some parenchyma cells which form rays. These rays provide a means of transport between the central and outer regions of the stem. These new rays replace the original rays between the vascular bundles.

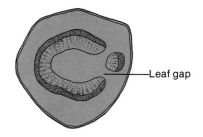

Leaf gap

Fig. 15-35 Leaf gaps. A disruption of the vascular system occurs where strands of vascular tissue enter a leaf.

Fig. 15-36 Vascular cambium in dicot stems. The vascular cambium becomes continuous between the vascular bundles.

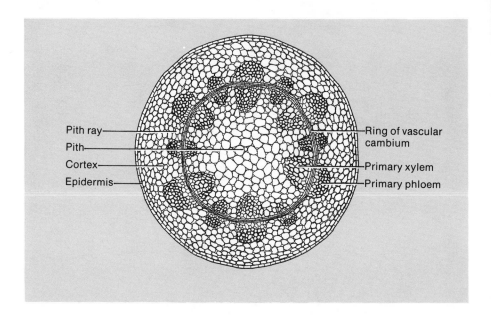

The rays produced by the vascular cambium are quite narrow, usually two to three cells wide. Cell division in the vascular cambium produces secondary xylem on the inside of the cambium and secondary phloem on the outside. By the end of the first year, these are both distinctly visible as rings of tissue when viewed in cross section (Fig. 15-37).

During each season of growth, successive layers of xylem and phloem are formed by the vascular cambium. Much more xylem than phloem is produced each year, creating a large woody core. This is necessary because the xylem, or wood, is the chief means of support for the stem. The new xylem gradually crushes the original pith in the centre of the trunk.

The epidermis and cortex outside the wood are not able to expand indefinitely as the stem enlarges. New protective layers are produced on the outside by a second meristem called the **cork cambium**. This layer comes from parenchyma cells in the cortex during the first year of growth. Mitosis in the cork cambium produces **cork** cells to the outside and parenchyma to the inside. The walls of cork cells contain an oily substance called **suberin**. The cells die soon after they mature, but the suberin in the dead cell walls protects the underlying tissue from insects, bacteria, fungi, and water loss. Layers of cork build up over time to form the outer bark of a tree. Cracks that develop as the diameter keeps increasing are filled in by new cork tissue. The oldest layers are shed from the outside as the new layers form inside.

Here and there in the cork tissue of the outer bark, one can often see long raised marks on the outer surface. These are called **lenticels** (Fig. 15-38). They consist of loosely arranged pockets of parenchyma cells with many spaces between the cells that allow air in and out of the trunk. They are produced by the cork cambium. They may occur singly or in rows. The pattern of lenticels on twigs of cherry trees and the stems of birch trees makes them very attractive.

Just beneath the outer bark is a white pulpy layer called the **inner bark**.

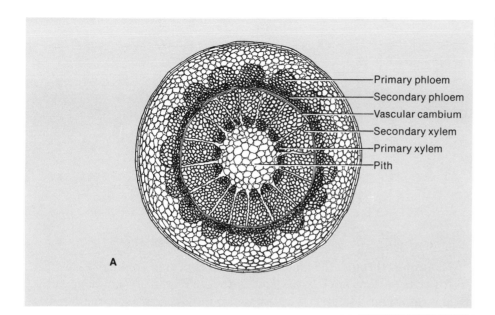

A

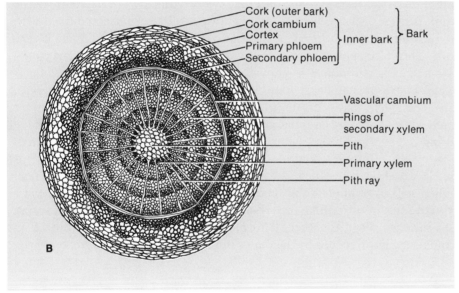

B

Fig. 15-37   A: one-year-old dicot stem. Continuous cylinders of secondary phloem and secondary xylem have been produced by the vascular cambium.

Primary phloem
Secondary phloem
Vascular cambium
Secondary xylem
Primary xylem
Pith

Cork (outer bark)
Cork cambium
Cortex   } Inner bark } Bark
Primary phloem
Secondary phloem

Vascular cambium
Rings of
secondary xylem
Pith
Primary xylem
Pith ray

Fig. 15-37   B: three-year-old dicot stem in cross section. Three layers of xylem form the woody core of this trunk. The outer bark that protects the tissues is composed of cork.

Fig. 15-38   Lenticels in the bark of cherry twigs.

It is made of the cork cambium, the cortex, and the phloem tissue. The cortex may gradually disappear as it gets crushed by the expanding core of the tree. The phloem farthest from the vascular cambium also gets crushed. However, the inner phloem continues to transport food materials through the trunk. If a complete ring of bark is removed from a tree, the phloem transport system is destroyed. This removal is called **girdling**. It sometimes happens to young trees in the winter when rabbits chew off the bark. When trees have been girdled, they gradually die because their roots starve from lack of food. The food may collect in the bark just above the girdle, producing a bulge as shown in Figure 15-39. Many birch trees in parks and forests die because they have been girdled by thoughtless people who peel off the attractive white bark.

Fig. 15-39   Girdling of trees. Girdling cuts off the flow of food to the roots because it removes the phloem tissue of the inner bark.

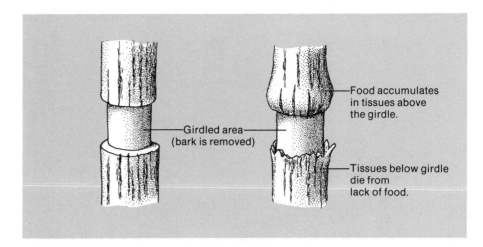

Food accumulates in tissues above the girdle.

Girdled area (bark is removed)

Tissues below girdle die from lack of food.

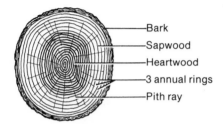

Bark
Sapwood
Heartwood
3 annual rings
Pith ray

Fig. 15-40   Heartwood and sapwood. The central heartwood is filled with resins and other deposits.

Transport through the phloem always occurs in the phloem tissue that is farthest from the outer bark. This is the youngest phloem next to the vascular cambium. The newest xylem is also the layer closest to the vascular cambium. After several years, the older xylem tissue toward the centre of the trunk is no longer needed for water transport. Its cells become filled with resins, tannins, and other deposits that darken the wood and make it very hard. This hard central core is called **heartwood**. The outer band of young xylem that is still active is known as **sapwood** (Fig. 15-40). Sapwood is lighter in colour than heartwood. It is not as durable when cut for lumber.

The xylem that is produced during the spring usually has much larger tracheids and vessels than that produced later in the summer. Growth in the spring is faster and more vigorous. Thus more water transport is required. This results in more numerous, larger cells with relatively thin walls. This **spring wood**, as it is called, is easily distinguished from the smaller, thick-walled cells of **summer wood**. The contrast between the two types produces **growth rings** in the trunk (Fig. 15-41). One **annual ring**, made up of spring and summer wood together, is formed each year. The total width of an annual ring is influenced by the yearly rainfall, the width being greater in wet

Fig. 15-41   Annual rings: Cross section of a woody trunk. The large cells of the spring wood alternate in bands with the small cells of the summer wood, creating annual rings.

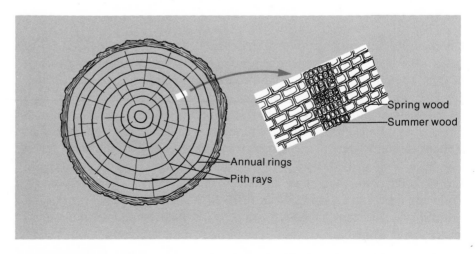

Spring wood
Summer wood

Annual rings
Pith rays

years. By counting the annual rings, you can accurately determine the age of a tree.

The pattern of growth rings and pith rays in various species of trees is known as **graining**. The graining in lumber differs with the type of tree and with how the wood is cut. Attractive graining patterns make many woods very valuable for furniture, panelling, and floors. Where branches were once attached to a tree, the graining pattern is disturbed and a **knot** is created. A knot is simply the base of an old branch that gets buried deep in the wood as the tree grows in diameter. Many people find the patterns created by knots very attractive in wall panelling and flooring. Knotty pine and cedar are both used in homes and cottages.

Review

1. Define the terms herbaceous and woody.
2. What features of the epidermis help to conserve water in a stem?
3. In what three ways is the cortex of a stem different from the cortex of a root?
4. What is a stele?
5. **a)** Describe the arrangement of the stele in a herbaceous dicot stem.
   b) Describe the arrangement of the stele in a monocot stem.
6. What is the function of pith rays?
7. Describe the cell arrangement of a typical vascular bundle of a herbaceous dicot.
8. How do leaf gaps arise?
9. Why do only a few types of monocots grow into trees?
10. As the vascular cambium produces the woody core of a trunk, what happens to the pith, pith rays, cortex, and epidermis?
11. **a)** What is the function of cork?
    b) What characteristics of cork cells help them perform this function?
12. What are lenticels?
13. What would happen to a tree if you scraped off the white pulpy layer underneath the outer bark? Why?
14. What are two differences between heartwood and sapwood?
15. Why are old trees whose trunks have been hollowed out by disease or rot able to continue growing?
16. Explain the formation of annual rings.
17. Why does lumber have knots in it?

15.7 Investigation

Stem Structure

You are now aware that stems of flowering plants are rather complicated structures with many types of cells. Cutting the stems into thin sections that can be seen through a microscope is a difficult technique. It takes practice to be able to cut them thin enough to see the tissues clearly. For this reason, it is easier to use prepared slides from biological supply companies for your investigations. The slides are stained with biological dyes that make the different types of cells easily visible.

Materials

prepared slide of a young dicot stem (buttercup, sunflower, coleus, alfalfa, or geranium)
prepared slide of an older, woody dicot stem (basswood, oak, geranium, elderberry, or tulip tree)
prepared slide of a monocot stem (corn, lily, or wheat)
compound microscope
cross section of a large branch or the trunk of a tree

Procedure A   Herbaceous Dicot Stem

a. Examine the slide of a young dicot stem under low power on the microscope. Locate the epidermis, cortex, vascular bundles, xylem, vascular cambium, phloem, and pith by moving the slide around under the field of view.

b. Make an outline drawing from low power. Label each layer of cells. Switch to medium and high power and draw in accurately a few cells of each type.

c. For each tissue, note the type of cells, their location, and the relative width of the layer. Copy and fill in the following table from your observations.

TISSUE	TYPES OF CELLS	LOCATION	RELATIVE WIDTH OF THE LAYER
Epidermis			
Cortex			
Vascular bundle			
—xylem			
—phloem			
—vascular cambium			
Pith			

Procedure B   Woody Stem

a. Examine a slide of a woody stem under low power. Move the slide around until you are familiar with the location of the various tissues.

b. Start at the centre and locate the pith (if any), xylem layers, vascular cambium, phloem, cortex, cork cambium, and cork. The two cambial layers may be difficult to find.

c. Observe the xylem in detail. Count the annual rings. On medium and high power, look for differences in the cell structure of spring and summer wood.

d. Compare the cells of the xylem with those of the phloem. Note the size of the cells and the thickness of the cell walls in each tissue.

e. Sketch and label the whole section to show the thicknesses of the layers. For a pie-shaped section of your sketch, draw in the cells of each layer as they appear in the microscope.

Procedure C   Monocot Stem

a. Observe the slide of a monocot stem under low power. Locate each type of cell listed in the chart in part A. Which one is absent?

b. Describe the arrangement of the vascular bundles. Focus on one bundle. Look for xylem, phloem, and any sclerenchyma fibers that may be present. Make a large drawing of one vascular bundle. Label the various tissues.

c. Draw a sketch of the stem, indicating the arrangement of the vascular bundles and the location of the other tissue layers. Do not draw the cells in detail.

Procedure D   Tree Trunk

a. Observe the specimen of a tree trunk or branch and count the annual rings. Look carefully at the annual rings. Determine the relative thicknesses of each year's growth.

b. Locate and describe the appearance of the outer bark, inner bark, sapwood, and heartwood. Determine if any pith rays are visible in the wood.

Discussion A

1. What type of tissue makes up the bulk of a young dicot stem?
2. **a)** What type of cell in the stem has the thickest walls?
   b) Why would this be an advantage?
3. What is the function of the cortex?
4. What differences have you observed between the stele of a stem and the stele of a root?
5. **a)** What part of a vascular bundle conducts food?
   b) What part of a vascular bundle conducts water and minerals?
6. Explain how the individual vascular bundles develop into a continuous cylinder as the stem ages.

Discussion B

1. What type of cell makes up the bulk of a woody stem?
2. What changes have occurred in the stele compared to the young dicot stem in part A?
3. How old was the woody stem you observed?

4. **a)** What differences were there between the cells of spring wood and summer wood?
   b) What differences did you observe between the cells of xylem and phloem?
5. What happens to the pith as woody stems grow older?
6. **a)** Describe the location of the vascular cambium.
   b) Of what type of cell is it composed?
7. **a)** Describe the location of the cork cambium.
   b) Why is a cork cambium necessary in woody stems?

Discussion C

1. What are two differences between young dicot and monocot stems?
2. What cell layer is missing in the stems of monocots?
3. How is the arrangement of vascular bundles different from that of dicots?
4. What would be the function of sclerenchyma fibers in the vascular bundles of monocots?
5. **a)** What tissue makes up the bulk of a monocot stem?
   b) What is the function of this tissue?

Discussion D

1. How old is the section of tree that you observed?
2. Describe the climate pattern for the years this tree was alive.
3. How is the bark of a tree formed?
4. **a)** What tissues make up the outer bark?
   b) What tissues make up the inner bark?
5. **a)** What is the difference between heartwood and sapwood?
   b) Which type of wood is more abundant?
6. **a)** What are pith rays? What is their function?
   b) How can you tell where pith rays were located in the living wood?

15.8 Investigation

Twigs

The growth of a stem is easy to understand if you observe the structure of a dormant twig from a deciduous tree. Leafless twigs can tell you how the leaves were arranged in the previous season and where the growth will start in the spring. Since the twigs of each species are different in structure, you can even identify a tree from just its bare twigs.

By observing closely, you should be able to find the scars of old leaves at the nodes of the twig. These **leaf scars** may be single, in pairs, or in whorls around each node (Fig. 15-42). If the leaves were arranged alternately on the twig, there will be one scar per node. Opposite leaves result in two scars per

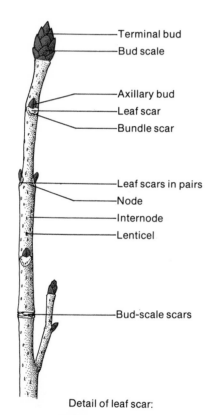

Terminal bud
Bud scale

Axillary bud
Leaf scar
Bundle scar

Leaf scars in pairs
Node
Internode
Lenticel

Bud-scale scars

Detail of leaf scar:

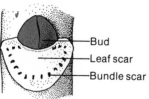

Bud
Leaf scar
Bundle scar

Fig. 15-42   A woody twig.

node. Whorled leaves result in more than two. The arrangement of the leaves is a good clue to start with when identifying the tree.

On the surface of each leaf scar are small corky dots called **bundle scars**. They are the ends of the vascular bundles that once extended into the leaf. The number and arrangement of the bundle scars is also a clue to the identity of the tree.

Twigs also have **buds**. The largest bud is usually right at the tip and is called the **terminal bud**. It is a meristematic region that will develop into the next year's leafy shoot. It is covered with overlapping rows of **bud scales**. These protect the bud from injury and from drying out during the winter. The number of bud scales varies with the type of tree.

In spring, the bud scales open and fall off, leaving an area of ring-like scars on the twig called **bud-scale scars**. The distance between each set of bud-scale scars is the length the twig grew during one season. By counting the scars from the terminal bud down, you can determine the age of the twig.

Other buds are also present below the terminal bud. Just above each leaf scar are the **axillary buds**. They are called axillary buds because they sit in the axils of last year's leaves. Their structure is the same as that of the terminal bud. They develop into lateral branches that are arranged in the same pattern (alternate, opposite, or whorled) as the leaves were in the previous season.

Since most buds develop into leafy shoots, they are called **leaf buds**. Both terminal and axillary buds may be leaf buds. Sometimes flowers are produced instead of leaves, and the buds are called **flower buds**. Some plants such as lilac and apple have **mixed buds** that produce both shoots and flowers from the same bud (Fig. 15-43).

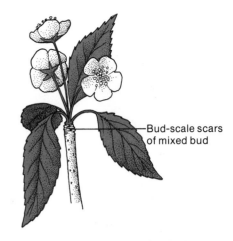

Fig. 15-43   Mixed bud of apple. Both leaves and flowers develop from a mixed bud.

Materials

dormant winter twigs of deciduous trees such as maple, oak, beech, poplar, birch, horse chestnut, tree of heaven, and sumac
hand lens

Procedure

a. Choose one twig from those available and observe its general appearance. Record its colour, thickness, and roughness. Look for lenticels in the bark on the internodes. Describe the shape, size, and colour of the lenticels.

b. Locate the terminal bud at the tip of the twig. On some species there will be a cluster of buds at the tip instead of only one. Some lack terminal buds completely. If there is a terminal bud, describe its shape, colour, and size in your notebook.

c. Look at the bud scales of the terminal bud closely with a hand lens. Describe their arrangement, number, and shape. Note whether they are smooth, scaly, dotted, or hairy.

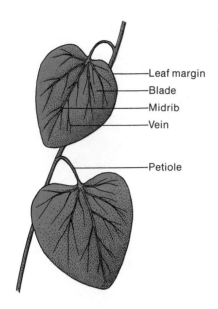

Fig. 15-44 Leaves with thin broad blades photosynthesize large amounts of food.

Fig. 15-45 Sessile leaves. The leaf blade attaches directly to the stem with no petiole.

d. Observe the internode of the twig below the terminal bud. Locate the rings of bud-scale scars that were left by last year's terminal bud. Describe their appearance. Use these scars to determine the age of the twig.

e. Look for leaf scars along the twig. Decide if the leaf scars are alternate, opposite, or whorled. Describe their shape and colour.

f. Choose one leaf scar and look closely at the bundle scars inside it. Record the number and the arrangement of the bundle scars. Sometimes they form straight lines, curves, or triangles. Note their colour compared to the leaf scar.

g. Just above the leaf scars you should be able to see axillary buds. Describe the axillary buds in terms of size, colour, shape, and number of bud scales.

h. Make a large labelled drawing of at least two years' growth of the twig. Label the terminal bud, bud scales, nodes, internodes, lenticels, bud-scale scars, leaf scars, bundle scars, and axillary buds.

i. Choose a second twig of a different species and repeat steps a to h for that twig. Compare the features of the two twigs by means of a chart in your notebook.

j. If there is time, your teacher may ask you to make up an identification key for all the twig specimens available. There may be a key already made up that you could use to identify your specimens. Ask your teacher about this.

Discussion

1. What are the three possible arrangements of leaves on a twig?
2. **a)** How do bundle scars form?
   b) How do bud-scale scars form?
3. What function do bud scales serve?
4. How can a twig grow in length if it lacks a terminal bud?
5. What is the function of lenticels?
6. How do the axillary buds of a twig compare to the terminal bud in shape, size, colour and number of bud scales?
7. What characteristics seem to be the most important in identifying one species of twig from another?

15.9 The Leaf

Although leaves exist in thousands of different shapes and sizes, they are all designed for one major function: photosynthesis. It is in the leaf that all the raw materials for photosynthesis are brought together. Water arrives through the xylem in the veins. Carbon dioxide enters from the air through small holes called stomata. Chloroplasts in the leaf cells capture the light energy required to produce food. Leaves are very efficient photosynthetic organs. Their tissues are organized for maximum food production. They are also designed to maximize the exchange of gases with the external environment. You will learn more about leaf functions as you read on in this chapter.

Form and Types

Form

Most leaves of flowering plants have thin, broad **blades** (Fig. 15-44). The blades contain chlorophyll and are therefore green. The blade of a leaf is supported by a network of branching veins, the largest of which is called the **midrib**. The midrib may extend below the blade as a short stalk called a **petiole**. The petiole attaches the leaf to the stem. Leaves that lack petioles are called **sessile** leaves (Fig. 15-45). The leaves of a few families of plants have extra appendages called **stipules**. Stipules are green leafy bracts found where the petiole joins the stem. Figure 15-46 shows the stipules of clover.

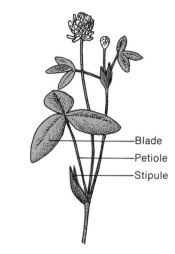

Fig. 15-46   Stipules of clover.

Venation

The leaf veins are arranged in different patterns in different kinds of plants. In most monocots, the veins are parallel and do not branch (Fig. 15-47). Corn, grasses, tulips, daffodils, and lilies have **parallel venation**. In dicots, the veins divide repeatedly, forming a network. Therefore, dicots are said to have **net venation**. If the main veins of the net all originate at the base of the midrib, the leaf is called **palmately net-veined**. Figure 15-48,A shows the palmate venation of a maple leaf. Note how the veins spread out like fingers from the palm of a hand. Net-veined plants whose veins originate at intervals up the sides of the midrib are called **pinnately net-veined**. Oak, cherry, and willow are examples of this type of leaf (Fig. 15-48,B).

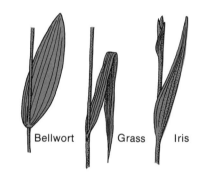

Fig. 15-47   Leaves of monocots. Monocot leaves have parallel veins.

Size and shape

The blades of leaves vary greatly in shape and size. Some palm leaves measure over a metre in length. Blades may be oval, rounded, elongate (long

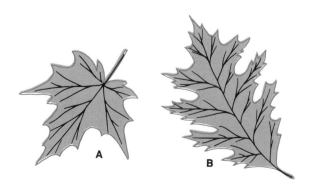

Fig. 15-48   Net venation. A: palmate net venation of maple. B: pinnate net venation of oak.

Fig. 15-49 Leaves. Although varied in appearance, all green leaves photosynthesize.

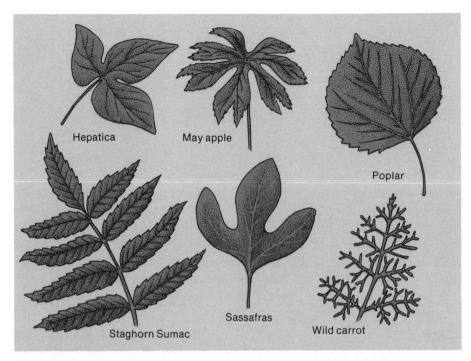

Hepatica

May apple

Poplar

Staghorn Sumac

Sassafras

Wild carrot

Fig. 15-50 Simple leaves.

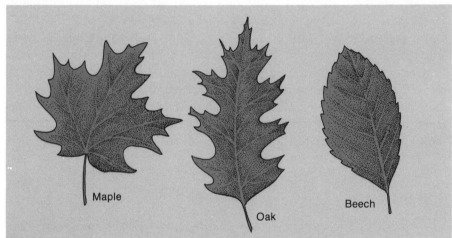

Maple

Oak

Beech

Walnut

Ash

Fig. 15-51 Pinnately compound leaves.

and narrow), or almost any other shape. Their margins may be toothed, smooth, or wavy. Figure 15-49 shows a variety of common leaf types. Characteristics such as the shape, size, and margin of the blade are used by botanists to distinguish one plant species from another.

Simple and compound leaves

Leaves in which the blade is all in one piece are called **simple** leaves. Maple, oak, and beech leaves are all simple leaves (Fig. 15-50). If the blade is divided into several parts, the leaf is **compound**. Each part of a compound leaf blade is called a **leaflet**. Leaflets may be arranged either pinnately or palmately. If the leaflets are attached along the sides of the midrib, the leaf is **pinnately compound**. Walnut and ash trees have pinnately compound leaves (Fig. 15-51). **Palmately compound** leaves are fan-shaped, with the leaflets all attached at one point. Figure 15-52 shows the palmately compound leaves of the horse chestnut tree and a vine called Virginia Creeper.

Deciduous and evergreen leaves

An individual leaf may live from a few days to several years. Young leaves produced on cacti in dry deserts fall off almost as soon as they form, leaving the fattened stem to photosynthesize. The leaves of annual plants last one season only.

Many trees such as maples and oaks lose their leaves each fall and are called **deciduous**. Trees whose leaves last more than one year are called **evergreen**. Pines, spruces, yews, and cedars are all evergreens with needle-like or scale-like leaves (Fig. 15-53). Their leaves last two to five years. They eventually drop off a few at a time instead of all together. In temperate climates, most evergreens have needle-like leaves because broad flat leaves cannot survive the winters. There are only a few broad-leaved evergreens in Canada. Christmas ferns and wintergreen are two of them (Fig. 15-54).

Deciduous trees in temperate climates often become brightly coloured

Horse Chestnut

Virginia Creeper

Fig. 15-52 Palmately compound leaves.

Red pine   Yew   White cedar

Fig. 15-53 Evergreens often have needle-like or scale-like leaves.

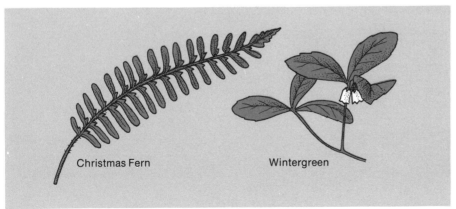

Christmas Fern   Wintergreen

Fig. 15-54 Broad-leaved evergreens.

Fig. 15-55   Leaf abscission.

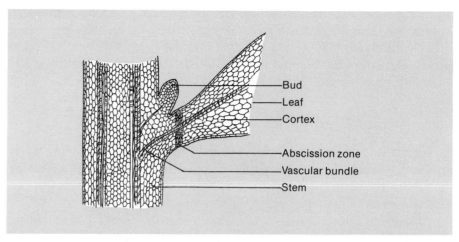

before they drop their leaves. Fall colours are caused partly by a loss of green chlorophyll triggered by falling temperatures. With the cold nights of early autumn, chlorophyll breaks down, making visible yellow and orange pigments that are masked all summer by the green. Yellow pigments are known as xanthophylls, and orange ones as carotenes. They are responsible for the beauty of beech, white birch, and poplar trees in the fall. Red colour is caused by another pigment called anthocyanin. Production of anthocyanin is triggered by cold temperatures, and is greater when there are many cool sunny days. The leaves of maples, sumac, and poison ivy turn brilliant red due to the presence of anthocyanin in the vacuoles of their cells.

With the falling temperatures and shorter days of autumn, the chemical activities in deciduous leaves change and the leaves fall off. Leaf fall is called **abscission**. It occurs when the parenchyma cells at the bases of the leaf petioles separate from each other, forming an **abscission layer** (Fig. 15-55). The leaves remain attached by their vascular bundles only, and are easily knocked off by wind or rain. Where the petiole falls off, a layer of cork forms over the exposed stem to protect it from drying out. The corky areas form the **leaf scars** that you observed on dormant winter twigs in Investigation 15.8.

Fig. 15-56   Cross section of a leaf.

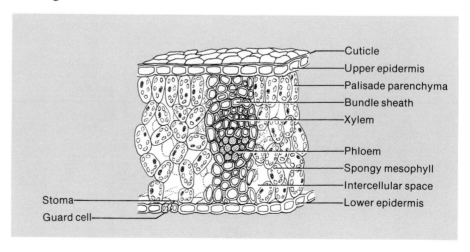

Leaf Structure

Compared to roots and stems, leaves are relatively simple in structure. They originate as young leaves from the tip meristem of a shoot. Mitotic divisions in the young leaves soon produce leaf cells that differentiate into the various tissues of the mature leaf.

Development of compound leaves is similar to that of simple leaves. The leaflets develop separately from localized meristems in the young leaf.

Leaves increase in both length and width by meristematic activity until they reach a size characteristic of the species. Growth often stops earlier at the tip than at the base. This produces leaves that are wider below than above. Some monocot leaves, including many grasses, remain meristematic *at the base of the blade* indefinitely. Such leaves will continue to grow, even when the tips are cut off.

If you cut out a very thin section of tissue from a leaf blade and put it under a microscope, you can see several layers of cells (Fig. 15-56). On both the upper and lower surfaces is a single layer of **epidermal cells**. As in a stem, the epidermis is a protective layer for the inner tissues. The cells of the epidermis appear rectangular in cross section. However, they are often very irregularly shaped in surface view (Fig. 15-57). They do not contain any chloroplasts.

The epidermis of most leaves is covered on the outside with a waxy waterproofing layer called the **cuticle**. The cuticle protects the leaf from losing too much water to the air. It is secreted by the epidermal cells, and is usually thicker on the upper epidermis than the lower. The thick cuticle on the upper surfaces of leaves like ivy makes them appear very shiny.

Although the epidermal cells of leaves do not have water-absorbing root hairs projecting from them, they often have other kinds of hairs. The epidermal hairs on many leaves make them appear woolly or downy (Fig. 15-58). Like the cuticle, the hairs help reduce water loss. They slow down the movement of air at the leaf surface. This forms pockets of non-moving air and reduces the evaporation of water from the leaf. Hairs also help to protect the leaf from damage. Sometimes they also contain glands that secrete oils into the leaf surface.

The epidermis of a leaf is perforated here and there by pores called **stomata**. A single **stoma** is formed by two specialized curved cells called **guard cells** (Fig. 15-59). The curved surfaces of the two guard cells face each other, forming a slit-like opening through which gases pass in and out of the leaf. Guard cells are the only epidermal cells that contain chloroplasts. Because they are responsible for opening and closing the stomata, they control gas exchange in the plant. The functioning of guard cells and the whole topic of gas exchange are discussed later in this chapter.

The number of stomata per leaf varies greatly from species to species and even from leaf to leaf on the same plant. The distribution is variable, too. Some plants have no stomata on the upper leaf surfaces, while others have them on both upper and lower leaf surfaces. The stomata of floating

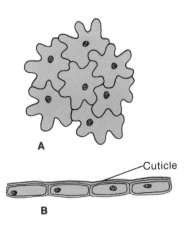

Fig. 15-57 Epidermal cells. A: surface view. B: cross section.

Fig. 15-58 Mullein. Many leaves are woolly like these mullein leaves because of epidermal hairs. Mullein is a common roadside weed, with very soft hairy leaves and yellow flowers.

Fig. 15-59   Stoma and guard cells. A: surface view. B: cross section.

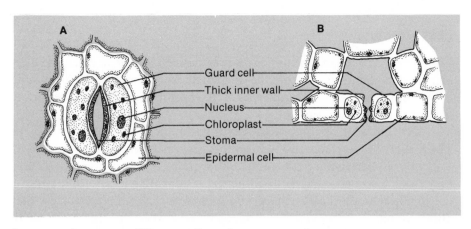

Guard cell
Thick inner wall
Nucleus
Chloroplast
Stoma
Epidermal cell

leaves such as water lilies are all on the upper surfaces. The arrangement of stomata on a leaf surface appears to be random in many broad-leaved plants. However, the needle-like leaves of many evergreens have regular parallel rows of stomata extending from base to tip.

The main part of a leaf blade between the upper and lower epidermis is called the **mesophyll**. Mesophyll is made of parenchyma cells that contain chloroplasts. The mesophyll is frequently differentiated into two layers. The upper part consists of elongated cells placed side by side just below the upper epidermis (see Fig. 15-56). Because the arrangement of these cells resembles the logs of a palisade wall around a fort, they are called **palisade** parenchyma. The palisade tissue may be one or several layers deep. Its cells are full of chloroplasts and act as the main light collectors of the leaf.

Below the palisade tissue is a zone of irregularly-shaped parenchyma cells called the **spongy mesophyll**. The cells are large and contain fewer chloroplasts than the palisade layer. The cells are separated by many large **air spaces** with which they exchange gases. Carbon dioxide, oxygen, and water are transferred from cell to cell through these intercellular spaces.

Supporting the leaf blade is a fine network of **veins** that forms a continuous transport system with the vascular tissue of the stem. From the stem, one or more **vascular bundles** enter the leaf. The bundles branch repeatedly, forming an interconnecting system that reaches nearly every cell. When viewed in cross section, the veins usually appear about half way between the upper and lower epidermis. Each consists of thick-walled **xylem** cells on the upper side and **phloem** on the lower (see Fig. 15-56). The xylem and phloem of most veins are surrounded by a layer of compact parenchyma cells called a **bundle sheath**. All materials that move in or out of the vein must pass through this sheath. Water and minerals diffuse from the xylem out through the sheath to the mesophyll. Food moves inward from the mesophyll to the phloem. Sometimes the sheath extends to the epidermal layers of the leaf, providing extra support for the flimsy blade. These strengthening tissues often rise above the surface of the leaf, forming ribs.

There are endless variations of the basic leaf structure just described. Leaves that grow in full sunlight often have more palisade layers than those in the shade. Shade leaves are usually much larger than sun leaves, and have more stomata. In moist climates leaves are very thin and delicate, and tend to wilt easily. Succulent plants that grow in deserts have thick and fleshy leaves that hold a great deal of water in the mesophyll. On some cacti the leaves are reduced to thin spines to prevent water loss. Although their structures vary, all leaves are equipped to carry out photosynthesis.

Review

1. Why could it be said that leaves are perfectly designed for photosynthesis?
2. Explain the difference between a sessile leaf and a petiolate leaf. Give an example of each.
3. **a)** Distinguish between simple and compound leaves.
   b) What is the difference between palmately net-veined and palmately compound leaves?
4. Define deciduous and give an example of a deciduous plant.
5. Explain how leaves turn colour and fall off in the autumn.
6. What differences would you expect between the epidermis of a leaf growing in a moist shady forest and that of a leaf growing in a dry, sunny field? Why?
7. Why might it be advantageous for a leaf to be hairy?
8. **a)** What is a stoma?
   b) How is it formed?
   c) What is its function?
9. Describe the two types of mesophyll that make up the bulk of a leaf.
10. **a)** Where are intercellular spaces most abundant in a leaf?
    b) What is their function?
11. **a)** How does the vein of a leaf differ from a vascular bundle of a stem?
    b) What are two functions of the bundle sheath?

15.10 Investigation

Leaf Structure

The purpose of this investigation is for you to become familiar with the various tissues of a leaf. Your teacher may give you prepared stained slides to observe, or might get you to make your own slide from a fresh leaf. You may need a bit of practice in slicing off sections that are thin enough to show the cells clearly.

Materials

prepared slide of the cross section
   of a leaf of lily, privet, lilac or
   any other readily available type
compound microscope

Procedure

a. First examine the cross section on low power. Scan the leaf by moving the slide slowly from side to side until you are familiar with the shape of the section. It will look something like a bat in flight. Arrange the slide so that the upper epidermis is toward the top of the field of view.

b. With low power, make a large sketch to show the general shape of the section. Do not draw the individual cells. Indicate the location of the midrib and some of the smaller veins. Label the upper and lower surfaces.

c. Switch to high power and focus your attention on the upper epidermis. Locate the cuticle. Look for any stomata that might be present. Describe the cuticle and stomata in your notes. Draw a few epidermal cells on your sketch from step b.

d. Repeat step c for the lower epidermis.

e. Next, focus on a small area of palisade tissue. Examine a few cells closely. Draw a few palisade cells on your diagram. Label the cell walls, chloroplasts, and cytoplasm.

f. Repeat step e for a small area of spongy mesophyll. Look for the intercellular spaces. Label them on your sketch.

g. Focus with high power on one of the smaller veins. Locate the thick-walled xylem cells and describe their position. Look for phloem cells next to the xylem, and the bundle sheath cells surrounding the vein. Determine if the sheath extends to either epidermal layer. Draw the entire vein carefully and label its parts.

Discussion

1. a) Was a cuticle present on either leaf surface?
   b) What is the function of a cuticle?
2. Where are the most stomata in this leaf?
3. a) What shape are the palisade parenchyma cells?
   b) How are they arranged?
4. How does the shape and size of the spongy parenchyma cells compare to the palisade parenchyma cells?
5. Where in the leaf are the most chloroplasts located?
6. a) In what part of the leaf are the most air spaces found?
   b) What gases would be present in these spaces in a living leaf?
7. What is the purpose of the leaf veins?
8. a) Is there a bundle sheath surrounding the veins of this leaf?
   b) Does it extend to the epidermis?
   c) What is the purpose of the bundle sheath?

15.11 Investigation

Leaf Epidermis

The epidermis of a leaf can be separated from the rest of the leaf tissue and easily observed under a microscope. The species suggested below work well, although your teacher may have others for you to try.

Materials

leaf from a well-watered fleshy
   plant that has been in bright sun
   or under fluorescent lights
   for several hours (Mexican
   Hat Plant (*Bryophyllum*),
   Peperomia, or Purple Heart
   (*Setcreasea*) are recommended)
microscope slides

forceps
cover slips
eye droppers (3)
iodine solution
salt solution
compound microscope
lens paper
paper towel

Procedure

a. Hold one leaf in your hands with its lower surface facing up. Fold the leaf and tear one half over the surface of the other half. This should leave a thin transparent piece of epidermal tissue sticking out from one of the torn edges.

b. With the forceps, carefully peel off a small piece of the epidermis. Mount it on a slide in a drop of water. Cover the mount with a cover slip.

c. Observe the mount under low then medium power. Scan the slide until you find some stomata. In your notes, describe the arrangement and shape of the epidermal cells and the guard cells of the stomata.

d. Count and record the number of stomata you can see in the low power field at one time. Move the slide to a different position and count the number of stomata again. Record the number of stomata for five different fields.

e. Switch to high power and carefully observe a few epidermal cells. Describe their shape, colour, and contents. Make a drawing of five or six.

f. Focus on a single stoma on high power. Find out if it is open or closed. Describe the shape, colour, and contents of the guard cells. Make a large labelled drawing of one stoma with its guard cells and a few surrounding epidermal cells.

g. Gently add iodine solution to the mount by first putting a drop of it on the slide at one edge of the cover slip. Then place a piece of paper towelling along the other edge of the cover slip to draw the iodine solution through the specimen. Do not remove the cover slip during this procedure. Once the iodine has been added, examine the stomata and record any changes in the guard cells.

h. Remove the slide from the microscope and discard the specimen. Make a second slide using the upper epidermis of the same leaf. Count the number of stomata that are visible in the low power field of view and record your count.

i. Make a third slide from a freshly picked leaf and find an open stoma under high power on the microscope. Without moving the slide, put a drop of concentrated salt solution at one edge of the cover slip. Draw it through the specimen using paper towelling as you did with the iodine solution. Observe the stoma and record any changes that occur.

j. Make one drawing of the stoma and guard cells as they appear in water and a second drawing as they appear in the salt solution. Label both drawings clearly.

Discussion

1. **a)** What shape were the epidermal cells of the plant you observed?
   b) What cell contents were visible in the epidermal cells?
   c) What is the function of epidermal cells?
2. **a)** What shape were the guard cells?
   b) What cell contents were visible?
3. **a)** Which leaf surface had the most stomata?
   b) What advantage might this give to the plant?
4. What differences did you observe after adding iodine to the slide?
5. Why is it important that the plant be well watered and receive bright light before this experiment?
6. **a)** What effect did salt solution have on the guard cells?
   b) Explain why this occurred. (Hint: the process of osmosis is involved.)

15.12 Transport and Gas Exchange in Plants

You have learned that roots, stems and leaves have vascular tissue designed for efficient transport of materials from place to place. Water, food, minerals, hormones and other chemicals are constantly being transported. It is because of efficient transportation that vascular plants have been able to grow to tremendous sizes in terrestrial environments.

The movement of water from the soil up to the top of the tallest trees has puzzled plant physiologists for years. Water is absorbed by the roots and moves upwards through the plant body by means of the xylem. Absorption by root hairs appears to be purely an osmotic process. However, the upward movement is regulated by the opening and closing of the stomata in the leaves by the guard cells.

The Guard Cells and Gas Exchange

To understand water transport, you must first learn a little about gas exchange and the operation of the guard cells. Stomata act as pathways for the entry and exit of carbon dioxide, oxygen, and water vapour. Beneath the stomata in a leaf are cavities into which gases diffuse from the surrounding cells and intercellular spaces of the mesophyll. The humidity inside the leaf is almost 100%. This keeps the cell walls moist and allows the diffusion of materials across them. Carbon dioxide, for instance, enters the open stomata, moves through the intercellular spaces, and dissolves in the wet surfaces of the cells. It then diffuses into the cells and is used for photosynthesis during daylight hours. The water needed for photosynthesis diffuses into the cells from the veins. Oxygen produced diffuses out of the cells and moves out through the stomata in the same manner that carbon dioxide moves in.

The leaf cells are constantly using oxygen and giving off carbon dioxide and water in the process of respiration. During daylight hours, the carbon dioxide is used in photosynthesis. At night when the stomata are usually closed, the carbon dioxide from respiration builds up in the cells because photosynthesis stops.

When a plant is photosynthesizing, the stomata are open to allow entry of the necessary carbon dioxide. How do stomata open and close? We know that the turgidity of the guard cells is responsible for the opening and closing. **Turgidity** means the fullness of a cell and it occurs when water moves into the cell. When guard cells absorb water and become turgid, the stomata open. This is because the walls of the guard cells are very thick along the inner side facing the stoma and quite thin elsewhere. As they expand, the thin walls are pushed outward and the thickened inner walls are pulled apart, creating an opening. When guard cells lose water and shrink, the inner walls collapse against each other and the stomata close (Fig. 15-60).

If stomata open when the guard cells take up water and become turgid, what makes them absorb water? No one knows for sure. There are several theories, all of which are quite complicated. More research has to be done to completely understand the opening and closing of stomata.

A Theory for Guard Cell Action

The most widely accepted theory is based on the amount of carbon dioxide in the guard cells. It has been shown that the concentration of carbon

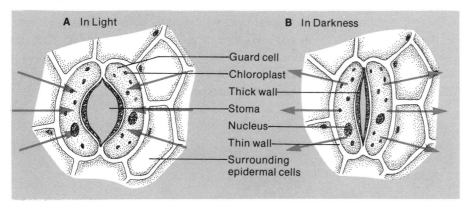

A In Light
B In Darkness

Guard cell
Chloroplast
Thick wall
Stoma
Nucleus
Thin wall
Surrounding epidermal cells

Fig. 15-60 The opening and closing of stomata. The coloured arrows indicate the flow of water into the guard cells during the daytime (A) and out of the guard cells at night (B).

dioxide indirectly causes water to move in and out of the guard cells. If the concentration of carbon dioxide is low, the guard cells become turgid and the stomata open. This happens during the morning when the leaf starts photosynthesizing. The daylight causes the chloroplasts in the guard cells to photosynthesize. This uses up the carbon dioxide. As the carbon dioxide is used up, the acidity of the guard cells decreases. This is because carbonic acid in the cells changes into carbon dioxide and water, leaving less acid in the cells. Under these conditions, starch molecules stored in the guard cells are converted into a soluble sugar. With more sugars in the cells, the osmotic balance of the cells and their surroundings changes. The increased sugar causes water to move in by osmosis. This makes the guard cells turgid, and the stomata will open.

At night, the whole process reverses. Without light, photosynthesis stops. The carbon dioxide builds up in the guard cells because of cellular respiration. This increases the acidity of the guard cells, because some of the carbon dioxide is converted into carbonic acid. In acidic conditions, sugars are converted to insoluble starches. Again the osmotic balance is changed. With fewer soluble sugars in the guard cells, water moves out of them by osmosis. The cells become less turgid and the stomata close. The closing may be complete or partial, depending on how much water is lost.

The carbon dioxide theory explains why stomata are usually open during the day and closed at night. However, the stomata will not open during the day if the plant tissues do not contain enough water. If the soil is dry, the root hairs will not be able to absorb enough water into the plant. All the cells will shrink and the plant will start to wilt. The stomata will remain closed to conserve water until more is available in the soil. In this case, changes in the carbon dioxide concentration will not cause the stomata to open. Therefore, photosynthesis stops in a wilted plant when all the available carbon dioxide in the leaves is used up.

Transpiration

The fact that the stomata have to be open during daylight hours for photosynthesis causes a tremendous loss of water from plant tissues. *Evaporation of water from the stomata is called* **transpiration**. It is the greatest on hot afternoons when the air is dry. The rate of transpiration is affected by humidity, temperature, wind, and light intensity. Leaves exposed to bright sun on hot days lose large quantities of water. This loss cools the leaf tissues because heat is used up in changing liquid water into water vapour. Of all the water absorbed by roots, only about one per cent is used for photosynthesis; the rest is transpired.

The rate of transpiration is controlled by the stomata. When too much water is lost, the stomata close. Unfortunately this prevents photosynthesis because carbon dioxide is kept out. There is a constant conflict in the plant between the need for carbon dioxide and the need to conserve water.

Cohesion-pull theory

The water that is lost through the stomata has to be replaced by water absorbed from the soil. Tremendous pressure would be required to push water up to the tops of tall trees. How the force of gravity is overcome is not exactly known. But the most widely accepted explanation is the **cohesion-pull theory**. As water transpires from the stomata and the spaces below them, osmotic pressure is set up that causes water to flow into the spaces from the surrounding cells. They, in turn, draw water from the xylem cells adjacent to them. From there the water molecules form a continuous column in the xylem all the way down to the roots. Water molecules have a strong cohesive force on each other, and are believed to "pull" themselves up the xylem because of the osmotic pressure created in the leaves. The tension created in the continuous column of water is sufficient to overcome the force of gravity and draw the water up the tree. The pressure of water in the soil may also be a driving force that "pushes" water into the root. The whole topic of water transport still needs much investigation.

Review

1. What are the two main functions of stomata?
2. Trace the path of a carbon dioxide molecule from the air into the mesophyll cell in a leaf.
3. **a)** What is turgidity?
   b) Explain how turgidity causes stomata to open and close.
4. According to the carbon dioxide theory, why do the stomata open during the day?
5. How is the rate of transpiration controlled?
6. What environmental factors affect the rate of transpiration?
7. Why would it be an advantage for a leaf to have all its stomata on the lower epidermis and none on the upper?
8. Why is transpiration necessary?
9. How does a plant prevent excessive water loss in hot dry weather?
10. Explain the conflict between photosynthesis and water conservation in leaves.
11. Explain briefly the cohesion-pull theory of transpiration.
12. Turn to Section 6.10, Homeostasis, on page 134. Review this section carefully. Now explain why the functioning of the guard cells is a good example of homeostasis.

15.13 Investigation

Transport of Water Through Shoots

Water is believed to be pulled up through a plant as water vapour evaporates from the stomata of the leaves. This movement can be observed by experimenting with red dye and stem cuttings of coleus plants.

Materials

stem cuttings of coleus, each about 15 cm long (6)
scalpel
small beakers (6)

water
eosin stain or red ink
hand lens
vaseline

Procedure

a. Put several centimetres of water into each beaker and insert one coleus cutting into each. Cut off 2 cm from the lower end of each cutting while the end is underwater. Do not remove the cuttings from the water. This step removes any air bubbles from the stem that may interfere with the experiment.

b. Coat the surfaces of all the leaves of one cutting with vaseline. Number the untreated cuttings from 1 to 5, and the one with vaseline number 6.

c. Add the same amount of eosin or red ink to each beaker so that the water becomes deeply coloured.

d. Put cutting 1 in a dark drawer or cupboard and cuttings 2 to 6 under a bright light. Leave cuttings 1, 5, and 6 for 24 h.

e. After 15 min, examine the stem and leaves of cutting 2. Make cross sections up the stem at regular intervals. Examine them with a hand lens. Record the distance the stain has travelled (if any). Make a cross section of one of the lower leaf petioles. Determine if any stain is present.

f. Thirty minutes after adding the dye to the beaker, examine cutting 3 in the same manner as in step e.

g. Repeat step e after 45 min with cutting 4.

h. The next day, examine cuttings 1, 5, and 6 in a similar fashion. Record all your observations together in a table in your notes.

Discussion

1. Describe any differences among cuttings 2, 3, and 4 on the first day.

2. Why were five of the cuttings put in bright light for this experiment?

3. Had cutting 5 absorbed more dye after 24 h than 4 had after 45 min? Explain.

4. a) What effect did vaseline have on the uptake of water?
b) Why would this be so?

5. What effect did darkness have on the uptake of water? Explain.

6. What causes water to move up the stems of plants?

Highlights

The tremendous success of seed plants on land has been mainly due to the specialization of plant tissues. With specialization, the functioning of the plant body as a whole has become more and more efficient. The various plant organs each contribute in different ways to the survival of the entire organism.

Roots, stems, and leaves have all evolved into many shapes, sizes, and forms. Roots are the primary absorbers and anchors of terrestrial plants. Absorption of water is accomplished passively by osmosis from the soil into root hairs. The water moves across the tissues of the root to the main conducting tissue, the xylem.

Stems transport materials from place to place. They also support the leaves, flowers, and fruits.

Leaves are highly efficient food factories that are adapted to trap light and exchange gases with the atmosphere. On sunny days leaves produce great quantities of sugars that are transported via the phloem to all parts of the plant. Leaves also keep the plant cool, by regulating the rate of transpiration through the stomata. Neither the operation of the stomata nor the mechanisms of transpiration are understood completely. Much research remains to be done on these topics.

16 Plant Growth And Regulation

The growth of plants is such a common event that it is often taken for granted. Yet growth is a complex process involving cell division, elongation, and differentiation. It results in both an increase in size and an increase in complexity. Growth occurs in green plants when they produce new protoplasm from the products of photosynthesis. It is genetically controlled. But the pattern of growth is greatly influenced by the environment. Both hereditary and environmental factors determine the final shapes and sizes of plants.

16.1 Development

Development of an organism is the series of changes that occurs until the organism reaches its final form. Most plants have a continuous period of development throughout their lives. This period lasts only one season in annual plants. However, it may extend through several thousand years as in the ancient bristle-cone pines of California.

During its lifetime, a plant continues to grow new leaves, stems, and roots and to produce flowers and fruits. Development begins with the seed, whose embryo has all the basic structures needed to produce the adult plant. Sometimes development is temporarily stopped when a seed or plant goes into a period of dormancy. Dormancy may be caused by something within the plant or seed. It may also be caused by changes in the environment, such as low temperatures or lack of moisture.

You are already aware that the complex tissues of a mature plant all develop from embryonic cells called meristems. Each cell contains all of the genetic information it requires to develop into any type of specialized cell. Why then do the outer cells of a stem always develop into epidermal cells and not into xylem cells or some other type of cell? The activation of certain enzymes and the suppression of others inside the cell is one factor. The relative position of the newly formed cells and the chemical environment immediately surrounding them also play a part. There is evidence that plant hormones have definite effects on cell development. Hormones are believed to coordinate the enzyme activities in all the cells of a plant, thereby controlling the plant's development and its reactions to the environment. Finally, the external environment affects the chemical activities of the cells and the action of the hormones. The whole process of growth and development is very complex because all the factors interact. It remains one of the great challenges of biology to completely understand the mechanisms that control plant development.

Why do the embryonic tissues in an acorn always develop into an oak tree and not into a maple? For the most part, the DNA blueprint inherited from the parents determines the final form of the offspring. The genetic information also controls the production of hormones that affect the development. As well, the appearance of the mature oak is dependent on the immediate environment in which it developed.

Review

1. What is meant by the term "growth"?
2. What steps are involved in the development of a plant to its mature state?
3. What is dormancy?
4. What three factors affect the differentiation of cells?
5. What determines the final form of a plant?

16.2 Plant Hormones

The coordination of the cell activities of complex organisms is largely controlled by hormones. Plant hormones are organic molecules that are produced in actively growing plant parts. They are transported to other regions where they influence cell activity. Hormones are active in very low concentrations. They are involved in the responses of plant parts to changes in the environment and in the development of flowers and fruits.

There are three known groups of hormones that regulate the development of flowering plants, although there may be others that are still not discovered. The known hormones are called auxins, gibberellins, and cytokinins. All affect the growth, division, and differentiation of cells. Also, they probably interact with each other to control development.

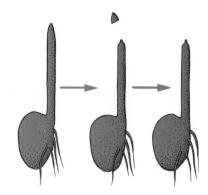

Fig. 16-1 If the tip of an oat coleoptile is removed, the shoot stops growing.

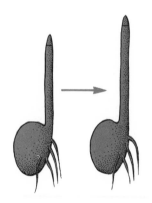

Fig. 16-2 Replacement of the tip of the coleoptile causes growth to resume.

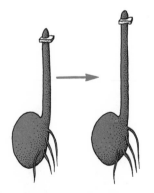

Fig. 16-3 With agar between the tip and the base, the shoot still grows.

Auxins

Growth hormones called **auxins** have been known since the 1930s. At that time, the chemical indoleacetic acid (IAA) was first discovered in a fungus. Since then it has been found in many different plants and is believed to be widely distributed throughout the plant kingdom. IAA is synthesized (produced) in the tips of roots and shoots, in young leaves, and in developing seeds. It is transported down the stem or up the root of a plant, where its primary effect is on the elongation of cells. Many other auxins have been discovered since the 1930s. They have been shown to have similar effects.

The effects of auxins have been studied most closely in oat seedlings. The shoot that develops from the seed of oats is covered with a protective sheath called a **coleoptile**. If the tip of the coleoptile is cut off, the shoot below it will stop growing (Fig. 16-1). If the tip is replaced, growth of the shoot resumes (Fig. 16-2). This shows that a growth substance is secreted from the tip and transported to the growing cells below, causing them to elongate.

To discover what type of substance is secreted, the tip of a coleoptile can be cut off and a layer of jelly-like substance called agar can be placed between the tip and the rest of the shoot (Fig. 16-3). When this is done, the shoot still grows. This shows that the growth substance is a soluble material that can diffuse through agar.

If a coleoptile tip is placed on an agar block for several hours and then the agar alone is placed on the shoot, the shoot also grows (Fig. 16-4). Untreated agar alone has no effect. If the treated agar block is put on one side of the shoot, an even more startling result occurs: the shoot bends away from the side with the agar on it (Fig. 16-5).

All this shows that a soluble chemical produced in the tip regulates the growth of the shoot by stimulating cell elongation. The bending of the coleoptile in Figure 16-5 must be due to faster growth on one side than the other. The chemical substance secreted by the coleoptile tip is called auxin.

The primary effect of auxins in shoots has been shown to be a stimulation of cell elongation. Plants growing on windowsills always bend toward the light because of the distribution of auxins in the cells (Fig. 16-6). For some unknown reason, light energy causes the auxins in the shoot to move to the shaded side of the stem. There, they cause more cell elongation than on the lighted side. The longer cells cause the stem to bend toward the light.

The actions of auxins in plants are not restricted to the stimulation of cell elongation. Their effects are varied and complex. Auxins produced in the terminal buds of leafy shoots and woody twigs inhibit the growth of axillary buds below them. This effect is called **apical dominance**. It is very noticeable in many plant species. The degree of apical dominance determines the growth habit of all plants by regulating the number of lateral branches that develop. Cutting off the terminal bud makes most plants more bushy because the inhibiting auxins are removed.

Auxins both inhibit and stimulate plant development. Different plant parts react differently to the same concentration of auxin. For example, a certain amount of auxin may stimulate the growth of stems and inhibit the growth of roots on the same plant. Auxins secreted from developing seeds inside an ovary stimulate the formation of fruits. Auxins produced in the root tip move through root tissues and promote the development of lateral roots from the pericycle. Auxin production in leaves during the growing season prevents the development of the abscission layer that causes leaf drop in the autumn. Auxins are also active in the initiation of flowering.

Many synthetic (human-made) auxins are now available for use in agriculture and horticulture. In high concentrations they inhibit growth, but in low concentrations they usually promote it. The two most common synthetic auxins are IBA (indolebutyric acid) and 2,4-D (2,4-dichlorophenoxyacetic acid). IBA is used on potato tubers to keep them from sprouting during storage. It inhibits the growth of the buds ("eyes") on the tubers into leafy shoots. 2,4-D is a selective auxin that kills broad-leaved plants but has little or no effect on grasses when used in low concentrations. For this reason, it is used to kill weeds in lawns and along roadsides. 2,4-D acts as a stimulant to cell growth. It causes the plants to grow rapidly, exhaust themselves, and die in a very short time.

The observation that auxins produced by seeds cause the development of fruits has led scientists to experiment with synthetic auxins on horticultural crops. Auxins sprayed onto flowers of some species before pollination result in the development of fruits that contain no seeds. Seedless tomatoes, cucumbers, and watermelons have been successfully grown this way.

The presence of auxins in fruits such as apples, peaches, and pears prevents them from falling off the trees before they are mature. When the

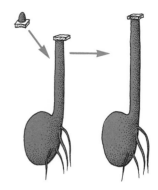

Fig. 16-4 Agar exposed to the coleoptile tip for several hours will cause elongation of the shoot in the absence of the tip.

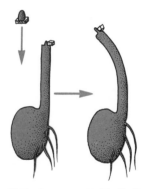

Fig. 16-5 Agar exposed to the tip and put onto one side of the shoot causes the shoot to bend.

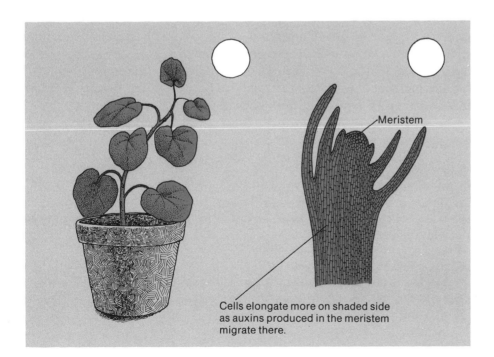

Meristem

Cells elongate more on shaded side as auxins produced in the meristem migrate there.

Fig. 16-6 Windowsill plants bend toward the light because of an unequal distribution of auxins in their cells.

fruits ripen naturally, a decrease in auxin level as the seeds mature causes an abscission layer to form in the stalk. This results in fruit drop. To make harvesting easier, farmers now spray or dust their crops with low concentrations of synthetic auxins to prevent early drop. This allows them to harvest all the fruit at one time and reduce crop loss.

Synthetic auxins are also used in horticulture to propagate leaf and stem cuttings. When the cut ends are dipped into a powder containing auxin, they root much faster than cuttings that are not treated. **Rooting hormone**, as the powder is called, is available in most nurseries and garden centres for use by both amateur and professional gardeners.

Gibberellins

Another group of plant hormones that are found in most complex plants is the gibberellin group. Gibberellins were first studied in Japan in the 1920s. They were extracted from a fungus that caused young rice stems to elongate and fall over. The substance in the fungus that caused this "foolish seedling disease" was called gibberellin. Many gibberellins have since been found in other plants. They all stimulate cell elongation, and are produced in the young leaves.

Dwarf plants have remarkably low concentrations of gibberellin. They will grow tall if a drop of gibberellin is added to the shoot tip. Rosette-type plants can be made to "bolt", or produce a central flower stalk, with synthetic gibberellin. Because gibberellins promote flowering in some species, they can be used to induce early blooms.

Cytokinins

The third group of plant hormones is the cytokinin group. Cytokinins stimulate cell division in the presence of auxins. They are very concentrated in young developing fruits. Their interaction with auxins seems to stimulate the development of leaves, shoots, and buds as well. Much research remains to be done on this group of hormones in plants.

Review

1. **a)** What is a hormone?
   b) What are three types of plant hormones?
2. Where are auxins produced in a plant?
3. What four characteristics of auxins were discovered by experiments with oat seedlings?
4. Explain how auxins cause plants to bend toward the light.
5. **a)** What is apical dominance?
   b) How is the effect of auxin different in apical dominance from its effect in the oat coleoptile experiments?
6. **a)** What is 2,4-D?
   b) What is 2,4-D used for and how does it work?

7. Explain how seedless cucumbers are produced.
8. List three advantages of using synthetic auxins in horticulture.
9. Explain what causes "foolish seedling disease" in rice plants.
10. What use can be made of synthetic gibberellins?
11. What effect do cytokinins have on plant development?

16.3 Investigation

Plant Hormones

Experiments with hormones have shown that plant tissues are quite sensitive to very low concentrations of these chemicals. Both auxins and gibberellins are used by horticulturalists to promote growth of roots, shoots, and flowers. You will be able to observe for yourself the effect of commercial rooting powders and gibberellins in the following two experiments.

Materials

similar cuttings of *Coleus, Peperomia,* or other available houseplants with non-woody stems (10)	bean seeds (50)
	large germination trays (2)
	gibberellic acid
rooting powder for "softwood cuttings"	eye droppers (2)
vermiculite	ruler
small pots, trays, or other containers suitable for cuttings (10)	labels
	water
pencils or sticks	plastic

Procedure A Effects of an Auxin

a. Fill the 10 containers to within 1 cm of the top with vermiculite. Soak them thoroughly with water.
b. Label 5 of the containers "untreated". Insert one stem cutting halfway down into the vermiculite in each. Each cutting should have 3 or 4 pairs of leaves if *Coleus* is used, or 4 or 5 leaves of *Peperomia*. Pack the vermiculite down around the cuttings so that they stand upright.
c. Mark the other 5 containers "treated". Dip the ends of 5 cuttings into the rooting powder before planting them. The ends should be wet first. Tap off any excess powder. Then plant the cuttings in the same manner as the untreated ones.
d. Cover each set of containers with a sheet of clear plastic propped up with pencils or sticks. Keep the cuttings in a warm bright location for two weeks (not full sun).
e. Three days after planting, pull up one untreated cutting and one treated one from the vermiculite. Observe any root development. Note the density, thickness, and length of the roots (if any). Record any

differences between the two cuttings. Make comparison sketches of their roots.

f. Repeat step e with another two cuttings (one untreated, one treated) after another 3 d (6 d from planting). Continue at three-day intervals with the remaining cuttings. Organize your observations in a table in your notebook.

Procedure B   Effects of Gibberellic Acid

a. Soak 50 bean seeds overnight and separate them into two groups of 25. Fill two trays with vermiculite and plant 25 seeds in each. Keep the trays well watered in a bright location for one week or until the seedlings are 5 to 6 cm tall.

b. Label one tray "treated" and one tray "untreated". In each tray, select 10 seedlings about the same height. Cut off the remaining seedlings at the base of the stem. Without disturbing the plants too much, measure the heights of the 10 plants left in each tray. Record the information in a table in your notebook.

c. For each of the 10 seedlings in the tray marked "treated", add one drop of gibberellic acid to the tip of the shoot.

d. To the 10 seedlings in the tray marked "untreated", add one drop of water to the shoot tip. Be sure to use a different eye dropper.

e. Leave both trays under similar light and temperature conditions for one week. Each day during that week, measure the heights of all the seedlings. Record the measurements in the table you started in step b.

f. Calculate the average height of the treated seeds and the average height of the untreated seeds for each day of the experiment.

Discussion A

1. What effect did rooting powder have on the growth of roots by your cuttings?

2. How long did it take for a difference in root development to occur between the treated and untreated cuttings?

3. What was the purpose of the plastic in this experiment?

4. Explain how auxins in rooting powders affect the growth of roots.

Discussion B

1. Why is it necessary to start with 50 bean seeds if all you use is 20?

2. Why should the seedlings all be about the same height at the beginning of this experiment?

3. Why was a drop of water added to each plant in the "untreated" tray?

4. a) What changes occurred in the untreated seedlings over the 5 days?
   b) What changes occurred in the treated seedlings?
   c) Describe any differences between the treated and untreated seedlings on the fifth day.

5. Why was it necessary to keep both trays under the same light and temperature conditions?
6. What effect does gibberellic acid have on bean seedlings?
7. Predict what might happen if the seedlings are left to grow for several more weeks.

16.4 Plant Movements

Have you ever thought that plants cannot move? Animals move from place to place, but plants usually do not. However, they do move in response to environmental stimuli such as light, touch, gravity, and temperature. Because the movements are slow, you may never have noticed them.

Plant movements can be grouped into three types. **Tropisms** are slow growth movements in response to stimuli such as light and gravity. The direction of movement is determined by the location of the stimulus. **Turgor movements** are faster reactions, that are independent of the direction of the stimulus. Finally, **circadian rhythms** are movements that occur approximately every twenty-four hours. They appear to be independent of the environment altogether.

Tropisms

Phototropism

Tropisms are named according to the type of environmental stimulus that produces them. *The bending of plants toward the light is called* **phototropism**. The bending of plants on windowsills allows all the leaves to receive the maximum amount of light available for photosynthesis. It occurs because the cells grow at different rates on either side of the plant due to different auxin concentrations. In some unknown way, the light causes more auxins to collect on the shady side of the stem. As a result, more cell elongation occurs there. Because tropisms are caused by the growth of cells, they are not reversible.

Phototropism can be observed outdoors in many situations. Trees often have more leaves on their sunny side. The leaves of vines growing on fences are usually arranged to face the sun for maximum photosynthesis.

Geotropism

Geotropism *is the response of plants to gravity.* Since plants never grow upside down, they must have some mechanism for detecting which way is up. If you turn a growing plant on its side, after a while the shoot will bend up and the roots down. The movement of a plant toward a stimulus is called a **positive tropism**, and movement away from it a **negative tropism**. Which is negatively geotropic—the shoot or the root?

As in phototropism, changes in auxin concentrations cause the

Fig. 16-7 Tendrils of pea plants (A) and watermelon vines (B) will curl around a support.

geotropic responses. More auxins go to the lower side of the shoot and cause that side to elongate faster, bending the shoot up. More auxins also go to the lower side of the root. Then why do the roots bend down? It seems that the concentration of auxin that causes cell elongation in shoots *inhibits* cell growth in roots. Therefore, the lower side of the root grows more slowly and the root curves downward. Experiments with many different concentrations of auxin have proven that roots are much more sensitive than shoots. They also prove that excess amounts of auxin inhibit cell growth.

Other Tropisms

Other tropisms are probably also controlled by auxins. **Hydrotropism** *is the growth of plant parts toward water.* It may be seen in the roots of willows and cottonwoods along riverbanks. **Chemotropism** *is the response to chemicals.* **Thigmotropism** *is the response to contact with solid objects.* Some climbing plants have modified leaves or branches called **tendrils** that will curl around a support (Fig. 16-7). Peas, grapes, watermelons, and cucumbers all have curving tendrils that twist around objects. The twisting is permanent because it results from more cell elongation on the side away from the stimulus. The mechanism for it is not well understood.

Of what importance are tropisms? Consider the growth of root tips through the soil. The path taken by a root is affected by gravity, water, and soil chemicals. Geotropism, hydrotropism, and chemotropism all interact to affect the development of the root system and the survival of the plant.

Turgor Movements

The opening and closing of flowers at certain times of the day and the folding up of leaves at night are two examples of turgor movements. Both are caused by changes in the **turgor**, or water pressure, in the cells. Turgor movements occur in response to light, temperature, touch, and wind. They are reversible when conditions change. Unlike tropisms, they are not directional responses.

Dandelions and tulips are two flowers that close at night and open again the following morning. Evening primrose flowers open in the evening, stay open during the night, and close the next day (Fig. 16-8). Morning glories open in the morning (Fig. 16-9). Both light and temperature may be factors in this type of behaviour. Daily changes also occur in the leaves of clover and oxalis. The leaves are horizontal during daylight but they droop down at night. How the environment causes the changes in cell turgor that result in these movements is unknown.

The classic example of turgor movement is the Sensitive Plant, *Mimosa pudica. Mimosa* is a small plant with compound leaves composed of two or four sets of leaflets (Fig. 16-10). The leaflets respond almost instantly to touch, heat, or wind by folding up. At the same time, the petiole droops. The reaction is due to a sudden loss of turgor in the cells at the base

Fig. 16-8 Evening Primrose flowers open in the evening.

of each leaflet and the base of the petiole. The leaves recover after 15 min or so, as the turgor pressure returns to normal. Hundreds of studies have been done on *Mimosa* to discover what causes the reaction. However, it is still poorly understood.

Circadian Rhythms

Certain movements in plants occur on a daily cycle, even when the environment is kept absolutely constant. In some plants the leaves close up at specific times although the light and temperature remain unchanged. This seems to be due to some internal time-measuring system, or "**biological clock**", in the plant itself. Changes that occur on a twenty-four hour cycle are called circadian rhythms (from the Latin words "circa", meaning nearly, and "diem" meaning day). How plants measure time and how their responses are regulated remain mysteries to be solved by future research.

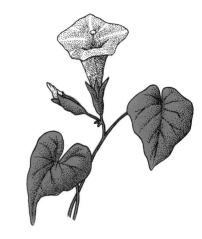

Fig. 16-9   Morning Glories open in the morning.

Review

1. What is the difference between a tropism and a turgor movement?
2. What is phototropism?
3. **a)** Why do people frequently turn the plants growing on windowsills?
   b) Could this turning have any bad effects on the plants? Explain.
4. Explain why roots are positively geotropic and shoots negatively geotropic.
5. **a)** What are tendrils?
   b) Explain how cucumbers climb up a trellis.
6. Why are turgor movements reversible?
7. **a)** What happens to the Sensitive Plant when you touch its leaves?
   b) Explain what happens to the cells when this occurs.
8. **a)** What is a circadian rhythm?
   b) Give as many examples as you can of circadian rhythms.

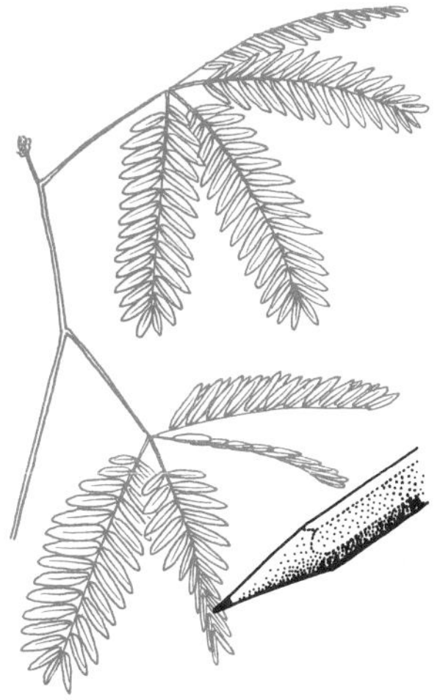

16.5 Investigation

Study of a Tropism

The growth responses of plants to environmental factors are slow but permanent. You can investigate tropisms by controlling the environmental stimuli and watching for responses in the growth patterns of experimental plants. In this investigation you will study one tropism.

Fig. 16-10   The Sensitive Plant, *Mimosa pudica*.

Materials

sphagnum moss or vermiculite
germinated seedlings of corn
   or pea, with roots about
    1 cm long (4)

large beaker
paper towelling
grease pencil
razor blade

Procedure

a. Line a beaker with several thicknesses of moistened paper towelling. Fill the inside with damp vermiculite or sphagnum moss to keep the towelling pressed against the sides of the beaker.

b. Measure and record the length of the root of one seedling. Place the seedling between the towelling and the side of the beaker, about halfway down, with its root pointing downward. Label the outside of the beaker, above the seedling, "Seed 1".

c. Take a second seedling, measure its root, and place it on the opposite side of the beaker with its root pointing upwards. Label this seedling "Seed 2".

d. Measure and place a third seedling between the other two with its root pointing sideways. Label it "Seed 3".

e. Cut 1 mm off the root tip of the fourth seedling with a razor blade. Measure the root and insert the seedling with its root downward into the beaker in the vacant fourth side. Label it "Seed 4".

f. Make a sketch of each seedling, showing the direction of the roots as they were placed in the beaker. Be sure to label the sketches.

g. Keep the seedlings moist and in a warm location for the duration of the experiment. After 2 d, observe and sketch each seedling. Note any changes in the length and direction of the roots in your sketches. Do not remove the seedlings from the beaker.

h. Repeat step g after another 2 d. You should have 3 sets of sketches altogether. On the fourth day, measure and record the length of each root.

i. Complete a table like Table 16-1 in your notebook.

SEEDLING	INITIAL ROOT LENGTH (cm)	FINAL ROOT LENGTH (cm)	INITIAL DIRECTION OF ROOT GROWTH	FINAL DIRECTION OF ROOT GROWTH
1				
2				
3				
4				

TABLE 16-1

Discussion

1. a) What were the effects of removing the root tip of seedling 4?
   b) From your knowledge of root structure, explain these results.
2. What changes occurred in the roots of seedlings 1, 2, and 3?
3. What effect does the orientation of the seed have on the length of the root?
4. To what stimulus are the roots responding in this experiment?

5. Does this experiment show a positive or negative tropism? Explain.
6. What benefit is this tropic response to the seedling?
7. Predict what might happen if you planted tulip bulbs upside down.

16.6 Investigation

Phototropism

In Investigation 16.5 you observed the growth of roots in response to gravity. The variables of light, temperature, and moisture were identical for each seedling. But the seedlings were placed in different positions. In other words, you controlled all the variables except one in order to study its effects. You observed the growth of the roots and compared the results in a table.

You now have a chance to investigate phototropism, another tropic response of plants. You know that this is the response of plants to light. The problem to be investigated is how different light conditions affect plant growth. You can probably make a few hypotheses right now. Your task is to design an experiment to test one or more of your hypotheses. Your teacher will supply either young seedlings or larger growing plants. Check what is available before you design your experiment. Remember to state your hypotheses and record the procedure you will follow step-by-step. You must decide how long to run the experiment and how you will control other variables in order to obtain useful results. When you have performed your experiment, record all your observations in a systematic way and state your final conclusions.

16.7 Factors Affecting Plant Growth

Investigations 16.5 and 16.6 have shown that environmental factors such as light and gravity affect plant growth. The availability of oxygen, carbon dioxide, minerals and water also affect plant development. Temperature and humidity are other important factors. All these things interact to control growth by affecting the metabolism of each individual cell in an organism. A change in only one factor can change the metabolism enough to cause stunting, weak growth, or poor flower development. Of all the environmental stimuli, light and temperature are probably the most important.

Temperature

Temperature affects all plant processes. As the temperature increases, the rates of photosynthesis, respiration, and transpiration rise. This results in faster growth, up to a certain point. Very high temperatures stop development and may cause injury to the cells. For each species of plant there is an **optimum temperature** at which growth is at a maximum. For most plants, this temperature is somewhere between 10°C and 40°C. Many plants will not grow at all below a certain temperature, but will remain alive. Garden vegetables planted too early in the spring will not put out any new growth until the temperature rises.

Temperature can also affect flowering. A period of low temperatures is required for flower production in some varieties of wheat and rye. Tulip bulbs need a cold period to bloom. Biennial plants that form rosettes in their first year must be exposed to cold before they will flower. In temperature climates, many woody plants go into a period of dormancy at the end of the growing season. Low winter temperatures are required to break the dormancy and start new growth. Branches from fruit trees or lilac bushes will not grow indoors in the autumn, but can be forced into bloom indoors in late winter after they have been exposed to cold. Florists have learned to treat many plants with artificial periods of cold to start flowering at unusual times of the year.

Light

The main effect of light on plants is its control of photosynthesis. The intensity of the light, its quality, and its duration all affect the rate of photosynthesis and, therefore, the rate of growth. Plants grown in darkness become tall, spindly, and yellowish. Their leaves do not expand. Plants in this condition are said to be **etiolated**. Lack of light causes etiolation by changing the rate of elongation of the internodes. It also prevents photosynthesis, and causes plants to die from lack of food.

Plants grown in shady places are usually taller than similar plants grown in full sun. Again, the amount of light affects the elongation of the stem. The leaves of shade plants are also larger and thinner, and contain fewer palisade cells. They also have a thinner cuticle and fewer veins than sun leaves in which water loss has to be more closely controlled. These differences are all due to variations in the intensity of the light received by the plants.

The duration of light is also important. The response of plants to varying periods of light and darkness has been studied a great deal. Some plants will bloom only when the days are short and others only when the days are long. The flowering of plants in response to the length of day is called **photoperiodism**. Photoperiodism is easy to observe in temperate climates where the day length changes from season to season. In Canada, the days gradually increase in length from December 21 to June 21, and decrease from June to December. Spring flowers such as tulips and daffodils are called **short-day plants**, because they bloom early in the season when the days are short. Hollyhocks, spinach, and lettuce flower only during the long days of summer. Therefore they are called **long-day plants**. Autumn flowers such as chrysanthemums are short-day plants like the tulips. There are other plants that are not affected by the length of day, or photoperiod, and will bloom right through the growing season. These plants are called **day-neutral**. Marigolds, roses, snapdragons, zinnias, and tomatoes are all day-neutral plants.

How is flowering triggered in plants that are sensitive to the photoperiod? It has been shown by experiments that it is actually the period of darkness that is critical. Short-day plants flower only when the dark period is greater than some critical value (often 14 h) and is uninterrupted by any flashes of light. In other words, short-day plants are really long-night plants! If they receive longer periods of light, or brief exposures to light during the night, they produce no flowers. The Christmas poinsettia is a plant that must be exposed to short days for several weeks in the fall in order to bloom and become red for Christmas. Florists grow chrysanthemums in greenhouses under artificially controlled photoperiods to produce blooms for Easter, Christmas, and other occasions.

Long-day plants produce flowers only when the period of daylight in each twenty-four hour cycle is greater than some critical value (approximately 14 h). Long-day plants will not bloom in the spring or fall unless the photoperiod is extended with artificial light.

The length of the photoperiod seems to be detected in the young leaves of plants by the use of a pigment called **phytochrome**. Phytochrome somehow affects the production of a hormone in the leaves. The hormone is transported to the shoot tip where it stimulates flower production. The interaction of light and phytochrome and its effect on flowering is not well understood as yet. Neither are the other effects of phytochrome on plant development. It is known that phytochrome is involved in stem elongation, chloroplast development, leaf expansion, and seed germination. But the exact mechanism is still a mystery.

Review

1. In what two ways does temperature affect plants?
2. Will tulip bulbs bloom if you keep them indoors over the winter and plant them outside in the spring? Explain.
3. Why is dormancy a useful adaptation for woody plants in the Canadian climate?
4. **a)** What is etiolation?
   b) What causes it?
5. What differences would you expect between two geraniums if you grew one in full sun and the other in the shade?
6. Distinguish between photoperiodism and phototropism.
7. Explain why chrysanthemums will not bloom in the summer and why hollyhocks will not bloom in the fall.
8. If someone complained to you that a Christmas poinsettia would not bloom again the second year, what would you tell that person to do?
9. How do plants "recognize" the length of daylight in any twenty-four hour period?

Highlights

Both hereditary and environmental factors affect the growth and development of plants. Differentiation of embryonic cells into the various tissues of a mature plant is a complex, poorly understood process. It involves environmental influences, the action of hormones, and the genetic make-up of the cells themselves.

Hormones are organic molecules that coordinate plant activities and their reactions to external stimuli. Three major groups of hormones, called auxins, gibberellins and cytokinins, are known to exist in plants. Auxins have been widely studied and are used in agriculture to control weeds, propagate (grow and reproduce) house plants, and produce seedless fruits.

The movements of plants in reaction to external stimuli have also been well studied. Fast movements that can be reversed are called turgor movements. Slow growth movements that are irreversible are called tropisms. Tropic responses to various stimuli are caused by the distribution of auxins in the plant cells.

The environment in which a plant grows has a great effect on its development. Of all the environmental stimuli, light and temperature are probably the most important. They affect the rates of photosynthesis, transpiration, and respiration in plant cells. They also promote or inhibit flowering in plants that are sensitive to them.

UNIT FIVE

Animal Form And Function

In Units Three and Four you studied the form and function of organisms in four kingdoms of living things: Monera, Protista, Fungi, and Plantae. In this unit you will investigate the form and function of organisms in the kingdom Animalia.

17 Animal Diversity

About 1 200 000 species of living things have been classified in the kingdom Animalia. Countless thousands of animals still remain to be discovered and classified. Most are probably insects and other arthropods.

Zoologists do not agree on exactly how animals should be classified. However, most modern classifications divide the animals among about 20 phyla. Table 17-1 shows the approximate number of species in each of the eleven main phyla of animals. The phyla are listed in order from the simplest to the most complex animals.

PHYLUM	COMMON NAME	NUMBER OF SPECIES
Porifera	sponges	4 500
Coelenterata	coelenterates	9 500
Platyhelminthes	flatworms	6 500
Nematoda	roundworms	10 000
Rotifera	rotifers	2 000
Bryozoa	bryozoans	3 000
Mollusca	molluscs	75 000
Annelida	segmented worms	10 000
Arthropoda	arthropods	1 000 000
Echinodermata	echinoderms	5 000
Chordata	chordates	45 000

TABLE 17-1 **The Main Phyla of Animals**

This chapter will describe the common characteristics of the organisms in the kingdom Animalia. It also gives an overview of the eleven main phyla. Then, the remaining chapters of Unit Five will describe more fully the forms and functions of some typical organisms in most of these phyla.

17.1 Characteristics of the Kingdom Animalia

The organisms in the kingdom Animalia share most of the following characteristics:

1. All animals are many-celled. The cells are differentiated to form specialized tissues. In most animals, the tissues are grouped into organs and the organs into organ systems.
2. Most animal cells have a nucleus and organelles such as mitochondria, Golgi bodies, lysosomes, and ribosomes.
3. Animal cells are bounded by the plasma membrane. They have no cell wall. Therefore, they tend to be flexible.
4. Animals are **heterotrophic**. They obtain their energy for life processes by eating other organisms. Animals that eat plants are called **herbivores**. Those that eat other animals are called **carnivores**.
5. Most animals are motile. That is, they can move from place to place.
6. All animal species reproduce sexually. During sexual reproduction, a male gamete (sperm) and a female gamete (egg) unite to form a zygote. The gametes have the haploid (n) number of chromosomes and the zygote has the diploid (2n) number of chromosomes. A few of the simplest animal species can reproduce asexually by processes such as **budding**.

    Most animal species have separate sexes. That is, the male and female gametes are produced by separate individuals. In other words, male and female animals exist for those species. However, some animal species are hermaphrodites. That is, both the male and female gametes are produced by the same individual.

Review

1. **a)** How do organisms in the kingdom Animalia differ from organisms in the kingdom Plantae? (See Section 12.1, page 311.)
   b) How are animals and plants alike?
2. **a)** The 2-kingdom system of classification places the protozoan protists in the kingdom Animalia. What justification is there for doing this? (See Section 10.1, page 245.)
   b) How do animals and protozoan protists differ?

17.2 A Survey of the Main Animal Phyla

Phylum Porifera: Sponges

About 4 500 species of animals are classified as sponges. These are the simplest animals. Most species do not have symmetrical shapes. However, a few species have **radial symmetry**. That is, they have similar parts arranged around a common central point.

The cells of sponges are grouped together but do not form true tissues. The body has many pores, canals, or chambers through which water flows.

Sponges have no organs, no movable parts, and no appendages. Most species have an internal skeleton that supports the body. The skeleton is made of silicates (glass-like substances) in some species, calcium carbonate (limestone) in other species, and proteins (spongin) similar to fingernails in still other species.

Sexual reproduction occurs in all species. As well, some sponges reproduce asexually by budding. Some species are hermaphrodites. Others have separate sexes.

Spongia
(Bath sponge)

Spongilla

Scypha

Microciona

Phylum Coelenterata: Coelenterates

About 9 500 species of animals are classified as coelenterates. These include hydroids, jellyfish, corals, and sea anenomes.

Coelenterates have radial symmetry about a central opening, or "mouth". That is, they can be cut along their central axis in any direction to give two equal halves. The mouth is surrounded by tentacles. Food is taken in and undigested particles are expelled through the mouth. Digestion occurs in a large body cavity.

Some coelenterates are sessile. That is, they live attached to some object or to each other (example, the corals). Other coelentrates, such as jellyfish, swim or float in the water.

Their body consists of two layers of tissue. A skeleton of limestone or protein may be present. These animals have no circulatory, breathing, or excretory systems. They have a few nerve cells in the body wall, but no organized nervous system.

Reproduction may be asexual by budding, or sexual. Some species are hermaphrodites, while others have separate sexes.

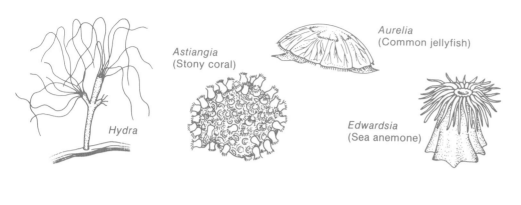

Astiangia
(Stony coral)

Aurelia
(Common jellyfish)

Edwardsia
(Sea anemone)

Hydra

Physalia
(Portugese
man-of-war)

Phylum Platyhelminthes: Flatworms

About 6 500 species of animals have been classified as flatworms. These include free-living flatworms such as the common planarian. The flukes, which are parasites of most animals, and the tapeworms, which are intestinal parasites of vertebrates, are also flatworms.

The body of a flatworm is flattened dorso-ventrally (from top to bottom), as the name suggests. It has no segmentation.

Flatworms have **bilateral symmetry**. That is, the body can only be cut along the central axis in one direction for the right and left halves to be mirror images of each other.

The body consists of three layers of tissue. All animals above this level also have three layers of tissue. The body of the flatworm lacks skeletal, breathing, and circulatory systems. The digestive system is incomplete. In other words, these animals have a mouth, but no anus. Food enters and waste particles leave by the same opening. A simple nervous system also exists.

Some parasitic species have suckers or hooks, or both, for attachment to the host on which they live. Muscle layers are well-developed in these animals. Most flatworms are hermaphrodites.

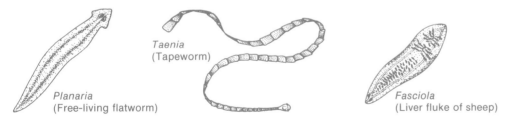

Planaria
(Free-living flatworm)

Taenia
(Tapeworm)

Fasciola
(Liver fluke of sheep)

Phylum Nematoda: Roundworms

About 10 000 species of animals have been classified as roundworms (nematodes). Among these are some worms that are scavengers. This means they feed on decaying organic matter. However, most species are parasitic. Many parasitize the roots of plants. Still others parasitize animals. In fact, over 30 species live in humans. Roundworms vary in length from about 0.5 mm to over 1 mm. Most are under 10 mm in length.

Roundworms have bilateral symmetry. The body consists of three layers. It is slender and cylindrical, with no segmentation or appendages.

The digestive system is complete. It consists of a long tube that runs from the mouth at one end of the animal to the anus, at the other end.

The body wall has longitudinal muscles only (muscles that run the length of the body). Therefore these animals move with a snake-like, or S-shaped, motion. They cannot move along like an earthworm.

Roundworms have no circulatory or breathing systems. Some species have simple excretory organs. The nervous system is fairly well developed. Most roundworm species have separate sexes. The male is normally smaller than the female.

Ascaris (Roundworm of pig)

Trichinella (A parasite of many animals)

Ancylostoma (Hookworm)

Phylum Rotifera: Rotifers

About 2 000 species of animals have been classified as rotifers. You will likely see these animals whenever you look at pond water, aquarium water, or any other fresh water habitat, through a microscope.

Rotifers have numerous beating **cilia** (hair-like projections) at their anterior ends. These sweep food into the mouth. Their method of beating makes them look like rotating wheels. Therefore they are called rotifers.

Rotifers have bilateral symmetry and three body layers. The bodies are generally cylindrical, with a ciliated region at the anterior end. At the posterior end is a forked projection. Although many-celled, the rotifers are microscopic organisms.

Many species have a complete digestive system. All have excretory organs. The nervous system is fairly well advanced, but no nerve cord is present.

Rotifers have separate sexes. However, in some species, males have never been observed. The females of all species can reproduce by **parthenogenesis**. This means their eggs can develop into mature animals without being fertilized by sperm.

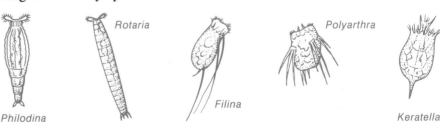

Philodina

Rotaria

Filina

Polyarthra

Keratella

Phylum Bryozoa: Bryozoans

About 3 000 species of animals are classified as bryozoans, or "moss animals". (You may recall that the *bryo*phytes are the *moss* plants.)

All bryozoans are aquatic, and most species live in salt water habitats. These animals are colonial and most species are sessile. They live in colonies, attached to solid objects. Although the colonies can generally be seen with the unaided eye, a microscope must be used to see the individual animals clearly.

Bryozoans have bilateral symmetry and three body layers. The digestive system is complete, and has a mouth surrounded by tentacles. No circulatory, breathing organs, or excretory organs are present. A simple nervous system exists. Most species are hermaphrodites.

Urnatella

Bugula

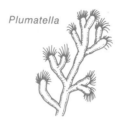

Plumatella

Phylum Mollusca: Molluscs

With 75 000 different species, the phylum Mollusca is the second largest phylum of animals. Among the molluscs are chitons, snails, whelks, conchs, slugs, clams, mussels, scallops, oysters, squids, and octopuses. Most species are aquatic. Of these, most live in sea water.

Many species have bilateral symmetry. All have three body layers. The body is soft and is often enclosed in a shell. The shell is very small or missing in some species.

Most species have a well-developed head region. All have a ventral muscular "foot" that is used for crawling, burrowing, or swimming.

The digestive system is complete. Most species have a radula, or

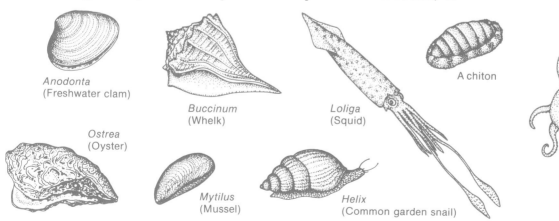

Anodonta
(Freshwater clam)

Ostrea
(Oyster)

Buccinum
(Whelk)

Mytilus
(Mussel)

Loliga
(Squid)

Helix
(Common garden snail)

A chiton

Octopus

tongue-like projection, that is equipped with a rough surface for grinding off food.

The circulatory system has a heart and a few blood vessels. Breathing occurs through gills, the epidermis, or a simple lung. Excretion is carried out by one or more simple "kidneys". The nervous system has many organs for smell, taste, touch, and sight.

Most species have separate sexes.

Phylum Annelida: Segmented Worms

About 10 000 species of animals are classified as annelids, or segmented worms. The most familiar annelids are terrestrial. You have likely seen several species of earthworms. However, most annelids are aquatic. Some saltwater species actually reach a length of 1 m. Giant Australian earthworms reach a length of more than 3 m!

Annelids have bilateral symmetry and three body layers. The body is elongated. It is cylindrical in some species and flattened dorsoventrally in others. The segmentation of the body is clearly visible in most species.

The body has a moist cuticle (covering) over the epidermis ("skin"). It also has both longitudinal and circular muscles, which allow these worms to move in their typical fashion.

The tubular digestive system is complete, running from a mouth at one end of the worm to an anus at the other end.

The circulatory system is well-advanced. Blood circulates through a closed set of blood vessels, with the help of five pairs of "hearts". The blood contains haemoglobin.

In most species, breathing takes place through the moist epidermis. Some marine species that live in burrows have gills. Excretion is carried out by a number of simple excretory organs. The nervous system is well-advanced, with a simple "brain", main nerve cord, and sensory cells and organs for touch, taste, and detection of light.

Most annelids are hermaphrodites, although some have separate sexes.

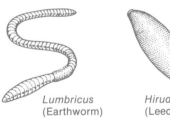

Lumbricus
(Earthworm)

Hirudo
(Leech)

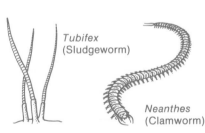

Tubifex
(Sludgeworm)

Neanthes
(Clamworm)

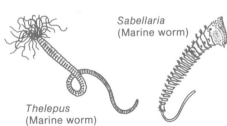

Sabellaria
(Marine worm)

Thelepus
(Marine worm)

Phylum Arthropoda: Arthropods

About 1 000 000 species of animals are classified as arthropods, making this phylum the largest. In fact, most animals are arthropods. The name Arthropoda comes from two Greek words, *arthros,* meaning "joint" and *podos,* meaning "foot". All arthropods have jointed appendages (legs, mouthparts, wings, antennas).

The phylum Arthropoda includes crustaceans such as shrimps, crayfish, lobsters, crabs, and barnacles. It also includes insects, spiders, scorpions, millipedes, centipedes, and a host of other animals.

Arthropods have bilateral symmetry and three body layers. The body is externally segmented and jointed. Arthropods have three body divisions: head, thorax, and abdomen. While these divisions are segmented, they may be fused to one another in various ways.

Arthropods have a hard **exoskeleton** (outer skeleton) that is made of chitin. Movement is made possible by muscles that are attached to the inside of this skeleton.

The digestive system is complete. The mouth is equipped with specialized mouthparts. The circulatory system is open. That is, the blood does not travel always within blood vessels. Instead, it is pumped by a heart through vessels into large spaces where it bathes the organs and tissues. Then it drains back to the heart without the aid of blood vessels.

Breathing in arthropods is varied. Aquatic species use gills. Terrestrial species use tracheae (air tubes). Still others use their body surface. Excretion is by means of simple excretory organs. The nervous system includes nerve centres, or a "brain", in the head, along with two nerve cords. Among the sensory organs are antennae, hairs, compound eyes, and simple eyes.

Arthropods have separate sexes. In some cases the males and females show differences in external appearance and food habits.

The illustrations show members of the main classes of the phylum Arthropoda.

Class Crustacea

All crustaceans have two pairs of antennae, and breathe with gills.

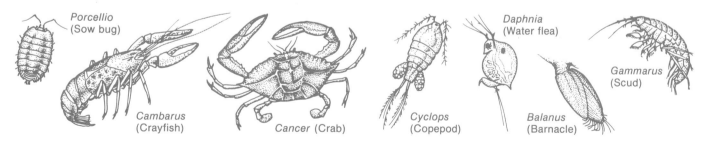

Class Arachnida

All arachnids lack antennae and have four pairs of legs. The head and thorax are fused, making these animals appear to have only two body regions.

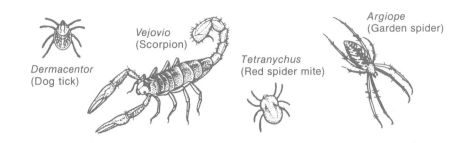

Dermacentor
(Dog tick)

Vejovio
(Scorpion)

Tetranychus
(Red spider mite)

Argiope
(Garden spider)

Class Chilopoda

The centipedes are elongated, visibly segmented, and flattened dorsoventrally. They have one pair of long antennae. Each segment of the body, except the first one and last two, bears *one* pair of legs. Centipedes are predators, feeding on insects, other small arthropods, and worms. They prey mainly at night.

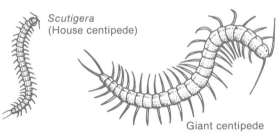

Scutigera
(House centipede)

Giant centipede

Class Diplopoda

The millipedes, like the centipedes, are elongated and visibly segmented. However, they are rounded, instead of flattened. Each body segment bears *two* pairs of legs. Millipedes are scavengers, feeding mainly on dead leaves and other decaying vegetation.

Spirobolus

Julus

Class Insecta

Most arthropods are insects. These animals have distinct head, thorax, and abdomen regions. The thorax consists of three segments. Each segment bears a pair of legs. Insects usually have two pairs of wings. However, some have only one pair and others have none. The wings are on the thorax.

The class Insecta is divided into 26 orders. Examples from the main orders are shown here.

Order	Examples
Odonata Dragonflies and damselflies	Damselfly — Dragonfly
Orthoptera Grasshoppers and crickets	Cricket — Grasshopper
Isoptera Termites	
Trichoptera Caddisflies	Adult — Larva
Diptera True flies	Housefly — Mosquito
Hymenoptera Ants, bees, and wasps	Bee — Wasp — Ant
Ephemeroptera Mayflies	Nymph — Adult
Plecoptera Stoneflies	Adult — Nymph
Hemiptera True bugs	
Lepidoptera Moths and butterflies	Moth — Butterfly
Coleoptera Beetles	

Phylum Echinodermata: Echinoderms

About 5 000 species of animals are classified as echinoderms. Among these are starfish, brittle stars, sand dollars, sea urchins, and sea cucumbers. All echinoderms live in salt water habitats.

Echinoderms have radial symmetry and five-part bodies. Each part usually has numerous projections called "tube feet" for locomotion, feeding, and breathing. Three body layers are present. However, these animals lack heads, brains, and segmentation. The body is covered by a thin epidermis that stretches over a hard **endoskeleton** (inner skeleton) of movable plates.

The digestive system is simple, but complete. Breathing is by tiny gills on the body surface and by the "tube feet". The circulatory and nervous systems are simple. Most echinoderms have separate sexes.

Asterias
(Starfish)

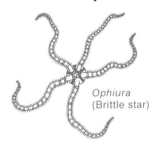

Ophiura
(Brittle star)

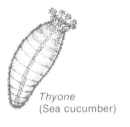

Thyone
(Sea cucumber)

Echinarachnius
(Sand dollar)

Strongylocentrotus
(Sea urchin)

Phylum Chordata: Chordates

You belong to one of the 46 000 species that make up the phylum Chordata. All **chordates** share the following characteristics:

1. Chordates have a **notochord** in the early embryo stage of their development. A notochord is a thin rod of cells containing a jelly-like material. It is surrounded by fibrous supporting tissue and usually extends the length of the body. In some of the lower chordates, the notochord lasts to the adult stage. In other words, it becomes the main axis of support for the organisms.

   About 45 000 species of chordates are in the **sub-phylum Vertebrata**. Fish, amphibians, reptiles, birds, and mammals are **vertebrates**. In these organisms, the notochord is surrounded or replaced by a **vertebral column** in the adult stage. A vertebral column is made up of many bones called **vertebrae** (singular: **vertebra**).

2. Chordates have a **dorsal nerve cord** in the early embryo stage of development. In vertebrates, this nerve cord differentiates to form the brain. Invertebrates (animals without a backbone or notochord) have a *ventral* nerve cord.

3. Chordates have **gill pouches** on the sides of the pharynx in the early embryo stage of development. (Your pharynx is at the very back of your mouth cavity.) In fish, the gill pouches develop into gill slits with gills. In amphibians, they develop into gill slits with gills in the tadpole but disappear when the tadpole undergoes metamorphosis to an adult. In reptiles, birds, and mammals, the gill pouches never become functional.

   The following illustrations show representative animals from the six main classes in the sub-phylum Vertebrata of the phylum Chordata.

Class Chondrichthyes

About 600 species belong to the class of **cartilaginous** fish. These fish have skeletons made of cartilage instead of bone. **Cartilage** is a stiff tissue that can give reasonable support, yet still be somewhat flexible. For example, your nose and ears are supported by cartilage. The sharks, skates, and rays belong to this class. All species in this class breathe with gills. The gill slits are clearly visible on the side of the pharynx.

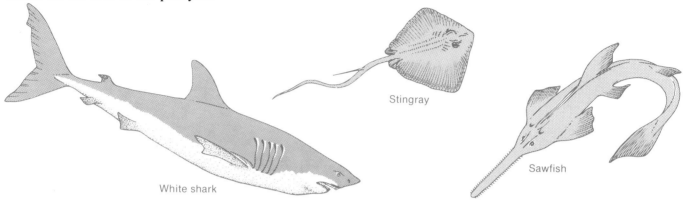

Stingray

White shark

Sawfish

Class Osteichthyes

About 20 000 species belong to this class of vertebrates. They are commonly called **bony fishes**. All species in this class have skeletons that are made chiefly of bone. Also, all species in this class breathe with gills. However, some species called lung fish also have a primitive lung and can breathe air.

Carp

American eel

Rainbow trout

Brown bullhead

Northern pike

Class Amphibia

About 3 000 species make up this class. As the name implies, most species of **amphibians** live partly on land and partly in the water. As a result, they have

Red-backed salamander

American toad

Newt

Leopard frog

Mud puppy

legs and lungs in the adult stage. However, the larvae are aquatic and breathe with gills. Frogs, toads, salamanders, newts, and mud puppies are amphibians. Amphibians have moist skin.

Class Reptilia

About 7000 species of animals are **reptiles**. These animals have dry skin, usually with scales growing out of it. They breathe with lungs at all stages of their development. Turtles, lizards, snakes, alligators, and crocodiles are reptiles. These animals lay eggs that are covered with a soft shell.

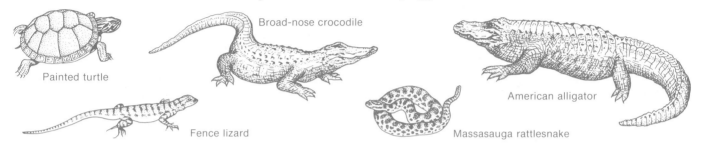

Painted turtle

Broad-nose crocodile

Fence lizard

Massasauga rattlesnake

American alligator

Class Aves

About 8600 species of animals are **birds**. These animals have smooth dry skin with feathers growing out of it. Birds breathe with lungs and lay eggs with hard shells. All birds have wings, though some species cannot fly. Unlike fish, amphibians, and reptiles, birds are **homeothermic**, or "warm-blooded". That means they have a constant body temperature, regardless of the outside temperature.

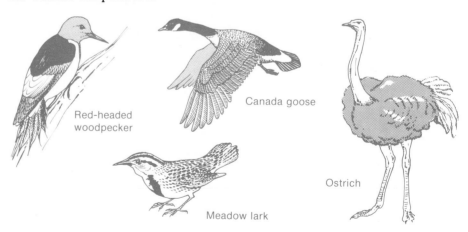

Red-headed woodpecker

Canada goose

Meadow lark

Ostrich

Class Mammalia

About 5000 species of animals are **mammals**. These animals have smooth dry skin with hair growing out of it. Some mammals are covered with hair.

Others, like whales, have very little. The young are fed with milk produced by the mammary glands of the female. Like birds, mammals are homeothermic.

Cat

Platypus

Squirrel

Seal

Elephant

Human

Review

1. Your teacher has prepared several stations around the room. At each station there is an animal that belongs to one of the eleven phyla discussed in this section.

   Visit each station and study the animal closely. Decide which phylum it is in. Record your reasons for placing it in that phylum.

18

Nutritional Requirements, Feeding, And Digestion

In Chapter 2 we discussed the characteristics of living things. We discovered that all living things must be able to obtain and use a constant supply of energy and matter. Energy provides the "power" for life processes such as movement, reproduction, and growth. Matter provides the new "parts" necessary for growth and repair of an organism.

Animals are heterotrophs. They rely on other living organisms. These organisms are sources of raw materials that contain both a usable form of energy and the molecules required to build and maintain the animals. What are the raw materials required by animals? How do they obtain these materials? How do animals use the food that they acquire? In this chapter we will discuss answers to these questions.

18.1 Nutritional Requirements of Animals

Nutrients *are the raw materials which provide the essential matter and energy for life.* Nutrition *refers to all the processes by which living organisms obtain and use these nutrients.* Some nutrients are called organic nutrients since they contain carbon and are made by living organisms. Organic nutrients include carbohydrates, fats, proteins, and vitamins. Other nutrients such as water and minerals are called inorganic nutrients.

Carbohydrates

Carbohydrates are chemical compounds made of carbon, hydrogen, and oxygen. They contain hydrogen and oxygen in the same proportion as in water, which is two to one. Carbohydrates are a major energy source for animals. Sugars, starches, and cellulose are carbohydrates.

Some sugars, you may recall, are monosaccharides (simple sugars). Glucose, fructose, galactose, and mannose are examples. They require no processing in order to be absorbed by a living organism.

Other sugars are disaccharides. These consist of two simple sugar units that are joined. Some examples are sucrose (one molecule of glucose linked to one molecule of fructose), lactose (one molecule of glucose linked to one molecule of galactose), and maltose (two molecules of glucose linked together). Before disaccharides can be used by living systems, they must be broken down into their simple sugar units.

Polysaccharides such as starch and cellulose contain many simple sugar units linked together. Starch must be broken down into its simple sugar units before it can be taken in and used by a living organism. Cellulose cannot be digested by humans and many other animals. However, it does provide essential roughage or bulk which aids in digestion. Such roughage stimulates the intestinal wall to contract, thereby forcing the food along the digestive tract. In herbivores, the digestive system contains millions of microscopic organisms which digest the cellulose of plant cell walls.

Fats

Fats are similar to carbohydrates in that they also contain only carbon, hydrogen, and oxygen. They also serve as sources of energy. However, the proportions of the three kinds of atoms are different. There is a much higher proportion of hydrogen in fats. A gram of fat may produce over two times as much energy as a gram of carbohydrate. A fat molecule consists of one molecule of glycerol and three molecules of fatty acid. During digestion, fats are broken down into these molecules.

Besides being a valuable energy source, fats may be stored beneath the skin and around internal organs. There they serve as an insulating material and as an energy storehouse. In addition, fats are solvents for certain vitamins.

Proteins

Proteins are large complex organic molecules that are made of amino acid units linked together. All amino acids contain carbon, hydrogen, oxygen, and nitrogen. Some proteins also contain sulfur, phosphorus, and even iron. During digestion proteins are broken down into their component amino acid units.

All living organisms contain proteins. Proteins have a number of functions in living organisms. Some are enzymes that aid processes such as digestion. Others are parts of many structural components of cells, such as cell membranes. Proteins can also serve as an energy source. During fasting, proteins in muscles may provide energy once the carbohydrate and fat stores have been used up.

Most animals do not have the ability to manufacture all their own amino acids. Those which they cannot produce on their own must be obtained from the food that they eat. Such amino acids are called essential amino acids. About twenty kinds of amino acids are commonly found in the various proteins of living organisms. About eight are essential for almost all animals.

Vitamins

In addition to carbohydrates, fats, and proteins, vitamins are also important organic nutrients for living organisms. However, they are usually needed in only very small amounts. Vitamins do not have to be broken down before they can be used. The vitamin requirements in diets vary from animal to animal. For example, ascorbic acid (vitamin C) is required by man but not by most other mammals. Most mammals can make their own ascorbic acid from other nutrients.

Table 18-1 lists some important vitamins and their roles in the human organism. Not all vitamins are present in any one food source. One must have a well-balanced diet in order to get all the vitamins. When a particular vitamin is not present in a person's diet, that person may suffer from a deficiency disease. Also, too much of a particular vitamin may prove harmful. Vitamin A is one vitamin which, in humans, must be supplied in just the right amounts. Too little can result in retarded growth, improper tooth formation, and night blindness. Too much can cause bone fragility, enlargement of the liver and spleen, and loss of hair. A carefully balanced diet is needed in order to prevent both deficiency diseases and diseases resulting from excessive amounts of certain vitamins.

VITAMIN	SOME KNOWN FUNCTIONS	SOME COMMON SOURCES
Fat soluble vitamins		
Vitamin A	—aids in maintaining resistance to infection —stimulates new cell growth —delays senility	—leafy and yellow vegetables —egg yolk —milk —liver —cod and halibut liver oil
Vitamin D (Sunshine vitamin)	—aids in building and maintaining bones and teeth —aids in absorption of calcium and phosphorus from intestinal tract	—eggs —milk —cod and halibut liver oil —fresh fish
Vitamin E	—essential for normal reproduction —maintains cell membranes	—corn, soybean, cottonseed, and peanut oil —wheat germ —leafy vegetables
Vitamin K	—essential for normal blood clotting	—spinach —alfalfa —soybeans —pork liver
Water soluble vitamins		
Vitamin B_1	—necessary for growth, fertility, and lactation —essential for normal nerve function	—whole grain cereals —yeast —milk —green vegetables —egg yolk
Vitamin B_2 (Riboflavin)	—plays a role in light adaption of eyes —essential for carbohydrate metabolism —essential for healthy normal skin	—lean meat —wheat germ —yeast —milk
Vitamin C (ascorbic acid)	—plays a role in tooth and bone formation —essential for normal capillary function	—citrus fruits —green vegetables —strawberries

TABLE 18-1 Some Vitamins Important to Humans: Functions and Sources

Water

A large portion of all living matter is water. For example, about 70% of your body mass is water. Water has many functions in living organisms. The main ones are:

1. It aids in the digestion and absorption of food.
2. It is a medium of transport within an organism.

3. It acts as a lubricant.
4. It helps in the excretion of harmful by-products of metabolic processes.
5. It aids in the regulation of heat loss. (Evaporation from the surface of a body causes cooling.)

Most organisms get much of the water they need by taking in water directly. They also get it indirectly from foods that they eat, since most foods contain water. They may also get some water as a by-product of certain chemical reactions that occur in living cells, such as cellular respiration.

Minerals

The four most common elements found in organisms are carbon, hydrogen, oxygen, and nitrogen. However, a number of other elements are needed in small amounts to maintain life functions. These important inorganic minerals are often classified into two groups, major and trace minerals. The **major minerals** include calcium, phosphorus, potassium, magnesium, sodium, chlorine, and sulfur. **Trace minerals**, or those present in very small amounts in living organisms, include iron, iodine, zinc, copper, and manganese. At least 14 trace minerals are necessary for the complete nutrition of most vertebrate animals.

The functions performed by some of the minerals are well known. For example, calcium is an important bone and tooth building mineral in vertebrates. Many invertebrate structures, such as coral and the shells of clams are made mainly of calcium carbonate. Calcium also plays an important role in muscle and nerve function. Milk, eggs, cheese, and vegetables such as beans, peas, and asparagus contain significant amounts of calcium readily available for human nutritional needs.

Phosphorus is important in the bone building process as well. It also forms part of a number of high energy organic compounds important in metabolism. Sources of this mineral include milk, eggs, liver, and various grains.

Iron is part of a number of substances that control cell activities such as cellular respiration. It is also found in hemoglobin, the oxygen-carrying material in the red blood cells of many animals. A deficiency of iron in humans results in a lack of energy and, in some cases, **anemia**. Such deficiency diseases are rare, since iron is not easily eliminated from the body but, instead, is recycled. Growing animals need additional iron and it must be supplied through their diets. For human dietary needs, iron is present in large amounts in such foods as organ meats (liver, heart), whole wheat, raisins, spinach, and lettuce.

The functions of some other minerals found in living systems are not well known. For example, lithium, rubidium, aluminum, boron, beryllium, and bromine have been found in living organisms. They may have some functional importance, but there is a lack of any definite evidence on the subject. Also, scientists do not yet know whether or not all animals have exactly the same mineral requirements.

In summary, carbohydrates, fats, proteins, vitamins, minerals, and water are the basic nutrients needed by all animals. Both the form in which they are obtained and the relative amounts required vary greatly from one animal species to another. However, one thing is clear: every animal must be able to meet its specific nutritional needs or it will die.

Review

1. **a)** Define the terms heterotroph, nutrient, and nutrition.
   b) What nutrients are referred to as organic nutrients? Why?
2. **a)** Why are carbohydrates important to animals?
   b) How do disaccharides and polysaccharides differ? How are they alike?
   c) Raffinose contains 1 molecule of glucose, 1 molecule of fructose, and 1 molecule of galactose. What class of nutrient is it? Hint: Read the section on carbohydrates. Then make up a name for this class of nutrient. Consult books in the library to find the source of raffinose.
3. **a)** What basic molecules do fats consist of?
   b) What role do fats play in animals?
4. **a)** What are the basic building blocks of proteins?
   b) What role do proteins play in animals?
5. Why is it necessary to have close to the right amount of vitamins in a diet?
6. Why is water so important to animals?
7. **a)** Describe the functions of two major minerals.
   b) Describe the functions of one trace mineral.
8. Oils are present in living organisms. Visit the library and find out what they are. How are they classified? In what kinds of living organisms are they found? What is their nutrient role?
9. Why would a loss of blood result in a deficiency of iron?

18.2 Investigation

Tests For Nutrients

In this investigation you will test for the presence of various nutrients in some foods.

Part A   Test for Starch

Materials

dry cracker	test tubes (3)	mortar and pestle
potato	test tube rack	adjustable clamp
apple	iodine solution	Bunsen burner

CAUTION: Wear safety goggles during this investigation.

Procedure

a. Crush a dry cracker in a mortar containing about 20 mL of water.

b. Pour the mixture into a test tube. Heat the mixture gently until it begins to boil.

c. Cool the test tube. Then add 3 drops of iodine solution (Fig. 18-1). Observe and record the resulting colour.

d. Repeat steps a to c, first with the potato and then with the apple. Use about 1 cm³ of each food.

Discussion

1. Iodine reacts with starch to produce a blue colour. (Sometimes the colour may be deep purple or almost black.) Which substances contain starch?

2. Why was the mixture boiled before the test was done?

Part B   Test for Simple Sugars

Materials

10% glucose solution
10% maltose solution
10% sucrose solution
test tubes (3)
Benedict's solution
hot water bath
test tube holder

CAUTION: Wear safety goggles during this investigation.

Procedure

a. Pour about 10 mL of the 10% glucose solution into a test tube. Then add 5 mL of Benedict's solution.

b. Place the test tube in the hot water bath. Observe and record any colour changes that occur within the next 5 min.

c. Repeat steps a and b, first with the 10% maltose solution and then with the 10% sucrose solution.

Discussion

1. Benedict's solution reacts with a simple sugar to produce a reddish-orange colour. Which solutions contained a simple sugar?

2. If a food sample contained a disaccharide would it give a reddish-orange colour when boiled with Benedict's solution? How do you know?

Fig. 18-1   The starch test. Fairly reliable results are also obtained by adding iodine solution directly to foods (no crushing and boiling). You may have iodine solution in the medicine cabinet at home. If so, test bread and other foods that you think contain starch.

Part C   Test for Protein (Biuret Test)

Materials

10% glucose solution   test tubes (3)
egg white   test tube rack
distilled water   Biuret reagent

CAUTION: Wear safety goggles during this investigation.

Procedure

a. Place 15 mL of 10% glucose solution in a test tube. Add 10 mL of the Biuret reagent. Turn the test tube around a few times to mix the contents.
b. Place the test tube in the test tube rack. Observe and record the resulting colour.
c. Repeat steps a and b, first using the egg white and then using the distilled water.

Discussion

1. This test is called the Biuret Test for Proteins. The Biuret reagent reacts with a protein to produce a violet colour. Which test substances contained protein?
2. Why was distilled water used as one of the test substances?

Part D   Test for Fats and Oils

Materials

crushed peanuts   vegetable oil
potato   filter paper

Procedure

a. Rub a small quantity of the crushed peanuts on a piece of filter paper. Wait 2 min and then hold the paper up to a light. Observe and record what you see.
b. Repeat step a, first using the potato and then the vegetable oil.

Discussion

If a translucent (semi-transparent) spot forms on the filter paper where the food was rubbed, the food contains a fat or oil. Which of the foods contain fat or oil?

Part E   Test for Vitamin C

Materials

indophenol solution	C-plus orange drink	distilled water
fresh orange juice	orange soda pop	medicine dropper
0.5% ascorbic acid	test tubes (3)	
solution	test tube rack	

Procedure

a. Place 10 drops of indophenol solution in a clean dry test tube.

b. Add 0.5% ascorbic acid solution to this test tube 1 drop at a time until no further colour change occurs. Record the resulting colour change. Record also the number of drops of ascorbic acid solution it took to bring about this colour change.

c. Repeat steps a and b, first testing fresh orange juice, then C-plus orange drink, then orange soda pop, and finally, the distilled water.

Discussion

1. a) What colour does indophenol produce in distilled water? This colour shows that no Vitamin C is present.

  b) In the presence of Vitamin C the indophenol becomes colourless. Which substances contain Vitamin C?

2. The greater the number of drops of test substance needed to make the indophenol colourless, the less Vitamin C is present. Which substance contained the most Vitamin C? Why?

Part F   Test for the Presence of Minerals

Materials

potato	deflagrating spoon
sugar cube	Bunsen burner
apple	

CAUTION: Wear safety goggles during this investigation.

Procedure

a. Place about 1 cm³ of potato in a deflagrating spoon. Heat the spoon and potato until no further change occurs. Describe the contents of the spoon. Record your results.

b. Repeat step a, first using the sugar cube, then 1 cm³ of apple.

Discussion

1. The ash that remains after intense heating consists of minerals. Which substances contained minerals?
2. **a)** What other nutrients besides minerals are present in potatoes? in apples? in sugar?
   b) Why do only minerals remain after intense heating?

For Further Investigation

1. An unknown substance gave the nutrient test results shown in Table 18-2.

REAGENT	RESULT
iodine	red brown colour
indophenol	became colourless
Benedict's reagent	green colour
Biuret reagent	violet colour

TABLE 18-2 Results of Nutrient Tests

   a) According to the tests, what nutrients does the unknown substance contain?
   b) Would these be the only nutrients that the substance contains? Explain your answer.
2. Your teacher may choose to give you some time in the laboratory to test some of the foods you eat. If so, bring samples of two or three foods that you eat to the laboratory for testing.

18.3 Feeding

As you know, the form and relative amounts of nutrients that organisms require vary with the type of organism. The methods by which these nutrients are obtained by organisms are also quite different. In this section we will study several feeding mechanisms.

The term **feeding** generally refers to *the process by which organisms obtain their food.* It includes the method by which the food is captured and the method by which the food is taken into an organism's body. *The taking in of food is also called* **ingestion**.

The methods that organisms use to get food are classified into three major groups according to the form in which the food is found:

1. Methods for ingesting very small particles;
2. Methods for ingesting large food masses;
3. Methods for ingesting fluids or soft tissues.

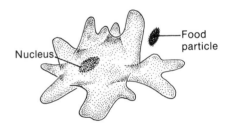

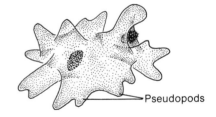

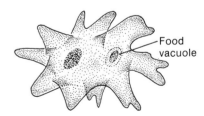

Fig. 18-2 Feeding by an amoeba. Pseudopods surround the food particle. Eventually a food vacuole is formed and ingestion is complete.

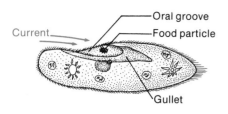

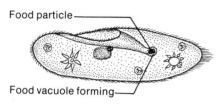

Fig. 18-3 Feeding by a parmecium. Food is carried into the oral groove by currents created by ciliary action. The food eventually enters into the paramecium within a food vacuole.

Methods for Ingesting Very Small Particles

Some aquatic organisms feed on bacteria, microscopic algae, and small invertebrate animals. For example, the amoeba engulfs food particles in the surrounding water. Its protoplasm forms **pseudopods** (false feet) which simply flow over and around the food until it is captured in a food vacuole (Fig. 18-2). Eventually, the food is digested or broken down in this vacuole. The required nutrients are then absorbed into the amoeba's single-celled body.

Several species of organisms use hair-like projections called **cilia** to create currents that carry food particles into a "mouth". For example, Figure 18-3 shows how the paramecium uses ciliary currents to sweep food particles into the oral groove ("mouth") and along the gullet. Food vacuoles form at the end of the gullet.

Many species of animals use filtering devices to trap food. These animals are called **filter feeders**. Suspended or free floating food particles are strained out of the water by a special filter. They are then passed along to the animal's mouth for ingestion. The crustacean *Daphnia* and molluscs such as

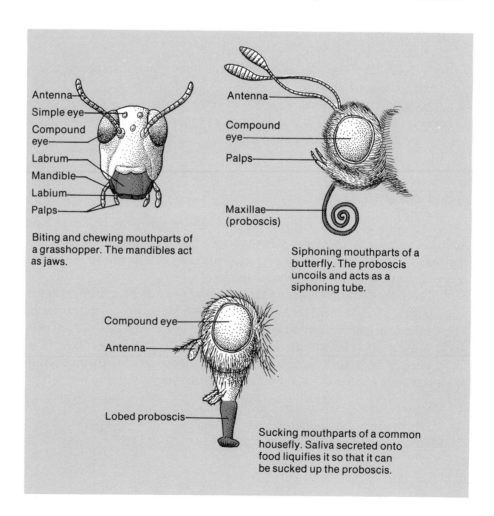

Biting and chewing mouthparts of a grasshopper. The mandibles act as jaws.

Siphoning mouthparts of a butterfly. The proboscis uncoils and acts as a siphoning tube.

Sucking mouthparts of a common housefly. Saliva secreted onto food liquifies it so that it can be sucked up the proboscis.

Fig. 18-4 Mouthparts of various insects. The structures are adapted for different types of feeding.

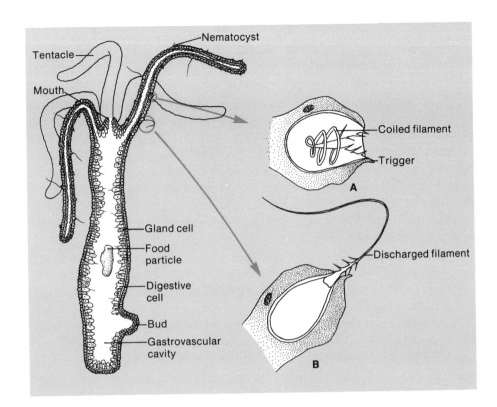

Fig. 18-5 The hydra: a longitudinal section showing detail of cellular structure. Note the detail of the nematocyst before discharge (A) and after discharge (B).

Labels in figure: Nematocyst, Tentacle, Mouth, Coiled filament, Trigger, Discharged filament, A, B, Gland cell, Food particle, Digestive cell, Bud, Gastrovascular cavity

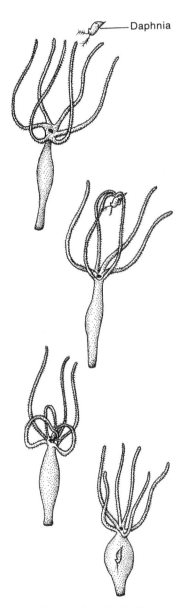

Label: Daphnia

Fig. 18-6 Feeding by a hydra. Here a *Daphnia* is trapped by the tentacles. It is then drawn into the hydra's gastrovascular cavity.

clams feed in this manner. A number of chordates are also filter feeders. Basking and Whale sharks feed only on plankton which are trapped in comb-like filters attached to their gills. Basking sharks may grow to be 10 m long. Obviously this method of feeding is quite efficient. A few birds are also filter feeders. Flamingos have beaks that strain small organisms from the water. One mammal, the Blue whale, filter feeds only on plankton. It is the largest living animal.

Methods of Ingesting Large Food Masses

Animals show a wide variety of methods for ingesting large particles or masses of food. Some animals ingest large quantities of material such as soil. They then digest the living or dead organic matter found in it. The digested organic matter meets the nutrient requirements of such animals. Annelids (segmented worms) such as the earthworm feed in this manner. The earthworm literally eats its way through the soil in which it lives. It gets nutrients from the soil, and leaves behind a trail of casts. You may have seen this undigested waste matter around earthworm burrows.

Many plant-eating invertebrates or herbivores feed by scraping, chewing, or boring into large plant masses. This method of feeding is characteristic of many insects. Biting and chewing mouthparts are part of their feeding mechanism. Grasshoppers, crickets, and leaf-hoppers are examples. Termites can bite off small portions of wood from tree trunks and wooden beams. Powerful and sturdy mouthparts enable termites to feed in

Fig. 18-7 To aid in feeding, a frog has a long tongue attached to the front of its mouth. The tongue can be flicked out to grasp the prey. Note that the two flexible lobes of the tongue grasp the prey and hold it until both prey and tongue are back in the mouth.

this manner. Compare the biting and chewing mouthparts of the grasshopper with the mouthparts of the butterfly and housefly (Fig. 18-4).

Most carnivores have specialized structures for capturing and swallowing prey. For example, the hydra makes use of **nematocysts** or "stinging cells" to paralyze its prey (Fig. 18-5). The nematocysts are located in specialized cells along the outer cell layer of the hydra's tentacles. They are thread-like filaments coiled within a capsule with a hair-like trigger sticking outward. When the prey touches the trigger, the nematocyst ejects the coiled filament. Some nematocysts are barbed and can penetrate the prey's body. Others are whip-like and can wrap around the victim. A paralyzing poison is then injected into the victim through the filament. Tentacles seize the prey and carry it to the hydra's mouth. There it is drawn into the **gastovascular cavity** by contractions of the body wall (Fig. 18-6).

Other than the hydra, very few invertebrates capture and swallow their prey whole. However, many vertebrate carnivores feed in this way. Included are most carnivorous fish, all adult amphibians (Fig. 18-7), many reptiles, many birds, and a few mammals.

In some cases the prey, once seized, is first mechanically broken down by chewing before swallowing. Many animals have teeth and powerful jaws and mouths for this purpose. Other animals lack teeth but have a special compartment called a **gizzard**. The gizzard has muscular walls which mechanically break up the food substance by strong contractions. The earthworm has a gizzard (Fig. 18-8). Birds also have a gizzard. Before the food gets to the gizzard, it enters the crop. There it is temporarily stored and moisturized before it enters the gizzard (Fig. 18-9). Fine gravel is sometimes swallowed by birds. It remains in the gizzard to aid in the grinding of hard food materials such as seeds.

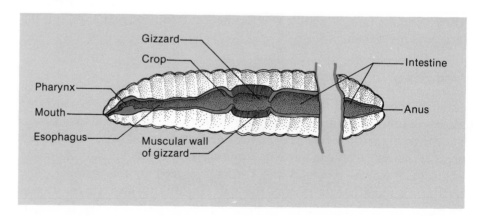

Fig. 18-8 The earthworm's digestive system. Mechanical breakdown of the food occurs in the gizzard.

Methods for Ingesting Fluids or Soft Tissues

Nutrients are not always in the form of small particles or large solid masses. They may also be in the form of a fluid or very soft tissue. Many animals get their nutrients by feeding on nutrient substances in these forms. Such animals are called **fluid feeders**.

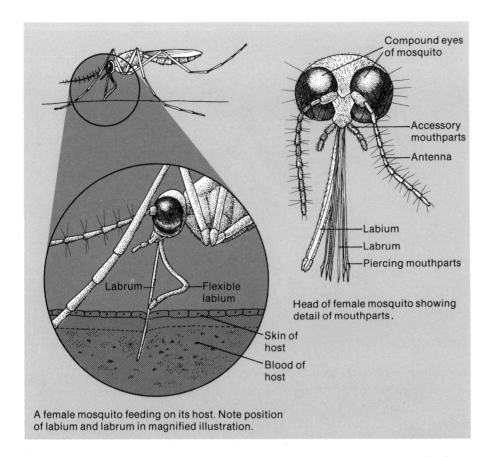

Compound eyes of mosquito

Accessory mouthparts

Antenna

Labium

Labrum

Piercing mouthparts

Head of female mosquito showing detail of mouthparts.

Labrum — Flexible labium

Skin of host

Blood of host

A female mosquito feeding on its host. Note position of labium and labrum in magnified illustration.

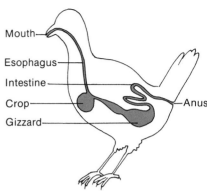

Mouth

Esophagus

Intestine

Crop

Gizzard

Anus

Fig. 18-9 Digestive system of a bird. A bird has no teeth. Thus food is swallowed, temporarily stored in the crop, and eventually broken down mechanically in the gizzard before it enters the intestine.

Fig. 18-10 Feeding by a female mosquito. Mouthparts pierce the skin of the prey and blood is pumped up the hollow labrum.

The mosquito is the fluid feeder most familiar to you. It usually feeds on plant juices. It has mouthparts which include a needle-like tube that can penetrate plant tissue. Juices containing nutrients are then sucked up the tube. The female mosquito often gets needed protein from the blood of animals. The mouthparts of the female are specially adapted for this type of feeding (Fig. 18-10). A grooved, flexible **labium** encloses a number of needle-like structures which pierce the skin of the animal to enable feeding. One of these structures, the **labrum**, is hollow. It serves as a "straw" through which the victim's fluids can be drawn into the mosquito. Once the mosquito has penetrated the skin, a liquid is injected into the animal to prevent the blood from clotting. Clotting would plug up the labrum. It is this fluid that causes the itching of a mosquito bite.

Spiders often feed on very large prey with hard outer body coverings. To do so, spiders simply pierce the prey's body and inject digestive juices into it. The juices liquify the inner tissues of the prey. The resulting fluid is then sucked up into the spider's body.

The phylum Platyhelminthes ("flatworms") consists mostly of fluid and soft tissue feeders. The planarian and the tapeworm both belong to this phylum. The planarian has a tube-like branched digestive system called a **gastrovascular cavity** (Fig. 18-11). The mouth opening is located on the ventral (lower) surface about half way between the head and the tail region. During feeding, a tube-shaped, muscular **pharynx** is extended through the

Fig. 18-11 The digestive system of a planarian. The sucking action of the pharynx breaks up the food into very small pieces. These, along with fluids, are taken into the gastrovascular cavity. They are then engulfed by cells lining the cavity.

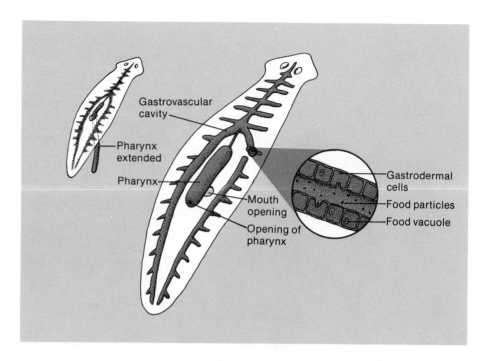

mouth opening onto soft tissue food substances. The food is torn into microscopic pieces by sucking movements of the pharynx. These small food particles, along with tissue fluids are then sucked into the gastrovascular cavity.

The tapeworm is another specialized fluid feeder belonging to the phylum Platyhelminthes. It has no digestive tract of its own. Instead, it absorbs already digested nutrients across its entire body surface from the intestine of the animal in which it is living.

You have seen that animals can obtain their food in a variety of ways. Yet, in most cases, the food they ingest is not in a form that can be used by the animal. Further breakdown is required to reduce the ingested food into basic units such as simple sugars and amino acids.

Review

1. a) What is the difference between feeding and ingestion?
   b) Methods of feeding are grouped into three major categories. On what basis are these groupings determined?
2. a) Describe the mechanism by which an amoeba feeds.
   b) What are cilia? What role do they play in feeding?
   c) Using an example, describe how filter feeders obtain nutrients.
3. a) How does the earthworm get nutrients?
   b) What are nematocysts? What role do they play in feeding?
   c) Using an example, describe the role a gizzard plays in the feeding process.
4. a) What are fluid feeders?
   b) How is the tapeworm adapted for feeding according to its way of living?

18.4 Investigation

Feeding and Digestion in the Paramecium

According to the 5-kingdom system of classification, the paramecium is not an animal. It is included here because it is a very simple organism. Therefore it is easy for you to observe feeding and digestion.

All heterotrophic organisms, single-celled or multicellular, must be able to obtain food from their environment. They must also be able to digest the food so that essential nutrients in the food become available to the organisms. As we have seen, the feeding mechanisms that organisms use to obtain food depend on the form the food is in. Once ingested, the food is digested or broken down into basic units which are small and soluble. They can then be absorbed into the organism and put to their necessary use.

In this investigation you will observe the feeding mechanism used by the paramecium, a single-celled organism. You will also observe the process of digestion that occurs within the food vacuoles of the paramecium following ingestion of the food. The food will be yeast that is coloured with a dye called Congo red.

Materials

paramecium (living culture)
yeast, in Congo red solution
microscope
microscope slide (depression type)

cover slip
medicine droppers (2)
1.5% methyl cellulose
toothpick

Procedure

a. Using a medicine dropper, place 2 drops of the paramecium culture into a depression slide.
b. Using a second medicine dropper, add a drop of methyl cellulose to the culture. The methyl cellulose will slow down the movement of the paramecia, making them easier to observe.
c. Add a small amount of yeast that has been dyed with Congo red to the culture. Add only the amount you can get on the end of a toothpick. Cover the depression with a cover slip.
d. Place the prepared slide under low power of the microscope. Find a paramecium that is not moving too much. Once you have located a paramecium switch to medium, then high power.
e. Locate the cilia, oral groove, and gullet (Fig. 18-12).
f. Observe the feeding process and the formation of a food vacuole. The dyed yeast cells will be visible within the vacuole. BE PATIENT.
g. Observe a newly formed food vacuole for several minutes. You should observe a colour change in the yeast cells within the food vacuole. How long does the colour change take to occur?

Fig. 18-12 Paramecium. Note the food vacuole forming at the end of the gullet.

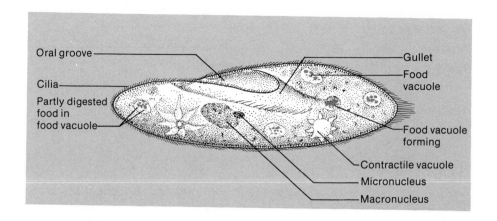

Oral groove
Cilia
Partly digested food in food vacuole
Gullet
Food vacuole
Food vacuole forming
Contractile vacuole
Micronucleus
Macronucleus

Discussion

1. How is the food (yeast) moved into the oral groove of the paramecium?
2. **a)** What is the likely reason for the colour change that occurred in the food vacuole? (Note: Congo red turns blue in the presence of an acid.)
   b) Do further colour changes occur?
3. **a)** What will eventually happen to the ingested yeast?
   b) What will eventually happen to the food vacuoles?
4. You will recall that food vacuoles are bounded by a membrane. How do required nutrients resulting from digestion within the vacuole get into the cytoplasm of the paramecium where they can be used?
5. Is the paramecium's digestive process intracellular or extracellular? Explain. If you cannot answer this question, read the next section.

18.5 Digestion

Digestion *is the process by which large food molecules are chemically broken down into simple basic nutrient units.* Required basic nutrient units are then absorbed into the body. There they can be used as building blocks for new living matter. They can also be used as sources of energy for the many functions of an organism. Digestion is accomplished with the aid of organic compounds called **enzymes**. These cause the breaking of certain bonds that join the basic nutrient units in the large food molecules.

Intracellular and Extracellular Digestion

Digestion may be **intracellular**. In this case, a food particle is acted on by digestive enzymes within a cell. For example, in the amoeba and paramecium, ingested food is enclosed in a food vacuole where digestive enzymes chemically break it down. Digestion is going on within the cell.

Extracellular digestion is a characteristic of most multicellular animals. In these animals, food is broken down in a digestive cavity of some kind. The resulting basic unit nutrients are then absorbed into the body and eventually into all cells.

Intracellular digestion is clearly limited to very small food particles. They must be small enough to be ingested by individual cells. Extracellular digestion enables an organism to get food from relatively large food particles.

In the paramecium, digestion occurs intracellularly. Food particles are swept into the oral groove (see Fig. 18-12). They collect in the gullet and eventually are enclosed in a food vacuole. The resulting vacuole moves within the paramecium's body, through the cytoplasm, toward the anterior end of the cell. Digestion of the food within the vacuole occurs when enzymes are secreted into the vacuole. Required basic nutrient units such as sugars and amino acids are formed. These diffuse across the vacuolar membrane into the cytoplasm of the cell. Unused food fragments are released from the food vacuole, once it reaches the anal pore and bursts open.

The hydra (see Fig. 18-6) ingests larger food particles. For example it may ingest whole, although small, crustaceans. Once the food is in the gastrovascular cavity, digestive enzymes secreted into the cavity, by specialized cells lining it, begin to break down the food particles. This extracellular digestion does not result in basic nutrient units. Instead, special cells called **gastrochemical cells** engulf the food fragments in an amoeba-like way. Food vacuoles are formed and digestion is completed intracellularly. Unused remains are discharged into the gastrovascular cavity. These are then expelled from the hydra through the mouth.

You may recall that, during feeding, a planarian ingests microscopic particles of food into its gastrovascular cavity. These particles are so small that they can be taken directly into cells lining the cavity in much the same manner that an amoeba takes in food. As a result, almost no digestion takes place within the gastrovascular cavity. Instead, intracellular digestion takes place in the food vacuoles formed within the cells lining the cavity. Once digestion is complete, usable nutrients diffuse to all cells of the body. Undigestible and unusable food particles are egested (expelled) from the planarian through the single opening in the pharynx.

Alimentary Canal Systems

Most complex animals have a one-way digestive tube or **alimentary** canal system (Fig. 18-13). This system has an opening, the **mouth**, for ingestion of food at one end. It also has a separate opening, the **anus**, for egestion of undigested remains at the other end. An alimentary canal system has specialized sections within it. These sections absorb, physically break down, chemically digest, and store food.

The earthworm shows this specialization within its alimentary canal system (see Fig. 18-8). Soil is sucked in through the mouth by the muscular **pharynx**. It then moves to the **esophagus** and into the **crop** where it is temporarily stored. The food then enters the **gizzard** and undergoes physical breakdown. Next it is mixed with water and forced along the **intestine**. In the intestines, enzymes are secreted onto the food, and chemical digestion takes

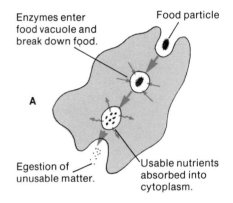

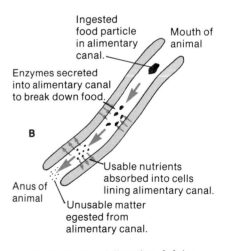

Fig. 18-13 Types of digestion. A: intracellular digestion in an amoeba. B: extracellular digestion in an animal with a one-way digestive tube or alimentary canal.

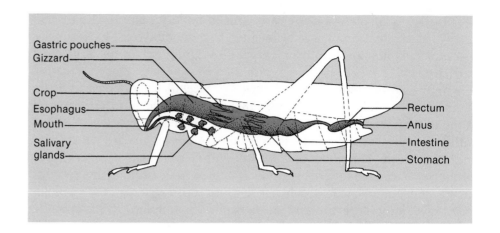

Fig. 18-14   The digestive system of a grasshopper.

Gastric pouches
Gizzard
Crop
Esophagus
Mouth
Salivary glands
Rectum
Anus
Intestine
Stomach

place. Required basic nutrient units are absorbed through the intestinal wall into the earthworm's blood system. Unused material is expelled from the canal by way of the **anus**. This undigested material forms worm casts at the entrance to burrows.

The grasshopper also has an alimentary canal system (Fig. 18-14). Specialized mouth parts mechanically break down the food into smaller pieces during feeding. **Salivary glands** secrete saliva which is mixed with the food in the **mouth**. Saliva aids in the passage of food along the **esophagus**. It also contains enzymes which begin chemical digestion of the food. Food passes along the esophagus into a temporary storage organ called the **crop**. The crop enables the grasshopper to eat large quantities of food at one time and then digest it later. The "brown juice" that a grasshopper spits when disturbed is really a mixture of partly digested food and saliva forced out of the crop.

In the muscular **gizzard**, food is further broken down by the shredding and grinding action of teeth-like plates. The food then enters the stomach. Six pairs of **gastric pouches**, located outside the stomach wall, manufacture a digestive juice containing enzymes. These enzymes are secreted into the **stomach**. Once mixed with the food, they complete the digestive process. Usable digested nutrients are absorbed into blood vessels surrounding the stomach wall. The remaining food passes into the **intestine**, through the **rectum**, and out the **anus**.

Vertebrate Digestive System

All vertebrates have the same basic design for an alimentary canal or digestive system:

1. a mouth for ingestion;
2. an esophagus, which may include a storage area or crop;
3. a region for initial digestion;
4. a region for final digestion and nutrient absorption;

5. a region for water absorption;
6. a region for waste elimination.

We have already dealt with the mouth region in the section on feeding. Variations occur in mouth structure, depending on the animal's method of feeding. In some vertebrates such as birds, the esophagus contains a storage area, the **crop**. In most others it is simply a connecting tube between the mouth and the region of initial digestion.

In many animals such as mammals, initial digestion occurs in a structure called the **stomach**. The food is mixed by muscular contractions. During mixing, enzymes are secreted by the stomach wall to begin chemical breakdown of the food.

Final digestion occurs in the **small intestine**. Absorption of required nutrients also occurs in this region.

Water absorption occurs in the **large intestine**. Wastes are formed into **feces**. The feces collect in the **rectum** and are eventually eliminated from the canal through the **anus**.

Rabbits and many other small herbivores digest cellulose in a large sac called the **caecum** which projects from the large intestine. Of special note are a certain group of large grazing mammals which are also able to digest and use cellulose as a nutrient. The group includes deer, sheep, goats, giraffes, moose, buffalo, camels, and cattle. All chew their cud (rumination) and are therefore called **ruminants**. These animals swallow a great deal of plant matter at one feeding. Then, over a long period of time, they regurgitate the swallowed food and chew it thoroughly before swallowing it again.

The ruminant stomach is divided into four distinct compartments: the rumen, the reticulum, the omasum, and the abomasum (Fig. 18-15). When first swallowed, most of the food enters the **rumen**. This is the largest of the four compartments. In it, food is acted upon by bacteria and protozoa that normally live there. These organisms get their food energy by breaking down the cellulose in the swallowed food matter. The food is not chewed much when it is first swallowed. Therefore the ruminant brings rumen contents back into the mouth and thoroughly chews the solid food matter before reswallowing. This behaviour is leisurely and may occupy up to 8 h of every day. Once thoroughly chewed, the reswallowed food enters the **reticulum** where it undergoes further digestion. The smaller fragments of food enter the **omasum**. Some usable nutrients are absorbed there. The remaining food moves into the **abomasum**. Further enzyme action completes the digestive process within this compartment. The food matter now passes into the **intestines**. Once all usable nutrients are absorbed, waste is eliminated from the digestive tract through the **anus**.

In the next section we will study the structure and function of the human alimentary canal system. We will also study the process of digestion as it applies to humans.

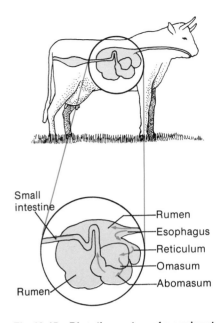

Fig. 18-15 Digestive system of a ruminant animal. The stomach is divided into four compartments: the rumen, reticulum, omasum and abomasum. The arrows indicate the path of food through the stomach.

1. **a)** What is digestion?
   b) What is the function of an enzyme in digestion?
   c) Using an example, explain the difference between intracellular and extracellular digestion.
2. **a)** Explain how a paramecium digests its food.
   b) Digestion in the hydra can be referred to as partly extracellular and partly intracellular. Explain why.
3. **a)** What is an alimentary canal system?
   b) Describe the alimentary canal system of an earthworm. Outline the role of each section in the process of digestion.
4. **a)** What are the basic parts of vertebrate digestive systems?
   b) What role does each part play in terms of digestion?
   c) Using an example describe how the structure of the digestive system in vertebrates is adapted for specific types of food eaten.
5. **a)** What is a ruminant?
   b) Describe fully the relationship that exists between the bacteria and protozoans in the rumen and the ruminant itself. What term describes this type of relationship?

18.6 The Human Digestive System: An Introduction

Your digestive system is a food processing system (Fig. 18-16). It is a healthy system if proper nutrients travel from outside your body into the bloodstream. These nutrients can then be carried to each cell in your body. Clearly, the bloodstream cannot transport large chunks of food. Thus the function of the digestive system is to break down food into particles small enough to be absorbed and transported by the bloodstream.

Most of the food processing in humans occurs in a 10 m muscular tube. This "food tube" runs through the middle of the body between openings much less than 1 m apart. This means that the food tube in humans, unlike the earthworm's, is not straight. Instead, it is coiled and folded back upon itself at least five times. Other names for the food tube are alimentary canal, digestive tube, and gastrointestinal tract.

Associated with the alimentary canal are other structures that help with food digestion. Some parts, like the tongue and teeth, help by mechanically moving, mixing, and grinding the food. Others, like the liver and pancreas, add enzymes and salts to help chemically digest the food.

Structural Regions

The alimentary canal of humans can be divided into six regions that are very different in structure. Most animals with alimentary canals have similar regions. These six structural regions are:

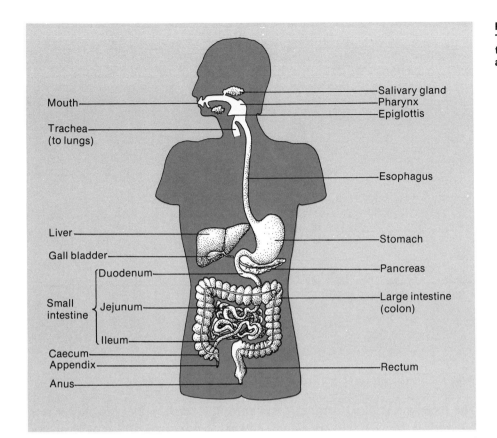

Fig. 18-16 The human digestive system. The food tube is specialized for the functions of mechanical and chemical digestion, and nutrient absorption.

Mouth
Trachea (to lungs)
Liver
Gall bladder
Duodenum
Small intestine
Jejunum
Ileum
Caecum
Appendix
Anus

Salivary gland
Pharynx
Epiglottis
Esophagus
Stomach
Pancreas
Large intestine (colon)
Rectum

1. the mouth
2. the esophagus
3. the stomach
4. the small intestine
5. the large intestine
6. the anal region

Some structural regions of the digestive system have more than one function. For example, the small intestine is a region of both digestion and absorption. Some functions are found in more than one region. For example, the stomach, small intestine, and large intestine all are absorbing regions, although they absorb different nutrients.

Review

1. Table 18-3 summarizes the relationships between the structures of the human digestive system and their functions. Study the table as you examine Figure 18-16. Then write a general description of what happens to a mouthful of food as it passes from the mouth region to the anal region.

STRUCTURE	FUNCTION					
	RECEIVING	CONDUCTING	CHEMICAL DIGESTION	NUTRIENT ABSORPTION	WATER ABSORPTION	ELIMINATION
Mouth	X		X (starch)			X (vomiting)
Esophagus		X	X (starch)			
Stomach		X	X	X (alcohol)		
Small intestine a. Duodenum		X	X			
b. Jejunum		X	X	X		
c. Ileum		X	X	X		
Large intestine		X		X (vitamin)	X	
Anal region		X				X

TABLE 18-3   **The Structure and Function of the Human Digestive System**

18.7 The Mouth Region

The receiving region, or mouth, includes lips, teeth, upper jaw, lower jaw (mandible), tongue, salivary glands, and hard and soft palates. The receiving region functions chiefly to mechanically break up bulk food into groups of molecules by:

1. grinding and crushing bulk food into smaller particles;
2. dissolving soluble substances in water;
3. chemically splitting some starch molecules.

The mouth region also helps as follows:

1. The tongue rolls the food into individual portions, each one called a **bolus**, before swallowing.
2. Saliva lubricates each bolus of food with a protein called **mucin**. This reduces the wearing away of and damage to the innermost layer of cells lining the throat and esophagus.
3. Sense receptors monitor the characteristics of the food in terms of texture, temperature, and taste.
   The mouth has other functions not related to digestion. These include its involvement in speaking and in expressing emotion.

Crushing and Grinding

The teeth and jaws, aided by the tongue, are responsible for separating large groups of food molecules from one another. Look into a mirror and locate on your lower jaw the following teeth. Use Figure 18-17 as a guide.

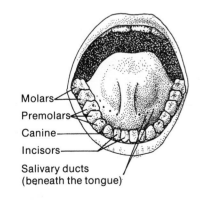

Molars
Premolars
Canine
Incisors
Salivary ducts
(beneath the tongue)

Fig. 18-17 Tooth arrangement. Determine how many people in your class have the third molars (wisdom teeth).

1. Four flattened, sharp **incisors**. Incisors are for cutting food.
2. Two single-pointed **canines**. These are for piercing and tearing food.
3. Four **premolars** that are used for crushing food.
4. Four or six **molars** that are used for grinding food.

You should have a total of 14 to 16 teeth in the lower jaw. Check that you have the same number of incisor, canine, premolar, and molar teeth on your upper jaw. You may have 28 to 32 teeth, depending on whether or not the third molars, or "wisdom teeth", have come in.

Prove to yourself that the shape or structure of the tooth determines its best use, or function. Try to chew some bread with your incisors only. Or try to split a sunflower seed between your molars. Good luck!

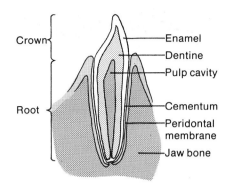

Fig. 18-18   Tooth structure. This is the vertical section of a canine tooth.

Tooth structure

All teeth consist of a visible **crown** and one or more roots. There is a hard white coating over the crown called enamel. It protects a softer, yellow substance called **dentine**. The dentine surrounds the **pulp cavity**, which contains blood vessels, lymph, and nerve fibres. The protective enamel coating is not needed deep in the jaw where the roots are found. A different substance, cementum, is found on the surface of the root. As you can guess by the name, this layer helps hold the tooth in the jaw. Fibres attached to the cementum hold the tooth to the surrounding **peridontal membrane** (Fig. 18-18).

Tooth hygiene

The most familiar problem with teeth is **dental caries** or "cavities". Lactic acid produced by bacteria and yeasts is responsible for the gradual dissolving of the protective enamel layer. If unnoticed, the dentine can disintegrate and infection of the pulp chamber can occur. Pain can be severe, since this is where the nerves are. A dentist cleans the damaged area of all bacteria and fills the remaining hole with a mixture of mercury and silver.

Some people may have very few fillings and others may have several. Resistance to infection varies from one individual to another since it is an inherited characteristic. However, the chances of infection can be reduced by either strengthening the enamel or by making it difficult for the acid-producing bacteria to survive. Enamel becomes harder if some fluoride ions are added to the tooth as it is growing. Using fluoridated toothpastes and drinking fluoridated water while the teeth are still young are two ways of obtaining the small amounts of fluoride required.

Regular brushing of teeth removes bacteria and the food particles they can use for their own nutrition. Brushing also removes the fuzzy-textured layer that forms over teeth between brushings. This layer, called **dental plaque**, is a carbohydrate produced by the bacteria. It helps to keep the acid from being washed away and provides a barrier to mouthwashes and toothpastes that can neutralize acids. The combination of alkaline (acid neutralizing) toothpastes and brushing, reduces the likelihood of cavities.

Fig. 18-19 The upper end of the alimentary canal. The head contains many specialized tissues assisting digestion.

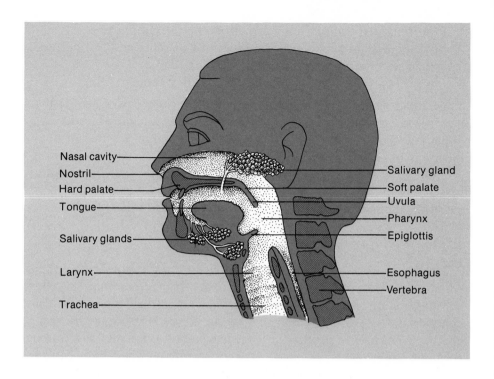

Nasal cavity
Nostril
Hard palate
Tongue
Salivary glands
Larynx
Trachea

Salivary gland
Soft palate
Uvula
Pharynx
Epiglottis
Esophagus
Vertebra

Dissolving, Digestion, and Lubrication

Your mouth produces up to 1.5 L of liquid every day in order to dissolve, digest, and lubricate your food. This liquid is called **saliva**. Saliva is produced when food is being tasted or chewed. The mere thought of biting into a lemon (a warm, yellow, juicy, sour lemon!) can sometimes cause your mouth to water.

Saliva is a solution of three main substances: **water**, the enzyme **amylase**, and the protein **mucin**. These substances come from three areas of your mouth. The inner lining of your cheeks has many separate mucin-producing glands, called **mucous glands** Underneath your tongue are the entrances to tiny tubes called **salivary ducts** which lead to the salivary glands (Figs. 18-17 and 18-19). These glands produce water, mucin, and the enzyme amylase. Opposite each second upper molar tooth is the entrance to a duct from other salivary glands. These two glands, found on each side of your face in front of your ears, produce water and amylase. The painful swelling of these glands is a symptom of the viral disease called mumps.

Water helps to moisten dry food and dissolve any soluble nutrients. Water is also essential for tasting food. The protein amylase is an enzyme that helps break starch molecules apart into shorter molecules of maltose sugar. Most vertebrates do not have any enzyme in their saliva. The protein mucin is produced as a lubricating and protective coating throughout the mouth, nasal cavity, and trachea. Mucin serves a very important non-digestive function by keeping all internal surfaces lubricated and protected from drying out.

Monitoring

The entire mouth region is very sensitive to touch, temperature, and taste. Your past experiences help you decide if you should keep eating or spit out a food of unusual texture, temperature, or taste.

Your lips have many nerve endings that are very sensitive to touch. There are other **touch receptors** throughout the inside of your mouth. The nerve endings in your front incisors are especially pressure sensitive. The signals from these nerves tell much about the texture of the food.

Temperature-sensitive nerve endings called **temperature receptors** are concentrated on the roof of the mouth (the **hard palate**). When you "burn" your mouth, it is the hard palate that feels hottest. There are temperature receptors scattered throughout the mouth region as well. Your tongue can have that "burnt" feeling. Scientists have shown that "hot" foods like mustard, curries, and pepper taste hot because they have chemicals that stimulate the temperature receptors. You interpret the stimulation as a hot feeling more than as a taste or smell.

Your tongue does the tasting in your mouth. The surface of the tongue is covered by many tiny bumps called **taste buds**. Each taste bud is most sensitive to either saltiness, sweetness, sourness, or bitterness (Fig. 18-20). Because the nerve endings are deep inside the taste bud, part of the food must be dissolved before it can be tasted. Dissolved molecules of food can get to the nerve endings, but solid particles cannot. There is some evidence to suggest that the shape of individual food molecules also determines how food will taste.

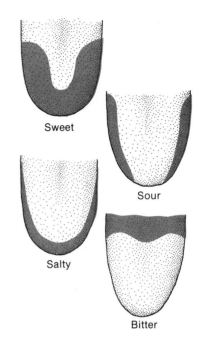

Fig. 18-20 Tastebud distribution. Try to verify these diagrams using sugar, lemon, salt, and tonic water solutions. Take a cotton swab, dip it in one solution and dab it on your dry tongue. Can you taste sugar at the back of your tongue? Can you taste bitter tonic water at the tip? Can you taste salt at the back and sour in the centre?

Swallowing

The muscular tongue helps to separate a mouthful of food into bite-size chunks, each one called a **bolus**. Pressure between the tip of the tongue and the hard palate (roof of the mouth) squeezes the bolus to the back of the mouth toward the **pharynx**, or throat (Fig. 18-21).

There is no hard palate at the rear of the mouth. However, though the bone that forms the roof of the mouth comes to an end, the tissue continues toward the back. This tissue is called the **soft palate**. As the tongue presses the bolus up, the soft palate is pushed upward by the bolus. This action closes the air passageway from the nasal cavity. The **uvula**, the fold of skin hanging down from the soft palate, helps cover the nasal passageway. This prevents food from being pushed up into the nasal airway and causing a blockage.

Once the bolus reaches the back of the throat, you lose voluntary control of it. A swallowing reflex is started. Only with great effort can you stop swallowing once the bolus has reached the **esophagus**.

Fig. 18-21   Mechanism of swallowing.

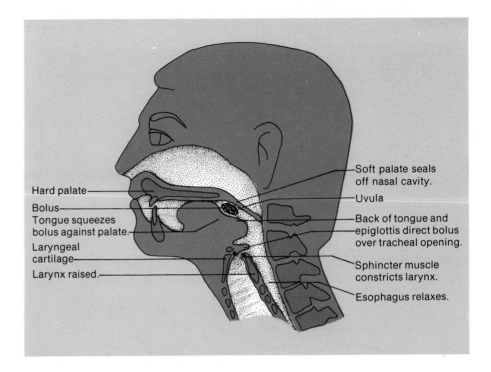

Hard palate

Bolus

Tongue squeezes
bolus against palate.

Laryngeal
cartilage

Larynx raised.

Soft palate seals
off nasal cavity.

Uvula

Back of tongue and
epiglottis direct bolus
over tracheal opening.

Sphincter muscle
constricts larynx.

Esophagus relaxes.

Review

1. **a)** What is saliva?
   b) How is mucin useful in the digestive system?
2. **a)** Explain why brushing teeth with toothpaste helps prevent dental caries.
   b) Why is fluoride put into many toothpastes and water supplies?
3. In what ways is food analysed in the mouth before it is swallowed?

18.8 The Esophagus

The section of the alimentary canal that moves the bolus through the thoracic (chest) cavity is called the **esophagus**. The esophagus is a muscular tube about 25 cm long and 2.5 cm across. Your esophagus is collapsed most of the time. It expands to a circular shape only when there is a bolus of food in it. The region that you call the throat is properly called the **pharynx**. Here the nasal passage from the nose meets the trachea and the esophagus.

Food is prevented from backing up into the nasal cavity by the flap action of the soft palate and uvula. A similar mechanism prevents food from entering the lungs. Nerves in the pharynx detect the bolus. An automatic reflex causes part of the windpipe, or **trachea**, to rise against a special cartilage called the **epiglottis**. This upward movement can be seen in mature males as a movement of the "Adam's apple". You can feel it yourself by placing your fingers over your throat and swallowing. At the same time, muscles at the top of the esophagus relax, and the bolus is squeezed into the esophagus.

Food is moved through the esophagus and the entire alimentary canal by muscular contractions. The esophagus wall contains an inner layer of **circular muscle** and an outer layer of **longitudinal muscle**. A bolus is literally squeezed forward by a coordinated contraction (shrinking) of the circular muscle behind it. This contraction must be coordinated with a relaxation of the circular muscle and a contraction of the longitudinal muscle in front of the bolus (Fig. 18-22). This coordination occurs automatically.

Peristalsis is *the scientific name for the coordinated contractions and relaxations of the esophagus.* Peristalsis also occurs in the intestines. Peristalsis explains why someone doing a headstand can still drink a glass of water successfully.

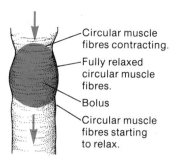

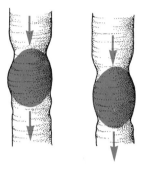

Circular muscle fibres contracting.

Fully relaxed circular muscle fibres.

Bolus

Circular muscle fibres starting to relax.

Review

1. **a)** How is food prevented from entering the nasal cavity?
   b) How is food prevented from entering the lungs?
2. Explain how someone doing a headstand can swallow a glassful of water successfully.

Fig. 18-22 Peristalsis. The shrinking of the alimentary canal behind the bolus, combined with relaxation of muscles in front of it, move the bolus along. Food moves by peristalsis throughout the length of the alimentary canal. Not shown are the longitudinal muscle fibres. Where would they have to contract?

18.9 The Stomach

Peristalsis carries the bolus down to the **stomach**. The stomach is an expanded section of the alimentary canal. It has special structural features that allow rather violent chemical digestion to occur. Yet the stomach itself remains undigested!

When empty, the stomach appears to be no more than a thickened section of esophagus. However, it can expand to form a J-shaped sac holding a total volume of up to 2 L. The lining layer is often quite wavy and ridged. The **cardiac sphincter** is found at the junction of the esophagus and stomach. A **sphincter** is a thickened ring of circular muscle that acts as a valve. The **pyloric sphincter** is found where the stomach joins the duodenum of the small intestine. Other sphincters along the alimentary canal are the oral and anal sphincters. You have some conscious control over them. They regulate the supply and elimination rates of your digestive system. Sphincters help to keep food in one section of the digestive system. Otherwise, peristalsis would move the food through too quickly. Many nutrients would not have a chance to be digested and absorbed.

The pyloric sphincter remains closed for most of your life. The circular muscle relaxes, opening the tube, only to let a single squirt of well-digested material into the duodenum.

When a bolus enters the stomach through the cardiac sphincter, two main processes are started.

1. Rhythmic contractions of the stomach muscles churn the contents (Fig. 18-23). This mixes and separates the food particles. Rhythmic contractions of an empty stomach are felt as "hunger pains".
2. Many glands in the lining start secreting all sorts of digestive chemicals.

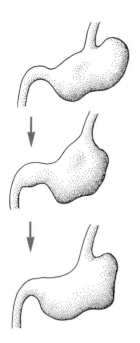

Fig. 18-23 The active stomach churns food back and forth by coordinated contractions of the muscular wall.

The glands are called **gastric glands**. The collective name given to the different chemicals they produce is **gastric juice**.

Gastric juice contains a protein-digesting enzyme, hydrochloric acid (HCl), mucin, and water. It is the coating of mucin that protects the stomach lining from the acid and enzymes. Up to 2 L of gastric juice is added to your meals every day by your stomach. The resulting mixture of acid, enzyme, water, mucin, and partially digested material is called **chyme**. After a meal has been churned and digested for 2 to 3 h, spurts of the chyme are released by the pyloric sphincter into the duodenum. Further digestion and absorption occur here.

Review

1. **a)** What is a sphincter? What is its function?
   b) Name the two stomach sphincters and describe their locations.
2. What two processes are involved in the digestion of food in the stomach?
3. Describe the composition of gastric juice.
4. Why does the stomach not digest itself?

18.10 The Small Intestine

The **small intestine** is the longest single structure of the alimentary canal (Fig. 18-24). It is almost 7 m long and is folded and coiled inside the abdominal cavity. It is called the "small" intestine because its diameter (3 cm) is less than that of the "large" intestine (7 cm). The first 30 cm leading from the stomach is called the **duodenum**. The duodenum receives fluid secretions from the **liver** and the **pancreas**. The remainder of the small intestine is divided into two sections. The **jejunum** is about 2.5 m long. It is followed by the 4.5 m **ileum**.

Chemical digestion of the chyme continues in the small intestine. The entire lining of the small intestine has many **intestinal glands** (Fig. 18-24). These glands produce a solution of digestive enzymes called **intestinal juice**. In addition, a concentrated solution of digestive chemicals are added to the duodenum from the **liver** and **pancreas**. The liver produces a concentrated fat-emulsifying solution called **bile**. The pancreas produces **pancreatic juice** which contains digestive enzymes. Both the bile and pancreatic juice contain sodium bicarbonate (baking soda) which neutralizes the acid chyme coming from the stomach. Tubes from the liver and the pancreas, called the **bile duct** and the **pancreatic duct**, enter the duodenum about 10 cm from the pyloric sphincter.

The great length of the small intestine ensures that nutrients can be fully digested and have a good chance of being absorbed. Food is not really "in" your body at all until the nutrients have gone through the lining cells separating the chyme from the bloodstream. The lining of your alimentary canal is just a specialized part of your outer body surface!

Molecules have a good chance of being absorbed if they are touching a

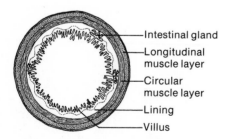

— Intestinal gland
— Longitudinal muscle layer
— Circular muscle layer
— Lining
— Villus

Fig. 18-24 Cross-section of the small intestine. Villi and muscle layers are shown clearly.

lining cell. The lining of the small intestine is not smooth. Rather, it is gathered in large folds. Most of these folds are covered with many millions of finger-shaped projections called **villi**. Each one is called a **villus**. All together the villi give the lining a velvety appearance. A villus is about 1 mm long and 0.2 mm in diameter. The surface of each villus contains hundreds of cells. And each single cell membrane has even smaller projections called **microvilli**. Each microvillus is about 0.000 05 cm long and 0.000 005 cm in diameter. Figure 18-25 shows the great difference the folding, villi, and microvilli make to the surface area of the small intestine. About 600 times more absorbing area is created than if the small intestine had a smooth lining.

Food moves through the small intestine by peristalsis. Pairs of contracted circular muscles 6 cm apart carry the digesting chyme forward and backward over the villi at the rate of 3 cm/min. In addition, the finger-like villi contract and wave from side to side. This helps nutrients circulate from the centre of the alimentary canal to the edge where they are absorbed.

The chyme spends about 4 to 6 h in the small intestine. After most of the nutrients have been absorbed, the remaining mixture is passed into the large intestine through a sphincter at the end of the ileum.

Review

1. Calculate the minimum surface area of the small intestine. Assume that it is a smooth cylinder 3 cm in diameter and 7 m long. How does your answer compare to the value given in Figure 18-25?
2. What are villi?
3. **a)** Describe the structure of the small intestine.
   b) How does this structure aid in the absorption of nutrients by the body?
4. Name the sources of the three digestive fluids that help in chemical digestion in the small intestine.

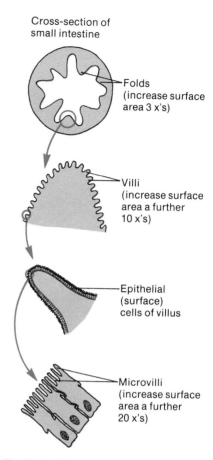

Cross-section of small intestine

Folds (increase surface area 3 x's)

Villi (increase surface area a further 10 x's)

Epithelial (surface) cells of villus

Microvilli (increase surface area a further 20 x's)

Fig. 18-25 Folds, villi, and microvilli combine to increase the surface area by 600 times. The intestinal surface area is about 200 m².

18.11 Large Intestine (Colon)

The digestive glands secrete 5 to 6 L of water on your food every day. This water comes from the sources shown in Table 18-4.

REGION	SECRETION	DAILY VOLUME OF WATER
Mouth	Saliva	1.5 L
Stomach	Gastric juice	1.5 L
Liver	Bile	0.8 L
Pancreas	Pancreatic juice	0.8 L
Small intestine	Intestinal juice	1.0 L
Total Volume of Secretions		5.6 L

TABLE 18-4 Sources of Water Used in Digestion

The water is added to dissolve the enzymes and to lubricate or dissolve the nutrients. Most animals have a region of their alimentary canal that reabsorbs water into the bloodstream. Therefore, rather than drink 5.6 L of replacement water every day, you automatically recycle all but 0.2 L.

The **large intestine** or colon, is the special section of the alimentary canal that all mammals have for reabsorption of water. Many dissolved minerals are absorbed into the bloodstream along with the water. The ileum joins your large intestine about 5 cm above a pouch-like **caecum**. Projecting from the caecum is the 10 cm **appendix**. The caecum and appendix are usually in the lower right of your abdomen. The large intestine leads upward along the right side, then curves across the front just below the liver and stomach. It then turns downward on the left side toward the **rectum**. The large intestine is about 7 cm in diameter and 1.5 m in length.

The harsh chemical conditions of the stomach and small intestine prevent most bacteria from growing. However, the large intestine provides ideal conditions for bacterial growth. It is dark and warm, with a steady supply of "leftovers" from the small intestine. The bacteria living with you are quite useful to you, for they manufacture some of the B vitamins, along with vitamin K. Some diseases, like cholera, will over-run the normal types of bacteria found in your intestine. Herbivores like the horse and rabbit depend on the bacteria in their enlarged caecum to help digest the cellulose in grasses. Bacteria of the **coliform** group assist the villi of the large intestine in the reabsorption of water.

The reabsorption of water and the addition of large numbers of bacteria make the undigested material much more solid. The solid matter is moulded back into separate boluses by the peristaltic contractions of the large intestine. The solid waste is called **feces**, or **fecal matter**. Up to 50% of the feces consist of dead bacteria. Even if no food is eaten, 7 to 8 g of solid waste is produced from the dead bacteria and dead lining cells every day. If normal peristalsis through the large intestine is speeded up, not enough water can be absorbed. **Diarrhea** results. If peristalsis slows down, or ceases, water absorption is very thorough. This results in **constipation**.

Review

1. **a)** How much water is supplied by your body to aid chemical digestion in the alimentary canal each day?
   b) How much of this water is reabsorbed?
   c) Why is the reabsorption of water so important?
2. **a)** Describe three ways in which the large intestine is suitable for bacterial growth.
   b) How are some of these bacteria useful?
   c) How can some be harmful?
   d) How much of your solid waste is bacteria?
3. **a)** Suggest how a drug that stops constipation might work.
   b) Suggest how a drug that stops diarrhea might work.
4. The relationship between your body and the coliform bacteria in your large intestine is a symbiotic relationship called **mutualism**. Refer to the index to find out what mutualism is. Explain this example of mutualism.

18.12 The Anal Region

Your body temporarily stores feces in the **rectum**. This final section of the alimentary canal is about 20 cm long. It ends at the **anal sphincter** which guards the opening called the **anus**. The rectum is usually empty except when defecation occurs. Feces accumulate in a final section of the large intestine. Defecation occurs when the longitudinal muscles of the large intestine contract and shorten. At the same time, a peristaltic contraction of circular muscle behind the fecal matter pushes it down into the rectum. Voluntary relaxation of the anal sphincter allows the feces to pass out of the body through the anus. (Toilet training of young children helps overcome the involuntary relaxation of the anal sphincter.) Once outside your body, the fecal matter is completely decayed by bacteria, and the nutrients are returned to the environment.

Review

1. Describe the location and function of the human rectum.
2. Explain how fecal matter is moved from the large intestine to the exterior.

18.13 Chemical Digestion

Why is Chemical Digestion Needed?

Chewing, churning, and mixing of food with digestive juices can only separate molecules from each other. These actions cannot split up molecules. However, some individual nutrient molecules are too large to pass through cell membranes. If you are going to receive any benefit from eating proteins, carbohydrates, and fats, the molecules must be split apart in your digestive system. Then each smaller molecule can pass through the lining cells and into the bloodstream. Molecules of water, minerals, and vitamins are already small enough that they do not have to be broken apart like the other three essential nutrients. *Splitting large molecules into smaller ones is called* **chemical digestion**.

Chemical digestion of proteins, fats, and carbohydrates is needed for three reasons. First, these molecules are too large to pass through a cell membrane. Second, fat and some carbohydrates are insoluble in water. Before nutrients can enter the blood, they must be dissolved in water. Third, the proteins, fats, and carbohydrates that you eat are not usually the same proteins, fats, and carbohydrates that your body uses. In fact, your body rearranges plant and animal proteins, fats, and carbohydrates into "people" proteins, "people" fats, and "people" carbohydrates.

How Does Chemical Digestion Occur?

Most chemical digestion in your body uses a water molecule to break the bonds between the parts of the larger molecules. This process is called

hydrolysis. Specific enzymes may help speed up the hydrolysis of the large molecules by positioning the water molecule in just the right place for the chemical reaction to occur. Enzymes are often identified by the molecule they act upon with the ending **-ase**. For example, proteinases (often shortened to proteases) help hydrolyze proteins, lipases help hydrolyze lipids (a synonym for fat), and carbohydrases help hydrolyze carbohydrates.

Chemical Digestion in the Mouth

Humans are unusual animals. Our saliva contains a digestive enzyme called **amylase**. The only carbohydrate that this enzyme helps to hydrolyze is starch. Many starch molecules are found in grains, potatoes, and many other foods. The starch molecules are very large and insoluble in water. Therefore, they must be broken apart before they can be absorbed into your body.

Amylase helps water molecules split starch apart in a very particular way. Hydrolysis of a starch molecule produces hundreds of disaccharide molecules of maltose. Amylase is produced by the salivary glands in the mouth. Thus some starch digestion takes place in your mouth. But you do not keep food in your mouth for very long, so not much maltose is produced. However, you may be able to taste maltose sugar if you chew an unsalted soda cracker for a few minutes. As soon as the swallowed bolus reaches your stomach, the hydrochloric acid there destroys the amylase activity, and the brief period of starch digestion stops.

Chemical Digestion in the Stomach

Gastric juice contains hydrochloric acid, water, mucin, and a proteinase called **pepsin**. The combined action of these substances furthers the chemical digestion of food.

The hydrochloric acid in your stomach assists digestion in four ways. It helps dissolve insoluble minerals, it kills many bacteria taken in with the food, it aids in the digestion of starch, and it provides the acidity needed to keep the pepsin enzyme working. The acid is very strong, strong enough to dissolve metals. The stomach lining is prevented from being dissolved by the mucin it produces. The mucin itself is continually being digested by the acid and pepsin. Therefore mucin must be continually produced. If not enough mucin is produced, the acid does start dissolving the lining and the muscle layers. This painful event is known as an **ulcer**.

"Heartburn" is another acid-related condition. Hydrochloric acid bubbles through the cardiac sphincter and irritates the lower esophagus. Taking an antacid tablet neutralizes the acid in the esophagus so that the acid-sensitive nerve endings there are no longer stimulated. Unfortunately, most antacid remedies end up neutralizing some of the acid in the stomach as well. This slows down normal protein digestion, and also stimulates the gastric glands to produce more replacement acid. Therefore you could end up with an even more acidic stomach.

Pepsin also helps water molecules split proteins apart. Hydrolysis of a protein by pepsin produces several shorter chains of amino acids. These shorter chains are further split up into single amino acids in the small intestine.

After a few hours, the pyloric sphincter relaxes, and the chyme squirts into the duodenum. It now contains a mixture of sugars, including maltose, short chains of amino acids, proteins unaffected by pepsin, undigested fats, minerals, water, and vitamins. As soon as the acid chyme touches the wall of the duodenum, the pyloric sphincter closes tightly. It will only open again when the duodenum is neutral or alkaline.

Chemical Digestion in the Small Intestine

There are three main sources of chemicals that assist digestion in the small intestine. These are the liver, the pancreas, and the intestinal lining.

Liver

The liver produces no digestive enzymes. Rather, it makes complex mineral salts in a solution called **bile**. Some animals, including you, have a special sac, the **gall bladder**, where the bile salts can be stored and concentrated until needed. Other animals, including horses and rats, do not have a gall bladder. Such animals are continually producing bile and putting it into the duodenum via the bile duct. The special bile salts act like a detergent on the fat droplets in the chyme. Fats and oils do not dissolve in water. Bile salts break up large drops of fat into many smaller droplets. The fat is said to be emulsified. Bile salts also prevent the small droplets from going back together again. The liver has a further function. It stores glucose in the form of starch-like molecules called **glycogen**.

Pancreas

The pancreas produces **proteinases**, **amylase**, and the fat-hydrolyzing enzyme **lipase**. These enzymes and the bile from the liver enter into the duodenum through the same opening. The breaking down of fat droplets caused by the bile salts allows the lipase to hydrolyze fat molecules much faster. Hydrolysis of fat molecules produces smaller molecules of glycerol and fatty acids. The amylase that the pancreas makes is a different enzyme from the amylase in saliva. But it acts in the same way to produce maltose from starch. Therefore, even hastily eaten french fries will be digested. The proteinases, like pepsin, secreted by the pancreas, split proteins into shorter chains of amino acids.

Intestinal lining

The lining of the small intestine contains many tiny intestinal glands (see Fig. 18-24). These glands produce enzymes that split the shorter chains of

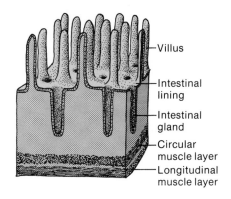

Villus

Intestinal lining

Intestinal gland

Circular muscle layer

Longitudinal muscle layer

Fig. 18-26 Profile of a section of small intestine. Note the many openings for the intestinal glands interspersed among the villi. Not shown is the network of blood and lymph vessels. Can you calculate the magnification of this diagram?

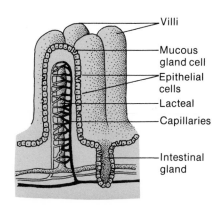

Villi

Mucous gland cell

Epithelial cells

Lacteal

Capillaries

Intestinal gland

Fig. 18-27 Absorption by the villus. Fats are absorbed into the lacteal. Amino acids, minerals and glucose are absorbed by the capillaries.

amino acids into individual amino acid molecules. The glands also produce enzymes that help hydrolyze disaccharides like maltose, lactose, and sucrose into simple sugars like glucose and fructose. Mucin is the other major substance produced by glands of the lining. Recall that the mucin helps lubricate and protect the intestinal lining.

Absorption of nutrients

The structural features of the small intestine make it the chief organ of nutrient absorption (Fig. 18-26). Within each villus is a **capillary bed** of the bloodstream and a **lacteal** of the lymphatic system (Fig. 18-27). Amino acids, simple sugars, minerals, and vitamins pass through the outer row of epithelial cells on the villus and into the capillary. From there, blood flow carries the nutrients into the portal vein leading to the liver. Excess sugars can be stored there in the form of glycogen. Excess vitamins and amino acids can be stored or transformed in the liver into other needed molecules. Microscopic fat droplets, glycerol, and fatty acids are all carried across the cells of each villus into the lacteal. These fats and digested fat products move into the thoracic lymph duct and are carried to your upper left shoulder area and enter the bloodstream.

Most nutrients are carried across the cells of the alimentary canal to the bloodstream against the concentration gradient (opposite to the direction of diffusion). In other words, **active transport** of nutrients is required. (Recall that cells use up some energy to do this.) However, the overall gain to the organism of these nutrients is worth the use of the energy required for transport.

Vitamins, minerals, and water are absorbed throughout the small and large intestines. Fat soluble vitamins, like A, D, and K, are carried into the lacteals along with fat droplets. Reabsorption of water in the large intestine is by the process of osmosis.

Review

1. Why is chemical digestion of starch, proteins, and fats necessary?
2. Why do most minerals not need chemical digestion?
3. Why are fats emulsified before their digestion?
4. How is the absorption of fat different from the absorption of amino acids and glucose?
5. What is glycogen?

ENZYME	SOURCE(S)	SUBSTANCE ACTED ON	PRODUCT(S) OF DIGESTION
Amylase	Saliva and ?	?	?
Pepsin	?	?	?
Other proteinases	?	?	?
Lipase	?	?	?

TABLE 18-5   Enzymes in the Digestive System

6. Copy Table 18-5 into your notebook. Review your knowledge of chemical digestion by filling in the blanks.
7. Describe the changes in a cheese sandwich as it passes through the digestive system. (A cheese sandwich is basically fat, protein, and minerals between two sheets of starch, protein, and minerals.)
8. Describe the advantages in digestion of a) chewing food, b) enzyme action, c) a small intestine of great length.
9. When you eat a bowl of oatmeal, why does the oatmeal turn into you, and not you into oatmeal?
10. Why can cows live by eating grass while you cannot?
11. Why would you not expect to find amylase in the saliva of a carnivore?

18.14 Investigation

Why is Chemical Digestion Necessary?

The sizes of individual molecules affect how easily they can be absorbed into the body. In this investigation you will determine how easily salt, glucose, and starch molecules can travel across an artificial membrane.

Materials

10 cm section of dialysis tubing	250 mL beaker
starch suspension	wooden stick
glucose solution (corn syrup solution)	string
sodium chloride (table salt) solution	depression slide
warm distilled water	test tube and holder
iodine	Bunsen burner
Benedict's solution	medicine dropper
silver nitrate solution	

CAUTION: Wear safety goggles during this investigation.

Procedure

a. Wet the dialysis tubing to make it flexible. Then roll one end between finger and thumb to open it. Tie one end of the dialysis tubing to make a bag.
b. Add to the dialysis bag about 4 mL each of starch, glucose, and salt solutions.
c. Tie the open end tightly. Rinse the outer surface. Then use string to suspend the bag in 50 mL of warm distilled water in the beaker as shown in Figure 18-28.
d. Wait 5 min. Then do the 3 tests outlined in steps e, f, and g.
e. Test the water in the beaker for the presence of starch as follows. Place 1 drop of the water in the depression of the depression slide. Then add 1

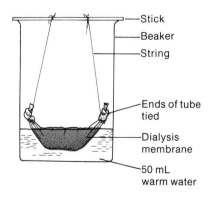

Fig. 18-28 The beaker should be set up like this. Keep the ends of the membrane out of the water.

Stick
Beaker
String
Ends of tube tied
Dialysis membrane
50 mL warm water

drop of iodine solution. (A blue or purple colour will form if starch is present in the water.)

f. Test the water in the beaker for the presence of salt as follows. Wash the depression slide and wipe it dry. Place 1 drop of the water in the depression. Then add 1 drop of silver nitrate solution. (A white precipitate will form if salt is present in the water.)

g. Test the water in the beaker for the presence of glucose as follows. Place 5 mL of the water from the beaker in a test tube. Add 5 drops of Benedict's solution. Heat the mixture carefully to the boiling point. (A reddish-orange colour will form in about 1 min if glucose is present.)

h. Repeat the 3 tests after 20 min.

i. Set the beaker aside for 1 d in a warm location. Repeat the 3 tests.

j. Record all your results in a table similar to Table 18-6. If there is no change after a particular test, write "No change".

| | REACTION WITH WATER IN BEAKER | | |
TEST SOLUTION	AFTER 5 min	AFTER 20 min	AFTER 1 d
Iodine solution			
Silver nitrate solution			
Benedict's solution			

TABLE 18-6   Observations on Membrane Permeability

Discussion

1. a) Describe the test result that shows the presence of starch.
   b) Can starch molecules pass through a dialysis membrane? Describe the observations that support this conclusion.
2. a) Describe the test result that shows the presence of a glucose-like sugar.
   b) Can glucose molecules pass through a dialysis membrane? Describe the observations that support this conclusion.
3. a) Describe the test result that shows the presence of salt. (Actually, this test only shows the presence of the chloride ions from the dissolved sodium chloride. But you can regard this test as one for salt in this investigation.)
   b) Can salt ions pass through a dialysis membrane? Describe the observations that support this conclusion.
4. Which observations suggest that some particles travel across membranes easier or faster than others?
5. a) A single starch molecule may have thousands of atoms, a glucose molecule has 24, and the ions of dissolved salt are the size of single atoms. Using your knowledge of diffusion across selectively permeable membranes (Chapter 6), explain the observations of this experiment.

b) Why is the chemical digestion of starch necessary?

6. Does this investigation show that starch molecules cannot move across living membranes? Explain.

7. **a)** Why would a careful student want to test the starch solution for the presence of starch, the glucose solution for the presence of glucose, and the salt solution for the presence of salt using the same test chemicals as used in this investigation?

   b) Why is the membrane rinsed off after it is filled and tied?

   c) Why is the membrane not placed in 500 mL of water (assuming you used a 600 mL beaker, of course!)

   d) Why does using warm water speed up the results of this investigation?

8. Describe the tests you would do to see if a) lipids (fats and oils) and b) proteins could travel across a dialysis membrane.

18.15 Investigation

The Effect of Amylase on Starch

In Investigation 18.2, you learned how to test for the presence of starch and glucose molecules in food. The enzyme amylase in your saliva provides a remarkable connection between these two carbohydrates.

Materials

dilute starch suspension
saliva (the source of amylase)
clean, new rubber band
drinking straw
thermometer
water bath equipment (stand, ring clamp, wire gauze,
   Bunsen burner, 250 mL beaker two-thirds full of water)
iodine solution
Benedict's solution
dilute hydrochloric acid
test tubes in test tube rack (5)
250 mL beaker with 150 mL of water at $37°C \pm 2.0°C$

CAUTION: Wear safety goggles during this investigation.

Procedure

a. Set up the water bath and get the water boiling.

b. Label 5 test tubes IC, IS, ISA, BC, BS. See Table 18-7 for the meanings of these labels.

LETTER	MEANING
C	Control
S	Saliva
I	Iodine solution
B	Benedict's solution
A	Acid (hydrochloric)

Example: The test tube labelled ISA contains iodine solution, saliva, and acid.

TABLE 18-7 **Code for Test Tube Labels**

c. Chew the rubber band to stimulate saliva production. Using the straw, put 2 mL of saliva into the ISA, IS, and BS test tubes.
d. Put 10 mL of starch suspension into each of the 5 test tubes.
e. Add 1 drop of dilute hydrochloric acid to test tube ISA.
f. Add 2 drops of iodine solution into test tubes IC, IS, and ISA.
g. Sit the 3 test tubes in the 37°C water bath. Observe what happens over the next 10 min.
h. Record your observations in a table similar to Table 18-8.

TEST TUBE	INITIAL COLOUR	FINAL COLOUR	SIGNIFICANCE OF TEST
IC			
IS			
ISA			
BC			
BS			

TABLE 18-8 **Effect of Amylase on Starch**

i. While you are waiting for the results from step h, prepare the other two test tubes as follows: Place 2 mL of saliva in BS. Then add 5 drops of Benedict's solution to both BC and BS.
j. Place the two test tubes in the 37°C water bath for 10 min. Now place BS and BC into the boiling water bath. Note any colour change after 5 min. Record your observations in the table.

Discussion

1. Complete your observation table by describing the significance of each test result.
2. a) What substance disappears from the saliva/starch suspension mixture?
   b) What new substance appears in the saliva/starch suspension mixture?
3. What effect does saliva have on starch?

4. What is the effect of hydrochloric acid on the saliva/starch suspension reaction?
5. Why are test tubes IC and BC needed in this investigation? Explain fully.
6. Think up an experiment that could show whether or not the amylase is used up during starch digestion.
7. Why is it not surprising that carnivores like dogs and cats have no amylase in their saliva? (If you have a dog, you may try to verify this fact.)
8. How could a plant make use of an enzyme like amylase?
9. **a)** Suggest the effect of placing 8 mL of starch solution along with 2 mL saliva in a dialysis membrane immersed in water as in Investigation 18.14.

   b) Do this experiment with your teacher's permission. Be sure to design your experiment carefully.

18.16 Investigation

Effect of Temperature on a Stomach Enzyme

Rennin is an enzyme found in the stomach of new-born mammals. It helps coagulate (clump) casein, which is the main protein found in milk. By observing how long coagulation takes at different temperatures, you can discover the best temperature for rennin activity. *Note*: **Coagulate** *means to change from a fluid into a curd-like or solid mass.*

Materials

rennin enzyme solution (1 cheese rennet tablet in 200 mL
   of water or 1 junket tablet in 50 mL of water)
skim milk
test tubes in a test tube rack (6)
medicine dropper
watch or clock
water bath equipment (mat, stand, ring clamp,
   wire gauze, Bunsen burner)
250 mL beakers (3)
thermometer
ice or snow for water bath

CAUTION: Wear safety goggles during this investigation.

Procedure A

a. Set up a water bath and start heating 150 mL of water in one of the beakers.
b. Label 6 test tubes as follows: 0°C (E), 0°C (C), 37°C (E), 37°C (C), 100°C

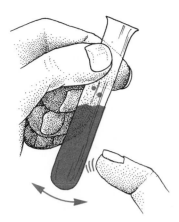

Fig. 18-29 Mixing liquids in a test tube. If you hold the top of the test tube firmly, the bottom springs back and forth as it is flicked by your index finger.

(E), 100°C (C). Note that all temperatures are Celsius. "E" stands for "Enzyme" and "C" stands for "Control".

c. Put 10 mL of skim milk in each test tube. (You do not have to be exact. But be sure that all test tubes have the same amount of milk.)

d. Prepare a 150 mL ice bath (at about 0°C) in one beaker.

e. Prepare a 150 mL 37°C water bath in the other beaker.

f. Do not proceed until the water in the third beaker is boiling.

g. Now put 1 mL (about half a dropper) of rennin enzyme solution into each of the 3 test tubes marked "E". Mix by flicking your finger against the side of the test tube as shown in Figure 18-29.

h. Place the two "100°C" test tubes in the boiling water, the two "37°C" test tubes in the 37°C water; and the two "0°C" test tubes in the ice water.

i. Note the time.

j. Use the thermometer to check the temperature of the water surrounding the test tubes. Adjust as necessary to keep the temperatures at their original levels.

k. Once each minute, lift and tilt each test tube. Note the time at which coagulation occurs. Stop after 10 min. If no change has occurred, write "No change" in your observations.

Procedure B

a. After the 10 min, remove the 37°C test tubes from the water bath. Place the 0°C (E), 100°C (E), and 100°C (C) test tubes into the 37°C beaker to warm or cool for 5 min.

b. After the 5 min, add 1 mL of enzyme solution to the 100°C (C) test tube.

c. Again, lift and tilt each test tube once a minute for 10 min. Note the time taken for coagulation to occur in each test tube.

d. Before you clean up, stir the coagulated milk. You should be able to identify the separate curds and whey phases of the milk.

e. Record all your observations in a table similar to Table 18-9.

Time started: _____

Part A

TEST TUBE LABEL	TIME FOR CHANGE TO OCCUR
0°C (C)	
0°C (E)	
37°C (C)	
37°C (E)	
100°C (C)	
100°C (E)	

5 min at 37°C

Part B

TEST TUBE LABEL	TIME FOR CHANGE TO OCCUR
0°C (E)	
100°C (C)	
100°C (E)	

TABLE 18-9 Temperature and Enzyme Activity

Discussion

1. What does rennin do to milk at 37°C?
2. **a)** At what temperature is rennin's effect the fastest?
   b) Why do you think this temperature was chosen? Which other Investigation have you done using this temperature?
3. Does cooling rennin to about 0°C destroy the enzyme? Which observations show this?
4. **a)** Does boiling rennin at 100°C destroy the enzyme? How do you know this?
   b) Is milk protein completely destroyed at 100°C? How do you know this?
5. What assumption do you make about the temperature of the milk and the temperature of the surrounding water?
6. **a)** What is the difference between the "C" and "E" test tubes?
   b) Why are the "C" test tubes necessary in this experiment? Could you have guessed the results found for the "C" test tubes? How?
7. **a)** Predict the effect of boiled amylase on starch.
   b) Predict the best temperature for amylase activity.
   c) Predict the best temperature for pepsin, lactase, and lipase activity.
   d) Predict the best pH (acid, neutral, or base) for rennin activity.
8. **a)** Suggest why you produced rennin when you were born, but now no longer produce that particular enzyme.
   b) Which protein-digesting enzyme does your stomach produce now? What is its function?
9. How is rennin used commercially?

Highlights

Since animals are heterotrophs, they must obtain nutrients from other living organisms. These nutrients provide the matter and energy needed for life processes. Nutrients may be organic or inorganic. The main organic nutrients are carbohydrates, fats, proteins, and vitamins. The main inorganic nutrients are water and minerals such as calcium.

Organisms feed in a variety of ways. The one-celled amoeba ingests its food by using pseudopods. The paramecium sweeps food into its one-celled body using cilia. Many small crustaceans and bivalves are filter feeders, as are a few large mammals such as the Blue Whale. Earthworms ingest large masses of organic laden material and digest from it the required nutrients. Hydra use nematocysts on their tentacles to capture food. The food is then transferred to a mouth and, finally, into a gastrovascular cavity. Some animals chew their food to break it up before it is swallowed. Still other animals are fluid feeders. They have specially adapted mouthparts for sucking up fluids or soft tissues.

Digestion is the process by which large food molecules are broken down or hydrolyzed into simple basic nutrient units. It may be either intracellular or extracellular. During intracellular digestion, the food is acted upon within a cell. During extracellular digestion the food is acted upon within a digestive cavity. The digestive cavity has only a single opening in the simpler animals. In higher animals the digestive cavity, or alimentary canal, has both a mouth and an anus.

The human digestive system consists of different structural modifications of the basic alimentary canal. The alimentary canal is lined in certain regions with a layer of lubricating, enzyme-producing cells. The lining layer is surrounded by a muscular layer.

Teeth, tongue, and hard palate assist the mechanical digestion of food into smaller particles. The smaller particles are more easily chemically digested with the help of enzymes such as amylase, pepsin, and lipase. These enzymes hydrolyse or break down starch, proteins, and fats into smaller, more easily absorbed molecules.

The liver and pancreas produce chemicals that assist in chemical digestion. Absorption of nutrients occurs in the small intestine. Water is reabsorbed by the large intestine. Unabsorbed or undigestible food and bacteria are expelled from the body to be recycled in the environment.

Gas Exchange And Breathing

19

Most animals are **aerobic**; that is, they need oxygen. Carbon dioxide also affects animals, since too much of it can be harmful. The exchange of these two gases between living cells and their environment is the topic of this chapter. How does such an exchange occur? How does biological size complicate gas exchange? What problems does an aquatic environment present to gas exchange? What problems does a land environment present? How do aquatic and terrestrial animals solve these problems?

19.1 The Need for Gas Exchange Systems

We learned in Chapter 7 that most cells require oxygen to release energy from organic materials (food). This process is called **cellular respiration**. A

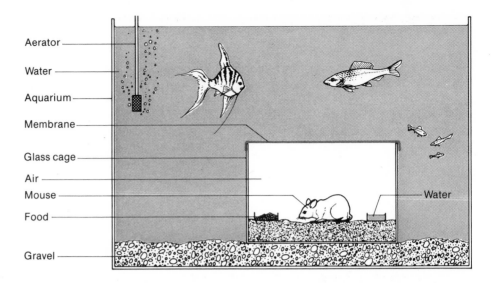

Fig. 19-1 The membrane in this illustration is permeable to gases but impermeable to water. The mouse continues to live in this solution. How is this possible?

Aerator
Water
Aquarium
Membrane
Glass cage
Air
Mouse
Food
Gravel
Water

harmful by-product of this process is another gas, carbon dioxide. The intake of oxygen and the elimination of carbon dioxide is a basic problem for most organisms (Fig. 19-1).

Oxygen and carbon dioxide are exchanged by diffusion across the membranes of most living cells. These gases must be dissolved in a liquid in order to diffuse across the membranes. Gaseous exchange is usually a simple process in single-celled and small multicelled organisms. All the cells in such organisms are in direct contact with or close to the external environment.

Gas exchange is a more complicated process in larger multicelled animals. Such animals may have many cells deep within their body, far from the external environment. Diffusion alone is too slow at moving gases across a large number of cells. They must have some other mechanism for carrying the gases to every cell. Therefore special regions for gas exchange are present in larger animals.

Gas exchange will only occur across a moist membrane. For terrestrial animals, especially, this presents some serious problems.

Although there are a variety of devices in large animals for solving the problems of gas exchange, they are all designed to meet the same general needs.

1. Each animal must have a gas exchange surface of adequate size.
2. Each animal must have some means of transporting the gases between the gas exchange surface and every body cell, no matter how far the cell is from this surface.
3. Each animal must have some way of keeping the gas exchange surface moist to allow diffusion.

The gas exchange surface of an animal may involve the entire body surface of the organism. It may also consist of infoldings or outfoldings of the body surface. As you study the gas exchange systems of various animals, try to determine the type of gas exchange surface involved in each case.

Respiration and Breathing

You are familiar with **cellular respiration**, the process by which cells release energy from foods. **Breathing** *is the means by which respiratory gases are exchanged between the entire organism and its environment.* Breathing is simply a mechanical process. In mammals, breathing causes air containing a high concentration of oxygen to enter the lungs and causes air containing a high concentration of carbon dioxide to leave the lungs. Animals must have some way of transporting gases between the breathing region and the cells throughout their bodies. Diffusion is involved and, in some cases, an internal transport system is required.

Review

1. Describe how gas exchange occurs in single-celled and simple multi-celled organisms.
2. Why does an increase in the size of an animal make the process of gas exchange more complicated?
3. What general requirements must all large animals meet in order to solve the problems of gas exchange?
4. Would you expect that terrestrial animals would have gas exchange surfaces with infoldings or with outfoldings? Why?
5. **a)** Explain the difference between respiration and breathing.
   b) Why is the term "artificial breathing" more correct biologically than the commonly used term "artificial respiration"?

19.2 Investigation

Gas Exchange, The Size Factor

One factor that determines how much oxygen an animal needs is its size. In general, the larger an animal is, the more oxygen it requires. The animal's gas exchange surface must permit oxygen intake at a rate adequate to meet its demands. Also, the surface must permit an adequate rate of carbon dioxide release. In this investigation you will study the effect of an increase in size upon the ratio of surface area to volume. Then, you will try to determine the implications this has for gas exchange in large multicelled animals.

Materials

sheet of metric ruled graph paper (1)

Procedure

a. Look at Figure 19-2. Calculate the surface area and the volume of a cube in which *l*=2.0 cm. Then calculate the surface area to volume ratio for this

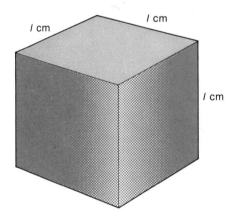

Area of one side = *l* cm²

Total surface area of cube = 6 *l* cm²

Volume of cube = *l* cm³

$$\frac{\text{Surface area}}{\text{Volume}} = \frac{6 \ l \ \text{cm}^2}{l \ \text{cm}^3}$$

Fig. 19-2 The surface area of the cube represents the area of an organism's gas exchange surface. The volume of the cube represents the volume of the organism. How does the surface area to volume ratio change as the value of *l* is increased? Or, how does the ratio of the area of the gas exchange surface to the volume of the organism change as *l* is increased? What problems does this cause for large organisms?

cube. Record all your results in a table similar to Table 19-1. To serve as an example, the table has been completed for l=1.0 cm.

b. Repeat step a for l=3.0 cm, 4.0 cm, 5.0 cm, 6.0 cm, 7.0 cm, and 8.0 cm.

c. Plot a graph with the length of the side on the horizontal axis and both surface area and volume on the vertical axis.

LENGTH OF SIDE (cm)	SURFACE AREA (cm²)	VOLUME (cm³)	SURFACE AREA TO VOLUME RATIO (cm⁻¹)
1.0	6.0	1.0	$\dfrac{6.0}{1.0} = 60$
2.0			
3.0			
4.0			
5.0			
6.0			
7.0			
8.0			

TABLE 19-1   Surface Area to Volume Ratio

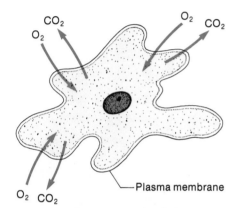

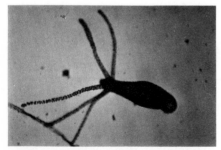

CO₂   O₂   CO₂
O₂
O₂
O₂   CO₂
——— Plasma membrane

Fig. 19-3   Gas exchange in an amoeba. Single-celled aquatic organisms exchange gases with the surrounding water by diffusion across the entire plasma membrane. Remember: It is *dissolved* molecules of oxygen (O₂) that diffuse, not the oxygen (O) of water molecules (H₂O).

Fig. 19-4   A hydra. Direct diffusion of gases is sufficient for this small animal. Would it be sufficient for the much larger squid and octopus which look much like the hydra?

Discussion

1. a) Use your calculations and your graph to describe what happens to the ratio of surface area to volume as the cube gets larger.

b) What do you think will happen to the ratio of the area of the gas exchange surface to the volume of the organism as the organism gets larger?

2. a) The results of this investigation suggest that a serious problem could arise as an animal increases in size. What is this problem?

b) How might an animal grow larger, yet avoid this problem?

19.3 How Aquatic Animals Solve the Problem of Gas Exchange

Direct Diffusion

Single celled aquatic organisms such as the amoeba do not require any special gas exchange system (Fig. 19-3). Dissolved oxygen in high concentration in the surrounding water diffuses across the plasma membrane into the organism's body. Carbon dioxide in high concentration in the cell diffuses outward into the surrounding water. The oxygen requirements of such organisms are so low that direct diffusion can provide all the gas exchange needed.

Many small multicelled animals such as the hydra (Fig. 19-4) also do not require a special gas exchange system. All cells in a hydra are in direct contact with the surrounding water or the water in the gastrovascular cavity (see Fig. 18-6). Therefore diffusion is adequate for gas exchange. The planarian's flattened shape also permits adequate gas exchange by direct diffusion (Fig. 19-5). No cell is very far from the body surface.

Use of Gills

Most larger aquatic animals have structures which increase the surface area available for diffusion. In addition, such animals require a faster method than direct diffusion for transporting gases within their bodies. Therefore they have specialized gas exchange structures called **gills**. Gills allow dissolved gases to come into close contact with the animal's internal transport system (circulatory system). Although they may differ somewhat from animal to animal, all gills usually have these same characteristics: they provide a large surface area for gas exchange and they contain a rich supply of blood vessels.

In aquatic animals, gills are usually feathery or leaf-like to provide a large surface area. Oxygen diffuses from the water across the thin moist gill

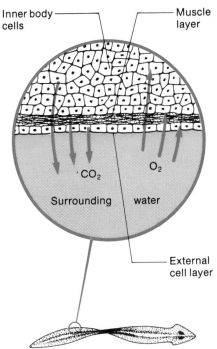

Fig. 19-5 Gas exchange in a planarian (flatworm). Its flattened shape enables all cells to exchange gases by direct diffusion.

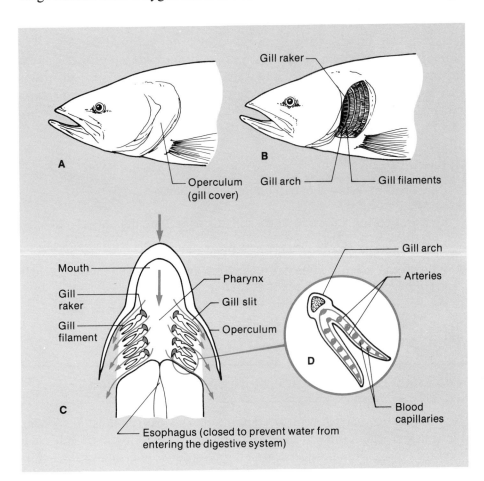

Fig. 19-6 Gills of a fish. A: operculum covers gills for protection. B: operculum cut away to show a gill. C: path of water through the gills. D: cross-section through a gill filament.

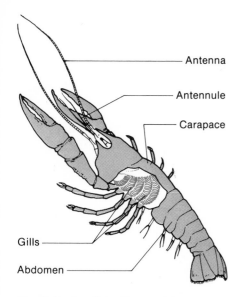

Fig. 19-7 Crayfish with part of the carapace removed to show the feather-like gills.

Labels on figure: Antenna, Antennule, Carapace, Gills, Abdomen

surface and into the blood transport system where it is carried to cells throughout the body. Carbon dioxide is transported in the blood to the gills from the body cells and diffuses outward across the membranes of the gill cells. Gills are quite delicate and, as a result, must be protected. Some animals have coverings over the gills, such as the **operculum** (gill cover) of a fish (Fig. 19-6) and the **carapace** of a crayfish (Fig. 19-7).

Fish gills

Most fish have four pairs of gills which are located on the sides of the pharynx. **Gill slits** separate each gill and permit water to flow past them. Each gill consists of a support made of cartilage called the **gill arch**. A double row of feathery **gill filaments** is attached to each arch. Blood vessels called **capillaries** are found within the thin walled filaments. The inner surface of the gill arch has numerous projections, the **gill rakers**, which act as straining devices. They prevent large particles from passing over the delicate gill filaments. Water enters through the mouth into the pharynx. The mouth is then closed and the floor of the pharynx and mouth is raised. This action forces the water out through the gill slits and past the gill filaments. Gas exchange takes place across the thin membranes of the gill filaments.

The problem of gas exchange in aquatic animals is complicated by the fact that very little oxygen is dissolved in water (around 0.005%). The amount decreases as the temperature of the water increases. Furthermore, the rate of diffusion of oxygen into water is very slow. Therefore, organisms that require large amounts of oxygen must have some means of constantly moving water over the gills. Swimming and the action of the pharynx do this for a fish.

Review

1. Describe how gas exchange occurs in the amoeba.
2. **a)** What are gills?
   b) Describe the structure of the gill of a fish. State the function of each part.
   c) Explain the breathing process in fish.
3. Why does a fish die when it is taken from the water, even though there is more oxygen in the atmosphere than in the water?

19.4 Investigation

The Effect of Water Temperature on Fish Respiration

Fish require oxygen for cellular respiration. This oxygen must come from the water in which the fish lives. It is present there as dissolved molecular oxygen, O_2. Carbon dioxide, a harmful by-product of cellular respiration, must be given off into the water. The gills carry out this gaseous exchange. In this investigation you will observe the operation of fish gills and study the effect of temperature on their action.

Materials (per team)

goldfish in an aquarium (1)
beaker (1 000 mL)
crushed ice
hot water
thermometer
dip net

Procedure

a. While the fish is still in the aquarium, observe its head region for a few minutes. Describe the movement of its mouth and gill covers (opercula).

b. Fill the beaker three quarters full of aquarium water. Then use the dip net to transfer the fish to the beaker.

c. After a few minutes, determine the fish's breathing rate by counting the number of times the opercula open and close in one minute. Take the temperature of the water. Record the breathing rate and the temperature in a table similar to Table 19-2.

WATER TEMPERATURE (°C)	BREATHING RATE (OPERCULUM MOVEMENTS PER MINUTE)

TABLE 19-2   **Temperature and Breathing Rate**

d. Add crushed ice to the beaker a little at a time. Observe the temperature and, with each 5°C drop in water temperature, record the fish's breathing rate. (As you add ice, pour enough water out of the beaker to keep the water level constant.)

e. When the temperature approaches 0°C, start adding hot water in small portions. The water should be poured into an area of the beaker away from the fish so that the fish will not be harmed. The water should then be stirred gently to equalize the temperature throughout. Determine and record the breathing rate with each 5°C change in temperature until the original water temperature is reached.

f. Return the fish to its original aquarium.

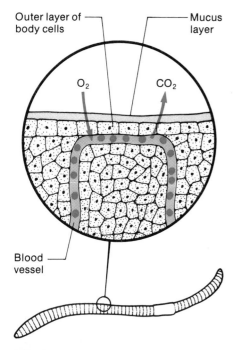

Fig. 19-8 Gas exchange in an earthworm. The moist skin acts as a gas exchange surface. A blood circulatory system carries gases to and from this surface.

Discussion

1. What relationship exists between the breathing rate of a goldfish and the water temperature? Account for this relationship.
2. Table 19-3 shows the solubility of oxygen gas in water when air at standard pressure is in contact with the water. If an industry or power plant pollutes the water with heat, what effect will this have on fish? Use both the data given in Table 19-3 and the results of the investigation to answer the question.

TEMPERATURE (°C)	SOLUBILITY OF OXYGEN (µg/g)
0	14.6
5	12.7
10	11.3
15	10.1
20	9.1
25	8.3
30	7.5

TABLE 19-3 Solubility of Oxygen Gas

19.5 How Land Animals Solve the Problem of Gas Exchange

Compared to water, the land environment is quite rich in oxygen (See Table 19-4). However, air is a very dry environment. Since a moist membrane is required for gas exchange, land animals must have some means of keeping gas exchange membranes from drying out.

Fig. 19-9 Insect tracheal systems. A: the flea. B: the cockroach. Note the extensive system of air-filled tubes or tracheae which carry gases to and from all parts of the body. The spiracles connect the tracheal system to the outside air.

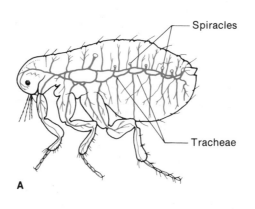

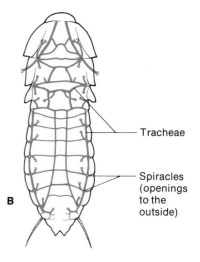

GAS	PERCENT BY VOLUME
nitrogen	78.09
oxygen	20.94
argon	0.93
carbon dioxide	0.034

TABLE 19-4   **The Main Gases in Air**

A Simple Gas Exchange System

The earthworm does not have a complex gas exchange system, yet it is well adapted to its land environment. It lives in moist soil. Numerous small blood vessels are located just beneath the outer layer of body cells. This surface must be kept moist by a secretion of mucus. Gas exchange occurs readily through the thin moist skin. Oxygen diffuses into the blood and carbon dioxide diffuses outward from the blood and into the soil (Fig. 19-8). If an earthworm is removed from its natural moist earth environment, its skin will likely dry out and it will probably die from lack of oxygen.

Tracheal Systems

Tracheal systems are characteristic of land arthropods such as insects. Air enters the organism through small openings called **spiracles** (Fig. 19-9). The spiracles lead to a system of branched tubes called **tracheae** (sing. **trachea**). These tracheae subdivide into smaller and smaller tubes called **tracheoles** which reach all parts of the insect's body (Fig. 19-10). The tracheoles contain a fluid. This fluid provides the moist surface needed for gas exchange by diffusion.

Air is brought directly to the body cells by the system of tracheal tubes. Some larger insects move the air along the tracheae by means of muscular contractions. The circulatory (blood) system does not play a part in gas transport in animals with tracheal systems. No cell is very far removed from a tracheole. However there is a limit to the rate at which such a system can supply the oxygen demands of the animal. No doubt this has been a factor in limiting the size of insects.

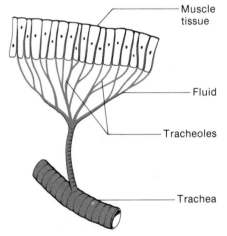

Fig. 19-10   The tracheal system carries oxygen to all cells (in this case, the cells of muscle tissue) so that cellular respiration can take place to produce energy.

One interesting adaptation for gas exchange is shown by the water beetle *Dytiscus* (Fig. 19-11). This insect comes to the water's surface, traps an air bubble under its forewings, and dives beneath the surface. The bubble does contain some oxygen, but other gases occupy a large part of the volume (see Table 19-4). As the insect uses up the oxygen, more diffuses into the bubble from the surrounding water. In a similar manner, carbon dioxide passes from the insect into the bubble and then into the water. Thus the bubble acts as a "physical lung".

Most of the spiracles of aquatic insects are non-functional. Usually only two at the tip of the abdomen are open to the external environment.

Fig. 19-11   The aquatic beetle, *Dytiscus*, traps an air bubble and uses it as a physical lung.

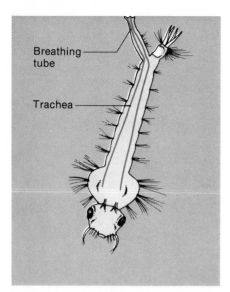

Fig. 19-12 The breathing system of a mosquito larva. A breathing tube allows air to enter the tracheae, making gas exchange possible. Although it is not shown here, the tracheal system branches to all parts of the body.

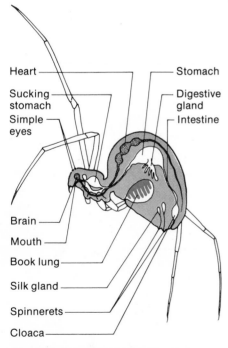

Fig. 19-13 A spider, showing book lungs. Air is drawn into the lung chamber and expelled by muscular action. Gas exchange occurs across the book lung membranes.

Thus, some aquatic insects can exchange gas with the atmosphere by forcing the tip of the abdomen up out of the water and into the air. Mosquito larvae carry out gas exchange in this manner (Fig. 19-12).

Lung Systems

Lungs are specialized gas exchange organs that are located in a certain region of the animal body. Closely associated with the lungs is a blood transport system. Gases are transported between the lungs and the internal body cells by the blood. Lungs are most characteristic of the vertebrate animals. Some fish, most amphibians, and all reptiles, birds, and mammals have lungs. However, some invertebrates have developed simple lungs.

Spiders

The spider, an arthropod, has a rather unique pair of lungs for gas exchange. Each lung has a number of leaf-like folds which resemble the pages of a book. As a result, they are called **book lungs**. Air enters the book lungs through slit-like openings on the ventral (lower) surface of the abdomen (Fig. 19-13). Gas exchange occurs across the membranes of the folds. A blood system carries gases dissolved in the blood to and from all parts of the spider's body. Sometimes tracheae are present; however, these are usually located only in a small area of the abdomen and are really of little significance.

Amphibians

The lungs of an amphibian, such as the frog or salamander, consist of a pair of thin-walled sacs (Fig. 19-14). Air enters the mouth cavity through the nostrils as the floor of the mouth is lowered (Fig. 19-15). The nostrils are then closed and the mouth floor raised, forcing the air into the lungs. The air is then forced out of the lungs by a contraction of muscles in the body wall. Small blood vessels line the lungs. These absorb oxygen from the air in the lungs and release carbon dioxide into that air. The lungs have numerous infoldings on their inner surfaces to increase the surface area.

When hibernating during the winter in the mud bottom of a pond, a frog is able to obtain sufficient oxygen by diffusion through its skin. In fact, at all times of the year, the skin is the major breathing surface when the frog is underwater. The mucous lining of its mouth is also an organ for gas exchange. The lungs of reptiles are similar to those of amphibians, except that the inner surfaces of reptilian lungs have many more infoldings.

Birds

The gas exchange system of a bird begins with two nostrils at the base of the beak (Fig. 19-16). Through these, air moves into the **trachea** or windpipe. A shorter tube called the **bronchus** (plural: **bronchi**) goes from the trachea to each of the two large lungs. Each bronchus branches into numerous fine tubes called **bronchioles** within each lung. Leading off from each lung are five **air sacs** that are located in the body cavity. Some of these air sacs extend into hollow bones. The complete breathing system of a bird occupies about one-fifth of its body volume.

To inhale, a bird lowers its **sternum** or breast-bone. Air rushes into the lungs and air sacs. Gas exchange occurs across the thin membranous bronchioles of the lungs and the walls of the blood vessels that line the bronchioles. No gas exchange occurs in the air sacs since they are not supplied with these blood vessels.

To exhale, a bird raises its sternum. The stale air is forced out of the lungs. At the same time, the fresh air in the air sacs rushes out through the lungs, providing additional oxygen for the bird. Thus a bird's lungs get fresh air both when it is inhaling and when it is exhaling.

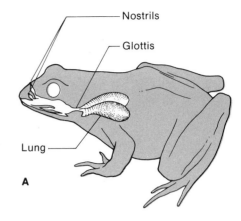

A

Mammals

The gas exchange system of mammals also involves lungs. Passageways connect the internal lungs to the external air. These passageways consist of many fine ducts leading into a vast number of tiny blind air sacs or **alveoli** (sing. **alveolus**). The actual gas exchange occurs in these air sacs. Their surfaces are covered by numerous blood vessels. Mechanical means are used to get air to enter and leave the lungs. A muscle layer called the **diaphragm** is a characteristic part of mammalian lung systems. It performs an important role in the process of breathing. The structure and function of mammalian lungs will be studied in detail in Section 19.6, using the human gas exchange system as an example.

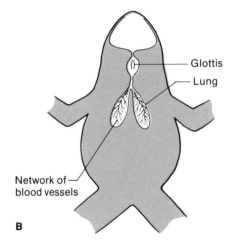

Review

1. How does gas exchange take place in an earthworm?
2. **a)** What are tracheae?
   b) What are lungs?
   c) What are the advantages of tracheal and lung systems for gas exchange in a terrestrial environment?
3. In contrast to insects, fish do not have a system of air tubes to carry the oxygen rich water to the different tissues of the body. How does oxygen get to these tissues?
4. A frog can obtain oxygen three ways: by diffusion through the skin, by mouth breathing, and by lung breathing.
   a) Which method do you think can supply oxygen at the fastest rate? Why?
   b) A frog that is hibernating during the winter requires only diffusion through the skin to keep it alive. Why?
   c) A frog that is resting on the edge of a pond may use only mouth breathing. That is, air is brought into the mouth but is expelled before it gets into the lungs. In contrast, a frog that is hopping must use lung breathing as well. Why is this so?
5. Reptilian lungs have a greater number of infoldings than amphibian lungs. Explain why this is an advantage.

B

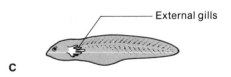

C

Fig. 19-14   The lungs of a frog. A: as seen from the side. B: as seen from the top. C: in the early tadpole stage when gills are external. (External gills are later replaced by internal gills and finally by lungs in the adult).

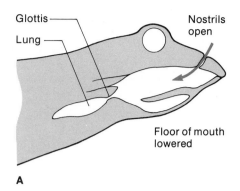

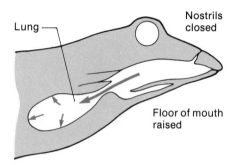

Fig. 19-15 Breathing movements of a frog. A: air enters the mouth cavity as the nostrils are opened and the floor of the mouth is lowered. B: air is forced into the lungs as the nostrils are closed and the floor of the mouth is raised. The elastic lungs expand.

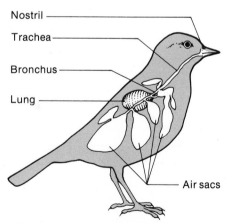

Fig. 19-16 Lung system of a bird. Air sacs adapt the bird for flight. They provide the continuous supply of oxygen-rich air that is needed to release the large amount of energy that flight requires.

6. **a)** Describe the gas exchange system of a bird.
   b) Explain how the bird's gas exchange system allows fresh air to pass through the lungs both on inhalation and exhalation.
   c) What advantage does this adaptation provide for the bird?
7. If a *Dytiscus* were placed in water that had just been boiled then cooled, it might die. Why?

19.6 The Human Gas Exchange System

Your breathing system is similar to that of most large land animals (Fig. 19-17). The basic breathing unit is the lung. However it is really the capillary network surrounding the air spaces of the lung that is the key part of most animals' breathing systems. This section deals first with the air passage and then with the lungs.

The Air Passage

The route that air travels to and from the lungs is called the **air passage**. It consists of three main sections: the nasal cavity, the pharynx, and the tracheal passage.

Nasal cavity

As air moves through your nasal cavity, it is filtered, warmed, moistened, and analysed. The hairs in your nose filter out the larger particles of foreign material. The smaller particles may stick to the moist layer of mucus that protects all the internal surfaces of the air passage. **Mucus** is a mixture of the protein mucin and water. In addition to cleaning the air, the mucus also moistens it. If the incoming air were not moistened, it would tend to dry out the lungs and decrease the rate of gas exchange. (Recall that a membrane must be moist for gas exchange to occur through it.) Mucus must be produced continuously to replace mucin lost by sneezing and water lost by evaporation.

Before entering the lungs, the air is warmed as it passes over the nasal membranes and **turbinate bones** (Fig. 19-18). If this did not happen, cooling of the blood in the lung capillaries would result. Chilling of the lung surface would increase the chances of infection.

Special areas in the nasal passage have many nerve cells, called **receptors**, that are sensitive to odours. The sense of smell is an important protection to an animal, and to the breathing system in particular. Many dangerous pollutants such as ammonia and hydrogen sulfide are easily detected. Your body may warn you to leave the environment before harm is done. Unfortunately, many pollutants such as carbon monoxide are odourless.

Fig. 19-17   The human breathing system.

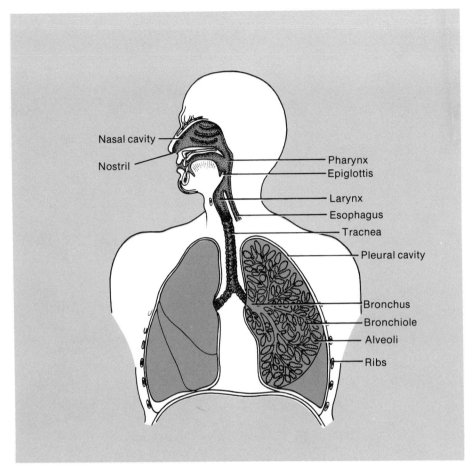

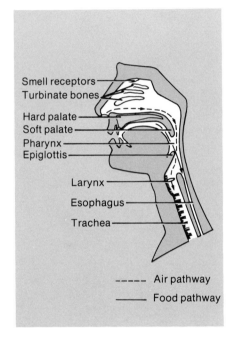

Pharynx

The air coming through your nasal passage enters the top of your pharynx (throat). Here this airstream may join with air inhaled by mouth. Air inhaled by mouth travels much faster than air inhaled by nose. As a result, it cannot be warmed as much, nor can it be as well filtered or moistened. Small particles will stick to the sides and roof of your mouth, and water will evaporate into the airstream. Breathe through your open mouth for a minute or two without rewetting your mouth. As the water evaporates, the sticky mucin remains.

Tracheal passage

At the back of the pharynx, the airstream passes over the **epiglottis** into the **trachea** (windpipe). You may recall from Chapter 18 that the epiglottis prevents food from entering the trachea. Why does air not go down the esophagus into the stomach?

The trachea is a tube 10 cm in length (Fig. 19-19). It is prevented from collapsing by several horseshoe-shaped segments of cartilage (Fig. 19-20).

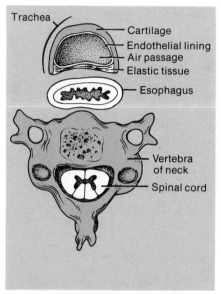

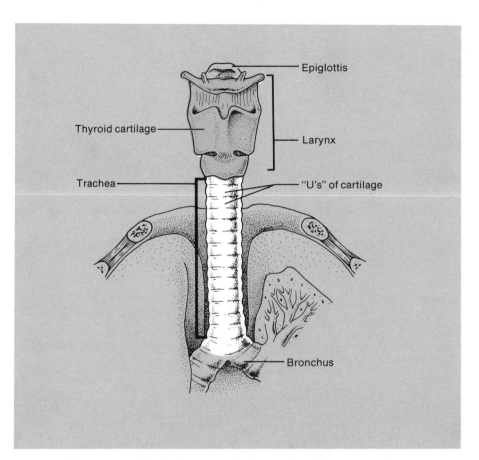

Fig. 19-20   Notice that each cartilage of the trachea does not completely encircle the air passage. What must happen when food is swallowed down the esophagus?

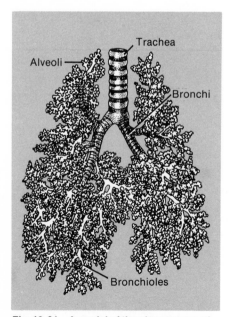

Fig. 19-21   A model of the air passages of the lungs. Such models can be made by filling all the air passages of a dead animal with a rubbery liquid. It is allowed to harden, and the tissues are dissolved away.

The trachea is lined with cells that have tiny hairlike **cilia** on their surface. The waving action of the cilia continually sweeps particles of foreign matter upward to the pharynx. The cartilage at the top of the trachea is modified to form the **larynx** (voicebox), where exhaled air is used to produce sound. The bottom of the trachea branches into two **bronchi** (singular: **bronchus**), each leading to a lung unit.

Structure of the Lung

Within each lung, the bronchus branches and rebranches into smaller and smaller air tubes called **bronchioles**, or "small bronchi" (Fig. 19-21). Like the trachea, both the bronchi and bronchioles are lined with ciliated cells. Also, their walls are supported by complete rings of cartilage. The bronchioles end in clusters of microscopic air sacs called **alveoli** (sing. **alveolus**). Each alveolus is surrounded by a network of **capillaries** (Fig. 19-22). Only at the alveolus is the tissue thin enough to let oxygen move into the blood and to let carbon dioxide move into the airspace. The large number of alveoli (approximately 3×10^8, or three hundred million) give your lungs a spongy texture. The capillaries give them a pink colour.

Gas Exchange in Lungs

The function of the breathing system is to absorb oxygen into the body and to expel carbon dioxide from the body. Let us see how this occurs.

You may recall from Chapter 6 that diffusion is the process by which molecules move from a region of high concentration to a region of low concentration of that molecule. Dark red blood, poor in oxygen and rich in carbon dioxide, is pumped by the heart into the **pulmonary arteries** (See Fig. 19-22). Eventually this blood ends up in a capillary surrounding the airspace of an alveolus (Fig. 19-23). Fresh inhaled air in the airspace is about 21% oxygen. This is a very high concentration compared to what is in the blood. Therefore oxygen molecules diffuse from the airspace into the blood stream through the thin lining of the alveolus (Fig. 19-24). The now bright red, oxygen-rich blood moves through the capillary into the **pulmonary vein** and back to the heart. It is continually replaced by oxygen-poor blood. Thus the diffusion of oxygen never stops so long as the flow of blood continues.

At the same time that oxygen diffusion is occurring, carbon dioxide is diffusing from the blood into the air space of the alveolus. Fresh air contains only 0.034% carbon dioxide, which is a lower concentration than that in the

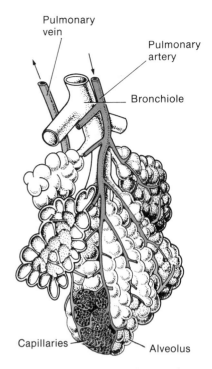

Fig. 19-22 The grape-like clusters of alveoli are found at the very ends of the bronchioles.

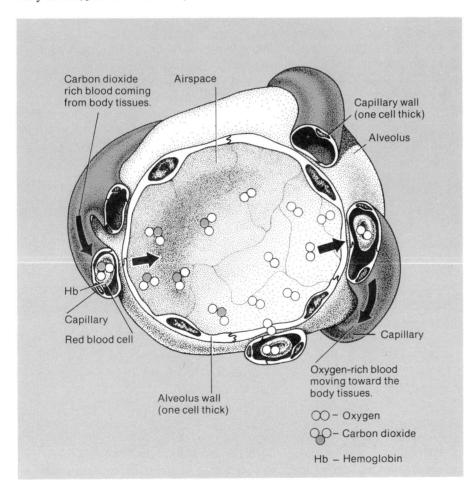

Carbon dioxide rich blood coming from body tissues.

Airspace

Capillary wall (one cell thick)

Alveolus

Hb

Capillary

Red blood cell

Alveolus wall (one cell thick)

Oxygen-rich blood moving toward the body tissues.

Capillary

○○ – Oxygen

⊘○ – Carbon dioxide

Hb – Hemoglobin

Fig. 19-23 Gas exchange. Carbon dioxide is expelled and oxygen absorbed by diffusion.

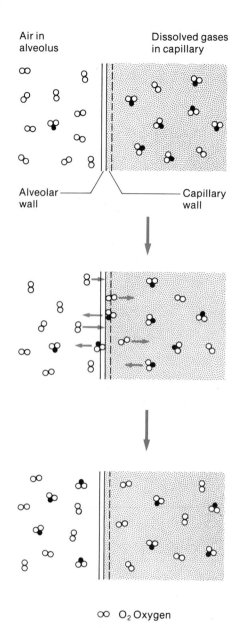

O₂ Oxygen
CO₂ Carbon dioxide

Fig. 19-24 Gas exchange. As blood moves through a capillary of the lung, its carbon dioxide concentration decreases while its oxygen content increases. A thin layer of water over the alveolar wall is necessary for diffusion to occur.

blood. Therefore diffusion of carbon dioxide occurs from the blood into the airspace. If you hold your breath you can stop the diffusion; the amount of carbon dioxide in the air space will increase until it eventually equals the concentration of carbon dioxide in the blood. The effect of gas exchange on fresh air is summarized in Table 19-5.

	PERCENT OXYGEN	PERCENT CARBON DIOXIDE	PERCENT WATER
Inhaled air	21	0.034	1
Exhaled air	16	4	6

TABLE 19-5 Effect of Gas Exchange on the Air You Breathe

Review

1. What is the function of the breathing system?
2. a) What effect does the nasal cavity have on inhaled air?
   b) What effect does it have on exhaled air?
3. a) What disadvantage does mouth breathing have?
   b) What advantage does mouth breathing have?
4. List, in order, the structures a microscopic camera would photograph if it were inhaled through a nostril and travelled to an alveolus.
5. How are oxygen molecules delivered to body cells from the alveolus?
6. a) Why are alveoli called "the sites of gas exchange"?
   b) Why does carbon dioxide leave the bloodstream at an alveolus, yet enter the bloodstream from the body tissues?
7. a) Design a "human" in which there is no crossing of the air and food pathways (see Fig. 19-17). Make a sketch of the head and neck regions of your new human. What advantage is there in the present design?
   b) Find out how the breathing system of whales is structured. Is it separate from the digestive system?

19.7 Investigation

Measuring Air Inhalation Time

Air rushes into the lungs to fill an expanding chest volume. Determine your inhalation time by trying this short investigation.

Materials

stop watch (1)

Procedure

a. Exhale until you can force no more air out of your lungs.
b. Have your partner measure the time taken for a full inhalation through just your nose.

c. Repeat steps a and b, this time inhaling by mouth only.
d. Change roles with your partner and repeat the entire experiment.
e. Record your data in a table similar to Table 19-6.

BREATHING PATH	INHALATION TIME (SECONDS)
Nose breathing	
Mouth breathing	

TABLE 19-6   **Inhalation Time**

Discussion

1. Account for the different inhalation times.
2. **a)** What advantages does mouth breathing have?
   b) What advantages does nose breathing have?
3. How does the size of the air passage affect the breathing rate?
4. Explain why coughing is so effective at expelling irritating particles from the trachea and pharynx. Refer to Figure 19-20 in your answer.

19.8 Investigation

Do You Produce Carbon Dioxide?

Proving experimentally that you produce carbon dioxide is more complicated than you may think. Be sure you understand the design of this investigation. The test for carbon dioxide uses bromthymol blue indicator. If carbon dioxide is present, the blue indicator will turn yellow.

Materials

bromthymol blue indicator
    solution
drinking straw
50 mL beaker

flasks (2) with stoppers, glass
    and rubber tubing, a "Y" tube,
    and a clean mouthpiece
    (see Fig. 19-25)

Procedure A

a. Place about 20 mL of bromthymol blue indicator solution in the beaker.
b. Blow gently through the straw into the solution.
c. Continue blowing until a colour change occurs in the indicator solution.

Procedure B

a. Set up the apparatus as shown in Figure 19-25. Be sure to obtain a clean mouthpiece.

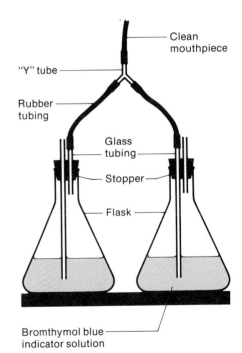

Fig. 19-25   **Do you produce carbon dioxide? How will air flow through this apparatus?**

b. Exhale gently into the mouthpiece. Note the flask in which bubbling occurs. Record your observation in a table similar to Table 19-7.

c. Now inhale through the same mouthpiece. Note the flask in which bubbling occurs. Record your observation.

d. Continue to breathe through the apparatus until there is a difference in colour between the 2 flasks. Record your results in a table similar to Table 19-7.

	FLASK A	FLASK B
Effect of exhaling		
Effect of inhaling		
Appearance after 1 min		

TABLE 19-7   Testing for Carbon Dioxide

Discussion

1. a) What is the effect of bubbling your breath through the indicator solution in Procedure A?
  b) What is the significance of this observation?
  c) Why does Procedure A not prove that you produce carbon dioxide? Where else could the carbon dioxide have come from?

2. a) Which flask bubbles when you exhale? Explain why this happens.
  b) Which flask bubbles when you inhale? Explain why this happens.

3. a) In which flask is the indicator solution not affected by the bubbling?
  b) Why is this observation so important?

4. a) What else besides carbon dioxide is added to exhaled air from your body? Design an experiment to prove this.

5. Design an experiment to show that another animal produces carbon dioxide. Fish, frogs, insects, gerbils, and hamsters may be good subjects. Pay careful attention to keeping the animal supplied with oxygen.

6. Design an experiment to show that a burning candle produces carbon dioxide.

7. How could you show that oxygen is taken out of the air inhaled into your lungs?

19.9 Investigation

How Much Oxygen Do You Need?

To find out how much of the world's oxygen you use up each day, you must find several different numbers. You need to know how many breaths you take during one day. Also, you need to know the size of each breath. Finally, you need to know how much oxygen is actually absorbed from all the air that goes in and out of your lungs.

Materials

500 mL (at least) jar (1)
500 mL graduated cylinder
overflow trays and sink
flexible tubing

watch or a pendulum timer
  that swings at 1 s intervals
glass plate

Procedure

a. Make a table similar to Table 19-8 in your notebook. Record all your measurements and calculations in it.

b. Count the number of breaths (an inhalation followed by an exhalation) that your partner takes in 1 min while sitting quietly. Record this number.

c. Multiply this number by 60 and then by 24 to find out the number of breaths taken in one day.

d. Fill the large jar with water. Invert it in the overflow tray as shown in Figure 19-26. Use the glass plate so that you do not lose any water from the jar.

e. Have your partner exhale 3 breaths through the tubing into the jar, breathing at the same rate that you measured earlier.

f. Cover the mouth of the jar with the glass plate. Remove the jar from the tray and set it upright on the desk. Determine the volume of air in the jar by adding water with the graduated cylinder until the jar is full.

g. Calculate the average breath volume by dividing the volume of air by 3.

h. Calculate the total amount of air that goes in and out of your body in one day using this formula:

Total daily volume = Number of breaths in one day × Volume of breath

i. Use the data in Table 19-5 to calculate the percentage of the air volume that is actually oxygen absorbed into the body.

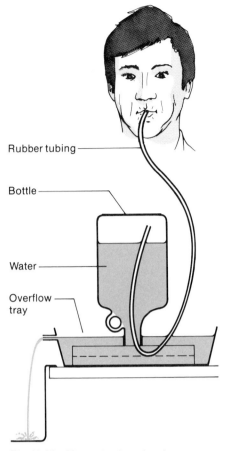

Fig. 19-26   Measuring breath volume.

Breaths taken in 1 min	
Breaths taken in 1 d	
Volume of 3 breaths	_____mL
Volume of 1 breath	_____mL
Volume of air taken in during 1 d	_____mL
Percent oxygen in inhaled air	_____%
Percent oxygen in exhaled air	_____%
Percent of air volume absorbed	_____%
Volume of oxygen absorbed in 1 d	_____mL

TABLE 19-8   **Oxygen Requirement**

Discussion

1. Compare your answer with those of your classmates. How do you compare? Calculate the class average. Try to determine why your value differs from those of other students.

2. a) The average breath volume of a resting person is 0.5 L. The average breathing rate is about 14 breaths/min. Using this information, calculate the average daily oxygen consumption.

    b) How does the value that you calculated in a) compare with the value obtained in the investigation? Account for any difference in the values.

3. a) How should exercising affect your oxygen consumption?

    b) Where will the effect of exercising show up in the calculations?

4. Calculate how much oxygen the world's four billion (4 000 000 000) people use up each day. Why does the oxygen supply of the world not get used up?

5. Suggest a method of measuring your breath volume using a plastic bag. Give two advantages this method has over the method that you used here.

19.10 The Breathing Mechanism

It is very important to realize that the lungs are attached to the body only by the bronchi. They are supported by, but not attached to, the diaphragm and ribs. There is a thin lymph-filled space, the **pleural cavity**, between the lungs and the inner wall of the chest cavity. Each lung is contained in an outer membrane that permits almost frictionless rubbing of the lungs inside the chest cavity. Sometimes chafing of the lungs occurs as they continually expand and contract against the supporting surfaces. This painful condition is called **pleurisy**.

How Breathing Occurs

Very carefully try to inhale and exhale three or four breaths of air without moving your ribs. Can you do it? Did you feel your abdomen moving in and out? How must your diaphragm move to push your abdomen out? What does moving your diaphragm have to do with getting air into your lungs?

    Now, keep your abdomen normal, and inhale and exhale three or four times moving only your rib cage. How do your ribs move when you inhale? And when you exhale? What does moving your ribs have to do with getting air in and out of your lungs?

    Summarize your results in a table similar to Table 19-9.

	MOVEMENT OF DIAPHRAGM	MOVEMENT OF RIB CAGE	EFFECT ON SIZE OF CHEST CAVITY
Inhalation			
Exhalation			

TABLE 19-9 **Breathing Movements**

Inhalation

As you can see from Figure 19-27, a contracting diaphragm and rising ribs can cause an increase in the volume of the chest cavity. The lungs expand to fill the available space. This happens because the larger chest volume has reduced the pressure of the air in the airspaces of the alveoli. The outside air pressure is higher, and pushes air into the lower pressure area of the chest cavity, thereby inflating the lungs. The surface tension of the lymph between the lungs and rib cage also helps pull the surface of the lungs outward along with the expanding rib cage.

Exhalation

Not as much muscular effort is required to exhale as to inhale. Relaxation of the diaphragm and the rib muscles is normally sufficient to push air out of the lungs. Gravity plays an important role in pulling the rib cage downwards. The lungs themselves are normally quite elastic, and contract from their forced expansion readily.

Forced exhalation is possible by contracting some sets of muscles. Contracting your abdominal ("stomach") muscles squeezes your abdominal contents upward. This reduces the volume of the chest cavity. Contraction of another set of rib cage muscles can pull the rib cage down further.

Whatever method is used, the effect is the same. The chest cavity volume is reduced. The air in the alveoli is compressed, just for an instant. It is then at a higher pressure than the outside air. The air then moves from the area of high pressure to the area of lower pressure by the easiest route: up the bronchi, trachea, larynx, and out the mouth or nasal passages.

First Aid for Choking

A useful lifesaving technique, the **Heimlich Maneuver**, relies on the high pressure air formed by contraction of the chest cavity. Many people have died by suffocation from choking on food. Often food can be removed from the throat with the fingers. However, if the food has become stuck below the epiglottis, it usually cannot be reached with the fingers. In this case the air passage can often be cleared easily by the Heimlich Maneuver (Fig. 19-28).

Procedure for the Heimlich Maneuver

This procedure can be done with the victim lying on his or her back on the floor, or by standing behind the victim.

STEP 1 Place your fist between the victim's "belly-button" and rib cage. Place your other hand on top of the fist.

STEP 2 Press forcefully into the victim's abdomen with a quick, upward thrust.

STEP 3 Repeat steps 1 and 2 if necessary.

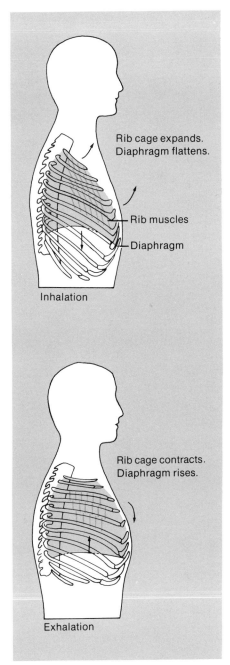

Fig. 19-27 The mechanism of breathing. Contraction of the rib muscles raises the rib cage. The chest falls as they relax. Describe the movements of the diaphragm.

Fig. 19-28   The Heimlich Maneuver.

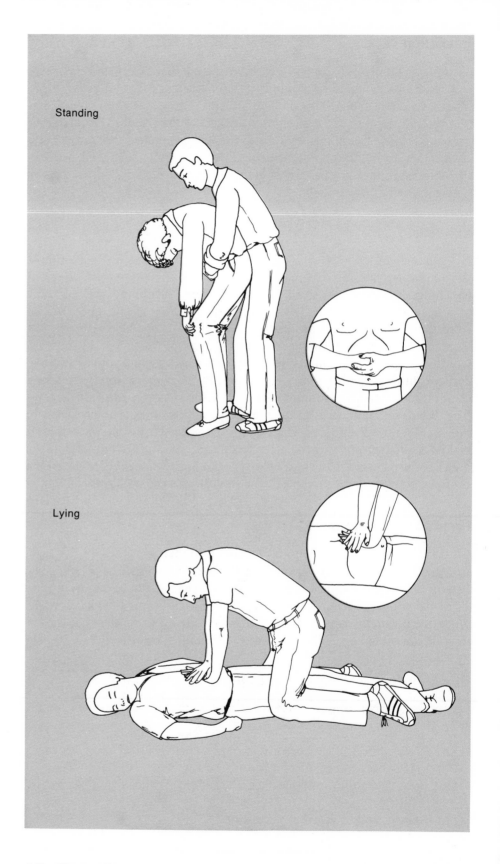

Standing

Lying

Be sure you realize you are reducing the chest volume by pushing the stomach and diaphragm upward. Do not press on the ribs. Very high pressures are produced in the chest cavity using this method. Chunks of food have shot out over 6 m from the victim.

Artificial Respiration

Some outdated methods of artificial respiration relied on the same principles as normal breathing. Air could be drawn into the lungs if the rib cage was expanded or if the diaphragm forced down. But the **mouth-to-mouth method of pulmonary resuscitation** is much more efficient and versatile to use (Fig. 19-29). Air is forced into the lungs under pressure; gravity and the natural elasticity of the lungs pushes it back out. The airway must be clear and the tongue away from the back of the throat. Make sure you know how to give mouth-to-mouth resuscitation properly.

Frogs use much the same principle in their breathing. They draw air into their mouth by enlarging the mouth cavity. They then compress the air by shrinking the mouth. Air is forced into the lungs because the air in the mouth is at a higher pressure than the air in their lungs just as in mouth-to-mouth resuscitation. Similar pressure changes between the mouth and the vocal sac allow frogs to croak even under water!

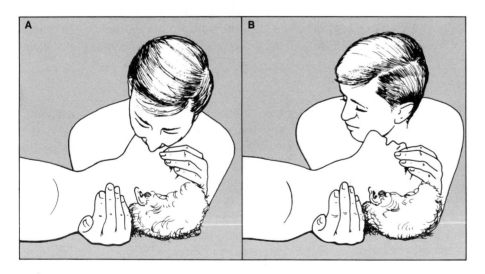

Fig. 19-29   Mouth-to-mouth resuscitation. A: nose pinched, airway cleared and straight. Watch for rise in chest. B: listen and watch for passive exhalation; then ventilate victim's lungs again.

Review

1. **a)** Describe how the diaphragm and rib muscles assist breathing during inhalation and during exhalation.
   b) Why does air move into the chest cavity during inhalation?
   c) Why does air move out of the chest cavity during exhalation?
2. Explain how the Heimlich Maneuvre works.
3. Why is "artificial respiration" not a technically correct term? ("Pulmonary resuscitation" is the term now gaining popularity.)

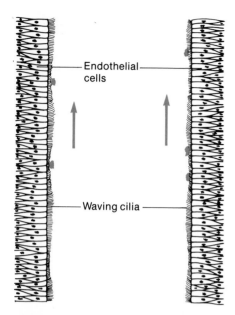

Fig. 19-30 Healthy nasal passages, trachea, bronchi, and bronchioles are lined with ciliated cells. Particles are swept upward by the cilia toward the pharnyx.

Endothelial cells

Waving cilia

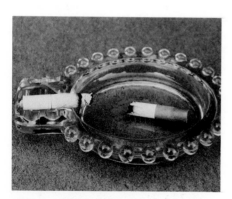

Fig. 19-31 The problem: Voluntary short-circuiting of the breathing defenses.

19.11 Health of Your Breathing System

Your lungs are the largest tissue area in your body that is exposed to the outside environment. Every kilogram of body cells needs $1 m^2$ of lung surface to absorb the necessary amount of oxygen. (What is the surface area of your lungs?) Lungs also present the thinnest possible barrier between "inside" you and whatever is "outside".

Your body has several different structures that help protect the lungs. These include the nasal passages, the pharynx, and the ciliated cells of the nasal passage, trachea, bronchi, and bronchioles (Fig. 19-30). Even the cleanest "country air" contains 40 000 particles in every breath. Only the smallest particles of dust, bacteria, and viruses ever reach the alveoli.

Smoking

Smoking is the most direct way to "short circuit" the breathing system's defenses (Fig. 19-31). Inhaling by mouth obviously prevents any nasal filtering. Hot vapourized gases enter your cooler mouth, pharynx, and trachea, and condense into "tars". A heavy smoker can inhale 2 L of tars per year! Cigarette tar is a mixture of several different chemicals, many of which are dangerous to humans. The tars condense between and over the cilia that line the trachea, bronchi, and bronchioles. Carbon particles from the smoke collect on the sticky surface. The gradual build-up of tars eventually "gums up" the cilia. They can no longer sweep particles up and out of the system.

Lung cancer

The chemical called nicotine is the stimulant that is best known and desired in tobacco smoke. Unfortunately, tobacco smoke contains other chemicals. Benzopyrene is just one of several known cancer-causing chemicals that are deposited with the tars. Lung cancer is very hard to detect. The spongy lung tissue can hold the growing mass of cancerous cells without discomfort. Often the mass is not found until it is too late to operate successfully. Growing tumours can interfere with the circulation of the entire body as they press on major veins and arteries near the heart (Fig. 19-32). Spreading of the cancer throughout the body is made easy by the rapid circulation of blood through the lung tissue. Smokers have a significantly higher chance of getting lung cancer than non-smokers. In 1974, lung cancer was the third-ranked cause of death in Canada.

Emphysema

The disease called emphysema is rapidly becoming a major killer among our ever-growing elderly population. Air pollutants, including cigarette smoke, gradually wear away the walls separating one alveolus from another. The effect is a reduction of lung surface area (Fig. 19-33). Deeper and faster breathing is required to balance the loss of surface area. The problem is

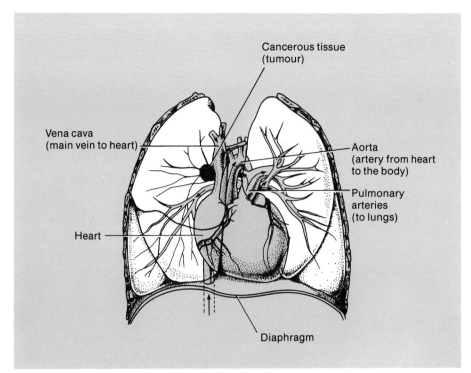

Fig. 19-32 A swelling mass of cancerous cells (tumour) can reduce the flow of blood along major veins and arteries. This extra tissue in the pleural cavity can also block off bronchioles and reduce breathing efficiency.

made worse by the loss of elasticity of the lungs. This means that more muscular effort is needed to expand and contract the chest cavity. In the advanced stages of emphysema, the person often lacks the strength in the lungs to blow out a burning match!

Chronic bronchitis

A closely related disease is chronic bronchitis. Irritation of the inner membranes of the bronchi and bronchioles causes swelling. Mucus accumulates and further irritates the surrounding tissues. The inner diameter of the bronchi and bronchioles is reduced. Airflow is restricted. What effect would this have on breathing rate and depth?

Both chronic bronchitis and emphysema are often found in the same patient. This condition is given the term "chronic obstructive pulmonary disease". These diseases occur much more frequently in smokers than in non-smokers.

Damage to the Breathing System

"Collapsed lung" can result if a chest wound lets outside air into the chest cavity. Even if the chest cavity expands, the lungs will not inflate. Instead, more outside air will fill the space between the lungs and inside rib surface. Without a pressure difference between outside air and the chest cavity, the elastic lung tissues will not expand. Sealing of the chest wound and use of mouth-to-mouth resuscitation are successful first aid procedures.

The same condition can affect careless scuba divers. If divers come up

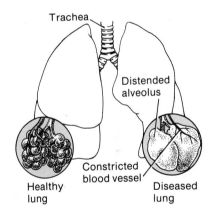

Fig. 19-33 Effects of emphysema on lung.

to the water surface holding their breath, the air in their lungs expands. A weakened alveolus can rupture and let air out into the chest cavity. Once again, the lung will no longer inflate because there is no longer any pressure difference between outside air and the chest cavity.

Review

1. How does smoking affect the defenses of the breathing system?
2. How can lung cancer affect the body?
3. **a)** Describe the effects of bronchitis and emphysema occurring in the same person.
   b) Why are older people especially affected by bronchitis and emphysema?
4. Explain why a chest wound must be covered as part of first aid procedure.
5. Design and carry out a survey to determine the background and the attitudes toward smoking of both smoking and non-smoking students.

Highlights

All animals must have a moist gas exchange surface. Gas exchange in simple animals can take place by direct diffusion across such a surface. However, complex animals must have some means such as a tracheal system or circulatory system to transport the gases between this gas exchange surface and every body cell.

Cellular respiration is the process by which organic materials (food) is oxidized in cells to release energy. Breathing is the mechanical process by which oxygen and carbon dioxide are exchanged between an organism and its environment.

All animals must have a gas exchange surface of adequate size. As the size of an organism increases, the surface area to volume ratio decreases. Therefore larger organisms generally have some special organs such as gills or lungs that increase the surface area for gas exchange.

PROBLEM	SOLUTION
1. Maximization of oxygen absorption and carbon dioxide expulsion	Large lung surface area; hemoglobin in blood carries oxygen
2. Reduction of heat and water loss and chilling of the lungs	Restricted opening to lung; preheating of inhaled air
3. Regulation of gas exchange	Nervous and chemical control of breathing rate
4. Protection of system	Lungs protected by ribs; trachea, bronchi, and bronchioles held open by cartilage; filtering of air by nose and air passage membranes; ciliated cells lining passageways

TABLE 19-10 **Problems Faced by Terrestrial Mammals**

The structure of the human breathing system deals efficiently with the four problems faced by warm-blooded terrestrial animals. Table 19-10 summarizes these four problems and shows how the structure of the human breathing system solves each problem.

20

Internal Transport

Chapter 18 described how animals get the nutrients required for their life functions. But how are these nutrients transported within the animal? How are harmful byproducts that are produced by life functions transported within the organism? What special means do single-celled organisms and small multicelled animals have for transporting substances? What special problems do larger multicellular organisms have? How do they solve these problems?

In this chapter we will discuss the need for internal transport systems. We will also examine some of the methods by which materials are transported within organisms.

20.1 The Need for Internal Transport Systems

Every living cell must be able to perform all the functions associated with life. It must, then, be able to carry out its own metabolic activities. It must be able to do this whether it is an independent, single-celled organism or part of a multicellular organism. You know the importance of raw materials such as food and oxygen in maintaining life functions. It is also essential that each cell rid itself of the harmful byproducts of its metabolic activities. For example, carbon dioxide is a byproduct of respiration. Every cell must be able to get rid of this substance. Therefore, all cells must be in contact with a medium from which they can get required raw materials and into which they can excrete their harmful byproducts (sometimes called wastes).

Single-celled organisms and some simple multicellular animals have no problems in this respect. Every cell is either directly surrounded by this medium or very close to it. In more complex and larger animals, many cells are far removed from this medium. Therefore, in such animals a greater division of labour occurs. As a result, many functions such as digestion and gas exchange take place in specialized body areas. Materials must be taken to and from these specialized areas by an internal transport system.

A further need for an internal transport system results from the fact that certain specialized cells produce important substances that must reach every cell. These substances help to control and coordinate life processes within the animal. For example, a chemical messenger called a **hormone** may be produced by a group of specialized cells in one part of an organism. Yet it controls the actions of some cells in another part of the organism, far removed from the source of production.

There is a need, then, for an internal transport system to move materials between specialized areas and the rest of the individual cells in larger animals.

Review

1. What kinds of materials are exchanged between a cell and its environment during metabolic activities?
2. Why is there a need for an internal transport system in larger multicellular organisms?

20.2 Transport in Structurally Simple Organisms

Diffusion and Cyclosis

Diffusion plays a key role in the transport of materials in single-celled organisms and in some small multicellular animals. Diffusion carries materials from where they enter or are formed in these organisms to where they are used or disposed of by the organisms. However, diffusion is a very slow means of transport. Often it cannot, by itself, transport materials fast enough to meet an organism's needs (Fig. 20-1). Therefore, in some organ-

Fig. 20-1 Every living cell requires certain substances from its environment and must get rid of certain harmful by-products. What problems would the coloured cell have in this respect? Could the problems be lessened if the organism had a different shape?

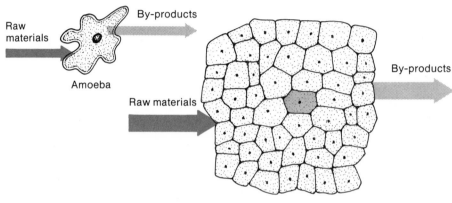

isms, the cytoplasm within the cells moves. Materials in the cytoplasm get carried along with it. Such movement can be seen in an amoeba when it is observed under high power on an ordinary laboratory microscope. This movement of the cytoplasm is called **cyclosis** or **cytoplasmic streaming** Materials can be transported by this means much faster than by diffusion. It is not entirely clear how this process works. However, it is known that energy is required during the process. Therefore the cells in which cyclosis occurs must be living and able to use energy to carry out this means of internal transportation.

Transport in Hydra

In a small single-celled organism, diffusion and cytoplasmic streaming are adequate means of transportation. They are also adequate means in some small multicellular animals in which the individual cells are not too far from the external environment. For example, the hydra has a body wall that is two cells thick. As you can see from Figure 18-5 (page 457), all its cells are in direct contact with water. Water is the medium from which the hydra receives the raw materials required for metabolism and into which it excretes its byproducts. The inner layer of cells is exposed to water because of the presence of a **gastrovascular cavity**. A branch of this cavity extends into each tentacle. Specialized cells with **flagella** provide a sweeping action that creates a current in the water. Specialized muscle cells in the body wall

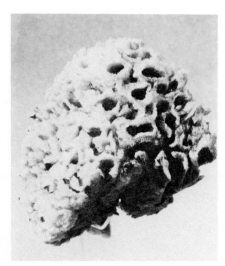

Fig. 20-2 This sponge is one of more than 5 000 known species of sponges. Most sponges are marine; that is, they live in salt water.

enable various parts of the hydra body wall to contract. Such contractions and currents cause circulation of water in the cavity. This ensures that a constant supply of water, rich in raw materials, enters the cavity. Furthermore, waste-laden water is forced away from cells and eventually out of the cavity altogether.

Even though all cells of the hydra do not actually take in food, no cell is very far from specialized ones that do. Food is taken in by specialized endoderm cells that line the digestive cavity. It is then circulated within these cells by diffusion and cyclosis. The endoderm cells pass soluble food substances to the ectoderm cells by active transport through the plasma membrane. However, each cell does take in its own oxygen and get rid of its own wastes, such as carbon dioxide.

Transport in Sponges

The sponge is another structurally simple animal that has adaptations which aid in the transport and distribution of materials within its body (Fig. 20-2). Sponges have sac-like bodies which contain numerous microscopic pores and one or more large openings (Fig. 20-3). The inner body wall is lined with specialized **collar cells** which have **flagella** attached to them. Movement of these flagella creates a current that draws water carrying food and oxygen into the body cavity through the **incurrent pores**. This fresh supply of nutrient-rich water forces waste-laden water out of the cavity through the large openings. Each large opening is called an **excurrent pore**, or **osculum**. This mechanism of water circulation is so efficient that even a small sponge may draw in up to 60 L of water a day.

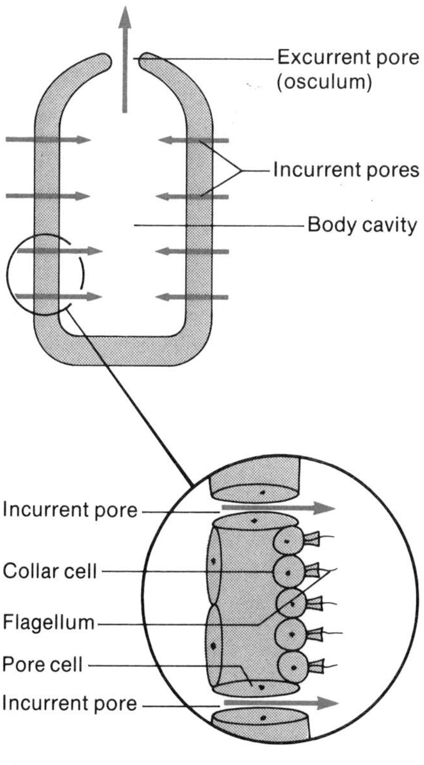

Transport in Other Simple Animals

Planarian

The planarian is another structurally simple animal with an interesting adaptation for transport (Fig. 20-4). Its **gastrovascular cavity** branches into all parts of the body. Therefore it serves as a primitive transport system.

Jellyfish

The body of a jellyfish contains a series of **radial canals** that link the **digestive cavity** to a **circular canal** which is just inside the outer body wall (Fig. 20-5). Contractions of the body wall cause the liquid contents of this system to flow through the canals. Digested materials are then distributed to all parts of the body.

Nematode

Nematodes have a fluid-filled cavity between an inner gut (digestive tube) and the body wall (Fig. 20-6). The fluid is permanently contained in this

Excurrent pore (osculum)

Incurrent pores

Body cavity

Incurrent pore

Collar cell

Flagellum

Pore cell

Incurrent pore

Fig. 20-3   Sponge body plan. The body of a sponge has numerous pores. Water enters the body cavity through the incurrent pores and is expelled through the excurrent pore.

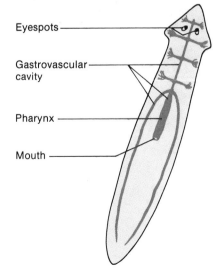

Eyespots

Gastrovascular cavity

Pharynx

Mouth

Fig. 20-4   The planarian. The gastrovascular cavity serves as a transport system.

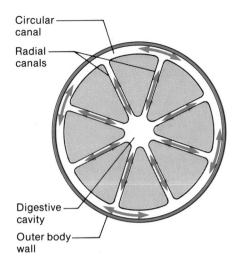

Fig. 20-5 The jellyfish. Note that the digestive cavity is linked to a circular canal, forming a transport system.

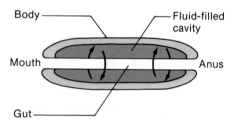

Fig. 20-6 Nematode body plan. Materials dissolved in the fluid in the body cavity are circulated to cells by means of muscular contractions.

cavity. Materials such as digested nutrients, which diffuse into this fluid, are distributed by contractions of the body wall. This represents a rather limited inner circulatory system.

Review

1. **a)** What is diffusion?
   b) What is cyclosis?
   c) What roles do diffusion and cyclosis play in transportation in structurally simple organisms?
2. **a)** Describe the body structure of the hydra.
   b) How is the hydra structurally adapted to ensure that all cells are in contact with a constant supply of nutrient-rich water?
3. What special adaptations do sponges have to aid in the transportation of raw materials and waste materials to and from all cells?
4. How do materials get from one place to another in planarians and jellyfish?
5. What special adaptation to transportation do nematodes have?

20.3 Transport Systems in More Complex Animals

Blood Systems

The majority of multicellular animals have specialized circulatory systems for the transport of materials. In such animals, a body fluid, usually called **blood**, circulates through the body by a system of tubes. Such a system is called a **blood system**. Blood systems generally contain:

1. A transport medium, the blood;
2. A pumping device called a heart to make the blood circulate;
3. Valves to control the direction of blood flow;
4. Tubes or vessels through which blood travels to all parts of the body and then back to the heart;
5. A means by which all body cells can get required raw materials from the blood and get rid of their waste byproducts.

Blood systems have the ability to regulate the flow of blood to tissues depending on their varying requirements. There is not enough blood in any one animal to completely fill every blood vessel at the same time. As a tissue or organ becomes active, vessels supplying blood to it will increase in diameter or **dilate** to increase the blood flow to meet the demand. At the same time, vessels leading to other parts of the body will decrease in diameter or **constrict**, thereby decreasing the flow of blood to those parts. An example of this is evident after one eats a full meal. The digestive organs are very active and require a great deal of blood. Therefore, blood is directed to them. Drowsiness is then experienced because the flow of blood to the brain is somewhat restricted.

All blood systems perform the same basic function. They ensure that the materials which they transport can be freely exchanged with any cell in the body. In various tissues, substances are exchanged across thin membranes by means of diffusion or active transport. Hearts, valves, and blood vessels are simply structures which help in this function.

Open and Closed Systems

In some animals, blood is pumped through vessels into large open spaces known as sinuses. As a result, tissues and organs within these sinuses are bathed in blood, and materials are exchanged by diffusion directly through the cell membranes. The blood eventually gets transported out of the sinuses by means of another set of vessels, thereby completing its circuit. This type of system is called an open transport system. It is characteristic of invertebrates such as arthropods (including insects) and the majority of molluscs.

In other animals, blood flows within vessels throughout the *entire* circuit. This type of system is called a closed transport system. The cells of tissues are bathed in tissue fluid. Substances in the blood capillaries pass out into the tissue fluid and then into the cells of the tissues. The reverse process also takes place. The tissue fluid forms a vital link in the closed transport system. Annelids, some molluscs (squid and octupus), and all vertebrates have this type of closed system. With a closed system, materials can be transported to a given location much more rapidly than with an open system because the blood flows more rapidly. Therefore, very active organisms have their demands met more efficiently by a closed system.

Review

1. **a)** What are the parts of a blood system?
   b) What is the function of a blood system?
   c) Blood vessels can dilate and constrict. Of what value is this to an animal?
2. **a)** Compare open and closed transport systems.
   b) What type of transport system do you have?
   c) Why is it an advantage for you to have such a system?

20.4 Investigation

Circulation—The Capillaries

Although interest in the study of circulation in animals began as early as the time of Aristotle (about 300 B.C.), it was not until the 17th century that the pattern of blood circulation, as we know it today, was discovered. William Harvey, an English doctor, studied the pumping action of various animal hearts. In addition, he observed the presence of vessels carrying blood to the

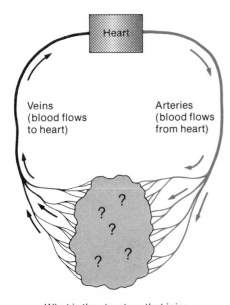

What is the structure that joins the arteries and veins to complete the circuit?

Fig. 20-7 Harvey's hypothesis and the question he could not answer.

heart (**veins**) and of vessels carrying blood away from the heart (**arteries**). Harvey's observations led him to hypothesize about the existence of a closed transport system in which the same blood that left the heart through one type of vessel returned to it through another type. This would make a complete circuit (Fig. 20-7).

Harvey did not have a microscope and thus never saw the structure of the fine vessels connecting the arteries and the veins. Just after Harvey's death, Marcella Malpighi, an Italian anatomist, did see them with the aid of a microscope. Malpighi examined the lung tissue of a frog and observed very fine vessels connecting the pulmonary (lung) arteries and veins. He called these hairlike vessels **capillaries**.

In this investigation you will observe, in the tail fin of a fish, capillaries and the flow of blood through them.

Materials

goldfish in aquarium	microscope
net	eye-dropper
petri dish (one half only)	glass slide
cotton	

Procedure

a. Using a dip net, obtain a small goldfish from the aquarium. If you follow the procedures quickly and carefully, the fish will not be harmed. After 5 min it must be returned to the aquarium.

b. Place the goldfish in a petri dish that has been moistened with water. Try not to touch the fish with your fingers. Handling a fish often removes the mucus coating that protects the fish from fungal and bacterial infections (Fig. 20-8).

c. Wrap the fish in wet cotton, leaving the mouth and tail exposed.

d. Place the petri dish on the microscope stage and focus on the tail fin with low power.

e. Observe the movement of blood through the vessels. The smallest vessels you see are capillaries. Switch to medium power to see the blood cells flowing through the capillaries.

f. Compare the rate and method of blood flow in the arteries, capillaries, and veins. Record your observations in a table similar to Table 20-1.

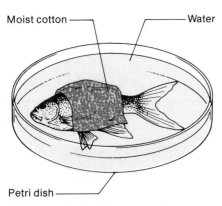

Fig. 20-8 Preparation of a goldfish for examination of capillary circulation in the tail fin.

TYPE OF VESSEL	RATE OF BLOOD FLOW	METHOD OF BLOOD FLOW

TABLE 20-1 **Comparison of Blood Flow**

g. Estimate the diameter of the arteries, capillaries, and veins.

h. Note the shape of the blood cells.

i. Sketch an outline of the fish tail and draw in and label the various vessels you observed. Indicate the direction of blood flow.

j. Return the goldfish to the aquarium (no later than 5 min from the time it was removed).

Discussion

1. In which vessel(s) does the blood flow most rapidly?

2. Compare the diameters of the various kinds of blood vessels.

3. Capillary walls are very thin in comparison to the walls of arteries or veins. Is there a reason for this?

4. How is the shape of the red blood cell adapted for movement through the capillaries?

5. Does blood always flow in the same direction through a particular capillary, or does the flow change directions?

6. Would you expect to find a capillary network in the gills of the goldfish? Why?

7. Certain chemicals can influence circulation by changing the diameter of some of the blood vessels. The effects of four substances on capillaries are shown in Table 20-2. Study these effects and then answer the questions that follow.

EFFECT	NICOTINE	LACTIC ACID	ALCOHOL	ADRENALIN
Diameter of capillaries increases and blood flow rate increases		√	√	
Diameter of capillaries decreases and blood flow rate decreases	√			√

TABLE 20-2 **Effects of Chemicals on Capillaries**

a) Why is it likely that, on a cold day, the temperature of the extremities (fingers and toes) of a smoker will be lower than those of a non-smoker?

b) Lactic acid builds up in muscles during long periods of activity. What influence will this substance have on the blood circulation to these muscles?

c) Would you expect a person who is drunk to be more susceptible or less susceptible to exposure from cold? Why?

d) Adrenalin is released into the blood by the body when one is frightened or excited. Would you expect such a person to turn pale or blush? Explain your answer.

8. How would an increase and a decrease in water temperature affect the circulation of blood in the arteries, veins, and capillaries of a fish? Why?

20.5 Invertebrate Circulatory Systems

A brief examination of two invertebrate circulatory systems will further clarify the differences between open and closed systems.

The Grasshopper

The grasshopper has an open circulatory system (Fig. 20-9). A single **dorsal blood vessel**, running through the insect's abdomen and thorax, carries a colourless blood. The posterior (rear) portion of this vessel functions as a **heart**. The anterior (front) end is called an **aorta**. Blood is pumped from the aorta, forward through the vessel and out the open anterior end into the body sinus. There it bathes the internal organs. As the blood moves through the body sinus from the anterior to the posterior end, it distributes nutrients to all tissues and picks up waste products from them. The blood finally reaches the heart and enters it through a one-way series of valves called **ostia** (sing. **ostium**). As you know, the circulatory system of a grasshopper does not transport oxygen and carbon dioxide. This role is carried out by a system of tracheal tubes.

Fig. 20-9 An open circulatory system. The body of an insect contains a body sinus into which the blood flows.

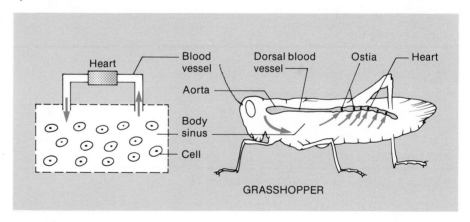

The Earthworm

The earthworm has a closed circulatory system (Fig. 20-10). A **dorsal blood vessel** carries the red blood forward to the anterior end of the earthworm. The blood returns to the posterior end through a **ventral blood vessel**. Small, thin-walled vessels called **capillaries** connect the dorsal and ventral vessels throughout the earthworm. Exchange of materials between the blood and body cells takes place through the capillary walls.

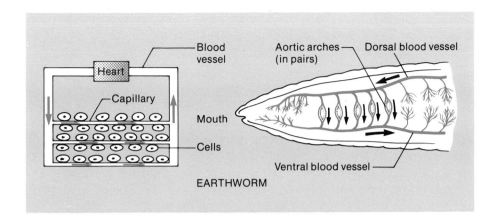

Fig. 20-10 A closed circulatory system. Blood remains within a system of vessels; it is not pumped into a body sinus. Capillaries connect the large vessels but, because of their small size, they are not shown in the earthworm diagram.

In the anterior end of the earthworm, five pairs of muscular tubes called **aortic arches** function as hearts. They pump the blood from the dorsal to the ventral vessel. Nutrients, respiratory gases, and waste products are all transported within this system.

Review

1. **a)** Briefly describe the circulatory system of a grasshopper.
   b) Why is it unnecessary for a grasshopper's blood to carry oxygen?
   c) Name five other animals that have open circulatory systems.
2. **a)** Briefly describe the circulatory system of an earthworm.
   b) Where and how would food and oxygen enter the earthworm's circulatory system? (If necessary, refer back to Chapters 18 and 19 for information.)
   c) Name five other animals that have closed circulatory systems.
3. What is the basic difference betwen a closed and an open circulatory system?

20.6 Transport Systems in Vertebrates

A General Description

The closed circulatory system of a vertebrate consists of a pumping device called a heart and blood vessels of three basic kinds: arteries, veins, and capillaries. **Arteries** carry blood away from the heart; **veins** carry blood toward the heart. A series of **valves** prevents the backward flow of blood through the veins. **Capillaries** are very small blood vessels which connect the arteries and veins. Together these vessels and the heart form a closed circuit. The heart pumps blood into the muscular thick-walled arteries which, in turn, carry the blood to all parts of the body. The arteries branch again and again, finally becoming tiny thin-walled capillaries. It is through the thin walls of these capillaries that the exchange of materials such as oxygen and carbon dioxide takes place between the blood and other tissues. Blood

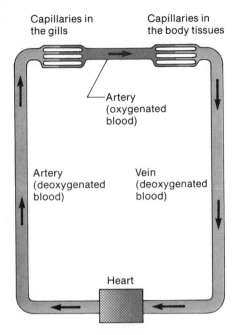

Fig. 20-11 A diagram of a single circulatory system. Blood passes through the heart only once during each circuit.

Fig. 20-12 A diagram of a double circulatory system. Blood passes through the heart twice during each circuit. Oxygenated and deoxygenated blood are kept separate in the heart.

moves through the capillaries into small veins which, in turn, join to form larger veins. The thin-walled, lightly muscled veins complete the circuit by carrying the blood back to the heart.

Single and Double Systems

If blood passes through the heart once each time it completes a circuit, the arrangement is called a single circulatory system. This type of system is characteristic of fish. Deoxygenated blood (blood without oxygen in it) from the tissues travels to the heart which, in turn, pumps it to the gills. On its way through the gills it becomes oxygenated and then travels to all parts of the body (Fig. 20-11). In birds and mammals the blood passes through the heart twice each time it completes a circuit. This is called a double circulatory system (Fig. 20-12). In a double circulatory system the left side of

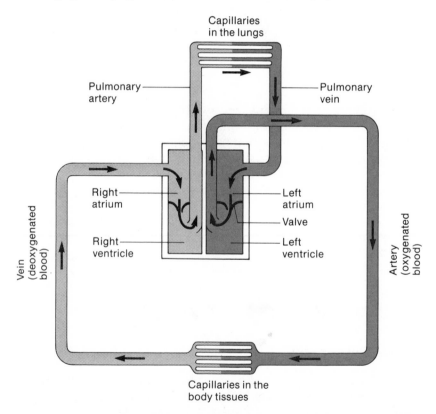

the heart always receives oxygenated blood from the lungs and pumps it to all body tissues. The right side of the heart always receives deoxygenated blood coming from the body tissues and pumps it to the lungs.

Vertebrate Hearts

A comparison of the structures of the hearts of the five vertebrate classes is shown in Figure 20-13.

Fish

The heart of a fish is two-chambered. It consists of *one thin-walled atrium* which takes deoxygenated blood from the veins and delivers it to the *thicker-walled ventricle.* The muscular ventricle pumps the blood into the arteries toward the gills. The blood must pass through at least two sets of capillaries before returning to the heart. The single pump of the fish heart must provide enough pressure for this to happen.

Amphibian

The amphibian heart has three chambers, *two atria* leading into a *single ventricle.* The left atrium receives oxygenated blood from the lungs; the right atrium receives deoxygenated blood from the other tissues of the body. The two atria pump in unison and oxygenated and deoxygenated blood are forced into the ventricle where they mix. This mixture is then pumped to all other body tissues.

Reptiles

Reptiles have a heart with *two atria* and *two ventricles.* In most species *the two ventricles are not completely separated.* Thus some mixing of oxygenated and deoxygenated blood may occur.

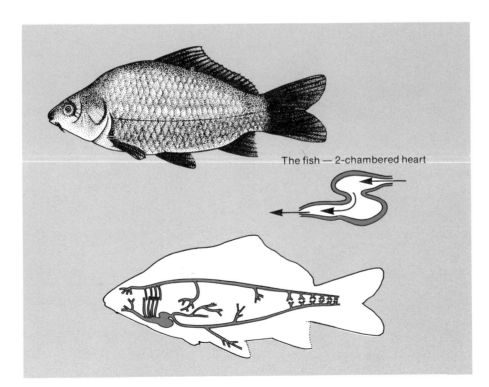

The fish — 2-chambered heart

Fig. 20-13   A comparison of the structure of the hearts of the five classes of vertebrates.

A: heart structure of a fish.

Fig. 20-13   B: heart structure of an amphibian.

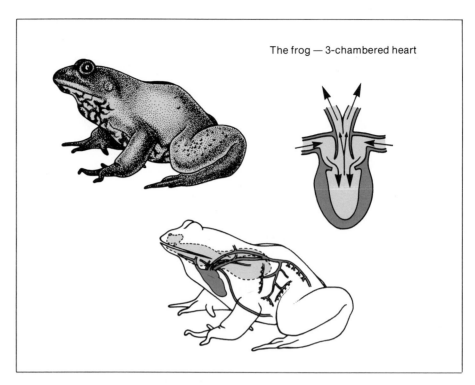

The frog — 3-chambered heart

Fig. 20-13   C: heart structure of a reptile.

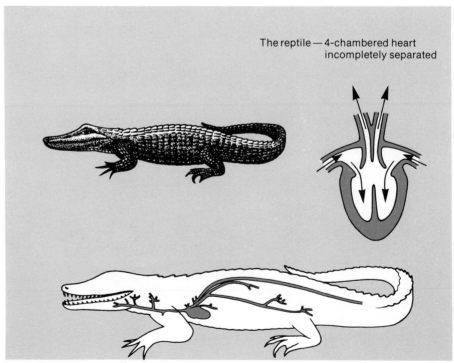

The reptile — 4-chambered heart incompletely separated

Fig. 20-13   D: heart structure of a bird.

The bird — 4-chambered heart
completely separated

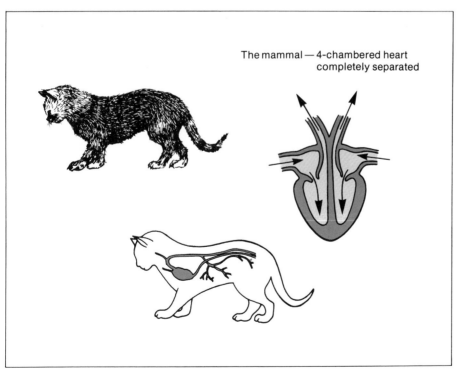

Fig. 20-13   E: heart structure of a mammal.

The mammal — 4-chambered heart
completely separated

Birds and mammals

These animals have *four-chambered hearts in which both atria and ventricles are completely separated* in the adult. Oxygenated blood, found only in the left atrium and ventricle, is completely separated from deoxygenated blood in the right atrium and ventricle.

Birds and mammals are warm-blooded or **homeothermic** animals. Their body temperatures and metabolic rates remain relatively high and constant despite changes in environmental temperatures. As a result, a highly efficient transport system is required to maintain both a high temperature and a high metabolic rate in these vertebrates. (*Note*: The term *homeothermic* means *"constant temperature"*.)

Fish, amphibians, and reptiles are cold-blooded or **poikilothermic** animals. Their body temperatures and metabolic rates fluctuate with changes in the environmental temperature. These animals do not have a circulatory system as efficient as that of homeotherms. (*Note*: The term *poikilothermic* means *"variable temperature"*.)

Review

1. **a)** Distinguish among arteries, veins, and capillaries.
   b) What function do valves serve in circulatory systems?
2. Use examples to distinguish between single and double circulatory systems.
3. **a)** Compare the structures of the hearts of fish, amphibians, reptiles, birds, and mammals.
   b) Why is a mammalian system considered more advanced than that of other vertebrates?
4. **a)** Distinguish between homeothermic and poikilothermic animals.
   b) What advantages are there in being poikilothermic? Homeothermic?
   c) Give an example which shows that the terms cold-blooded and warm-blooded can be misleading. (Hint: Is a frog's blood always "cold"?)

20.7 Investigation

The Heart in Action

Background

Daphnia is a small freshwater crustacean commonly called a waterflea. It has a transparent body wall that allows one to see the internal organs at work when viewed through a microscope (Fig. 20-14). In this investigation you will observe the action of the heart as you change the animal's environment.

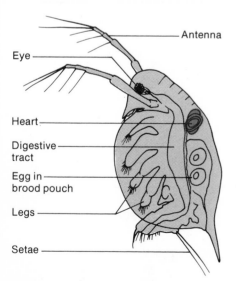

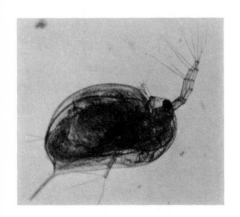

Antenna

Eye

Heart

Digestive tract

Egg in brood pouch

Legs

Setae

Fig. 20-14 Daphnia. A photograph (A) and a diagram (B) showing internal organs. Note the position of the heart.

Materials

microscope
Daphnia (4-8 specimens in 150 mL
  beaker of aquarium water)
microscope slide (depression type)
medicine dropper
thermometer
stop watch or watch with second
  hand
beaker (500 mL)—large enough
  to hold *Daphnia* beaker

hot water
crushed ice
graph paper
chlorpromazine (a tranquilizer)
dexedrine sulfate (a stimulant)
methyl cellulose or cotton fibres

Procedure

a. Using a medicine dropper take one *Daphnia* from the small beaker and transfer it to a depression slide. Record the temperature of the water in the beaker. Place just enough water in the depression with the *Daphnia* to barely cover it. Place the slide under low power on the microscope. (Methyl cellulose or cotton fibres may be added to the depression to restrict the *Daphnia*'s movement and to prevent it from moving out of the field of vision.)

b. Observe the action of the heart. Count the number of beats in one minute for three successive minutes. Average these three readings and record the result in a table similar to Table 20-3.

	WATER AT ROOM TEMP. _____ °C	COLD WATER TEMP. _____ °C	HOT WATER TEMP. _____ °C	WATER AT ROOM TEMP. AND CHLORPROMAZINE	WATER AT ROOM TEMP. AND DEXEDRINE SULFATE
Heart rate (beats per minute)					

TABLE 20-3   **Heart Rate of *Daphnia***

c. Next, using a medicine dropper place a drop of chlorpromazine in the depression with the *Daphnia*. Again determine the number of heartbeats in one minute for three successive minutes. Record the average of these readings. Rinse the dropper thoroughly and return the *Daphnia* to its original beaker.

d. Select a second *Daphnia* and place it on a clean depression slide. Remove as much water as possible. Determine its normal heart rate. Then add a drop of dexadrine sulfate to the depression. Again, determine and record the average heart rate over a three minute period as in step c. Rinse the dropper and return the *Daphnia* to its original beaker.

e. Place the beaker of *Daphnia* into a larger beaker containing ice and water. Stir the water in the *Daphnia* beaker and, after 2 min, record its temperature. Quickly, place a *Daphnia* from this beaker into a depression slide. Determine and record its average heart rate per minute over a three minute period as you did in step c. Return the *Daphnia* to its original beaker.

f. Place the beaker of *Daphnia* into a larger beaker containing water at 80°-90°C. Again, stir the *Daphnia* water and, after two minutes, record its temperature. Quickly place a *Daphnia* from this beaker into a depression slide. Determine and record its average heart rate per minute as you did in step c. Return this *Daphnia* to its original beaker.

Discussion

1. a) What is the effect of chlorpromazine on the heartbeat rate of *Daphnia*? Why is chlorpromazine called a tranquilizer?
   b) What is the effect of dexedrine sulfate on the heartbeat rate of *Daphnia*? Why is dexedrine sulfate called a stimulant?
2. What effects do changes in temperature have on the heartbeat rate of *Daphnia*? Why do you think this happens?
3. Examine the data in Table 20-4. On the same sheet of graph paper, plot the data for each individual, placing heartbeat rate on the vertical axis and temperature on the horizontal axis. Draw the lines in a different colour for each individual.
   a) What relationship does the graph show between temperature and heartbeat rate for *Daphnia*?
   b) Account for the variations in rate shown by individual *Daphnia* in the graph. That is, why do all *Daphnia* not have the same graphs?
   c) Would you expect similar effects of temperature on the heartbeat rate of a vertebrate? Explain your answer.

DAPHNIA INDIVIDUALS	TEMPERATURE					
	0°C	10°C	20°C	30°C	40°C	50°C
A	30	44	81	140	201	0
B	29	45	80	143	200	42
C	31	46	80	141	200	0

TABLE 20-4 Influence of Temp. on Heartbeat Rate of *Daphnia* (heartbeats per second for various individuals at temp. intervals of 10°C)

20.8 The Human Circulatory System

Waste, nutrients, and water are carried throughout your body by a closed transport system that is called the **circulatory system**. As you know, a closed system is one in which the blood is always enclosed in tubular blood vessels; it does not come into direct contact with tissue cells. By means of such a

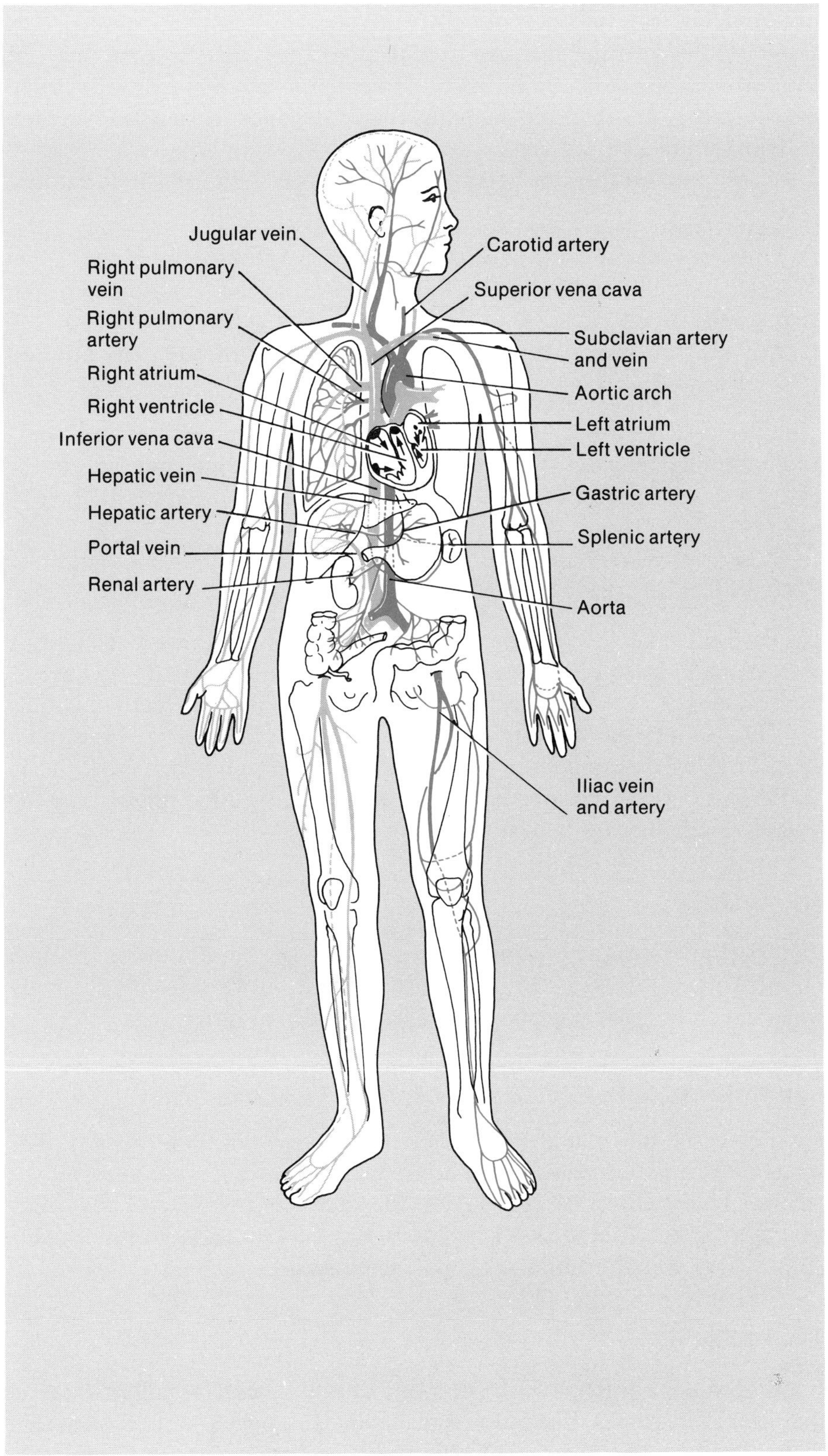

Fig. 20-15 The circulatory network. What would a diagram showing all the capillaries look like?

Jugular vein

Right pulmonary vein

Right pulmonary artery

Right atrium

Right ventricle

Inferior vena cava

Hepatic vein

Hepatic artery

Portal vein

Renal artery

Carotid artery

Superior vena cava

Subclavian artery and vein

Aortic arch

Left atrium

Left ventricle

Gastric artery

Splenic artery

Aorta

Iliac vein and artery

system, blood can be piped directly to the organs where it is needed. The amount flowing to an organ can be regulated. The return flow is rapid. The large size of the human body and its homeothermic nature make the closed system more suitable for humans than the open system of smaller, poikilothermic invertebrates.

The closed transport system of the human has four basic structures:

1. A fluid tissue (the blood and lymph);
2. A network of tubing (the veins and arteries) to carry the blood;
3. Specialized tubing (the capillaries) to allow diffusion of molecules to and from the blood;
4. A pump (the heart) to keep the blood moving through the arteries, capillaries, and veins.

These structures all serve the function of transporting, distributing, and collecting the gases, nutrients, wastes, and regulating chemicals of the body.

Parts of the Circulatory System

The pathway taken by the blood through the body must let every cell in your body come in close contact with the blood. Considering that you have to keep about 6×10^{13} (sixty trillion) cells alive, this is some feat!

The path taken by a single blood cell is always through the heart to an artery, through a capillary network, and then through a vein back to the heart. This cycle is repeated again and again. Figure 20-15 shows only the largest vessels through which the blood flows.

The heart

The heart provides the pressure needed to keep the blood flowing through the network of tubing. The fact that you bleed outward from your body when cut shows that your blood is always under pressure.

The arteries

Arteries carry blood under the highest pressures (around 16 kPa). The structure of an artery must be able to handle both high pressure and the changes in pressure that result from the rhythmic pumping of the heart. Therefore arteries are thick-walled tubes, wrapped with elastic muscle tissue (Fig. 20-16). Small arteries are called **arterioles**.

The capillaries

Capillaries are a finely divided network of tiny tubes as small as $10\,\mu m$ in diameter (Fig. 20-17). Each tube is made up of a single layer of cells so thin that the cell nuclei bulge into the bloodstream. Individual capillaries often

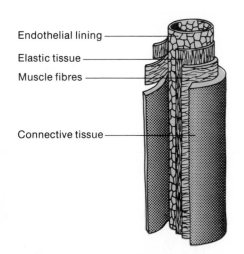

Endothelial lining

Elastic tissue

Muscle fibres

Connective tissue

Fig. 20-16   Cross-section of an artery. Notice the layer of elastic muscle fibres. Why must the walls of an artery be so thick?

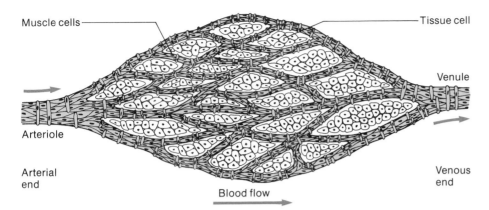

Muscle cells

Tissue cell

Venule

Arteriole

Arterial end

Venous end

Blood flow

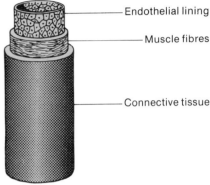

Endothelial lining

Muscle fibres

Connective tissue

A

Cross-section of a vein. Why do veins not need walls as thick as those of arteries?

have tiny rings of muscle that can control the bloodflow through a tissue. Capillaries "leak" nutrients from the bloodstream to all cells in the body. Cell wastes diffuse through the capillary back into the blood. The walls of arteries and veins are too thick to allow such diffusion of molecules to and from the bloodstream.

The veins

Veins provide a return system for blood under lower pressure than the blood in the arteries. Therefore the walls of veins do not need to be as thick or elastic as those of arteries (Fig. 20-18,A). The pressure of the blood coming out of the venous end of some capillaries often is not high enough to push the blood all the way back to the heart. For this reason, some veins in the legs and arms have one-way valves to ensure that the blood travels only toward the heart (Fig. 20-18,B). The valves help support the column of returning blood. Leaky valves can produce **varicose veins**. The unsupported blood column stretches the veins. Varicose veins near the skin surface are noticeably larger and darker because of the stretching of the venous walls. Small veins are called **venules**.

The Pathway

Blood returning to the heart from the arms, head, abdomen, and legs has little oxygen left in it. As a result, the pathway that the blood follows should put more oxygen into the blood and then direct the re-oxygenated blood back out to the body tissues. The part of the circulatory pathway that re-oxygenates the blood is called the **pulmonary (lung) circuit**. The pathway that circulates blood through the rest of the body systems is called the **systemic circuit**. Each of these circuits requires its own pump. The pulmonary circuit uses the right half of the heart; the systemic circuit uses the left half. Follow Figures 20-15 and 20-19 as you read the following descriptions of the two circuits.

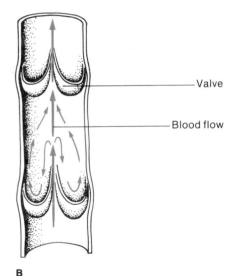

Valve

Blood flow

B

Valves in veins. These ensure returning blood flows in one direction only.

Fig. 20-18   Structure of veins.

Fig. 20-19   A schematic diagram of the blood circuit.

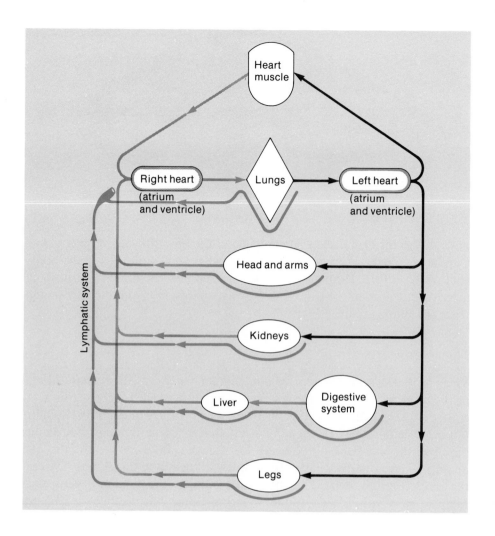

Pulmonary circuit

Veins from the arms, head, abdomen, and legs all unite and feed oxygen-poor and carbon dioxide-rich blood into the top right chamber of the heart called the **right atrium**. When the lower right chamber, called the **right ventricle**, is ready to be filled up with more blood, a one-way valve between the right atrium and right ventricle opens. The muscles of the atrium contract, forcing blood into the ventricle. When the ventricle is full, the much stronger muscles of the ventricle contract strongly. This pushes the blood through another one-way valve and into the **pulmonary artery**. (Recall that arteries are designed to withstand high and varying pressures.) The pulmonary artery carries blood toward the lungs.

The pulmonary artery branches to each of the lungs. It divides and redivides into smaller and smaller tubules called **arterioles** that connect to the **capillaries** of the lung. The lung capillaries cover the small alveoli or air sacs found at the end of the bronchioles (Fig. 20-20). Within each capillary,

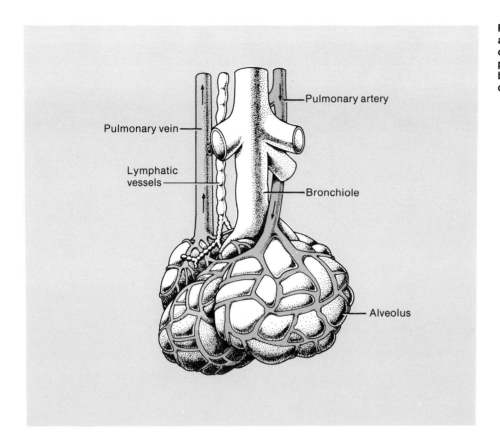

Fig. 20-20 Capillaries, surrounding each air sac (alveolus) of the lung, exchange carbon dioxide for oxygen. Arteries of the pulmonary circuit carry oxygen-poor blood, unlike the arteries of the systemic circuit.

Pulmonary artery

Pulmonary vein

Lymphatic vessels

Bronchiole

Alveolus

carbon dioxide diffuses from the blood to the alveolus, and oxygen diffuses from the alveolus into the blood. The blood pressure created by the right ventricle keeps the blood moving through the capillaries into tiny veins called **venules** that unite and reunite into larger veins. A total of four **pulmonary veins** feed the oxygen enriched blood into the top left chamber of the heart, called the **left atrium**.

Systemic circuit

When the **left atrium** contracts, blood is squeezed through a one-way valve into the **left ventricle**. When the left ventricle is completely full of blood, the muscle fibres surrounding it contract. This pumps the oxygenated blood through yet another one-way valve into a large artery called the **aorta**. The **coronary arteries** branch off from the aorta right on to the surface of the heart (Fig. 20-21). Blood carried in these arteries nourishes the heart muscle itself. The aorta is an upside down "J" shape. It arches behind the heart and carries blood to the lower abdomen and legs. Arterial branches at the top of the arch carry blood to the head and arms. In the abdomen, smaller arteries branch off to serve the organs and muscles of the digestive system. The last major branching of the aorta is into arteries that carry blood into the legs and feet.

Fig. 20-21 The coronary circulation. Major veins and arteries servicing the heart muscle are shown. Blockage of blood supply in the heart muscle results in a "coronary" heart attack.

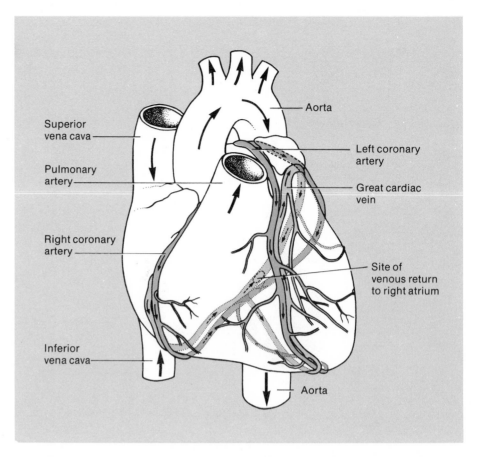

Superior vena cava

Pulmonary artery

Right coronary artery

Inferior vena cava

Aorta

Left coronary artery

Great cardiac vein

Site of venous return to right atrium

Aorta

Exactly as they do in the lungs, the larger arteries divide and redivide into smaller and smaller tubules, or capillaries, that are only 10 μm in diameter and a single cell thick. Again, a high pressure is needed to force blood through these tubules. A network of capillaries supplies all tissues in your body, with a few exceptions such as the lens in your eye. But, unlike the lungs, oxygen diffuses out of the blood and into the tissues. Carbon dioxide, the waste product of cell respiration, is absorbed into the blood. Nutrients, originally absorbed from the small intestine, are also exchanged for cell wastes. After being squeezed through a capillary, the blood enters tiny veins, or venules, that unite and reunite into larger veins. The largest veins, called **venae cavae** (singular: **vena cava**), carry the blood back to the **right atrium** of the heart.

There is one atypical instance in which blood from a vein goes through a second capillary network before it goes back to the heart. Capillaries surrounding the small and large intestines absorb nutrients. The veins from these capillaries all lead to the **liver**. Here a second capillary system allows excess nutrients to be stored in the liver before general distribution via the systemic circulation. Nutrients stored in the liver may be added to the blood if required. The liver also removes harmful substances such as bacteria and poisons from the blood. The blood may have absorbed such substances from the intestines.

The Lymphatic System

As blood is pushed through the capillaries, the liquid portion of the blood, called **plasma**, is squeezed through the capillary walls. Plasma that is not reabsorbed into the capillaries builds up among the body cells as **tissue fluid**. A complete, separate drainage system, the **lymphatic system**, absorbs excess tissue fluid (Fig. 20-22). Vessels called **lymph ducts** absorb and carry away the extra tissue fluid that builds up around body cells. They return the fluid to the blood stream near your left and right shoulders. **Lymph** is the name given to the fluid in the lymph ducts. It is largely plasma, but also contains white blood cells important for protection from infection. Certain sections of the lymph ducts are enlarged to form **lymph nodes**. Infectious organisms such as bacteria are destroyed in the lymph nodes. The nodes sometimes swell up and become painful. The tonsils are lymph nodes located in the throat.

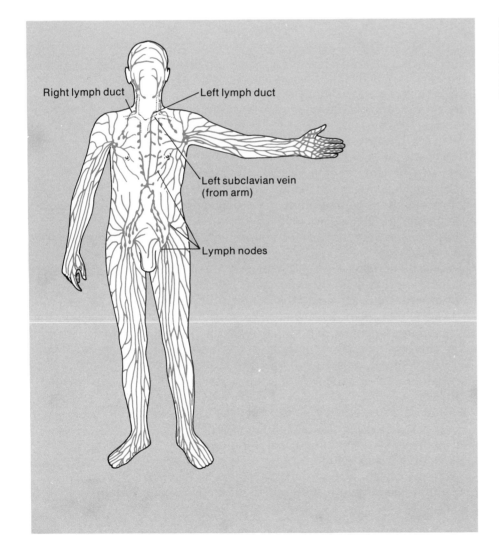

Right lymph duct

Left lymph duct

Left subclavian vein
(from arm)

Lymph nodes

Fig. 20-22   The lymphatic system. This network of lymph vessels returns tissue fluid squeezed out of capillaries to the blood stream. Lymph vessels from the small intestine carry absorbed fats. Lymph nodes trap foreign matter, including bacteria.

CARRIES OXYGEN-POOR BLOOD		CARRIES OXYGEN-RICH BLOOD	
LOCATION	NAME	LOCATION	NAME
away from the brain	jugular vein	toward the brain	
away from the left arm		toward the right arm	
away from the stomach		toward the stomach	
away from the liver		toward the liver	
away from the left kidney		toward the right kidney	
away from the intestines		toward the spleen	
away from the right leg		toward the left leg	
into the right atrium		out of the left ventricle	
away from the heart		into the left atrium	
toward the liver		toward the liver	
toward the left lung		away from the left lung	
away from the heart muscle		into the heart muscle	

TABLE 20-5 Functions of Blood Vessels

Review

1. What is the function of the circulatory system?
2. **a)** What is a closed circulatory system?
   b) What are three advantages of a closed circulatory system over an open system?
3. Why is a powerful heart needed in a closed circulatory system?
4. What are the differences in structure among arteries, capillaries, and veins? What are the differences in their functions?
5. Why are capillaries found almost everywhere in your body? Name three parts of your body that do not have capillaries. Why are capillaries not necessary for those parts?
6. Describe the path that a blood cell would follow from the right atrium to the aorta.
7. How is blood kept flowing in only one direction?
8. Where are nutrients, freshly absorbed from the intestines, carried to? What purpose is served by this route?
9. Using Figure 20-15 as a guide, complete Table 20-5 by naming the correct artery(ies) or vein(s) that serve the listed function.
10. Describe the circulatory network of the liver, the lungs, and the heart. For each organ, be sure that you describe the following:
    a) nature of the blood entering the organ (e.g. nutrient-rich, oxygen-poor);
    b) source of the blood entering the organ;
    c) how the blood is changed in the organ;
    d) pathway by which blood leaves the organ.
11. **a)** What is tissue fluid?
    b) Where does it come from and what is its function?
12. Describe the structure and function of the lymphatic system.

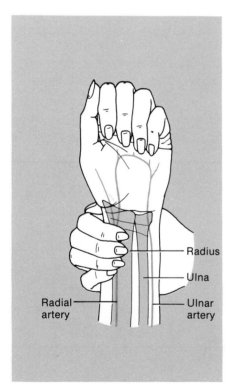

Fig. 20-23 Finding your pulse. Use the first three fingers of one hand to gently press the radial artery. The best place is where the radius bone narrows away from the wrist. Can you find your ulnar pulse?

Radius

Ulna

Radial artery

Ulnar artery

20.9 Investigation

Finding Your Pulse

Each time your ventricles contract, 70 mL or more of blood are forced into the arteries. Every time this happens, the pressure is raised throughout the arterial system. However this arterial pressure falls as the blood is pushed through the capillaries. Then a new ventricular contraction repressurizes the system. This rhythmic change in arterial pressure is called the **pulse**. The pulse can be felt almost anywhere an artery can be squeezed by a finger.

Materials

a stethoscope is useful but not necessary

Procedure

a. Find your **radial pulse** as shown in Figure 20-23. If you have difficulty finding it, change the angle of your hand.
b. Find the radial pulse of the other arm.
c. Check for a pulse at your temple, jaw, neck, arm, and foot. Use Figure 20-24 to help you find the proper locations. The pulse at your ankle may be found easiest by first crossing one leg over the other and changing the angle of the foot.
d. Use one hand to find the ankle pulse. At the same time, use the other hand to pick out the pulse of the carotid artery. Concentrate carefully. Do they occur at the same time?
e. If you have a stethoscope, determine whether the pulse corresponds to the first or second heart sound. Is the radial or ankle pulse exactly in time with the heart beat?

Review

1. a) Which occurs first, the ankle or carotid pulse?
   b) Which pulse location is closest to the heart?
   c) Explain why the carotid and ankle pulses are not in time with each other.
2. a) What is a pulse?
   b) Is your pulse rate the same as your heartbeat rate?
   c) Why can your pulse be used to measure heartbeat rate.
3. Explain how you can sometimes hear your pulse in your head.

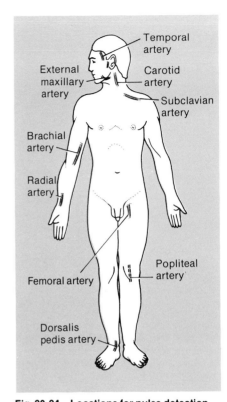

Fig. 20-24 Locations for pulse detection. These locations are often called "pressure points". Firm pressure can actually drop blood flow to the "downstream" tissues.

20.10 The Human Heart

The heart is a very muscular section of the circulatory system. The contraction of the heart muscle is strong enough to squirt a stream of blood 1.5 m into the air. Of course, the 16 kPa pressure produced by the left ventricle is used to squeeze blood through the capillaries of the body. This pressure overcomes the friction of the blood against the walls of the capillaries. Sufficient pressure is left over to return the carbon dioxide-rich blood to the right atrium and the pulmonary circuit.

Structure of the Heart

Figure 20-25 shows the structure of the human heart. The walls of the heart are made of **cardiac muscle**. Figure 20-26 shows the bands of cardiac muscle that help squeeze blood out of the ventricles. Cardiac muscle is unlike other muscles in your body. Only cardiac muscle can repeatedly contract with very little relaxation time in between. Cardiac muscle cells are able to get rid of their toxic wastes much more rapidly than other muscle cells.

Fig. 20-25   The human heart. Note the four chambers. Why are the muscular walls of the left ventricle so much thicker than those of the right ventricle?

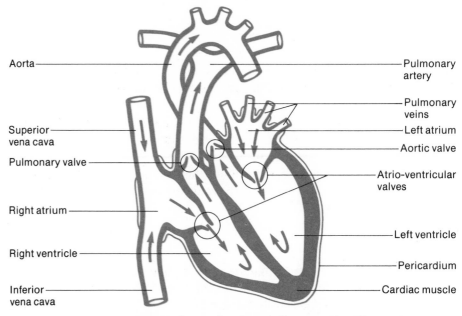

Special nerve cells are found in the cardiac muscle. These **pacemaker cells** control the heartbeat, and ensure that the correct cardiac muscle cells contract at the same time.

Two sets of **valves** are found in the heart. These valves ensure that blood flows only from atrium to ventricle, or from ventricle to artery. Heart muscle has its own blood supply, the **coronary circulation**. Surrounding the heart is a tough double membrane called the **pericardium**. Fluid between the two layers in the pericardium makes sure that the constant beating of the heart does not cause friction. Rubbing of the beating heart on the breastbone or the lungs could cause dangerous irritation inside the chest cavity.

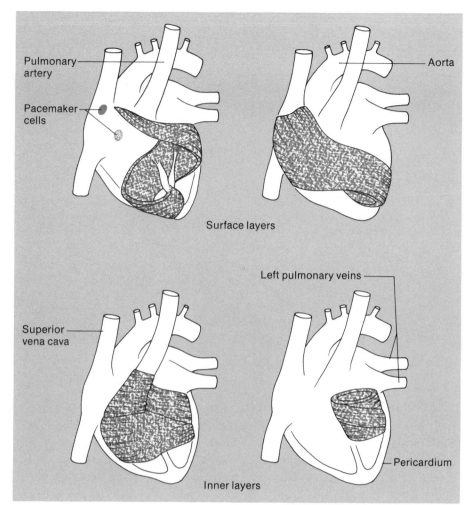

Pulmonary artery

Pacemaker cells

Aorta

Surface layers

Superior vena cava

Left pulmonary veins

Pericardium

Inner layers

Fig. 20-26 The muscles that squeeze the heart. Recall that muscles only contract. Which ventricle is wrapped with the most muscle? Why?

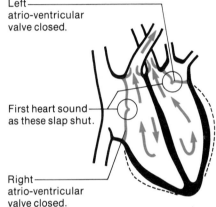

Left atrio-ventricular valve closed.

First heart sound as these slap shut.

Right atrio-ventricular valve closed.

The Heartbeat

Contraction and relaxation of the cardiac muscle occurs a little faster than once a second. For adults, the average heart rate is 72 beats/min; young children may have a resting rate of up to 100 beats/min. Difficult or prolonged exercise can raise the rate to 170 beats/min. A long distance runner may have a resting heart rate as low as 40 beats/min.

A beat consists of the simultaneous (at the same time) emptying of the left and right ventricles, followed by their refilling. The familiar "lub-dup . . . lub-dup . . . lub-dup" noise of a beating heart is caused by the closing of the heart valves (Fig. 20-27). The "lub" sound is the noise of the atrioventricular valves (between each atrium and ventricle) closing when blood is forced into the arteries. During the "dup" sound, the empty ventricles are refilling, and the valves between the ventricles and arteries close. These valves prevent the blood that was just squeezed out of the ventricles from flowing back to where it came from.

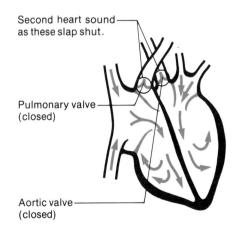

Second heart sound as these slap shut.

Pulmonary valve (closed)

Aortic valve (closed)

Fig. 20-27 The heartbeat. Why do the valves open and close?

Fig. 20-28 The location of heart sounds. The aortic valve closes slightly before the pulmonary valve. Why are the sounds heard best to the side of the valves. Where would you listen for the first "lub" sound?

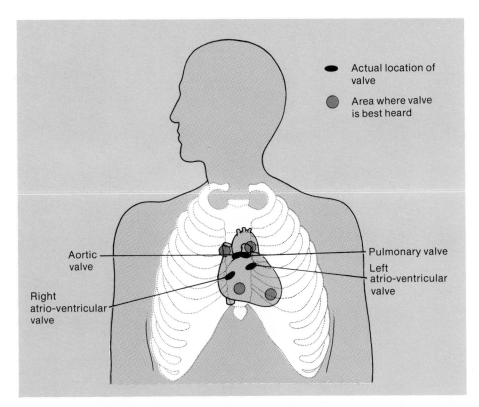

Aortic valve

Right atrio-ventricular valve

Actual location of valve

Area where valve is best heard

Pulmonary valve

Left atrio-ventricular valve

A doctor does not keep a stethoscope in one place when listening to your heart. By moving it to certain places on the chest cavity, the sounds of the closing valves can be heard separately. Figure 20-28 shows the key locations for listening to each set of valves. The listening locations are not directly over the actual valves. The heart, lung, and bone tissue all muffle and deflect these sounds elsewhere. If you have a chance to use a stethoscope, check these locations. Also try to show that the first "lub" sound is really coming from the closing of the atrioventricular valves.

Two pumps in one

The heart is really two pumps in one. One is for the pulmonary circuit, and the other is for the systemic circuit. More pressure is needed to push the blood through the capillaries of the body than through the capillaries of the lungs. Therefore it should not surprise you to see that the wall of the left ventricle is much thicker than the wall of the right ventricle (see Fig. 20-25). What is interesting about the pump action of the heart is that one muscle contraction causes pumping through both circuits.

External Cardiac Massage

Heart massage is an emergency means of pumping blood through the body. Heart stoppage can be detected by the widening that it causes in the pupils of the eye. This sign, along with a blueness of the lips and the absence of any pulse in the neck (carotid) artery, indicate lack of oxygen to the brain. Brain damage occurs after 4 min of oxygen starvation; so time is extremely important.

During cardiac massage, the heart is squeezed between the breastbone and backbone at the rate of one squeeze per second. The chest is depressed 3 to 5 cm on an adult, less in children. The valves in the heart keep blood flowing in the right direction. Mouth-to-mouth resuscitation MUST be applied at the same time as cardiac massage. Pumping blood with no oxygen around the body is of no value. The combination of cardiac massage and mouth-to-mouth resuscitation is called **cardio-pulmonary resuscitation**.

Once begun, massage must be continued without interruption until the patient has recovered or until death has been confirmed by a medical doctor. *Therefore you must know what you are doing before you dare attempt this extreme lifesaving method.* Patients have recovered after 4 h of cardio-pulmonary resuscitation.

Review

1. What metabolic wastes would be produced by a living cardiac muscle cell?
2. Why are valves necessary in the heart?
3. What is the function of the pericardium?
4. About how many times has your heart beat in your life so far? Show your calculations.
5. **a)** What is happening in the heart when the "lub" sound is made?
   b) What is happening as the "dup" sound is made?
6. Describe what would happen if the valve between the left atrium and left ventricle did not seal properly.
7. **a)** Which half of the heart is the pump for the pulmonary circuit?
   b) Which half is the systemic circuit pump?
   c) Explain why the wall of the left ventricle is thicker than the wall of the right ventricle.
   d) Suggest why more pressure is needed to push blood through the body than through the lungs.
8. What advantage is there in having the heart pumps side by side rather than separated as shown in Figure 20-19?
9. Describe the effect of a hole between the left and right ventricles.
10. What is a pacemaker? Find out why artificial pacemakers are used.
11. **a)** Suggest two abnormalities of the heart that a doctor could find with a stethoscope. For each, suggest what noise(s) might be heard.
    b) How else is a stethoscope useful for diagnosis of illness?

20.11 Investigation

Determination of Your Heart Performance Score (HPS)

If you exercise vigorously, your heartbeat rate, or pulse rate, increases. The amount that it increases depends on how fit you are. A strong heart can pump blood more efficiently than a weak one. Therefore the pulse rate of a physically fit person will not increase as much with exercise as will that of a physically unfit person. In this investigation you will determine how fit your heart is by calculating your Heart Performance Score (HPS).

Materials

watch or timer with second hand

Procedure

Note 1: For all pulse rate measurements, take the radial pulse for 20 s; then multiply by 3 to get the pulse rate (in beats per minute).
Note 2: Record all data in a table.

a. Lie down for 2-3 min. Then determine your pulse rate while lying down.
b. Stand up, wait for 10 s, then determine your pulse rate again.
c. Calculate the difference between your pulse rate while lying down and standing.
d. Perform vigorous exercise as follows: Obtain a chair about 0.5 m high. Have your partner hold it securely. Then, raise your left leg and step up on the chair. Stand right up on it with both feet. Then get off the chair using your right leg. Switch legs after every 5 complete steps. (A complete step is an up and down motion.) Continue this exercise for 60 s. You must perform about 15 complete steps in the 60 s.
e. Immediately determine your pulse rate.
f. Rest for 60 s; then determine your pulse rate again.
g. Calculate your HPS by adding together the 4 pulse rates that you measured *plus* the difference that you calculated in step c.

Discussion

1. Determine your general fitness level by examining Table 20-6.
2. a) Calculate the average HPS for your class, the average HPS for just the girls of your class, and the average HPS for the boys of your class.
   b) Compare your HPS to the class average and to the average for your sex. Try to account for any differences.
3. a) What is the effect of exercise on the heart?
   b) Suggest how the heart "knows" to react in this way.
   c) How does the body benefit from the change in heartbeat rate when the body is being exercised?

d) Why does the heart itself need more oxygen during exercise?

e) Describe the effect of resting after exercise on the heartbeat rate.

4. a) Each ventricle pumps out 70 mL of blood with every beat. How many beats are necessary to pump 4.2 L of blood through the heart?

b) How long would this take at the resting heartbeat rate?

c) How long would this take when the body is hard at exercise?

HPS	GENERAL FITNESS LEVEL
200-250	Endurance athlete
250-300	Athletic
300-325	Very good
325-350	Good
350-375	Fairly good
375-400	Fair
400-450	Poor
450-500	Very poor

TABLE 20-6   HPS and Fitness Level

20.12 Blood

Functions of Blood

Blood is a unique body tissue. Since it is fluid, it can circulate throughout the body. Cells far away from the digestive system and lungs must obtain nutrients. They must also get rid of the poisonous wastes that they produce. Only a circulating fluid tissue such as blood can keep a large multicelled animal alive.

Blood also transports hormones throughout the body. Hormones are chemicals that help regulate body processes. In addition, heat from cellular metabolism is absorbed by the blood. The circulatory system helps maintain a relatively constant body temperature by distributing the warm blood to colder tissues. Increased circulation to the skin also helps to release excess body heat. Finally, blood helps protect the entire body by fighting bacterial infections. Blood clotting is a defense against both blood loss and the entry of bacteria.

Physical Nature of Blood

You have at least 4 L of blood in your body. Blood is a fluid in two phases: a liquid phase consisting of water and dissolved materials, and a solid phase consisting of living cells suspended in the liquid phase (Fig. 20-29). The liquid phase is called **plasma**. You may have seen this clear, straw-coloured substance oozing through a reopened cut. Plasma is over 90% water; therefore it is easily forced out of capillaries. Plasma that bathes the body cells is

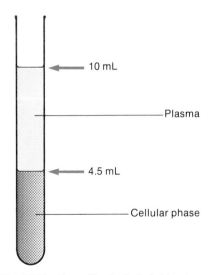

Fig. 20-29 Every 10 mL of whole blood contains 4.5 mL of cells — red cells, white cells, and platelets. Plasma is largely water, containing dissolved proteins and minerals.

called **tissue fluid**. Tissue fluid that is not reabsorbed into the capillaries drains into the lymph vessels.

Blood Cells

There are two types of living cells carried in the plasma. These are the red blood cells and white blood cells. Cell fragments called platelets are also suspended in the plasma. Figure 20-30 shows the size, shape, and number of each of these parts of the blood. These cells make up 45% of your blood.

Red blood cells

The red blood cells of mammals, including humans, are unlike most other single cells you have studied; they have no nuclei. (Red blood cells in most other vertebrates, such as the frog, do have nuclei.) Yet each red cell functions well for almost four months. The cells usually maintain the shape shown in Figure 20-30. This shape is called biconcave, since both sides of the disc-like cell are concave. (Each cell looks like a donut that did not turn out quite right!)

Fig. 20-30 Human blood cells.

Type of cell	Size and shape	Number in 4L of human blood
Red blood cell	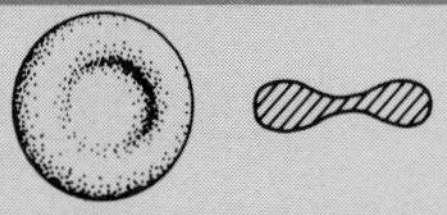	2000×10^{10}
White blood cell	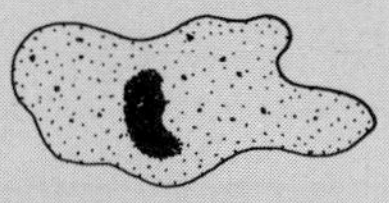	3×10^{10}
Platelet	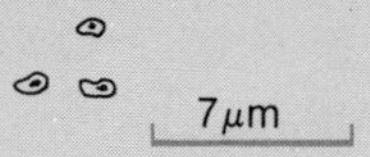 7 μm	120×10^{10}

The colour of red blood cells is related to their function of carrying oxygen. Red blood cells contain molecules of **hemoglobin**, which is dull red in colour. When these molecules come in contact with oxygen in the lung capillaries, they combine with the oxygen to form molecules of **oxyhemoglobin**. Oxyhemoglobin gives blood a bright red colour. When the red blood cells are carried to tissues that are low in oxygen, the oxyhemoglobin molecules break up into oxygen molecules and hemoglobin molecules. The oxygen molecules diffuse out of the red blood cells into the tissues, leaving molecules of hemoglobin behind in the blood. The dull red hemoglobin of red blood cells found in the veins give blood a much darker red colour. Veins appear blue due to several layers of skin cells covering the dull red blood.

Not all the red blood cells that you have are circulating in your body. Many are stored as reserves in your **spleen**, ready to be released into the bloodstream when more are needed (Fig. 20-31). Loss of blood from an accident, or a heavy need for oxygen can cause the spleen to pour the extra cells into the circulatory system. Aging red blood cells are destroyed by both the spleen and the liver. The liver excretes the used hemoglobin in the bile. The iron atoms that were part of the hemoglobin are recycled to make new red cells in the bone marrow.

White blood cells

White blood cells are similar to amoebas. They have nuclei; they are colourless; they can change shape, and can move in any direction. There are several types of white blood cells. Some wander outside of the capillaries, in the tissue fluid between the body cells. White blood cells are present in the lymph vessels and lymph nodes. Like amoebas, white blood cells engulf particles, including bacteria, by **phagocytosis**. In this way they act as scavengers and also protect the body from disease.

Platelets

Platelets are not true cells. They have no nucleus, and are only 2 or 3 μm in diameter. They contain enzymes and other chemicals enclosed in a cell membrane.

Platelets speed up the clotting process in the blood. When any vertebrate is wounded, many reactions take place to stop the loss of blood. Muscles tighten around the damaged vessels, narrowing their diameter. Also, the blood itself gets thicker near the wound and forms a clot. A **clot** is a network of long proteins that is fine enough to prevent blood cells from escaping. Why does this protein network usually form only when a blood

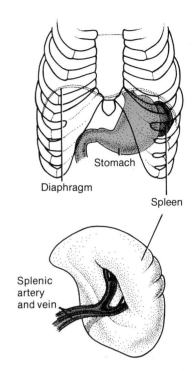

Fig. 20-31 The spleen. The spleen stores red blood cells. It also destroys the old red blood cells after their 120 d lifetime. What other organ could take over this job if the spleen is removed in surgery?

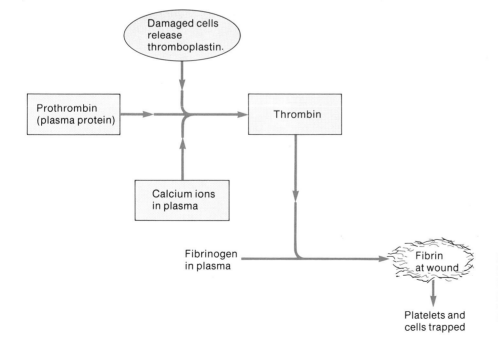

Fig. 20-32 Clotting. Fibrin fibres only form if thrombin is present. Thrombin will only be formed if both calcium and thromboplastin are present. Thus thromboplastin controls the entire sequence of steps leading to the clotting of blood.

vessel is cut? Surprisingly, it is not contact with air that causes clotting. Damaged cells start the clotting process. If platelet membranes are broken, a chemical called **thromboplastin** is released. This chemical triggers a series of reactions that cause the long protein called **fibrin** to form (Fig. 20-32). The fibrin molecules form a dense network that keeps white and red cells from leaking through. Only the pale yellow plasma can filter through. Even it is soon stopped by the plug of fibrin at the end of the cut blood vessel.

Dissolved Substances in Plasma

The water portion of the plasma is an excellent solvent. Dissolved in it are many substances that are essential to the healthy functioning of the body. The plasma contains varying amounts of cell wastes and nutrients. It also contains fairly constant amounts of special proteins, hormones, and minerals.

Nutrients and wastes

All nutrients and poisons that are absorbed by the body are carried in the blood. Thus in the blood, one finds nutrients such as glucose, amino acids, fatty acids, glycerol, vitamins, and minerals from intestinal absorption. Also, one finds some poisonous chemicals that are absorbed into the bloodstream from the skin surface. Skin absorption of cancer-related PCB's (polychlorinated biphenyls) has occurred in countless school laboratories from careless use of the lens immersion oil used with microscopes. Many airborne pollutants are absorbed into the bloodstream from the lungs and nasal passageways. Cell wastes such as urea, ammonia, and carbon dioxide are carried to the kidneys, sweat glands, and lungs by the blood.

The kidneys and liver are largely responsible for maintaining the correct waste and nutrient levels in the blood.

Special proteins

Proteins of many different types are circulating throughout the body. Some proteins help protect the body from foreign particles such as bacteria. Some, like prothrombin and fibrinogen, help blood to clot. Others help transport insoluble chemicals safely through the blood vessels. All plasma proteins, including albumin, help keep the blood from losing too much water to the surrounding tissues. Because many of the protein molecules are too large to diffuse or be squeezed through capillary walls, water is reabsorbed by osmosis at the venous end of a capillary.

A special group of protective proteins are called **antibodies**. They can react with and inactivate proteins that are foreign to your body (Fig. 20-33). Since bacteria have a protein membrane, they are destroyed by the reaction with antibodies. Antibodies are also related to your **blood type**. Red blood cells that you receive in a blood transfusion could be destroyed by your own antibodies if you are given the wrong blood type.

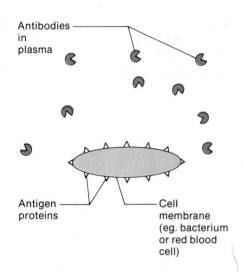

Antibodies in plasma

Antigen proteins

Cell membrane (eg. bacterium or red blood cell)

Inactivated cell

Fig. 20-33 Antibody protection. Bacteria or foreign red blood cells may have protein-like antigens in their surface. Some antibodies in the plasma "fit" onto the antigen surface. The different "texture" or chemistry of the membrane may allow dangerous cells to be engulfed by white blood cells.

Antibody proteins are also responsible for immunity against diseases. Usually a small amount of protein from dead disease-producing cells is injected into the blood. Antibodies that react with that protein are soon formed. They generally remain for several years, and actively protect the body from a real infection of that same disease.

Antibodies also determine the success of tissue transplants. A transplant will "take" only if the recipient's antibodies and other immune defenses do not react with the new tissue. Kidney and heart transplants have only become possible through careful research into antibody and immune reactions.

Hormones

Hormones are a special group of chemicals that regulate different body processes. The fast circulation time of the blood (less than 60 s) allows control of many different parts of the body by small amounts of these chemicals. Many hormones are produced in a section of the brain called the **pituitary**. Hormones produced here diffuse into the bloodstream, and are circulated throughout the body.

Adrenalin, produced by adrenal glands near the kidneys, is likely the most familiar hormone to you. It causes the reactions of fright: faster, shallower breathing, racing heartbeat, and the cold sweat of fear.

Minerals

Almost all essential minerals can break up in water to form soluble ions (charged atoms or groups of atoms). The common ions carried by the blood include hydrogen, sodium, chloride, calcium, and potassium ions. Ions such as sulfates, phosphates, and bicarbonates are also dissolved in the plasma, or are carried by plasma proteins.

The overall effect of these dissolved minerals is to give plasma a salt content of almost 1%. Minerals are needed for the normal functioning of your body cells. For example, you know that calcium is needed for proper clotting of blood. The kidneys help maintain constant concentrations of dissolved minerals in the blood (see Chapter 21).

Review

1. Give five functions of blood in your body.
2. How much of the plasma is water? How much of the plasma is dissolved material?
3. How much of the blood is cellular? How much of the blood is plasma?
4. **a)** Blood has about the same density as water (1 g/mL). What is the mass of 4.2 L of blood (the volume in an average person)?
   b) About 8.75% of your body mass is blood. What is the mass of all your blood cells?
   c) Calculate the volume of blood in your body.

d) Re-do Discussion question 4 of Investigation 20-11 (page 547), using your blood volume instead of 4.2 L.

5. **a)** Where does blood change colour from dark red to bright red?
   b) Where does it change from bright red back to dull red?
   c) Account for the changes in colour described in a) and b).

6. **a)** Where are red blood cells stored? When are they released?
   b) Why are blood transfusions sometimes necessary? What danger can there be in a blood transfusion?

7. **a)** How does a blood clot form?
   b) Find out how "serum" differs from "plasma".
   c) Find out the causes of internal blood clotting. How could an un-attached clot, or **thrombus**, that formed in a vein affect circulation in the lung?

8. **a)** Name three nutrient molecules carried by the blood.
   b) Name three waste molecules carried by the blood.
   c) What two organs control the waste and nutrient levels in the blood?

9. **a)** Name five functions of blood proteins.
   b) How is tissue fluid reabsorbed into the venous end of a capillary?

10. **a)** Transplant patients are often given drugs to reduce antibody reactions. What benefit would this have? What could be wrong with giving too much of these drugs?
    b) Patients with burned areas of the face and hands are sometimes given tissue transplants called skin grafts, using skin from their back or legs. Do you think these transplants would require antibody suppression drugs? Why?

11. Make a copy of Table 20-7 in your notebook. Complete it and give it an appropriate title.

12. Briefly describe the cause, symptoms, effects, and treatment of each of the following diseases of the circulatory system: leukemia, anemia, thrombosis, arteriosclerosis. You should be able to find this information in an encyclopedia.

NAME	SKETCH	FUNCTION
Red blood cell		
White blood cell		
Platelet		
Artery		
Capillary		
Vein		

TABLE 20-7

20.13 Investigation

Preparation and Examination of Blood Cells

It is easy to see red blood cells under the microscope. Their natural red colour makes them stand out clearly against the background. However, platelets and white blood cells are not so easily seen. They must be stained for greater visibility. Obtaining good results in this investigation requires good, careful, and clean laboratory habits.

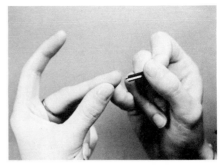

Fig. 20-34   Making a puncture with a lancet. Prop the wrists of both hands on the edge of the desk. The lancing hand should dart firmly toward the middle finger of the opposite hand. Use the lancet only once!

Materials

prepared blood slides
Wright's stain
microscope slides (2)
distilled water (or pH 6.8 phosphate buffer)
microscope (with known diameter of the high power field)
sterile, disposable blood lancet
rubbing alcohol and cotton batting
medicine dropper

CAUTION: Use the lancet once only. Discard it after using. Reusing it can spread diseases.

Procedure

a. Thoroughly clean and dry the 2 microscope slides. Handle them by their edges only. Any oils from your fingers will ruin your results.

b. Wipe the end of a middle finger with the alcohol. Move blood towards your finger tips by shaking your hand for 10 s. Make a small puncture in the fleshy part of the finger tip with the lancet (Fig. 20-34). **DISCARD IT AFTER USING.**

c. Place a drop of blood near the end of one slide. Use the other slide to produce an even smear across the first slide as shown in Figure 20-35.

d. Allow the smear to dry completely to a dull finish.

e. Cover the entire smear with a few drops of Wright's stain.

f. After 2 min, add drops of distilled water (or buffer) until a gold sheen appears.

g. Let the slide sit undisturbed for 13 min. While you are waiting, set up the microscope and observe the prepared slides as described in steps i, j, k, and l.

h. After the 13 min, gently wash the slide with distilled water (or buffer). Wipe the bottom of the slide dry and allow the top to air dry.

i. Inspect the slide at low, medium, and high power. A coverslip is not needed unless you use oil immersion.

j. Identify and describe the red blood cells and the white blood cells. You should see at least two types of white blood cells, differing in nuclear

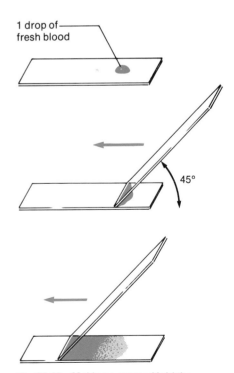

1 drop of fresh blood

45°

Fig. 20-35   Making a smear. Hold the second slide at 45°, and push it *away* from the drop of blood.

shape and amount of cytoplasm. Look for platelets between red and white cells.

k. Count how many red blood cells would fit across the diameter of the field of view at high power. Knowing the diameter of the high power field, calculate the diameter of a red blood cell. From this figure, estimate the diameter of a white blood cell.

l. Make a sketch of a red blood cell, and the two kinds of white blood cells. Make them larger than they appear in the microscope, but keep their sizes in proper relation to each other. Calculate the magnification of your sketches.

m. Record your observation in a table similar to Table 20-8.

CELL TYPE	DESCRIPTION	ESTIMATED DIAMETER
Red blood cell		
White blood cell		
White blood cell		

TABLE 20-8   Observations of Blood Cells

Discussion

1. What is the purpose of a stain?
2. a) Why is the very tip of the finger not punctured?
   b) Why is your finger tip wiped with alcohol?
   c) Why do you not share your used lancet with someone else?
3. Describe how to make an even smear of blood over a slide.
4. Compare red and white cells under these headings: colour, size, nuclei, cytoplasm, shape, number.
5. Explain why platelets may have been difficult to see even after staining.
6. Why are the centres of red cells so much lighter in colour?
7. a) Compare your blood preparation to the permanent preparation. Are there as many cells? Are they the same shapes? Are they the same colours? Explain how any differences could have occurred.
   b) Do all peoples' blood cells seem to look the same?
8. The number of red blood cells that can fit across a field of view of 200 μm is 29. What is the diameter of one red cell? A single red cell is sketched with a diameter of 3.5 cm. What is the magnification of the sketch?
9. You may observe the natural colour of blood cells by first placing one drop of blood in 20 mL of 0.85% salt solution. Then place a few drops of the mixture in a depression slide. Once you know the natural colour, describe the effect of Wright's stain on the different blood cells.
10. Calculate how many red blood cells could fit in the entire high power field of view of your microscope. Use this number to estimate the relative number of white blood cells to red blood cells.
11. Use a reference source to find the names and shapes of the different types of white blood cells.
12. You may want to see for yourself why the staining procedure in this investigation is so complicated. Carefully change the procedure to determine the effect of staining time, and write a report of your results.

20.14 Investigation

Blood Typing

In Investigation 20-13 you saw that all human red blood cells look the same. However, different people actually have different substances in their red blood cell membranes. Two of these substances are called **Type A** and **Type B** molecules. Whether or not your red cell membranes have these protein-like molecules determines your blood type. A person whose red blood cells have only Type A molecules has **Type A blood**. A person whose red blood cells have only Type B molecules has **Type B blood**. Someone who produces red cells with both substances in the membrane has **Type AB blood**. People who have neither of these antigens in their red cell membranes have **Type O blood**.

Your plasma contains many different kinds of antibodies. Recall that antibodies are proteins that help protect you from materials that do not belong in your body. Two of the antibodies you may possess react specifically with the Type A and Type B red cell proteins. These antibodies are given the names **anti-A** and **anti-B**. You will determine your blood type by seeing if these antibodies react with a drop of your blood.

Materials

microscope slides (2)
anti-A and anti-B typing sera
toothpicks
disposable, sterile blood lancets
rubbing alcohol and cotton batting
marking pen

CAUTION: Use the lancet once only. Discard it after using. Reusing it can spread diseases.

Procedure

a. Clean and dry each slide thoroughly. Label one as "1" and the other as "2".
b. Wash your hands. Then rub alcohol on the tip of the middle finger of your least-used hand. Swing your arm for 10 s to move blood toward your fingertips. Make a small puncture in the fleshy part of the finger tip with the lancet (See Fig. 20-35). **DISCARD THE LANCET.**
c. Discard the first drop of blood. Then place one drop in the centre of each slide.
d. Add 1 drop of anti-A serum beside the blood on slide 1. Add 1 drop of anti-B serum beside the blood on slide 2.
e. Use a toothpick to mix the two drops on slide 1 together. Use a different toothpick to mix the two drops on slide 2 together.

f. You should have any reactions within 2 min.

g. Record your observations in a table similar to Table 20-9. Wash off the slides when finished.

SLIDE	ANTI SERUM USED	SKETCH AFTER 1 MIN
1	anti-A	
2	anti-B	

TABLE 20-9 **Blood Typing**

Discussion

1. Why were the slides not labeled "A" and "B"?

2. a) Why is the fleshy part of the finger tip punctured?

b) Why is the finger tip wiped with alcohol?

c) Why do you not share your used lancet with someone else?

3. a) Determine your blood type using Table 20-10. Share your result with the class.

b) Forty-six percent of Canadians have Type O blood, 42% have Type A, 9% have Type B and 3% have type AB. Compare these figures with the values for your class. Why are there differences?

IF THE REACTION WITH ANTI-A PRODUCED	AND	IF THE REACTION WITH ANTI-B PRODUCED	YOUR BLOOD TYPE IS
no change		no change	O
no change		black dots	B
black dots		no change	A
black dots		black dots	AB

TABLE 20-10 **Blood Type Determination**

4. a) The reaction of anti-A antibodies with the Type A antigen produces a characteristic clumping together of the red cells. Suggest what might happen in your body if clumps of red cells travelled to a capillary.

b) How do you know that people with Type A blood do not have anti-A antibodies in their plasma?

c) Table 20-11 describes the antibodies present in people of different blood types. Explain why clumping occurs if a Type O person is given a transfusion of Type B blood by accident.

BLOOD TYPE	ANTIBODIES PRESENT
O	anti-A and anti-B
A	anti-B
B	anti-A
AB	neither anti-A nor anti-B

TABLE 20-11 **Antibodies and Blood Type**

5. **a)** Verify the information in Table 20-11 by mixing 1 drop of your blood with 1 drop of a different type of blood.

   b) Why is blood typing so important?

   c) Suggest a hypothesis to explain how someone's Type A blood can cause clumping of someone else's Type A blood.

6. Define: antigen, antibody, and blood type.

7. Suggest one source of the anti-A serum you used in the investigation.

8. **a)** Another antigen important to blood typing is one called the **Rh factor**. You can test whether you have this antigen (**Rh-positive**) or not (**Rh-negative**) with anti-Rh serum. The reaction, if any, occurs at 37°C. Perform the test if time permits.

   b) Find out why the Rh factor is of such importance to an Rh negative mother bearing an Rh positive baby.

20.15 A Comparison of Transport Systems

Now that you have studied the transport systems of several animals, a summary may help you to sort out the similarities and differences.

A Comparison of Structures

Some invertebrates, such as insects, have an open transport system with a single muscular tube that directs blood into the body spaces. This tube ("heart") is similar to a section of muscular artery with a one-way valve in it. The open transport system of the insect is not used for oxygen transport; therefore a more complicated system is not really necessary.

Earthworms have a closed transport system. Their "hearts" consist of ten muscular tubes that circulate blood through the body. Gas exchange occurs directly through the skin. There is no need for a separate pulmonary circuit.

Vertebrates have chambered hearts. The fish heart has one atrium and one ventricle. The single ventricle can exert enough pressure to force blood first through the capillaries of the gills and then on through the capillaries of the body.

The amphibian heart has two atria and one ventricle. Blood is forced by the single ventricle to all body regions. Oxygenated blood from the lungs returns to the left atrium. The blood is then forced into the ventricle and out to all body regions again. Oxygen-poor blood from the body tissues enters the right atrium and also flows into the single ventricle. Of course, the ventricle cannot separate the oxygen-rich and oxygen-poor blood. Therefore some oxygen-poor blood recirculates through the body, and some oxygen-rich blood recirculates through the lung and skin capillaries.

Mammals and birds have four-chambered hearts. The pulmonary, or oxygen-securing, circuit is completely separate from the systemic circuit. This separation ensures that oxygenated blood flows throughout the body with every ventricular contraction.

Comparative Blood Chemistry

Oxygen is carried in the blood of most animals by **hemoglobin** molecules. Some animals, such as the earthworm, have the hemoglobin molecules freely dissolved in the plasma; they are not carried by the red blood cells.

There are molecules other than hemoglobin that can transport oxygen. **Hemocyanin** is the most common alternative oxygen carrier. Unlike hemoglobin, which contains iron atoms, hemocyanin contains copper atoms. Hemocyanin is found in the blood of some molluscs and crustaceans.

Terrestrial insects get their oxygen by means of hollow air-filled vessels called tracheae. Oxygen transport by blood is not significant. Heart stoppage in an insect is not a serious event, since cells can survive much longer without nutrients than without oxygen.

It is interesting to look at how the oxygen-carrying ability of blood changes from one animal class to another. Table 20-12 shows how blood volume, oxygen content, and oxygen available to 100 g of tissue vary with the class of animal.

	BONY FISHES	AMPHIBIANS	REPTILES	BIRDS	MAMMALS
Blood volume as % of body	2%	6-9%	6-9%	6-9%	7-10%
Oxygen content (mL of oxygen/ 100 mL of blood)	6-15	8	10	11-20	15-29
Total oxygen in 100 g of animal (mL)	0.1-0.4	0.5-0.7	0.6-0.9	0.8-1.8	1.1-2.9

TABLE 20-12   A Comparison of Oxygen-Carrying Ability of Blood

Highlights

Animals continually exchange materials with their external environment. Within an animal various means are used to transport such materials. Structurally simple organisms rely on diffusion to transport materials to the individual body cells. Efficiency in transport is aided by cyclosis and various simple structural adaptations such as the flattened body in flatworms.

In more complex animals, exchange of materials takes place in specialized structures. A transport system links the cells with the environment in these organisms. Such transport systems generally consist of a fluid (blood), a pumping device (heart), and a network of tubes (blood vessels). The efficiency of these systems varies; it is determined by the structural adaptations that each animal has.

The closed circulatory system of humans is typical of mammals. Separate pulmonary and systemic circuits are present. This allows for efficient gas exchange between the blood and the air in the lungs.

Contraction of the ventricles of the heart forces blood into arteries that are built to withstand the high pressures. Passage of molecules into and out of the blood occurs through the thin-walled capillaries. Veins return the blood under lower pressure to the atria of the heart.

Blood is a unique fluid tissue. It consists of several types of cells suspended in a water solution of dissolved minerals and proteins. As well as transporting nutrients, several components of the blood help protect the body from infection. The internal transport system of humans shows many similarities in structure to other animals, since the basic problem of the transport of materials is common to all animals.

21 Excretion

Through metabolic activities, living organisms obtain nutrients, assimilate materials for growth, maintenance, and repair, and release energy to carry out life functions. However, these same metabolic activities also produce a number of by-products. For example, cellular respiration produces two main by-products: carbon dioxide and water. The metabolism of proteins results in the formation of some nitrogen-containing by-products, called **nitrogenous wastes**. In addition, the ingestion of food substances often results in a build-up of mineral salts in an organism.

When allowed to build up in excess amounts, these by-products have a poisonous, or toxic, effect on an organism. Therefore, a living organism must be able to rid itself of such harmful by-products. Unless it is able to do so, death will eventually result.

The process by which metabolic by-products are eliminated from an organism is called **excretion**. How do animals excrete these potentially harmful substances? You have already seen that carbon dioxide is eliminated from living systems by diffusion across a gas exchange surface. In this chapter you will study the **excretory systems** by which animals rid themselves of nitrogenous wastes and, in some cases, regulate their salt and water content.

21.1 Investigation

Problems Encountered by a Cell in Different Osmotic Environments

A living cell interacts constantly with the environmental medium that surrounds it. The plasma membrane surrounding a cell is a living, selectively permeable structure. It helps to regulate which materials enter and leave the cell. Both the cytoplasm of a cell and its external environmental medium consist mainly of water. You may recall that the plasma membrane is permeable to water. Therefore, water enters or leaves a cell through the process of osmosis.

A cell's environmental medium is called hypotonic if it contains a greater water concentration than the cell itself. If a cell's environmental medium has a lower water concentration than the cell, it is said to be hypertonic. An environmental medium that has the same water concentration as the cell is called isotonic.

What might happen to a cell that is placed in an environmental medium that has a different water concentration than the contents of the cell? In this investigation, you will study the effects of various osmotic environments on living human red blood cells.

Materials

3% sodium chloride (salt) solution   cover slips (3)
0.9% sodium chloride solution   lens paper
distilled water   paper towel
alcohol   sterile lancet
compound microscope   absorbent cotton
microscope slides (3)

CAUTION: Dispose of the lancet after use. It must not be reused, since reuse can spread diseases.

Procedure

a. Label 3 microscope slides as follows: 3% sodium chloride solution, 0.9% sodium chloride solution, distilled water.
b. Sterilize the tip of one of your fingers, using absorbent cotton soaked with alcohol. Then prick the tip of this finger with the sterile lancet. **DISCARD THE LANCET.**
c. Place a drop of blood on each of the 3 slides. Place cover slips on the slides. Then place the slide labelled 3% sodium chloride on the stage of the microscope.
d. Examine the mount under low, medium, and high power. Describe the appearance of individual red blood cells.

SOLUTION ADDED TO SLIDE	OBSERVATIONS	SKETCH OF RED BLOOD CELLS
3% sodium chloride		
0.9% sodium chloride		
distilled water		

TABLE 21-1   **Effects of Different Osmotic Media on Red Blood Cells**

e. Add a drop or two of 3% sodium chloride to the edge of the cover slip. Draw the solution under the cover slip by using the paper towel. Observe the red blood cells until no further change occurs. Record your observations and sketch a diagram of an individual red blood cell as it appears in a 3% sodium chloride solution. Record your results in a table similar to Table 21-1.

f. Repeat procedures d and e using the remaining two slides and the appropriate solutions.

Discussion

1. a) Which medium was hypertonic? How do you know?
   b) In what kind of medium does shrinking of red blood cells occur? What causes it?

2. a) Which medium was hypotonic? How do you know?
   b) In what kind of medium does swelling and bursting of red blood cells occur? What causes it?

3. a) Explain what happens when the cells are placed in 0.9% sodium chloride solution.
   b) Which solution is isotonic in comparison to the contents of a red blood cell?

4. a) What must be the osmotic relationship between human red blood cells and blood plasma?
   b) What do you think is the concentration of salts in human blood? Why?

5. Describe what would happen if you transferred some red blood cells from the 3% sodium chloride solution into distilled water. Explain your answer.

21.2 Nitrogenous Wastes

During digestion, proteins are broken down into their amino acid units. These basic units are then absorbed into the cells of the animal. Some of these amino acids are used by the cells to make new proteins required for growth and repair. The rest are broken down and used as an energy source. Amino acids cannot be stored by cells.

During amino acid breakdown, ammonia (NH_3) is formed. In high concentrations, ammonia is toxic to living organisms. Therefore, it must be excreted.

In small aquatic animals, excretion of ammonia is a relatively simple process. The surrounding water dilutes the ammonia while the ammonia is still in the animal's body. The resulting dilute solution diffuses across the cell membrane to the external water environment. The water environment further dilutes the ammonia. It also aids in removing the ammonia from the area around the animal. In summary, small aquatic animals have little problem excreting ammonia because of their small size and the abundant supply of surrounding water.

Amphibians, some terrestrial animals, and some large aquatic animals have a problem excreting ammonia. First, their large size makes simple diffusion of ammonia to the environment almost impossible. Second, these animals simply cannot take in enough water to dilute the ammonia enough. As a result, such animals usually convert ammonia into a less toxic form instead of trying to excrete it. For example, adult amphibians and mammals convert ammonia to **urea** by combining the ammonia with carbon dioxide (Fig. 21-1). Urea is formed in the liver, then transported in the blood to the kidneys. From there, it is excreted in the **urine**. A great deal of water is used in excreting the urea. But, as we shall see, such animals have special excretory structures that recycle much of this water during the excretion process. For these animals, then, the problem of excess water loss during the removal of nitrogenous waste is solved.

Insects, reptiles, and birds solve the problem in another way. Within their livers, they convert ammonia into a relatively harmless substance called **uric acid** (Fig. 21-1). Uric acid is insoluble. Therefore, it can be easily removed from internal fluids and excreted with very little water loss (see Table 21-2).

Uric acid formation provides another interesting advantage for these animals. They all produce eggs that are enclosed in a shell. Thus the embryos would not be able to excrete nitrogenous wastes as either ammonia or urea. Such substances would be trapped within the shell and eventually poison the developing embryo. Fortunately, these animals are able to convert the ammonia into relatively harmless uric acid. In this form the nitrogenous wastes can be excreted from the embryo and yet safely stored within the shell.

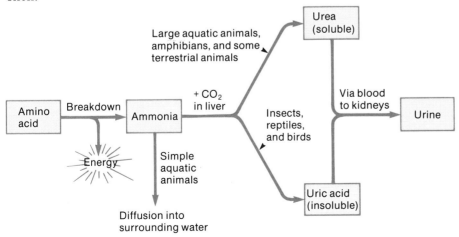

Fig. 21-1   Formation of nitrogenous wastes.

For most animals, the end product of excretion is a solution called urine. Depending on the animal, urine contains various proportions of excess water, nitrogenous wastes, and excess salts taken from body fluids.

Review

1. **a)** Into what basic units are proteins broken down during digestion?
   b) What are these basic units used for in cells that have absorbed them?
2. **a)** What happens to excess amino acids in the body?
   b) What three substances may be produced from excess amino acids in an animal?
3. Why is ammonia a suitable form of nitrogenous waste for small aquatic organisms?
4. Why is ammonia not a suitable form of nitrogenous waste for terrestrial organisms?
5. Why is uric acid a suitable form of nitrogenous waste for birds and reptiles?
6. Describe the part played by the human liver and kidneys in the excretion of nitrogenous waste.

FORM	RELATIVE TOXICITY	ORGANISMS THAT PRODUCE THIS FORM
Ammonia	Very toxic in high concentrations (soluble)	Most aquatic (marine and freshwater) animals, including crustaceans, bony fish, the tadpole stage of amphibians, alligators, and aquatic insects
Urea	Less toxic, even in high concentrations (soluble)	Cartilaginous fish (sharks and rays), many amphibian and terrestrial animals such as adult frogs, turtles, and mammals
Uric acid	Almost harmless (insoluble)	Most terrestrial insects, many terrestrial egg-laying animals (reptiles and birds)

TABLE 21-2   Nitrogenous Waste Excreted by Animals

21.3 Salt and Water Regulation in Animals

An excretory system enables an organism to rid itself of harmful nitrogenous wastes. However, in many organisms this is not its only function. An excretory system may also enable an organism to maintain a constant internal fluid medium, regardless of the osmotic conditions of the surrounding environmental medium. It does so by a process called osmoregulation. During this process, the excretory system continually adjusts to selectively eliminate or retain water and the dissolved salts that make up the organism's internal fluid medium. Therefore, it is probably more fitting to refer to such an excretory system as a regulatory system.

Marine, freshwater, and terrestrial environments all impose different sets of osmotic conditions on organisms living in them. Let us examine animals that live in each of these environments. We will look at both the osmotic problems they encounter and the methods they use to solve these problems.

Marine Animals

Invertebrates

A number of marine invertebrates are isotonic with their external salt water medium. That is, their body fluids contain dissolved salts and water in about the same proportion as does the sea water in which they live. Water balance is usually no problem for such animals. They neither take in nor lose a large amount of water by osmosis. But, what would happen to such animals should they move to an **estuary**, the area where a river enters the sea? In an estuary, fresh water and sea water mix, forming a diluted medium called **brackish water**. Brackish water is hypotonic to the body fluids of these invertebrates. Therefore, water tends to diffuse *into* these animals, diluting their body fluids.

Some marine invertebrates, such as the spiny spider crab *Maia*, are not able to regulate the concentration of their body fluids. As a result, they soon die in brackish or freshwater environments. Others, such as the shore crab *Carcinus* and the marine annelid *Nereis*, can regulate their internal fluids to a great extent. Their body fluids are isotonic to their normal sea water environment. However, they become hypertonic as they move into brackish water. Salts are removed from the brackish water by the crab's gills and by the annelid's body surface. The salts are then deposited in their body fluids. Excess water is eliminated by excretory organs called the **antennal glands** (Fig. 21-2). This ability to osmoregulate enables these animals to adapt to and live in a marine environment where the osmotic concentration is continually changing.

Bony fish

The body fluids of marine bony fish, such as herring and cod, are hypotonic with respect to their external sea water environment. Therefore, they tend to lose water by osmosis. An impermeable outer covering of skin and scales partially solves the problem. But diffusion of water can still occur across the exposed gills of these fish. To replace the water they lose, these fish drink sea water continuously. Obviously this would also result in the intake of a large quantity of salts. However, special gland cells in the gills excrete the excess salts to keep the body fluids hypotonic to the sea water (Fig. 21-3).

Cartilaginous fish

Marine cartilaginous fish, such as sharks and stingrays, solve the problem of living in sea water in a unique way. They have an unusually high tolerance

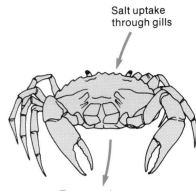

Salt uptake through gills

Excess water excreted through antennal glands

Carcinus (shore crab)

Salt uptake through the body surface

Excess water excreted through antennal glands

Nereis (marine annelid)

Fig. 21-2 Osmoregulation in two marine invertebrates living in brackish waters. Normally their internal body fluids are isotonic to sea water. But, as they move from sea water into brackish water, their body fluids become hypertonic.

Fig. 21-3 Osmoregulation in marine and freshwater bony fish. Freshwater fish are hypertonic to their environment. Active uptake of salts and excretion of dilute urine compensates for water gain and salt loss by osmosis. Marine bony fish are hypotonic to their environment. They tend to lose water by osmosis. Water is gained by drinking, and excess salts are excreted from the gills.

FRESHWATER BONY FISH (CARP) Hypotonic medium

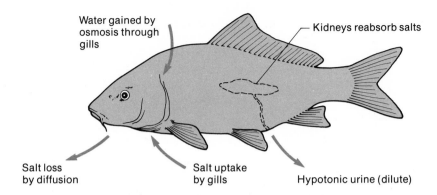

Water gained by osmosis through gills

Kidneys reabsorb salts

Salt loss by diffusion

Salt uptake by gills

Hypotonic urine (dilute)

MARINE BONY FISH (COD) Hypertonic medium

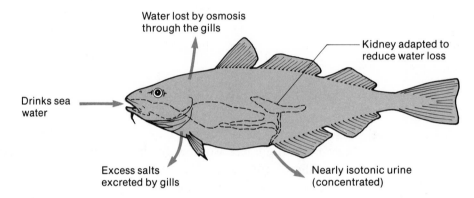

Water lost by osmosis through the gills

Kidney adapted to reduce water loss

Drinks sea water

Excess salts excreted by gills

Nearly isotonic urine (concentrated)

for urea and can retain a high concentration of it in their blood. Their kidneys are adapted to enable them to do this. As a result, their body fluids are slightly hypertonic in relation to sea water. Therefore, unlike bony fish, cartilaginous fish tend to gain water. Excess amounts of water are removed from the body fluids by the kidneys and excreted from the body (Fig. 21-4).

Reptiles and birds

Some marine reptiles, such as the Green turtle, have special salt-secreting glands near the corners of their eyes. A concentrated salt solution is excreted from these glands. Salt glands allow these animals to rid themselves of excess salts taken in from the sea water they drink. This osmoregulatory process gives these animals the appearance of crying or shedding tears when they are observed on land laying their eggs. Crocodiles also have such glands to rid themselves of excess salts. One can observe what appear to be large transparent tears coming from the eyes of a crocodile. For quite some time this was interpreted as mourning or weeping, presumably over the prey the crocodile had just eaten. As a result the term "crocodile tears" came to be used to describe the false tears accompanying a person's insincere show of sorrow. Marine birds, such as gulls and penguins, have similar salt-secreting

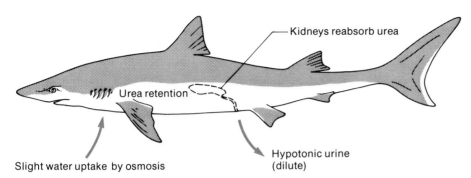

MARINE CARTILAGINOUS FISH (SHARK)   Slightly hypotonic medium

Kidneys reabsorb urea

Urea retention

Slight water uptake by osmosis

Hypotonic urine (dilute)

Fig. 21-4   Osmoregulation in marine cartilaginous fish. Unlike marine bony fish, the shark does not drink sea water. Instead, it retains urea. Internal fluids are therefore, slightly hyptertonic. Excess water that enters by osmosis is excreted in a dilute urine.

glands that enable them to maintain a salt and water balance by excreting excess salts from body fluids.

Mammals

Marine mammals, such as whales and seals, do not usually drink sea water. They get their water by eating fish and other marine life. The body fluids of the fish they eat are hypotonic to sea water. Therefore, they contain a high concentration of water. Marine mammals also eat marine invertebrates, which are rich in protein and salts. To balance this input, the excretory organs of marine mammals are able to excrete high concentrations of nitrogenous wastes and salts.

Freshwater Animals

The body fluids of freshwater animals are hypertonic to their external freshwater environment. As a result, water always tends to diffuse into these animals. Therefore, they must have a means of eliminating it. Protozoans, you will recall, rid themselves of excess water by means of contractile vacuoles. However, most freshwater animals use another mechanism. Their excretory organs also help in osmoregulation. We will study these mechanisms more closely in Section 21.5.

To illustrate the problems encountered by freshwater animals, let us examine the freshwater bony fish (Fig. 21-3). The body fluids of freshwater fish, such as trout and carp, are hypertonic to the water in which they live. Unlike marine fish, freshwater fish do not need to drink the water. Instead, water diffuses through their gills and into their body fluids. To get rid of excess water, these fish excrete large quantities of it, along with their metabolic wastes. Their kidneys act as both excretory organs and as osmoregulation organs. Some salt is also lost in the process. However, this is replaced either through diet or through active intake by specialized cells in their gills.

Most aquatic animals are adapted to live in a specific osmotic environment, either sea water or fresh water. However, some have osmoregulatory mechanisms that enable them to withstand small changes in the concentration of their watery environment. For example, animals that live in estuaries can survive the varying osmotic concentrations of brackish waters. However, very few aquatic animals are capable of living in both sea water and fresh water. The Atlantic salmon is one that can. Salmon are basically marine fish. Yet they migrate to freshwater streams to reproduce. Salt glands in their gills, and excretory organs called **kidneys** are specially adapted to maintain a balance of salt and water within the salmon. These organs enable the salmon to selectively excrete or retain salts and water, depending on the external osmotic environment. As a result, the Atlantic salmon can migrate either from salt water to fresh water or from fresh water to salt water.

This process of osmoregulation requires the use of energy. During spawning, a great deal of energy is spent on migration and preparation for reproduction. As a result, little energy is available for osmoregulation. Therefore, in going from salt to fresh water, the salmon body fluids become diluted with fresh water. The Atlantic salmon is able to correct this imbalance after spawning. Most other salmon cannot, and therefore die after spawning. Thus the Atlantic salmon is able to return to the sea and spawn again during its lifetime.

Terrestrial Animals

Terrestrial animals are faced with a problem very different from that of aquatic animals (Fig. 21-5). They must conserve water rather than prevent

Fig. 21-5 Water balance problems of a terrestrial animal. To prevent dehydration, water input must equal water loss.

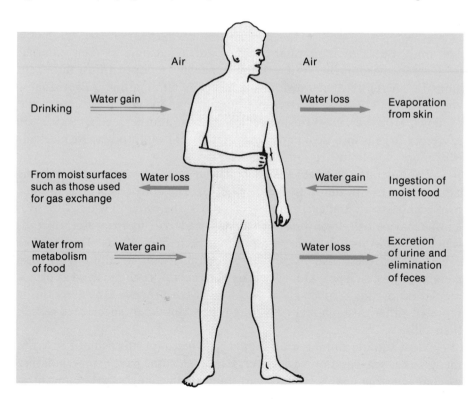

its entry or excrete an excess. In their relatively water–scarce environment, they are in constant danger of drying up. Water is lost to the environment by evaporation from exposed moist membranes such as respiratory surfaces. Water is also used to dilute nitrogenous wastes. Thus it is often lost in excretion. Land animals replace lost water by drinking it or by eating foods containing it. Some water is also formed during the metabolic breakdown of foods during cellular respiration.

Invertebrates

Insects and many other terrestrial invertebrates have a protective outer covering to lessen water loss due to evaporation. Also, you may recall that insects convert nitrogenous wastes to uric acid. Uric acid is insoluble. Therefore no water is lost during its excretion.

Reptiles

Snakes and lizards also excrete nitrogenous wastes in the form of uric acid. A waterproof covering of scales also aids in water conservation.

Birds and mammals

Birds and mammals have protective coverings too. For example, the feathers of birds and the hair of mammals reduce the amount of water lost from body fluids to the dry, terrestrial environment. The kidneys of birds and mammals enable them to reabsorb large quantities of water that would have been lost during the excretion of nitrogenous wastes.

Desert inhabitants have some other unique adaptations to the problem of water balance. For example, kangaroo rats do not drink water (Fig. 21-6). They live on seeds and other dry plant material that contain a high fat concentration. Water is formed during the breakdown of the fats during cellular respiration. They have no sweat glands. They are active only at night, when it is cool and more humid. They lose very little water during excretion. All these conservation measures enable the kangaroo rat to keep its water loss in balance with its water gain.

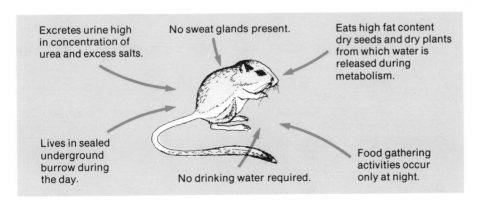

Excretes urine high in concentration of urea and excess salts.

No sweat glands present.

Eats high fat content dry seeds and dry plants from which water is released during metabolism.

Lives in sealed underground burrow during the day.

No drinking water required.

Food gathering activities occur only at night.

Fig. 21-6 The kangaroo rat is specially adapted to acquire and conserve water while living in a dry desert environment.

Amphibians

An amphibian, such as the frog, spends part of its life in water and part on land. To cope with this changing environment, osmoregulation in the frog is rather complicated. When in water, the frog tends to gain water through its skin. Its kidneys excrete excess water and retain needed salts. On land, the frog is subject to water loss. However, while on land, the frog carries with it a supply of water. The frog's kidneys remove excess water from the body fluids and store it in the bladder and lymph spaces. This stored water serves as a reserve to help maintain a constant water balance in the body fluids.

Humans

We are terrestrial animals. As such, we must also cope with the problem of salt and water balance. As you will see in Section 21.6, our kidneys can, under normal conditions, regulate the salt and water levels in our body fluids. To prevent water loss from our cells, our body fluids must contain a constant salt level of about 1%. Our kidneys are also excretory organs. Therefore they enable us to get rid of harmful nitrogenous wastes.

Think about what water balance problems you would face if you were shipwrecked on a raft at sea. If you were without fresh water, how could you replace the water you normally lost from your body fluids? Sea water has a salt concentration of about 3.5%. If you drank it, the salt level of your body fluids would increase to about 3.5%. However, your kidneys cannot excrete solutions that are over 2% salt. Therefore, you would have to use additional water from body fluids to excrete all the excess salt. Soon your body tissues would dehydrate, and you would die.

If you attempted to survive by eating fish, you would face another problem. Fish contain a large amount of protein. This would result in a large amount of nitrogenous waste being produced. This waste must be diluted with water in order to be excreted. In the process, you would lose more water from your body fluids than you would gain. Thus, unless you were able to find a supply of fresh water to replace the water you lost, you would not be able to survive for very long.

Review

1. Why is it more fitting to refer to an excretory system as a regulatory system?
2. What is meant by the term osmoregulation?
3. **a)** What is an estuary?
   b) Brackish waters pose an osmotic problem for marine animals. Describe this problem. Then, using a specific example, show how marine animals that live in these waters solve the problem.
4. **a)** Compare the osmotic concentrations of body fluids in marine cartilaginous fish and marine bony fish.
   b) What problems must each solve in order to survive in a marine environment?

c) How do cartilaginous fish solve the problem of osmoregulation?

5. How do freshwater fish maintain a salt and water balance?

6. Some aquatic animals such as the sea eel are able to live equally well in salt or fresh water. Describe the osmotic problems that these animals must overcome in order to do this.

7. a) What problems do terrestrial animals face with respect to salt and water balance?

   b) Give an example of how one such animal solves these problems.

8. a) If a kangaroo rat is fed a diet consisting only of seeds high in protein, it will soon die. Explain why.

   b) Study the information given with Fig. 21-7. Then answer the question that is asked.

9. a) It is usually recommended that workers in hot environments take salt tablets to increase their salt intake. Why?

   b) Too much salt in the diet can raise the blood pressure of some people. Why might this be so?

10. If you were on a raft in the ocean, you could consider it to be a desert. Explain.

Fig. 21-7 The dromedary camel (*Camelius dromedarius*) is well adapted to the arid conditions of the Sahara desert. Although its ability to do without water is exaggerated, it can survive severe water loss. The camel's hump is a deposit of fat, not a water storage tank. A camel can draw on this fat deposit as a temporary source of food and water. How could a reservoir of fat be a source of food and water?

21.4 Investigation

Contractile Vacuoles in Paramecia

Many single-celled and simple multi-celled organisms lack special excretory mechanisms. Harmful metabolic by-products simply diffuse directly across their cell membranes into the surrounding water environment. However, a number of protists have specialized excretory devices called contractile vacuoles. Protists that have contractile vacuoles usually live in water that is hypotonic to their internal body fluids.

Contractile vacuoles may excrete some metabolic by-products such as nitrogenous wastes. However, their main function is to rid the organism of excess water.

The paramecium has two contractile vacuoles, one near each end of the organism (see Fig. 10-11, page 256). Each vacuole has a series of canals radiating from it into the surrounding cytoplasm. These canals fill with water from the cytoplasm and then empty their contents into the vacuole. The vacuole gradually enlarges as its water content increases. It eventually contracts, ejecting its contents into the surrounding external environment (Fig. 21-8). This process requires energy.

The paramecium can normally rid itself of a volume of water equal to the volume of its cytoplasm in about half an hour. The two vacuoles usually fill and contract at regular intervals. What would happen to this rate of contraction if the osmotic concentration of the paramecium's external environment was changed?

In this investigation, you will observe and determine the rate of contraction of a contractile vacuole of a paramecium. You will also determine the effects of different environmental conditions on this rate.

Fig. 21-8 Contractile vacuole cycle in a paramecium. Water enters the radiating canals from the cytoplasm (A). As the canals force water into the vacuole, it gradually expands (B,C). Eventually, it contracts and forces water out into the external environment (D). This process is repeated again and again.

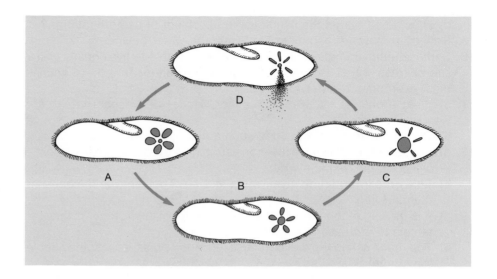

Materials

paramecium culture	cover slip
1.5% methyl cellulose solution	medicine droppers (4)
distilled water	compound microscope
3% salt solution	lens paper
microscope slide (depression type)	timing device

Procedure

a. Place two drops of paramecium culture in the depression on the slide. Add a drop of methyl cellulose solution to the depression. Place the cover slip over the depression.

b. Using low power, observe the mount and focus on a paramecium. Switch to medium, and then high power. Note the location of the two contractile vacuoles.
Note: Turn the fine adjustment until the paramecium is slightly out of focus. The vacuole should then stand out in contrast.

c. Observe one of the contractile vacuoles. Count the number of times it contracts during a one minute period. You should examine at least two other paramecia and repeat the procedure. Average the counts and record your results in a table similar to Table 21-3.

TYPE OF ENVIRONMENT	RATE OF CONTRACTION (PER MINUTE)
Normal culture medium	
Distilled water	
3% salt solution	

TABLE 21-3 **Effects of Different Osmotic Environments on the Rate of Contraction of Contractile Vacuoles in Paramecia**

d. Prepare a second slide by repeating procedures a to c. This time, add two drops of distilled water to the depression. Record your observations.

e. Prepare a third slide, again repeating procedures a to c. This time, add two drops of the 3% salt solution to the depression. Record your observations.

Discussion

1. How does water normally enter the body of a paramecium?
2. What effects do the different external environments have on the rate of contraction of the paramecium's contractile vacuoles? Give an explanation for the results.
3. Why would protists that normally inhabit hypotonic environments tend to have contractile vacuoles?
4. Potassium cyanide is a substance that causes paralysis of contractile vacuoles. If a solution of potassium cyanide were added to the paramecium culture, what would happen to the paramecia? Why?
5. In what sort of osmotic environment would you expect to find a protist without contractile vacuoles? Explain.
6. Fluid which is found in the radiating canals of a paramecium's contractile vacuoles is almost 100% water. The cytoplasm, in comparison, is hypertonic. Explain how this water could possibly enter the radiating canals from the paramecium's cytoplasm.
7. Some protists that live in sea water also have contractile vacuoles. However, the rate of contraction of their vacuoles is very slow. They contract only two or three times an hour. Explain what would happen if such organisms were placed in a freshwater environment.

21.5 Excretory Systems of Animals

As you have seen, one main function of excretion is to rid an organism of harmful metabolic by-products. These include carbon dioxide and nitrogenous waste. In addition, excretion gets rid of excess water and salts. Some organisms are designed so that most of their cells are in direct contact with the external environment. In such organisms, each cell carries out excretion on its own. Unwanted substances are excreted directly into the external medium, either by diffusion or active transport. Animals such as the hydra and the sponge excrete substances in this manner. However, in most large animals, harmful metabolic by-products from individual cells diffuse into a tissue fluid that bathes the body cells. The by-products then diffuse into the blood which is then processed by the animal's excretory system. This system removes the harmful nitrogenous wastes from the body fluids. You may recall that, in most cases, the excretory system also performs an osmoregulatory function.

All excretory systems use one or more of the following three processes:

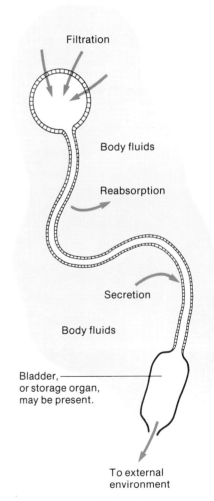

Fig. 21-9 **The basic function of an excretory system. Living cells carry out one or more of the processes of filtration, reabsorption, and secretion in every excretory system.**

Filtration

Body fluids

Reabsorption

Secretion

Body fluids

Bladder, or storage organ, may be present.

To external environment

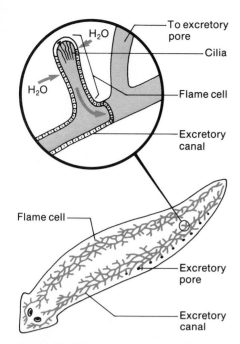

Fig. 21-10 The flame cell system of a planarian. Magnified portion shows detail of a flame cell. This system is mainly a water regulating system.

filtration, reabsorption, and secretion. **Filtration** takes place across a special membrane of the excretory system. This membrane is in contact with the body fluids. Substances in the body fluids are filtered out into the excretory system. Then the filtrate is forced through the rest of the excretory system. Various useful substances, such as salts or water, may be selectively **reabsorbed** into the animal's body fluids through specialized cells lining the system. Also, additional excess substances may be actively **secreted** into the filtrate at some location in the excretory system. The final product is eventually released into the external environment (Fig. 21-9). Let us look at a few examples of animal excretory systems to see how they solve waste removal problems.

Flame Cell System

Flatworms such as the planarian have a primitive excretory system called a **flame cell system** (Fig. 21-10). It consists of a series of tubes called **excretory canals** that run the length of both sides of the body. Side branches from these canals contain **flame cells**. Water and some dissolved wastes diffuse into the flame cells from the surrounding tissues. Each flame cell contains a number of **flagella**. These hair-like projections move continuously and create a current along the excretory canal. Fluids that collect in the flame cells are forced into the canals. From there, the fluids pass out of the planarian's body through **excretory pores**. To some observers, the motion of the flagella resembled a flickering flame. Therefore the cells were called flame cells.

Due to the planarian's body design, harmful metabolic by-products such as carbon dioxide and ammonia can diffuse rapidly into the surrounding water environment. No special structures are needed to remove these substances from the planarian body. As a result, the flame cells are mainly water-regulating structures. Since the body fluids of a planarian are hypertonic to its external fresh water environment, water is constantly diffusing into the animal. The flame cells "bail out" any excess water.

Nephridial System

Several groups of invertebrates have an excretory system called a **nephridial system**. Annelids such as the earthworm have this type of system. The body of an earthworm is made up of a series of distinct segments. Most segments contain a pair of excretory organs called **nephridia** (Fig. 21-11). Each **nephridium** has a funnel-shaped opening called a **nephrostome**. Cilia surround the nephrostome. The nephrostome is connected to a coiled **tubule** that extends into the next body segment. A **bladder** is located at the end of this tubule. The bladder is a storage structure. It empties to the outside through an opening called a **nephridiopore**. The earthworm also has a circulatory system. A network of capillaries surrounds the tubule of each nephridium.

Fluids enter the nephrostome from the body cavity of the earthworm.

Movement of the cilia creates a current that carries these fluids along the tubule. Some useful substances are reabsorbed from the fluid through special cells in the tubule wall. These substances move into the surrounding blood capillaries. In addition, some wastes may be secreted into the tubule from the blood in these capillaries. Then the waste materials pass into the bladder where they are temporarily stored. Eventually, the waste fluids are emptied to the outside through the nephridiopore. These wastes consist of urea and ammonia, as well as excess salts and water. Carbon dioxide is carried in the bloodstream to capillaries near the surface of the earthworm. From there, it diffuses into the environment.

Malpighian Tubules

Insects excrete carbon dioxide through trachea that form part of their gas exchange system. The rest of an insect's harmful metabolic by-products are excreted by means of organs called **Malpighian tubules**.

In the grasshopper, these excretory structures are attached to the intestine (Fig. 21-12). They are actually blind sacs that project outward as extensions of the intestine into the blood sinuses of the grasshopper's open transport system. Nitrogenous wastes, salts, and water enter the tubules from the blood. The tubules then allow water and some salts to be reabsorbed into the blood. The remaining nitrogenous waste, in the form of uric acid crystals, continues on through the intestine. This solid material is eventually excreted through the anus along with undigested materials. The conversion of nitrogenous waste to uric acid and the reabsorption ability of the Malpighian tubules enable the grasshopper to conserve water while getting rid of harmful wastes.

Kidneys

As you have seen, invertebrate excretory systems generally consist of simple tubules that lead to the external environment from various parts of the body. Certain metabolic by-products in the form of fluids are filtered into

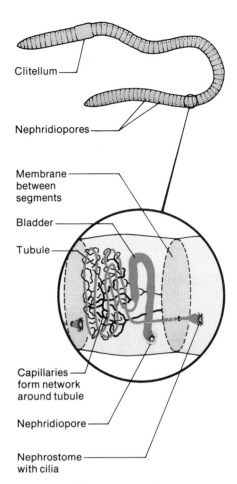

Fig. 21-11 The excretory system of an earthworm. A segment is magnified, showing the structure of the excretory organ, the nephridium.

Fig. 21-12 The excretory system of a grasshopper. Malpighian tubules are the major excretory organs.

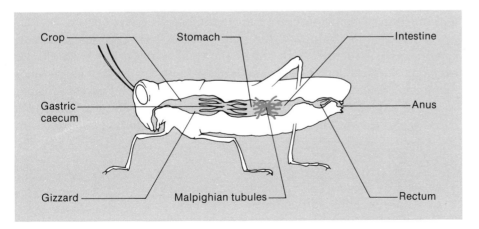

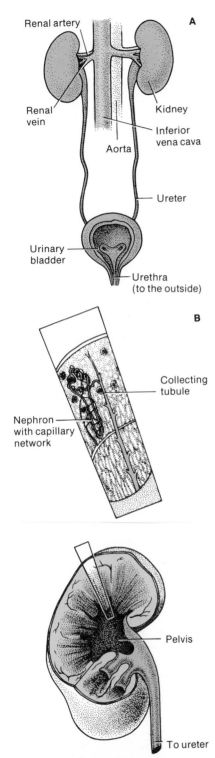

Renal artery

Renal vein

Kidney

Aorta

Inferior vena cava

Ureter

Urinary bladder

Urethra (to the outside)

A

B

Collecting tubule

Nephron with capillary network

Pelvis

To ureter

Fig. 21-13 The mammalian excretory system. A: General structure of system. B: Enlarged view of a kidney section, showing location of one of the functional units, a nephron.

these tubules. Fluids move along the tubules, perhaps aided by the movement of cilia. Cells lining the tubules reabsorb from or secrete substances into these fluids. This is a means of regulating both the content of the fluid to be excreted and the concentrations of the internal body fluids.

In vertebrates, the main excretory organs are the **kidneys**. The exact structure and function of kidneys varies with each class of vertebrates. Aquatic vertebrates, you may recall, live in a variety of osmotically different environments. The body fluids of most marine vertebrates are generally hypotonic to sea water. Those of fresh water vertebrates are usually hypertonic to their environments. Terrestrial organisms, of course, must conserve water. These factors affect the form in which various metabolic by-products such as nitrogenous wastes are excreted. Ammonia and urea require dilution; uric acid does not. Since vertebrates inhabit a variety of environments, their requirements with respect to excretion will vary. The difference in kidney structure and function from one class of vertebrates to the next permits all these animals to solve their individual excretory problems.

In general, each vertebrate has a pair of kidneys. A circulatory system brings blood to the kidneys where it is filtered. Useful materials are reabsorbed from the filtrate, and various unwanted substances are secreted into it. The kidneys perform both excretory and osmoregulatory functions. They differ from other types of excretory organs mainly because of greater efficiency.

Each kidney consists of thousands of kidney tubules called **nephrons** (Fig. 21-13). These nephrons are in close contact with a network of capillaries. The tubules of the many nephrons join together to form **collecting tubules**. These, in turn, empty into a central cavity of the kidney called the **pelvis**. A single duct called the **ureter** leads from the pelvis to a temporary storage chamber. In mammals and some other vertebrates, this chamber is called the **urinary bladder**. The urinary bladder drains to the outside through a duct called the **urethra**.

The major difference between vertebrate nephrons and the excretory tubules of other animals is the degree to which they are associated with the circulatory system. This close association is the main reason for the kidney's superior efficiency. We will examine, in Section 21.6, the complex structure and function of the nephron of the human excretory system.

Although the kidney is the main part of the vertebrate excretory system, other structures also aid in the total process. For example, gas exchange surfaces such as gills and lungs excrete carbon dioxide. Also, in mammals, sweat glands in the skin aid in the secretion of excess salts and water. In any one animal, many structures may be involved in collectively carrying out the very important life processes of regulation and excretion.

Review

1. **a)** What is a flame cell? Describe its structure.
   b) What functions do flame cells perform?
2. Describe the excretory system of an earthworm.

3. How do the Malpighian tubules of insects function in excretion?
4. **a)** What is the major excretory structure in vertebrates?
   b) How do vertebrate excretory systems differ from invertebrate systems?
5. Vertebrates do not depend entirely on kidneys to carry out the excretory process. Explain this statement.
6. In general, describe the processes involved in animal excretion and regulation.
7. Figure 21-14 shows the excretory system of a crayfish. Study the diagram carefully, then answer these questions:
   a) A crayfish is a freshwater arthropod. Its body fluids are hypertonic to its environment. Would you expect a crayfish to excrete a diluted or a concentrated urine? Why?
   b) Describe the excretory system of a crayfish.
   c) Describe how this excretory system functions.

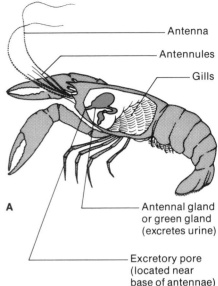

A

- Antenna
- Antennules
- Gills
- Antennal gland or green gland (excretes urine)
- Excretory pore (located near base of antennae)

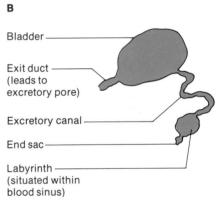

B

- Bladder
- Exit duct (leads to excretory pore)
- Excretory canal
- End sac
- Labyrinth (situated within blood sinus)

Fig. 21-14   The excretory system of a crayfish. A: Crayfish with carapace removed, to show the location of one of the two excretory organs, the antennal or green glands. B: A green gland enlarged, showing structural detail.

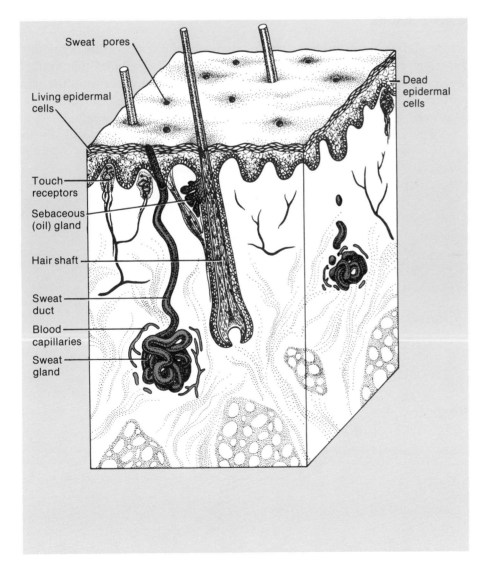

- Sweat pores
- Living epidermal cells
- Dead epidermal cells
- Touch receptors
- Sebaceous (oil) gland
- Hair shaft
- Sweat duct
- Blood capillaries
- Sweat gland

Fig. 21-15   Sweat glands in the skin. A dilute solution of salt, urea, and carbon dioxide enters the coiled sweat gland from the surrounding capillaries. This sweat rises through the duct and escapes through the pore on the skin surface.

21.6 The Human Excretory System

Your body produces a large amount of metabolic waste each day. If this waste builds up in your body, it becomes poisonous. Your circulatory system acts as a "waste collection service". It carries wastes to "dump sites". These include your lungs, liver, sweat glands, and kidneys. This process is called **excretion** and the structures involved are the **excretory organs**.

The wastes produced by your body include carbon dioxide and water from cellular respiration. These wastes are carried directly by the circulatory system to the excretory organ. Carbon dioxide is disposed of by the lungs. Water is excreted by the lungs and the sweat glands. Other wastes, like the amino acids from broken-down proteins, are carried to the liver. The liver converts most amino acids to urea. The urea is put back into the bloodstream and is later filtered out by the kidneys and sweat glands.

Sweat Glands

Sweat glands perform both a waste removal and a temperature regulating function (Fig. 21-15). These glands consist of coiled tubules that extract water, salts, and urea from the blood. Evaporation of water from the skin surface produces a cooling effect. This can be refreshing on a hot day. However, it can produce excessive chilling in cool, windy weather.

The salts and urea left behind by the evaporating water nourish the

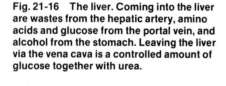

Fig. 21-16 The liver. Coming into the liver are wastes from the hepatic artery, amino acids and glucose from the portal vein, and alcohol from the stomach. Leaving the liver via the vena cava is a controlled amount of glucose together with urea.

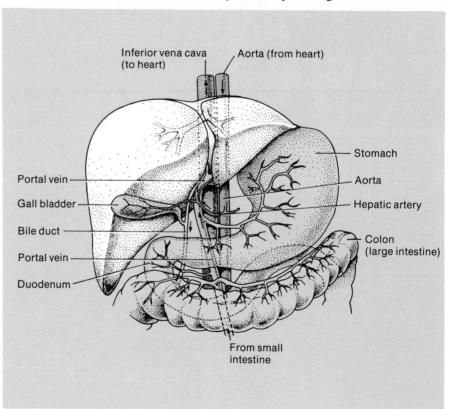

large number of bacteria already on the skin. Body odour is caused by the waste substances produced by these bacteria, not by sweat itself.

The Liver

This organ is perhaps the most important regulating organ in the body (Fig. 21-16). It can affect the amounts of both waste and nutrient substances circulating in the blood. It stores excess glucose as glycogen. It makes vitamin A, proteins, and some amino acids. Also, as mentioned before, the liver plays an important role in the excretion of the nitrogenous waste, urea. It converts waste amino acids into urea. The urea is put back into the bloodstream, and is eventually extracted by the kidneys. In addition, the liver breaks down the hemoglobin from damaged and worn-out red blood cells. The hemoglobin forms waste **bile pigments**, which pass out in the **bile**. The bile fluid collects in the **gall bladder** where it is temporarily stored. The bile then passes through the **bile duct** to the duodenum. Besides the waste bile pigments, bile also contains bile salts. These salts are important in fat digestion.

The Kidneys

The kidneys are the main excretory organs of the body (Fig. 21-17). All your blood is filtered through the kidneys once every 4 min. The kidneys remove

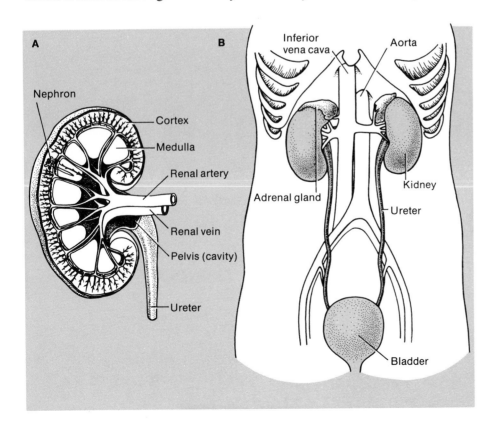

Fig. 21-17 The kidneys: these two organs both filter and regulate substances in the blood.

Fig. 21-18   A single nephron.

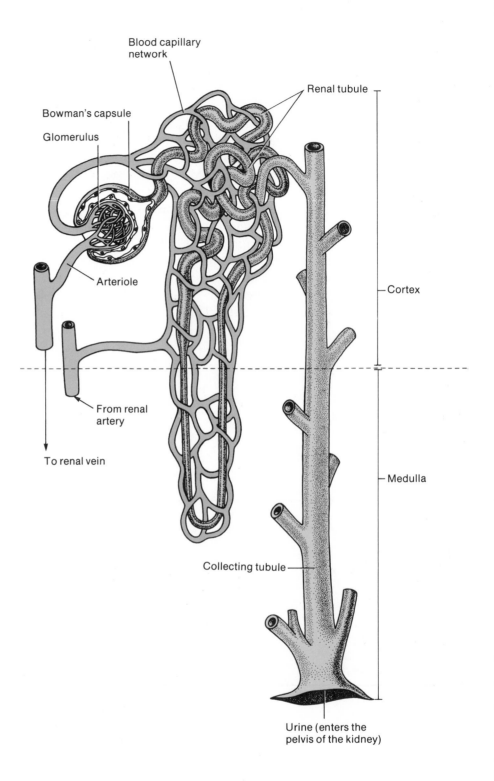

Blood capillary network

Renal tubule

Bowman's capsule

Glomerulus

Arteriole

Cortex

From renal artery

To renal vein

Medulla

Collecting tubule

Urine (enters the pelvis of the kidney)

excess minerals and urea from the blood. At the same time, they regulate the water content and the pH (acidity) of the blood.

A human kidney, like all vertebrate kidneys, contains functioning units called **nephrons**. A human kidney has about one million of these. Each nephron consists of two main structures, a **blood capillary network** and a **renal tubule** (Fig. 21-18). Wastes are removed from the blood by each nephron in two steps: filtration and concentration. The resulting waste fluid is called **urine**.

Filtration

Arterial blood under pressure is squeezed through a knot of capillaries about 200 μm in diameter. This special capillary bed is called a **glomerulus**. Small particles such as water, glucose, urea, and salts, are forced through the capillary walls. Large protein molecules and red blood cells are too big to be squeezed out of the capillaries.

About 180 L of liquid are squeezed through the capillaries into renal tubules every day. The cup-shaped end of the renal tubule is called **Bowman's capsule**. It collects the filtrate (the water and dissolved substances) from the glomerulus. Obviously no person passes out 180 L of urine each day. In fact, about 1.5 L of urine leave the body in a 24 h period.

Concentration

The great length of the renal tubule allows almost all the water, glucose, and amino acids to be reabsorbed into the blood. Most of the urea stays in the tubules.

Reabsorption of water into the blood occurs partly because the capillary surrounding the renal tubule is carrying the same flow of blood that was filtered. This blood has a high concentration of proteins. It therefore has a relatively low concentration of water. Thus diffusion of water from the tubule back into the capillary takes place. The capillary membrane is permeable to water but not to the large proteins.

Diffusion, alone, does not explain the reabsorption of many of the dissolved materials in the filtrate. For example, glucose that is filtered into the renal tubule is completely reabsorbed (Fig. 21-19). If passive diffusion was the only way reabsorption of substances could occur, the concentration of glucose inside the renal tubule would equal the concentration in the capillaries. But no glucose is left in the tubule at all. Glucose is reabsorbed into the blood by active transport. Cells lining the tubule use energy to carry glucose out of the tubule into the blood. Energy is required because the glucose is carried to an area of higher glucose concentration. Many salts and amino acids are also actively transported from the tubule back into the blood against the existing concentration.

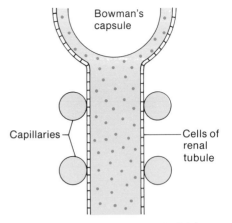

Filtrate from Bowman's capsule is rich in glucose.

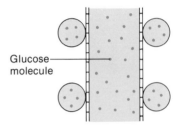

If only passive diffusion occurred, glucose concentrations would become equal.

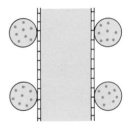

Instead, no glucose is found in the collecting tubule. This indicates active transport of glucose by the renal tubule.

Fig. 21-19 Active transport of glucose. Only glucose molecules are shown.

Urine

The renal tubules from the nephrons connect with longer **collecting tubules** which empty into the pelvis of the **kidney**. A single **ureter** leads from the pelvis of each kidney (see Fig. 21-17). The two ureters lead to the **bladder**. About 200 mL of the liquid waste, called urine, can be stored in the bladder. Urine is about 96% water, 2% urea, and 2% other wastes. The urea has been concentrated almost 100 times by the action of the kidneys. The other wastes in the urine consist mainly of sodium and chloride ions (salt solution).

Chemical analysis of urine is often carried out by doctors to detect unusual amounts of protein or glucose. If these molecules are present in urine, it indicates an unusual and possibly dangerous condition of the excretory or circulatory system.

Kidneys as Regulators

As well as acting as a blood filter, the kidneys act as regulators of blood chemistry. That is, they have mechanisms that can control how effectively they extract substances from the blood. For example, if you are losing considerable water and salt by sweating, or if you eat large amounts of protein, the concentrations of water, salt, and urea in the blood change. The kidneys act to keep these concentrations from getting too high or too low in the blood.

The total amount of blood that is filtered can be controlled by reducing or increasing the flow of blood into each glomerulus. The body does this by changing the diameter of the arteriole that carries blood into each glomerulus. Recall that an arteriole has muscular walls. Contraction of circular muscles reduces the diameter of the arteriole. Therefore less blood enters the glomerulus to be filtered.

Substances such as glucose and sodium ions that are carried by active transport have their return to the blood regulated by cells of the renal tubule. Certain drugs can actually block the active transport mechanisms, and thus affect the blood and urine concentrations of these substances.

Regulation of the water content of blood is more complicated. Special nerve cells in the brain are sensitive to concentrations of dissolved substances. If solute concentrations go up, these cells stimulate the production of a regulating hormone called vasopressin. Vasopressin circulates throughout the body. Cells toward the end of the renal tubule are sensitive to this hormone. Vasopressin causes these tubule cells to become very permeable to water. Therefore more water diffuses back into the blood, and the solute concentrations in the blood are reduced. Vasopressin production then stops. Vasopressin has no effect on most other body cells.

Kidneys and Health

The total blood volume is filtered once every 4 min by the kidneys. Damage to the capillaries or tubules of a kidney may make it completely useless. If only one kidney is actually regulating and removing wastes from the blood, it takes about 8 min for a complete filtering of the blood. A healthy kidney can handle the extra blood volume. However, wastes such as urea often accumulate in the blood faster than they can be removed. For this reason, kidney machines and kidney transplants have been developed.

Kidney machines

The dialysis membrane (tubing) that you used in the diffusion experiments of Chapter 6 is a direct result of research on artificial kidney machines. This material is the filtering tube in these machines. However, there is an important difference between a dialysis membrane and a renal tubule. A dialysis membrane acts solely as a selectively permeable membrane through which diffusion takes place. Because it is not alive, it cannot carry out active transport.

Figure 21-20 shows, in a simplified way, how urea is removed from the blood while, at the same time, bicarbonate ions are added to the blood. The composition of the surrounding bath is crucial. For example, even though glucose is not to be added to the blood, it must not be removed. For this reason, the bath must contain glucose and other substances in exactly the concentrations that they are found in the blood.

A further problem was found during the development of the kidney machine. Blood travelling through the dialysis tubing would often start clotting. The friction of the blood on the sides of the membrane was enough to release clotting agents such as thromboplastin. Membranes are now treated with a chemical to prevent clotting.

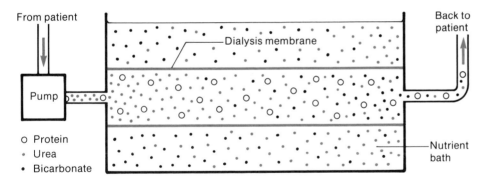

Fig. 21-20 Principle of the kidney machine. Urea leaves the blood while bicarbonate is added by passive transport. Why can proteins not leave the bloodstream?

Kidney transplants

For people with kidney damage, the once or twice weekly trip to the hospital to be hooked up to the kidney machine is inconvenient and expensive. Smaller units for home treatment have been developed. However, the long-term solution to kidney failure problems is a kidney transplant. Careful analysis of donor kidneys is made to make sure that a potential transplant will not be rejected by the patient's body. Fig. 21-21 shows how a donated kidney is placed in a patient's body. Many driver's licences now include a section granting use of body parts including kidneys for transplant purposes.

Fig. 21-21   A kidney transplant. Notice how the donated kidney is "hooked up" into the circulatory system. Find out the names of the artery and vein used.

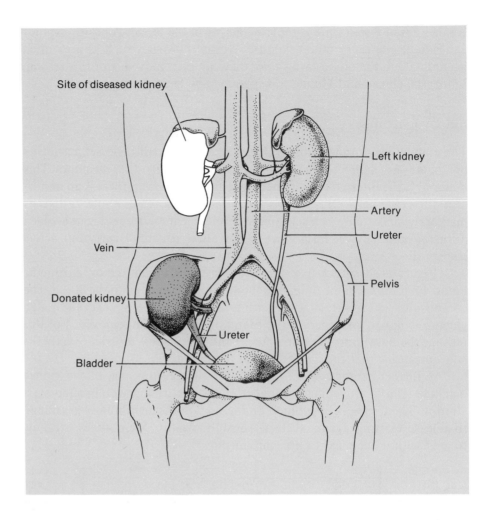

Site of diseased kidney

Left kidney

Artery

Ureter

Vein

Pelvis

Donated kidney

Ureter

Bladder

Review

1. Name three waste substances produced by living cells.
2. Why must cellular wastes be removed from the body?
3. Where are wastes excreted from the body?
4. **a)** Why do unwashed feet and underarms usually have an unpleasant odour?
   b) Why do bacteria survive so well on your underarms and feet?
5. Describe two ways that the liver functions as part of the excretory system.
6. Describe the structure of a nephron.
7. Describe how urea from the bloodstream is concentrated into urine. Be sure to include the functions of the glomerulus and renal tubules.
8. Describe how an artificial kidney machine works.
9. When are kidney transplants necessary? What precautions are taken before such an operation?
10. Suppose that a large amount of protein is found in a person's urine. Why might this be evidence of kidney damage?
11. How do you think a longer renal tubule would affect the amount of water found in urine? Why?

21.7 Investigation

Absorption in the Kidneys

Table 21-4 describes what happens to five substances found in the blood when the blood is filtered through the kidneys. The figures given apply when 1 L of urine is formed. In this "dry lab," you will learn about the functioning of the kidneys as you perform certain calculations.

SUBSTANCE	COLUMN A AMOUNT INITIALLY FILTERED OUT OF BLOODSTREAM	COLUMN B AMOUNT RETURNED TO BLOODSTREAM	COLUMN C AMOUNT IN URINE	COLUMN D PERCENT REABSORBED BY BLOOD
Water	100.0 L	99.0 L		
Chloride ion	370.0 g	364.0 g		
Glucose	70.0 g	70.0 g		
Urea	30.0 g	10.0 g		
Uric acid	4.0 g	3.5 g		
Calcium ion	10.0 g	9.85 g		

TABLE 21-4 Absorption in the Kidney

Analysis

1. For each substance, calculate the amount left in the urine. Amount in urine = Amount initially filtered out – Amount returned to blood.

2. a) For each substance, calculate the percent reabsorbed into the blood.

$$\text{Percent reabsorbed} = \frac{\text{Amount returned to bloodstream}}{\text{Amount initially filtered out}} \times 100\%$$

   b) On a percentage basis, which of these five substances is *removed* from the bloodstream in the greatest quantity?

   c) On a percentage basis, which two substances are *returned* to the bloodstream in the greatest quantity?

3. a) Suggest how the percentage of certain substances in urine might be changed by a low salt diet. Justify your answer.

   b) Suggest how the percentage of certain substances in urine might be changed by a high protein diet. Justify your answer.

4. a) How does the body benefit by reabsorbing so much water and so much glucose back into the bloodstream?

   b) All glucose is normally returned by the kidneys to the bloodstream. Why, then, do you need to keep eating foods that are a source of new glucose?

5. a) Suggest how scientists obtained a sample of the initially filtered liquid (Column A).

b) Suggest how scientists obtained a sample of blood right after filtering by the kidney (Column B).

c) Suggest how scientists obtained a sample of the urine (Column C).

6. a) What might the presence of glucose in urine show about levels of glucose in the blood?

b) What disease could cause this condition?

Highlights

Excretion is the process by which metabolic by-products are eliminated from an organism. These by-products include carbon dioxide and nitrogenous wastes. Ingestion of food often allows excess salts to build up in an organism. These, too, must be excreted. Most animals have an excretory system that eliminates nitrogenous wastes and, in many cases, regulates the salt and water content of the body fluids.

Nitrogenous wastes are formed when amino acids are broken down by an organism. The first waste formed is ammonia. Small aquatic animals excrete it directly by diffusion into the surrounding water. Many large aquatic animals and terrestrial animals convert the ammonia to urea before it is excreted. Still other animals excrete the nitrogenous waste in the form of uric acid crystals.

The excretory system of most animals is also responsible for osmoregulation. That is, it maintains a constant concentration of materials in the body fluids, regardless of the osmotic conditions of the surrounding environment. Marine animals, freshwater animals, and terrestrial animals all have unique methods of osmoregulation.

Many simple animals have no excretory system. Each cell carries out its own excretion. A wide variety of excretory systems occur in more complex animals. These include the flame cell system of the planarian, the nephridial system of the annelids, the Malpighian tubules of insects, and the nephrons in the kidneys of vertebrates.

The lungs, sweat glands, liver, and kidneys are structures that function to excrete cell wastes from the human body. Kidneys are the most important excretory organs. Blood is filtered through a capillary network called a glomerulus. Essential minerals, water, and glucose are reabsorbed in renal tubules. The remaining urine contains higher concentrations of urea and salts. This liquid is stored in the bladder for a time before it is eliminated.

Reproduction: The Perpetuation of Life

22

Chapter 8 discussed the reproduction of cells. It explained mitosis and its role in the growth, development, maintenance, and repair of living things. It also described meiosis and its role in the maintenance of genetic continuity during sexual reproduction. You should review mitosis and meiosis before you begin this chapter, if you do not clearly remember what they are.

This chapter deals with the reproduction of organisms. In this chapter, reproduction refers to the formation of new separately-existing individual organisms from organisms already in existence.

Reproduction is a characteristic of all living things. Its importance to a species is obvious. All organisms have a limited life span. Death eventually results from such causes as accidents, starvation, disease, predation, and old age. A species would quickly become extinct if individuals of that species were not able to reproduce new individuals. Also, there must be some means of ensuring the successful development of newly reproduced individuals into organisms which can themselves reproduce.

How do animals reproduce? What mechanisms are there that ensure successful reproduction in animals? In this chapter we will discuss the general features of reproduction in animals. We will also examine some specific patterns of reproduction in both invertebrates and vertebrates. The final section of this chapter discusses the process of reproduction in humans.

22.1 General Features of Reproduction in Animals

The methods by which organisms reproduce their own kind can be grouped into two categories: **asexual reproduction** and **sexual reproduction**. A great deal is known about each method. In order to more fully understand the processes, it is necessary to become familiar with some of the basic terminology that is part of the language of this area of biology. In this section we will review some terms that you already know and introduce some new ones.

Asexual Reproduction

Asexual reproduction is the formation of one or more new offspring without the fusion, or joining, of two genetically different cells. In asexual reproduction, the parent organism, by the process of mitosis, produces a new cell or cells which develop into offspring. Since the process of mitosis is involved, the offspring are identical to the parent. This method of reproduction does not allow for any variations in the offspring. Minor changes could result from mutations. However, major mutations are rare and usually cause death. As a result, asexual reproduction practically ensures that the hereditary charcteristics remain the same, generation after generation.

Binary fission

Many single-celled organisms, such as the amoeba and bacterium, reproduce asexually by undergoing mitosis and simply dividing into two. This process is called **binary fission** (see Fig. 8-9, page 181 and Fig. 9-13, page 211). The parent ceases to exist by becoming two distinct and identical individuals.

Budding

Budding is another type of asexual reproduction. In this process, new offspring develop as bud-like outgrowths from the parent organism. As the bud grows, it differentiates into a completely new individual, identical to the parent. Eventually the bud breaks away, becoming independent of the parent organism. In budding, the parent organism maintains its identity and eventually dies. Hydra reproduce by budding. We will discuss reproduction in hydra in Section 22.4.

Fragmentation

Another type of asexual reproduction in animals involves the breaking off, or **fragmentation**, of a part of an organism. The fragment develops into a new organism. For example, sponges can be cut up into small pieces and each piece will develop into a new, separate individual. The flatworm, planaria, often pinches off the posterior portion of its body. Each part then **regenerates** the missing structures (Fig. 22-1).

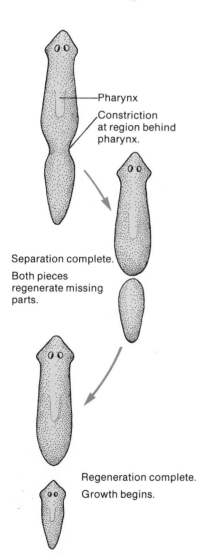

Pharynx

Constriction at region behind pharynx.

Separation complete.

Both pieces regenerate missing parts.

Regeneration complete.

Growth begins.

Fig. 22-1 Asexual reproduction in the planarian. Fragmentation occurs and regeneration results in two complete planarians from a single parent organism.

Sexual Reproduction

Even though reproduction in many animals can occur asexually, most animals do not depend entirely on this method. They also reproduce sexually. In fact, sexual reproduction is more common among animals than asexual reproduction.

Gametes and fertilization

Sexual reproduction involves the formation of offspring as the result of the fusion, or joining, of two individual cells. Sexual reproduction almost always involves two parent organisms. Each parent produces specialized reproductive cells called **gametes** by the process of meiosis. Then, a gamete produced by one parent unites with a gamete produced by the other parent. This fusion of gametes is called **fertilization**. Fertilization results in the formation of a single cell called a **zygote**. The zygote eventually develops into a new organism. Each gamete contains half the hereditary information of each of its parents. As a result, a zygote develops into a new individual with hereditary information from both parents. Unlike asexual reproduction, sexual reproduction can result in the formation of new individuals that may differ from their parents in a variety of ways (Fig. 22-2).

Fig. 22-2 Asexual versus sexual reproduction. What advantages and disadvantages would there be in each method?

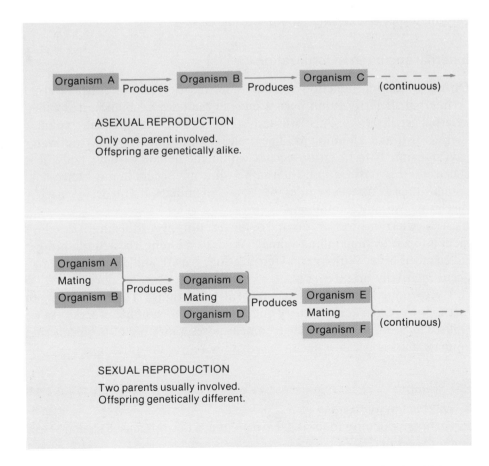

In most sexually reproducing species of animals, there are two distinctly recognizable kinds of individuals that produce gametes. These are called **males** and **females**. Males produce gametes called **sperm**. Females produce gametes called **ova** (singular: **ovum**), or **eggs**. In sexual reproduction, a sperm and an ovum unite to form a zygote. Ova are usually larger than sperm. The size difference is due to the presence of stored food materials in the ova. The stored food in an ovum is used to nourish the **embryo** that develops from the zygote. Sperm are smaller and more numerous than ova. They are also motile. They swim to the female gametes by means of tail-like flagella. In almost all sexually reproducing animals, gametes are formed in special organs called **gonads**. Gonads that produce sperm are called **testes**. Those that produce ova are called **ovaries**.

Hermaphrodites

In most animal species, testes and ovaries are each found in different parent organisms. However, in some species, they are found in the same parent organism. For example, in the hydra, planarian, and earthworm, each individual organism has both ovaries and testes. Such organisms are called **hermaphrodites**. **Self-fertilization** (the fusion of ova and sperm from the same individual) does not usually occur in hermaphroditic animal species. Most often, **cross-fertilization** occurs. Two individual members of a hermaphroditic species exchange sperm during mating.

External and internal fertilization

During sexual reproduction in animals, fertilization may be either external or internal. If fertilization occurs outside the parents' bodies, it is called **external fertilization**. This is usually the case with many aquatic animals such as fish and amphibians. They release their gametes into their water environment. If fertilization occurs within the female parent's body, it is called **internal fertilization**. Internal fertilization occurs in most terrestrial animals such as insects, reptiles, birds, and mammals. It also occurs in a few species of fish and amphibians.

Once fertilization has occurred, the resulting zygote begins its development into a fully functioning animal. As it is developing, the new offspring is called an embryo. With internal fertilization, growth and development may occur internally or externally.

Reproduction is a characteristic of all living things. There is a variety of methods by which animals carry out this very important life process. As we will see in Section 22.2, the environment poses a number of problems that animals must solve in order to successfully reproduce offspring.

Review

1. What is binary fission?
2. Compare budding to fission. How are these two processes similar? How do they differ?

3. Would asexual reproduction be an advantage or a disadvantage to a species in a constantly changing environment? Explain your answer.
4. How is a zygote formed?
5. What is a hermaphrodite?
6. Self-fertilization does not occur in most hermaphrodites. Would this be an advantage or disadvantage? Explain.
7. Self-fertilization is quite common in internal parasitic hermaphrodites such as the tapeworm. Why do you think this would be so?
8. What advantages would there be for an organism to carry out reproduction sexually? What disadvantages would there be?

22.2 Problems in Sexual Reproduction

In order for sexual reproduction to be successful, two conditions are necessary. First, sperm and ova must unite. That is, fertilization must occur. Second, the resulting zygote must become an embryo and then develop into a mature organism. Sexually reproducing animals must overcome a number of problems to ensure that these two conditions are met. Since animals live in a variety of different environments, many of the problems vary from species to species. However, there are a number of problems that are common to all sexually reproducing animal species. In this section we will discuss some of these common problems and the means by which different animal species solve them.

Problems Associated with Fertilization

Three problems are associated with fertilization:

1. A liquid medium is required to allow movement of sperm to the ovum.
2. The release of gametes must be timed to ensure that they meet one another.
3. Some sort of protection for gametes is required prior to fertilization.

Transportation medium

An ovum has no means of locomotion. On the other hand, a sperm is capable of movement. A tail-like flagellum propels it in a swimming motion from its point of release at the testes to an ovum. However, this form of locomotion requires a liquid medium of some sort. Animals in which fertilization takes place externally, usually live in or near water. Thus the water acts as the liquid medium. The gametes are simply released into the water.

In internal fertilization, sperm are released into a confined space, the female reproductive tract. A liquid produced by the testes of a male is released into the female tract along with the sperm. This liquid enables the sperm to swim along the tube-like tract until they meet an ovum which has been released by an ovary. Then fertilization occurs within the female reproductive tract.

Since the tract is a confined space, the chances of a sperm meeting an ovum are much greater than if the gametes were released externally. As a result, animals that undergo internal fertilization produce relatively few ova.

Proper timing

Once gametes are released by the gonads, they have a very short life span. Sperm generally lack a food supply. Therefore, unless fertilization occurs shortly after they are released, they die. Since they lack food, they also have a limited amount of energy. Therefore they can only travel short distances. Mature ova can be fertilized for only a short period of time after their release. The length of this fertile state of ova varies from species to species. In humans it lasts for about one day. Therefore, in order to ensure that fertilization can occur, the release of gametes must take place, not only in the same general location, but also at about the same time.

In animals that undergo external fertilization, proper timing is often achieved through various behaviour patterns. It appears as though hormonal control plays an important role in many of the **courting**, or **mating behaviour**, patterns of these animals. Such behaviour patterns stimulate each sex to release gametes into the water in the same place and at the same time. As an example, let us look at reproduction in the frog.

Fertilization of frog ova occurs externally. Body contact between the two adults is an essential part of the mating behaviour. The male grasps the female. This contact stimulates the female to release ova into the water. The male releases sperm at the same time. Fertilization occurs soon after. As you can see, this behaviour pattern ensures the proper timing and place of release of both types of gametes. Clearly this increases the chances of fertilization.

Behavioural patterns may also play a role in solving the problem of timing in animals that undergo internal fertilization. The two different sexes must be brought into close contact for fertilization to occur. Therefore, part of the mating behaviour may serve as an advertisement or signal to attract the opposite sex when they are ready to mate. This behaviour occurs when gametes have matured and are fertile.

Hormonal control is often involved in the proper timing of gamete release in more complex animals. We will discuss the role of hormones in human reproduction in Section 22.5.

Some animals, such as insects, solve the problem of timing in another way. Females are often able to store sperm within their reproductive systems after they are deposited there by the male during mating. The sperm are stored in special sacs. Later, the sperm are released into the reproductive tract at the same time that ova are released from the ovary. Queen bees are able to store fertile sperm for several years. However, biologists do not know exactly how this is done.

Protection of gametes prior to fertilization

Gametes that are simply released into a water environment during external fertilization face a number of hazards. Predators may eat them. Certain abiotic factors, such as the availability of dissolved oxygen and the temperature of the water, may also present some problems. With the exception of a jelly-like coating around some types of ova, the gametes generally have no physical means of protection against such environmental hazards. Furthermore, external fertilization is a relatively uncertain process. Even though a sperm is quite often attracted to an ovum by chemicals that the ovum produces, the two do not always meet. As a result, fertilization does not always take place. However, animals that carry out external fertilization usually produce and release large numbers of gametes. Though this may seem rather wasteful, it does increase the chances that fertilization will occur.

The reproductive behaviour patterns of many aquatic species help to solve some of these problems. For example, mating generally occurs during a definite season. Ova are usually not produced and released when environmental conditions such as water temperature are not suitable.

Terrestrial animals live in a very dry environment, and generally undergo internal fertilization. Gametes produced by these animals are quite sheltered, and not subject to the dangers of an external environment.

Problems Associated with the Developing Embryo

In animal species that reproduce by external fertilization, the developing embryo generally receives little or no protection. Under such circumstances, most of the fertilized ova, or zygotes, will perish. However, as we have already seen, these animals produce and shed large numbers of gametes. The

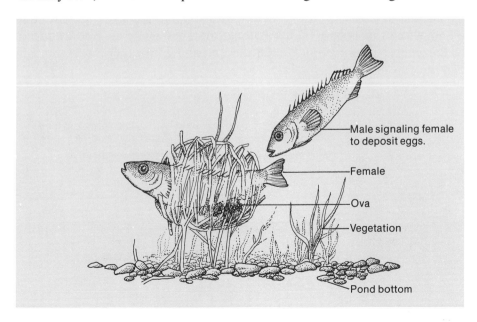

Fig. 22-3 Reproductive behaviour of the ten-spined stickleback. After a complex courting behaviour, the female sheds her mature ova into a nest previously prepared by the male. Once the eggs are released, the male enters the nest and releases sperm over them. The male then guards the developing embryos and young offspring until they are capable of living independently.

Male signaling female to deposit eggs.

Female

Ova

Vegetation

Pond bottom

few zygotes that do survive are usually enough to maintain the species.

Some animals, such as frogs, shed ova that are surrounded by a jelly-like coating. This coating provides some protection for the ova prior to and after fertilization. Usually a small amount of food reserve is also present in the ova. This provides some nourishment to the embryo in early stages of development. The eggs may also be coloured in such a way as to camouflage them from predators.

Certain species of catfish are **mouthbreeders**. A parent places fertilized eggs in its mouth until development is complete. When the offspring are able to survive on their own, they are released. Other fish species, such as the stickleback, are **nest builders** (Fig. 22-3). The female sheds ova into a tunnel-like nest that has been built by the male. The male then sheds sperm on the ova. After fertilization, the embryos develop within the nest. The male guards the nest during embryonic development. He also creates water currents that carry fresh supplies of dissolved oxygen to the embryos.

Animals that undergo internal fertilization usually provide much more protection and nourishment to the developing embryo than do animals that undergo external fertilization. Reptiles and birds enclose the zygote in a porous **shell** (Fig. 22-4). The shell encloses a **yolk** encased in a membranous **yolk sac**. The yolk is the food supply for the developing embryo. Other membranes are also present. The **amnion** forms a fluid-filled sac around the embryo. This keeps the embryo in a wet environment and protects it from shock. The **allantois** forms a sac-like structure that is actually an outgrowth from the digestive tract of the embryo. It contains numerous blood vessels that allow gas exchange. Also, many metabolic by-products are excreted into and stored within this sac. A fourth membrane, the **chorion**, is an outer membrane that encloses the embryo and all the other membranes. It is a thin membrane situated just beneath the porous shell. It is involved in the exchange of respiratory gases. Oxygen passes through the shell from the outside air. It then diffuses across the chorion and allantois, and into the blood vessels that supply the developing embryo. Carbon dioxide produced by the embryo is transported to the allantois and diffuses across the chorion and out through the shell.

Fig. 22-4   Cross section of a chicken egg, showing the developing embryo and the relative positions of the embryonic membranes.

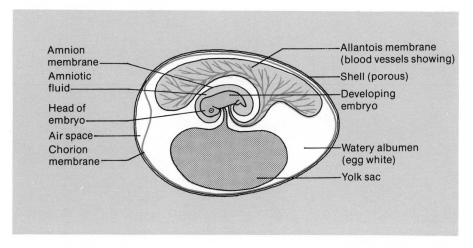

In summary, the shell and its associated membranes protect the developing embryo. They also assist in such life processes as gas exchange, excretion, and nourishment of the developing embryo.

Some shark and reptile species retain the developing embryo within a special cavity in the female's body. The embryo has its own stored food supply, is enclosed by membranes, but is not encased in a shell. The female's body cavity offers protection. When fully developed, the young emerge as miniature adults.

In most mammals, development of the embryo occurs within the female's body. Some, such as the kangaroo and opossum, are called **marsupials**, or pouched mammals (Fig. 22-5). In marsupials, following internal fertilization, the embryo begins its early development within the female's body. Only a small amount of yolk is present in the ova. It is not enough to provide nourishment for complete development of the embryo. Once the food supply is used up, the immature offspring are expelled from the female's body. They find their way into a special pouch located on the female's abdomen. Here, they attach themselves to the nipples of the mammary glands to receive a nutrient milk substance. The offspring remain in this pouch until development is complete.

In most other mammals, complete internal development takes place with the aid of a special structure called the **placenta** (Fig. 22-6). This structure consists of a network of capillaries. Nutrients and wastes are exchanged across this network between the developing embryo and the female's body. Once development is complete, the offspring is born as a miniature adult. It is usually nourished by milk from the female's mammary glands until it can find its own food. Humans are placental mammals. So are cats, dogs, horses, cows, and elephants.

Fig. 22-5 The wallaby is a marsupial. Mammary glands within the female's "pouch" supply nourishment to the developing young.

Fig. 22-6 A mammalian embryo and its placenta. Food and oxygen are supplied to the developing embryo. These substances in the mother's blood diffuse across the placental membranes into the embryo's blood. Harmful metabolic by-products diffuse from the embryo's blood into the mother's blood.

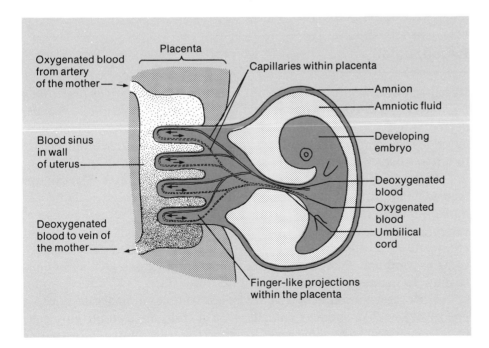

1. What two conditions must be met in order to ensure successful reproduction in animals?
2. What three problems must animals deal with to ensure that fertilization does occur?
3. Explain why proper timing of gamete release is essential to successful fertilization.
4. How do aquatic animals generally solve the problem of protection of gametes prior to fertilization?
5. What function do "mouth breeding" and "nest building" serve in sexually reproducing aquatic animals?
6. a) List and explain the functions of the four major membranes associated with a reptile or bird embryo.
   b) What do you think would happen to a developing embryo within a shell if the shell was coated with wax? Explain.
7. a) What are marsupials?
   b) How is the developing offspring protected and nourished in a marsupial animal?
8. What role does the placenta play in the development of the human embryo?
9. Prepare a report on the courting behaviour of the stickleback and the grunion. How do these two species of fish overcome the problems associated with successful reproduction?
10. The platypus is a rather unusual animal. Find out about its pattern of reproduction. How does the platypus solve the problems related to successful reproduction that are mentioned in this section?

22.3 Investigation

Regeneration

Regeneration *is the process by which lost or damaged parts of an organism's body can be repaired or replaced.* All organisms possess this ability to some degree. In some invertebrates, small fragments of the organism's body may develop into a whole animal. In more complex animals, regeneration occurs to a lesser degree.

Pieces of sponge may regenerate into whole new sponges. Starfish can regenerate lost arms. Some crabs, crayfish, and insects can regenerate lost limbs. An earthworm can regenerate a new head or tail. Several groups of vertebrates can also regenerate lost parts. Some salamanders and lizards can regenerate a tail lost by accident or during combat. However, in more complex vertebrates, regeneration is limited to processes such as wound-healing. In mammals, a lost limb will not regenerate.

How regeneration occurs and why less complex animals show it to a greater degree than complex animals is not yet fully understood. However, there does appear to be some general pattern to regeneration. Once a part is lost, epithelial cells migrate out from the cut edges and cover the wound. This is part of the normal healing process. If a new part is to be replaced, cells from tissues near the wound develop into a mass of unspecialized cells called a **blastema**. These cells undergo mitosis. The mass grows, and gradually the new cells differentiate to form the replacement body part. Recent evidence shows that the nervous and chemical (hormonal) control systems play an important role in the regeneration process.

In this investigation you will attempt to induce regeneration in two simple invertebrates, the planarian and the hydra.

Materials

planaria culture	microscope slides (2)
hydra culture	small petri dishes
single-edged razor blade or	eye droppers
sharp fine scalpel	ice cubes
dissecting microscope or	small, artist's brush
hand lens	

Procedure A   Regeneration in Planaria

a. Using the artist's brush, place a planarian in a drop of culture water on a microscope slide. Examine the animal closely under a dissecting microscope.

b. Allow the planarian to extend itself. Then place the slide on top of an ice cube. The cold temperature will slow the animal's movements. Cut off the "head" with a razor blade or scalpel. Place the "head" section in a petri dish with some culture water. Label this dish "head section". *Note*: The cut should be perpendicular to the animal's body as shown in Fig. 22-7,A.

c. Next, cut off a "tail" section as shown in Fig. 22-7,A. Place this section in a separate petri dish labelled "tail section".

d. Place the rest of the planarian in another petri dish and label this dish "middle section". Add some water from the original planarian culture to each of the petri dishes.

e. Using two more planaria, repeat the above procedures. Place all "heads" in the petri dish labelled "head section", all tails in the petri dish labelled "tail section", and all middle pieces in the petri dish labelled "middle section".

f. Cover all three petri dishes. Then place them in a cool, dark place. Do not place the dishes in a refrigerator. No food should be supplied while the planaria are regenerating.

g. Observe the petri dishes every 2 to 3 d for a three week period. Note any changes that occur in the cut sections. Record your observations in the form of a sketch and a brief description.

h. The water in the petri dishes should be changed every third day. Any dead sections should be removed.

Fig. 22-7 Three patterns for sectioning a planarian. Note: cuts should be made so that complete sectioning results.

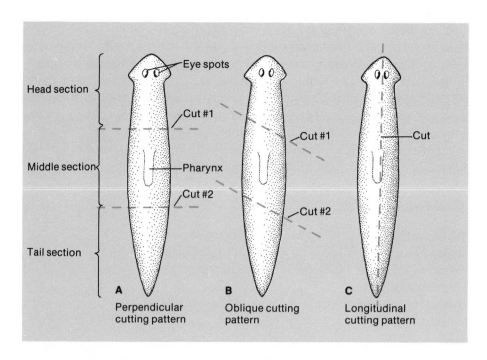

Eye spots

Head section

Cut #1

Cut #1

Cut

Middle section

Pharynx

Cut #2

Cut #2

Tail section

A
Perpendicular cutting pattern

B
Oblique cutting pattern

C
Longitudinal cutting pattern

Fig. 22-7 Three patterns for sectioning a planarian. Note: cuts should be made so that complete sectioning results.

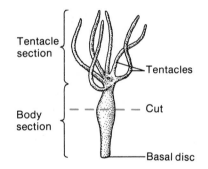

Tentacle section

Tentacles

Body section

Cut

Basal disc

Fig. 22-8 Pattern for sectioning a hydra. Note: the cut should be made so that complete sectioning results.

Procedure B Regeneration in Hydra

a. Using an eye dropper, place a hydra in a drop of culture water on a microscope slide. Examine the animal closely under a dissecting microscope.

b. When the hydra extends itself, cut it into two pieces using a scalpel or razor blade. The cut should be made approximately midway between the hydra's tentacles and basal disc as shown in Fig. 22-8.

c. Place the tentacle portion in a petri dish with some culture water. Label this dish "tentacle section".

d. Place the rest of the hydra in a separate petri dish with some culture water. Label this dish "body section".

e. Using two more hydras, repeat the above procedures and place the cut sections in the appropriate petri dishes.

f. Cover both petri dishes and place them in a location designated by your teacher. Again, no food should be supplied while the hydra are regenerating.

g. Observe the petri dishes every day for the next two weeks. Sketch and briefly describe any changes that occur.

Discussion

1. Is there any pattern in the tendency for dissected sections of a planarian to form heads or tails? Explain.

2. Is there any pattern in the tendency for dissected sections of a hydra to form tentacle (head) ends or body ends? Explain.

3. Why should you not feed the cut sections during regeneration?

4. Why was more than one planarian and one hydra used in each case?

For Further Research and Discussion

1. Repeat the set of procedures as outlined for the planarian by cutting obliquely as shown in Fig. 22-7,B. Observe and record the results. Compare them with those obtained using a perpendicular cut.
2. Again, repeat the investigation using planaria, but this time cut the animal longitudinally as shown in Fig. 22-7,C. Observe and record the results.
3. Design and perform an experiment that might produce a planarian with two heads attached to the anterior end.
4. Design and perform an experiment that might produce a planarian with a head at both the anterior and posterior end.
5. Design and perform an experiment to determine how small a section of a hydra is capable of regenerating a whole new hydra.
6. Place a hydra on a microscope slide. Place a cover slip over the hydra and gently mash it. Put the mashed hydra in a petri dish along with some culture water. Observe the mashed hydra daily for two weeks. Watch for and describe any signs that the cells are reassociating to form a new hydra.

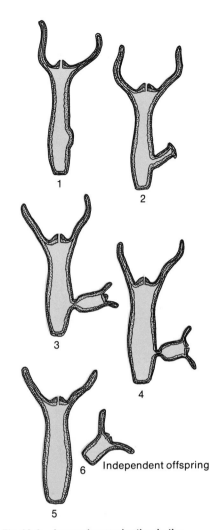

Fig. 22-9   Asexual reproduction in the hydra by budding. A: A hydra with developing buds. B: Longitudinal section of hydra, showing successive stages of the developing bud.

22.4 Examples of Reproductive Patterns in Animals

This section describes several examples of reproductive patterns in animals. As you read through the examples, keep in mind the problems referred to in Section 22.2. Then try to determine how the animals mentioned here solve these problems.

The Hydra

The hydra can reproduce both asexually and sexually. Asexual reproduction occurs by budding (Fig. 22-9). It is the major method of reproduction. It usually occurs during the warmer months of the year. Buds form as outgrowths of the body wall, and occur about one-third of the way up from the base of the hydra. As cell division continues, the buds gradually increase in size. In 2 to 3 d, they resemble the parent hydra, complete with tentacles and mouth opening.

During development, the gastrovascular cavity of the bud is continuous with that of the parent. Therefore, the developing bud can obtain nourishment from the parent. Once the mouth of the bud becomes functional, a constriction forms at the base of the bud. Soon the bud pinches off from the parent and begins its life as an independent organism. A single parent may produce as many as six buds at once. Also, it is not uncommon for a bud to develop its own buds before it detaches from the parent. Three generations can actually be seen attached at the same time.

Sexual reproduction of hydra usually takes place in the autumn (Fig.

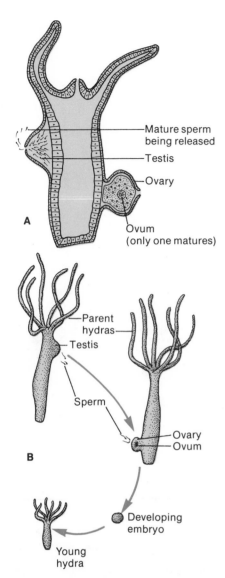

A

B

Fig. 22-10 Sexual reproduction in the hydra. A: Longitudinal section showing temporary gonads. B: Pattern of reproduction in the hydra. Note: self-fertilization seldom occurs.

22-10). The gonads are not permanent organs. They develop from cells in the ectoderm (outer layer) only during the season of sexual reproduction. They disappear immediately afterwards. **Testes** develop as swellings near the tentacles. The testes produce a large number of **sperm** cells. **Ovaries** develop as swellings near the base of the hydra. Each ovary produces a single **ovum** or **egg** cell. Once they have matured, the flagellated sperm are discharged into the surrounding water. Then they swim toward an ovum. Eventually one sperm penetrates the ovum. Fertilization takes place and a zygote is formed.

A mature ovum has a stored supply of food that is used later to nourish the developing embryo. The zygote divides many times and eventually forms a ball-like mass of cells, the embryo. A thick protective coat forms around the ball of cells at this stage. Then the embryo breaks away from the ovary. The encased embryo drops to the bottom of the pond or stream, where it remains dormant through the winter. When favourable conditions return in the spring, the dormant embryo resumes development. Eventually an independent hydra is formed.

Most species of hydra produce both testes and ovaries on the same individual. Therefore they are hermaphrodites. However, in such individuals, the sperm cells can survive only for a few hours. Since they are usually released a couple of days before the ova have matured, self-fertilization rarely occurs.

The Earthworm

Earthworms are terrestrial animals. As such, they must somehow protect both the gametes and the developing embryo from the drying conditions of their land environment. As you will see, their pattern of reproduction solves this problem in a rather unique way.

Earthworms are **hermaphrodites**. Each individual produces both the male and female gametes required for sexual reproduction. Since it is bisexual, you might think that each earthworm is capable of fertilizing its ova with its own sperm. However, this does not happen. Both the structure of the earthworm's reproductive system and its pattern of reproduction prevent self-fertilization.

The gonads are located in the anterior end of each earthworm (Figs. 22-11 and 22-12). The ova, or eggs, are formed in a pair of small ovaries located in segment 13. During the mating season, mature ova are released by the ovaries and are picked up in the funnel-like ends of two **oviducts**. An egg sac attached to each oviduct stores the mature eggs until mating occurs. The oviducts lead to the ventral surface of the worm and open to the outside as two small pores on segment 14.

Sperm are produced in two pairs of testes, located in segments 10 and 11. The testes are enclosed in large sac-like structures called **seminal vesicles** (sperm chambers) that extend from segments 9 to 13. These store the sperm until they are mature. A **sperm duct** receives mature sperm through funnel-like structures within the seminal vesicles. Sperm are transported along the sperm ducts to openings in segment 15 on the ventral surface of the earthworm.

Two other structures are involved in reproduction in the earthworm. Two pairs of seminal receptacles, located in segments 9 and 10, serve as storage sacs for sperm received from another earthworm. Finally, the clitellum, a swollen ring of tissue containing gland cells, surrounds the earthworm, just behind the region of the sex organs.

Mating, or copulation, between two mature earthworms usually occurs on the surface of the ground (Fig. 22-13). The two worms come together, ventral surface to ventral surface, with their anterior ends pointing in opposite directions. The clitellum of one earthworm is opposite segments 9 to 11 of the other. Gland cells in the clitellum secrete a mucous slime tube around each earthworm from about segment 13 to segment 31. These slime tubes prevent the mixing of the sperms from the two earthworms during sperm exchange. In the region of each clitellum a common slime tube is secreted to hold the two worms closely together. Sperm from the seminal receptacles are discharged through the external openings in segment 15 of each earthworm. The sperm are carried in the sperm grooves along the ventral surface of each earthworm, to seminal receptacles of the other worm where they are stored. Having exchanged sperm, the two earthworms separate.

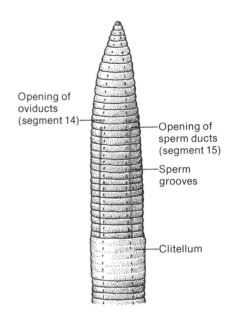

Opening of oviducts (segment 14)

Opening of sperm ducts (segment 15)

Sperm grooves

Clitellum

Fig. 22-11   Ventral view of the anterior end of an earthworm, showing external features of the reproductive system.

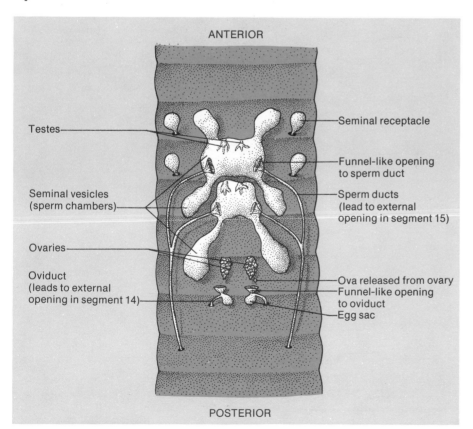

ANTERIOR

Testes

Seminal receptacle

Funnel-like opening to sperm duct

Seminal vesicles (sperm chambers)

Sperm ducts (lead to external opening in segment 15)

Ovaries

Oviduct (leads to external opening in segment 14)

Ova released from ovary
Funnel-like opening to oviduct
Egg sac

POSTERIOR

Fig. 22-12   The reproductive system of the earthworm. The diagram shows the general structure (dorsal view) of the system as it would appear in a dissection. The seminal vesicles are drawn as if they were transparent, to show the location of the testes and sperm duct openings.

Fig. 22-13 Mating of earthworms. Diagram shows pathway of exchange of sperm from sperm ducts of one worm to the seminal receptacles of the other. Fertilization occurs much later.

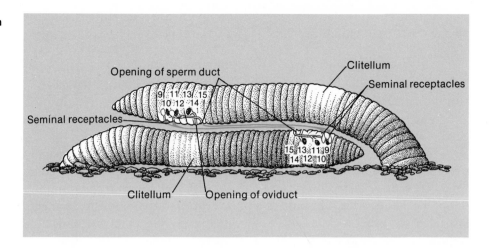

Once the ovaries have produced mature ova, gland cells in the clitellum secrete a mucous sleeve. A form of food reserve to nourish the developing embryos is also deposited within the mucous sleeve. The sleeve slips forward over the front of the earthworm. As it passes segment 14, it receives ova from the oviducts. And, as it passes segments 9 and 10, sperm are released into it

Fig. 22-14 Reproductive organs of a male crayfish. A: Lateral view of left side showing relative position. B: Dorsal view, showing details of structure.

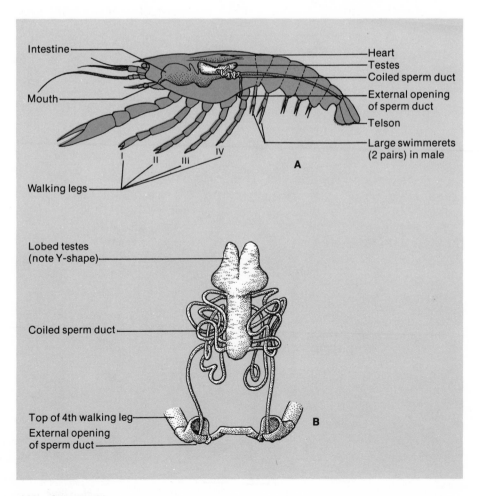

from the seminal receptacles. Fertilization takes place within the mucous sleeve. Once the sleeve slips off the earthworm, its ends seal, forming a capsule called a cocoon. Only one zygote continues to develop within the cocoon, which becomes lodged in the soil. The earthworm embryo reaches full development in a few weeks. Then the independent young worm escapes from the cocoon.

The Crayfish

In the crayfish, the sexes are separate and easily distinguished by external characteristics. The male is identified by four large, stiff, sharp-pointed swimmerets that form the first two of six pairs located beneath the abdomen (Fig. 22-14). In the female, the first two pairs of swimmerets are much smaller in size (Fig. 22-15).

The gonads of each sex are found in the thorax, above the intestine and just below the heart. In both sexes, they have the same general shape. They consist of three lobes arranged much like the shape of the letter Y. In each sex, a pair of ducts extends from the gonads to external openings on the walking legs. The openings of the oviduct are at the top of the second pair of walking legs in the female. The opening of the sperm duct is at the top of the fourth pair of walking legs in the male. The female also has a seminal receptacle between the bases of the third and fourth pairs of walking legs.

During mating, the female is held on her back. Then sperm released through the opening of the male's sperm duct are guided into the seminal

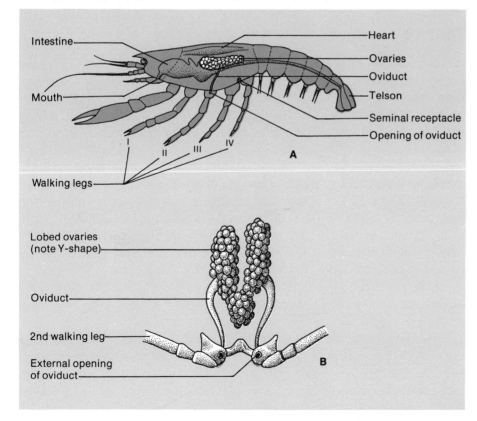

Fig. 22-15 Reproductive organs of a female crayfish. A: Lateral view of left side showing relative position. B: Dorsal view showing details of structure.

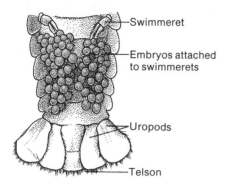

Fig. 22-16 Ventral surface of abdomen of a female crayfish "in berry".

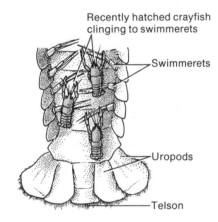

Fig. 22-17 Ventral surface of abdomen of a female crayfish with "hatchings".

receptacle of the female. The male's first two pairs of swimmerets aid in the sperm transfer. Sperm are stored in the seminal receptacles of the female until mature eggs are produced by the female. Sperm deposited in the seminal receptacle in the fall might not fertilize eggs until the next spring.

Once the ova have matured, the female lies on her back, with her telson curled over her walking legs, and releases the eggs through the opening of the oviduct. As they are released, sperm cells from the seminal receptacle are also released. Fertilization takes place and the resulting embryos, coated with a sticky secretion, become attached to the female's swimmerets. The attached embryos resemble a bunch of berries. As a result, the female is said to be "in berry" (Fig. 22-16). The swimmerets are constantly waved back and forth in the water. This action circulates fresh, oxygen-rich water over the developing embryos.

Full development of the embryos into young crayfish usually takes from five to eight weeks, depending on the water temperature. When they hatch, the young crayfish remain attached to the mother's swimmerets for another few weeks until they are able to survive on their own (Fig. 22-17). The young crayfish reach adult form by going through a series of **moults**. During each moult, the crayfish sheds its exoskeleton and swells up to a larger size by absorbing water. Gradually, a new exoskeleton forms. The crayfish is practically defenseless without its exoskeleton. Therefore it usually hides during the moulting process. Moulting most often occurs about seven times during the first year and about twice a year thereafter. Crayfish have an average life span of three to four years.

The Frog

Although adapted for life partly on land, the adult frog must return to the water to reproduce. Separate sexes, external fertilization, and external development are all characteristics of the reproductive pattern of this amphibian. **Metamorphosis** is also a characteristic of the frog. That is, during development from fertilized egg to adult, a series of changes in form take place.

In the frog, the reproductive system is closely associated with the excretory system. As a result, the two are often referred to as the **urinogenital system** (Fig. 22-18). The male has a pair of oval, creamy-white testes. Each testis is found on the ventral surface of a kidney. Flagellated sperm, produced in the testes, pass into the kidney through sperm ducts. The mature sperm then enter the **ureters** and move along to **seminal vesicles** (sperm chambers). Here they are stored until mating occurs. Note that, in the male, the ureter is used for passage of both urine and sperm. Sperm are discharged from the seminal vesicles, through the **cloaca**, and out the **cloacal opening** during mating.

The ovaries of the female are large, lobed structures. One is found on the ventral surface of each of the two kidneys. During breeding season, eggs, which are half black and half white, are produced and released by the ovaries into the female's body cavity. The eggs are then swept into the ciliated

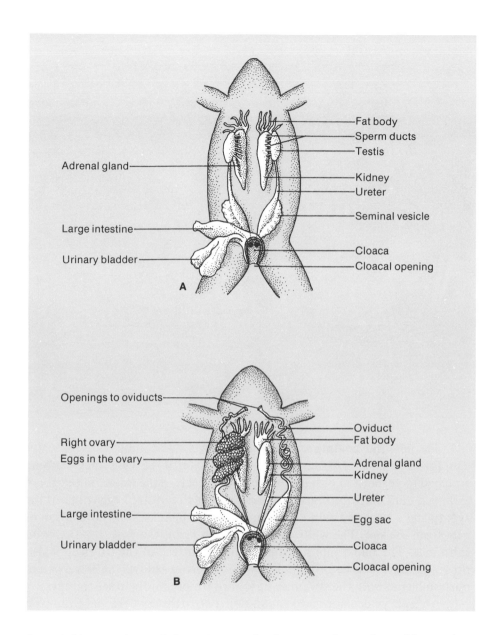

A labels:
Fat body
Sperm ducts
Testis
Adrenal gland
Kidney
Ureter
Seminal vesicle
Large intestine
Urinary bladder
Cloaca
Cloacal opening

B labels:
Openings to oviducts
Oviduct
Fat body
Right ovary
Eggs in the ovary
Adrenal gland
Kidney
Ureter
Large intestine
Egg sac
Urinary bladder
Cloaca
Cloacal opening

funnel-like openings of the oviducts. As they pass down the oviducts, they are coated with a jelly-like substance that is secreted by the walls of the oviduct. The coiled oviducts lead to an enlarged egg sac. Here, the eggs are stored until mating occurs. Then they are released into the cloaca and out the cloacal opening.

Frogs generally mate in the spring and usually at night. The female, body swollen with mature eggs, seeks out a male. The male perches on the back of the female and clasps his forelimbs around her body (Fig. 22-19). Swollen pads on the male's index digit help him to secure his grip. The pressure of his grip also forces eggs out through the female's cloaca and into the surrounding water. As the eggs enter the water, the male releases sperm over them and fertilization takes place.

Fig. 22-19   The frog life cycle.

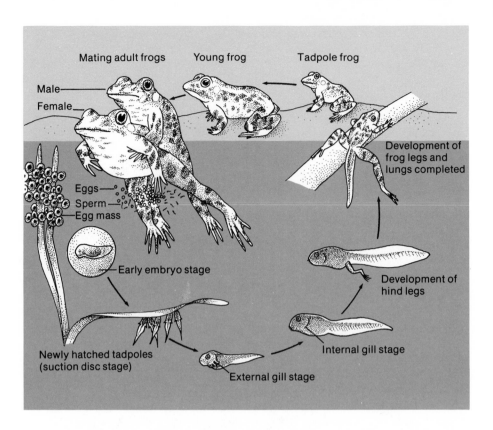

The jelly-like coating around the eggs swells on contact with the water and forms a large egg mass. This mass of fertilized eggs may become attached to vegetation or it may simply float freely in the water. The eggs are arranged dark side up. The dark portion of each fertilized egg contains the zygote and a dark pigment. This dark pigment increases the absorption of heat from the sun. The white portion of each egg is the **yolk**, a stored food substance. The half-white half-black colour pattern of the frog's eggs also provides some degree of camouflage. Seen from below, the lower white portion blends with the sky, making it hard for predators to see the eggs.

An embryo develops into a **tadpole** in two to three weeks, depending on the species and the weather conditions. A young tadpole has a fish-like tail and breathes through external gills. In this independent form, it feeds on algae and other plant matter. Its digestive tract is long and coiled and adapted to digesting the vegetable matter that it eats. During its development, further changes occur. External gills are replaced by internal gills. Then these, in turn, are replaced by lungs. The mouth broadens and teeth develop. Hind legs appear, closely followed by the development of front limbs. Changes take place in the digestive tract that permit the frog to feed on insects as well as plants. The tail is gradually absorbed and serves as a nutrient source for the developing tadpole. Eventually, development is completed and the frog is capable of living a terrestrial life. Approximately three months are required for metamorphosis from egg to adult to occur in the leopard frog.

During the winter months the frog hibernates. It buries itself in the mud at the bottom of a pond. Its body temperature drops to the temperature of the cold mud and its metabolism slows down. What little oxygen it requires can be obtained from the moisture surrounding it. Dissolved oxygen simply diffuses across its outer skin. Food reserves stored in the liver and the fat bodies above kidneys provide the frog with all the nutrition it requires. As the temperature of the environment warms up in the spring, the adult frog becomes active again.

Review

1. Describe the pattern of asexual reproduction in the hydra.
2. How is cross-fertilization of a hydra ensured during sexual reproduction?
3. What possible advantages or disadvantages might there be in each of the two methods by which a hydra reproduces?
4. **a)** Describe the pattern of sexual reproduction in an earthworm.
   b) How does this pattern overcome the problems related to reproduction which were discussed in Section 22.2?
5. **a)** Briefly describe the reproductive systems of the crayfish.
   b) What features of the reproductive pattern of crayfish ensure a high percentage survival rate of the zygotes?
   c) After young crayfish leave the mother, their survival rate falls greatly. What factors would lead to this fall?
6. Compare the reproductive systems of male and female frogs.
7. **a)** Describe the changes that occur in the metamorphosis from fertilized egg to adult in the frog.
   b) Describe how the physical features of the various stages during metamorphosis adapt the frog to the environment in which it lives.
8. **a)** What is hibernation?
   b) How does a frog obtain nourishment during hibernation?

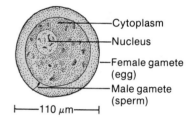

Fig. 22-20 The human gametes. Estimate the length of the sperm cell.

22.5 Human Reproduction

An Overview

Mammals, including you, share many characteristics with other animals that reproduce sexually. Recall that the distinguishing characteristic of sexual reproduction is the uniting of separate male and female gametes to produce a zygote. In mammals, the female gamete, or egg, is much larger and less motile than the male gamete, or sperm (Fig. 22-20). As a result, the sperm must move from the male into the female for fertilization to occur (Fig. 22-21).

Fertilization of the egg occurs in the female oviduct (Fallopian tube). The zygote then becomes implanted in a special muscular organ called the uterus. Different mammals have slightly different structures for their uterus (Fig. 22-22). The lining of the uterus is well supplied with blood. Thus the uterus is able to nourish the developing embryo until birth. The total time

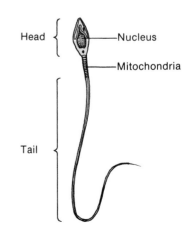

Fig. 22-21 The male gamete. What is the function of the long tail?

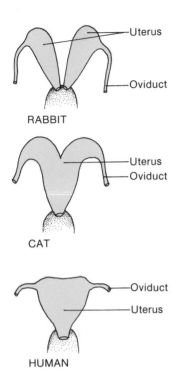

Fig. 22-22 Various uteruses. Which animals usually produce more than 1 young at birth?

for the development from zygote to birth is called gestation time. As you can see from Table 22-1, this time varies from mammal to mammal.

MAMMAL	AVERAGE GESTATION TIME (DAYS)
Elephant	645
Horse	336
Cattle	282
Human	267
Goat	151
Sheep	147
Dog	63
Cat	63
Rabbit	31
Mouse	21

TABLE 22-1 Gestation Time for Some Mammals

After birth, further nourishment is obtained from milk produced in the mammary glands of the mother. In fact, milk production by the female is a distinguishing characteristic of all mammals.

The Male Reproductive System

The male reproductive system of a human has three functions (Fig. 22-23). First, it produces the male gametes, or sperm. Second, it produces the male hormone testosterone. And, finally, it places the sperm into the female reproductive tract.

The main reproductive organs are the testes or testicles. They produce the male gametes, or sperm. The testes are located outside the body cavity in a sac of skin called the scrotum. The cooler temperature outside the body cavity is needed for proper sperm production. Each testis contains about 600 coiled tubules. The sperm are produced by cells in these tubules. The sperm then move down the tubules to the epididymis, where they are stored. When the sperm are released, they travel down the sperm duct, or vas deferens. This duct carries the sperm to the urethra, for ejaculation from the body. As the sperm pass along the sperm duct and urethra, fluids are added to them by three glands. These are the seminal vesicles (sperm chambers), prostate gland, and Cowper's gland. The sperm combines with the fluids from these three glands to make up a mixture called semen. This is the fertilizing fluid that enters the female reproductive system.

Note that the semen leaves the penis through the urethra. The same duct also carries urine from the bladder during urination.

The testes also produce the male sex hormone testosterone. Cells in between the sperm-producing tubules produce this chemical. Testosterone is absorbed into the blood and circulates throughout the body. Continual

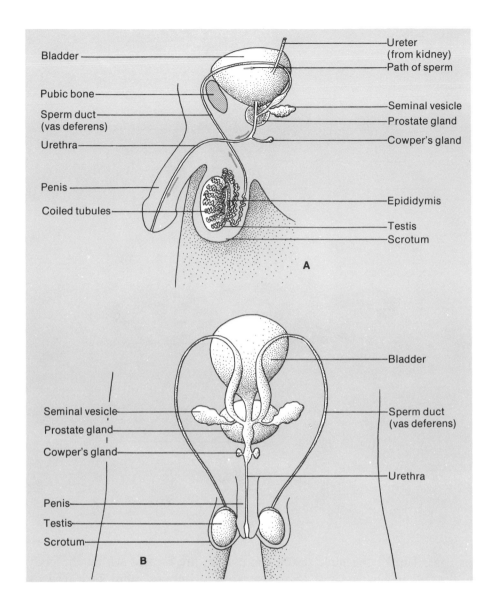

Fig. 22-23   The male reproductive system.
A: side view. B: front view.

Bladder

Pubic bone

Sperm duct
(vas deferens)

Urethra

Penis

Coiled tubules

Ureter
(from kidney)

Path of sperm

Seminal vesicle

Prostate gland

Cowper's gland

Epididymis

Testis

Scrotum

A

Bladder

Seminal vesicle

Prostate gland

Cowper's gland

Penis

Testis

Scrotum

Sperm duct
(vas deferens)

Urethra

B

production of testosterone maintains the typical characteristics of the
mature male. These characteristics include thick facial and body hair, a deep
voice and heavy muscular development. Interestingly, testosterone injected
into a female produces many of these male characteristics in the female.

The Female Reproductive System

The female reproductive system of a human has four functions (Fig. 22-24).
First it produces the female gametes, or eggs. Second, it produces the female
hormones estrogen and progesterone. Third, it provides the place for the
fertilization of an egg by a sperm. Finally, it nourishes the developing
embryo during the nine month gestation period.

   The main reproductive organs are the ovaries. They produce the ova,

Fig. 22-24 The female reproductive system. A: side view. B: front view.

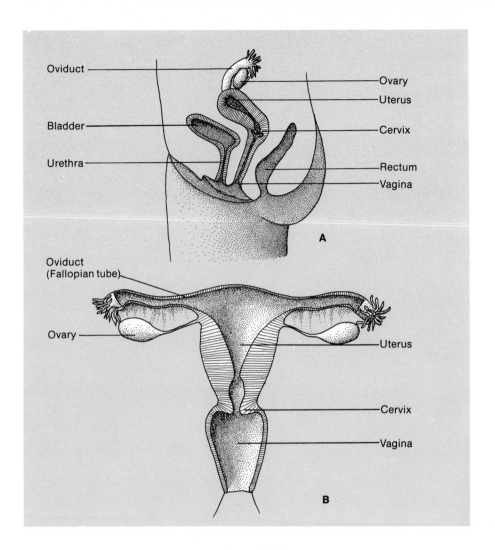

Oviduct

Bladder

Urethra

Ovary

Uterus

Cervix

Rectum

Vagina

A

Oviduct (Fallopian tube)

Ovary

Uterus

Cervix

Vagina

B

or **eggs**. Unlike the male testes, the ovaries are located within the body cavity. When a female is first born, each ovary contains about 300 000 partly developed eggs. Each egg is surrounded by a single layer of cells called a **follicle**.

The ovaries are not connected directly to the **oviducts (Fallopian tubes)**. However, the oviducts are lined with cilia. Therefore, when an egg is released by an ovary, the beating of the cilia draws the egg into an oviduct. The egg then moves down the oviduct to the **uterus**. If the egg is not fertilized, it passes through the neck of the uterus, called the **cervix**. From there it enters the **vagina** and is eventually discharged from the body.

Note that, unlike the male, the urethra is separate from the reproductive system.

The ovaries produce two hormones called **estrogen** and **progesterone**. These two hormones are absorbed into the blood and circulate throughout the body. Estrogen maintains the typical characteristics of the mature female. These characteristics include the broad pelvis, developed breasts and rounded body contours. Estrogen injected into a man produces many of these same characteristics. Progesterone helps prepare the uterus (or womb) for nourishment of the developing embryo.

Cyclic Changes in the Female Reproductive System

An egg should be produced only when the uterus is ready to nourish a new life. Therefore, egg development and development of the uterus are closely coordinated. This coordination is controlled by several hormones. Two of these hormones are follicle stimulating hormone (FSH) and luteinizing hormone (LH). They are secreted by the pituitary gland which is near the

Fig. 22-25   Regulation of egg production. The three hormones, FSH, estrogen, and LH, are transported in the blood to their target site.

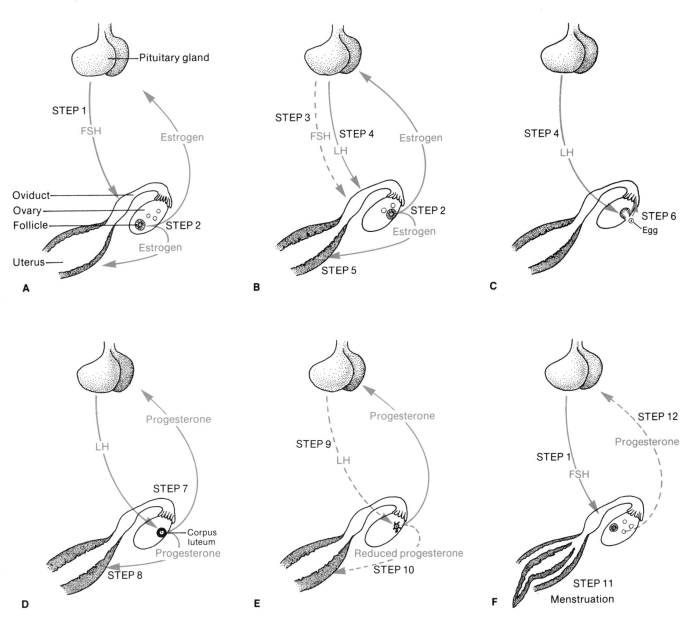

brain. A releasing hormone, gonadotropin releasing hormone (GnRH) from the hypothalamus region of the brain directs the production of the two pituitary hormones.

Once every 28 d or so, the quantity of FSH reaches a high level in the bloodstream. FSH causes a follicle in one of the ovaries to start developing (Fig. 22-25, Step 1). Growing follicles produce the hormone estrogen (Step 2). The increasing estrogen levels cause three changes. First, FSH production drops to a very low level (Step 3). Therefore no more follicles mature. Second, LH production increases (Step 4). And third, the lining of the uterus starts thickening (Step 5).

The increasing level of LH triggers the release of the egg from the follicle. This is called ovulation (Step 6). The single egg is immediately drawn into the open end of the oviduct by the sweeping action of cilia. If sperm cells are present in the oviduct, fertilization may occur. If fertilization does not occur, the egg loses its ability to be fertilized and continues its journey to the uterus, to the vagina, and out of the body. About 14 d are required for this journey.

The LH circulating through the blood stimulates the burst follicle on the ovary to develop further. A collection of cells called the corpus luteum forms (Step 7). The corpus luteum produces the other sex hormone, progesterone. Progesterone causes further development of the lining of the uterus (Step 8). An extensive network of capillaries forms in preparation for nourishing an embryo.

The increased amount of progesterone in the blood stops LH production by the pituitary (Step 9). If fertilization has not occurred, the lower amount of LH causes the corpus luteum to break down. Therefore less progesterone is formed, and further preparation of the lining of the uterus stops (Step 10). Soon the lining detaches from the uterus. In most animals, the lining is reabsorbed. However, in humans, this capillary-rich lining leaves the body. This process is called menstruation (Step 11). About 4 d is needed to rid the uterus of the old lining before preparation of a new one begins.

Eventually the corpus luteum disappears and no more progesterone is formed (Step 12). With no estrogen or progesterone to stop FSH production by the pituitary, FSH levels begin to build up again (Step 1). Therefore a new follicle is stimulated into maturing and the cycle repeats itself. Figure 22-26 summarizes this repeating cycle that consists of four stages: the growth of the follicle, ovulation, the corpus luteum stage, and menstruation.

Fertilization and Development

After the semen has been deposited by the male in the vagina of the female, the sperm begin to move up the reproductive tract of the female. Fertilization usually occurs in one of the Fallopian tubes (oviducts). At that point, one sperm enters an ovum (egg), forming a zygote. As soon as the zygote is formed, a fertilization membrane forms around it. This membrane prevents other sperm from entering the ovum.

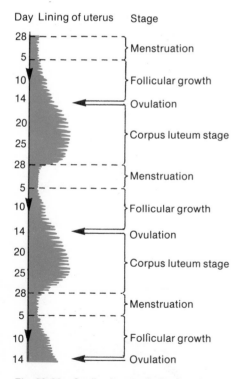

Fig. 22-26 Cyclic changes in the female reproductive system. Variations in the length of the cycle occur from one individual to another.

When fertilization occurs, the corpus luteum does not break down. Instead, it continues to secrete progesterone. The progesterone, in turn, causes the uterus to thicken with many blood vessels and glands. The uterus is now ready to receive the zygote.

It takes about 6-8 d for the zygote to move from the Fallopian tube to the uterus. As it moves, the zygote begins to grow by mitotic divisions. It first divides into two, then four, then eight cells and more (Fig. 22-27). Eventually a hollow sphere of about 100 cells is formed. It is called a blastocyst. About 4 d after fertilization, this sphere enters the uterus and lodges itself in the nutrient-rich wall of the uterus. A special tissue, the placenta, forms where the blastocyst lodges in the lining. The placenta represents an area of close contact between capillaries of the mother and capillaries of the developing embryo. Oxygen and nutrients diffuse from the uterine lining into the circulatory system of the embryo. Wastes, including carbon dioxide, diffuse from the embryo to the circulatory system of the mother. Nine months of internal development is required to produce a baby that will survive in the outside environment (Fig. 22-28). In the later stages of development, the embryo is called a fetus.

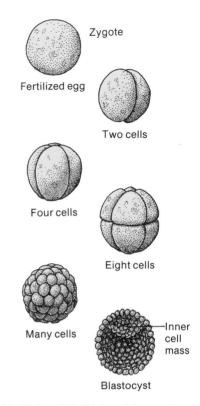

Fig. 22-27   Cell division of the zygote.

Hormonal Control of Sexual Development

Almost all humans are incapable of sexual reproduction until at least 10 years after birth. Until then, boys and girls have much the same bone structure and features. Then the pituitary gland starts producing more and more FSH and LH under the influence of GnRH from the hypothalmus. These two hormones produce different effects in young men and women. Table 22-2 summarizes these different effects.

SEX	APPROXIMATE AGE (YEARS)	EFFECT OF FSH	EFFECT OF LH
Male	13-16	causes maturing of testes, leading to sperm production	causes production of testosterone by testes
Female	10-14	causes maturing of follicles, leading to egg and estrogen production	induces ovulation and development of corpus luteum, leading to progesterone production

TABLE 22-2   **Effects of FSH and LH on Young Men and Women**

In males, testosterone production starts the growth of thicker hairshafts over the entire body, but particularly over the face, armpits, genitals, and legs. Growth of the larynx produces longer vocal cords and thus a deeper voice. Rapid growth of muscle tissue also occurs.

In females, estrogen and progesterone levels usually induce thicker hair growth only over the armpits and genitals. (Males and females have about the same number of hairs. However, the hormones affect the thickness of each shaft.) Fat cells under the skin of the female body grow larger, round-

Fig. 22-28   Development of the embryo. The embryo develops inside a fluid-filled amniotic sac that stays intact until just before birth.

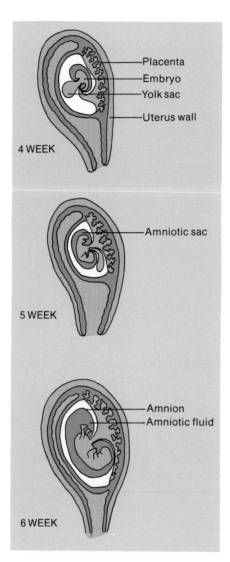

4 WEEK
—Placenta
—Embryo
—Yolk sac
—Uterus wall

5 WEEK
—Amniotic sac

6 WEEK
—Amnion
—Amniotic fluid

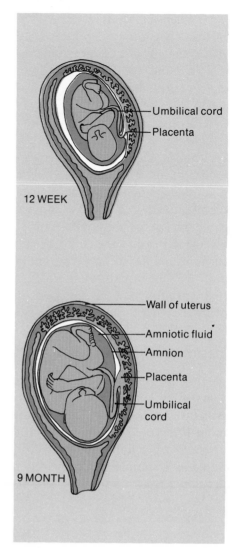

12 WEEK
—Umbilical cord
—Placenta

9 MONTH
—Wall of uterus
—Amniotic fluid
—Amnion
—Placenta
—Umbilical cord

ing the female figure. In particular, the breasts develop, as fat storage cells increase in number and size. The shape of the pelvis changes, and the hips broaden. In both males and females, the hormones cause personality changes. The terms **puberty** and **adolescence** describe the changes in physical appearance and mental outlook that occur in maturing humans.

Review

1. **a)** What is the name given to the male gamete and the female gamete?
   b) Compare the size of the male gamete to that of the female gamete.
   c) How does this size difference affect the mechanics of fertilization?
   d) Where does fertilization of the egg occur?
   e) What is the function of the uterus?
2. **a)** What is meant by "gestation time"?

b) How does gestation time seem to change with the size of the mammal?

c) Suggest a hypothesis to explain the observation in 2 b).

d) How is a gestation time determined?

3. a) What are three functions of the male reproductive system?

b) Where are sperm produced?

c) Why are the testes not inside the body cavity?

d) What does testosterone do in the body?

e) Describe the path by which sperm leave the male body. In addition, outline how semen is formed along this path.

4. a) What are four functions of the female reproductive system?

b) Where are the eggs, or ova, produced?

c) Describe an ovarian follicle.

d) How does estrogen affect the body?

e) What is the function of the oviducts?

5. a) Why must egg production be carefully timed?

b) What is the effect of FSH on the female?

c) Where does estrogen come from?

d) What is the effect of estrogen on the female?

6. a) What is ovulation?

b) What two effects does LH have on a mature follicle?

7. a) What hormone is produced by the corpus luteum?

b) What is menstruation?

8. a) Define the terms blastocyst and placenta.

b) What is the function of the placenta?

9. a) How is it possible to have twins?

b) What is the difference between fraternal and identical twins?

10. What effects would you expect to see if an adolescent male produced:

a) FSH but no LH?

b) LH but no FSH?

11. a) What would be the effects of testosterone injections into a mature woman?

b) What would be the effects of estrogen injections into a mature man?

Highlights

Many animals can produce both asexually and sexually. However, sexual reproduction is more common among animals. During sexual reproduction, a male gamete (sperm) unites with a female gamete (egg, or ovum) to form a zygote. This process is called fertilization. The zygote develops into an embryo and, finally, into a mature animal.

In order for fertilization to occur, a liquid medium must be present to allow transportation of the sperm to the egg. Also, the gametes must be released at the proper times and must be protected from the environment. Internal fertilization solves these problems most efficiently.

Male animals produce sperm in gonads called testes. Female animals produce eggs, or ova, in gonads called ovaries. Some animals are hermaphrodites. That is, both male and female gonads develop in the same animal.

The reproductive systems of the human male and female are similar to those of most animals. The testes produce the smaller, motile male gametes, or sperm. The ovaries produce the larger female gametes, the eggs, or ova. Fertilization occurs when a single sperm unites with an egg in the female oviduct. Further development occurs in the uterus, ending in the birth of a new human nine months after conception.

Hormones control the development of the secondary sexual differences between male and female. Hormones also regulate the continual cycle of egg production and preparation of the uterine lining in the human female.

UNIT SIX

Ecology

Ecology deals with the inter-relationships between organisms and their environments. In this unit you first learn the basic principles of ecology. Then you apply those principles to the study of aquatic and terrestrial ecosystems. In addition, you investigate the adaptations and behaviour of organisms. With an understanding of ecology, perhaps you can help your generation avoid the mistakes made by earlier generations and find solutions to our environmental problems.

23

Interdependence of Living Things

Ecology *is the study of the relationships among organisms and between organisms and their environments.* In this chapter you will study the basic principles of ecology. From your studies, you should develop a good understanding of the interdependence of all living things. You will apply that understanding in the remaining chapters of Unit 6.

23.1 Ecology and Levels of Biological Organization

In the preceding units of this text, you studied organisms from all five kingdoms. You learned a great deal about many of these organisms. You studied them at various levels of biological organization: cellular, tissue, organ, organ system, and individual. However, no organism lives completely on its own. Each organism is dependent upon other organisms and upon the environment in which it lives. Therefore it is important to study organisms at levels above the individual level. These levels are the popula-

tion, community, biome, and biosphere levels. Ecology is the field of biology that deals with the study of organisms at these levels.

First, we must define population, community, biome, and biosphere.

Fig. 23-1   The Canada goose population on a section of the shoreline of Lake Ontario.

Population

A population is a group of individuals of the same species, living together in the same area. Examples of populations are the white pine trees in a woodlot, the leopard frogs in a pond, and the Canada geese on a lake (Fig. 23-1).

Community

A community is a naturally occurring group of organisms living together in the same area. A community consists of several populations. For example, a woodlot community might consist of the beech tree population, the maple tree population, the trillium population, the racoon population, the great-horned owl population, the earthworm population, the mosquito population, and hundreds of other populations. The woodlot community includes the populations of all living things in the woodlot.

Biome

A biome is any large geographical region with a characteristic climate. A biome has a characteristic geography. That is, it has a characteristic **climate** (precipitation and temperature) and a characteristic **topography** (land shape). Therefore, it will be dominated by characteristic plant and animal populations. These populations may occur in many communities throughout a biome. For example, a desert is a biome. It may consist of a sand plain community, a sand dune community, an oasis community and others. Each community will have its own populations of desert organisms. The tundra of northern Canada is a biome. It consists of bog communities, coastal communities, lake communities, and others.

Biosphere

The **biosphere** *is the region on earth in which life exists* (Fig. 23-2). Organisms live in the lower regions of the atmosphere. They also live in almost all the bodies of water on the earth. Organisms live on the soil and in the first metre or two of the soil. This thin layer from the lower atmosphere to the upper part of the soil makes up the biosphere. The biosphere is made up of many biomes. Among them are desert biomes, coniferous forest biomes, deciduous forest biomes, grassland biomes, marine biomes, and others. You will learn more about biomes in Chapter 25.

Fig. 23-2 Levels of biological organization that are involved in ecological studies. Note how the five levels are related to one another.

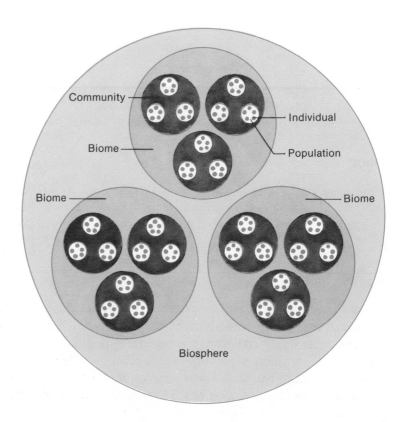

1. **a)** Define ecology.
   b) Define population, community, biome, and biosphere.
   c) Why do ecologists generally study organisms at the population, community, biome, and biosphere levels?
2. Identify each of the following as a population, community, or biome:
   a) The porcupines in a woodlot.
   b) The deer in a meadow.
   c) The grasses and other living things in a meadow.
   d) All the living things in a forest.
   e) The living organisms in a pond.
   f) The brook trout in a stream.
   g) The prairies of western Canada.
   h) The hemlock trees in a forest.
   i) The earthworms and other living things in a certain section of soil.
3. Ecologists sometimes speak of a plant community and an animal community. What do you suppose they mean by these terms?

23.2 The Ecosystem Concept

What is An Ecosystem?

An ecosystem is an interacting system that consists of groups of organisms and their non-living or physical environment. An ecosystem consists of two main parts, a biotic community and an abiotic environment. The **biotic** community includes all the living organisms in the ecosystem—animals, plants, fungi, protists, and monerans. The **abiotic** environment includes all the non-living aspects of the ecosystem—water, carbon dioxide, soil minerals, light, humidity, temperature, wind, and other physical factors.

There are no fixed limits to the size of an ecosystem. Any community of organisms interacting with its environment forms an ecosystem. A woodlot is an ecosystem. It consists of a community of living things plus abiotic factors such as wind speed, light intensity, and temperature. A lake is also an ecosystem, as is a meadow, a city park, and even a classroom aquarium.

The most important thing to remember about an ecosystem is that all its parts, biotic and abiotic, are highly interrelated (Fig. 23-3). Each part is affected by all the other parts. Therefore, if one part is changed in any way, all the other parts will be changed also.

Suppose, for example, that a lumber company completely cleared a forested area of its trees. What effects would this have on the forest ecosystem? The trees would no longer add humus to the soil, since they would no longer drop leaves onto the ground. Snails, slugs, earthworms, bacteria and fungi thrive on leaf litter. Therefore, they would decrease in numbers. Some species might vanish completely. Animals that feed on these organisms would also be affected. Soil erosion might occur, since the leaf canopy would no longer be present to absorb the energy of a heavy rainfall. If the

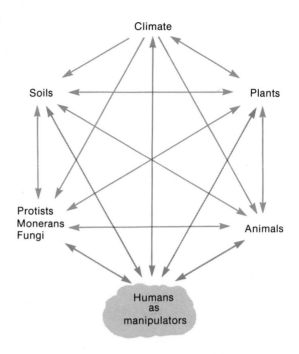

Fig. 23-3 Each part of an ecosystem, biotic or abiotic, is influenced by the other parts.

soil was washed away, many plant species would disappear. The animals that eat these plants would move away or die of starvation. Plants like ferns, that require abundant shade and moisture, would die. Broad-leafed plants like Jack-in-the-pulpit and trillium cannot live in direct sunlight. Therefore they, too, would die.

On the other hand, many species of plants that require bright sunlight might then be able to grow in the area. Grasses, goldenrod, and other sun-tolerant plants might gradually become established. Shrub and tree species that could not grow in the shade of the forest might appear. New insect populations might be established. New bird populations might appear. Mammal species, such as white-tailed deer, might increase in numbers. But the original ecosystem would be gone, perhaps forever.

The chain of events that occurs when one factor in an ecosystem is changed is long and involved. However, these events are certain to occur. Try now to imagine further changes that would occur in a forest that has been cleared. Do you think that neighbouring ecosystems would be affected? Would a nearby stream, pond, or meadow change in any way as a result of lumbering activities in the forest?

Humans and Ecosystems

Humans are animals. Therefore, like other animals, we fit into the interaction shown by the green arrows in Figure 23-3. However, unlike other animals, we also manipulate or influence ecosystems as shown by the gray arrows. That is, we change them to our own advantage. For example, we remove the natural grasses of the prairies and replace them with grasses that we want, such as wheat and other grains. We replace deer, groundhogs, and other natural animals of a meadow with cattle, sheep, and other domestic animals. We use herbicides to destroy plants that we do not want. We till the soil so that it will grow what we want it to grow. We clear mountain slopes to produce ski areas. We bulldoze natural areas to make room for new homes, shopping plazas, schools, and roads. We even change the climate by polluting the air and by cloud seeding.

Often humans are unaware of the broad ecological consequences of changing ecosystems for their own advantage. We should always remember that we are part of an ecosystem. We are dependent upon *all* the other parts. Therefore, for our own survival as a species, we must develop an ecological awareness of our actions.

Review

1. **a)** What is an ecosystem?
   b) Distinguish between the biotic and the abiotic parts of an ecosystem.
   c) Name three ecosystems that are not named in this section.
2. Suppose that a farmer owns a farm of which half is flat land under cultivation. The other half is hilly and wooded, with a river flowing through it. The farmer decides to cut down the trees and cultivate the hilly half. Make a list of the possible consequences of this action.

23.3 Structure of an Ecosystem

Habitat and Niche

Habitat

The **habitat** *of an organism is the place in which it lives.* An ecosystem, such as a woodlot, has many habitats. For example, the habitat of an earthworm is the rich woodlot soil. The habitat of a land snail is the moist leaf litter. The habitat of a porcupine is a hollow tree. The habitat of a bluejay is the branches of the trees. Habitats may overlap. For example, the porcupine may seek out a meal of bark in the branches that are also the habitat of the bluejay. However, since these animals do not eat the same food, no problems result from the overlap of their habitats.

Niche

The **niche** *of an organism is its total role in the community.* For example, the niche of a deer is to feed on grass and other plants, to become food for wolves, to provide blood for blackflies and mosquitoes, to fertilize the soil

Fig. 23-4 What is the habitat and niche of this animal?

with nutrients, and so on. The niche of a frog in a pond is to feed on insects, to become food for snakes and other animals, and many other things.

Many people confuse habitat and niche. It may help you to remember the difference if you think of the habitat as the "address" of the organism and the niche as its "occupation", or "job" (Fig. 23-4).

If two species have the same habitat and similar niches, they will compete with one another. For example, mule deer and elk often live in the same mountain valley. That is, they have the same habitat. Both species eat grass and other plants. Both are preyed upon by wolves and are attacked by many of the same parasites. That is, they have similar niches. Clearly they will compete for available space and food in the valley.

Trophic Levels

As you know, all the organisms in an ecosystem depend on one another and on their physical environment. This interdependence shows up clearly when one studies the way in which various organisms in an ecosystem obtain their food.

Producer level

In Chapters 2 and 7 you learned that all living things need energy to support life processes. Almost all this energy originally comes from the sun. Green plants, many monerans, and many protists contain chlorophyll. These living things, through photosynthesis, store some of the sun's energy in the bonds of glucose molecules. They make their own food, or store their own energy, by converting light energy into chemical potential energy. *Organisms that produce their own food are called* **autotrophs** *("self-feeders")*. They are also called **producers**. Biologists say that they occupy the **trophic level** ("feeding" level) of producer.

Consumer levels

Organisms that depend on other organisms for food are called **heterotrophs** ("other feeders"). All ecosystems have heterotrophs. Most heterotrophs are animals that feed on plants or on other animals. These organisms are called **consumers**. They occupy the trophic level of consumer.

Those animals that feed directly on producers are called **first-order consumers**, or **herbivores** ("plant eaters"). Deer, rabbits, and cows are herbivores. Animals that eat other animals are called **carnivores** ("flesh eaters"). Those carnivores that feed on herbivores are called **first-order carnivores**, or **second-order consumers**. The wolf and fox are first-order carnivores, since they eat deer and rabbits respectively. If a mountain lion kills and eats a wolf, it is a **second-order carnivore**, or **third-order consumer**. Ecosystems also have what is called a **top carnivore**. What do you think is meant by this term?

Some animals are both herbivores and carnivores. For example, a fox eats berries as well as mice and rabbits. Such animals are called **omnivores** ("all eaters"). Are you a herbivore, carnivore, or omnivore?

Animals which feed on live organisms are called **predators**. The organisms that are eaten are called **prey**. Some animals feed on dead organisms. These special consumers are called **scavengers**. Snails, crayfish, and some fish are scavengers of lakes and ponds. Crows, magpies, and vultures are also scavengers. Some animals may be predators sometimes and scavengers at other times, depending on the food available.

Decomposer level

The trophic level of decomposer is a most important level in all ecosystems. All organisms eventually die, and all animals produce waste products. If decomposers were not present, non-living organic matter would soon smother all life on earth. **Decomposers** break down non-living organic matter and return valuable nutrients to the ecosystem. Most decomposers are bacteria and fungi such as moulds and yeasts. Decomposers are **saprophytes**. This means they feed on non-living organic matter.

Summary

Producers, consumers, and decomposers are the necessary biotic parts of any ecosystem. Every ecosystem has these three main trophic, or feeding, levels. Energy and nutrients flow from level to level as shown in Figure 23-5. You will study this flow in Section 23.4.

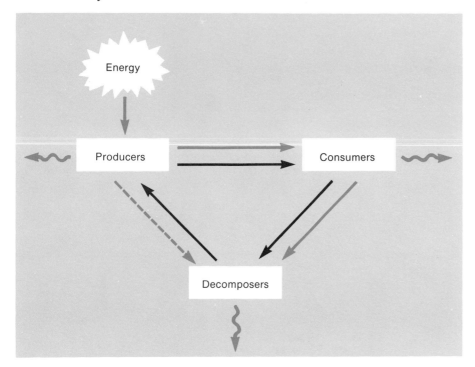

Fig. 23-5 The three main biotic parts of an ecosystem are producers, consumers, and decomposers. The green arrows indicate the flow of energy from trophic level to trophic level. The wavy green lines represent energy losses due to metabolic activities. The black arrows indicate the flow of nutrients. What is the meaning of the broken green arrow?

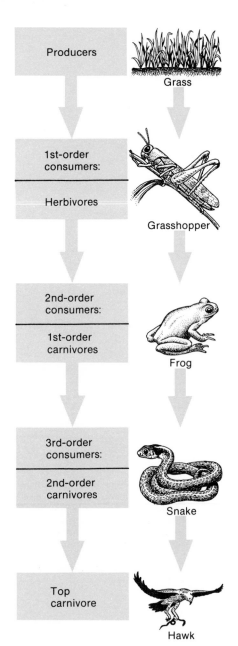

Fig. 23-6 The general pattern for a food chain, illustrated by one example.

Food Chains and Webs

A wolf is a predator and a deer may be its prey. The organisms in an ecosystem are generally linked together through predator-prey relationships in what is called a **food chain**. For example, clover is food for the rabbit. The rabbit, in turn, is food for the fox. This simple food chain may be summarized in the following way:

Clover ⟶ Rabbit ⟶ Fox

All food chains follow the general pattern shown in Figure 23-6. Of course, some food chains are longer than others. Some may have more carnivore levels than others.

Most organisms can occupy a trophic level in more than one food chain. In other words, they have more than one source of food. For example, a rabbit could be part of the following food chains and many others, as well:

Grass ⟶ Rabbit ⟶ Fox
Clover ⟶ Rabbit ⟶ Fox
Clover ⟶ Rabbit ⟶ Wolf
Grass ⟶ Rabbit ⟶ Owl

Food chains can be interconnected to form a **food web** (Fig. 23-7).

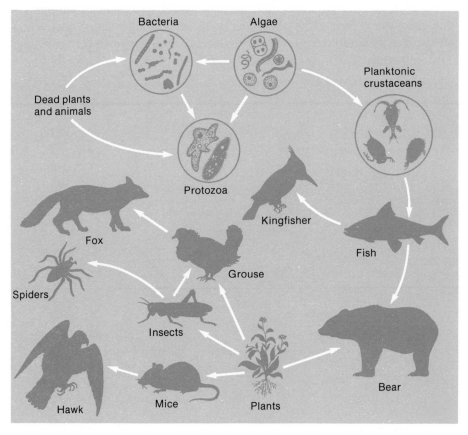

Fig. 23-7 A food web. Follow the arrows to find out who eats whom.

Ecological Pyramids

Pyramid of numbers

Many predators are limited to prey in a certain size range. They normally eat animals smaller than themselves. Therefore many food chains tend to go from very small organisms to larger and larger ones. Also, as the size of the organisms increases, the number of individuals involved in the food chain decreases. For example, consider this simple food chain:

Wheat ⟶ Mouse ⟶ Owl

A mouse that eats only wheat, must eat several kernels of it a day to stay alive. Yet, at the next trophic level, the owl needs only five or so mice a day. Such numerical relationships are represented by a **pyramid of numbers** (Fig. 23-8). As you can see, this diagram emphasizes the fact that a large number of organisms is required to support one organism at the next higher level in the food chain.

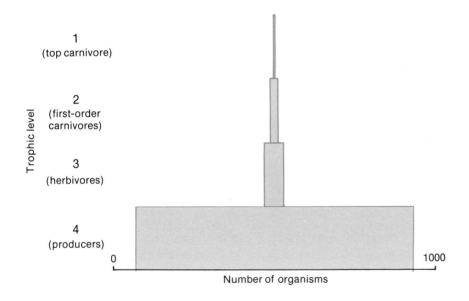

Fig. 23-8 A pyramid of numbers. Note how the number of organisms decreases as the trophic level increases.

Not all pyramids of numbers have such a regular shape. In fact, some are inverted. What would be the likely shapes of the pyramids of numbers that illustrate these food chains?

Dead fish ⟶ Moulds

Grass ⟶ Rabbit ⟶ Fox ⟶ Flea

Pyramid of biomass

A pyramid of numbers gives some interesting information. However, it is not of great importance in ecological studies because it treats all organisms

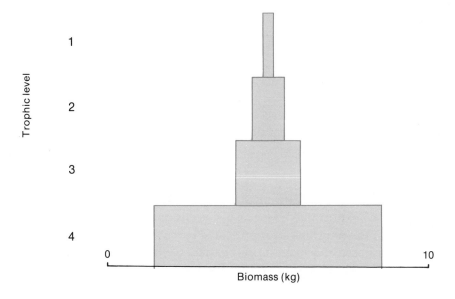

Fig. 23-9 A pyramid of biomass. Why is it more useful to ecologists than a pyramid of numbers?

as though they were the same. It ignores differences in size. Yet, to a hungry fox, size is important. One rabbit makes a better meal than one mouse! Therefore ecologists decided that a better pyramid would give the total mass of organisms, or the **biomass**, at each trophic level. If you think about it, you will agree. It makes more sense to equate 1 g of rabbit with 1 g of mouse than it does to equate one rabbit to one mouse.

Ecologists measured the total biomass at each trophic level in several food chains. They discovered that, in general, there was a decreasing biomass the higher the trophic level. This discovery can be summarized in a **pyramid of biomass** (Fig. 23-9). Let us illustrate the pyramid of biomass using a food chain of which you are a part:

$$\text{Grain} \longrightarrow \text{Chicken} \longrightarrow \text{Human}$$

Chickens eat grain, and humans eat chickens. About 3 g of grain are required to form 1 g of chicken. Yet that 1 g of chicken will form only a small fraction of a gram of human tissue.

Pyramid of energy

A pyramid of biomass has one basic weakness. It equates unit masses of all organisms. For example, it implies that 1 g of rabbit and 1 g of mouse will provide a fox with equal amounts of energy. Yet experiments have shown that this is not true. Different types of tissue have different energy contents. The most noticeable differences occur between plant and animal tissues. You can obtain about 20% more energy by eating 1 g of animal than you can by eating 1 g of plant.

Some animals eat plants to obtain energy for life processes. Other animals eat animals to obtain their energy. Therefore, the efficiency with which energy is passed along a food chain is more important than either the

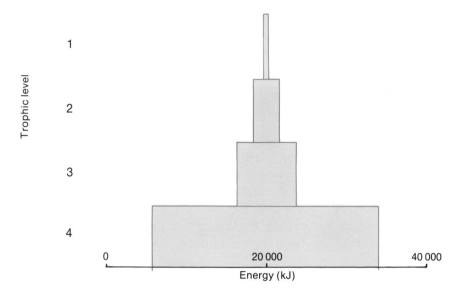

Fig. 23-10   A pyramid of energy. What advantage does it have over a pyramid of biomass?

numbers of organisms or their biomasses. Therefore, ecologists today direct most of their attention toward **pyramids of energy** (Fig. 23-10). A pyramid of energy is never inverted. There is always a decreasing total energy the higher the trophic level.

Symbiosis

Many unusual feeding relationships exist among organisms. Foremost among them is symbiosis. The word *symbiosis* means "living together". **Symbiosis** *is a relationship in which two different kinds of organisms live together in close association.* There are three kinds of symbiosis: parasitism, mutualism, and commensalism.

Parasitism

Parasitism *is a symbiotic relationship in which one organism benefits and the other suffers harm.* The organism that benefits is called the **parasite**. The organism that suffers harm is called the **host**. The parasite lives on or in the host. It gets its food and shelter from the host. However, the host is harmed and, in some cases, eventually killed. Fleas, ticks, and mosquitoes are parasites. Tapeworms are parasites in the human intestine. Many plant diseases such as rusts, smuts, and mildews are parasites. Most parasites do not kill their hosts. If they did, they would lose their food supply.

Mutualism

Mutualism *is a symbiotic relationship in which both organisms benefit.* For example, the relationship in a lichen is mutualism (see Section 11.9, page

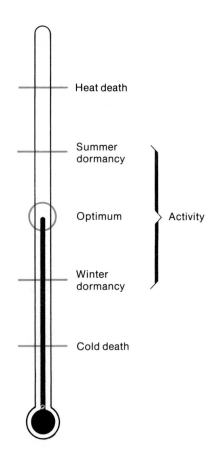

Fig. 23-11 The range of tolerance to temperature has limits at which dormancy occurs and at which death occurs. An organism lives best at the optimum temperature.

304). A lichen consists of an alga and a fungus, growing together in a close relationship. The alga is autotrophic. It manufactures foods through photosynthesis. It shares these foods with the fungus. The fungus, in turn, provides the alga with water, minerals and support. Another example of mutualism occurs when a bee pollinates a flower as it takes nectar from it.

Commensalism

Commensalism *is a symbiotic relationship in which one organism benefits and the other neither benefits nor suffers harm.* Spanish moss (a lichen) and oak trees have a commensal relationship. The lichen lives high in the branches of oak trees in the southern United States. The oak trees provide a home and exposure to sunlight for the lichen. Yet the oak tree receives neither harm nor benefit from the relationship.

Abiotic Factors in Ecosystems

As you know, the abiotic (non-living) parts of an ecosystem interact with the biotic (living) parts. You will study in detail the relationships between the abiotic and biotic parts of aquatic ecosystems in Chapter 24 and of terrestrial ecosystems in Chapter 25. A brief outline of these relationships follows.

Five main abiotic factors affect terrestrial ecosystems. They are temperature, moisture, light, wind, and soil characteristics. Each organism has a range of tolerance for each of these factors (Fig. 23-11). This range depends on the factor and on the organism. When the range is exceeded, in either direction, the organism suffers. Within each range of tolerance there is a point at which each organism lives best. This is called the optimum. However, conditions are seldom at the optimum. Organisms with the broadest range of tolerance generally survive best.

Many abiotic factors affect aquatic ecosystems. For example, a stream ecosystem may be affected by temperature, speed of the water, nature of the stream bottom and banks, light, the chemical properties of the water and many other factors. For example, fast cool water with a high oxygen content will support trout. However, sluggish warm water with a low oxygen content will not. A pond or lake ecosystem may also be affected by temperature, light, depth, the nature of the bottom, the transparency of the water, the chemical properties and other factors. The life in the pond or lake is determined, in part, by these factors. The life, in turn, influences these factors.

Review

1. a) Distinguish between habitat and niche.
   b) Distinguish between niche and trophic level.
2. a) Distinguish between an autotroph and a heterotroph.
   b) Name and describe the three main trophic levels that occur in all ecosystems.

3. Complete the following food chains:
   a) In a meadow: grass → crickets → . . .
   b) In a mixed forest: saplings → deer → . . .
   c) In a lake: diatoms → microscopic animals → . . .
   d) In an oak woodlot: acorns → squirrels → . . .
   e) In a garden: seedlings → earthworms → . . .
   f) On a mountain: grass → mountain goat → . . .
4. **a)** What is the difference between a food chain and a food web?
   b) Are predator-prey relationships involved in all food chains? Explain.
5. **a)** What is a scavenger? Give three examples of scavengers.
   b) Name three scavengers that are not named in the text.
6. **a)** What is a saprophyte? Give three examples of saprophytes.
   b) Why are decomposers classified as saprophytes?
7. **a)** What is a pyramid of numbers?
   b) What is the major weakness of a pyramid of numbers?
8. **a)** What is a pyramid of biomass?
   b) What is the major weakness of a pyramid of biomass?
9. Why do ecologists direct most of their attention toward pyramids of energy?
10. **a)** What is symbiosis?
   b) Distinguish among parasitism, mutualism, and commensalism.
   c) Give one example of each form of symbiosis that is not mentioned in the text.
11. **a)** State the five main abiotic factors that affect terrestrial ecosystems.
   b) Each of these five main factors can be subdivided into two or more related factors. For example, temperature can be subdivided into air temperature and soil temperature. Divide the remaining four factors into as many subdivisions as you can.
   c) Make a list of the abiotic factors that you think might affect the marine ecosystems that occur in the tidal zone of a rocky shore on the Atlantic coast.

23.4 Energy Flow and Nutrient Cycling

If you spent a few hours in the middle of a wood in late spring, you would likely be surprised by the constant activity around you. Squirrels seem to dart everywhere. Birds are constantly searching for food. If you are quiet, you might even see a deer nibbling on the saplings or a fox chasing its prey. Activity is the essence of life. And, in order to have activity, energy and nutrients are required. Ecologists measure the functioning of an ecosystem by the rate of energy flow and nutrient cycling.

Energy Flow

The ultimate source of energy is the sun. The producers of an ecosystem convert light energy from the sun into chemical potential energy in the bonds of glucose molecules. In so doing, the producers support, directly or indirectly, all the other organisms of the ecosystem. Herbivores lose heat

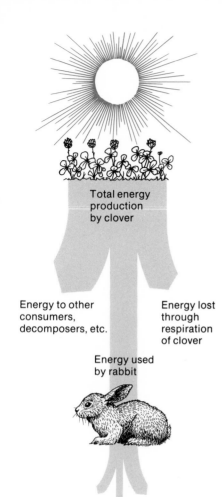

Total energy
production
by clover

Energy to other
consumers,
decomposers, etc.

Energy lost
through
respiration
of clover

Energy used
by rabbit

Energy to other
consumers,
decomposers,
etc.

Energy lost
through
respiration
of rabbit

Energy used
by fox

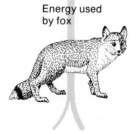

Energy to other
consumers,
decomposers,
etc.

Energy lost
through respiration
of fox

Fig. 23-12 Energy flow along a simple food chain.

energy as they graze on producers. Similarly, carnivores lose heat energy as they prey on herbivores. *Thus energy is gradually lost along a food chain* (Fig. 23-5). As a result, the flow of energy through an ecosystem is one-way. Little, if any, of the energy trapped by producers is returned to them.

Here is an example that explains the decrease in total energy at higher trophic levels. The example considers this food chain:

$$Clover \longrightarrow Rabbit \longrightarrow Fox$$

Study Figure 23-12 carefully as you read this description. Not all the chemical potential energy stored by the clover is available to higher trophic levels. Much of it is lost as heat through cellular respiration by the clover. When the rabbit eats the clover, the rabbit obtains some of the stored energy. However, it does not pass all this on to the fox. Like the clover, it loses much of its total energy intake as heat during cellular respiration. The fox can use only a part of the energy provided by the rabbit. For instance, bones are not digestible. Much of the energy intake of the fox is lost as heat during cellular respiration. Parasites and decomposers also use some of the fox's energy. In the end, little, if any, energy remains to be passed on to higher trophic levels.

The energy lost as heat at each trophic level cannot be recaptured by any organisms in the ecosystem. It is lost forever to that ecosystem. Thus energy flow is one-way. *For an ecosystem to keep operating, energy must always enter it from the sun.*

Humans are as dependent as any other organism on the flow of energy through ecosystems. Our very existence depends on it. However, we can control energy flow and divert it for our own use. For example, we can destroy a natural ecosystem like a forest and replace it with a grain crop that will yield more energy for us. As the human population increases, more crops will be needed to supply the larger food demands. Therefore, more natural ecosystems will be destroyed. Fewer suitable habitats and niches will be available for many organisms. As a result, they will be threatened by extinction. However, alternatives do exist. What are they?

Nutrient (Biogeochemical) Cycles

Although an ecosystem cannot function without the input of energy, an ecosystem requires more than this input of energy in order to function. Over 20 different elements must be present to sustain life processes in an ecosystem. Ecologists call these elements **nutrients**. The main elements required are carbon, hydrogen, oxygen, nitrogen, phosphorus, and sulfur.

Unlike energy, these elements are recycled within an ecosystem. There need not be a constant input of the elements. Figure 23-5 illustrates this recycling of elements. As you can see, producers pass the elements on to herbivores. Herbivores pass them on to first-order carnivores. And so the elements pass along the food chain. How do they get back to the producers to complete the cycle? **Decomposers** perform this task. They break down animal excretions and dead organisms into simpler parts that can be taken in and used by producers. Often microorganisms called **transformers** must act on the materials formed by the decomposers. They break these materials down into particles small enough to be absorbed by producers.

Ecologists call these cycles of elements **nutrient cycles**, or **biogeochemical cycles**. As the name suggests, a biogeochemical cycle involves living things (**bio**), the earth (**geo**), and of course, **chemicals**. Let us see how these cycles operate by considering three of the basic cycles, those of water, carbon, and nitrogen.

The water cycle

Water vapour enters the atmosphere through transpiration from vegetation and by evaporation from bodies of water and the soil (Fig. 23-13). In the cool upper atmosphere this vapour condenses, forming clouds. In time, enough water collects in the clouds to cause precipitation. When this occurs, some of the water falling on the ground is absorbed. Some of it runs along the surface of the ground to a stream, pond, or other body of water. The amount of water absorption and surface runoff depends on the nature of the soil and the amount of precipitation. Some of the water in the soil is absorbed through the roots of green plants. Thus animals can obtain water by eating green plants. Of course, they can also obtain it directly by drinking it from a body of water. When plants and animals die, they decompose. During this process, the water present in their tissues is released into the environment. Some of the soil water seeps to ground water level, only to return to a body of water.

Fig. 23-13   The water cycle.

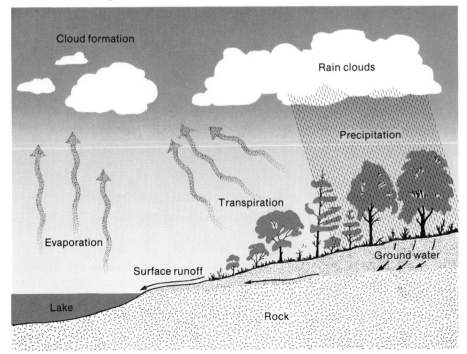

The carbon cycle

At one time, the carbon cycle was considered a perfect cycle (Fig. 23-14). That is, carbon was returned to the atmosphere as quickly as it was removed. Lately, however, the increased burning of fossil fuels has added carbon to the atmosphere faster than producers can remove it.

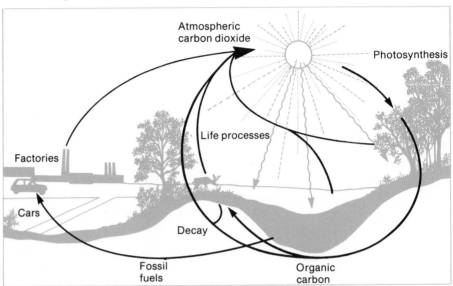

Fig. 23-14   The carbon cycle.

Carbon is present in the atmosphere as carbon dioxide. Producers such as green plants use carbon dioxide to make glucose and other organic compounds during photosynthesis. Animals cannot make their own organic compounds this way. Thus they obtain them by eating producers or other animals. Some carbon dioxide is returned to the atmosphere as a by-product of the cellular respiration of living things. A major contribution is the respiration of decomposer organisms during the decay of non-living organic matter. Some organic matter does not decompose easily. Instead, it builds up in the earth's crust. Oil and coal result from the accumulation of plant organic matter in the distant past.

The nitrogen cycle

All living things need nitrogen to make proteins (Fig. 23-15). Although almost 78% of the atmosphere is molecular nitrogen (N_2), neither plants nor animals can use this form of nitrogen directly. The nitrogen must be in the form of a nitrate before it can be absorbed by the roots of plants. Lightning flashing through the atmospheric mixture of nitrogen and oxygen can cause this conversion to nitrates. The bacterium *Rhizobium* lives on nodules on the roots of legumes like clover (see Fig. 9-14, page 217). It, too, can convert molecular nitrogen to nitrates. Many other bacteria and blue-green algae can also do this.

Plants use the nitrates that they absorb to make plant proteins. Animals get the nitrogen that they need to make proteins by eating plants or other animals.

Fig. 23-15 The nitrogen cycle.

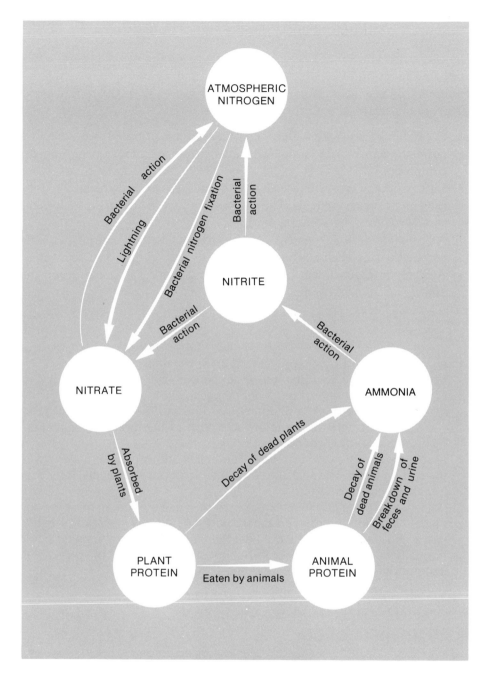

When plants and animals die, bacteria convert their nitrogen content to ammonia. The nitrogen in the metabolic wastes (urine) and fecal matter of animals is also converted to ammonia by bacteria. Ammonia, in turn, is converted to nitrites and then to nitrates by bacteria, thus completing the main part of the cycle. Bacteria convert some nitrites and nitrates to molecular nitrogen to complete the total cycle. The nitrogen cycle need not and often does not involve this last step.

Summary of Structure and Functioning of Ecosystems

By this point you should have a clear understanding of the general structure and function of ecosystems. Test your understanding by reading this summary of key ideas. If you do not understand any of them, reread the appropriate sections.

1. Ecosystems may vary widely in species composition. However, all require the same three biological parts: producers, consumers, and decomposers.
2. Energy flow in ecosystems is one-way. Energy is progressively lost along food chains in ecosystems. Thus an outside source of energy, such as the sun, is required by all ecosystems.
3. All ecosystems require essentially the same nutrients. These nutrients are recycled within each ecosystem.
4. A highly interdependent relationship exists among all parts, biotic and abiotic, of an ecosystem.

Review

1. **a)** What is meant by energy flow in an ecosystem?
   b) Why is energy flow one-way in an ecosystem?
2. **a)** What is a nutrient, or biogeochemical, cycle?
   b) How does nutrient flow differ from energy flow in an ecosystem? Explain your answer by referring to one of the three cycles that are discussed in this section.
3. An oxygen cycle exists in nature. Ecologists often consider it along with the carbon cycle. They call the result the **carbon-oxygen cycle**. Why does it make sense to do this?
4. It has been suggested that, to make the best use of food on this crowded planet, we should all become herbivores. Therefore, instead of eating cattle, pigs, and fowl, we would eat the plants that these animals would normally have eaten. Debate this suggestion in the light of your knowledge of pyramids of energy and energy flow.

23.5 Investigation

A Classroom Ecosystem

This is a long-term experiment that will allow you to study the many aspects of an ecosystem that you have read about in earlier sections. In this investigation you will build a model ecosystem and study its structure and function.

Materials

large bottle or jar with top (preferably about 3-4 L)
table lamp with 60 W bulb
khuli loach or another small herbivorous fish
sprigs of an aquatic plant (3 or 4)
pond snails (8-10)
clean gravel and/or sand

Procedure

a. Place sand or gravel to a depth of 2-3 cm in the jar.

b. Fill the jar with water. If you use tap water, let the jar stand with the top removed for 48 h in order that the chlorine can leave the water.

c. Add a few strands of an aquatic plant such as *Cabomba*.

d. Add 8-10 pond snails to the water.

e. Place a khuli loach in the water.

f. Put the top on the jar and seal it tightly.

g. Place a table lamp with a 60 W bulb in it close to the jar as shown in Fig. 23-16.

h. Place the setup in a location away from windows and other places where light and temperature conditions change greatly during the day.

i. Leave the lamp on 24 h per day or place it on a timer that provides at least 16 h of light per day. Do not depend on your memory to turn the light off and on!

j. A healthy ecosystem will have a pale green colour in the water. This colour is caused by algae, which are food for the fish and snails. As the days pass, move the light closer if the water does not develop a green colour. Move it further away if the green becomes intense.

k. Observe your ecosystem closely from time to time for several months. Make careful notes of any changes that occur.

Fig. 23-16   A classroom ecosystem.

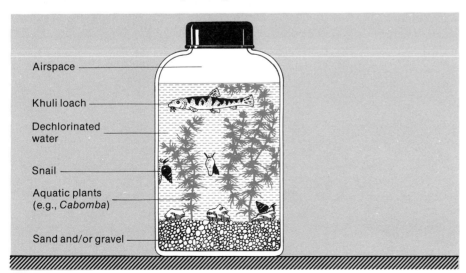

Airspace

Khuli loach

Dechlorinated water

Snail

Aquatic plants (e.g., *Cabomba*)

Sand and/or gravel

Discussion

Many months may pass before you can answer all these questions.

1. **a)** Your ecosystem is a closed ecosystem. How does this differ from natural ecosystems?
   b) What advantages are gained by studying a closed ecosystem?
2. **a)** State the trophic level of each species of organism in the ecosystem.
   b) List any food chains that exist in the ecosystem.
3. **a)** Describe the preferred habitat of the khuli loach.
   b) Describe the niche occupied by the khuli loach.
4. **a)** Are decomposers present in the ecosystem? How do you know?
   b) Name the three main biological components of all ecosystems.
   c) Name the organisms in your ecosystem that make up each of these components.
5. What nutrient cycles are taking place in your ecosystem? How do you know?
6. What evidence do you have that the energy flow in your ecosystem is one-way?

23.6 Ecological Succession

You have probably noticed that a vacant lot or field, left untouched, does not remain in its original state for long. Although it may have started out as a grassy area, it soon becomes overgrown with weeds. As time passes taller weeds dominate the shorter ones. In a few years shrubs may appear. Several years later small trees may begin to colonize the area. If you could sit down under one of these trees and observe the area for a few hundred years, you would continue to witness one of nature's most remarkable phenomena—ecological succession. What is succession? Why does it occur?

A well-defined succession occurs on sand dunes. Let us examine this environment to gain an understanding of the meaning and causes of succession.

Fig. 23-17 Formation of a sand dune.

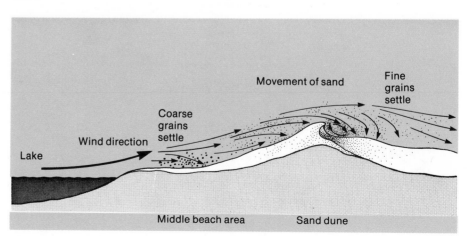

Succession on Sand Dunes

Sand dunes commonly form on the leeward side of large lakes (the shore toward which the wind is blowing) and on ocean shores. Wave action piles sand on the beach. The sun dries out the sand. Then winds blow it inland, creating large mounds called sand dunes (Fig. 23-17).

Such an area is a harsh environment for any organism. Lack of moisture, few available nutrients, intense light, strong winds, shifting sands, high day temperatures, and low night temperatures are just a few of the problems that confront any organism that attempts to live here. Yet some organisms do establish communities on sand dunes. And, as you know, their presence modifies the environment. The modified environment supports different organisms. These organisms further modify the environment. And on it goes. The interaction of biotic and abiotic factors makes a sand dune a region of change. Let us take an imaginary trip through time to see exactly what happens. Imagine that you are sitting on a sand dune that has no life on it, other than yourself (Fig. 23-18). This sand dune is located on the shore of one of the Great Lakes. You are going to sit there for several hundred years to observe and study succession.

As time passes, some dead organic matter from the lake is sure to be swept onto the dune. This small amount of organic matter enriches the sand enough for patches of a **pioneer plant**, sand grass, to begin to grow around you (Fig. 23-19). This hardy plant is well adapted to the dry conditions of a sand dune. It has an extensive root system that absorbs the small amount of water available and anchors the plant in the shifting sand. The plant's narrow leaves bend against the force of wind-driven sand. The sand grass tends to **stabilize** the dune with its large branching root system. Also, it traps drifting sand, making the dune even larger.

Fig. 23-18 Sand dunes along Lake Ontario.

Fig. 23-19   Sand grass.

As sand grass dies and decays, humus is added to the soil. This modified soil supports the growth of shrubs like the sand cherry. The presence of these shrubs stabilizes the dune still further.

Next cottonwood trees begin to shoot upward from among these shrubs. These trees are the **index plants** of this stage in succession. The shade cast by the shrubs and cottonwood trees helps the soil to retain its moisture longer. In addition, decaying leaves add further to the organic content of the soil. Ants and beetles move busily among the sand grass plants. Birds feed on the sand cherries. The digger wasp, an **index animal** of the cottonwood stage, burrows into the sand at your feet. You notice that the sand is darker in colour because of the added organic matter.

The enriched soil now enables pine seedlings to become established in the area. Eventually they become the **dominant species**. Therefore they are called the index plants of this stage. As the pine trees develop, they drop their needles. The soil becomes still richer. But now, a strange thing happens. Pine trees are sun-loving plants. But the large pine trees cast a dense shade on the soil. As a result young pine trees do not receive enough light to develop. The pines, through living, have brought extinction to their species! As the adult trees mature and die, no young pines replace them.

However, black oak seedlings grow well in the shade of pine trees. Therefore, if a squirrel or blue jay drops an acorn from a black oak into the pine forest, it will germinate and grow. Eventually black oaks dominate the area. However, the trees are not very large or closely spaced because nutrients and water are still in short supply. This stage may continue for centuries, as the black oaks slowly add humus to the soil in the form of leaves, bark, and fallen branches.

As further humus accumulates, red and white oaks come to the area. They grow well in the shade of black oak trees, whereas young black oaks do not. After many years, still more shade-tolerant trees invade the area—ash, basswood, and hickory. But they, too, cast a shade too dense for their young to survive. However, young maple and beech trees are very shade-tolerant. They can develop in the shade of parent trees. Therefore, they thrive in this environment. Thus, after the passage of a great deal of time, they dominate the area (Fig. 23-20). As a result, the community becomes self-perpetuating. This means young trees are always ready to replace dead ones of the same species. Such a community is called a climax community.

You have observed a hot, dry, bright environment change to a cool, moist, shade one. The most obvious biological changes were the changes tree species. Looking carefully, you can see still other important biological features. Remnants of previous stages can be seen throughout the climax forest—an occasional oak and basswood tower among the beech and maple trees. In moist areas, hemlock, an evergreen, forms part of the climax community. Also, as the succession of plant communities took place, a succession of animal communities accompanied it. Invertebrates like earthworms, insects, millipedes, centipedes, and snails became more abundant as the climax was approached. Toads, salamanders, and a host of mammals and birds appeared.

Now, let us return to the original question. What is succession? Why does it occur? In summary, living organisms modify their environment. In doing so, they make the environment less favourable for themselves but more favourable for another community of plants and animals. Each stage in succession, except the climax, brings about its own downfall. As succession continues, species diversity, population numbers, and niche availability increase. Also, total biomass and organic matter increase. All these add to the complexity of the community. This complexity increases the stability of the community. The presence of many plant and animal species mean more food webs will be formed. Thus there is less chance of the entire community collapsing because of the extinction of one species.

Fig. 23-20 A climax forest, dominated by maple and beech trees.

Fig. 23-21 Succession on a sand dune in the Great Lakes regions.

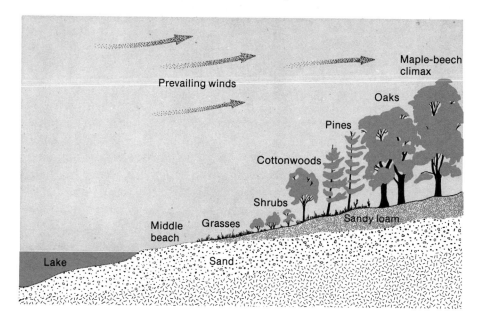

Obviously it is not possible to wait in one place for several hundred years to observe succession. Fortunately, you do not have to. You can see all the stages within a period of a few minutes by walking inland from the water's edge (Fig. 23-21).

As a dune enlarges and becomes covered with vegetation, a new dune forms closer to the water. As it becomes covered with vegetation, a still younger dune forms between it and the water. Meanwhile the original dune has advanced further in succession. This process can happen again and again. As a result, a young dune with little or no vegetation exists closest to the water. The next dune inland will be further on in succession and the next dune still further on. Thus you can see in space, as you walk inland, the stages of a succession that occurred in time.

Types of Succession

Primary succession

Succession that begins in an area that has not supported life within recent geological times is called primary succession. A sand dune succession is in this category. So, too, is the succession that occurs on bare rock. Lichens are the pioneers of a rock succession. They attach themselves to rocky surfaces and extract nutrients from the rocks. This weakens the rock surface. Weathering contributes to the breakup of the rocks. Dead lichens are decomposed which adds humus to the rock fragments. This forms soil. Mosses can now colonize the area. They, too, add humus to the soil as they die and decay. They contribute greatly to succession by trapping wind-blown earth and organic matter. The resulting soil can support plants like ferns and grasses. Later, other herbaceous plants and shrubs invade the area. From this point on the succession follows a course similar to that on sand dunes. Thousands of years may be required before the climax is reached.

Secondary succession

Succession that begins in a region that once supported life is called secondary succession. An abandoned meadow and a forest destroyed by fire or lumbering will undergo this type of succession. Secondary succession is generally more rapid than primary succession, since soil is already present. Also, some forms of life are already in the area. Grasses usually dominate the early years of secondary succession. But tall weeds like wild asters, fireweed, and goldenrod soon follow. Shrubs like hawthorns quickly invade the area. They are followed closely by sun-loving trees like poplars and pines. Succession then proceeds along the usual path to a climax community.

Heterotrophic succession

Both primary and secondary succession are examples of autotrophic succession. In both cases succession is dominated by green plants—they are largely responsible for the modification of the environment. However, many interesting examples of heterotrophic succession also exist. For example, a fallen log undergoes succession. Bacteria, fungi, and a host of invertebrates succeed one another as they inhabit and feed on the log. The climax is reached when the log has been converted to soil. The climax community is the resulting soil community. Similarly, a rotting carcass undergoes succession. Bacteria are the first to colonize the carcass. Worms follow. Other invertebrates such as flies, beetles, and wasps follow to complete the decomposition of the carcass.

The stages in succession where you live may not be the same as those described here. For example, in the St. Lawrence River valley, the climax consists largely of white spruce and balsam fir. Minnesota has a climax community of maple and basswood. The giant redwoods of the California coast and the Douglas firs of British Columbia dominate the climax forests of their respective regions. But, wherever you live, you are never far from an example of succession.

Fig. 23-22   What species have likely been used to reforest this dune?

1. Give the meanings of these terms: pioneer plant, index plant, index animal, dominant species, and climax community.
2. **a)** List, in order, the index plants of a typical succession in the area where you live.
   b) Explain carefully why a succession of plant communities occurs. Be sure to mention all the abiotic factors involved.
3. Why do earthworms and other invertebrates become more abundant as succession approaches the climax stage?
4. Foresters make use of their knowledge of succession when they are selecting tree species to reforest a sand dune area that has lost its tree cover by fire, wind, or lumbering. Fig. 23-22 shows an area that has been replanted. What tree species do you think have been planted here? Why did the foresters select them?
5. **a)** Distinguish between primary and secondary succession.
   b) Distinguish between autotrophic and heterotrophic succession.
6. Many operators of trail bikes, dune buggies, and 4-wheel-drive vehicles like to drive over sand dunes. It is a challenge to see if you can get over the dunes without getting stuck. What do you think of this practice? Why?

23.7 Population Ecology

As you know, *a* **population** *is a group of individuals of the same species, living together in the same area.* In this section you will study the growth of populations and the factors that affect population growth.

The Growth of Populations

Each population has a population size. *The* **population size** *is the number of individuals in the population.* For example, if 80 muskrats live in a certain lake, the population size for muskrats in that lake is 80.

A factor closely related to population size is population density. *The* **population density** *is the number of individuals in a certain area.* Suppose that the lake containing the 80 muskrats has an area of ten hectares. Then the population density of muskrats in that lake is:

$$\frac{80 \text{ muskrats}}{10 \text{ ha}} = 8 \text{ muskrats/ha.}$$

On the average, each hectare of that lake contains 8 muskrats.

All species have a tendency to reproduce. This tendency is called the **biotic potential**. The biotic potential of a species tends to increase the population size of that species. Clearly, other factors must work against this tendency. If they did not, the population size would eventually become so large that the species would overrun the earth. The factors that limit the biotic potential are called the **environmental resistance** of the ecosystem. You can probably guess what some of these factors might be.

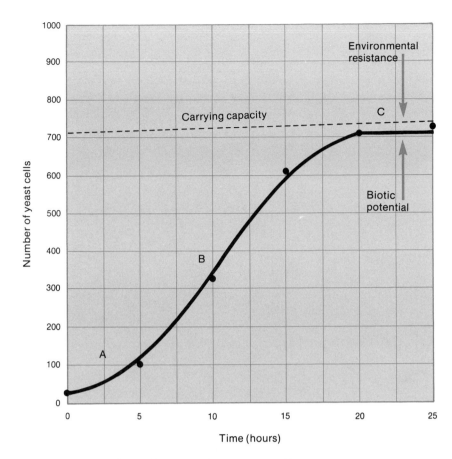

Fig. 23-23 Population growth curve for a yeast population.

In order to investigate the factors that affect population size, ecologists often study population growth. **Population growth** *is the change in the size of a population with time.* Examine Figure 23-23. In the study that resulted in this graph, a scientist started a yeast culture by placing a few yeast cells in a suitable medium. The scientist then counted the yeast cells in a certain volume of medium every 5 h for a total of 25 h. She graphed the results and obtained this **population growth curve**. It is also called a **sigmoid curve**, or **S-shaped curve**, because of its shape. Note that the graph has three main regions. For the first 5 h, the population size increased slowly (Region A). It apparently took some time for the yeast cells to reach their optimum biotic potential. However, once this biotic potential was reached, the population size increased rapidly for the next 10 h (Region B). But, after 20 h, the population size stopped increasing. In this plateau region (Region C), environmental factors limited the biotic potential. These environmental factors make up the environmental resistance of the ecosystem.

In the plateau region the **natality rate** (birth rate) equals the **mortality rate** (death rate). When this point is reached, ecologists say that the ecosystem has reached its **carrying capacity**. The final size of the population is determined by the balance between the biotic potential and the environmental resistance.

The growth of most populations follows an S-shaped curve similar to

Fig. 23-24 What could cause the sudden change in the population growth curve at A?

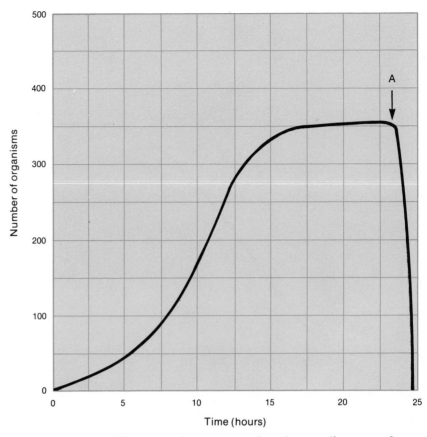

that of the yeast. For example, suppose that the natality rate of trout in a lake is greater than the mortality rate. The population size of the trout will increase. However, the larger population will soon have difficulty finding enough food. Weak and sick trout will starve to death. For a while, the mortality rate will exceed the birth rate. Soon the remaining trout will be able to find enough food. Then the population size will level off at the carrying capacity.

Under some circumstances the environmental resistance may not check the biotic potential soon enough. The population size may overshoot the carrying capacity to the point where recovery is impossible. In that case, the population may die out completely or suffer a great drop in population size (Fig. 23-24). What factors could cause this to happen?

Factors Affecting Population Density

The population density, or the number of organisms in a certain area, is affected by many factors. Some of these factors are **density dependent**. In other words, they depend on the population density. One such factor is predation. If the population density of the prey is high, the rate of predation will increase. The predators will have less trouble finding prey. Other density dependent factors are natality rate, mortality rate, disease, parasitism, and availability of food.

Other factors that affect population density are density independent. They do not depend on the population density. One such factor is temperature. A very low temperature, such as a sudden frost, can reduce the population density of certain organisms to zero, regardless of the density of the organisms. Whether there are two organisms or two hundred organisms per hectare, all will die if the temperature becomes too low. Other density independent factors are relative humidity, light conditions, and sudden weather changes such as a hurricane.

Interspecific competition

Suppose that two species of organisms, A and B, live in the same area. Suppose, further, that both species require exactly the same food and that, because of other environmental factors, species A thrives and reproduces more rapidly than species B. Clearly the two species will compete for the food. Thus this competition itself may become a limiting factor to population density. In the long run, species A will out-compete species B because of sheer numbers. Species B will likely vanish from that area.

*Competition between populations of **different** species is called interspecific competition*. When the competition is for a limited resource such as food, one competitor will be eliminated. Complete competitors cannot live together in the same place. Or, simply stated, *two populations cannot occupy the same ecological niche*.

Intraspecific competition

*Competition between organisms of the **same** species is called intraspecific competition*. This type of competition is density dependent. The greater the population density, the greater the competition (Fig. 23-25).

Some intraspecific competition is good for a species. Weak or otherwise unfit organisms are "weeded out" by the competition. The healthy organisms remain to carry on the species. However, too much intraspecific competition can be harmful. Overcrowding often leads to stress among members of the population. Aggressive behaviour may increase. Natality and mortality rates could be affected by this behaviour. For example, crowded rats often eat their young or abandon them. The stress also frequently results in smaller litter size and poorly built nests.

Fig. 23-25 These pine trees show intraspecific competition. Because of overcrowding, they are competing for light, space, nutrients, and water.

Review

1. Define the following terms: population, population size, population density and population growth.
2. **a)** To ecologists, population density is generally a more useful measurement than population size. Why?
   b) Fifteen deer were counted in a total area of 1 000 ha. What are the population size and the population density of deer in that area?
3. **a)** Explain how the biotic potential and the environmental resistance of

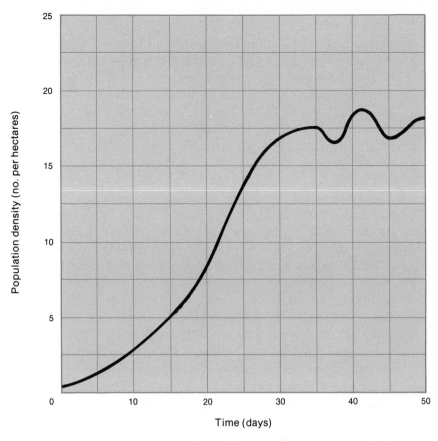

Time (days)

an ecosystem produce the plateau region in a population growth curve.

b) The plateau region of a population growth curve is generally not as flat as indicated in Figures 23-23 and 23-24. Instead, it is usually wavy, as shown in Figure 23-26. Why is this so?

4. Do you think that the biosphere (world ecosystem) has a carrying capacity for the human population? Defend your answer by discussing the biotic potential and the environmental resistance of this ecosystem.

5. a) Distinguish between density dependent and density independent factors that affect population density.

b) Distinguish between interspecific competition and intraspecific competition.

6. a) What kind of competition occurs between the grasses and the weeds in a lawn?

b) Without human help, the weeds normally out-compete the grasses. Why is this so?

Highlights

Ecology is the study of the relationships among organisms and between organisms and their environments. Ecologists study organisms at the population, community, biome, and biosphere levels.

The most important ecological concept is the ecosystem concept. An ecosystem is an interacting system that consists of groups of organisms together with their non-living or physical environment. An ecosystem is a biotic community interacting with an abiotic environment.

The habitat of an organism is the place in which it lives. The niche of an organism is its total role in the community. Ecologists also speak of the trophic level of an organism. The main trophic levels are producer, consumer, and decomposer. Food chains and food webs show how the trophic levels of an ecosystem are interrelated. Three kinds of ecological pyramids also summarize these interrelationships. Some unusual relationships, called symbiosis, exist in nature. There are three kinds of symbiosis: parasitism, mutualism, and commensalism.

Energy flow in an ecosystem is one-way. Energy must always be provided to keep an ecosystem functioning. However, nutrient flow is cyclic. Nutrients are repeatedly recycled through an ecosystem.

Ecological succession is the replacement of one community by another as time passes. A climax community normally results from succession. Succession can be classified as primary or secondary. It can also be classified as autotrophic or heterotrophic.

The population density in an ecosystem is regulated by a balance between the biotic potential and the environmental resistance. The point of balance is called the carrying capacity of the ecosystem.

24 Aquatic Ecosystems

Over 70% of the earth's surface is covered by water. Most of this, of course, is made up of the oceans. In fact, 97% of the world's water is salt water and only 3% is fresh water. Of that 3%, 98% is frozen in the ice caps of Antarctica and Greenland. The 2% of fresh water that is not frozen is in ponds, lakes, and rivers (Fig. 24-1). In this chapter you will study the abiotic and biotic factors that interact in the various kinds of aquatic ecosystems on earth.

24.1 Abiotic Factors in Aquatic Ecosystems

The abiotic factors that operate in aquatic ecosystems can be divided into two groups, chemical and physical. Among the **chemical factors** are oxygen, carbon dioxide, and phosphate concentrations. Among the **physical factors** are temperature, speed of the water, and turbidity (cloudiness). This section discusses the main abiotic factors and explains their effects on living things.

Fig. 24-1 This river is just one of many kinds of aquatic ecosystems. Is it, in any way, related to other aquatic ecosystems such as ponds, lakes, and oceans? Do any of them affect it? Does it affect any of them?

Oxygen

Most aquatic organisms depend on dissolved oxygen for survival. Thus the determination of the dissolved oxygen content of water is one of the most important tests that you can perform to measure the quality of a body of water (Fig. 24-2). In general, an acceptable environment for aquatic life must contain no less than $5 \mu g/g$ of oxygen (five micrograms of oxygen in one gram of water). Variations from this figure are wide, of course. The amount of oxygen required varies with the organism, its degree of activity, the water temperature, the pollutants present, and many other factors. Here are some examples of these variations.

The trout species, *Salmo trutta*, requires, under normal conditions, an oxygen concentration greater than $10 \mu g/g$. The chub, *Leuciscus cephalus*, requires only $7 \mu g/g$. The carp, *Cyprinus carpio* can remain alive in water containing as little as $1-2 \mu g/g$ of oxygen.

One species of trout uses about 55 mg of oxygen per hour for each kilogram of body mass when it is resting at a temperature of $5°C$. At $25°C$, under similar conditions, the consumption rises to about 285 mg/h. In contrast, one species of goldfish has an oxygen consumption of only 14 mg/h for each kilogram of body mass when it is resting at $5°C$. However, the consumption increases ten-fold to 140 mg/h when the temperature is raised to $25°C$. Clearly the rate of oxygen consumption by fish depends on both temperature and species.

Fig. 24-2 Pond water being tested for its dissolved oxygen content.

Solubility of oxygen in water

Table 24-1 gives the solubility of oxygen gas in water when air (21% oxygen) is in contact with the water at standard pressure. For example, at 10°C, a maximum of 11.3 μg of oxygen will dissolve in 1 g of water if air is the only source of the oxygen.

TEMPERATURE OF WATER (°C)	SOLUBILITY (μg/g)
0	14.6
5	12.7
10	11.3
15	10.1
20	9.1
25	8.3
30	7.5

TABLE 24-1 **Solubility of Oxygen in Water (When air containing 21% oxygen at standard pressure is the only source of the oxygen)**

Oxygen as a limiting factor

As you know, the oxygen concentration can be a limiting factor in determining the species present in a body of water. It is generally not a limiting factor in rapidly flowing, clean rivers. In such rivers, the concentration of oxygen usually stays at the saturation point. Why is this so? In sluggish or polluted streams, however, the oxygen concentration may be well below saturation. Respiration by microorganisms such as bacteria uses up much of the oxygen in the water. Therefore, the oxygen concentration may be a limiting factor.

Lakes and ponds get some of their oxygen from the atmosphere. The remainder comes from photosynthetic organisms in the lake or pond. Sometimes oxygen is used by the respiration of organisms in the body of water faster than it is replaced. This often occurs in old or polluted lakes and ponds that have dead organisms or sewage on the bottom. In such a case, an oxygen deficit results that limits the species that can live in that body of water. For example, sewage keeps the central basin of Lake Erie below 10% oxygen saturation. Thus fish, like trout, that prefer the cool deep water of a lake, yet require high oxygen levels, cannot live in that part of the lake.

On a sunny day, a pond or lake that contains many photosynthetic organisms will often have oxygen levels well above those given in Table 24-1. These organisms add pure oxygen to the water as they photosynthesize.

Carbon Dioxide

Air is only about 0.034% carbon dioxide by volume. Therefore, when air is the only source of carbon dioxide for a body of water, very little carbon dioxide ends up in the water. For example, at 0° C, the solubility of carbon dioxide in water is only $1\,\mu g/g$, when air is the only source. However, in most natural ecosystems, the air is just a minor source of carbon dioxide. Respiration by living organisms is the major source. As a result, the surface water of most bodies of water can contain up to $10\,\mu g/g$ of carbon dioxide. If the bottom of the body of water contains decaying organic material, much higher levels of carbon dioxide will be found in the deeper regions. This results from the respiration of the decomposers, chiefly bacteria.

Water that contains over $25\,\mu g/g$ of carbon dioxide is harmful to most organisms. In fact, concentrations of $50\text{-}60\,\mu g/g$ will kill many species.

pH

The pH of a body of water is a measure of its acidity. It is normally measured on a scale that runs from 0 to 14 (Fig. 24-3). On this scale, 7 is neutral, below 7 is acidic, and above 7 is basic. In other words, the acidity of a solution decreases and its basicity increases as the pH goes from 0 to 14.

The pH range from 6.7 to 8.6 supports a good fish population. As long as the pH is within this range, it appears to have little negative effect on the life processes of most species of fish. In fact, most species can tolerate, for a limited time, pH values beyond this range. However, only a very few species can tolerate pH values lower than 5.0 or greater than 9.0. Further, little animal life of any kind is found outside of the range from 4.0 to 9.5.

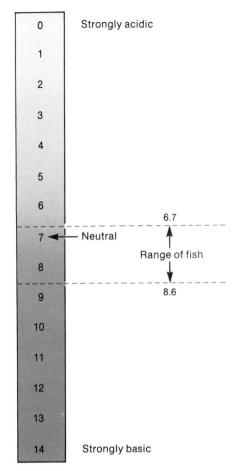

Fig. 24-3 The pH scale for acidity.

Alkalinity

Alkalinity is defined as the ability of water to neutralize acids. In natural waters, the substances that make up most of the alkalinity are bicarbonates and carbonates. These are the same substances that are used in many antacid preparations that people take to relieve excess stomach acid. How do the bicarbonates and carbonates get into the water in the first place? They are formed by the reaction of carbon dioxide with water. When carbon dioxide dissolves in water, carbonic acid is formed. This acid is present in the soda water used to make most soft drinks. The fizz that occurs when you take the top off a bottle of "pop" is due to the escape of some of the carbon dioxide. Some of the carbonic acid breaks up to form the bicarbonates and carbonates. The following word equation summarizes these reactions:

carbon dioxide + water ⇌ carbonic acid ⇌ bicarbonates + carbonates

The arrows indicate that all the reactions are reversible. In other words, bicarbonates and carbonates can form carbonic acid and, finally, carbon dioxide. In fact, during sunny days when carbon dioxide is being used for photosynthesis, the bicarbonates and carbonates replace it, thereby keeping the process of photosynthesis going.

Water with an abundant source of carbon dioxide will likely have a high alkalinity. Normally, such water contains decaying organic matter. The carbon dioxide is produced by the respiring bacteria that feed on the organic matter.

An alkalinity of 50 μg/g is considered low. In fact, one might even say that water with an alkalinity below 50 μg/g is "too clean". It contains little carbon (as bicarbonates and carbonates) to feed the process of photosynthesis. A lake with such a low alkalinity will likely contain little life. In contrast, an alkalinity of 200 μg/g is quite high. Unless you know of a natural source of high alkalinity in a body of water you might suspect dumping of domestic sewage or other organic waste into that water.

Hardness

Hardness in water is caused mainly by calcium and magnesium ions. In regions of sedimentary rocks such as limestone (calcium carbonate) and dolomite (magnesium calcium carbonate), the hardness is usually quite high. Calcium and magnesium ions are picked up by the water as it runs over these minerals.

Some hardness in water is necessary for living things. All living things need calcium and magnesium ions. However, very hard water is undesirable for many reasons, most of which are economic. For example, calcium and magnesium ions react with soap to form curds which hinder effective washing. Also, these ions often precipitate out when the water is heated, forming the familiar tea kettle scale and the costly boiler scale that torments industry. Because of such problems, synthetic detergents were developed that do not form curds. Also, water softeners are installed in many homes and industries.

Table 24-2 shows the commonly accepted standards for degrees of hardness.

DEGREE OF HARDNESS	TOTAL HARDNESS (μg/g)
Soft	0-60
Moderately hard	61-120
Hard	121-180
Very hard	Over 180

TABLE 24-2   Scale of Hardness

Water with a total hardness around 250 μg/g is thought to be the best for drinking. People who drink soft water for extended periods of time are more likely to get cardiovascular diseases (diseases of the heart and circulatory system). Yet, water with a total hardness above 500 μg/g can make you very ill if you are not accustomed to it. It often causes diarrhea. In your large intestine are billions of coliform bacteria. These bacteria help the villi of the large intestine extract water from the waste food material. The villi then put this water back into the bloodstream. However, a sudden change in the hardness of the water causes the coliform bacteria to stop their mutualistic relationship with your body. Thus the water stays in the waste food material and you get diarrhea.

Nitrogen

The element nitrogen is present in all proteins. Thus it occurs in all living things and is an essential part of all ecosystems.

Ammonia

Nitrogen occurs in the compound ammonia (NH_3). Ammonia is a by-product of the decay of plant and animal proteins and fecal matter. It is also formed when metabolic wastes such as urea decompose. Thus the presence of ammonia can be a sign of sewage entering the water. It may, of course, be of natural origin as well. Since many fertilizers contain ammonia and ammonium compounds, runoff from farms can also add ammonia to streams and lakes.

Nitrate

The nitrogen cycle shows you that ammonia is converted to nitrite and then to nitrate (see Fig. 23-15, page 635). Thus a high nitrate concentration could indicate the presence of decaying organic matter such as sewage or fertilizer runoff. Nitrates themselves are present in most fertilizers. Thus a high nitrate concentration could also mean that nitrate-bearing fertilizers are leaching from farm soil.

Most beef cattle are raised today in beef feedlots. Hundreds of cattle are placed in a very small area and force-fed until they are ready for market. The cattle urine and fecal matter produced can create a serious nitrogen pollution problem for nearby waters.

Ecologists generally agree that, on the average, a lake should not have over 0.30 μg/g of total nitrogen in it. If it does, it could have an algal bloom, if other nutrients are present in adequate amounts.

Phosphorus

Phosphorus, like nitrogen, is an element of great significance to living things. Many important organic molecules within cells contain phosphorus atoms. For example, adenosine triphosphate (ATP) is a phosphorus-bearing compound found in every living cell. It plays a key role in energy storage and supply.

Like other nutrients, phosphorus follows a cycle in nature. Natural waters normally contain some dissolved phosphates. These phosphorus compounds are absorbed by plants and other producers, and are used by them to make molecules like ATP. When herbivores eat the producers, they obtain much of this phosphorus. When the producers and animals die, decomposers return the phosphorus to the water. Additional phosphorus is returned to the water when decomposers break down animal waste matter. The phosphorus eventually ends up as phosphates, ready to repeat the cycle.

Phosphorus enters water from many of the same sources as nitrogen: sewage, animal wastes, agricultural runoff, and decaying plants and animals. Although fertilizers usually contain high concentrations of phosphorus, runoff from fields generally contains very little phosphorus. Apparently many types of soil can "fix" or hold on to phosphate ions. The phosphate ions are attracted to the surfaces of soil particles. Therefore, unless soil erosion occurs, the phosphate ions do not wash away to streams. Human sewage and beef feedlots are major contributors of phosphorus to water. Most detergents still contain phosphorus, although the concentrations have been reduced considerably in the last few years.

Most ecologists feel that phosphorus is a limiting factor in the production of algal blooms in lakes. Therefore, they believe that by controlling the phosphorus concentration, they can control the formation of algal blooms. As a result, treatment units have been added to many sewage treatment plants to remove phosphorus from the water.

Ecologists generally agree that, on the average, a lake should not have over $0.015\,\mu g/g$ of total phosphorus in it. If it does, it could have an algal bloom, if other nutrients are present in adequate amounts.

Temperature

As you know, each species of organism has its own optimum or preferred temperature. For example, the optimum temperatures for three common species of fish are *Cyprinus carpio* (carp), $32°C$; *Perca flavescens* (perch), $24°C$; and *Salmo trutta* (trout), $15°C$. These and other aquatic organisms can stand some variation from the optimum temperature. However, if the temperature shifts too far from the optimum, the organism either dies or moves to a new location. With most species of fish, an increase of $5°C$ above the optimum can be quite harmful. This is particularly true if the increase is unexpected for that time of year. For example, if a stream has an average May temperature of $18°C$ and hot outflow from an industry raises the temperature to $25°C$, a heavy fish kill will probably occur. If, on the other hand, the temperature gradually rises to $25°C$ as a result of normal summer warming, a fish kill will probably not occur. The fish have time to migrate to cooler regions.

Why does an increase in temperature kill fish and other aquatic life? As the water temperature increases, the body temperature of any poikilothermic ("cold-blooded") animal in the water increases. This, in turn, results in an increase in the rate of metabolism in the animal. Of course, this increases the need for oxygen by the animal. Yet, as the temperature of the water goes up, its ability to hold oxygen goes down. The animal needs more oxygen, but less is available. Death by suffocation may result.

Review

1. **a)** What is the effect of an increase in temperature on the solubility of oxygen gas in water?
   b) What is the effect of an increase in temperature on the rate of oxygen consumption by a fish? Why is this so?
2. **a)** What is the main source of oxygen for fast streams?
   b) What is the main source of oxygen for most ponds and lakes?
   c) What removes oxygen from bodies of water?
   d) Why do many lakes have little oxygen near the bottom?
3. **a)** What is the main source of carbon dioxide for most aquatic ecosystems?
   b) What is an acceptable level of carbon dioxide for an aquatic ecosystem?
4. **a)** What is pH?
   b) Describe the pH limits for fish and other aquatic animal life.
5. **a)** What is alkalinity?
   b) Explain how carbon dioxide in water results in alkalinity.
   c) A lake with an alkalinity of 10 μg/g was stocked with trout. The lake had a proper pH and adequate oxygen. Yet all the fish eventually died. Why?
6. **a)** What causes hardness in water?
   b) Why is some hardness necessary in natural waters?
   c) Why is excessive hardness undesirable at times?
7. **a)** How does ammonia get into water?
   b) How do nitrates get into water?
   c) What is an acceptable level of total nitrogen in a lake? What could happen if this level was exceeded?
8. **a)** How does phosphorus get into water?
   b) What is an acceptable level of phosphorus in a lake? What could happen if this level was exceeded?
   c) Phosphates have been limited by law to less than 5% in laundry detergents sold in Canada. This has greatly helped to reduce the amount of phosphorus pollution entering the Great Lakes. Yet automatic dishwasher detergents are not restricted to that amount. They contain up to 45% phosphate. The main reason is that the phosphate makes the dishes dry spotless. What do you think of this fact?
9. Power generating plants pollute lakes with hot water. When accused of such pollution, the manager of one plant replied that the pollution could not harm fish. It only raised the temperature of the lake near the plant by a few degrees. How would you answer this person?

24.2 Biotic Factors in Aquatic Ecosystems

You saw in Section 24.1 that the chemical properties of water determine, to a large extent, the types of organisms that live in that water. You saw, also, that the organisms can affect the chemical properties of the water. Thus organisms, as well as chemical tests, can be used as indicators of water quality. This section describes just how that is done.

Index Species

A pollutant can be present in water in a concentration high enough to kill all living things in that ecosystem. More commonly, though, the pollutant kills only certain species without harming others. The few species that are not killed increase in numbers, since interspecific competition has been reduced. Thus, *when water becomes polluted, there tends to be a shift from many species of moderate population density to a few species of high population density. A decrease in the diversity of species present is probably the best biological indicator of pollution.* The species present in the highest population densities are called the **index species** or the **indicator organisms** of water quality. Biologists use five groups of organisms as indicators of water quality. These are fish, bottom fauna (animals that live on, in, and near the bottom of the body of water), algae, zooplankton, and bacteria.

Eutrophication

All lakes and ponds will eventually fill up and disappear due to a natural aging process called eutrophication. To see how this works, let us consider the history of a lake that resulted from the last ice age. When the glaciers retreated, they left behind a basin that filled with water and became a lake. The lake was a cold, clear body of relatively clean water. Such a body of water contains little life and is called a **oligotrophic** ("little nourishment") lake. However, over the years, the rivers that fed this lake brought nutrients like nitrogen and phosphorus into it. In addition, these rivers carried silt and organic materials into the lake. As a result, the lake became more shallow. Further, the added nutrients increased the productivity of the lake. More life appeared, resulting in still more organic material on the bottom. The combination of warmer water, decreased depth, and added nutrients greatly increased the amount of life in the lake. When this stage is well advanced, the lake is called a **eutrophic** ("adequate nourishment") lake. *The process of aging or increasing productivity is called* **eutrophication**. In between the oligotrophic and eutrophic stages, a lake is usually called **mesotrophic** ("middle nourishment").

When natural eutrophication is well advanced, the water may appear polluted to an untrained observer. The diversity of species is usually low and the population density of the few remaining species is high. Natural eutrophication is generally a slow process. However, pollution with organic material such as sewage can greatly speed up the rate of eutrophication. Addition of nutrients such as nitrogen and phosphorus can do the same.

Fish as Indicators of Water Quality

The dominant fish in oligotrophic lakes are those that need clean, cool water. Among these are lake trout, whitefish, walleye, lake herring, and char (Fig. 24-4). At the mesotrophic stage, most of these species remain, but in decreased numbers (Fig. 24-5). Perch, black bass, pike, and smelt become the dominant species. After further eutrophication, carp, sunfish, and catfish dominate.

The presence of perch, carp, and sunfish does not mean that the water is polluted. It may simply mean that the water is eutrophic. More specifically, it means that the oxygen concentration in the deep, cool water is low (Fig. 24-6). The decomposition of organic material in the bottom sediment uses up oxygen. This makes survival impossible for trout and other fish that require the cool water near the lake bottom. However, fish like carp that can live in the warm upper water thrive.

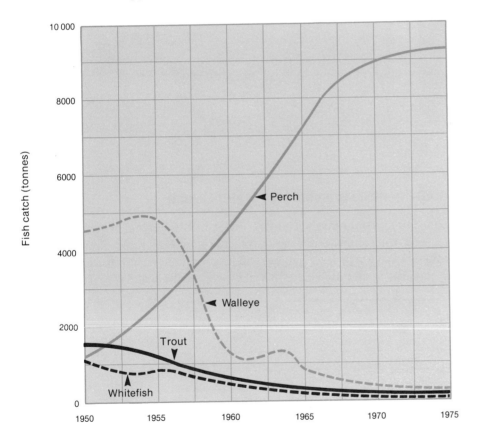

Fig. 24-4 **Fish as indicators of water quality.**

Fig. 24-5 **Changes in fish population numbers in Lake Erie.**

In a similar manner, the presence of carp and sunfish in a stream may simply mean that the water is generally too warm for trout and similar species. The water could be quite free of pollution. However, should you find a cool, deep stream that contains trout at its headwaters but mainly carp in the waters below a town, you should suspect the town of polluting the water.

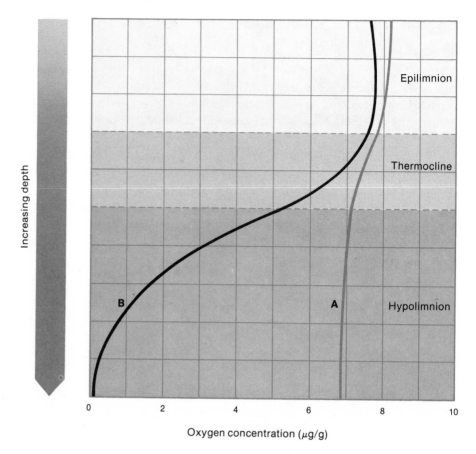

Fig. 24-6 Oxygen concentration versus depth for an oligotrophic lake (A) and a eutrophic lake (B), in mid-summer. Epilimnion means "upper lake". Hypolimnion means "lower lake". The thermocline in a region of rapid temperature changes with depth.

Epilimnion

Thermocline

Hypolimnion

Increasing depth

0   2   4   6   8   10
Oxygen concentration (μg/g)

Bottom Fauna as Indicators of Water Quality

Bottom fauna are animals that live on or in the bottom sediment, under rocks, and on submerged vegetation. The types of fauna found on the bottom of a stream or river depend to a large extent on the nature of the bed. This, in turn, is directly related to the speed of the water (see Table 24-3).

SPEED	NATURE OF BED
over 100 cm/s	rock
60-100 cm/s	rocks and heavy gravel
30-60 cm/s	light gravel
20-30 cm/s	sand
10-20 cm/s	silt
under 10 cm/s	mud

TABLE 24-3   **Speed and Nature of the Bed**

In rapidly flowing water, all but the larger rocks are gradually carried downstream. Gravel is deposited in slower regions and sand in still slower regions. Where there is little or no current, silt and mud are deposited on the bed (bottom).

As you know, the oxygen concentration plays an important role in determining the fauna in a stream. The water in a fast stream is usually turbulent. Therefore it will, as a rule, be highly oxygenated. Thus the bottom fauna in a fast area of a stream are those that require high oxygen conditions. They must also have adaptations that enable them to cling to the bottom. Otherwise, they would be swept downstream. Some of the dominant bottom fauna of this area of a stream are insects such as stoneflies, some species of mayflies, and some species of caddisflies (Fig. 24-7). All three are generally found under and around rocks and coarse gravel. If the stream is polluted to the point where its oxygen concentration drops greatly, these animals will not be found in the stream.

At the bottom of Figure 24-7 are organisms that live in mud-silt bottoms where the oxygen concentration is usually very low. The sludge worm is a segmented worm like the earthworm. It operates on the muddy bottom in much the same manner that an earthworm lives on the soil in a garden. It constructs a tube above a burrow extending down into the bottom ooze. While it feeds in the stream bottom, its tail sticks out of the tube, swaying back and forth, exchanging carbon dioxide for any available oxygen in the water. This swaying motion and the presence of blood vessels close to the body surface allow the sludge worm to get even the smallest traces of oxygen from the water.

The midge larva is another common bottom burrower that can tolerate low oxygen conditions. This larva occupies a tube that it makes from sludge material. Some species are red in colour and are called bloodworms. However, they are not worms. They are insect larvae. You have likely seen the adult midge fly. It is an annoying little creature that looks like a small mosquito without piercing mouthparts. Clouds of midge flies often gather around your head on a humid, windless day.

Leeches, isopods (aquatic sow bugs), and ostracods generally live in muddy areas with slightly higher oxygen levels than those in which sludge worms and midge larva live. Note that the fauna which live in mud-silt bottoms of streams are also among those that you will find on the bottoms of many lakes and ponds. Why is this so?

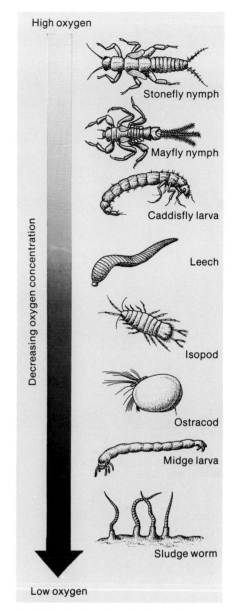

Fig. 24-7 Bottom fauna and oxygen requirements.

Review

1. **a)** What is an index, or indicator, species?
   b) What is the best biological indicator of water pollution?
2. **a)** Distinguish between an oligotrophic lake and an eutrophic lake.
   b) Define eutrophication.
3. **a)** A small lake is found to contain mainly sunfish, catfish, and other fish that can tolerate low oxygen conditions. Is the lake polluted? Explain your answer.
   b) Look closely at Figure 24-6. Account for the difference in the shapes of the two curves.
4. **a)** Four decades ago mayflies were so abundant in some towns and cities along Lake Erie that shovels were used to remove their bodies from streets and sidewalks. This problem no longer exists. Why?
   b) Sludge worms and midge larvae are usually more abundant at the mouth of a river than at the headwaters of the same river. Why?

A

B

Fig. 24-8 A pond ecosystem (A) and a lake ecosystem (B). How are they the same? How do they differ?

24.3 Freshwater Ecosystems

Freshwater ecosystems can be grouped into two categories, standing waters and flowing waters. A pond is an example of standing water. A stream is an example of flowing water. In some cases it is difficult to classify a body of water as standing or flowing. For example, some rivers flow so slowly near their mouths that they appear to be standing waters. In a similar manner, most ponds have inflowing water and outflowing water. As a result, they have a flow of water through them.

Standing Waters

Even a puddle of water is standing water. However, two types of standing water make up larger freshwater ecosystems. These are ponds and lakes (Fig. 24-8). There is not a sharp distinction between the two types. However, most limnologists (those who study standing waters) define a pond as a body of water in which light can reach the bottom at all places. In contrast, light cannot reach the bottom in the deeper regions of lakes. In general, then, ponds are smaller and more shallow than lakes.

Ponds

An interesting way to study a pond ecosystem is to view it with the concept of succession in mind. You may recall that succession is the replacement of one biological community by another as time passes (see Section 23-6, page 638). This process takes place in all ponds. Suppose that you sat in a boat at the centre of a freshly dug, deep pond. If you could remain there for a few hundred years, you would observe the complete succession of the pond. The following is what would likely happen.

Since the pond is quite deep, sufficient light cannot reach the bottom to promote plant growth. Thus the bottom consists of nothing but the parent earth material. This is the first stage in pond succession. It is called the pioneer stage, or bare-bottom stage. As the years pass, runoff from the surrounding land and the decay of dead organisms in the pond gradually add rich soil to the bottom of the pond. The pond gradually becomes more shallow. Eventually it is shallow enough that green plants can grow on the bottom. Soon the bottom is covered by submergent, or underwater, plants such as *Chara* (stonewort), *Elodea* (Canada waterweed), *Ceratophyllum* (coontail), and *Cabomba* (fanwort). This stage in succession is called the submergent vegetation stage. Over the years, the decay of these plants adds further humus to the bottom of the pond. In time, the pond is shallow enough that *floating-leafed* plants such as the water lilies can grow. These plants, in turn, make the pond shallow enough that it can support the growth of emergent plants such as cattails, rushes, and sedges. This stage in succession is called the emergent vegetation stage.

The decay of emergent vegetation fills the pond to the point where no large open expanse of water remains. This is the marsh stage. As the

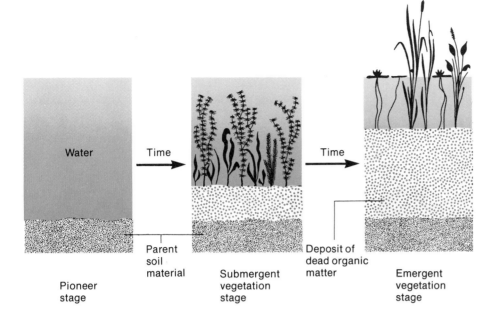

Water

Time

Time

Pioneer
stage

Parent
soil
material

Submergent
vegetation
stage

Deposit of
dead organic
matter

Emergent
vegetation
stage

"islands" of land become drier, trees may begin to grow on them. Ecologists call this the swamp stage. Whether trees appear or not, the pond eventually becomes so shallow that it dries up in the summer. It only contains water in the spring and late fall. It is completely dry in the summer and completely frozen in the winter. This is the temporary pond stage.

Given enough time, any pond or lake should fill in completely. However, the process of succession does not stop at this point. A whole series of terrestrial plant communities will continue the process. You studied terrestrial succession in Section 23-6.

Figure 24-9 summarizes the process of pond succession from the pioneer stage to the emergent vegetation stage. These are the stages that you will likely study if your teacher takes you on a pond field trip. Fortunately, you will not have to wait in a boat for a few hundred years to see these stages. A pond is generally deepest near its centre. Further, it gradually becomes more shallow towards the margin (edge). Therefore, you can usually see the first three stages of succession if you examine the pond along a line that runs from the deepest spot to the edge (Fig. 24-10).

The preceding discussion of pond succession dealt mainly with changes in vegetation. However, changes in other organisms accompany these changes in vegetation. Let us now look at these.

In the pioneer stage, most of the organisms are plankton. (The word "plankton" means small drifting organisms.) Most plankton are microscopic. A few are barely visible to the unaided eye. Some plankton are protists; some are animals. Some plankton are producers; others are consumers. The plankton that are producers are called phytoplankton ("plant plankton"). Most of these are algal protists such as diatoms and flagellates. Some are blue-green algae and green algae. The plankton that are consumers are called zooplankton ("animal plankton"). Some of these are

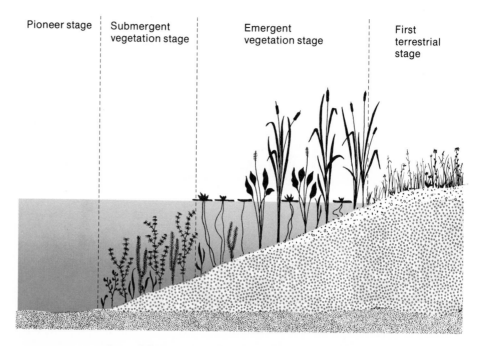

Fig. 24-10 Pond succession can be observed by examining the vegetation along a line from the deepest spot to the margin.

Pioneer stage | Submergent vegetation stage | Emergent vegetation stage | First terrestrial stage

protozoan protists. Many are animals such as rotifers and tiny crustaceans—cladocerans, copepods, and ostracods, to name a few. The zooplankton are mainly herbivores. They feed on the phytoplankton and, in turn, become food for higher order consumers such as aquatic insects and fish. The bottom of a pond contains large quantities of decaying organic matter. Decomposers such as bacteria, yeasts, and moulds break down this material, returning nutrients to the water. These nutrients promote the growth of phytoplankton, thereby "feeding" the food chains of the pond.

The submerged vegetation stage of succession is dominated by plants that grow entirely under water. In a similar manner, the animals that live in this vegetation are adapted to life in a submerged stage. Most are gill-breathing insect larvae such as mayflies, damselflies, and dragonflies. In most ponds, the submerged vegetation is an important source of oxygen and food for the pond ecosystem.

The emergent vegetation stage is dominated by plants that grow partly in and partly out of the water. It is interesting to note that many of the animals that live in this vegetation are also adapted to a life that is partly aquatic and partly terrestrial. Amphibians such as frogs and toads are one example. Reptiles such as snakes and turtles are another.

Lakes

A lake is a body of standing water in which light cannot reach the bottom at all points. Therefore, much of the bottom is in the pioneer stage of succession. As a result, submerged vegetation plays a lesser role in a lake ecosystem than it plays in a pond ecosystem. In fact, phytoplankton form the base of most food chains in a lake. The phytoplankton live mainly in the upper few metres of a lake. They require light for photosynthesis. However, consumers are found at all depths in most lakes. Those that are herbivores dwell near the surface where they graze on phytoplankton. Those that are carnivores spend their time at various depths. Some live at a depth of several hundred metres in certain lakes. Of course, these consumers need oxygen. Yet photosynthesis does not occur at such depths. How, then, does oxygen get to the bottom of a deep lake? It does so by an interesting process called **overturn**. Let us see how it happens.

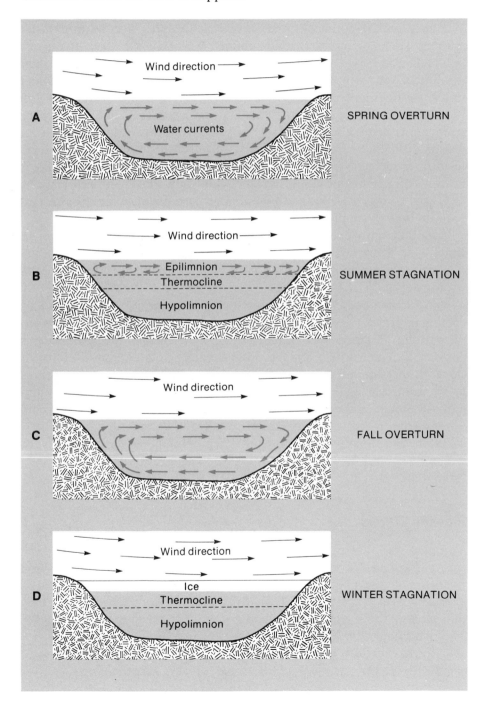

Fig. 24-11 Spring and fall overturn carry oxygen to the bottoms of many deep lakes.

As the ice on a lake melts in the spring, the cold water that forms sinks to the bottom. (You may know that water is most dense at 4° C.) Therefore the lake gradually fills from the bottom with cold water. At some time in the spring, the lake is the same temperature throughout. As a result, it also has the same density throughout. Such water mixes easily when the spring winds blow over it (Fig. 24-11,A). This mixing of the water carries oxygen to the bottom. It also brings nutrients from the bottom to the top, where they support the growth of phytoplankton. This process is called spring overturn.

As summer approaches, the sun warms the upper layer of water faster than the winds can mix it. By mid-summer, the lake usually has three layers in it (Fig. 24-11,B). The upper layer, or epilimnion, contains warmer water that circulates freely. The lower layer, or hypolimnion, contains colder water that does not circulate. In between is a layer of transition from warm to colder water. It is called the metalimnion, or thermocline. Usually the temperature drops about 1.0° C for every depth increase of 1.0 m. Because of density differences, the hypolimnion is cut off from circulation. It receives no further oxygen from the epilimnion. Limnologists say that the lake is now in summer stagnation. If the lake is eutrophic, much of the remaining oxygen is quickly used up. The hypolimnion becomes unsuitable for life that requires a large supply of oxygen.

As fall approaches, the epilimnion cools. Also, it gradually gets deeper. Finally the entire lake is a uniform temperature again. The fall winds are now able to cause another mixing of the water. This mixing is called the fall overturn (Fig. 24-11,C).

With the coming of winter, the upper water cools still further. This cooling increases the density of the water. The resulting cool water sinks to the bottom. This process continues as long as the temperature of the water is above 4° C. Water with a temperature below 4° C is less dense. Therefore it stays on the top and eventually freezes (Fig. 24-11,D). Once again three layers form. The epilimnion is now ice and water near 0° C. The hypolimnion is water at or above 4° C. The thermocline is reversed. It goes from cold water at the top to warmer water below. Again, because of density differences, the hypolimnion is cut off from the epilimnion. No additional oxygen will reach it until the next spring overturn. The lake is said to be in winter stagnation.

Flowing Waters

You likely call flowing waters by such names as brook, creek, stream, and river. In general, brooks and creeks are the smallest, streams larger, and rivers are largest. Creeks and brooks are often fed by springs (Fig. 24-12). They may also drain ponds and small lakes. However, sooner or later, they join together to form streams. Streams, in turn, join together to form rivers. The rivers carry the water to a lake or, in some cases, to the sea.

In Section 24.1, you learn that clean flowing waters are usually saturated with oxygen. As the water of brooks, creeks, and streams rushes over

Fig. 24-12 The girl has found one of the springs that feeds this brook. Rainwater percolated into the soil of the hills in the background. It then came back to the surface by way of the spring.

rocks, air circulates freely through it (aeration). Phytoplankton are not an important source of oxygen in fast waters. In fact, most phytoplankton are swept away by the current. However, a few producers do live in such waters. Clinging to the rocks are some blue-green algae, green algae, diatoms, and aquatic mosses. These organisms photosynthesize. As a result, they add some oxygen to the water. However, the amount added is very small compared to that which enters by aeration.

Photosynthesis by such producers also forms glucose. Thus these producers may support some food chains in the fast water. For example, herbivores such as snails may graze on these producers. The snails, in turn, may be eaten by carnivores such as fish. However, only a small portion of the total food supply of a fast water ecosystem comes from these producers. Most of it enters the water from the surrounding land. Leaves, grass, twigs and other organic matter fall into the water or are washed in by a rain. Scavengers feed on this organic matter. They, in turn, are eaten by carnivores. Also, insects and other small terrestrial animals often fall into the water and are eaten by fish and other predators.

Rivers are generally wider, slower, warmer, and more nutrient-rich than streams, brooks, and creeks. Some phytoplankton often live in slower rivers. Also, decaying organic matter builds up on the bottoms of such rivers. In fact, in some respects, a wide slow river has many of the properties of a pond. Submergent plants may even grow in it. Since the river contains more producers per square metre, it is said to have a higher productivity. As a result, it can support a larger number and variety of consumers.

The organisms that live in flowing waters show unique adaptations to their habitats. For example, the stonefly nymphs that live under the rocks in fast brooks and streams are streamlined in shape and have hooks on their feet. In contrast, the sludge worms that live in the bottom ooze of slow rivers show no adaptations for holding on to the bottom material. However, they do show adaptations to the low oxygen concentrations of their habitat (see Section 24.2, page 661).

Review

1. a) Distinguish between standing and flowing waters.
   b) Distinguish between a pond and a lake.
2. a) List the main stages of succession that a pond or lake may pass through as it changes from a young body of water to dry land.
   b) Give a brief description of each of the stages listed in a)
3. a) What are plankton?
   b) Distinguish between phytoplankton and zooplankton.
   c) Why are phytoplankton usually more important in a lake than in a pond?
4. a) What causes spring and fall overturn in a lake?
   b) What two useful functions does overturn perform for a lake ecosystem?
5. a) Describe the layering present in a lake at the time of summer stagnation.

b) How do the hypolimnions of an oligotrophic lake and an eutrophic lake differ at the time of summer stagnation?

c) How do summer stagnation and winter stagnation of a lake differ?

6. a) Compare the methods by which standing waters and running waters obtain oxygen.

b) Compare the methods by which standing water ecosystems and running water ecosystems obtain most of their food supply.

7. a) What would likely happen to a pond ecosystem if all the decomposers suddenly died? Why?

b) Trace the sequence of events that would likely occur in a pond if the owner fertilized it by adding large quantities of lawn fertilizer.

8. a) Construct three pond food chains. Begin each food chain with a producer.

b) Construct three stream food chains. Begin only one of these with a producer that lives in the stream.

24.4 Marine Ecosystems

The oceans and seas of the world are marine ecosystems. However, the tiny tidal pools that dot a rocky coast are also marine ecosystems. Big or small, marine ecosystems differ from freshwater ecosystems in one basic way. They contain much higher concentrations of dissolved minerals.

Salinity

The water in the oceans and seas is usually called sea water or salt water. It is about 3.5% minerals. Expressed in micrograms per gram, the concentration of minerals in sea water is about 35 000 μg/g. Compare this concentration to those that were discussed for freshwater ecosystems in Section 24.1.

The minerals in sea water are mainly salts. The most abundant salt in sea water is sodium chloride, or common salt. In fact, most sea water is about 2.7% common salt. As a result, the concentration of minerals in sea water is usually called the **salinity** of the water. Thus the salinity of sea water is about 3.5%, or 35 000 μg/g.

The salinity of a marine ecosystem varies from time to time and from place to place. For example, the salinity of a tidal pool may be only 0.5% after a rainfall. Yet it may be close to 3.5% after the tidal pool has been flushed with sea water by the tides. Also, the salinity of the surface water in a tidal pool is usually higher than the salinity of the water in the bottom. Evaporation of water by the sun concentrates the salts near the surface.

As you might expect, not all marine ecosystems have the same salinity as the oceans. For example, the Baltic Sea has a salinity of only 1.0%. In contrast, the Red Sea has a salinity of over 4.5%.

Life in the Open Ocean

The dominant producers in the ocean are diatoms and dinoflagellates. Diatoms tend to be more abundant in northern waters, whereas dinoflagellates tend to dominate warmer waters. The herbivores that feed on these producers are mainly copepods and other crustaceans (Fig. 24-13). Herring, other small fish, squid and many other animals feed on the crustaceans. Then tuna and other larger fish feed on the smaller fish. If you eat a tuna sandwich, you get some of the sun's energy that was originally trapped by the phytoplankton of the ocean.

Life on a Rocky Shore

A fascinating array of life exists on the shores where the ocean and the land meet. All shores—rocky, muddy, sandy, and marshy—have one thing in common. They are exposed to the tides. Each has an **intertidal zone**, or **littoral zone**. This is the zone between the high tide and the low tide marks. Within this zone, conditions are always changing. When the tide is in, the littoral zone is basically a sea water environment. However, when the tide is out, this zone is a moist, salty, land environment. The organisms that live here must be adapted to the harsh conditions that accompany the changing water levels. The organisms in the upper region of the littoral zone experience the greatest changes. They are out of the water, exposed to the sun and drying winds, for the longest time each day. Those in the lower region are out of the water for only a short time each day.

The various zones within the littoral zone have different abiotic conditions such as time under water and temperature. Therefore these zones contain different organisms. In fact, a **zonation** of organisms can easily be seen within the littoral zone. This zonation shows up well on a rocky shore. Figure 24-14 shows the zonation at one place along the North Atlantic coast, north of Cape Cod. The following is a description of that zonation. Keep in mind that all rocky coasts show a similar general pattern of zonation. However, because of temperature differences, the organisms making up the zones differ from place to place. For example, quite a different set of organisms occupy the zones on a South Carolina coast as compared to this North Atlantic coast.

A rocky coast has three main zones. Our attention will centre on the **littoral zone**. Above it is the **supralittoral zone**. Below it is the **sublittoral zone**. Let us begin our study of these zones by taking an imaginary walk from the supralittoral zone on the right of Figure 24-14 to the water's edge.

Supralittoral zone

The supralittoral zone is also called the **spray zone**. It is made up of two zones, the **land zone** and the **bare rock zone**. This region becomes wet only during times of intense wave action. At such times, spray soaks the region. However, most of the time, this region is dry. Thus it is essentially a non-marine environment and is dominated by lichens and land plants. From time to time, the rough periwinkle moves into this region to graze on the lichens.

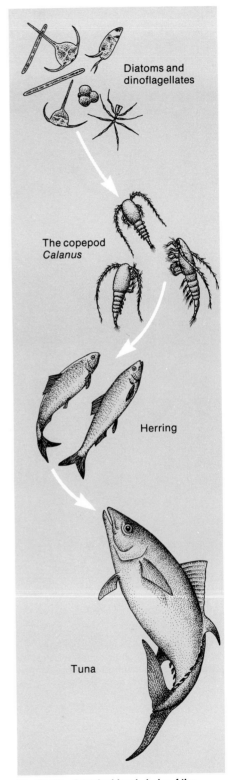

Diatoms and dinoflagellates

The copepod *Calanus*

Herring

Tuna

Fig. 24-13 A typical food chain of the open ocean.

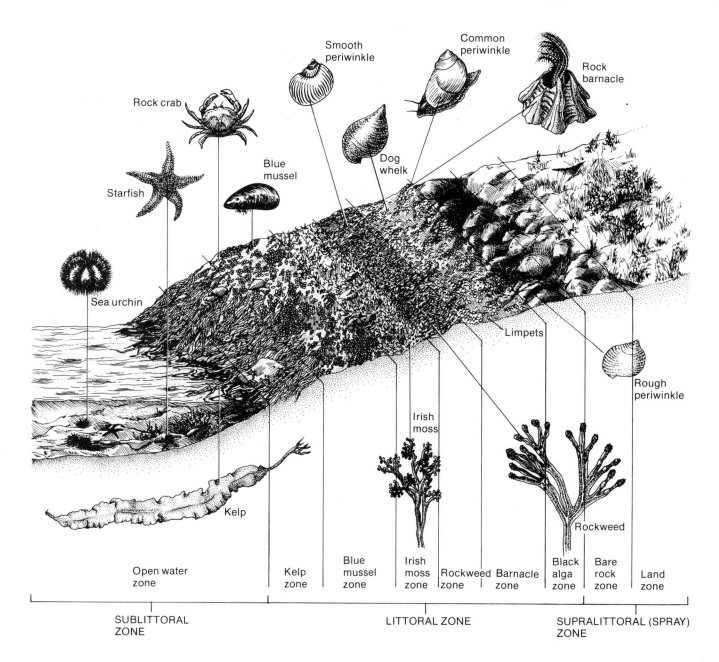

Smooth periwinkle

Common periwinkle

Rock barnacle

Rock crab

Blue mussel

Dog whelk

Starfish

Sea urchin

Limpets

Rough periwinkle

Irish moss

Kelp

Rockweed

| Open water zone | Kelp zone | Blue mussel zone | Irish moss zone | Rockweed zone | Barnacle zone | Black alga zone | Bare rock zone | Land zone |

SUBLITTORAL ZONE

LITTORAL ZONE

SUPRALITTORAL (SPRAY) ZONE

Fig. 24-14 Zonation on a rocky shore along the North Atlantic coast, north of Cape Cod. Even in this region, the zonation varies in detail from place to place.

Littoral zone

As you walk from the supralittoral zone toward the ocean, you will have little trouble telling when you have reached the littoral zone. The first zone of many within the littoral zone is the **black alga zone**. Every two weeks the spring tides cover this zone. Also, it is frequently wetted by wave action and spray. This zone stays wet enough to support the growth of certain blue-green algae and lichens. These organisms cling to the rocks and give them a characteristic black colour. The rough periwinkle is the main herbivore in this zone. It grazes on the algae and lichens. The black alga zone is often dry. Thus the rough periwinkle has to be very resistant to drying out.

As we walk further into the littoral zone, we enter the **barnacle zone**. This zone is covered and uncovered daily by the tides. Barnacles are crustaceans. They live stuck to the rocks and depend on the tides to bring them food. At high tide, the barnacles open up. Appendages emerge and sweep dinoflagellates, diatoms, and other protists into the barnacles' interiors. At low tide the barnacles close up to prevent water loss. Also found in the barnacle zone are limpets, dog whelks, common periwinkles, and, in some cases, blue mussels. Limpets move about, scraping algae from the rocks for food. Dog whelks are predators. They feed on barnacles, mussels, and periwinkles. The blue mussels in the barnacle zone are usually quite small. Apparently the tides do not bring these filter feeders enough food for maximum growth when they live this far up in the littoral zone.

Just below the barnacle zone is the **rockweed zone**. This zone is dominated by brown algae called rockweeds. These plants may grow to a length of 2 m in sheltered regions. However, they are usually about 30-40 cm long. The smooth periwinkle is a characteristic animal of this zone. These animals, along with limpets, graze on the rockweed.

Still closer to the water is the **Irish moss zone**. This red alga, along with other red algae, often forms a spongy carpet several centimetres thick. This zone varies in colour from purple to green and yellow. Rock crabs may hide in this vegetation, awaiting passing prey.

The **blue mussel zone** often overlaps into the Irish moss zone. The mussels are often so tightly packed together that no rock surface can be seen. They attach themselves firmly to the rocks. They do not move about like most bivalves. The blue mussel zone is under water most of the time. Even at low tide, wave action may keep it covered.

The lowest zone of the littoral zone is the **kelp zone**. This zone is dominated by a forest of the large brown alga, *Laminaria*, one of the kelps. The kelp zone may be uncovered once every two weeks during spring tides. However, even at that time, wave action often keeps it covered. The dense growth of kelp provides food and habitat for a wide range of animals. Limpets, mussels, isopods, amphipods, crabs, and starfish are just a few of the inhabitants. At high tide the starfish leave the kelp beds and move up into the blue mussel zone to prey on the mussels.

Sublittoral zone

Beyond the littoral zone is the **open water**, or sublittoral zone. This is a favourite spot for scuba divers. The bottom is covered with starfish, brittle stars, sea urchins and other fascinating animals. The exact nature of the community depends on the bottom material and other abiotic factors.

Review

1. **a)** What is the basic difference between a marine ecosystem and a freshwater ecosystem?
   b) What is meant by the term salinity?
   c) What is the average salinity of the oceans?

d) What portion of the salinity of the oceans is common salt?

2. Describe a typical food chain of the open ocean.

3. **a)** Name the seven zones of the littoral zone, beginning with the zone farthest from the water's edge.

   b) List two or three characteristic organisms from each of the zones that you named in a).

4. Select one marine organism, plant or animal, that inhabits the littoral zone of rocky shore. Research this organism's taxonomy, life cycle, feeding habits, and its morphological and behavioural adaptations to its environment. Prepare a written report of about 300-500 words on your findings.

24.5 Investigation

Oxygen, Temperature, and Metabolic Rate

You learned in Section 24.1 that temperature plays an important role in determining the dissolved oxygen content of water. You also learned that temperature affects the metabolic rates of poikilotherms ("cold-blooded" organisms). In this investigation you collect solubility data for oxygen in water. You also study the effect of temperature on the metabolic rate of a goldfish.

Materials

1 L beakers or other large
   containers (5)
aquarium aerators to service
   the 5 beakers
aquarium heater (optional)

ice cubes
hot plate
Hach oxygen test kit
goldfish of similar size (5)
$-10°C$ to $110°C$ thermometers (5)

Procedure A   Preparation for the Investigation

Procedure A will be started by your teacher or a small group of students at least 1 h before you begin this investigation. These steps provide water, saturated with oxygen, at a range of temperatures from about $0°C$ to about $40°C$.

a. Fill the 5 beakers with tap water. Begin aerating the water in all of them. Number the beakers from 1 to 5.

b. Maintain beaker 1 at about $0°C$ as follows: Add ice cubes to the water. Replace the ice as fast as it melts. Keep ice in the water at all times during the investigation.

c. Maintain beaker 2 at about $10°C$ as follows: Add one ice cube to the water. Let it melt completely. Then add another. Continue this process throughout the investigation.

d. Maintain beaker 3 at room temperature (about $20°C$).

e. Maintain beaker 4 at about 30°C by using an aquarium heater or by adding warm water from time to time.

f. Maintain beaker 5 at about 40° by using an aquarium heater or by placing it 1-2 cm above the surface of a hot plate set in the simmer position.

g. Place a goldfish in each beaker.

Procedure B The Investigation

Perform Procedure B after the beakers have been allowed to equilibrate at least 1 h.

a. Go to any one of the beakers. Record the temperature of the water. Determine the oxygen concentration using the Hach kit. Determine the breathing rate of the goldfish by counting the number of times the gill covers open and close in one minute. (This is a measure of the fish's metabolic rate.)

b. Now repeat step a with each of the other 4 beakers.

c. Record all your data in a table.

Discussion

1. Plot a solubility curve for oxygen in water. Put the temperature on the horizontal axis and oxygen concentration on the vertical axis. On the same set of axes, plot solubility data given in Section 24.1, page 652. Try to account for any differences between your data and the ideal data.
2. How does temperature affect the solubility of oxygen gas in water?
3. Describe the effect of temperature on the metabolic rate of goldfish, as measured by the rate of breathing.
4. Why can even a small temperature change have a significant effect on poikilotherms?

24.6 Investigation

Analyzing the Water in the Classroom Aquarium

You must be able to use water testing kits properly before you go on a field trip. This investigation gives you a chance to use the kits in a controlled situation. Also, it allows you to see how the data from water analysis can be used to draw conclusions regarding water quality.

Materials

an aquarium that has been in use for several months
thermometer (–10°C to 110°C)
selection of Hach water testing kits (oxygen, carbon dioxide, pH, alkalinity, hardness, ammonia, nitrate, phosphate)

Procedure

a. Obtain one of the water testing kits from your teacher. Perform the test on the water in the aquarium. Enter your data in the table that your teacher has prepared on the chalkboard.

b. Return the kit, obtain another, and repeat step a.

c. Continue tests until the teacher announces that time is up. Your teacher will see that each test is performed at least 3 times so accuracy can be checked.

d. Take the temperature of the water.

e. Record all the data in your notebook in a table.

f. Make careful notes on other abiotic properties of the aquarium ecosystem. These should include light intensity, clarity of the water, presence or absence of aeration, and the nature of any debris on the bottom.

g. Make careful notes on the biotic properties of the aquarium ecosystem. These should include a description of the types and abundance of organisms present, including fish, snails, plants, phytoplankton, and zooplankton.

Discussion

1. Account for the results of each test. For example, why was the oxygen concentration 8 $\mu g/g$ instead of some other value? Refer to Section 24.1 if you need help.

2. Explain the effects of each result on living organisms. For example, if you obtained a pH of 9, what effect will that pH have on living organisms? Can the aquarium support a wide range of fish species?

3. Make an overall judgment on the quality of the water in the aquarium. Will it support a wide range of species of organisms? Is it oligotrophic or eutrophic? Is it polluted?

24.7 Investigation

Effects of Organic Matter on Water Quality

In general, excess decaying organic matter has a negative effect on water quality. Discovering just what it does is the object of this investigation.

Materials

small aquaria or pails of at least 5 L capacity (2)
several sprigs of *Cabomba* or other aquatic plant
100-W lamps (2)
can of fish food (preferably the flaky type)
selection of Hach water testing kits (as for Section 28.6)
thermometer (–10°C to 110°C)

Procedure

a. Fill both containers to within 2 cm of the top with water.

b. Add the same quantity of *Cabomba* to each. About 10 sprigs, each 10-20 cm long, is sufficient for each container.

c. Place the containers side by side, each one under a 100-W lamp.

d. Dump the can of fish food (the organic matter) into one of the containers.

e. Wait at least 2 or 3 d. Then test the water in each container using the water-testing kits. Record your results in a table. Note, also, any changes in the appearance of the water. Record the temperature of the water in each container.

Discussion

1. No organic matter was added to one container. What was the purpose of that container?
2. Describe and account for the effect of organic matter on each abiotic factor that was investigated.
3. In general, how does excess organic matter affect water quality?
4. Should a well-balanced freshwater ecosystem have some decaying organic matter in it? Explain.

24.8 Investigation

Long-Term Study of a Mini-Ecosystem

In this investigation you prepare a small, balanced freshwater ecosystem in a jar. Then you change one or more of the abiotic factors in that jar and monitor the effects of that change on the ecosystem. Your teacher may ask you to do this investigation as a home project.

Materials

wide-mouthed jars, with a capacity of at least 1 L (2)
2 L of pond water
sprigs of *Cabomba* or other aquatic plant, each about 10-20 cm long (6)
pond snails (6)
chemicals, test kits, and microscopes as required

Procedure A Preparation of the Mini-ecosystem

a. Fill both jars with the pond water.

b. Add 3 sprigs of *Cabomba* to each jar. Put the same amount of *Cabomba* in each jar.

c. Add 3 pond snails to each jar.

d. Let the jars sit, side by side, in a bright location for 2-3 d. They should appear as similar as possible before you begin Procedure B.

Procedure B   Effect of Abiotic Changes on the Mini-ecosystem

a. Change one or more abiotic factors in one jar. For example, you could aerate the jar. Or, you could increase or decrease the light intensity on the jar. You could add a nutrient element such as nitrogen or phosphorus. You could even add a mixture of nutrients such as lawn fertilizer. You may wish to investigate the effects of oil pollution by adding a film of oil to this jar. Before you do any one of these, predict the effects that it will have on the ecosystem. Then discuss your prediction with your teacher. Your teacher will recommend the quantities of pollutants that you should add.

b. Monitor the effects that the change has on the ecosystem. Make careful notes of changes that occur in both jars. This is a long-term experiment. You may not see any changes for several days. Also, changes will continue to occur for weeks and even months. You may require test kits and a compound microscope during your studies.

Discussion

Prepare a paper on your investigation. It should clearly state the purpose, hypothesis, procedure, results, and conclusions of your investigation. You should also include a discussion of your results and conclusions, explaining how this investigation demonstrates the close relationship of abiotic and biotic variables in an ecosystem.

Highlights

A number of abiotic factors affect aquatic ecosystems. Among these are chemical factors such as oxygen concentration, pH, and phosphate concentration. There are also physical factors such as temperature, light intensity, and the nature of bottom material. If you measure several of these factors, you can often make a judgment on the water quality.

Many biotic factors also affect aquatic ecosystems. Among these are the fish, bottom fauna, bacteria, phytoplankton, and zooplankton. If you study the population densities of such organisms, you can often make a judgment on the water quality. Of course, you must understand the concepts of index species and eutrophication before such judgment can be made. The best judgment of water quality can be made by combining abiotic and biotic data.

Ponds and lakes will eventually disappear due to succession. This process begins with a planktonic community. This community is first succeeded by a submergent vegetation community, and then by an emergent vegetation community. A marsh community may follow and lead, ultimately, to dry land.

Lakes are, on the average, much deeper than ponds. Sufficient light cannot reach the bottom of a lake to cause plant growth, except near the margin. Oxygen gets to the bottom of a deep lake at the time of spring and fall overturn. Overturn also brings nutrients from the bottom of the lake to the surface.

A stream or river is not as self-contained an ecosystem as a pond or lake. Much of the energy that supports life in flowing waters enters by means of terrestrial organic matter.

Marine ecosystems are characterized by their salinity and other abiotic factors such as tides. The tides produce a remarkable zonation of organisms in the littoral zone of a rocky shore.

25 Terrestrial Ecosystems

Terrestrial ecosystems are those ecosystems that are based on land. They may be as small as a handful of earth or a fallen log; they may be as large as the Sahara Desert or the Arctic tundra.

This chapter begins with a study of the largest terrestrial ecosystems, called biomes. Then, in the latter part of the chapter, you will apply the ideas that you learned in the first part of the study of some local ecosystems.

25.1 What Are Biomes?

A **biome** *is a large geographical region of the earth with a characteristic climate and biota.* The **climate** of a region is created largely by temperature and precipitation patterns over the period of a year. The **biota** of a region consist of plants, animals, and other living things.

Usually, a biome is most easily recognized by its climax vegetation. For example, the prairie grasslands is a biome in which the climax vegetation is grass; the coniferous forest is a biome in which the climax vegetation is coniferous (cone-bearing) trees.

The major climatic regions that create the biomes encircle the earth in

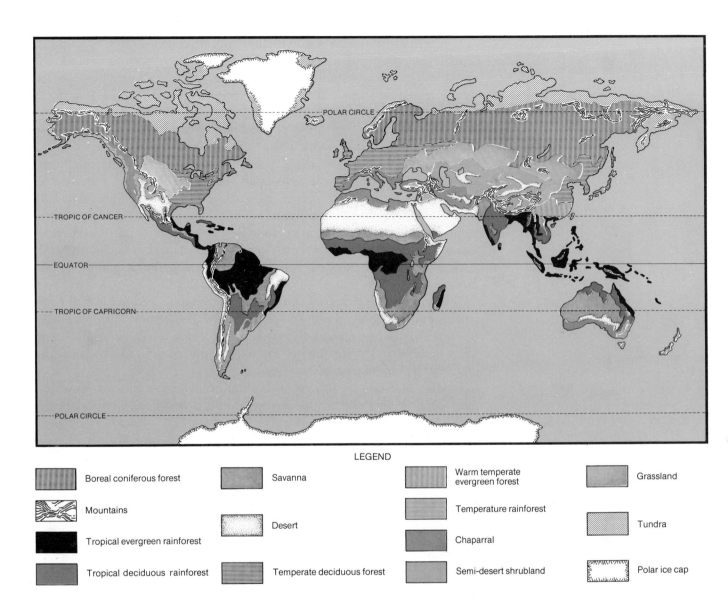

LEGEND

Boreal coniferous forest

Mountains

Tropical evergreen rainforest

Tropical deciduous rainforest

Savanna

Desert

Temperate deciduous forest

Warm temperate evergreen forest

Temperature rainforest

Chaparral

Semi-desert shrubland

Grassland

Tundra

Polar ice cap

wide bands. These bands are broken by oceans, mountains, and other surface features across each continent. As a result, the exact boundaries of biomes are very difficult to define. Geographical regions that are classed as biomes by some ecologists are classed as **ecotones** (transition zones between biomes) by others. Therefore Figure 25-1 should be considered as simply a general guide to the major biomes of the world.

Fig. 25-1 A general guide to the major biomes of the world.

Climatograms

Since climate is such an important property of a biome, scientists have developed what they call climatograms as a simple way of describing climate quantitatively. *A **climatogram** summarizes, in a graph, the monthly variations in temperature and precipitation* (Fig. 25.2).

A climatogram does not include any direct measures of sunlight (radiant energy), wind, humidity, or sky conditions. All these are important variables in an ecosystem. However, temperature and precipitation are clearly related to these other factors. Experience has shown that looking at temperature and precipitation is enough when accounting for the distribution and adaptations of most of the biota. Therefore climatograms are drawn with only temperature and precipitation data.

Many biomes are described in this chapter: tundra, coniferous forest, temperate deciduous forest, grasslands, desert, mountains, and tropical evergreen rain forest. In each case abiotic variables (mainly climate) are described. Then the biotic variables are discussed and related to the abiotic variables. Emphasis is placed on the climax vegetation and its associated animal life. Be sure you keep the ecosystem concept in mind as you study this chapter. Always think of the relationships between abiotic and biotic factors.

Climatograms are used throughout this chapter as a method of comparing biomes. Study Figure 25-2 carefully. Make sure you understand where the climatograms came from. Also, make sure you can interpret them. Then proceed to the Review section where you will find out if you really understand climatograms.

Most of our examples in this chapter are North American. However, the general principles that you learn here apply everywhere. Only the species are different.

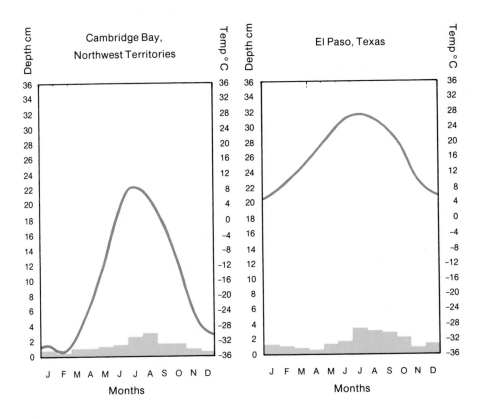

Fig. 25-2 Comparison of climatograms. The line graph represents average monthly temperatures (in degrees Celsius). The bar graph represents average monthly precipitation (in centimetres). Both these regions are classed as desert. Why?

1. Define biome, climate, biota, ecotone.
2. Why are the exact boundaries of biomes difficult to determine?
3. What one factor has the greatest influence on the nature of a biome? Why?
4. How might the vegetation affect the animal life found in a biome?
5. Using Figure 25-2 as a guide, draw the climatogram for each of the three sets of data in Table 25-1.

T=Temperature (°C)   P=Precipitation (mm)

		JAN.	FEB.	MAR.	APR.	MAY	JUNE	JULY	AUG.	SEPT.	OCT.	NOV.	DEC.
Biome A	T	26.7	26.9	26.7	26.7	26.4	25.8	25.3	26.1	26.7	26.7	27.2	26.7
(altitude 83 m)	P	262.0	196.0	254.0	269.0	305.0	234.0	224.0	183.0	150.0	175.0	183.0	264.0
Biome B	T	10.4	12.5	15.8	20.4	25.0	29.8	32.9	31.7	29.1	22.3	15.0	11.4
(altitude 330 m)	P	18.0	21.0	17.0	8.0	3.0	3.0	2.0	2.9	1.8	1.1	1.3	2.1
Biome C	T	-6.0	-5.6	-0.7	6.6	12.3	18.2	20.5	19.7	15.7	9.9	3.1	-3.6
(altitude 280 m)	P	76.0	65.0	72.0	78.0	75.0	81.0	81.0	73.0	79.0	74.0	83.0	87.0

TABLE 25-1   **Temperature-Precipitation Data for Three Biomes**

6. Climatograms only show the monthly changes in temperature and precipitation for a region. However, other information can be scientifically deduced. For each of the following factors, select the most likely biome from your climatograms. Be prepared to explain your choices. Which of the three biomes:
   a) lies closest to the equator?
   b) lies furthest from the equator?
   c) has seasonal changes in temperature?
   d) has the longest growing season?
   e) has the highest total yearly precipitation?
   f) has a measureable snowfall?
   g) has the highest humidity?
   h) has the driest soil?
   i) has the highest soil moisture content?
   j) has the fastest rate of decomposition?
   k) has the greatest abundance of plant growth?
   l) has the greatest variety of plant growth?
   m) has plants adapted for storing water?
   n) has plants which enter a seasonal period of dormancy?
   o) has the greatest variety of animal life?
   p) has animals adapted to reduce loss of body moisture?
   q) has animals which enter a seasonal period of hibernation?
   r) most closely resembles the biome you live in?
7. Try to identify each of these three biomes, using Figure 25-1.
8. Obtain from your local weather office or newspaper the monthly averages of temperature and precipitation for your area. Draw a climatogram. Compare it with those shown throughout this chapter. What biome do you live in?

Fig. 25-3 Repeated freezing and thawing of ice in the tundra soil produces the patterned land surface shown in this aerial photograph.

25.2 The Tundra Biome

Stretching beyond the northern forests to the edge of the Arctic ice cap lies a vast, treeless plain called the **Arctic tundra**. Figure 25-1 shows that this biome surrounds the north pole. Therefore, it is circumpolar. Although a similar climate exists in the far Southern Hemisphere, no corresponding biome is found there. Instead, oceans cover the comparable land area north of the icefields of Antarctica. Isolated tundra landscapes do appear, however, in the high alpine regions of mountains throughout the world.

Abiotic Factors

The tundra climate is extremely cold, windy and dry. Even the wettest months in summer have only about 2.5 cm of precipitation. The very light snowfall is swept by the wind into drifts. Hence the tundra is often called a frozen desert.

Tundra north of the Arctic Circle lies in the "land of the midnight sun". Here the seasonal changes in **photoperiod** (length of daylight) are extreme. This area has 24 h of daylight in midsummer. In contrast it has 24 h of darkness in midwinter when the sun lies too far south to rise above the Arctic horizon. South of the Arctic Circle, latitude determines the photoperiod. Although summer days are long in the tundra biome, the sun is never very intense. This is because the sun's rays strike this region of the earth's surface at a very low angle.

Winter lasts for nine months. The soil and shallow ponds freeze solid. Tundra lakes are covered with a thin blanket of ice. The spring thaw and Arctic summer are crowded into three months. The average growing seasons lasts only 60 d. Summer temperatures are cool and killing frosts are a constant threat to plant growth. The surface of the soil thaws to varying depths, ranging from a few centimetres to half a metre. Below this lies the **permafrost**, soil that never thaws. This frozen layer is 600 m deep in spots. It prevents proper drainage of spring meltwater. As a result, this water collects on the flat land surface, producing vast marshy areas called **muskeg**. This important water reservoir permits plant growth despite the low rainfall. The constant freezing and thawing of soil has moulded the tundra landscape into unique land patterns resembling a patchwork quilt (Fig. 25-3). Vegetative growth is greatly hindered by this steadily heaving, unstable surface.

Biotic Factors

Temperature alone greatly limits the number of organisms which can survive in this biome. Food chains and food webs are simple and easily determined.

Vegetation

During the Arctic summer, tundra vegetation flourishes. It undergoes characteristic habits of growth. Grasses, sedges, and herbs along with lichens and

Fig. 25-4 The summer landscape of the Arctic tundra resembles a grassy plain. Dwarf woody plants are scattered among grasses, sedges, mosses and lichens. Note the total absence of trees.

Fig. 25-5 This Arctic willow is a mature tree. Compare the growth with that of its Southern counterpart.

mosses dominate the biome (Fig. 25-4). Only a few of the woody shrubs such as birches, heaths, and willows, can survive. Their growth is greatly stunted by the chilling Arctic winds (Fig. 25-5). Buds can only grow on branches which spread laterally, close to the warmer soil surface. Grasses and sedges form a shallow, branching root network to anchor in the shifting soil and better absorb available moisture. Mosses and lichens grow hair-like rhizoids rather than true roots. Their ability to absorb and retain water enables them to grow on the driest surfaces. Yet they also thrive in the bogs. Arctic plants are adapted to the low light intensity, but they need long periods of daylight for sufficient photosynthesis. Because the cold climate greatly retards decomposition, tundra soils are nutrient-deficient. Wherever animal wastes provide these nutrients, plants thrive. Lush, green vegetation often marks animal burrows and the nesting sites of waterfowl.

Most tundra plants are **perennials**. They must grow for several seasons before they have stored enough energy to flower. To attract the limited number of insect and bird pollinating agents, large, colourful blossoms bloom in clusters. Since seeds have little chance of germinating, most Arctic plants reproduce by **vegetative propagation**.

Animals

During winter, both predator and prey have coats of white to blend with the snowy landscape. In spring, many of these animals, such as the Arctic hare, grey wolf, lemming, Arctic fox, and ptarmigan, change to darker colours for better camouflage during the summer (Fig. 25-6). To survive the cold

Fig. 25-6 The summer and winter plumage of a ptarmigan. Tundra species which rely upon camouflage must change colour with the seasons.

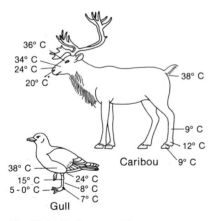

36° C
34° C
24° C
20° C
38° C
9° C
12° C
Caribou   9° C

38° C
15° C   24° C
5 - 0° C   8° C
7° C
Gull

Fig. 25-7 Body extremities such as legs and feet have little protection against snow, ice, or cold water. In most polar animals, these extremities function at low temperatures. Otherwise body heat would be rapidly lost.

winters, the larger mammals and permanent bird residents develop an insulating fat layer. They also have air pockets trapped within long, dense fur or feathers. Ptarmigans and snowy owls even grow extra feathers on their legs and feet.

Some animals reduce body heat loss by having two internal temperatures. The body core is kept at a normal temperature. However, exposed extremities, such as nostrils and feet, function at a temperature which varies with the external temperature (Fig. 25-7). A simple heat exchange system uses both veins and arteries. Blood from the heart, en route to the extremities, warms the cold blood returning from the extremities and is cooled in the process.

Since a large animal has less surface area per unit volume, it loses heat more slowly. A spherical shape also helps conserve body warmth. Thus Arctic species tend to be larger and more round than their southern relatives. Also, extremities such as ears, legs, and tails tend to be shorter, thus further reducing heat loss (Fig. 25-8).

Ptarmigans often roost and feed in snowbank tunnels. Smaller creatures such as lemmings and voles also escape both cold and predators by using tunnels connecting nests beneath the snowdrifts in sheltered hollows. They feed on plant growth or stored seeds. Shaggy muskox paw the snow to uncover a grassy meal. These heavy powerful creatures have a mass between 350 and 400 kg. Continuous summer grazing builds up winter stores of fat (Fig. 25-9). Many of these animals starve when a severe sleet storm leaves an impenetrable layer of ice over their winter food supply. The white Peary caribou searches the snow for reindeer moss and other lichens. Female polar bears shelter in dens where cubs are born in mid-winter. The only hibernating animal is the Arctic ground squirrel. It burrows wherever unusual drainage has left a mound of unfrozen soil beneath the snow. Since refuge below the frost line is almost impossible to reach, most invertebrates winter in a larval or pupal stage which is unharmed by freezing. Aquatic organisms such as rotifers and midgefly larvae remain frozen in ice for months or even years. Yet they resume activity as soon as they thaw!

Many animals migrate before the onset of winter. The autumn skies are filled with departing birds. The Barren Ground caribou move southward seeking shelter and vegetation below the tree line (Fig. 28-10). Water is often a barrier to herd movement; however, the caribou's warm, air-filled coat serves as a lifejacket. These animals swim faster and more easily than other deer.

In spring, this cycle repeats. Huge flocks of geese crowd the skies as they return to the many nesting sites beside freshwater ponds and lakes. Animal activity is geared to the short Arctic summer. Many birds mate before their arrival and begin nesting immediately. In the absence of perching trees, many birds must proclaim territories from stations high in the air. Northern birds tend to lay larger clutches of eggs and the young develop at a faster rate than do southern birds. The ground squirrel emerges from hibernation in May, ready to mate. Young ground squirrels, born in June, are self-sufficient within a month. They reach adult size and prepare for hibernation in late September.

Although tundra food chains lengthen in summer, the total number of species involved remains low. Reptiles and amphibians are rare. A plague of blackflies, mosquitoes, and deer flies emerges from the wet areas. But the diversity of insects is very limited. Bumblebees are plentiful, yet the highly adaptable ants are scarce. Songbirds feed on insects, seeds, and August berries. Waterfowl make up most of the tundra bird life. They feed on aquatic life, even though the limited period of thaw results in low productivity in the ponds and lakes. The largely migratory fish are most numerous in the rivers.

The most critical link in the tundra food chain is a little rodent called the lemming. A female brown lemming can produce a litter of 3-11 offspring monthly from April to September. The young are born in a grassy nursery lined with moss, feathers, and fox moult. They are weaned within two weeks and soon reach a length of about 15 cm. Lemmings lead a hazardous life. They are hunted by foxes, weasels, bears, wolves, and birds of prey. Even browsing caribou abandon their normal lichen meal to munch lemmings crushed beneath their broad hooves. Lemmings attempting to swim streams fall prey to large trout. As the lemmings flourish, the predator population grows (Fig. 25-11). However, nature seems aware of the dangers posed by unlimited animal activity in the fragile tundra ecosystem. The growth cycle peaks every three to four years. Then suddenly the balance is upset. Overcrowding leads to death and disease sweeping through the lemming colonies. Sometimes vast numbers of these rodents make a mass migration across the tundra, leaving a trail of devastated vegetation behind. In their search for living space, the lemmings swim across small ponds and streams. However, they cannot distinguish larger bodies of water like the sea from streams and ponds. Therefore thousands drown after plunging into the sea. Many even leap from high cliffs to enter the sea!

In the short tundra food chain a radical change at any trophic level has

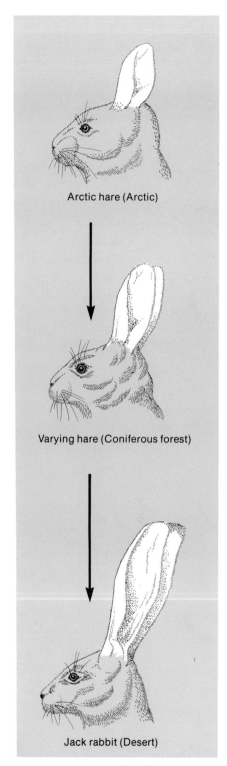

Arctic hare (Arctic)

Varying hare (Coniferous forest)

Jack rabbit (Desert)

Fig. 25-8 Species in colder climates have smaller body extremities to reduce heat loss.

Fig. 25-9 The Inuit name for the muskox is oomingmak, meaning "the bearded one". Dense, insulating wool lies beneath the hairy robe. Sleet storms can form an immobilizing coat of ice on the shaggy animals, making them an easy target for wolves. (Reproduction of art by Glen Loates courtesy of Setaol Incorporated.)

a drastic effect on other levels. Starvation now faces the predators. Northern shrikes, rough-legged hawks, snowy owls, and other birds fly south in search of food. The Arctic fox turns scavenger, following the hunting polar bear. Caribou vanish from ranges they have overgrazed while relieved of attack from wolves. Wolf numbers drop. Until the lemming population is renewed, many birds, particularly owls and hawks, will not breed. With fewer predators, lemming numbers increase and the cycle begins anew.

The Arctic tundra is a harsh environment for plant and animal alike. Yet life has adapted and thrives in seasonal abundance.

Fig. 25-10 Caribou means "shoveller", referring to the use of the splayed hoofs to dig through snow for winter forage. These hoofs act like snowshoes in winter and distribute the animal's mass on the soggy summer terrain.

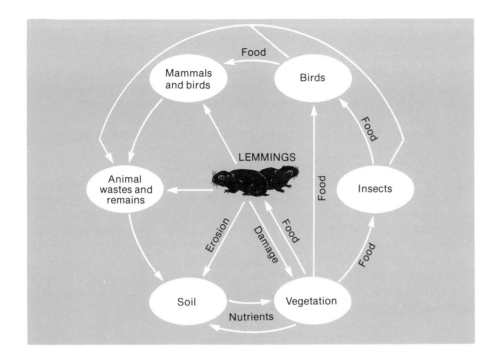

Fig. 25-11 The lemming population affects, directly or indirectly, the food supply of many animals in the tundra ecosystem. Lemming tunnels also modify soil structure and vegetation.

Review

1. Why is the tundra often called the Arctic Desert?
2. How do the hours of daylight (photoperiod) change with the seasons north of the Arctic Circle?
3. Why are summer temperatures low even when the sun shines 24 h a day?
4. **a)** How long is the growing season?
   b) What major problems would a gardener face trying to grow vegetables in the Arctic during the summer?
5. **a)** What is permafrost?
   b) Why is tundra soil unstable for plant growth?
6. **a)** What is muskeg?
   b) How does it form?
   c) Why is muskeg important to Arctic vegetation?
7. Why is tundra soil so nutrient-deficient?
8. How do tundra plants photosynthesize sufficiently in the low light intensity?
9. Explain the effects on Arctic plant growth of each of the following:
   a) cold winds
   b) low summer temperatures
   c) the short growing season
   d) nutrient-poor soil
   e) permafrost
   f) heavy, shifting soil.
10. State three ways in which different Arctic plants have adapted for reproduction in the Arctic environment.
11. **a)** Describe six different adaptations to the winter cold and snow that Arctic animals show.
    b) For each of the following Arctic species, state at least one of these adaptations: Arctic hare, caribou, ptarmigan, lemming, Arctic ground squirrel, muskox, polar bear, snowy owl, Arctic fox, grey wolf.
12. Why are reptiles and amphibians rare in the Arctic?

13. Soil organisms are essential for the recycling of plant nutrients.
    a) What type of decomposers can exist in the Arctic tundra?
    b) How do they survive the long, frozen periods?
14. Thousands of migrating birds return each spring to nest in the Arctic tundra.
    a) What type of birdlife finds the tundra most favourable? Why?
    b) State three special problems that these birds encounter.
    c) Why do many Arctic species lay a larger than average number of eggs?
    d) Explain two other adaptations of these birds.
15. In view of the growing food crisis, is farming the tundra feasible? Could grains, developed for a short growing season, be successfully cultivated in the Arctic? What major problems do you foresee? Discuss.
16. Conservationists throughout North America fear permanent ecological damage as a result of the development of oil and other natural resources in the Arctic tundra. What are the major problems created by surveying or drilling in the tundra? Why is an oil pipeline considered by many to be an environmental hazard? Compare this hazard to that of huge tankers travelling along the Arctic coastline.

25.3 The Coniferous Forest Biome

South of the Arctic tundra is a transition zone or **ecotone**. The physical factors gradually change as one moves from north to south in this ecotone. This change is accompanied by a gradual change in vegetation and animal life. Clumps of dwarf trees, scattered in sheltered nooks, gradually increase in size and numbers. Finally one reaches a fairly distinct **tree line** marking the edge of the **northern boreal forest**. This vast coniferous (cone-bearing) forest stretches across North America and Eurasia (Fig. 25-1). We will examine the features of the North American range.

Abiotic Factors

Northern Boreal Forest

Lying closer to the equator, this region receives more of the sun's radiant energy than the tundra. Average monthly temperatures are higher and the growing season, at 60 to 150 d, is longer. Summer days are shorter, but warmer than they are in the tundra. More important, the ground thaws completely. The winters are not as long or severe; few areas in this biome are without midwinter sun. Snowfall is heavier, but the total precipitation is still low. Summer rains provide most of the moisture.

Centuries ago, a massive glacier covered this region. This ice sheet scraped away topsoil, gouged out countless depressions in the land, and deposited tons of loose rock and earth when it melted. These scars have filled with water to produce the pattern of lakes and swamps which the Russians call **taiga**, meaning "swamp forest" (Fig. 25-12). The shifting ice cover altered the surface features and hampered the formation of an effective river

drainage system. Since low temperatures reduce evaporation, most surface moisture moves downward. As a result, the shallow boreal forest soil is usually water-logged. Melting snow produces spring flooding as rivers overflow.

In this cold, wet surface, earthworms are rare and the action of bacteria is slowed down. Conifer needles and other dead vegetation decompose slowly to form a peaty surface layer instead of mixing with the soil. Water, filtering through this decaying layer, becomes acidic and washes away plant nutrients such as calcium, nitrogen, potassium, and iron. This leaves a grey, acidic, nutrient-deficient topsoil called a **podsol**—from a Russian word meaning "ashes".

Fig. 25-12 Northern boreal forest, or taiga. The great northern woods abound with lakes and swamps.

Oceanic Coniferous Forest

Along the Pacific coast south of Alaska, climate and topography combine to produce an **oceanic coniferous forest** which differs greatly from the northern boreal forest. Prevailing westerly winds moderate the climate resulting in mean monthly temperatures from 2°C to 18°C. The ground is frost-free for a period of 120 to 300 d. In winter, moist westerly winds pass over the warm

Fig. 25-13 Comparison of climatograms. The line graph displays average monthly temperatures (in degrees Celsius). The bar graph displays average monthly precipitation (in centimetres). Note the moderate temperatures and heavy precipitation characteristic of the oceanic coniferous forest. This region extends along the Pacific coast from Southern Alaska to Oregon.

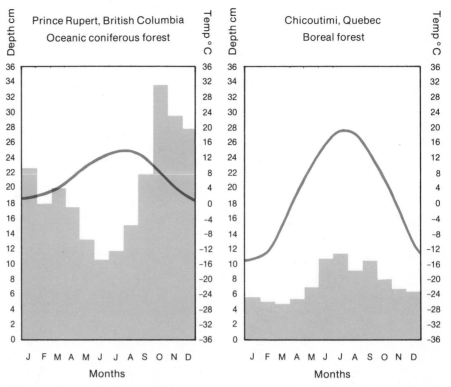

Prince Rupert, British Columbia
Oceanic coniferous forest

Chicoutimi, Quebec
Boreal forest

Fig. 25-14 Evergreen giants of the temperate rain forest. The forest floor is covered with tall ferns and deep cushions of moss.

Japanese current. As they move inland from the ocean, they are forced to rise abruptly over the coastal mountains. In the higher, colder atmosphere the condensing moisture is released as rain or snow. Hence the coastal forest receives as much as 635 cm of precipitation yearly! Compare the climatograms in Figure 25-13. In summer, the prevailing northwest winds are cooled by the northern seas. Although they carry little moisture, these colder air masses cause heavy fogs which soak the forest canopy. They contribute another 130 cm or more moisture to the soil. This abundance of moisture and high humidity in a warm climate nourishes this oceanic coniferous forest. The result is a rich growth of coniferous giants such as the Douglas fir and California redwood (Fig. 25-14).

Biotic factors (Northern Boreal Forest)

As in the tundra, the plant and animal communities of the boreal forest have more limited structures and relationships than those in warmer regions.

Vegetation

One major factor we should investigate is the dominance of coniferous trees throughout this biome. Boreal forest plants must cope with poor soil, low temperatures, and limited rainfall. Although the soil contains enough moisture to support tree growth, it is frozen during much of the year. Humidity is always low. Boreal vegetation must be able to tolerate long dormant periods and use to full advantage the available moisture. **Conifers** (cone-bearing

Fig. 25-15 The nature of the coniferous forest changes in different regions of the biome. Compare the needle structure of these common Northern American species.

Red pine

White pine

Hemlock

White cedar

Tamarack

White spruce

Black spruce

Balsam fir

Jack pine

Douglas fir

Sitka spruce

Lodgepole pine

trees) thrive in this environment because they are well adapted to dry conditions. Their leaves are modified into needles or scales wrapped in a thick cuticle (Fig. 25-15). This waxy outer skin greatly restricts water loss by

Fig. 25-16 The wolverine is a ravenous predator. Its common name, "glutton", is earned from its enormous appetite.

evaporation from the inner leaf cells. These needles can also withstand freezing. This adaptation helps conifers survive periods of frost and drought.

The shape and structure of coniferous trees are ideally suited for northern winters. Heavy masses of snow cannot collect on the small surface area of the needles. The flexible branches bend, causing clumps of snow to fall to the forest floor. The northern summer is too brief for most deciduous trees. They would lose more stored food when their large leaves were shed in the fall than they could replace during the short growing season. Also, deciduous leaves decay quickly. Therefore essential plant nutrients released by the decay would be leached from the soil before the trees could reclaim them for new leaf growth in the spring.

Most conifers are **evergreens**. In other words, they retain their leaves during the winter. Dead needles are gradually shed and replaced throughout the course of the year. The fallen needles decay slowly because their resin content and thick outer covering resist bacterial action. Thus, evergreens conserve valuable nutrients and are ready to photosynthesize whenever conditions permit. If a conifer's bark is damaged, a sticky resin produced by the tree covers the wound. This prevents attack by fungi or bacteria.

The dense evergreen canopy restricts forest floor growth to shade-loving plants such as ferns, mosses, and a few herbs. Fungi, the vital decomposers, grow throughout the needle-laden soil.

Animals

Boreal forest animals must survive the long, cold winter in which snow covers the frozen ground. Foxes, wolves, and moose have thick winter fur. The snowshoe hare turns snow-white and travels about on built-in "snow-shoes"—large tufts of fur which cover its feet. One of its predators, the

Fig. 25-17 The moose, symbol of the boreal forest, is the largest of all deer species. A grown moose must eat from 3 600 - 4 500 kg of vegetation to survive the harsh northern winter.

wolverine, is the swiftest mammal in the winter forest (Fig. 25-16). Spreading toes permit this stealthy hunter to run over deep drifts without sinking. Moose wade through deep snow on stilt-like legs (Fig. 25-17). Though normally solitary, moose gather in winter to trample snow so that they can reach tree shoots, brushwood and twigs. Starving moose will even chew conifer bark. By spring, the hungry survivors are thin from their winter ordeal.

Air spaces trapped in plant undergrowth between the snow and the soil create a microclimate in which the temperature never drops more than one or two degrees below freezing. Here lemmings, voles, and other rodents are active all winter long (Fig. 25-18). Chattering red squirrels search for hoarded supplies of conifer seeds, obtained by stripping away cone scales. Hungry squirrels often damage young trees by eating conifer buds. Porcupines chew away conifer bark, especially in the uppermost branches.

The woodchuck hibernates as long as eight months in a burrow or hollow log. Its metabolism is reduced to the minimum rate necessary to keep body cells alive. Heart rate, normally 200 beats/min, drops to a mere 4-5 beats/min. Breathing may only occur twice a minute, and stored body fat is consumed very slowly. Body temperature falls. However, if the outside temperature drops too low, nervous response awakens the animal. It resumes activity for a few hours, and normal blood circulation is restored. Chipmunks hibernate too, waking at intervals to feed on stored nuts and seeds. Bears also enter a winter sleep. However it is not considered true hibernation because the animal's body temperature falls only slightly.

The choice of food is greatly limited in any season. Only about 50 species of birds can feed on the tough resinous conifers. Most seed-eaters have strong jaw muscles which control short, sturdy beaks fitted with sharp cutting edges (Fig. 25-19). Conifer needles contain little nutrient value. Therefore, to survive on this food, birds such as the blue grouse and spruce grouse must eat constantly (Fig. 25-20). Grouse chicks cannot live more than a few hours without the insect pupae which they seek independently. These fragile balls of fluff often starve in cold, wet weather because they cannot leave the protecting warmth of the hen. Hungry chicks that do venture forth quickly perish from exposure. Some baby birds resist the spring cold by falling into a coma until the returning parent restores warmth to the nest.

Very few insect species can feed on coniferous growth but every part of a conifer is subject to attack by some type of insect. Forest areas dominated by one or two tree species are especially vulnerable. Outbreaks of spruce budworm and larch sawfly have wiped out vast areas of trees. In summer, the swarms of blackflies, deer flies, and mosquitoes that thrive in the northern forest greatly irritate larger animals. Moose escape by submerging in lakes and marshes where they browse on aquatic plants. Insects have two main seasons—a short active summer of reproduction and a long, dormant period when they winter as larvae or nymphs in bark crevices beneath the soil.

Fig. 25-18 The tiny shrew collects dormant insects from the soil litter. It must eat constantly to survive. A person with a matching appetite would have to consume 230 kg of food daily! The other rodent is a vole. It chews vegetation under an insulating blanket of snow.

Fig. 25-19 The crossbill's beak can shear through tough cone scales. The bird can then use its long tongue to reach the seeds inside. Penetrating the densest foliage, it hangs like a feathered acrobat while feeding.

Fig. 25-20 The spruce grouse dines exclusively on mature spruce trees. One bird may spend days robbing a single tree of its needles.

The birds of prey that patrol the forest sky include owls, hawks, falcons, ospreys, and eagles. They are equipped with keen vision, deadly talons, and strong, hooked beaks (Fig. 25-21). The smaller carnivores of the boreal forest—minks, martens, weasels, and wolverines—all belong to the weasel family. They hunt rabbits, rodents, birds, and insects. They do not hibernate, but grow protective winter coats (Fig. 25-22).

Like the lemming of the tundra, the snowshoe hare plays a major role in the boreal forest food chain. Over a ten-year period, the number of hares increases. Predators raise large families during this time of plenty. Suddenly, the hare population falls. The lynx is hardest hit because, in some areas, the hare provides 70% of its diet (Fig. 25-23).

One adaptation, not yet mentioned, is the well-developed hearing and vocal system of forest animals. In a habitat where trees limit visibility, vision is often a secondary sense. Most animal "music" such as birdsong is designed to attract mates or to proclaim territorial domains. But to the listener it conveys the magic and mystery of the northern woods.

Review

1. Describe how the boreal forest region differs from the Arctic tundra with respect to each of the following:
   a) average monthly temperatures,
   b) length of growing season,
   c) duration of sunlight throughout the seasons,
   d) intensity of sunlight,
   e) depth of snowfall,
   f) total yearly precipitation.
2. a) What is boreal forest?
   b) Why did it form in this part of the continent?
3. a) Describe the soil of this region.
   b) What major problems would farmers face in this region?
4. How does the climate of the oceanic coniferous forest zone differ from the climate of the boreal forest zone?
5. a) State three problems faced by boreal forest plants.
   b) Describe six features of conifers that enable these trees to thrive in the northern boreal forest.
   c) Give two reasons why deciduous trees are not suited to this region.
6. a) Explain two factors that restrict the growth of smaller plants under the canopy of a coniferous forest.
   b) What are the main decomposers of this region?
7. Explain how each of the following animals is adapted to winter in the coniferous forest: snowshoe hare, wolverine, moose, woodchuck, chipmunk, bear, insects, weasel.
8. a) What do each of the following animals feed on during the winter: moose, lemming, red squirrel, porcupine, spruce grouse, wolverine?
   b) Why is the food supply for most herbivores limited in the boreal forest at any time of the year?
9. a) What types of insects dominate this region during the summer?
   b) Why do they thrive in the taiga?

c) Why is a forest region consisting of only two or three different tree species very vulnerable to insect damage?

10. a) How are seed-eating birds adapted for feeding?
    b) Why are grouse chicks endangered by cold spring weather?
11. a) Explain how the snowshoe hare population affects the numbers of predators in the boreal forest.
    b) Which predator is affected most? Why?
12. Why do forest creatures need well-developed hearing and vocal systems?

25.4 The Temperate Deciduous Forest Biome

Along the southeastern fringe of the North American boreal forest, deciduous trees invade in ever-increasing numbers forming a **deciduous-coniferous ecotone**. This mixed growth gradually blends into the next major biome southward—the **temperate deciduous forest**. As Figure 25-1 shows, this biome is not continuous. It is also found in western Europe, eastern Asia, and in small areas of South America. We will examine the North American forest.

Abiotic Factors

This region is largely restricted to the eastern half of this continent where the average annual precipitation varies from 75 to 125 cm. This precipitation is fairly evenly distributed throughout four, well-defined seasons. The winter snow is not as deep or long-lasting as in the boreal forest. The climate is moderate. Relative humidity is high during the growing season which may last for more than six months.

The short winters are cold enough to greatly reduce both growth and photosynthesis. To increase efficiency, deciduous trees enter a dormant period and shed their leaves. One hectare of forest floor may be carpeted with more than 25 000 000 leaves each fall! These leaves and other matter rapidly decompose on the moist ground to produce a rich layer of humus on the "brown earth" soil. A balance between the downward drainage of rain

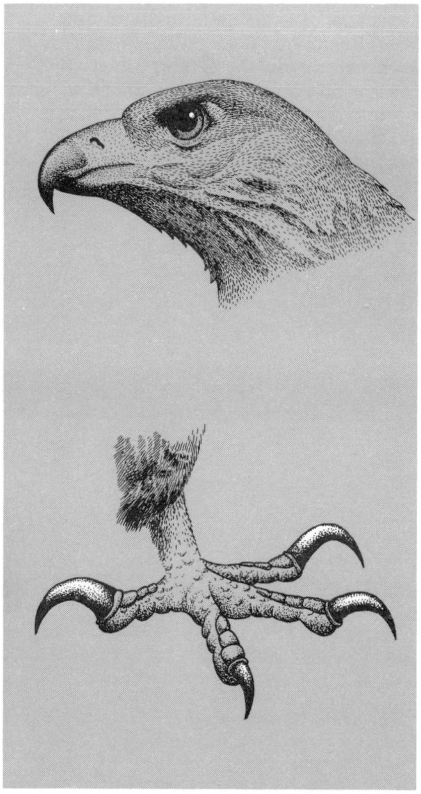

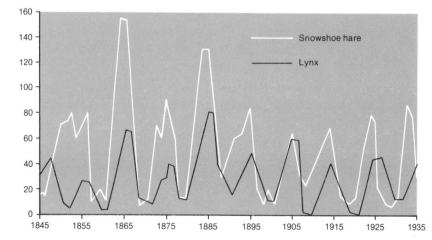

Fig. 25-21 This keen-eyed hunter has sharp claws for grasping prey, and a strong, hooked beak for tearing meat.

Fig. 25-22 The short-tailed weasel is called an ermine when its red brown summer coat turns to winter white. This little predator may claim a hunting territory of 340 000 m^2.

Fig. 25-23 Snowshoe hare populations peak every 10-11 years and then fall sharply. Records kept by Canada's Hudson Bay Company show that lynx populations have corresponding fluctuations.

Fig. 25-24 Some common deciduous trees. Beech-maple forest dominates the north central regions of this biome, while oak and hickory make up much of the western and southern forest. Florida and the Mississippi delta support a magnolia-oak forest.

American beech   Trembling aspen   Sugar maple

Red oak   Black willow

Sycamore   White birch   Shagbark hickory

Fig. 25-25 Stratification in a deciduous forest. How many levels can you observe?

water (**percolation**) and the upward movement of water (**capillarity**) circulates soil nutrients.

Biotic Factors

Since the arrival of the early settlers, this biome has been greatly altered. Only patches of the climax forest remain (Fig. 25-24). Large carnivores have been wiped out or forced to retreat into a few remaining pockets of wilderness.

Vegetation

Various plant species grow to different levels, or **strata**, within the forest (Fig. 25-25). The upper canopy of taller trees receives full exposure to the sun. Broad deciduous leaves permit maximum absorption of light energy. A small oak tree produces more than 100 000 leaves with a total surface area about the size of two tennis courts! Some sunlight filters through to the next layer, which contains smaller trees. Beneath these, a shrub layer grows. Finally, at ground level, ferns, mosses, and other small plants compete for the remaining light. The floor of an oak forest receives only about 6% of the noonday sunlight. Hence, most of the smaller plants flower very early in the spring before the new leaves emerge on the trees above (Fig. 25-26). By the time a new upper canopy has blocked off sunlight, these ground plants have stored photosynthesized food in roots or underground stems. After releasing seeds, these plants become dormant until the following spring.

Deciduous leaves are easily damaged by frost and dried out by winter winds. To preserve tree moisture, leaves are shed in the fall. Growth is retarded as the trees depend on food reserves stored in roots, trunks, and branches. The long growing season ensures new growth of leaves and seeds each year, although growing conditions affect the size of the annual seed crop. Some trees, such as the horse chestnut and apple tree, produce colourful, fragrant blossoms each spring to attract pollinating insects (Fig. 25-27). Others, like the beech and the oak, depend on wind-carried pollen to fertilize their inconspicuous flowers. The resulting bounty of fruits and seeds is harvested by many forest dwellers in preparation for winter. In autumn, many creatures feast upon the fruiting bodies of fungi growing on tree stumps and damp soil.

Fig. 25-26 Trilliums bloom on a sunny hillside in early spring. The upper canopy has not yet developed.

Fig. 25-27 These bright, fragrant chokecherry blossoms attract pollinating agents in the spring.

Fig. 25-28 The woodpecker uses two toes at the back of its four-toed foot to brace itself on the side of a tree.

Animals

Forest animals are adapted in structure, function, and behaviour to live among trees. Here they find shelter, protection, and nesting sites. From the branches they can sight enemies and proclaim territorial boundaries. Many dwell in the rich humus beneath the trees. Others simply take advantage of the shade, moderate temperatures, and high humidity of the forest.

Tree dwellers are well equipped for climbing. Woodpeckers have opposing toes to maintain balance (Fig. 25-28). Squirrels have sharp claws and some even have a built-in parachute as well as a bushy tail for balance. Tree frogs cling to the bark using suction discs on their toes (Fig. 25-29). Snails and slugs use a slimy "foot". White-tailed mice and opossums use their tails for climbing and grasping, much as monkeys do.

Unlike conifers, deciduous trees are a major food source for animals. The largest herbivores include the wapiti (elk) and deer. Deer prefer to feed on leaves in woods bordering meadowland rather than in deep forest (Fig. 25-30). They winter in small herds that break trails and form a "deer yard" for feeding. The buds and seeds of trees provide a year-round supply of concentrated protein. Berries and nuts are found far into the winter when other food is scarce. Rabbits and smaller rodents nibble at herbs, tree bark, and small plant growth. Beavers, the largest North American rodents, chew the inner bark of small branches of trees such as poplars, birches, and willows. In return for proteins and sugar, bees and other insects act as pollinating agents for blossoms. But two major groups of insects are highly destructive to deciduous foliage. Leaf chewers, such as caterpillars and beetle and fly larvae, use biting and chewing jaws to eat the leaf tissue (Fig. 25-31). Sap-suckers, such as aphids, have piercing and sucking mouthparts designed to suck fluids from plants. These insects provide food for birds, tree frogs, shrews, toads, and spiders. Larger carnivores include bobcats and members of the weasel family. Birds of prey such as owls and hawks are still widespread. Some omnivorous feeders will eat many types of food. The red fox (Fig. 25-32) is a crafty, nocturnal (night) hunter that prefers mice but will settle for birds, large insects, fish, eggs, and even berries and grass. Raccoons, skunks and opossums also have varied menus. But the most resourceful eaters are the intelligent, sociable crows. Their hardy digestive system can manage almost any form of food.

The greatest feeding activity occurs on the organically rich forest floor. One square kilometre of soil litter may be home to more than 120 different species of invertebrates, each one a specialized eater! The surface is alive with hungry spiders, beetles, snakes, toads, and small mammals—all links in complex food webs. Essential nutrients are recycled by fungi, earthworms, and bacteria.

This abundance of food disappears with the coming of winter. Some of the birds remain to feed on seeds, insect pupae, or insect eggs. However, most birds migrate south. Amphibians and reptiles become dormant. Forest activity will renew in the spring breeding season. Yet, at any time of year, this biome presents a complex study of the diversity and abundance of life.

Fig. 25-29 The tree frog uses suction discs on its toes to maintain balance in its leafy environment.

Fig. 25-30 An ecotone or transition zone between a deciduous forest and a meadow. Why would you expect to find a greater variety of plant and animal species here than in either the woods or the open meadow?

Fig. 25-31 Hundreds of tent caterpillars come from this nest to strip the host tree of its foliage. At night the gorged feeders return to the nest.

Review

1. Compare the temperate deciduous forest and the northern boreal forest with respect to each of the following factors:
   a) average monthly temperatures,
   b) length of growing season,
   c) intensity of sunlight,
   d) depth of snowfall,
   e) total yearly precipitation.
2. a) Why is photosynthesis greatly reduced in both evergreen and deciduous trees during the winter months?
   b) Explain two reasons why deciduous trees lose their leaves during winter dormancy.
   c) How does this leaf loss affect the forest soil?
   d) How do deciduous trees survive the winter period without their food manufacturing system?
3. a) List the four levels or strata of a deciduous forest.
   b) Why do many smaller woodland plants, such as the trillium, flower very early in the spring?
   c) How do the blossoms of an insect-pollinated tree differ from those of a wind-pollinated tree?
4. a) Describe three ways in which the plants of the deciduous forest benefit the animal life.
   b) How are each of the following animals adapted to tree life: woodpeckers, squirrels, tree frogs, snails, mice, opossums?
   c) What do each of the following forest dwellers feed upon: deer, rabbits,

Fig. 25-32 The red fox is a resourceful feeder. When eggs are abundant, foxes bury them only to dig them up months later when the food supply dwindles.

beavers, caterpillars, aphids, shrews, toads, foxes, skunks, crows? Define the niche of each.

5. **a)** Which level or stratum of the deciduous woods would you expect to support the greatest diversity of animal life? Explain.

   b) List three ways in which forest animals respond to the coming of winter.

6. **a)** Compare the temperate deciduous forest and the northern boreal forest as a summer animal habitat with respect to each of the following: availability of food, shelter from the elements, protection from predators.

   b) Compare the same factors during the winter period.

25.5 The Grasslands Biome

Moving westward in North America one notices that the temperate decidu-ous forest gradually thins and merges into a **savanna ecotone**. Here, the trees are scattered over an area dominated by grasses and sedges. Eventually all remnants of the forest disappear, leaving miles of rolling **prairies** or **grass-lands**. This type of biome is referred to as the "veldt" in South Africa, the "steppe" in Asia, and the "pampas" in Argentina (see Fig. 25-1).

Abiotic Factors

This region lies within the same latitudes as the deciduous forest. Hence, the seasonal changes and radiant energy supply are similar. However, both

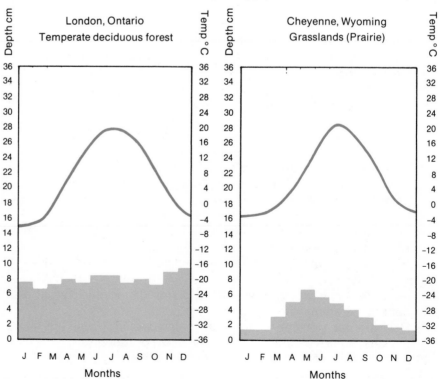

Fig. 25-33 Comparison of climatograms. The line graph represents average monthly temperatures (in degrees Celsius). The bar graph represents average monthly precipi-tation (in centimetres). These two regions both have a mean annual temperature of 7.5° C. Precipitation is the critical factor that determines the difference in vegetation.

seasonal and daily temperature fluctuations are more pronounced in the prairies. The critical factor that produces the drastically altered vegetation is the diminished rainfall. The continental pattern of air circulation from west to east produces *decreasing and irregular precipitation*, combined with an *increasing rate of evaporation* from the soil surface. The annual rainfall, 25-75 cm, is sufficient for many grass species, but is too low for tree growth (Fig. 25-33). Nor can trees survive the frequent droughts which can be severe and prolonged. Also, occasional fires kill seedling trees while the grasses quickly recover.

The **chernozem** (Russian for "black earth") soils, characteristic of the prairies, are the most fertile in the world. The rapid decay of short-lived grass plants forms a deep layer of dark humus. The upward movement and evaporation of water leaves the topsoil rich in nutrient deposits of calcium and potassium.

Biotic Factors

Until the nineteenth century, pronghorns and vast herds of bison wandered the prairies (Fig. 25-34). The buffalo, constantly on the move, seldom overgrazed an area. Instead, they moulded the prairies, killing trees by rubbing against them to remove shedding fur, and stripping off bark with their horns, or browsing on seedlings. Then white settlers, with their weapons,

Fig. 25-34 Enormous herds of bison — the largest North American animal — once roamed the heartland of this continent. They were slaughtered by the thousands when the white man invaded.

fences, and agriculture, changed the face of the grasslands. Less than 100 years after the first settlers came, the buffalo had all but vanished. Longhorns and then modern beef cattle overgrazed and upset the natural balance of the range. The plow produced a dust bowl in the drier plains. Animals such as the badlands-grizzly and white-plains-wolf were hunted to extinction. Today, even the prairie dog is endangered. Only a few small areas remain where the earlier grassland ecosystem can be seen and studied.

Vegetation

The difference in rainfall produces three distinct types of grassland (Fig. 25-35). Moderate rainfall makes the eastern prairies a **tall-grass zone**. Here the plants develop a solid mat over the ground. Further west, the drier central grasslands support **mid-grasses**. Most of these species are bunch grasses that grow in well-spaced clumps. In the arid western plains, a region of high winds and low humidity, the sod-forming **short grasses** grow. Their shallow roots absorb moisture from the upper soil layer. However they do not penetrate the permanent dry zone beneath. Herbs and nitrogen-fixing legumes flourish among the grasses. Tree growth, largely cottonwood, is generally limited to stream valleys or low mountain ranges.

There are three strata in grassland vegetation: the roots, the growth at

Fig. 25-35   From prairie to plain. Soil depth and rainfall combine to produce three distinct types of grassland between the eastern forest and the western desert.

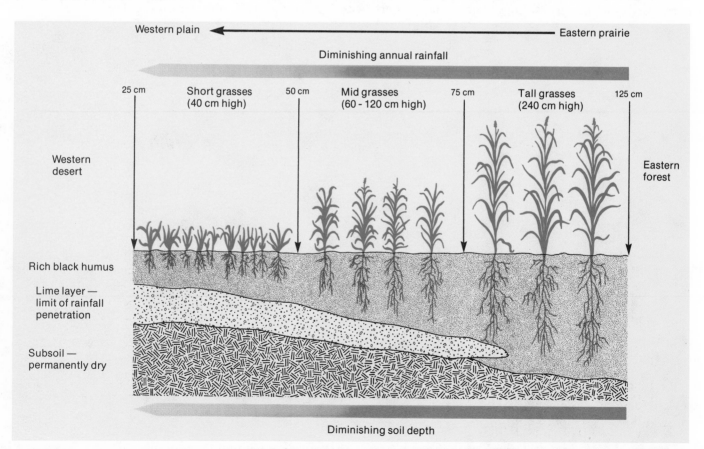

ground level, and the taller foliage. At least half of the total growth of each plant lies beneath the soil. In fact, the mass of the roots of healthy plants is much more than that of the portion above ground.

Animals

Grassland animals have many fascinating adaptations to open country. Long distance vision is very important for both predator and prey. The eyes of grazing animals are usually located well above the snout to enable the animal to see above the grass while feeding. Smaller creatures, such as the ground squirrel, stand up on their haunches to peer over the vegetation (Fig. 25-36). Others, such as the kangaroo rat hop up and down on well-developed hind legs. There are no trees to provide hiding places. Therefore some animals rely on camouflaging colouration for protection. Sensing danger, they remain motionless in the deep grass to escape notice. If the enemy approaches too closely, these creatures suddenly flee by running, hopping, or flying. By startling the predator with this flurry of motion, the would-be victim gains a head start. Many prairie species rely on speed for survival. Pronghorns (Fig. 25-37) have sturdy legs, large lungs and wind-pipe, and a heart double the expected size. They can race with bursts of speed reaching 100 km/h! Using 8 m leaps, jack rabbits bound across the prairie at 70 km/h, easily clearing obstacles more than 2 m high. Many creatures escape from predators by diving into underground burrows. These shelters also protect smaller, temperature-sensitive residents from the sur-face heat or cold.

Prairie dwellers must believe there is safety in numbers because they typically live within a herd or colony. Any alarmed member can alert the others to danger. One vast prairie dog "town" in Texas used to cover 65 000 km² and was home to more than 400 000 000 of these rodents! Such communities supported a host of predators before humans began an exter-mination campaign which destroyed large parts of the prairie dog nation (Fig. 25-38).

The large number of insects and seeds attract a wide variety of birds.

Fig. 25-37 Speed means survival for the pronghorn. When alarmed, the white rump hairs bristle, flashing a danger signal to the herd. (Reproduction of art by Glen Loates courtesy of Setaol Incorporated.)

Grassland birds must be strong fliers to fight the high winds which sweep the prairies. They hide their nests in the tall grass. More than a hundred types of grasshoppers thrived in the northern Great Plains when the first settlers arrived. When these grasshopper populations reach plague proportions, they cause extensive crop damage. In the drier grasslands, ants replace earthworms in the important role of mixing and aerating the soil.

Many animals leave the grasslands in winter, when strong prairie winds form deep snowdrifts. Bison used to move south with migrating birds. Pronghorns shelter in woodlands; many other animals hibernate below the frost line.

Although the grasslands are vitally important to us, probably no other biome has received greater abuse. Thousands of hectares are still being converted into barren desert through our failure to understand or respect the ecology of this area.

Review

1. **a)** Explain why the grassland biome supports a different type of vegetation than the temperate deciduous forest.
   b) State three factors that discourage tree growth in the prairies.
   c) Why are prairie soils darker and richer in nutrients than deciduous forest soils?
2. **a)** Compare the three types of grassland vegetation observed while moving from the eastern prairies to the western plains. What causes this change?

Fig. 25-38 A grassland food pyramid. The mass of primary consumers, feeding directly on producers, is far greater than that of succeeding trophic levels. The top predators, forming the peak, have a relatively small mass. Many animals occupy several levels. Try to identify the animals illustrated. Some are endangered species.

b) List the three strata found in grassland vegetation.

c) Where is the very limited tree growth found in the prairie region?

3. a) How did the vast herds of buffalo help to maintain the prairies?

b) Why are the modern beef cattle damaging their ranges?

c) State three ways in which human beings have affected the grasslands.

4. Explain eight different adaptations to life on the open range demonstrated by prairie animals.

5. a) Why are many birds attracted to the flat, open prairies?

b) What two major problems do they encounter?

6. a) Why are ants beneficial to the drier grasslands?

b) Which grassland creatures have most benefited from human development of the prairies? Why?

7. The true extent of human impact on prairie wildlife can best be understood by researching animal populations, past and present. Visit your school or community library and seek answers to these questions:

a) What is the role of the prairie dog in the grassland ecosystem? Why is it the target of extermination campaigns? How many interrelated species will vanish with it?

b) How were the ranks of the following species drastically reduced: prairie chicken, whooping crane, golden plover, trumpeter swan, and

Fig. 25-39 Most cacti have a rounded shape to minimize surface area exposed to the sun. Sharp spines protect fleshy parts from browsing animals. Some cacti are folded like an accordian and can quickly expand by soaking up rain water.

the Eskimo curlew? What is the greatest threat now facing prairie waterfowl? Why is the bobwhite quail declining in numbers?

c) Why did the white plains wolf and badlands grizzly bear become extinct before the turn of the century? Why are each of the following species on the verge of extinction today: the prairie falcon, the tule elk, the black-footed ferret, and the kit fox?

25.6 The Desert Biome

Along the western edge of the drier plains lies one of the most arid regions on earth—the **desert biome** of North America. Vast stretches of desert also extend through Africa, Asia, Australia, and the southern tip of South America (see Fig. 25-1).

Abiotic Factors

The range of latitudes produces two types of desert on this continent—the northern "cool" deserts of the Great Basin and the "hot" southwestern deserts. While summer temperatures in Death Valley have soared as high as 57° C (in the shade), the winter season brings snow and cold weather to the Great Basin desert of the United States. But the real impact of temperature lies in the tremendous fluctuations during each 24 h period. The desert sands receive almost 90% of total available solar radiation because there are no clouds, water vapour, or canopies of vegetation to absorb the sunlight. However, at night, temperatures drop rapidly as 90% of this accumulated surface heat is lost by radiation.

Yet lack of water, rather than heat, produces deserts. In North America, the deserts are located by the "rain shadow" of the western coast mountains. After a long drought, the unpredictable moisture supply for an entire year may fall during one short period, usually in the form of a thunderstorm or cloudburst. Since the sun-baked ground cannot absorb much moisture, most of it drains away in surface runoff. The relative humidity of desert air averages less than 30% at midday.

Steady winds erode rock into sand and stir up dust storms which scour the land surface. Soil organisms, such as earthworms, cannot endure the dryness. In addition, the scanty desert plant life does not provide much humus.

Biotic factors

Water is the key to desert life. Many plants and animals have survived because they are able to develop and reproduce rapidly during any period of rain. Dew formation provides an important source of moisture for some organisms.

Vegetation

There are three main types of desert plants—annuals, succulents, and desert shrubs. Since **annuals** live only for one season, each generation must produce enough fertile seeds to continue the species. The dormant seeds will only germinate when enough moisture is available to enable rapid plant growth. **Succulents**, or "juicy" plants, such as cacti, survive long droughts by storing water (Fig. 25-39). Cacti do not lose water through leaf stomata. Their green stems, which photosynthesize, are covered with a thick, rubbery cuticle to protect stored water. Spines help to shade the cacti from the direct rays of the sun. They also reduce surface air currents that cause water evaporation. The small, thick leaves of desert **shrubs** have sunken stomata and waxy leaf cuticles to reflect heat and retard water loss. The competition for water keeps desert plants well spaced (Fig. 25-40). The roots of some plants produce toxins to kill any competing plants that invade their growth sites.

Animals

Desert animals are adapted to conserving body moisture which may be lost in any of the following ways: evaporation from the body surface, exhalation from the lungs, and elimination through excretion of body wastes. Many desert species are **nocturnal**—they confine their activity to the cool desert nights. Some, such as the scorpions (Fig. 25-41), are physically equipped to dig burrows which provide cool storage sites for food and a retreat from enemies (Fig. 25-42). Reptiles and scorpions have a nearly impermeable

Fig. 25-40 Most desert plants send out shallow, widely-branching roots to rapidly soak up any traces of moisture. Others develop long tap roots which reach underground water sources more than 30 m deep. Animals digest desert fruits and eliminate the seeds in nutrient-rich droppings where new plants germinate.

Fig. 25-41 Desert scorpions have enlarged claws for digging and capturing prey. They often consume a daily ration of insects equivalent to their own body weight.

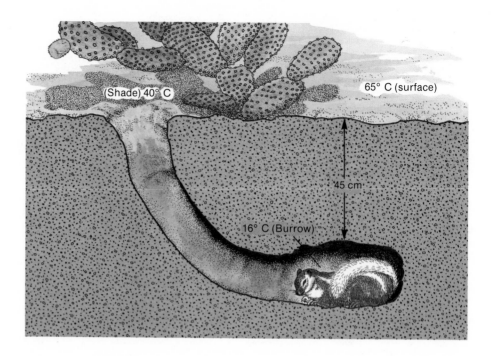

Fig. 25-42 The insulating air in the underground burrow maintains a fairly constant temperature, day or night. Moisture from the animal's breath raises the relative humidity, greatly reducing water loss from the body surface of the inhabitant. Animals that cannot burrow seek out any shady patch.

(Shade) 40° C

65° C (surface)

45 cm

16° C (Burrow)

Fig. 25-43 Some lizard species can stride rapidly for short distances across hot desert sands using only their hind legs. The long tail is raised to maintain balance.

outer covering to minimize surface evaporation. Lizards and snakes have no sweat glands. Many spiders and insects are protected by a waxy **exoskeleton**. All desert creatures try to minimize body contact with the hot sand (Fig. 25-43).

When **homeotherms** ("warm-blooded" animals) sweat or pant, they are simply using a built-in cooling system to keep body cells at or near the normal body temperature. Severe dehydration in hot, dry air causes "explosive heat death". Lost body moisture is replaced by water from the blood. Eventually the blood becomes so thick that it cannot circulate fast enough to transfer metabolic heat to the skin surface. The body temperature rises, causing rapid death. The body temperature of **poikilotherms** ("cold-blooded" animals) is regulated by their surroundings rather than by an internal mechanism. Although such creatures are often called "cold-blooded", a lizard basking on a desert rock measuring 38° C has anything but "cold" blood! To maintain a functional temperature, such animals must alternate from sun to shade. Ten minutes of direct exposure to the hot desert sun can kill a rattlesnake.

Birds have no sweat glands in their skin and feathers provide good insulation from the sun. Since most birds have higher body temperatures than other animals, they can withstand the heat more readily. Flight cools their bodies as they escape the hot surface in search of food or water (Fig. 25-44).

To conserve body moisture, the pocket mouse and ground squirrel enter a deep summer sleep called **estivation**. The blood vessels in the conspicuously large ears of many desert dwellers help radiate body heat (Fig. 25-45). Most desert animals pass body wastes in a highly concentrated form to further reduce loss of body moisture. The metabolism of the kangaroo rat actually breaks down its food to yield water for the animal!

The rainy season brings water and life to the face of the desert. It is a time for blossoms and birth as all life struggles to continue itself in this endless cycle of moisture and drought.

Review

1. **a)** What major climatic factor produces desert regions?
   b) Why are daily temperature fluctuations so extreme in the desert?
   c) Give three reasons why desert soils are so dry.
2. **a)** Name the three main types of desert vegetation. Explain how each is adapted to desert conditions.
   b) How does competition for water affect the growth of vegetation?
3. **a)** State three ways in which animals lose body moisture.
   b) Explain eight special adaptations that aid in survival, displayed by different desert creatures.
   c) Give four advantages provided by an underground burrow.
4. **a)** Explain why prolonged exposure to the desert sun can kill homeotherms and poikilotherms.
   b) Give four reasons why birds are the most active desert creatures during the day.
5. **a)** What is estivation?
   b) When is the normal season of birth in the desert? Why?
6. **a)** A brief rainstorm will not stimulate germination of desert annual wildflower seeds. Yet, more than 50% of these seeds begin to sprout after a heavy rainstorm. Furthermore, this water must come from above, not from beneath the soil. Why is this remarkable adaptation so critical to desert plants?
   b) Some desert plants, such as the night-blooming cereus, have evolved flower petals that open at night. The blossoms of such plants are usually white and highly fragrant. What purpose could this unusual behaviour serve?
   c) Many desert plants have a special form of photosynthesis. They absorb carbon dioxide through the stomata at night rather than during the day. Why is this an advantage?
7. Desert animals show many remarkable adaptations. Investigate the habits of the following and explain how they are suited to the desert.
   a) Why are kangaroo rats so important to life in the desert?
   b) What types of birds live in the desert? Where do they nest? Does the climate affect their breeding habits?
   c) How does the antelope ground squirrel remain active during the hot desert day?
   d) When are desert snakes most active? How do they locate and capture their prey?
   e) How do the wild pigs called "peccary" survive in the desert?

Fig. 25-44 **Birds of prey regain considerable moisture from their food. Desert speedster, the road-runner, preys on insects, lizards, and snakes. Prey is swallowed head first.**

Fig. 25-45 **Nocturnal hunters and their prey rely on their hearing for survival. Like the desert jack rabbit, the little kit fox displays large ears which are highly sensitive to sound.**

25.7 Mountain Biomes

A change in altitude can affect environment as much as a change in latitude. As altitude increases, the temperature falls as much as 1°C for every 150 m. Even in the Smoky Mountains of the southeastern United States, the temperature drops on the average 1.24°C for every 300 m increase in altitude. This is equivalent to a movement northward of about 640 km. Wind speed increases at higher altitudes. Mountain soil, eroded by rain, frost, and falling rock, becomes thinner and more mineral deficient at higher altitudes. As the environment changes, so does the natural life. Ecosystems resembling the circumpolar biomes of high latitudes are found on mountains (Fig. 25-46). An area less than 6.5 km up a mountainside at the equator provides as many different environments as an area 10 500 km along the surface between the equator and one of the poles.

Fig. 25-46 Altitude versus latitude. Compare the effect on vegetation in western North America.

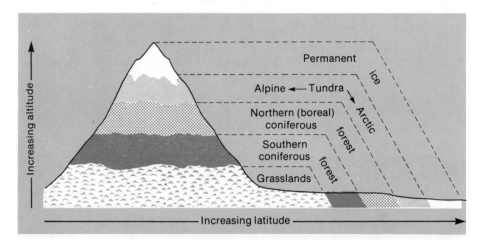

Fig. 25-47 The timberline marks the limit of tree growth on the mountainside. Alpine tundra lies beyond.

Northern species are found in mountain ridges where conditions match those of more northern latitudes. While these mountain zones may appear to resemble certain biomes we have studied, many mountain species have unique adaptations.

Abiotic Factors

As we have discussed, the polar regions are cold because the sun's rays strike these latitudes at a very low angle. Much of the radiant energy is absorbed during its long path through the earth's atmosphere. In the mountains, the solar rays have a much shorter route through a thinner layer of atmosphere. The density of mountain air is very low. As a result, most of the heat is radiated back into space instead of being absorbed. Yet the mountain surface absorbs heat energy readily. This results in a great difference between ground and air temperatures during the day. Since low-density air cannot retain much moisture, humidity is low. The clear, thin air permits more of the ultra-violet portion of the sun's rays to reach the surface. The high wind speeds increase evaporation, further reducing temperatures.

Fig. 25-48 A permanent snow belt cloaks the peaks of higher mountains. Life is scarce in this region of harsh conditions and limited food supply.

Biotic Factors

Forest creeps up most mountain slopes to the **timberline**—the uppermost limit at which trees can survive (Fig. 25-47). Above the timberline lies the **alpine tundra** and, in higher mountains, there is also a permanent snow belt with little food to support life (Fig. 24-48).

Vegetation

As in the Arctic, alpine tundra plants are small and stunted. However, mountain species are adapted to a shorter photoperiod with more intense sunlight. To survive the low temperatures and strong winds, alpine growth hugs the ground in a dense, compact mass. This tangle of foliage can absorb

Fig. 25-49 Stunted alpine vegetation grows close to the ground to avoid chilling winds and absorb heat energy.

and trap heat energy. In fact, the interior temperature of such growth might be twenty degrees higher than the surrounding air. This warm haven attracts pollinating insects. Some plants have a fuzzy coat of hairs to resist wind chill, trap heat and moisture, and also reduce the intensity of mountain sunlight. Too much ultraviolet radiation can damage plant cells. Many leaves are thick and waxy to resist both freezing and evaporation. Certain plants contain a red pigment which converts light into heat energy. Others have a rich cell fluid containing dissolved nutrients which acts as an anti-freeze. Alpine grasses have narrow leaves and stems that bend easily instead of breaking in the wind (Fig. 25-49).

Alpine plants use most of their energy just for survival. Hence growth and reproduction are very slow. Since mountain soil is dry much of the year, a large root system is needed to obtain sufficient moisture. Extensive roots also help secure plants against fierce winds and sliding rubble on the slopes. Most mountain species are perennials because the plants must be well established before flowering occurs. The first tiny blossom may appear after ten years of growth! Since insects are not abundant at high altitudes, many plants are wind–pollinated or self-pollinated. Normal seed germination is difficult in the harsh mountain conditions. The seeds of some plants sprout while they are still attached to the parent. Many mountain plant species rely upon vegetative propagation.

Animals

The warmth of spring melts the snow from the southern slopes first. Here, plant and animal activity gains a head start over the northern slopes. Summer bird residents nest on the southern edges of thickets where their young can best benefit from the sun's warmth. These nests are unusually compact, possibly for better insulation against the cold mountain air. Many birds feed and raise their offspring in sheltered crevices or holes. Rather than combat the strong mountain winds, the smaller birds avoid flying and remain close to the ground. When in the open, they usually face into the wind.

The low oxygen concentration affects mammals the most. Birds are adapted to fly at high elevations. Invertebrates and plants have lower metabolic rates and, hence, lower oxygen requirements. Increased heartbeat and respiration rate can temporarily help mammals, including humans, to adjust to the oxygen-poor atmosphere. However, animals that have permanently adapted to a high altitude have larger lungs and hearts. They also have more hemoglobin in their red blood cells. These adaptations increase the oxygen-carrying capacity of the blood.

Many alpine species have developed dark surface colours to absorb the increased ultraviolet radiation. This prevents damage to underlying tissues. Dark colouration also helps insects to absorb and retain more body heat during the day. They cannot remain active without the warmth of the sun. Since the mountain winds usually make controlled insect flight impossible, 60% of all alpine insects are wingless.

Fig. 25-50 The tiny mountain haymaker, the Pika, dries and stores winter food supplies. A single Pika can gather as much as 23 kg of hay — 150 times his own weight — to nourish him through the long winter months.

Mountain and Arctic wildlife share many of the same adaptations for withstanding low temperatures. Many alpine animals retreat to lower altitudes in winter. Some hibernate; others store food for nourishment through the long winter months (Fig. 25-50). Springtails (Fig. 25-51) are often seen eating conifer pollen on the snow. After freezing at night, these insects thaw out and become active during the day.

Snow meltwater is the main source of moisture in the mountains. At higher altitudes, where food supplies diminish, the population density of many alpine species is controlled by aggressive territorial behaviour and low reproductive rates.

Review

1. Describe how increasing altitude in a mountain region affects each of the following: temperature, wind, soil, and biotic community.
2. **a)** Why are the surface temperatures higher than air temperatures during the day?
   b) Why is humidity low?
   c) Why does more ultraviolet radiation reach the ground at higher altitudes?
3. **a)** What is the timberline?
   b) How do mountain tundra plants resemble Arctic tundra plants?
   c) How do they differ?
4. List six adaptations demonstrated by different alpine tundra plants and explain why each is an advantage.
5. **a)** Why is the rate of plant growth and reproduction so low in the mountain tundra?
   b) Give three benefits of a large root system to an alpine plant.
   c) Why are most mountain plant species perennials?
   d) How are many mountain plants pollinated? Why?
6. **a)** Where does the plant and animal activity first begin in the spring? Why?
   b) List four adaptations displayed by mountain bird life.
7. **a)** Why does the low oxygen concentration affect mammals more than birds or insects?
   b) How have mountain animals adapted to the oxygen-poor atmosphere?
8. **a)** State two advantages in the dark surface colouration many alpine species display.
   b) How are insects affected by the mountain climate?
9. **a)** How do mountain animals respond to the coming of winter?
   b) How are animal populations regulated at higher altitudes where food is scarce?

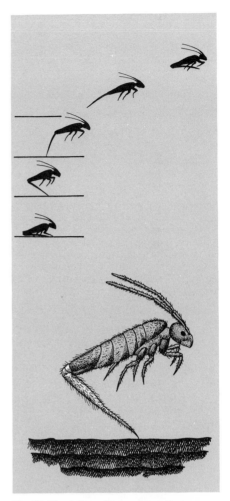

Fig. 25-51 A lever-like organ called a furcula snaps downward to propel the Springtail into the air, throwing the insect forward many times its own length.

25.8 The Tropical Evergreen Rain Forest Biome

We have completed our survey of the major biomes of North America. However, you should realize that many other biomes exist. Among them are the **chaparral**, the **tropical deciduous forest**, and the **tropical evergreen rain**

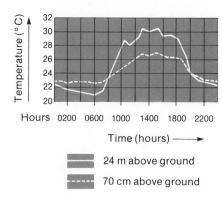

24 m above ground

70 cm above ground

Tropical Rain Forest Temperature — The upper canopy traps most of the sun's heat by day and is exposed to heat loss by radiation at nights. Since little sunlight penetrates the understory, the daily temperature change is less. Temperature varies by nearly 10°C at 24 m but by only 4°C at 0.7 m.

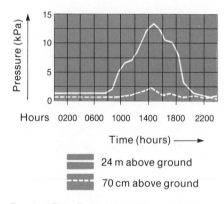

24 m above ground

70 cm above ground

Tropical Rain Forest Humidity — In the canopy, humidity nears saturation during the cooler night. The heat of day causes humidity to fall. In the shaded understory, the air is constantly near saturation with moisture. This graph measures humidity in kilopascals (barometric pressure)—as humidity increases, the reading decreases.

Fig. 25-52 Variations in Tropical Rain Forest temperatures and humidity.

forest. Each has its own array of plant and animal species. We will end our discussion of biomes with a brief look at the tropical evergreen rain forest.

This biome occurs in three separate areas around the equator (see Fig. 25-1). These moist, warm tropics provide the most favourable environment for terrestrial life. The supply of radiant energy continues throughout the year, since the noon sun is always within 23.5° north or south of the zenith. The daily rainfall keeps humidity high. Temperatures vary little within a 24 h period because the sky is often cloudy during the day and the high moisture content in the air reduces heat loss by radiation at night (Fig. 25-52). The tropical rain forest is dominated by broad-leaved trees. However, they are evergreen because of the constantly favourable growing conditions.

Although natural growth flourishes, tropical forest soil is infertile for agriculture. The rich humus quickly decomposes when exposed to sunlight. Also, the heavy rains wash away nutrients essential for agriculture. The natural growth of the tropical rain forest includes a great variety of plants that normally occur in five distinct strata within the mature forest (Fig. 25-53). This stratification increases the variety of niches available to animals (Fig. 25-54). Hence, species diversity is greater than in other land biomes. However, individual populations are often low in numbers and widely scattered. Since plant production never ceases, more animals can specialize in a particular food such as nectar or fruit. Non-flying animals must climb to reach most of their food. They tend to be small, agile, and well adapted for moving about in the trees.

As a result of the lack of seasonal variation, many tropical species have no precise breeding time. In the humid forest, some amphibians lay their eggs on land rather than in pools of water. This eliminates the larval stage of their life cycle. The high temperature and humidity favour insects and other invertebrates. Their constantly high metabolic rate stimulates growth, producing giant specimens of many species. Life cycles and hence reproduction are speeded up. More rapid evolution is possible since the greater numbers of new generations increase the chances of genetic variation. Many forest dwellers have keen colour vision and bright distinctive features. These markings help to attract a mate and discourage rivals.

A special animal habitat is provided by the epiphytes. These are plants which grow up or on other taller vegetation to reach sunlight. Many epiphytes have dangling roots which absorb moisture from the humid air. Debris accumulates around these roots, forming humus in which many invertebrates thrive. Some ferns and orchids are epiphytes.

Review

1. **a)** Describe the climate of the tropical rain forest.
   b) Why is the temperature fairly constant during each 24 h period?
   c) Why are higher levels of rainfall needed to support large tree growth in the tropical rain forest than in the northern coniferous forest?

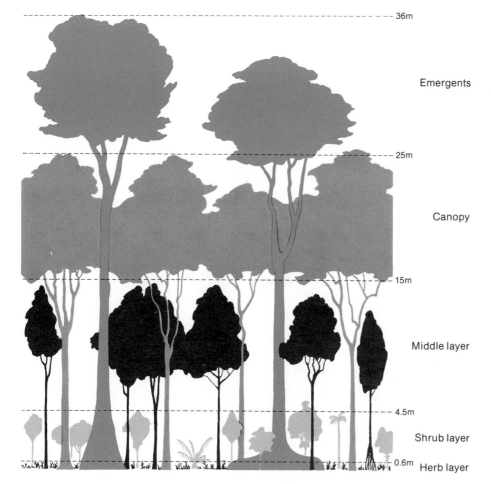

Fig. 25-53 The five main tropical rain forest strata. A: the tallest emergent trees. B: the continuous canopy. C: the middle layer of smaller trees with long narrow crowns for maximum use of poor light. D: a sparser layer of shrubs, and dwarf trees. E: a ground layer of herbs.

36m

Emergents

25m

Canopy

15m

Middle layer

4.5m

Shrub layer

0.6m Herb layer

2. **a)** Why is tropical rain forest soil not suitable for agriculture after clearing?
   b) Humus is organic matter that is resistant to decay. Decomposition of litter in tropical forest soils is so rapid that no appreciable humus layer forms. State two reasons why this decomposition is so fast.
3. **a)** Why is a tropical environment the most favourable for terrestrial life?
   b) Give two reasons, other than climate, why species diversity is greatest in the tropical rain forest.
4. **a)** Describe three characteristics of most non-flying animals in the tropical rain forest.
   b) Discuss three effects of the tropical climate on animal reproductive cycles.
   c) What is the function of colourful animal markings?
5. Explain how the tropical climate affects poikilothermic organisms with respect to: metabolic rate, growth, life cycles, and genetic variation.
6. **a)** What are epiphytes?
   b) Describe their role in the tropical forest ecosystem.

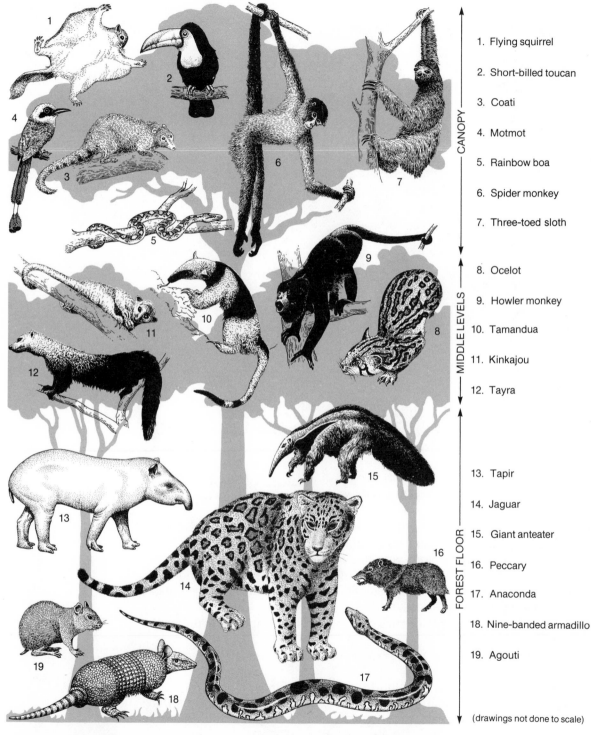

CANOPY

1. Flying squirrel

2. Short-billed toucan

3. Coati

4. Motmot

5. Rainbow boa

6. Spider monkey

7. Three-toed sloth

MIDDLE LEVELS

8. Ocelot

9. Howler monkey

10. Tamandua

11. Kinkajou

12. Tayra

FOREST FLOOR

13. Tapir

14. Jaguar

15. Giant anteater

16. Peccary

17. Anaconda

18. Nine-banded armadillo

19. Agouti

(drawings not done to scale)

Fig. 25-54 Each stratum seems to have its own characteristic animal life. Although most animals can easily move between strata, they tend to live and feed at one particular level. Why? How does this distribution affect competition?

25.9 Investigation

A Study of Two Terrestrial Sites

During your studies of biomes, you learned that, if the abiotic factors in an ecosystem change, the biotic factors change. In a similar manner, if the biotic factors change, the abiotic factors will change. Everything in nature is interdependent. If you could visit two different biomes, you could easily prove this, experimentally. However, there is not much chance that your teacher will take you to a desert today and then to the tundra tomorrow! Fortunately, however, you can perform a similar experiment in the schoolyard or a nearby park. In this investigation you measure and compare the abiotic and biotic factors in two contrasting sites.

Some Suggested Sites

a) Area dominated by long grass vs. area dominated by cut grass
b) Wooded area vs. meadow (grassy area)
c) Coniferous tree area vs. deciduous tree area
d) Low wet area vs. sunny area of the schoolyard
e) Area with organically rich soil vs. area with sandy soil
f) Sheltered area vs. windy area
g) South-facing slope of a hill vs. north-facing slope of a hill

Materials

The following is an ideal list. If your school does not have a certain piece of equipment, you can find substitutes. For example, a camera light meter can take the place of the light meter. Or, you can simply rate the light intensity on a relative scale such as: very light, bright, moderate, dim, very dim.

light meter
sling psychrometer
air thermometer
soil thermometer
soil test kit for pH (and, if possible,
   nitrogen, phosphorus and potassium)
soil sampler (or small shovel)

trowel
wind meter
soil sieves
tape measure (10 m)
sweep net
white tray
hand lens

Procedure

a. Select two sites for the comparison. They must differ in at least one obvious way. The sites need not be large. An area of 100 m² for each study plot is more than adequate. Stake out study plots of identical size and shape at the two sites.
b. Use the light meter to measure the average light intensity at each site.

c. Use the sling psychrometer to measure the relative humidity at each site.

d. Use the wind meter to measure the average wind speed at 3 levels at each site. Take readings 0.5 m, 1.0 m, and 2.0 m above the ground.

e. Measure the air temperature at each site. If the sun is shining, shade the bulb of the thermometer from the direct rays of the sun.

f. Use the soil thermometer to measure the soil temperature at depths of 2 cm, 7 cm, and 15 cm. Do this at both sides.

g. Determine the soil pH at each site. If possible, measure the concentrations of nitrogen, phosphorus, and potassium at each site.

h. Use the trowel to collect an "average" sample of soil from each site. Put the sample in the soil sieve and separate it by shaking the sieve. Note the approximate fraction of the total sample that ends up on each level of the sieve.

i. Use the soil sampler or shovel to determine the nature of the soil profile at each site. How deep is the litter layer? How deep is the topsoil?

j. Sweep each study plot thoroughly with the sweep net. Note the numbers of different species of organisms that are collected. Note, also, the relative abundance of each species. Is it abundant, frequent, occasional, or rare? Then note any adaptations to their environment that these organisms show.

k. Put a few scoops of soil in the white tray. Sort through it, looking for organisms. Again, note the numbers, relative abundance, and adaptations.

l. Examine the vegetation at each site. Note the numbers of different species, their relative abundance, and their adaptations.

m. Record all your findings in a table.

Discussion

1. Using the findings of this field study, write a paper which supports the ecological principle that everything in nature is interdependent. Pay particular attention to the differences in the abiotic and biotic factors at the two sites.

Highlights

A biome is a large geographical region of the earth with a characteristic climate and biota. When studying the climates of biomes and other regions, scientists often record their findings in climatograms. Climatograms graphically summarize monthly variations in temperature and precipitation.

Seven major biomes are found in North America. These are the tundra, coniferous forest, temperate deciduous forest, grasslands, desert, mountain, and tropical evergreen rain forest biomes. Each biome has a characteristic climate, a certain set of other abiotic factors, and characteristic plant and animal communities. Relationships always exist between the abiotic and the biotic factors in a biome. Of course, this is true for any terrestrial ecosystem and, for that matter, any ecosystem.

UNIT SEVEN

The Continuity of Life

No two individuals of the same species are exactly alike. Yet, for each species, the offspring resemble their parents in most respects. You know that during sexual reproduction each parent passes on to the offspring a basic plan that is characteristic of the species. In this unit you discover the nature of that plan and investigate many other aspects of heredity and the related topic, evolution.

26 Principles of Heredity

Fig. 26-1 Every organism inherits certain traits from its parents. What traits do you think the calf inherited from its mother?

How often have you been told that you "look just like your father" or that you "have your mother's eyes"? There is more to these remarks than just politeness. **Traits**, or characteristics, such as straight hair, length of long bones, and eye colour are inherited. They can be passed on from generation to generation, that is, from parents to offspring. This transfer of traits is called **heredity**, or **inheritance**. The study of heredity, in plants, animals, protists, fungi, or monerans, is called **genetics** (Fig. 26-1).

Parents and offspring do, of course, have differences between them. These differences are called **variations**. The study of these variations is part of the field of genetics. In this chapter and the rest of this unit, you will study several aspects of genetics. Let us begin with the first experiments in genetics.

26.1 Mendel's Experiments

The science of genetics began in the middle of the 19th century, as a result of the work of Gregor Mendel. Mendel was an Augustinian monk who taught school in a town in what is now Czechoslovakia. He lived in a monastery and, in his spare time, performed experiments in hybridization. **Hybridization** *is the process of cross-breeding between two varieties of the same species that are distinctly different in one or more traits, or cross-breeding between two closely related species.* The crossing of a black pig and a white pig is hybridization. Another example is the crossing of a short red tulip with a tall red tulip. The offspring of hybridization are called **hybrids**.

Mendel performed experiments with animals (mainly mice and bees) and with several species of plants. He studied how traits reappeared in several succeeding generations. His most important experiments were performed with the common garden pea. He chose this plant because he wanted one that was normally self-pollinating. In other words, he wanted a plant in which the pollen from the stamens was transferred to the pistil of the same flower. As you can see from Figure 26-2, the pea plant produces such a flower. Its stamens and pistil are located deep within a closed corolla. As a result, insects and the wind cannot easily carry pollen from other flowers to it. Further reasons for choosing pea plants for the experiments were that they are annuals, grow easily and mature quickly. But most important was the fact that pea plants show several pairs of obvious contrasting traits. For

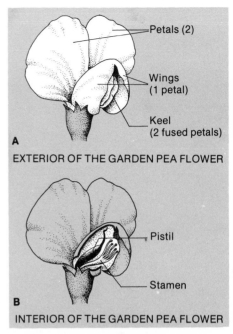

Petals (2)
Wings
(1 petal)
Keel
(2 fused petals)
A
EXTERIOR OF THE GARDEN PEA FLOWER
Pistil
Stamen
B
INTERIOR OF THE GARDEN PEA FLOWER

Fig. 26-2 The garden pea flower. The stamens and pistil are enclosed by the keel and the wings. How does this tend to make this flower self-pollinating?

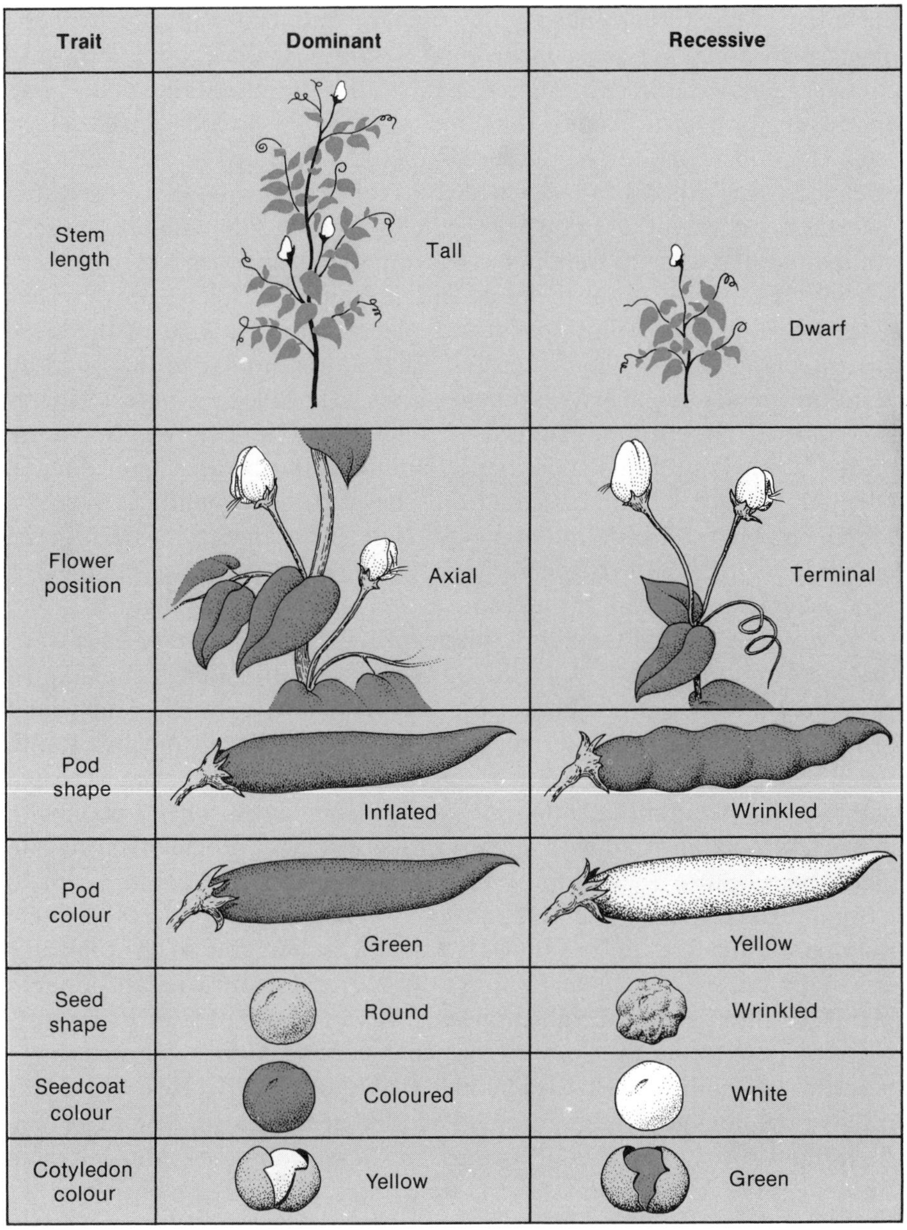

Trait
Dominant
Recessive
Stem length
Tall
Dwarf
Flower position
Axial
Terminal
Pod shape
Inflated
Wrinkled
Pod colour
Green
Yellow
Seed shape
Round
Wrinkled
Seedcoat colour
Coloured
White
Cotyledon colour
Yellow
Green

Fig. 26-3 Pairs of contrasting traits studied by Mendel.

example, some plants grow tall while others are dwarfs; some plants have smooth, inflated pods while others have wrinkled pods; some plants have green pods while others have yellow pods. Mendel found seven pairs of such contrasting traits that he wanted to study. These are shown in Figure 26-3.

Through many generations of natural self-pollination, the garden pea had developed into pure lines. A **pure line** is an organism that breeds true from generation to generation for a certain trait. In other words, any pea flower, if self-fertilized, produced seeds that grew into plants that were identical to the parent plant. To be sure this was the case, Mendel tested his seed for two years prior to his hybridization experiments.

Mendel had training in both mathematics and biology. Therefore he knew that it was important to perform a large number of identical experiments before making any conclusions. He also knew the value of having controls in his experiments. Therefore he studied the inheritance of only one contrasting pair of traits at a time, for example, green pods versus yellow pods. To do this, he mated purebred stock that differed in one trait only. Such a cross is called a **monohybrid cross** (one trait cross). Let us see how this was done when the trait being studied was pod colour. To prevent self-pollination, the stamens were removed from the plant which was to be seed-bearing. In this case, it was the plant with green pods. Pollen from the other plant (the one with yellow pods) was placed on the stigma of the seed-bearing plant. This method of cross-pollination is similar to that used by plant breeders today. The two parent plants were called the **parent generation**, or the **P generation**. The seeds that developed were allowed to mature on the vine. Some traits such as pod colour could be seen immediately. Other traits such as tallness could only be studied by planting the seeds. Mendel planted the seeds and observed the reappearance of the traits in the offspring plants. This first generation of hybrid offspring was called the **first filial generation**, or the F_1 **generation**. (*Filial* means son or daughter).

Mendel expected the hybrid plants of the F_1 generation to have traits that were intermediate between the traits of the P generation. For example, he expected that a cross between a tall plant and a dwarf plant would produce plants that were intermediate in height between tall and dwarf. But, much to his surprise, this was not the case. When a tall plant was crossed with a dwarf plant, all the offspring were tall. Similarly, when a plant with yellow pods was crossed with a plant with green pods, all the offspring had green pods. Since the traits of tallness and green pod colour seem to dominate the traits of dwarfness and yellow pod colour, Mendel called them **dominant traits**. Figure 26-3 shows the traits that he found to be dominant.

Why were all the F_1 generation tall? Why did no dwarf plants or even plants of intermediate height appear? Was the trait of dwarfism "lost" forever? To discover the answers to such questions, Mendel allowed the F_1 plants to self-pollinate. He then planted the resulting seeds. He observed the traits in the next generation, the **second filial generation**, or the F_2 **generation**. In the seven pairs of contrasting traits that he observed, Mendel found similar results. In each cross, the F_1 generation all showed one trait of the pair. In the F_2 generation, about ¾ of the plants showed the same trait. For example, in the case of tallness versus dwarfism, the F_1 generation all showed tallness. About ¾ of the plants in the F_2 generation showed tallness (Fig. 26-4). Mendel labelled the tall trait **dominant** over the dwarf trait because the tall trait was most often seen in the offspring. He called the dwarf trait **recessive** to the tall, since it was not expressed in the presence of the dominant trait.

Similarly, when he observed the results of crossing parent plants where one had green unripe pods and the other had yellow unripe pods, all the F_1 generation had green pods. Also, about ¾ of the F_2 generation had green pods. Mendel concluded that green pods were dominant to yellow. Or, to put it another way, yellow pods were recessive to green pods.

Mendel analyzed the information from all his crosses and formed general rules that summarized his results. These rules are discussed in Section 26.3.

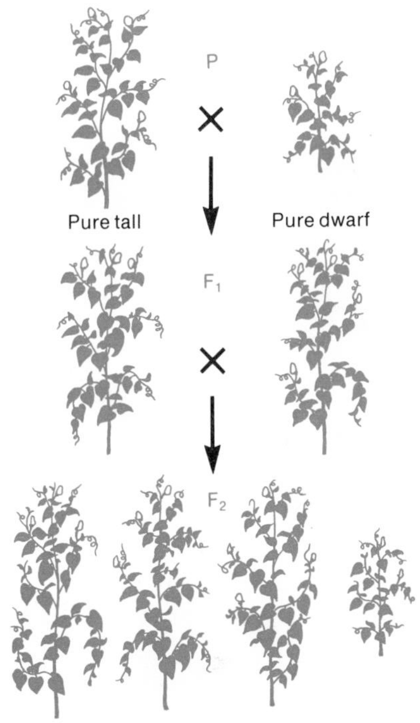

Review

1. Distinguish between heredity and genetics.
2. Why did Mendel choose the garden pea for his experiments?
3. Why is it important in research to repeat experiments many times?
4. What is the difference between dominant and recessive traits?
5. Suppose that a monohybrid cross has been performed with garden peas. The P generation includes a parent with round seeds and a parent with wrinkled seeds. The F_1 generation all have round seeds. The F_2 generation have mainly round seeds, but some are wrinkled. Seed shape is one of the traits that Mendel studied. Which trait, round seed or wrinkled seed, is dominant? Show how you arrived at your answer.
6. In garden peas, seed coat colour can be either coloured or white. White seed coat is recessive to coloured seed coat. Suppose the P generation includes one parent with the white seed coat trait and one with coloured seed coat trait. What results would you expect in the F_1 and F_2 generations?

Fig. 26-4   When the parents are a pure tall and pure dwarf plant, the F_1 hybrids are all tall. If the F_1 hybrids self-pollinate, the F_2 generation are ¾ tall and ¼ dwarf.

26.2 Investigation

The Transmission of Traits in Corn and Soybean

This investigation explores patterns of inheritance in corn seedlings and in soybean seedlings. The corn seedlings can be albino (colourless) or green. The soybean seedlings can be one of three colours—green, light green, or yellow.

After you get this investigation started, proceed with the study of Sections 26.3 and 26.4. They will provide knowledge that is required to complete this investigation.

Materials

enough containers to hold 16 seeds, planted 4-5 cm deep
  and 4-5 cm apart (old flats for flowers, old cake pans,
  margarine tubs with drainage holes, or flowerpots)
potting soil
corn seed representing the F_2 generation of a cross
  between albino and green corn
soybean seed, representing the F_2 generation of a cross
  between green and yellow soybean plants

Procedure

a. Fill the containers to a depth of 7-8 cm with potting soil.
b. Plant 8 corn seeds of the F_2 generation, about 5 cm deep and 5 cm apart.
  Label the containers.
c. Plant 8 soybean seeds of the F_2 generation, about 4 cm deep and 4 cm
  apart. Label the containers.
d. Water the soil whenever it appears to be drying out.
e. Observe the containers after 7, 10, and 14 d. Record the number of
  seedlings of each colour. Be sure to keep the corn and soybean results
  separate.
f. Pool the class results.

Discussion

1. From the class results, which trait is dominant in corn—albino or green
   leaf colour? To the nearest whole number, what is the ratio of green to
   albino colour?
2. How does the trait of leaf colour differ in its inheritance in soybean as
   compared to corn? To the nearest whole number, what is the ratio of
   green to light green to yellow leaf colour in soybean?
3. What would be the genotype of the F_1 generation of a cross between a
   green corn parent (GG) and an albino corn parent (gg)? Draw a Punnett
   Square of the F_1 cross of these two parents. What is the expected F_2 ratio
   of albino to green? Do the class results hold true to this? If not, why not?
4. The F_1 generation of a cross between dark green (GG) and yellow (gg)
   soybean parent plants is all light green in phenotype and genotype (Gg).
   Draw a Punnett Square of the F_1 cross, Gg x Gg. Then determine the
   expected F_2 ratio. Explain any difference between this and the class
   results.

26.3 Mendelian Laws of Inheritance

As Mendel recorded the results of his experiments, patterns began to show.
For each of the seven pairs of traits he studied, one of each pair was
dominant and the other was recessive. In the F_2 generation, about ¾ of the

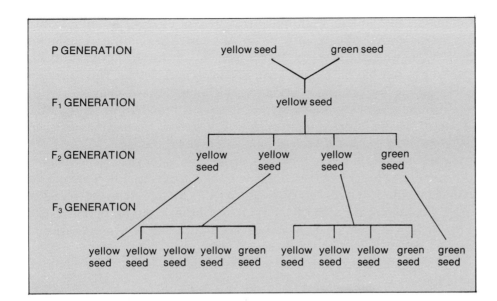

"grandchildren" showed the dominant trait (for example, tall) and ¼ showed the recessive trait (for example, dwarf). In other words, the ratio of dominant to recessive (tall to dwarf) was 3:1.

Mendel concluded that units of inheritance that he called "**factors**" caused this pattern. Today we call these units of inheritance **genes**. To test his theory, Mendel predicted what would happen if he let the F_2 generation self-pollinate and produce an F_3 generation (the offspring of the F_2 self-pollination). He knew that, if one parent of the P generation had yellow seeds and the other had green seeds, the F_1 generation would be all yellow. He knew, also, that the F_2 generation would be ¾ yellow and ¼ green. He predicted that the F_3 generation would show the following pattern: Of the F_2 yellow seeds, ⅓ would produce only yellow seeds and ⅔ would produce both yellow and green seeds, in the ratio of 3:1. The green F_2 seeds would produce only green seeds (Fig. 26-5). When he did the experiment, the ratios in the F_3 generation were almost exactly as he predicted. An explanation follows later in this section.

Mendel decided that each trait which he saw was controlled by a unit of inheritance which we now call a gene. For any given trait, different forms of the gene occur in pairs. We call these forms **alleles**. For example, the gene for seed shape has two alleles, an allele for round seed and an allele for wrinkled seed. Similarly, the gene for height has two alleles, a dominant tall allele and a recessive dwarf allele. When a tall plant is crossed with a dwarf plant, the resulting F_1 generation is all tall. Yet each F_1 plant has two alleles for height. The tall allele showed itself in all cases, even though the allele for dwarf was present. In other words, the dwarf allele did not express itself when the tall allele was there.

Mendel described this result as the Law of Dominance. This law is stated as follows: *Dominance occurs when one allele of a pair has the ability to express itself while the second is not expressed.*

The members of each pair of alleles segregated. That is, they appeared to separate into different reproductive cells. For example, a reproductive cell, or gamete, could have an allele for tall or an allele for dwarf, but not for both. Then, when the male and female gametes join during fertilization, the resulting zygote has the required pair of alleles for each trait. This pairing of alleles during fertilization occurs according to the "laws of chance".

From these observations Mendel formulated the **Law of Segregation:** *A pair of alleles for each trait occurs in any organism. The members of a pair are segregated (separated) when gametes are formed.*

Mendel used letters of the alphabet as symbols for genes. This method of representing genes is still used today. Capital letters (e.g. B) represent the dominant allele and lower case letters (e.g. b) represent the recessive allele. Let us look at one of Mendel's crosses, using letters for the alleles (see Fig. 26-6). You can see how this makes the results of crosses easier to understand.

The parents, each with two members of the two alleles, are symbolized as follows:

P generation: Yellow seed parent   X   Green seed parent
                          YY   yy

According to the Law of Segregation, the gametes of the yellow seed parent must all contain Y and those of the green seed parent must all contain y. When the gametes for each parent join, the F_1 generation is formed. The pair of seed colour alleles in each plant is Yy. All these plants have yellow seeds, because yellow seed colour is dominant. However, they contain two types of alleles, one for green seed colour and one for yellow seed colour. The appearance of the plant, which may or may not indicate all the genes that are present, is called the **phenotype**. The phenotype of the F_1 generation is yellow seed colour. The **genotype** of the plant is its gene makeup. In this case, the genotype is Yy. Here, the phenotype does not truly reflect the genotype. In the green seed parent, the genotype is yy. There the phenotype truly reflects the genotype.

The F_1 genotype, Yy, is self-pollinated to produce the F_2 generation. In this case, the cross is symbolized as follows:

Yellow seed colour   X   Yellow seed colour
       Yy   Yy

Here, the gametes for each parent can contain either Y or y (Fig. 26-6). In the F_2 generation, these gametes can join together in any manner, according to the "laws of chance", as long as one male gamete joins with one female gamete. Possible combinations can be YY, Yy and yy. These are all the genotypes possible. But what phenotypes are there? Since yellow is dominant, the Yy combinations and the one YY combination are all yellow seed colour. The yy combination is green seed colour. Therefore, the ratio of yellow seed colour to green seed colour should be about 3:1.

When Mendel produced an F_3 generation by self-pollination of the F_2 generation, he said that $\frac{1}{3}$ of the yellow F_2 should produce only yellow.

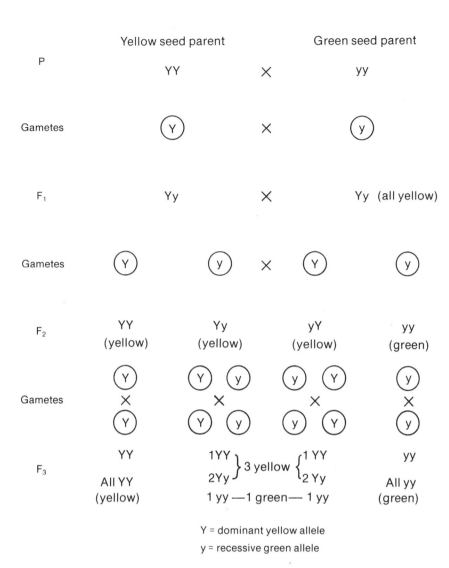

Yellow seed parent   Green seed parent

P   YY   ×   yy

Gametes   (Y)   ×   (y)

F_1 Yy   ×   Yy   (all yellow)

Gametes   (Y)   (y)   ×   (Y)   (y)

F_2

| YY | Yy | yY | yy |
| (yellow) | (yellow) | (yellow) | (green) |

Gametes

F_3

YY	1YY ⎫	1 YY ⎫	yy
	2Yy ⎬ 3 yellow	2 Yy ⎬	
All YY	1 yy —1 green— 1 yy		All yy
(yellow)			(green)

Y = dominant yellow allele

y = recessive green allele

Fig. 26-6   A Mendelian cross of yellow-seed and green-seed parents, using letter symbols for the alleles. Note the segregation of alleles when gametes are formed.

These would be the YY genotype of the F_2. Since the YY genotype is a pure line for yellow seed colour, it can only produce yellow seeds. Similarly, Mendel predicted that the green F_2 seeds (yy) should produce only green seeds in the F_3. Of the F_2 generation, $\frac{2}{3}$ of the yellow seed plants have the genotype Yy. These hybrids are similar to the F_1 hybrids and therefore produce an F_3 generation in a 3:1 ratio, yellow to green.

Review

1. Explain the difference between a gene and an allele.
2. Demonstrate why a yy genotype, which is green seed colour, would produce only green seeds when self-fertilized.
3. Assume that straight hair (c) is recessive to curly hair (C). Two curly-haired parents have a straight-haired child.
   a) What are the genotypes of the parents?

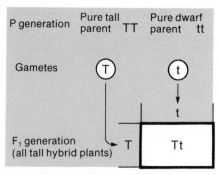

Fig. 26-7 Punnett square for the cross of a pure tall pea plant and a pure dwarf pea plant.

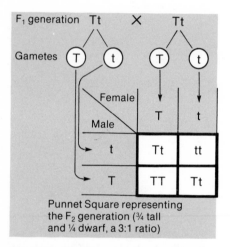

Punnet Square representing the F₁ generation (¾ tall and ¼ dwarf, a 3:1 ratio)

Fig. 26-8 The Punnet square for the self-fertilization of the F₁ generation of the cross represented in Figure 26-7.

b) What is the genotype of the child?
c) Give two genotypes for curly-haired people.
d) Using letter symbols, show two different crosses for curly hair x straight hair. Indicate the phenotypes of the offspring of both crosses.

4. Suppose that a cross had been performed with a garden pea, between a parent with round seeds (RR) and a parent with wrinkled seeds (rr). Outline the genotypes and phenotypes in the F_1 and F_2 generations resulting from this cross.

26.4 The Punnet Square

The Punnett Square is used to show the possible ways in which gametes may join. If a pure tall pea plant (TT) is crossed with a pure dwarf pea plant (tt), the Punnett Square appears as in Figure 26-7. The possible alleles found in each gamete are placed along the sides of the square. One side is used for the male parent gametes and the other side for the female gametes. All possible combinations are then found in the grid of the square. These combinations represent the members of the F_1 generation. Note that all the F_1 have the genotype Tt. If the F_1 are self-fertilized, the resulting Punnett square appears as in Figure 26-8. The F_2 generation is seen in the grid. It consists of 1 TT, 2 Tt, and 1 tt. As expected, there is a ratio of 3:1 of the phenotypes tall: dwarf.

Sometimes Mendel did experiments in which he crossed parent plants and observed the inheritance patterns of *two* characteristics at once. These are called **dihybrid** crosses. For example, a cross between a pure dwarf, pure round-seeded parent and a pure tall, pure wrinkle-seeded parent is represented as follows:

P generation: Pure dwarf; Pure round   X   Pure tall; Pure wrinkled
                          ttRR   TTrr

The gametes produced by one parent have tR alleles. The gametes of the other parent all have Tr alleles. Thus all the offspring in the F_1 generation will be TtRr. They are all tall and round-seeded. If these are allowed to self-fertilize, the cross would be TtRr x TtRr. The gametes possible from each parent are TR, Tr, tR, tr. The Punnett Square for this cross is shown in Figure 26-9. In the F_2 generation, $9/16$ of the plants are tall, round-seeded plants, $3/16$ are tall, wrinkle-seeded plants, $3/16$ are dwarf, round-seeded plants, and $1/16$ are dwarf, wrinkle-seeded plants. This ratio of the F_2 phenotypes of 9:3:3:1 is found in all dihybrid crosses which involve fully dominant and recessive alleles and in which pure lines of these traits are found in the P generation.

When an organism carries the same allele for both members of a gene pair, the organism is said to be **homozygous** in that gene. If two different alleles are present in a gene pair, the organism is **heterozygous** in that gene. For example a pure tall pea plant has a genotype of TT. It is homozygous. A heterozygous tall pea plant has a genotype of Tt.

Female Male	TR	Tr	tR	tr
TR	TT RR	TT Rr	Tt RR	Tt Rr
Tr	TT Rr	TT rr	Tt Rr	Tt rr
tR	Tt RR	Tt Rr	tt RR	tt Rr
tr	Tt Rr	Tt rr	tt Rr	tt rr

9/16 tall round

3/16 tall wrinkled

3/16 dwarf round

1/16 dwarf wrinkled

For pea plants, in all the paired traits (such as round versus wrinkled seeds), one trait was completely dominant or completely recessive to the other. The F_1 phenotype was always clearly the phenotype of one of the parent plants, even though the F_1 was heterozygous in its pair of alleles and the parent plant was homozygous.

Occasionally, both alleles of a pair are fully expressed without blending in a heterozygous organism. For example, in the ABO blood group, a person whose pair of alleles for blood type A and B, has an AB blood type. In this case, both alleles are expressed. These alleles are said to be **codominant**.

Sometimes a cross of two different homozygous parents will produce heterozygous F_1 offspring which are different from both parents. This is called **intermediate inheritance**. Quite often this heterozygote F_1 has a characteristic that is intermediate between the characteristics of each parent, as if they had been blended together. For example, in snapdragons, plants with broad leaves have the homozygous genotype BB. Plants with narrow leaves have the homozygous genotype B'B'. If these two types of plants are crossed, all the members of the resulting F_1 generation have the genotype BB' for the trait of leaf width. If this were an example of complete dominance, these plants would all have either broad leaves or narrow leaves, depending on which allele was dominant. However, in this case of intermediate inheritance, all members of the F_1 generation have intermediate-width leaves (Fig. 26-10). In the F_2 generation resulting from the self-fertilization of the F_1, the genotypes of the offspring are ¼ BB, ½ BB', and ¼ B'B'. The pattern is the same as that seen with completely dominant alleles. However, here the phenotypes take the form of one parental type: two intermediate types: one parental type. Clearly then, in intermediate inheri-

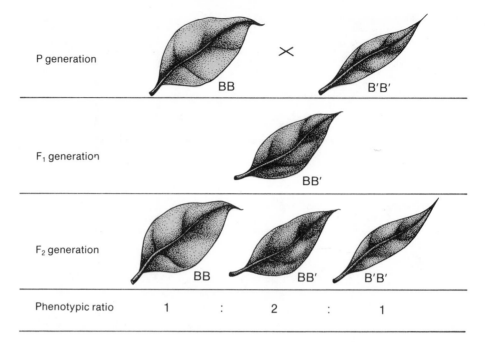

Fig. 26-10 Intermediate inheritance demonstrated in snapdragon leaf width.

| P generation | BB | × | B'B' |

| F₁ generation | BB' |

| F₂ generation | BB | BB' | B'B' |

| Phenotypic ratio | 1 | : | 2 | : | 1 |

tance, the phenotypes, or physical characteristics, of the F$_2$ generation truly reflect its genotypes, or gene makeup. The alleles B and B' do not actually blend together. Apparently they are transmitted separately to the F$_2$ generation.

Review

1. Use a Punnett Square to find the phenotypes and genotypes resulting from each of the following monohybrid crosses of pea plants:
   a) homozygous tall with heterozygous tall.
   b) heterozygous tall with homozygous dwarf.
2. Distinguish between a monohybrid and a dihybrid cross.
3. Find the F$_2$ phenotypes and genotypes of a dihybrid cross of a dwarf, yellow-podded pea plant (ttgg) with a tall, green-podded pea plant (TTGG). Use a Punnett Square.
4. Assume that curly hair (C) is dominant to straight hair (c). Albinism is a condition in which cells which normally produce pigment, do not do so. A person with this condition is called an albino. The allele for skin albinism (n) is recessive to the normal allele (N). A woman with curly hair and albinism and a man with straight hair and normal skin pigment have a child that has straight hair and is an albino. What are the genotypes of the parents?
5. Distinguish between codominance and intermediate inheritance.
6. The Four o'clock flower shows intermediate inheritance. The pure red-flowered plant (RR) is crossed with a pure white-flowered plant (R'R'). The intermediate phenotype is pink (RR'). Using a Punnett Square, find the phenotypes and genotypes of the F$_2$ generation. What offspring phenotypes would you expect if a pure red-flowered plant were crossed with a pink-flowered plant?

26.5 Investigation

The PTC Taste Test

Some people are able to taste the compound phenylthiocarbamide (PTC) in small amounts. Others are not able to taste it until high concentrations are reached. Those who taste its bitterness when it is in small amounts are called "tasters". Those who taste PTC only at high concentrations are called "non-tasters". This trait is inherited. The taster allele (T) is dominant to the non-taster allele (t). In the following investigation, you will test yourself and your family for the ability to taste PTC in small amounts. Then you will analyze the results.

Materials

PTC test paper

Procedure

a. Obtain a strip of PTC test paper for each member of your family who is to be tested. Test as many people as possible in your immediate family—mother, father, grandparents, brothers, and sisters.

b. Place the strip on your tongue, toward the back. This is where bitterness is usually tasted. There is no need to chew the strip. The results should be clear as soon as the strip is moist. Discard the strip.

c. Record the results for each person tested. Note the relationship of each person to you.

d. Share your results with your classmates.

Discussion

1. Is anyone in your family a non-taster? What must his or her genotype be?
2. Tasters can be one of two genotypes. What are they?
3. If your parents were both tested, draw some possible Punnett Squares for your immediate family. Remember, this is a monohybrid cross. Do the offspring fit the expected outcome?
4. Non-tasters of PTC are almost unknown among the Inuit and North American Indians. They are not common in North American and African Negroes. But they are fairly common in Caucasians (about 30%). What was the percent of tasters in your class?
5. In Section 26.4 the concept of codominance was introduced, using the ABO blood group as an example. The ABO blood group shows both codominance and the usual dominant-recessive pattern. The A and B alleles are codominant, and the O allele is recessive to them.

    In Section 20.14, page 555, you determined your blood type. You found out that you are A, B, O, or AB. These groups are the phenotypes.

a) What genotypes are possible for each blood type?
b) What are the possible genotypes for yourself?
c) Suppose that a father and mother both have Type A blood. Can they have a child with Type O blood? Explain your answer.
d) The most common blood types in North America are A and O. Each is found in 40 to 50% of the white population. The AB type occurs in about 3-4% of the population. What is the frequency of each blood group in your class?

Highlights

The transfer of traits from one generation to the next is called heredity. The study of heredity is called genetics. Gregor Mendel is known as the father of genetics because of his successful hybridization experiments with the common garden pea.

A monohybrid cross between a parent that is pure dominant in a trait and a parent that is pure recessive in the same trait produces an F_1 generation that shows the dominant trait. However, if the F_1 generation self-fertilizes, $\frac{3}{4}$ of the F_2 generation will show the dominant trait and $\frac{1}{4}$ will show the recessive trait.

Mendel attributed traits to "factors" in the parents. These factors are now called genes. Different forms of the gene for the same trait are called alleles. Having assumed that factors (genes) exist, Mendel went on to state his Law of Dominance and Law of Segregation. A knowledge of these laws and the use of a Punnett Square make it easy for us to predict and understand the results of many inheritance experiments.

Chromosomes, Genes, And The Genetic Material

27

In 1866 Mendel published a paper that carefully reported the results of each of his experiments. He also drew conclusions that led to his important laws of inheritance. However, he had no idea why things happened as they did. Chromosomes had not been discovered at that time.

Mendel's paper lay forgotten in a library for 34 years. Then, in 1900, it was rediscovered. By this time chromosomes were known to be present in cells. Biologists suspected that chromosomes carried the information for inheritance. Today we know that the genes are located on the chromosomes.

27.1 Meiosis and Mendel's Findings

In Chapter 8 you studied chromosomes and the process of meiosis. Let us first review some of the main ideas developed there. Then we will see how those ideas tie in with Mendel's discoveries.

Every cell in an organism has the same number of chromosomes. In fact, the body cells of all normal individuals of the same species have the same number of chromosomes. For example, all normal human cells have 46 chromosomes. The cells of the common garden pea each have 14 chromosomes and onion cells have 16. A crayfish cell has 200 chromosomes, and a fruit fly cell has 8. The chromosomes in each body cell always occur in pairs. The chromosomes in each pair are similar in form and in the way in which the genetic material that they carry is arranged. Such a pair of chromosomes is called a **homologous pair**. Thus the human has 23 homologous pairs of chromosomes. The garden pea has 7 homologous pairs. Each chromosome in a homologous pair is called a **homologue**. Each of the two homologues carries genetic information for the same trait as its partner. However, the specific information is often quite different. For example, one homologue may be coded for brown hair. The other homologue may be coded for black hair. The final hair colour of the offspring is determined by the combination of these two pieces of genetic information. Each homologous pair can carry genetic information for thousands of traits such as hair colour and eye colour.

The number of homologous pairs of chromosomes is represented by **n**. Therefore, the total number of chromosomes in a body cell is represented by **2n**. This is called the **diploid number** of chromosomes. Thus the diploid number (2n) for humans is 46.

A gamete, male or female, contains just one homologue of each homologous pair of chromosomes. That is, it has half the diploid number of chromosomes. This number is called the **haploid number**. It is represented by **n**. The haploid number (n) for humans is 23.

A human sperm (male gamete) has 23 chromosomes (one of each homologous pair). A human egg (female gamete) also contains 23 chromosomes (the homologues of the 23 in the sperm). During fertilization, the sperm and egg unite to form a zygote. The zygote will have 46 chromosomes, in 23 homologous pairs. Thus sexual reproduction has restored the diploid number of chromosomes. All your body cells came from that zygote by mitosis and cell division. Therefore, all your body cells have the diploid number of chromosomes. Note that one homologue of each pair came from one parent and the other homologue came from the other parent. As you know, the haploid gametes are formed in the sex organs by meiosis. If you have forgotten how meiosis performs this task, go back to Section 8.4, page 179, and review the process before you go on.

After the process of meiosis was discovered, Mendel's garden pea experiments acted as experimental proof for the theory that chromosomes carry genetic information. Mendel believed that the hereditary "factors" segregated (separated) during the formation of gametes (Law of Segregation). During Stage 1 of meiosis, the chromosomes segregate. This fact suggests that the chromosomes carry the hereditary "factors" of Mendel's experiments. Only one factor, for example, the factor for flower colour, appeared to be carried into the gamete. Therefore it seemed that a factor for any given trait must be found only on one particular chromosome and at just one spot. Thus a **hereditary factor**, or **gene**, came to be defined as *a unit of inheritance located at a fixed position on a chromosome*.

The movement of chromosomes during Stage 1 of meiosis also demonstrates a genetic law called the **Law of Independent Assortment**. It states that the distribution of one pair of genes is not controlled by other genes in chromosomes that are not homologous. For example, in fruit flies, the recessive gene for white eyes is in chromosome number 1. The recessive gene for black body is in chromosome number 2. Because these are not homologous chromosomes, they do not pair during Stage 1 of meiotic division. Therefore, the chromosome carrying the gene for black body may or may not go to the same pole as the chromosome carrying the gene for white eyes. In other words, these genes are "assorted" or grouped at the poles, independently of one another.

Review

1. A fruit fly normally has eight chromosomes in each of its body cells.
   a) What is its haploid number?
   b) What is its diploid number?
   c) How many homologous pairs of chromosomes are in each cell?
2. Explain how the events that occur during Stage 1 of meiosis support Mendel's Law of Segregation.
3. Explain how the events that occur during Stage 1 of meiosis support the Law of Independent Assortment.

27.2 Investigation

Heredity in Fruit Flies

The fruit fly, *Drosophila melanogaster*, has long been used for studies in genetics. It has a short life cycle of 10 d at 25°C. Also, it is small. Because of these two factors, large populations can be raised quickly in the laboratory.

   Here is a brief outline of the life cycle of the fruit fly. The eggs produce larvae. These grow and, after a few days, become pupae. The pupae develop into adults. Low temperatures lengthen the lifetime of survivors. However, fewer of the flies survive. Temperatures above 30°C result in sterile flies and death.

   In the cross that you will carry out in this investigation, you will put five male and five female flies in each culture bottle. Therefore you must be able to sex the flies. The following three differences are fairly easy to recognize (Fig. 27-1). The tip of the abdomen in the male is more rounded than that of the female. The male abdomen has a solid black tip. The female abdomen has seven segments and the male abdomen has five segments.

   When handling these flies, be careful to keep them inactive with a substance like ether. They should be kept alive when you are counting and observing them. They will be used in the F$_1$ cross. You can tell if they have died by looking at their wings. When a fruitfly dies, its wings go at right angles to the body rather than lying flat.

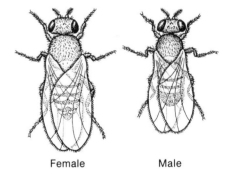

Female   Male

Fig. 27-1   Adult fruit flies (*Drosophila melanogaster*).

Materials

culture bottles (8) containing new
   medium (food)
stock cultures (2) of *D. melanogaster*,
   one of wild (normal) flies and one
   of flies with vestigial wings (wings
   which are small and crinkled)

ether
etherizer
soft paintbrush
low power microscope
white paper

CAUTION: Do not inhale ether. Make sure no flames are present in the room.

Procedure

a) Moisten the sponge of the etherizer with ether.

b) Thump the stock bottle with the wild flies in it against a soft object such as a sponge on the desk. This moves the flies to the bottom of the bottle.

c) Take the plug from the bottle. Then fit the top of the etherizer to the top of the stock bottle.

d) With the etherizer below the culture bottle, tap the whole unit on the desk several times. This should move the flies into the etherizer.

e) Remove the stock bottle and close the etherizer.

f) About 30 s after the flies stop moving, shake them out onto a sheet of white paper. They should stay inactive for 5 to 10 min.

g) Place 5 wild males in each of 2 culture bottles containing new medium. Place 5 wild females in each of another 2 culture bottles containing new medium. The flies will probably not be injured if you do this separation using a soft small paintbrush.

h) Repeat the etherizing procedure with the stock bottle of vestigial-wing flies.

i) Place 5 vestigial-wing males in with the 5 wild females. Then place 5 vestigial-wing females in with the 5 wild males. The 4 bottles should then contain 2 P (parent) crosses: wild male x vestigial-wing female; and vestigial-wing male x wild female. The bottles should be labelled as such. Include the date.

j) In about 7 d, when the F_1 larvae appear, release the P flies through a window. Then they will not be included in your F_1 generation observations.

k) When the F_1 flies emerge from the pupae, check the phenotypes of both males and females. Record them.

l) From each of the first 4 bottles, set up an F_1 cross of 5 males and 5 females in 4 new bottles containing medium. Note what the F_1 cross was for each bottle and label the new bottles accordingly.

m) Remove the F_1 flies when the F_2 larvae appear in about 7 d (see step j).

n) Count the F_2 flies as they emerge. Record both the phenotype and the sex of the flies from each bottle. Release any flies which you count so that none will be counted twice. You will need to count every 2 or 3 d for about a week, until all the F₂ flies have emerged.

o) Draw 2 Punnett Squares for each of your 2 crosses. One should predict the F₂ generation if the gene for vestigial-wing is sex-linked (see Section 27.3). The other should predict the F₂ generation if the gene is not sex-linked.

Discussion

1. What was the ratio of wild: vestigial-wing in your results?
2. Is the allele for vestigial wing dominant to or recessive to the wild-type wing? What ratio would you expect if the allele for vestigial wing were dominant?
3. Is vestigial-wing a sex-linked gene? Explain your answer.
4. Were the results in the F₁ generation all the same? If not, explain why they might not be.

27.3 Sex Determination and Sex Linkage

In organisms which are more advanced, reproduction is mainly sexual. In many plant species there are no recognizable sex chromosome differences. However, in many animals, special chromosomes are involved in determining sex. These **sex chromosomes** are called the **X** and **Y chromosomes**.

Humans have 46 chromosomes, forming 23 pairs. Of these, 22 are true homologous pairs. The remaining pair determines the sex of the person. This pair can be either the XX or XY combination of chromosomes. The XX person is female and the XY person is male. Figures 27-2 and 27-3 show human chromosomes as they appear under the light microscope during early mitosis. The chromosomes are in the double-stranded state. The pictures show the chromosomes before (Fig. 27-2) and after (Fig. 27-3) they have been karyotyped. A **karyotype** is a chart of the chromosomes of an individual. In making a karyotype, a biologist groups the chromosomes into their homologous pairs and numbers them according to their appearance. Number 1 is always the longest chromosome. The last chromosome is the shortest. Sometimes this pairing of chromosomes is made easier by treating them with chemicals so that special bands show on them. The karyotypes in Figure 27-4 are "banded" karyotypes.

The X chromosome carries the genes for female sex determination. The Y chromosome is made mainly of inactive material. In many species, the Y chromosome has few, if any, active genes. However, it appears to arrest the action of the X chromosome in some species. In fruit flies, it seems to carry a fertility factor. The Y chromosome usually has part of its length homologous to the X. The Y and X chromosomes will pair during meiosis. This means that each gamete will receive just one sex chromosome, rather than two or none, unless gamete formation is abnormal.

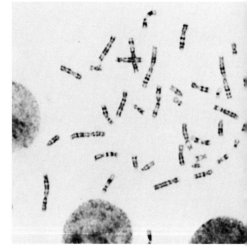

Fig. 27-2 A normal human male chromosome spread. The chromosomes have been chemically treated to produce "bands" called Giemsa — trypsin banding.

Fig. 27-3 Photograph A shows a normal human male karyotype using the chromosome spread of Figure 27-2. Photograph B shows a normal human female karyotype using Giemsa-trypsin banding.

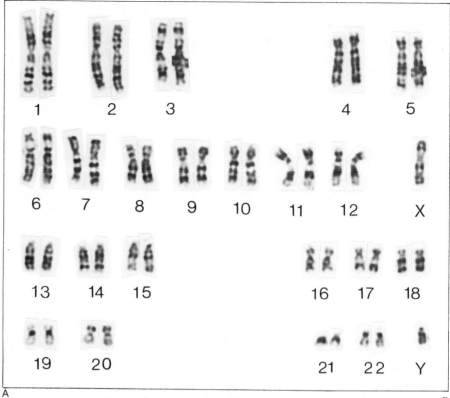

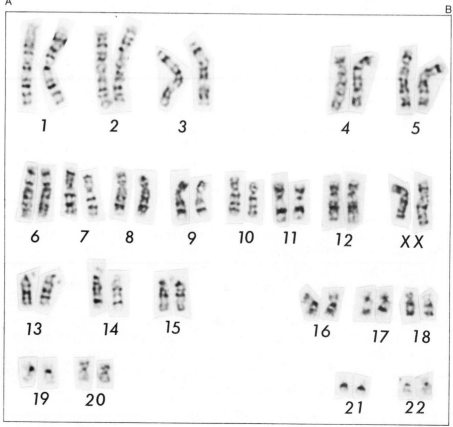

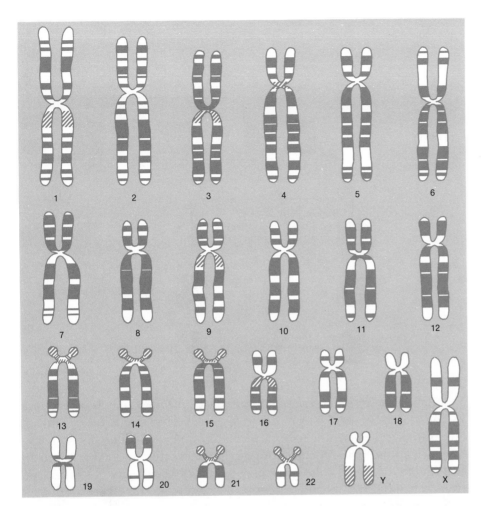

When gametes are formed in an XY system, again using the human as an example, all the eggs contain 23 chromosomes, one of which is an X chromosome. The sperm all contain 23 chromosomes. However, 50% of the sperm have an X chromosome and 50% have a Y chromosome. When the gametes meet at fertilization, the combination of sex chromosomes will be XX or XY. In this way, the sex of the offspring is determined. The XY pattern for sex determination is found in most higher animals and some plants.

Some organisms, such as birds and butterflies, have a reverse pattern. The male has the XX pair of sex chromosomes and the female has the XY combination of sex chromosomes.

When genes which are not alleles are in the same chromosome, they tend to stay together during meiosis. Therefore, they tend to be in the same gamete. This tendency to stay together is called **linkage**. For example, in the corn plant called maize, one gene for crinkly leaves and one gene for dwarf plants are close together on the same chromosome. They move together in meiosis and usually end up in the same gamete. Therefore, they are said to be "linked" genes. All the genes on a single chromosome are linked and tend to be inherited together.

Fig. 27-5 A human pedigree for red-green colour blindness. The genes for colour vision are carried on the X chromosome.

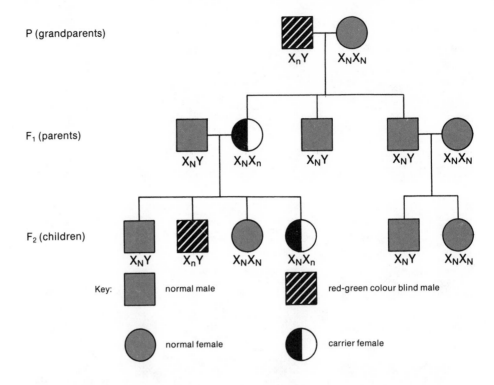

P (grandparents)

X_nY X_NX_N

F₁ (parents)

X_NY X_NX_n X_NY X_NY X_NX_N

F₂ (children)

X_NY X_nY X_NX_N X_NX_n X_NY X_NX_N

Key: normal male   red-green colour blind male

   normal female   carrier female

Genes for sex determination are in the X chromosome. If other genes are found to be in the X chromosome, these genes are said to be sex-linked. For example, the gene for red-green colour blindness in humans is a recessive allele found in the X chromosome. Therefore it is sex-linked. When an allele in the X chromosome is a recessive, it is expressed in the XY organism, even though there is only one recessive allele present. This is because there is no corresponding dominant allele for that gene on the Y chromosome. Remember, it is a small chromosome of mainly inactive material. A female who has an allele for red-green colour blindness on one X chromosome will not be red-green colour blind. The other X chromosome has a dominant normal allele to stop the red-green colour blindness from expressing itself fully. This female is called a **carrier** of the allele.

Let us consider two examples. First, with humans and many domestic animals, pedigrees are usually drawn to make it easier to study inheritance. A pedigree is a diagram of a family history (see Fig. 27-5). In this example the grandfather is red-green colour blind and his wife has normal vision. They have two sons who are not colour blind and one daughter who is not colour blind, but is a carrier of the allele. This means that, of her two X chromosomes, one has a recessive allele for colour blindness (X_n). The other has the dominant allele for normal vision (X_N). She has two sons. One is colour blind because he received a gamete with an X_n chromosome. The other has normal vision. His X chromosome has the allele for normal vision (X_N). Similarly, her two daughters each received a different type of X chromosome from her. Therefore, one daughter is normal and one is a carrier with normal vision.

The second example concerns fruit flies. One recessive allele (r) for white-eye colour in fruit flies is sex-linked. Therefore, it is on the X chromosome. If normal red-eyed females ($X_R X_R$) are crossed with white-eyed males ($X_R Y$), the F_1 generation are all red-eyed. This includes both females ($X_R X_r$) and males ($X_r Y$). The F_2 generation can be determined from a Punnett Square (Fig. 27-6). All the females and 50% of the males are red-eyed. The remaining 50% of the males are white-eyed.

Now, compare the two examples, humans and fruit flies. In both cases, the same pattern emerged. The original crosses produced all normal offspring in the F_1 phenotypes. However, all the females in this generation were carriers of the recessive allele. In the F_2 generation, all the females and 50% of the males had a normal phenotype. But 50% of the females were, in fact, carriers. Also, 50% of the males were affected by the recessive allele. This pattern of inheritance is standard for sex-linked genes. Note that the recessive sex-linked gene cannot pass directly from father to son. This is because male offspring receive their X chromosome from the mother.

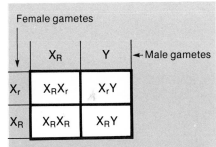

F2 phenotypes:
females all have normal vision (50% carriers); males are 50% normal, 50% colour blind.

Fig. 27-6 Punnett square of the F_2 generation of a cross of red-eyed female fruit flies with white-eyed male fruit flies.

Review

1. Show why, in humans, 50% of sperm contain an X chromosome and 50% contain a Y chromosome.
2. What is linkage? Explain how linkage and sex-linkage are similar. How do they differ?
3. A man and a woman both have normal vision. They have three offspring, all of whom marry people with normal vision. The three offspring and their children are as follows:
   a) A son with red-green colour blindness who has a daughter with normal vision.
   b) A daughter who has three sons with normal vision.
   c) A daughter who has one red-green colour blind son and one son with normal vision. Draw the pedigree for this family, indicating the affected people and the carriers.
4. In fruit flies, one white-eye colour allele is sex-linked. Red-eye colour is normal and dominant to the white-eye colour. Draw a Punnett Square for the F_2 generation of the following crosses:
   a) white-eye female x red-eye male
   b) female carrier of white-eye x red-eye male
   c) female carrier of white-eye x white-eye male

27.4 Investigation

The Pattern of Hemophilia in Queen Victoria's Pedigree

Queen Victoria of England is thought to have been a carrier of a sex-linked recessive gene for hemophilia. Hemophilia is a condition in which the affected person bleeds easily and for extended periods of time when injured. In other words, the blood does not clot normally.

This "dry lab" uses Queen Victoria's pedigree as it is now known. A few changes and omissions have been made. As you can see in Figure 27-7, the marks indicating who were hemophiliacs, who were carriers, and who were normal have been removed from the middle generations, that is, generations II and III.

Fig. 27-7 A partial pedigree of Queen Victoria's family.

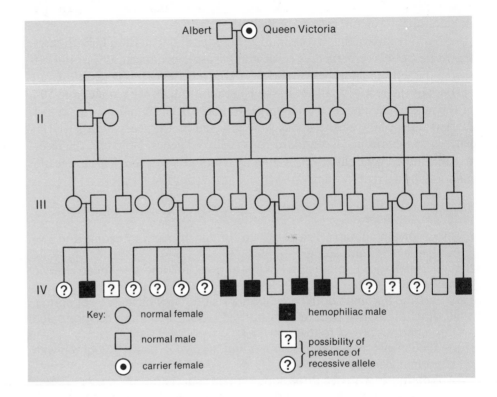

Materials

pencil
eraser
blank paper

Procedure

a) Draw the pedigree as it is shown in Figure 27-7.
b) Mark the intermediate generations until you have a reasonable pattern to connect Queen Victoria and her descendants. It is best to begin at generation IV and work back to generation III. Some members of generation III do not have offspring shown. Using the information in Section 27.3 and Fig. 27-5, estimate what percentage of the members should be marked as carriers or affected males. Mark them. Similarly, work back to generation II, estimate the percentage of carriers and affected males, and mark them.

Discussion

1. Check your version of this pedigree against the real one. It was published in Scientific American, Aug. 1965, pages 88-95. Your teacher has a copy of it. How close were you in your version?
2. If you were able to obtain the results of the PTC taste test (Section 27.5) for a few generations of your family, draw your own pedigree. Mark the "non-tasters", even though the gene is not sex-linked.
3. If red-green colour blindness is present in your family, or in the family of someone you know, interview the members of the family and draw up a pedigree indicating affected members of the family and possible carriers.

27.5 Chromosomal Variations

Most individuals of a species have the normal chromosome number, 2n. For example, most trilliums have 10 chromosomes, in 5 homologous pairs, in each cell. However, there are some individuals which have other than the usual 2n number of chromosomes. Also, individuals may develop with some of their chromosomes different in shape or size from the corresponding chromosomes in other individuals.

Differences in the number of chromosomes per cell usually occur because the chromosomes do not separate properly during cell division, be it mitosis or meiosis. For all species, the symbol, n, stands for the haploid number. In most species, the diploid number, 2n, is the normal chromosome

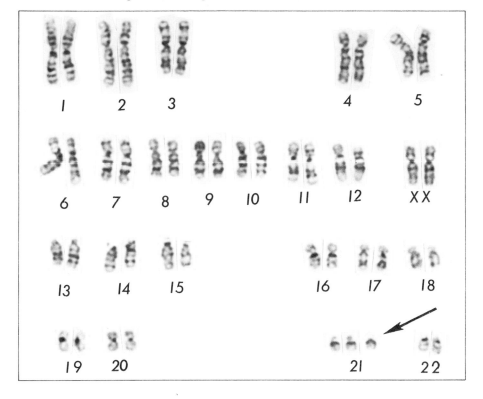

Fig. 27-8 A banded karyotype of a female with an extra 21st chromosome. This condition is called Down's Syndrome.

Fig. 27-9 Two ways in which a pair of homologous chromosomes might separate incorrectly and result in some abnormal gametes.

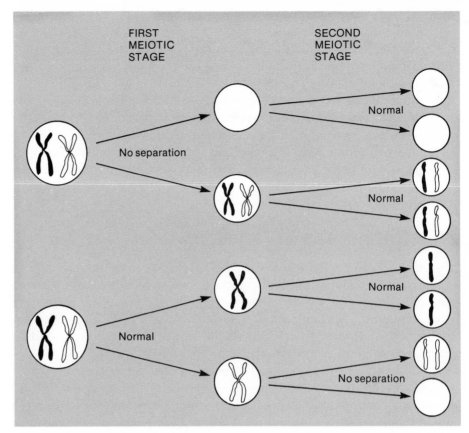

count per cell. However, some species normally have a chromosome number of 3n or 4n. These species are said to be **polyploid**. They have a multiple of n chromosomes, but it is greater than 2n. Polyploids are commonly plants. Most are 4n in chromosome number. Biologists believe that the development of polyploid species can occur in several ways. The simplest method occurs when the chromosomes of the organism fail to separate in the first division after fertilization. The new organism then has 4n chromosomes in every cell. Some kinds of barley, corn, and lettuce are polyploid plants that are thought to have formed this way. However, this simple method is not common. Almost all plant polyploids are formed by hybridization, the crossing of two species. Their chromosomes are often very different, and the 2n offspring are usually infertile. Any 4n offspring which might form by chance are usually fertile, and can continue on as a new plant form. Chromosome numbers higher than 4n are less common. However, New Zealand grass is approximately 38-ploid.

Polyploid animal species that reproduce sexually are uncommon. Their sex chromosomes are multiple (XXXX and XXYY). These individuals are usually sterile. When they are paired with a normal XX or XY individual, even more abnormal situations arise (for example, XXXY). Organisms with these chromosome patterns are usually either sterile or unable to survive.

Whenever the number of chromosomes is not an exact multiple of n,

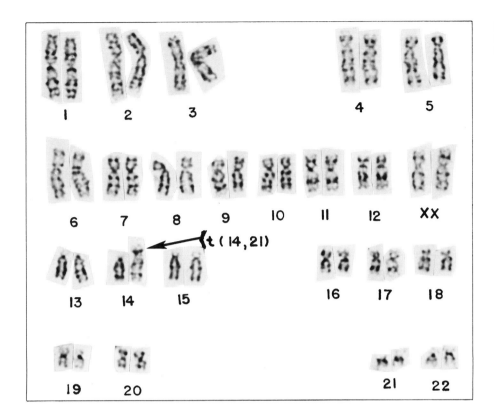

Fig. 27-10   A banded karyotype of a female showing a 21 chromosome fused to the end of a 14 chromosome (see arrow).

the organism produced is frequently abnormal. For example, some humans have 2n + 1 chromosomes. When the extra chromosome is number 21, the individual has Down's Syndrome and is mentally retarded. Compare the karyotype of a person with Down's Syndrome in Figure 27-8 with that of a normal person in Figure 27-3. Sometimes the number is 2n–1. For example, a female mouse can have XO rather than XX for sex chromosomes. Species of plants and animals vary in their ability to survive with chromosome numbers other than 2n or exact multiples of n. Petunia and corn tolerate extra chromosomes well. However, tomato does not.

Chromosome numbers other than 2n or exact multiples of n usually arise because separation of one or more homologous pairs of chromosomes in meiosis has not occurred properly. In the case of the XO mouse mentioned above, the egg which helped form that individual might have had no X chromosome. This would happen if, at the first meiotic stage, the XX which paired at the equator did not separate. Instead, both went to one pole and left the other pole with no X chromosome. Similarly, this might have occurred during the second meiotic stage rather than the first division. The chromatids might not have separated properly (Fig. 27-9).

Occasionally, an organism is found which has one chromosome less than the normal diploid number. In this organism, one small chromosome or part of a chromosome may have fused to another (Fig. 27-10). However, it still has the right amount of total chromosome material. Therefore such organisms are normal in phenotype, although their karyotype may be

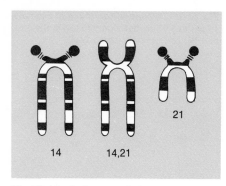

Fig. 27-11 A diagram of #14, #14/21, and #21 of Figure 27-10.

unusual. These organisms may produce gametes which are imbalanced. As a result, they may have abnormal offspring.

The ways in which chromosome abnormalities occur are now known to some extent. Chromosome fusion is one way. However, the reasons why chromosome abnormalities occur are largely unknown. Radiation can cause chromosome damage such as loss of parts of chromosomes or fusion of some chromosomes. Also, radiation may cause poor separation of chromosomes in cell division. Some viruses, such as the measles virus, can cause chromosome damage. Some chemicals, such as drugs, may also cause damage to chromosomes. Some theories suggest that certain genes can lead to poor separation of chromosomes in cell division. Unfortunately, little is definitely known about the causes of many chromosome abnormalities and variations.

Review

1. Figure 27-8 shows a karyotype of a person with an extra 21 chromosome. This abnormality is found in about $\frac{1}{3}$ of all cases of mental retardation. Figure 27-11 shows an unusual group of chromosomes. One #14 has #21 attached to one end. If this 14/21 were present with two other #21's, the person still gets Down's Syndrome. Why? If this person had one #21, one #14, and one #14/21 (assuming all other chromosomes were normal), would you expect him/her to be normal or retarded? Explain your answer.

2. The tree frog *Hyla versicolor* has a karyotype as diagrammed in Figure 27-12.
   a) What is the haploid number of chromosomes in this animal?
   b) Express the total chromosome number in terms of n.
   c) Illustrate with diagrams a simple way that this condition might have come about.

3. Figure 27-13 is a karyotype of a person with an abnormal number of sex chromosomes.

Fig. 27-12 Karyotype of *Hyla versicolor*, the tree frog.

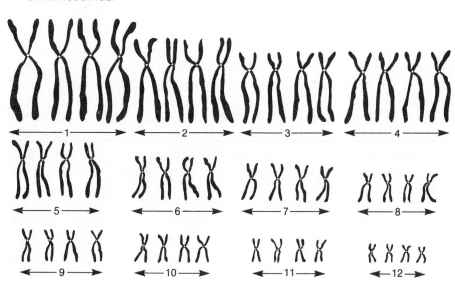

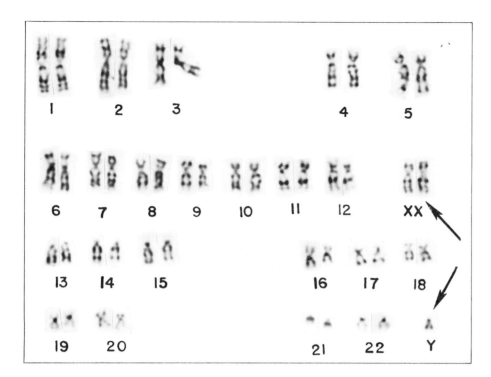

a) What is the sex chromosome makeup of this individual?
b) What are the possible combinations of sex chromosomes in the gametes which may have come together to produce this individual?

27.6 Control of Cellular Processes

The Genetic Material

Chromosomes contain a molecule called **deoxyribonucleic acid**, or DNA. This DNA contains the genetic code which controls the synthesis of proteins in the cell. In 1953, J.D.W. Watson and F.H.C. Crick, working at Cambridge University in England, proposed a model for the structure of the DNA molecule. They received a Nobel prize in 1962 for their research. This model led to a much better understanding of the nature of chromosomes and genes. The following is a simplified description of the model.

The model has the shape of a double helix. It looks like a ladder that has been twisted on itself (Fig. 27-14). Consider the ladder as if it were untwisted. The steps in the ladder are formed by chemical groups called **bases**. There are four types of bases. We will label these with the first letter of their full names—G, C, A, and T. Each step of the ladder is always formed by two bases plus their backbones of **sugar** and **phosphate**. These groups of sugar and phosphate form the sides of the ladder. The bases are made up in such a way that the G and C bases just join together and the A and T bases just join together (Fig. 27-15,A).

Replication of the chromosomes to form chromatids begins when a cell

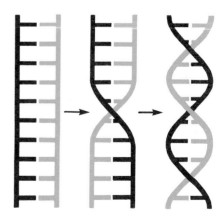

Fig. 27-14 The basic shape of DNA is illustrated best with a twisted ladder.

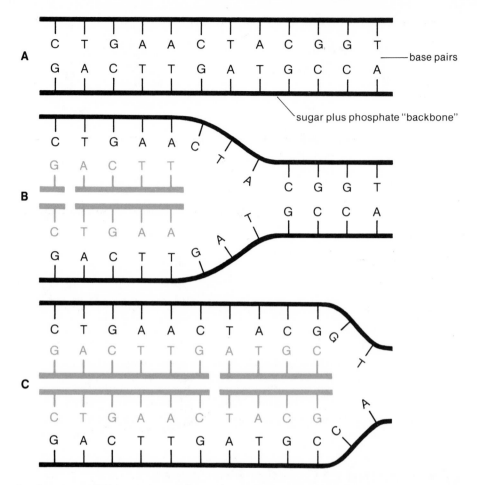

Fig. 27-15 DNA replication. A: the parent DNA molecule is untwisted. B and C: progressive stages of the replication process.

is about to divide and continues into the beginning of cell division. This replication of DNA must ensure that the DNA going into the daughter cells is exactly like the DNA of the parent cell. Figure 27-15, B and C, illustrates this replication process. The DNA untwists itself. Then the pairs of bases begin to separate along the DNA molecule. As these bases separate, other bases, with their attached sugars and phosphates, float into the area. These bases will pair with the exposed bases of the DNA strand, A with T and C with G. A pair of DNA molecules will be formed. This process requires the help of another molecule, **ribonucleic acid**, or **RNA**. If you look at Figure 27-15 carefully, you will see that the sequence of bases in the two new DNA molecules is exactly the same as in the parent molecule. However, knowing that two daughter molecules of DNA are the same as the parent molecule still does not tell us how DNA controls what goes on in the cell.

The molecule RNA, which assists in the replication of DNA, plays an important role in the control of the cell. RNA is known to exist in the cell in three forms. These are **mRNA** (messenger RNA), **tRNA** (transfer RNA), and **rRNA** (ribosomal RNA). As you have probably noticed, DNA (deoxyribonucleic acid) and RNA (ribonucleic acid) are very similar in name. They are also somewhat similar in structure. RNA differs from DNA in the sugar which is its "backbone". Also, RNA contains the bases C, G, and A, but contains a

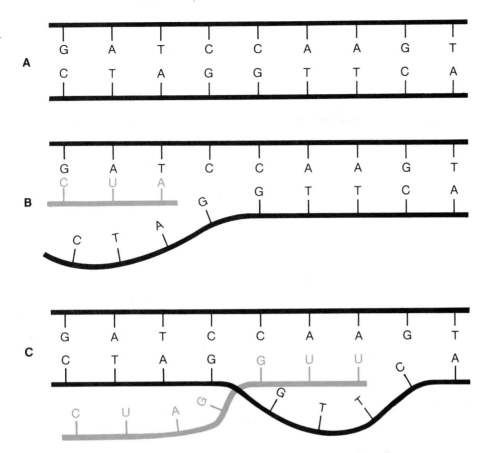

Fig. 27-16 Production of RNA from a DNA template. A: DNA template. B: RNA forms, using top side of ladder as its template. C: a later stage of RNA production.

A

Top strand: G A T C C A A G T
Bottom strand: C T A G G T T C A

B

Top strand: G A T C C A A G T
RNA: C U A
G T T C A

G
A
C T

C

Top strand: G A T C C A A G T
Bottom strand: C T A G G U U A
RNA: G G T T C
C U A G

base which we will label U, in place of the base T. In addition, RNA is a single strand rather than a double strand and is found in the cytoplasm. Most RNA is made using DNA as a **template**, or as a pattern for production. The DNA code of information is carried from the nucleus to the cytoplasm by mRNA. This is how the DNA indirectly controls what goes on in the cell. To produce RNA, one strand of a section of DNA is "unzipped" and "read". The RNA bases (C, G, A, and U) and their sugar-phosphates move into place and link up. When the DNA is being "read", it is always "read" in one direction. In Figure 27-16, the bases on the top of the ladder would read as GATCCA . . . The sequence would not be read as TGAACCT . . . Notice in the same figure that U takes up any spot that T would be in if DNA were being formed. This is controlled by molecules in the nucleus.

You can see here that the RNA that is produced is like a mirror-image of the DNA from which it was made, except that U replaces T. In other words, the sequence of the bases is kept, even though it is kept by the "partner" bases. Thus, the information of the DNA has a way to get out into the cytoplasm without the DNA having to leave the nucleus itself.

The RNA that is made from the DNA is in several forms. **Messenger RNA** (mRNA) carries the DNA code into the cytoplasm where it assists in the making of protein molecules. **Transfer RNA** (tRNA) is used to transfer amino acid molecules to the ribosomes in the cytoplasm. **Ribosomal RNA** (rRNA) is needed to form the ribosomes, the sites of protein synthesis.

1. **a)** If the pattern of bases in one side of a piece of DNA were AGGCTACATGA, what would the sequence of the bases be in the other side of the same section of DNA?
    b) If mRNA were produced from this same segment of DNA, what would its base sequence be?
2. How are DNA and RNA similar? How do they differ?
3. **a)** In what part of the cell is DNA mainly produced?
    b) Where is RNA mainly produced?
    c) What are the three main types of RNA? Where are they chiefly found in the cell?
    d) Describe the functions of the different types of RNA.

27.7 Making Protein in Cells

Proteins are very important molecules in the working of a cell. For example, many cellular proteins are **enzymes**. These protein molecules control how quickly products are made in the cell. As a result, they also control how much of a cell product is made. Since DNA directs what types of proteins are made, it controls how much of any given enzyme is in the cell. Since DNA controls which enzymes are produced, it indirectly controls the processes going on in the cell that are affected by enzymes. The making of sugars, fats, and hormones, and the transport of materials into and out of the cell are a few of the processes that are indirectly controlled by DNA. Some proteins which are produced under the control of DNA are not enzymes. Instead they are proteins needed in structures in the cell. Still other proteins are made to act as antibodies for the organism. As you can see, controlling the manufacture of protein means controlling the whole cell. DNA has that control (Fig. 27-17).

The actual production of protein occurs in several steps. First, the

Fig. 27-17 Some ways in which DNA controls cellular processes.

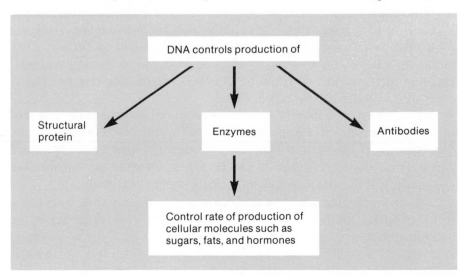

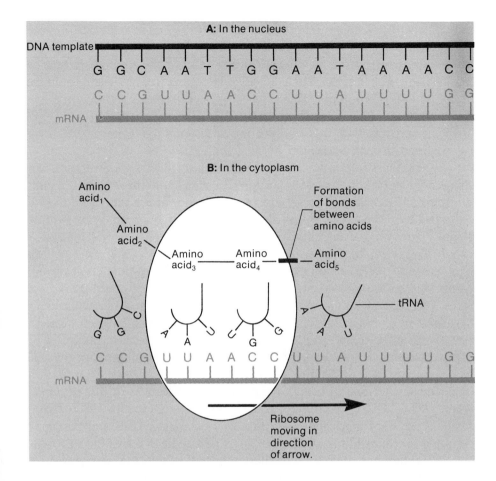

A: In the nucleus

DNA template

G G C A A T T G G A A T A A A A C C

C C G U U A A C C U U A U U U U G G

mRNA

B: In the cytoplasm

Amino acid₁

Amino acid₂

Amino acid₃ — Amino acid₄

Formation of bonds between amino acids

Amino acid₅

tRNA

C C G U U A A C C U U A U U U U G G

mRNA

Ribosome moving in direction of arrow.

Fig. 27-18   Stages in the synthesis of protein.

appropriate section of DNA unzips and one side of the ladder is "read" in the right direction. This process is controlled by enzymes. Messenger RNA is made from this template (Fig. 27-18,A). This mRNA enters the cytoplasm and becomes attached to a ribosome. The ribosomes contain rRNA and "read" the mRNA by reading its bases in groups of three. These base triplets are read in a straight line and do not overlap. They are important in the next step which involves transfer RNA.

Protein molecules are made up of one or more chains of molecules called **amino acids**. When several amino acids are linked together, they are called a **polypeptide**, or a polypeptide chain. Transfer RNA carries amino acids to the mRNA so they can be linked up to form a polypeptide. The order of the amino acids in the polypeptide is important in manufacturing a usable protein. The tRNA has three bases open at the bottom of its molecule. These three bases seem to determine what amino acid will be attached to the top of the tRNA. This "triplet" of bases can also pair up with partner bases which are "read" into the mRNA (Fig. 27-18,B).

As the ribosome moves along the length of the mRNA, the tRNA, with its attached amino acid, moves into the triplet of the mRNA which is being read. The tRNA then momentarily attaches to its partner bases. Notice that the triplets of bases at the bottom of the tRNA, when in sequence, are the same as the original DNA, except that U replaces T in the tRNA. As the amino acids fall in side by side at the tops of the tRNA, bonds form between the amino acids. The first tRNA is released and goes out free into the cytoplasm. There it picks up another amino acid. The joined amino acids have formed a polypeptide chain in the right sequence. When the chain takes on the proper shape, it may join with other polypeptides, or remain on its own, and form the required protein molecule.

When it was found that a section of DNA controlled the production of a single polypeptide chain, the definition of a gene had to be looked at again. A gene had been defined as a unit of inheritance at a fixed position in a chromosome. The smallest unit that DNA controls is a single polypeptide chain. Therefore, the current definition of a gene is: *A gene is the amount of DNA used to produce a certain polypeptide chain.*

When a gene, a certain piece of DNA, changes, a **mutation** has occurred. Mutations can be spontaneous. That is, they can happen naturally. This is rare. The role of spontaneous mutation in the process of evolution is discussed in Chapter 28. Scientists can make mutations happen. That is, they can induce them. Mutations can be induced by using radiation or chemical agents. How does a mutation appear in the DNA? Imagine that the base sequence in one side of some DNA is ATCCGTTAGCAAGT. A mutation might mean that one or more of these bases was lost, as in, for example, ATCCGT*AGCAAGT. Similarly, a base might be added to the original sequence, as in, for example, ATCCGTTAG*GCAAGT. Also, one base may simply be exchanged for another, as in, for example, ATCCC*TTAGCAAGT.

This changing of the original base sequence of the DNA may seem very simple. But its effect can be drastic. Consider the original base sequence in the triplets in which it would be read—ATC CGT TAG CAA. If one base were lost from the sequence, as shown, the triplets would be ATC CGT AGC AAG. As you can see, up to the spot where the base was lost, the amino acids would all be the same, since the triplets are the same. After that spot, most of the amino acids would be different because the triplets have changed. This means that the polypeptide chain that is made from the DNA with the mutation in it may not work properly. This is especially true if the mutation is an important part of the chain. For example, Queen Victoria probably carried a mutation that occurred spontaneously. The mutation involved one of the genes which controls a protein needed for blood clotting. This change in DNA was in the X chromosome. Therefore the mutated gene was X-linked. Because the mutated gene did not show up in Queen Victoria herself, it must have been recessive. Most mutated genes are recessive.

Throughout all this discussion, we have been looking at DNA which is active and controls protein production. Some DNA in chromosomes is inactive DNA. What decides which DNA will be active and which will be inactive is not known at this time.

1. Beginning with Mendel's definition of a gene, outline how the concept of the gene has evolved into its present form.
2. **a)** A section of DNA has a base sequence of AATGGCAAACCCTCCTCA. What will be the order of the bases in the mRNA made from this DNA?
   b) In Table 27-1, the base triplets of several types of tRNA are listed. What is the order of the amino acids in the final polypeptide chain using the DNA code in part a)?

ATTACHED AMINO ACID	tRNA BASE TRIPLET
B	AAA
D	GGC
E	UCC
F	UCA
H	AAU
J	CCC

TABLE 27-1

3. **a)** A scientist has found that, by using a certain chemical agent, he can cause mutations which affect a certain protein. Is this a spontaneous or an induced mutation?
   b) In part of the new protein, the amino acids are in the order LMNOPQRS. How many bases in the DNA are involved indirectly in the production of this piece of protein?
   c) The original protein had its amino acids in the order LMNOPQXY. How many triplets are different in the new DNA?

27.8 Investigation

Mutation Studies in Plants

Irradiation can induce mutations. To a certain extent, the amount of mutation induced is related to the amount of radiation to which the organisms are exposed. Some types of organisms are more resistant to radiation than others. In this investigation, you will be comparing the resistance of two types of seed, squash and marigold, to different levels of radiation.

Materials

potting soil
enough containers to hold 50 seeds, planted 1 cm deep and 5-8 cm apart
squash seed—non-irradiated control seed, and seed exposed to 4 different doses of radiation (for example, 5 000 R (röntgen), 10 000 R, 50 000 R, and 100 000 R)
marigold seed—non-irradiated control seed, and seed exposed to the same doses of radiation as the squash seed

Procedure

a. Form a group of 3 to 5 students.

b. Using potting soil, fill the containers to a depth of 8-10 cm.

c. Plant 5 squash seeds and 5 marigold seeds of each radiation dosage. Also, plant 5 control seeds of both squash and marigold. Plant the squash seed 1 cm deep and 5-8 cm apart. Plant the marigold seed 0.5 cm deep and 5 cm apart. Label the containers.

d. Water the containers whenever the soil appears to be drying out.

e. Observe the containers at intervals of 1, 2, 3, and 4 weeks. Record the appearance of all seedlings at each interval. Keep the marigold and squash results separate.

f. Pool your class results.

Discussion

1. Irradiated seeds frequently have mutations which are visible as colour differences and larger sizes of leaves or fruit. How did your plants, grown from the irradiated seed, compare with those grown from the control seed?

2. In what ways, if any, did the squash and marigold vary in the dosage level at which they were sensitive to radiation?

3. a) Did the number of mutations in either plant species vary with the amount of radiation?

   b) Was there a dosage level at which the seeds were so damaged that they did not germinate? If so, what was the dosage?

4. Why is it necessary to plant control seeds?

27.9 Current Issues in Genetics

Recombinant DNA

As biologists uncover more and more information about genes, problems begin to arise. Will humans be able to deal wisely with this information?

Some single genes can now be isolated in a reasonably pure state. For example, DNA can be separated from one virus. The virus DNA can then be placed in a host cell. The virus DNA joins or recombines with the host DNA. This new **recombinant DNA** is then replicated when the host cell divides. The new cells containing the recombinant DNA have different traits than the original host cells. The new cells can make proteins which they could not previously make. Will this ability to produce recombinant DNA be used to produce very infectious "germs" for "germ warfare"? Will we be able to ask for and get the exact kinds of babies we want?

People with diseases such as diabetes mellitus, in which not enough insulin is produced, could possibly be helped. Insulin is in limited supply.

Suppose biologists could make recombinant DNA which can indirectly produce insulin (a protein). Suppose, also, that they could develop bacteria which contain this DNA. Then insulin could be produced in larger amounts from bacterial culture. Perhaps bacteria could be cultured to produce blood-clotting factors for people with hemophilia. Perhaps bacteria could be developed to break down plastics which have been specially treated, helping to reduce one of our pollution problems. Bacteria might be designed to make food directly from the sun, like green plants do. Or strains of bacteria might be developed to split the water molecule. This would produce another energy source, something that we need right now.

All these possibilities suggest that, whether the intentions are good or bad, all the bacteria developed will be planned. But suppose that there is a slip-up which is not intended? A very harmful bacterium might escape from a laboratory and infect people.

Scientists in this field know about these risks. In 1974, they volunteered to stop certain kinds of experiments until guidelines could be made for them. Some experiments are allowed only in very protected areas. Perhaps with these limitations, the experiments which will benefit the world will continue and the chances of disaster will be kept very small.

Genetic Engineering

Discoveries in genetics have raised the issue of whether or not genes control behaviour. If they do, can behaviour be changed to something more useful and less harmful by **genetic engineering**, that is, by changing gene makeup? How do you decide what is unacceptable behaviour? Geneticists who feel that genes affect behaviour may use the Lesch-Nyhan syndrome as an example. This is a condition in humans in which there is mental retardation and several physical problems, as well as strange behaviour. These people bite themselves (self-mutilation). The condition arises from the lack of one particular enzyme. It develops from a defective gene. But does this prove that other behaviours are caused by gene action? The role of genes (and chromosomes) in determining behaviour requires much more research before accurate answers are obtained.

Genetic Counselling

Advances in genetics have increased the demand for genetic counsellors. They are needed to support and give information to families with problems that are inherited. Usually people who come to genetic counsellors are parents of a child with a genetic defect. They want to know whether or not another child of theirs might be affected. The counsellors and the parents discuss the likelihood of another child being born with the defect. They also discuss the burden that the defect will likely place on the child, the family, and on society. The parents can then decide whether to have another child. Also, they will be better prepared should the next child be defective.

Other people who may want counselling include: closely-related married couples who want to know about the risks for future offspring; women who are older and want to have children, but do not have accurate information on the risks involved; and adoptive parents being counselled about the child they are about to adopt. One of the main problems that the counsellor faces is the problem of accurately determining the risk of having a child with the same defect again. Scientists strongly suspect that many diseases may be caused by several defective genes rather than just one. It is therefore difficult to predict what undesirable traits offspring will inherit. Another problem is determining which people are carriers of defects. Unfortunately, for some of the more common disorders such as cystic fibrosis, reliable tests for carriers are not yet available.

Occasionally, a genetic counsellor can talk with some certainty. Some defects can be identified during pregnancy. A small amount of fluid can be drawn from the amniotic sac in which the baby is carried inside the mother's womb. Cells from this fluid are grown and tested for abnormal chromosomes and for any one of over 40 disorders in body chemical reactions. However, this procedure is not the final answer. Many genetic disorders have no test for them as yet. In these cases, the procedure would be useless, as well as an unnecessary risk to the mother and the unborn child. Any new, reliable tests for genetic defects in the unborn child are welcomed by genetic counsellors.

Review

1. Define recombinant DNA. What are some ways, other than those already mentioned, in which recombinant DNA could be used?
2. A husband and wife wish to know if the child she is carrying is a boy. They are told that the procedure of removing fluid from the baby's sac has some risk for both the mother and child. They still want the procedure done. Should the genetic counsellor have the necessary procedure done to see if they are having a boy? In your answer, consider whether people should be able to use this procedure for other than medical reasons.
3. A survey of newborn children is being run in a hospital. One of the infants is found to be XYY. Some researchers feel that there is evidence that an extra Y chromosome produces a male who is aggressive and prone to criminal behaviour. Others feel that some XYY's are normal and, in cases where parents have known about their child being XYY, the child has become aggressive because aggressive behaviour has been expected of him. Should the parents be told that the child may have behaviour problems? Should they be told at all that the child is XYY? Should the child be raised in a special way?

Highlights

Mendel's experiments with the common garden pea served as experimental evidence to support theories regarding the nature of chromosomes and genes. For example, Mendel's experiments resulted in his Law of Segregation. This law states that the hereditary factors separate during the formation of gametes. During Stage 1 of meiosis, the chromosomes segregate. Thus Mendel's findings supported the belief that chromosomes carry the hereditary material.

Humans have 46 chromosomes, of which 22 are true homologous pairs. The remaining pair can be an XX or an XY combination of sex chromosomes. An XX person is female and an XY person is male. Genes for factors other than sex may be found in an X chromosome. Such genes are called sex-linked, or X-linked. Red-green colour blindness is an example of a sex-linked factor.

Some individuals of a species may not have the normal 2n chromosome number. For example, Down's Syndrome is caused by a chromosome number of 2n + 1.

Cellular processes are controlled by DNA. DNA replicates itself during nuclear division. It also plays an important role in the control of cells. DNA directs protein synthesis. RNA plays an important role in this process. Enzymes are proteins. Therefore DNA controls enzyme synthesis. As a result, it also controls many important cellular processes such as the making of sugars, fats, and hormones, as well as the transport of material into and out of cells.

Recent studies in genetics deal with recombinant DNA, genetic engineering, and genetic counselling. Important moral decisions must accompany scientific work in these areas.

Genetics and Populations

Before many biologists were studying chromosomes and genetic material, biologists were studying how species in large groups, or populations, acquired variations. Much controversy arose over new theories about how populations change with time. Before long, it became obvious that two factors, heredity and environment, play a key role in many of the theories.

28.1 Heredity and Environment

The surroundings of an organism are its **environment**. The transfer of traits from generation to generation is **heredity**. Every organism inherits the potential to be something. But it will not necessarily reach that potential. Whether it reaches the potential or not depends upon how favourable the environment is for the organism or any given trait. For example, a snapdragon plant may inherit the trait of tallness from its parent plants. This is heredity. Whether it reaches its full height potential will depend upon whether there is sufficient water in the soil, whether the soil is fertile, whether there is enough sunlight, and so on. This is environment. Similarly, corn plants inherit the ability to make chlorophyll. As a result, most corn plants are green when they are full grown. However, if the corn seed germinates and begins to grow in the dark, the plant will not be green. In other words, the environment is not right. However, after these plants are exposed to light, they will turn green. They will reach their full genetic potential with regard to chlorophyll.

The Lamarck theory

One man who studied the interplay of heredity and environment was a French biologist called Lamarck. He published his theory in 1809. He felt that changes in the environment of a species could bring about an adaptation or a variation in that species. According to his theory, the environment brought about changes in the heredity of traits. For example, suppose that a part of an organism had much use or had use in a new way, and changed shape as a result. Lamarck believed that the newly-acquired shape would be inherited, not the old shape. In the past, according to Lamarck, a short-necked animal may have existed that fed on tree leaves. As the lower leaves of the trees in the area were eaten, the animal stretched its neck somewhat to reach the next level of leaves. Over a lifetime, this animal's neck became somewhat longer from the continued stretching. Over many generations, the species in this population got longer and longer necks. Finally an animal like our present-day giraffe was produced (Fig. 28-1).

This theory originally gained considerable support. However, some experiments began to show different results. One experimenter cut the tails off mice for many successive generations. According to Lamarck, this tailless population should have started to produce offspring with shorter tails because the tails were not used. Unfortunately for Lamarck, the mice in the more recent generations had tails the same length as those of the original mice. Lamarck's theory was discredited by this type of experimental data.

Fig. 28-1 Did the giraffe develop a long neck as a result of continual stretching?

The Darwin-Wallace Theory

Another theory about heredity and environment was put forth many years later, in 1859. This theory was published jointly by Charles Darwin and Alfred Wallace. It has two parts, the Theory of Organic Evolution and the Theory of Natural Selection. Darwin published much more material than Wallace on these theories. As a result, he is often considered to be the only person who thought of the ideas.

The **Theory of Organic Evolution** stated that today's living species are descendents of species that were on earth in the past. The present day species are somewhat modified from the earlier species. This theory gained some acceptance from the general population of the time. The second theory, the **Theory of Natural Selection**, tried to explain how the changes in species occurred. It was this theory which became so controversial. Darwin and Wallace both knew that, without pressure from the environment, most species would multiply very quickly. For example, a population which doubled itself in the first year, would have four times the original number in the second year, eight times the original number in the third year, and so on. Darwin and Wallace had also seen that, in nature, population sizes remained fairly constant. They varied only slightly from year to year. When they put these two observations together, Darwin and Wallace decided that not all the eggs and sperm produced by organisms became zygotes. They also decided that not all zygotes became adults, and not all adults survived to reproduce. There was obviously a "struggle for existence" in nature.

The Theory of Natural Selection had another side to it. Darwin and Wallace noticed that, in nature, not all members of a species were exactly alike. In other words, there was quite a bit of individual variation. They concluded from this observation that some variations gave some individuals of a species an advantage when the competition was high. The variations would help those individuals survive. Eventually individuals with those variations would be present in larger numbers and would produce offspring that made up a greater percentage of the total. For example, if certain birds of a species acquired beaks that could split seeds more easily, those birds would be better fed. As a result, they would survive in larger numbers. In summary, Darwin and Wallace said that the environment of a species was the main cause of the evolution of that species. The natural environment selected those individuals most adapted for survival.

Immediately, questions arose from both biologists and the general public. Where did these variations in species come from? How could this process create any new species? Natural selection was seen by many as a very negative, destructive force. Further, Darwin and Wallace could not answer any of these questions. Genetics was not yet a science. Yet only in genetics could answers be found to such questions. During the first half of the 20th century, the modern **Theory of Evolution** developed. Since it is still a theory, alternative theories continue to be presented for consideration.

Some people, scientists included, have studied the modern Theory of Evolution and found it lacking in some respects. The idea of the creation of the world and everything in it by a superior being is not included in the Theory of Evolution. Yet the modern Theory of Evolution and the Theory of Special Creation do not necessarily exclude one another. There are those who feel that the mechanism of evolution, as seen in modern-day terms, may have been set in motion by a superior being. In this way, both the creationist and the evolutionist have a middle road to follow. Perhaps this is the direction in which both camps should look in the future.

Review

1. **a)** Distinguish between heredity and environment.
   b) A set of twins were said to be "identical" at birth. They were separated and put into different adoptive homes shortly after birth. One twin was loved and well fed; it became a healthy child. The other was poorly treated and suffered from malnutrition. When they were compared at age three, one was 90 cm tall, thin, fair-haired, had poor teeth, and was unhappy. The other child was 100 cm tall, muscular, fair-haired, had good teeth, and was in good spirits. Was the diagnosis at birth of "identical" twins wrong? Explain.
2. Describe Lamarck's theory of how species acquire new characteristics. Use an example in your discussion.
3. Describe the difference between the Theory of Organic Evolution and the Theory of Natural Selection.

4. The peppered moth is a species that is active at night and rests on tree trunks in the daytime. In the 1890s, it was plentiful in England. About 50% of these organisms were pale with dark markings. They blended well with the light, clean tree trunks on which they rested. About 50% were found to be quite dark with a few light markings. These could be easily seen on the light tree trunks. By the 1950s, about 95% of the moths were darker in colour in parts of the country where the air was polluted. Only 5% of the moths were of the light variety. In unpolluted areas, about 85% of the moths were light-coloured and about 15% were dark-coloured. Over the 60 years, industry had developed in certain areas, filling the air with soot. The soot collected on the bark and leaves of the trees. The peppered moth is preyed upon during the day by birds which pick them off the tree trunks. How might the Darwin-Wallace Theory of Natural Selection explain the shift in the percentage of dark-coloured peppered moths in the polluted areas of England?

28.2 The Evolutionary Process

Since the time of Darwin and Wallace, the Theory of Evolution and the Theory of Natural Selection have been changed somewhat. This section outlines the modern forms of these theories.

The Gene Pool

Natural selection is usually discussed in terms of its action on a given population. This population is in one geographical area, and the individuals of the population are of one species. These individuals prefer to breed with one another. But, occasionally, there is breeding with individuals of neighbouring populations of the same species. Because of the interbreeding, a group of genes is spread throughout the whole population. *This total genetic content of a population is called its* **gene pool**. The gene pool of one population is connected to the gene pools of neighbouring populations by the occasional breeding between these neighbouring populations.

Evolution works through these gene pools by way of new variations. There are two ways that new variations which are inherited may occur. They may be caused by mutations. Or, the variations may be acquired during meiosis or the first mitotic division after fertilization. These chromosomal variations may be one of several types. For example, chromosomes may exchange unequal amounts of genetic material. You have already studied examples in humans where one chromosome has fused to another (see Fig. 28-10). Figure 28-2 illustrates another type of variation that could be inherited. This person has two extra chromosomes. There are a number of possible explanations for them. One is that these may be extra bits of chromosomes produced during DNA replication in a faulty first mitotic division after fertilization. These extra chromosomes will be replicated at all later cell divisions. Again, this is an inheritable variation which can be passed on to this person's offspring. In any generation, some individuals

Fig. 28-2 A chromosome spread of a normal, healthy person carrying 2 extra chromosomes (see arrows), making a total of 48 chromosomes.

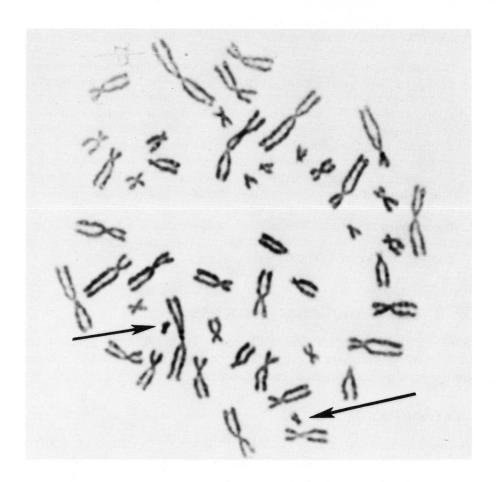

may appear with a new genetic variation. If these individuals survive and have offspring, a new variation will have been added to the gene pool. This variation may spread throughout the population.

Differential Reproductive Success

Whether a variation spreads or not depends upon **natural selection**, or **differential reproductive success**, as it is now known. This simply means that some individuals in the population have more surviving offspring than others. If an individual with a new genetic variation has differential reproductive success for several generations, the new variation is spread throughout the whole population. Consider a new variation that tends to give reproductive success to an individual. For example, the individual with the new variation leaves three offspring rather than the usual one. Eventually, the variation will spread throughout the population simply because of the numbers of offspring produced (Fig. 28-3).

Do the individuals that are best adapted to the environment leave relatively more offspring? Usually these well-adapted individuals are more successful. But this is not always so. A poorly-adapted individual may be very fertile. Also a very "fit" individual might be sterile and have no

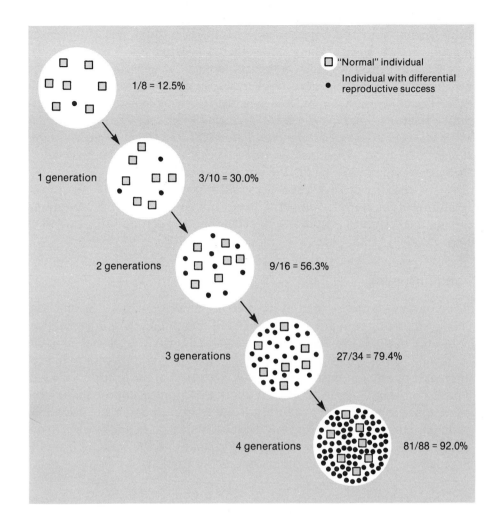

□ "Normal" individual

● Individual with differential reproductive success

1 generation   3/10 = 30.0%

2 generations   9/16 = 56.3%

3 generations   27/34 = 79.4%

4 generations   81/88 = 92.0%

1/8 = 12.5%

offspring. Therefore, the well-known principle of "survival of the fittest" is not always true. The factor that counts most in evolution is how many offspring live, not just their fitness or how well adapted they are to the environment.

When a variation appears in one individual of a population and spreads throughout the whole population, it is called a **unit of evolutionary change**. In order to have a new species, many units of evolutionary change must collect in a population. The new form of the population must be different from the original neighbouring populations. The differences can be physical, functional, or psychological. The formation of new species is discussed at greater length later in this chapter.

Hardy-Weinberg Law

The appearance of mutations and other genetic variations is random. This means that there is equal chance of them occurring in any individual of a species. The genetic variations which survive tend to produce better adapted individuals in a population. Natural selection spreads these better adapted traits throughout the population. Therefore, natural selection can be thought of as a positive thing rather than a negative one.

Over a period of time, biologists looked for a pattern in how the percentages of genes, or gene frequencies, in a population changed with time. The conclusion is now known as the **Hardy-Weinberg Law**. It states that, *if mating is random, if mutations do not occur, and if the population is large, the gene frequencies in a population will stay the same from generation to generation.* When the gene frequencies in a population stay the same, there is no evolution going on. Why?

The reason why evolution does take place is that all the "ifs" at the beginning of the Hardy-Weinberg Law are not always obeyed. Non-random mating does happen, mutations do occur, and populations are not always large.

Random mating

Random mating means that every member of a population has an equal chance of mating with any other member of that population. In actual fact, non-random mating is more common than random mating. Social structure in insects such as bees is a good example. Here, only certain groups of bees are mated at any given time. As a result, only part of the gene pool is used. This is non-random mating. Here is another example: suppose that a gene existed in a population in two forms, B and b. In each organism, it could be paired as BB, Bb, or bb. However, suppose that any organism with bb did not live to reproductive age. This removes many of the b alleles from the population. Therefore, random mating is not present.

Mutations

When a gene mutates, the mutation may have a good effect on a trait. It may also be harmful, or it may have a neutral effect in its present environment. Natural selection will either work for or against the mutation, if it is given enough time. The number of mutants will either increase or decrease in the population. If the mutation is dominant, its effect on the particular characteristic of the individual will be obvious at once. Natural selection will work either for or against it at once. However, mutations are usually recessive. Therefore it is often some time before their effect is seen. When a homozygous recessive individual finally appears in the population, natural selection will then act on it. Often, these recessive mutations spread through the population because they are linked to a dominant gene which has differential reproductive success. It is thought that "small" mutations (those which do not affect a trait drastically) tend to stay in populations longer than "large" mutations (such as major changes in the structure of the heart). Evolution is thought to happen as a collection of many small changes in characteristics rather than as a few large changes.

Population size

The Hardy-Weinberg Law also says that the population must be large. Because there are few matings in small populations, certain genes may be in high percentages by chance alone. For example, if you toss a coin several hundred times, approximately 50% will be heads and 50% will be tails. However, if you toss the coin only five times, by chance, they may all be heads. Similarly, consider the organisms in a population having a pair of genes that could be AA, Aa, or aa. By chance, few matings could produce all AA. This would eliminate the "a" allele from the population.

It is obvious that the Darwin-Wallace theories have changed greatly since their time. Newly-acquired knowledge in genetics and new techniques of study have made modification of their ideas possible.

Review

1. **a)** Define a gene pool.
   b) An individual in a population is born with a new genetic variation that gives it differential reproductive success. In the "normal" individuals, two offspring are left by each member of the population. The individuals with the new variation leave four offspring in the population. Show how the individuals with the new variation increase in percentage in the total population over the next four generations.
2. Compare the Darwin-Wallace Theory of Natural Selection to the modern-day view of natural selection.
3. **a)** State the Hardy-Weinberg Law.
   b) What is meant by non-random mating?
   c) Describe how non-random mating, mutation, and small populations affect the Hardy-Weinberg Law.
4. Some organisms reproduce only by asexual reproduction. How might they evolve?

28.3 Investigation

Analysis of Data Concerning Sickle-Cell Anemia

Sickle-cell anemia is an inherited blood condition. It is caused by a mutant recessive gene. In people with this disease, the red blood cells, which are normally disc-shaped, are C-shaped when oxygen levels are low. This occurs because the cells contain an unusual form of hemoglobin. Individuals who are homozygotes for the allele for sickle-cell hemoglobin tend to die early in life. Few survive to reproductive age. With this being the case, one would expect the mutant gene to be present in a low percentage in any population. However, the mutant gene is very common in some populations in Africa—up to 30%. Biologists have found that the mutation rate is low for the gene that controls hemoglobin production. The individuals carrying the gene for sickle-cell hemoglobin along with the normal gene must have a high survival rate in order that the frequency of the sickle-cell gene be high. These heterozygotes have a mild form of sickling of their red blood cells. Therefore they can be identified.

Materials

the information provided in Table 28-1

DISEASE	TOTAL NUMBER OF PEOPLE WITH THE DISEASE	% OF THE TOTAL WHO WERE HETEROZYGOTES FOR SICKLE-CELL GENE
1. Pneumonia	110	16
2. Malaria (mild form)	85	15
3. Malaria (severe form)	54	4
4. Tuberculosis	38	24
5. Diarrhea and vomiting	108	21
6. Typhoid Fever	25	33

TABLE 28-1 **Diseases Contracted in an African Population**

Procedure

Study Table 28-1 closely. It lists several diseases that were contracted by an African population. About 18% of the total population were heterozygotes for the sickle-cell gene. Is there any disease that appears not to have affected people carrying one sickle-cell gene? Do the heterozygotes have an advantage with regard to any diseases?

Discussion

1. What disease appears to affect very few carriers of the sickle-cell gene?
2. Investigate how this disease affects the body and what parts of the body it invades. Does it use red blood cells as part of its disease process?
3. The sickle-cell gene is only common where this disease is found in the population. Elsewhere, the gene is not as common. How does the modern Theory of Evolution explain why the sickle-cell gene has not disappeared from some African populations?

28.4 Investigation

Bacterial Resistance and Evolution

Bacteria are small, easily grown, and produce new generations very quickly. Therefore, they are often used in laboratory experiments in which many generations must be studied. In this investigation, the bacterium, *E. coli*, is used. *E. coli* is a bacterium that is usually found in the intestines of animals. It is often cultured in the laboratory on a solid material, or medium, called agar. The agar is melted. Then the nutrients that the bacteria require are added to the agar before it is solidified again. When *E. coli* is grown on nutrient agar, colonies are present after 24 h at 37° C. The colonies are round with smooth edges, and opaque. These colonies are relatively large, being up to 5 mm in diameter. They contain a large number of bacteria. Individual bacteria are not visible without a microscope. However, they are visible when present in large numbers in colonies.

The antibiotics, ampicillin and cephalothin, interfere with the normal function of the *E. coli* bacteria. When antibiotic (or sensitivity) discs are placed on nutrient agar, the antibiotic diffuses into the agar close to the disc. Usually *E. coli* colonies will not grow in this zone if ampicillin or cephalothin are used.

Materials

sterile petri dishes containing nutrient agar medium
marking pen
forceps (sterile)
Bunsen burner or alcohol lamp
inoculating needle
plugged test tube containing *E. coli* culture
sensitivity discs containing either ampicillin or cephalothin
incubator

Procedure

a. Hold the loop part of the inoculating needle in the flame of the burner or lamp until it becomes white hot.

b. Hold the loop out of the flame to cool. A loop that is too hot will kill the bacteria it touches. Do not blow on the loop to cool it. Also, do not set the needle down. Both of these actions will contaminate the loop.

c. Pick up the culture tube in your less dextrous hand (left hand if you are right-handed).

d. Remove the plug from the culture tube using the 5th digit of your dextrous hand. Do not set the plug down. This will contaminate the plug.

e. Pass the mouth of the culture tube through the flame. Then stick the inoculating loop into the tube. Do not touch the sides of the tube as the loop enters.

f. Gently remove some of the *E. coli* from the culture tube. If the *E. coli* is growing on an agar slant, try to avoid getting any agar on the loop.

g. Pass the mouth of the culture tube through the flame again. Then replace the plug.

h. With your less dextrous hand, remove the lid from the petri dish. Set it down so that the inside of the lid is open to the air.

i. Pick up the petri dish base. Gently "streak" the agar with the loop containing the culture. This means that you gently brush the top of the agar in straight lines. Do not puncture the surface of the agar with the loop. Try to have most of the surface of the agar covered by streaks. Do not be concerned if you cannot see the bacteria. They are there!

j. Put the loop in the flame until it is white hot. Then let it cool. Set it down.

k. Using the sterile forceps (if the discs are not in a dispenser), place a sensitivity disc in the centre of the streaked agar. Replace the lid immediately.

l. Label the petri dish with your name(s).

m. Incubate the petri dish at 37° C for 24 h. Then refrigerate if it is not going to be read at that time.

n. Record your observations.

Discussion

1. Are there any colonies present in the zone near the disc? If so, do they look like the colonies on the rest of the agar, or are they colonies resulting from contamination? Describe some ways in which colonies of *E. coli* might come to grow in this zone full of antibiotic.

2. How have the environmental conditions changed for all the *E. coli* near the disc? Suppose that some *E. coli* has mutated and is able to tolerate the antibiotic. Using the concept of differential reproductive success, describe how these results might be used to support the modern Theory of Evolution.

3. Has a "unit of evolutionary change" taken place in the case described in 2? Explain.

28.5 The Characteristics of Evolution

How do "units of evolutionary change" in a population give rise to a new species? This is the main question to be answered by followers of the modern Theory of Evolution. *A* species *may be defined as a group of populations sharing the same gene pool.* Within one gene pool, there is a free flow of genes. There is not a free flow between two gene pools. That is, a reproductive barrier separates two gene pools (two species). The problem of showing how new species come about is the same as the problem of showing how reproductive barriers form.

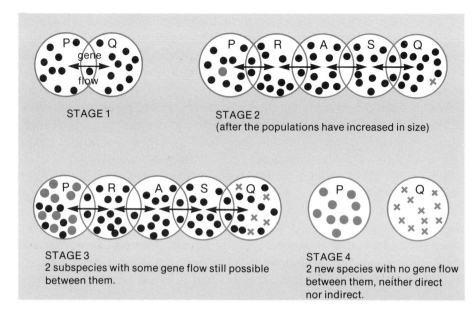

Fig. 28-4   How a new species can develop, according to the modern Theory of Evolution.

STAGE 1

STAGE 2
(after the populations have increased in size)

STAGE 3
2 subspecies with some gene flow still possible between them.

STAGE 4
2 new species with no gene flow between them, neither direct nor indirect.

Usually, barriers between populations develop geographically before reproductive barriers exist. The most effective geographic barrier is distance. For example, if a population of a species got larger and larger, the organisms would cover more and more territory. Eventually the populations P and Q, at the opposite ends of the territory, would be too far apart to have direct breeding between them (Fig. 28-4). There is still gene flow between the members of the populations in the middle of the territory, that is, between R, A, and S. As a result, there is still indirect gene flow between P and Q through the R, A, and S populations.

Suppose that, by chance, different genetic variations appear in both P and Q. Suppose, further, that these variations spread well in their own populations by natural selection, with differential reproductive success. This spreading by natural selection is easier if the environments in which populations P and Q are found are different. If the changes which happen in P and Q occur at a faster rate than the rate of flow of genes between P and Q through the other populations, the P and Q populations may become more and more different. Eventually, P and Q are different enough that they are **subspecies**. In other words, there are different traits in the two populations, but there can still be gene flow between them through the populations in between. The geographic barrier of distance has let differences develop between the two populations, P and Q.

Suppose the differences between population P and population Q became greater. For example, the gametes of neighbouring populations might form in such a way that they did not join together properly at fertilization. Or, perhaps a change in reproductive organs themselves make mating impossible, or the members of one population begin breeding earlier in the season than the other population. In these cases, gene flow would stop altogether. These are reproductive barriers. Because there is no gene flow between the two populations, they have become two different, new species.

The most common kind of geographic barrier is distance. However, there are other kinds—islands surrounded by water, a forest jutting into a prairie, and mountains. In time, geographic barriers give rise to new species. On the average, the estimated time taken for a new species to evolve is approximately one million years.

One characteristic of evolution is a process called **adaptive radiation**. In Figure 28-4, you saw that one parent species gave rise to two new species. This "branching" effect is found in the evolution of all organisms. Then the new species, in turn, becomes a possible ancestor of many different lines, which arise at about the same time. This evolving of different characteristics in closely related groups is called evolutionary **divergence**. The "branching" effect is shown in Figure 28-5. For example, a mammal-like ancestor gave rise to burrowing animals like moles, to tree dwellers like monkeys, to flying animals like bats, and to several types of aquatic animals like whales and seals, all at about the same time. It is this forming of lines of descent, all leading away from a common ancestor, that is called adaptive radiation. The old saying that "humans descended from the apes" is now known to be

Fig. 28-5 At the end of the paleozoic era and the beginning of the mesozoic era, conditions were such that a large number of new forms arose. This figure shows part of the "reptilian radiation" of that period and some of its members which persist to the present day.

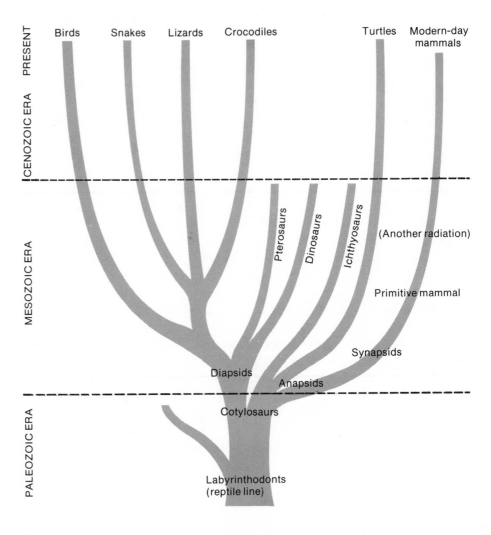

untrue. However, most biologists believe that humans and apes have a common ancestor.

Not all the branches of the "evolutionary tree" go all the way to the top (see Fig. 28-5). There are species that have become extinct. Why do some species become extinct? Environmental conditions change. Some species do not change fast enough to adapt to the different conditions in which they must live. These changes are not just physical. For example, suppose two species eat from the same supply of grass. However, one of them must eat only grass to survive, whereas the other eats meat as well. Clearly, the species which is very strict in its diet is more likely to die off if grass should become scarce. As the numbers of a species decrease, fewer potential mates are available. Eventually the species becomes extinct if environmental conditions remain the same.

The process of extinction allows for another process which is characteristic of evolution—**replacement**. Whenever a species becomes extinct, another group will eventually evolve to take its place in its environment. For example, ancient reptiles called ichthyosaurs were large, marine, fishlike animals that became extinct about one hundred million years ago. For about forty million years, their place in the environment was empty. Then the specialized mammals, dolphins and porpoises, evolved and filled the environmental spot left by the ichthyosaurs. This replacement took a long time. Sometimes the replacement is very rapid, as when several organisms evolve to fill a not-too-strict environment.

When two or more unrelated groups adapt to the same type of environment, whether they are replacing one another or not, they are evolving in the same direction. They may come to be alike in some ways. This is called **convergence**, or **convergent evolution**. For example, flightless birds like the emu, the ostrich, and the rhea, fill very similar roles on different continents. Similarly, the eyes of squid and of fish are much alike. Squid and fish are not directly related. However, they are both fast swimmers. Thus eyes made in a particular way are an advantage to both of these organisms in the way they live. Both these examples show how natural selection has brought about evolution in the same direction in two or more species.

As you can see from this chapter, the modern theory of evolution is dependent on new advances in genetics. As new information becomes available on the fine details of inheritance, the theory of evolution will be similarly modified. As a result, genetics and evolution should be considered constantly changing fields of study. They are not permanent and unchanging.

Review

1. **a)** Define a species.
   b) Describe how units of evolutionary change can result in the forming of a new species.
   c) The Labrador retriever and the dachshund are very different breeds of dog. Why are they considered the same species?

2. What is adaptive radiation? Give an example of this process.
3. Reproductive barriers arise between populations. How?
4. **a)** What causes extinction?
   b) Define replacement. Show how extinction and replacement are related.
5. Two birds, the African yellow-throated longclaw and the American meadowlark, look and act very much alike. Both prefer prairies and grasslands as habitats. Yet they are different species. Is this an example of convergence or divergence? Explain your answer.

Highlights

Both heredity and environment affect what an individual will become. According to the Lamarck Theory, the environment brings about changes in the inheritance of traits. This theory is no longer accepted by biologists. The Darwin-Wallace theory also deals with the relationship of heredity and environment. This theory has two parts, the Theory of Organic Evolution and the Theory of Natural Selection. The first part states that today's living species are descendents of species that were on earth in the past. The second part states that variations are responsible for evolutionary changes.

The total genetic content of a population is called the gene pool. Evolution works through a gene pool by means of variations. Variations may be induced by mutations, or by chromosomal variations during nuclear division. Whether a variation spreads or not depends on differential reproductive success, or natural selection.

The Hardy-Weinberg Law describes the effects of random mating, mutations and population size on gene frequencies in a population.

A species is a group of populations sharing the same gene pool. A reproductive barrier separates two gene pools. The barriers between populations sharing the same gene pool usually begin as geographical barriers.

Divergence is the evolving of different characteristics in closely related groups of organisms. Convergence occurs when two or more unrelated groups adapt to the same type of environment.

Many recent findings in genetics support the modern Theory of Evolution. However, this theory is still just a theory. It is not fact.

Index

of ruminants **465**
of vertebrates 465
Stomata (stoma) 390, 405-6, **406**
role in gas exchange of 410-11, **411**
and transpiration 412-13
Stoneworts 314, **315**, 333, **333**
Streams 666-67
bottom fauna of 660-62, **661**
Streptobacillus 207
Streptococcus 207, 208
Streptomyces griseus 237
Streptomycin 237
Structural formulas 80
Styles 357
Subclass 41
Suberin 392
Sublittoral zone 671
Subspecies
formation of 769
as taxonomic group 41 (table)
Substrate 156
Succession 638-43
definition of primary 642
definition of secondary 642
heterotrophic 643
in ponds 662-64, **663, 664**
plant, definition 306
of sand dunes 639-42, **641**
Sugars
see also names of individual
sugars
in DNA 747
types of 80-81
Sulfa drugs 238-39
Sundew **15**
Superclass 41
Supralittoral zone 669
Suspension, definition 73
Swallowing, mechanism of 471, **471**
Sweat glands **577**
in humans **577**, 578
Symbiosis 629-30
in ciliates 264
definition 217, 629
of lichens 304-5
Synapsis 184
Synura **269**, 270

T

Tadpole 606
Taiga 688, **689**
Tamarack 350
Tapeworm 460
Taproots 378, **378**
Taste buds 471, **471**
Taxa 39-41
see also Classification system
Taxonomy
see also Classification
definition 4, 33
Teeth
in human 468-69, **468, 469**
structure of 469, **469**
Teliospores 287

Temperature, in aquatic ecosystems 656-57
Terramycin 237
Testes 184, 590
of humans 608, **609**
of frog 604
Testosterone 608-9, 613
Tetanus 212, 214, 232-33
Thalli (thallus) 277, 278
of algae 322, **322**
Thermocline 666
Thigmotropism 424, **424**
Thromboplastin 550
Tissue
definition 110
types of **110**
Tobacco mosaic disease 226
Tobacco mosaic virus 227, **227**
Tolerance, range of 630, **630**
Tongue
functions of 468, 471
taste bud distribution on 471, **471**
Topography, definition 620
Toxoids 236-37
Trachea(e)
of insects **496**, 497, 558
of humans 472, 501-2, **502**
Tracheal systems **496**, 497-98
Tracheids 376, **376**
Transfer agent, see Coenzyme
Transfer RNA 748, 749, 751-52
Transpiration 412-13
Transport systems 517 ff.
see also Circulatory systems
comparison of 557-58
in complex animals 520 ff.
need for 517
of simple organisms 518-20
Tribonema **269**, 271
Trichocysts 256
Trillium 697, **697**
Trophic levels 624-25, **625**
see also Autotrophs; Consumers;
Herbivores
Tropical rain forest 713-16, **714, 715**
Tropisms 423-24
definition 423
Truffles 298
Trypanosoma 247, **248**
Trypanosomes **248**, 264
Tsetse fly 234, 264
Tuberculosis 214, 232
Tubers **386**, 387
Tundra **306**, 682-87, **682**
animals of 683-87, **684**
climate of 682
vegetation of 682-83, **683**
Turbine bones 500, **501**
Turgor movements 423, 424-25
Turgor pressure 128-29
Typhoid fever 214, 221, 232
Typhus 234

U

Ulcers 478
Ulothrix **326**, 327, **328**

Ulva 324, **325**
Urea 563, **563**, 564
Ureter 576, 582
Urethra 576, 608
Uric acid 563, **563**, 564
Urine 563-64, 582
Urinogenital system, of frog 604-7, **605**
Uterus 607, **607**, 610, **610**
Uvula 471

V

Vaccine 236
Vacuoles 94, 96
cell sap of 108
central 108
contractile 109
structure and function of 108-9
Vagina 610
Valves, of human heart 542, 543, **543**
Vapour, see Gas
Variable, definition of 9
Variation, as instrument of change 19
Varicose veins 535
Vas deferens 608
Vascular bundles 390, **390, 391**, 406
Vascular cambium 382, **382**, 390, 391-92, **392**
Vascular cylinder 379, 380
Vascular tissue 375-76, **376**
Vasopressin 582
Vaucheria 271-72, **272**
Veins 525
in humans 535, **535**
in leaves 401, **401**, 406
pulmonary 503, 537
Venae cavae 538
Venereal disease 233
Venules 535
Venus' flytrap **15**
Vertebrata, sub-phylum 442
see also Vertebrates
Vertebrates 443-45, **443, 444, 445**
circulatory system(s) of 521, 525-30
excretory systems of 576, **576**
hearts of 526-30, **527-29**
Vessels, blood, see Blood vessels
Vessel elements 376, **376**
Villi (villus) 475, **475**, 480, **480**
Vinegar 218
Viruses 193, 225-31, **227, 228**
as cause of disease 226, 230
discovery of 226-27
effects on cells 229, **229**
reproduction of 228-29
size and shape of 227-28, **228**
structure of 225, 228, **228**
Vitamins 448, 449
Vorticella 246

W

Walking fern 344, **344**
Wallace, Alfred 759-60